Brenda

Algebra: A Fundamental Approach

WILLIAM M. SETEK, JR.
Monroe Community College
Rochester, New York

1977 W. B. SAUNDERS COMPANY
Philadelphia • London • Toronto

W. B. Saunders Company: West Washington Square
Philadelphia, PA. 19105

1 St. Anne's Road
Eastbourne, East Sussex BN21 3UN, England

1 Goldthorne Avenue
Toronto, Ontario M8Z 5T9, Canada

Library of Congress Cataloging in Publication Data

Setek, William M

Algebra, a fundamental approach.

Includes index.

1. Algebra.   I. Title.

QA 152.2.S46   512.9   76–19611
ISBN   0–7216–8074–7

Algebra: A Fundamental Approach   ISBN 0-7216-8074-7

© 1977 by W. B. Saunders Company. Copyright under the International Copyright Union. All rights reserved. This book is protected by copyright. No part of it may be reproduced, stored in a retrieval system, or transmitted in any form or by any means, electronic, mechanical, photocopying, recording, or otherwise, without written permission from the publisher. Made in the United States of America. Press of W. B. Saunders Company. Library of Congress Catalog card number 76-19611.

Last digit is the print number: 9   8   7   6   5   4   3   2   1

For
Scott and Joe

preface

This text has been written for that increasingly large group of students who need a firm foundation in algebra as preparation for a variety of later courses, such as college algebra, elementary functions, and technical mathematics. It is assumed that students using this book have had a first course in algebra, but for a variety of reasons have forgotten or never mastered much of the material. Hence, all of the topics covered in this text have been developed completely. This enables students to review the topics of elementary algebra and provides them with background, skills, and confidence, so necessary for success in intermediate algebra. There is definitely a large audience for a text of this type. Virtually every community college and many general purpose colleges and universities offer an intermediate algebra course, a special need of today.

This text is based on an intuitive foundation, rather than an axiomatic one. The explanations will enable students to understand the concepts and techniques involved. The primary purpose of this text is for the student to obtain an intuitive understanding of these topics and develop the necessary skills so essential in mastering the topics of intermediate algebra. The voluminous number of illustrative examples and homework exercises will enable the student to follow the development of each chapter. The objectives and chapter review are designed to highlight the contents of each chapter. Each chapter ends with an exercise set that completely reviews the chapter's topics, which will serve as a check on the student's mastery of the material. Answers for all review exercises are in the back of the book, as are answers to selected homework exercises; answers to other exercises are in a separate manual.

I wish to express my thanks to the W. B. Saunders Company for the excellent editorial service provided, in particular by John S. Snyder. I would like to express my gratitude to Charles G. Denlinger of Millersville State College, Herbert A. Gindler of San Diego State University, Lynn H. Brown of Illinois State University, and David Rogachefsky, a student and friend, for their valuable suggestions for improving the manuscript. A special thanks to Pamela Dretto, whose excellent typing eliminated many aggravations. Finally, and most importantly, I wish to thank my wife Addie, for her understanding, patience, and assistance throughout this project. A special acknowledgement to my sons, Scott and Joe, who helped in their own special way.

I welcome any and all comments. Feel free to write and let me know your thoughts.

WILLIAM M. SETEK, JR.
Monroe Community College
Rochester, New York 14623

contents

Chapter 1

THE NUMBER SYSTEM 1

1.1	Introduction	2
1.2	Numbers	11
1.3	Natural Numbers and Whole Numbers.............................	18
1.4	The Integers	26
1.5	The Rational Numbers	38
1.6	Irrational Numbers and the Set of Real Numbers.............	53
1.7	Absolute Value..	63
1.8	Summary	67
	Review Exercises for Chapter 1	68

Chapter 2

FIRST DEGREE EQUATIONS AND INEQUALITIES IN ONE VARIABLE 71

2.1	Introduction	72
2.2	Solving a First Degree Equation	74
2.3	More First Degree Equations	85
2.4	Literal Equations...	96
2.5	Word Problems...	101
2.6	Inequalities...	112
2.7	More Inequalities	118
2.8	Summary..	125
	Review Exercises for Chapter 2	127

Chapter 3

EXPONENTIALS AND RADICALS 129

3.1	Introduction	130
3.2	Integral Exponents	137
3.3	Scientific Notation	144
3.4	Rational Exponents	151
3.5	Simplifying Radicals	157
3.6	Multiplication of Radicals...	164
3.7	Addition and Subtraction of Radicals	170
3.8	Multiplication of Radical Expressions	175
3.9	Division of Radical Expressions (Rationalizing a Radical Expression)...	181
3.10	Equations with Radicals	187
3.11	Summary..	190
	Review Exercises for Chapter 3	192

Chapter 4

POLYNOMIALS 195

4.1	Introduction	196
4.2	Sums and Differences of Polynomials	198

Contents

4.3	Multiplication of Polynomials	204
4.4	Special Products	209
4.5	More Special Products	214
4.6	Division of Polynomials	219
4.7	Synthetic Division	228
4.8	Summary	233
	Review Exercises for Chapter 4	235
	Cumulative Review	236

Chapter 5
FACTORING ... 239

5.1	Introduction	240
5.2	The Difference of Two Squares	247
5.3	Trinomials That Are Perfect Squares	251
5.4	Trinomials of the General Form $x^2 + bx + c$	254
5.5	Trinomials of the General Form $ax^2 + bx + c$	258
5.6	The Sum or Difference of Two Cubes	263
5.7	Summary	269
	Review Exercises for Chapter 5	270

Chapter 6
RATIONAL EXPRESSIONS (ALGEBRAIC FRACTIONS) ... 273

6.1	Introduction	274
6.2	Simplifying Fractions	275
6.3	Multiplication and Division of Rational Expressions	283
6.4	Addition and Subtraction of Rational Expressions	290
6.5	Complex Fractions	300
6.6	Solving Fractional Equations	307
6.7	Applications	314
6.8	Summary	323
	Review Exercises for Chapter 6	326

Chapter 7
COMPLEX NUMBERS AND QUADRATIC EQUATIONS ... 329

7.1	Introduction	330
7.2	Complex Numbers	336
7.3	Quadratic Equations	345
7.4	Solving Complete Quadratic Equations by Factoring	353
7.5	Solving Quadratic Equations by Completing the Square	363
7.6	The Quadratic Formula	371
7.7	Applications	381
7.8	Quadratic Inequalities	388
7.9	Summary	395
	Review Exercises for Chapter 7	398

Chapter 8
RELATIONS AND FUNCTIONS 401

8.1	Introduction	403
8.2	Relations and Functions	407
8.3	Linear Functions	420
8.4	The Slope of a Line	427
8.5	The Equation of a Straight Line	437
8.6	The Distance and Midpoint Formulas	451
8.7	Quadratic Functions	462
8.8	Variation	472
8.9	Conic Sections	483
8.10	Summary	494

Review Exercises for Chapter 8 497

Cumulative Review 499

Chapter 9
SYSTEMS OF EQUATIONS 503

9.1	Introduction	504
9.2	Solution of Systems of Equations by Substitution	508
9.3	Solution of Systems of Equations by Addition	516
9.4	Applications	528
9.5	Systems of Linear Equations with Three Variables	535
9.6	Determinants	545
9.7	Solution of Systems of Linear Equations by Determinants	552
9.8	Systems of Inequalities	561
9.9	Summary	567

Review Exercises for Chapter 9 570

Chapter 10
EXPONENTIAL AND LOGARITHMIC FUNCTIONS 573

10.1	Introduction	574
10.2	Logarithmic Functions	579
10.3	Properties of Logarithms	587
10.4	Common Logarithms	594
10.5	Interpolation	600
10.6	Computations Using Logarithms	605
10.7	Exponential Equations	615
10.8	Summary	620

Review Exercises for Chapter 10 623

APPENDIX

A. Useful Tables and Formulas 627
    1. Powers and Roots 628
    2. Four-Place Logarithms of Numbers 629
    3. Metric Conversion Factors 631
    4. Useful Formulas 633

B. Answers to Selected Exercises 635

Index 704

chapter 1

The Number System

As a result of this chapter, the student will be able to:

1. Describe a set as a collection of objects and name a set by listing its elements in roster form or by set-builder notation;
2. Determine the intersection of two sets A and B (denoted by A ∩ B), that is, the set of elements that are members of both A and B; and the union of two sets A and B, denoted by (A ∪ B), that is, the set of elements that are members of A, or members of B, or members of both A and B;
3. Distinguish between a number and a numeral and identify a numeral as a symbol or name for a number;
4. Identify a natural number as a number belonging to the set of numbers {1, 2, 3, ...} and identify a whole number as a number belonging to the set of numbers {0, 1, 2, 3, ...};
5. Identify an integer as a number belonging to the set of numbers consisting of the set of whole numbers and their additive inverses, that is, {..., −2, −1, 0, 1, 2, ...}, and add, subtract, multiply, and divide integers;
6. Identify a rational number as any number that can be written in the form $\frac{x}{y}$ ($y \neq 0$) where x and y represent integers; and add, subtract, multiply, and divide rational numbers;
7. Identify an irrational number as one that is not rational, i.e., a non-terminating and non-repeating decimal;
8. Identify the set of real numbers as the union of the set of rational numbers and the set of irrational numbers;
9. Recognize, identify and give examples of the properties that the real number system satisfies—closure, commutivity, associativity, identity elements, inverse elements, and the distributive property for the operations of addition and multiplication;
10. Simplify expressions that contain parentheses, brackets, and braces, and also evaluate expressions containing absolute value notation.

1.1 INTRODUCTION

What is algebra? Intuitively speaking, we can probably describe algebra as the area of mathematics that generalizes the basic concepts of arithmetic; that is, it is the study of the number system of arithmetic. In algebra, we study various types of numbers and their properties. Algebra is often referred to as a generalization of arithmetic because we investigate the general properties of the number system, as opposed to specific examples. Intermediate algebra involves a detailed study of some of the things you have previously learned in elementary algebra, plus study of the manner in which the properties of numbers can be extended.

The Number System

Numbers are basic ingredients of all branches of mathematics. Another essential ingredient is the use of "sets." Sets are often used to clarify ideas and terminology; they are also used to join together many different ideas. The idea of sets is the unifying concept that unites all of the different branches of mathematics.

We are all familiar with the use of sets in one way or another. We are exposed to sets in everyday living. For example, we use sets of dishes, sets of keys, sets of books, matching sets of luggage, and so on. In mathematics, we discuss sets of points, sets of lines, sets of numbers, and so on.

A *set* may be described as a collection of objects. The objects that make up the collection are called *elements* or *members* of the set. A set can be described by naming or listing its elements and when we list the elements of a set we enclose them with braces, { }. This is called the "roster form" of the set. In discussing the set of the first three letters of the alphabet, we would probably write it as

$$\{a, b, c\}.$$

Note that the elements are separated by commas. Capital letters are used to label or name sets and therefore we can write

$$A = \{a, b, c\}.$$

The order in which the elements of a set are listed is immaterial in the roster form. Consider

$$B = \{c, a, b\}.$$

We can say that A and B are *equal* or *identical* sets, that is, A = B because both sets contain exactly the same elements. The order of listing members of a set is not important except, sometimes, for convenience and clarity.

The symbol "$\in$" is used to indicate that an object is a member of a set. The symbol "$\in$" is read "is an element of" or "is a member of." The notation "$\notin$" is read "is *not* an element of." Using the example

$$A = \{a, b, c\},$$

we can state that $a \in A$, $b \in A$, and $c \in A$. We can also say that $z \notin A$, $q \notin A$, $4 \notin A$, and so on.

If a set contains a large number of elements, too many to list, then we may use the three dot notation "..." to indicate that there are other elements in the set, but they have not been listed. The three dots mean that the elements continue on in the indicated pattern. Note that when we use the three dot notation, we must list the elements in some logical order so that the pattern can be determined. The set of odd counting numbers would be written as

$$\{1, 3, 5, 7, \ldots\}.$$

If A = {1, 3, 5, 7, ...}, then we can say that 5 ∈ A, 7 ∈ A, 9 ∈ A, 99 ∈ A, 105 ∈ A, and so on. Similarly, we can also state that 2 ∉ A, 4 ∉ A, 100 ∉ A. We can also use the three dot notation in another manner. Consider the set of the letters in the English alphabet. It would take some time to list all 26 letters; a more efficient method would be to write the set as

$$C = \{a, b, c, \ldots, z\}.$$

This notation indicates the first element of the set, several succeeding elements, and the last element. Even though every element in set C is not listed, we can still determine the number of elements contained in C because we can count them. Set C is called a *finite* set because we can count the number of elements contained in it. Given that set A = {1, 2, 3, 4, ...}, we would describe this set as an *infinite* set because the number of elements in A is indeterminate. A set that does not contain a finite number of elements is said to be *infinite*.

EXAMPLE 1

Is set A = {1, 2, 3, ..., 100} an infinite set or a finite set?

Solution

A = {1, 2, 3, ..., 100} is a finite set. From the pattern, we would assume that set A is the set of counting numbers from 1 through 100, and there are 100 elements in this set.

EXAMPLE 2

Is set $B = \{2, 1, \frac{1}{2}, \frac{1}{4}, \ldots\}$ an infinite set or a finite set?

Solution

$B = \{2, 1, \frac{1}{2}, \frac{1}{4}, \ldots\}$ is an infinite set. It has an unlimited number of elements. Each succeeding element is one-half of the preceding element and the pattern will continue indefinitely.

EXAMPLE 3

Is set $C = \{1, \frac{1}{2}, \frac{1}{4}, \ldots, 0\}$ an infinite set or a finite set?

Solution

Set C is an infinite set. It does contain a last element, but it has an unlimited number of elements.

The Number System

Consider the following two sets:

$$A = \{1, 2, 3\} \quad B = \{1, 2, 3, 4, 5\}$$

Note that all of the elements in set A are also elements in set B. Therefore, set A is contained in set B and we can denote this by

$$A \subseteq B.$$

If all the elements of a given set A are contained in another set B, we say that A is a *subset* of B, denoted by $A \subseteq B$.

> Set A is a subset of set B ($A \subseteq B$), if every element in A is also an element in B.

Consider the following two sets: $C = \{1, 2, 3\}$, $D = \{3, 2, 1\}$. Note that C and D are equal sets and every element of C is also an element of D. Therefore, $C \subseteq D$. But we can also say that $C \subseteq C$, since every element of C is contained in C. Therefore, every set is a subset of itself.

If set $A = \{1, 3, 5, 7\}$ and set $B = \{1, 2, 3, 4, 5, 6, 7\}$, then A is a subset of B. Note that there are some elements in B not contained in A. We can better describe this situation by saying that A is a *proper subset* of B, and we denote this by $A \subset B$.

> If A is a subset of B, and there is at least one element in B not contained in A, then A is a proper subset of B ($A \subset B$).

The *empty set* (also called the null set) is a set that contains no elements; it has no members. Consider the set of months that contain 32 days. There are no months that have 32 days. If we are to list the elements in this set we would have { }, the empty set. Another notation also used to denote the empty or null set is ∅. Note that the empty set, ∅, and the set containing 0, {0}, are not the same. The empty set has no elements while the set containing 0 does have an element in it, namely the digit 0! Both sets are finite sets, but the empty set has no elements while the set containing 0 has one element.

Given that $A = \{1, 2, 3\}$, is the null set a subset of A? The answer is yes. Why? It satisfies the definition: every element of the null set is also an element of A. Since the null set contains no elements, there are no elements of ∅ that can fail to appear in A. Therefore, the null set is a subset of every set.

EXAMPLE 4

List all the possible subsets of {0, 1}.

Solution

From our previous discussion, we know that every set is a subset of itself and also that the empty set is a subset of every set. Thus far we have {0, 1} and { }. It appears that all the possible subsets of {0, 1} are

$$\{\ \}, \{0\}, \{1\}, \{0, 1\}.$$

EXAMPLE 5

List all the possible subsets of {5, 6, 7}.

Solution

We can list {5, 6, 7} (the set itself) and { } (the empty set). Next we shall list those sets containing one element, {5}, {6}, {7}. Now those with two elements, {5, 6}, {5. 7}, {6, 7}. It appears that the list is complete. All the possible subsets of {5, 6, 7} are

$$\{\ \}, \{5\}, \{6\}, \{7\}, \{5, 6\}, \{5, 7\}, \{6, 7\}, \{5, 6, 7\}$$

In order to determine the number of subsets of a set containing "n" elements, we may use the following rule:

> If a set contains "n" elements, then it has 2^n subsets.

The set {0, 1} contains two elements, therefore it has $2^2 = 2 \cdot 2 = 4$ subsets. The set {5, 6, 7} contains three elements, therefore it has $2^3 = 2 \cdot 2 \cdot 2 = 8$ subsets.

Consider the set A = {0, 1, 2, 3, 4, 5, 6, 7, 8, 9}. Upon inspection, we see that set A contains the digits 0 through 9. This technique of listing the elements of a set is called the *roster* technique or form (as stated earlier) because we have listed the elements. A roster is a list of items. It is often cumbersome and impractical to list the elements of a set. Consider the set of all people who are United States senators. We would not want to describe this set by the roster technique because of the large number of elements in the set.

Another way that sets can be described is by the *set builder notation*.

The Number System 7

An example of a statement in set builder notation is

$$A = \{x \mid x > 1\}.$$

This means that set A is the set of all x's such that x is greater than 1. The x after the first brace indicates that we are considering the set of all x's and the vertical line following the first x stands for "such that."

EXAMPLE 6

Interpret: $\{y \mid y < 0\}$.

Solution

The set of all y's such that y is less than 0.

$$\{y \mid y < 0\}.$$

The set of all y's   such that   y is less than 0.

EXAMPLE 7

Interpret: $\{x \mid x \neq 10\}$.

Solution

The set of all x's such that x does not equal 10. Note that this set has an infinite number of elements and it would be impossible to completely list all the elements.

INTERSECTIONS AND UNIONS OF SETS

Thus far, we have discussed sets and some of their various properties. Let us now define some operations on pairs of sets that will be helpful in later sections. We shall consider the *intersection* and *union* of sets. The intersection of sets A and B produces another set that contains all the elements that are contained in both A and B. The intersection of A and B is denoted by A ∩ B.

> Given sets A and B, then A ∩ B (the intersection of A and B) is the set of all elements that are members of both A and B.

Using the set builder notation, we can define the intersection of A and B as

$$A \cap B = \{x \mid x \in A \text{ and } x \in B\}.$$

EXAMPLE 8

Given $A = \{1, 2, 3, 4\}$ and $B = \{3, 4, 5, 6\}$, find $A \cap B$.

Solution

The elements contained in both A and B are 3 and 4. Therefore $A \cap B = \{3, 4\}$.

EXAMPLE 9

Given $A = \{1, 2, 3, \ldots\}$ and $B = \{1, 2, 3, \ldots, 10\}$, find $A \cap B$.

Solution

The elements that are common to both A and B are 1, 2, 3, 4, 5, 6, 7, 8, 9, 10. Therefore, $A \cap B = \{1, 2, 3, \ldots, 10\}$. Note that, for this example, we can also say $A \cap B = B$.

EXAMPLE 10

Given $A = \{1, 3, 5, \ldots\}$ and $B = \{2, 4, 6, \ldots\}$, find $A \cap B$.

Solution

There are no elements common to both A and B. Set A contains the odd counting numbers, while set B contains the even counting numbers. The intersection of A and B is empty! Therefore, $A \cap B = \emptyset$, or $A \cap B = \{\ \}$.

When the intersection of two sets is empty, that is, $A \cap B = \emptyset$, then we say that A and B are *disjoint* sets. They have no elements in common.

The *union* of two sets A and B produces another set that contains those elements that are members of A, or members of B, or members of both A and B. The union of sets A and B consists of the set of elements that are elements of at least one of the two sets. The union of A and B is denoted by $A \cup B$.

> Given sets A and B, then $A \cup B$ (the union of A and B) is the set of all elements that are members of A, or members of B, or members of both A and B.

The Number System

Using the set-builder notation, we can define the union of A and B as

$$A \cup B = \{x \mid x \in A \text{ or } x \in B \text{ or both}\}.$$

EXAMPLE 11

Given $A = \{1, 2, 3, 4\}$ and $B = \{5, 6, 7\}$, find $A \cup B$.

Solution

The elements contained in A are 1, 2, 3, 4 and those in B are 5, 6, 7. The union of A and B is the set containing all of these elements because each element is a member of set A or set B. Therefore, $A \cup B = \{1, 2, 3, 4, 5, 6, 7\}$.

EXAMPLE 12

Given $A = \{1, 2, 3, 4\}$ and $B = \{3, 4, 5, 6\}$, find $A \cup B$.

Solution

The union of A and B is a set that contains all of these elements because all of these elements are members of set A or set B, or both sets A and B. Therefore, $A \cup B = \{1, 2, 3, 4, 5, 6\}$. Note that 3 and 4 are elements of both sets, but we list each element only once in our solution.

EXAMPLE 13

Given $A = \{1, 3, 5, \ldots\}$ and $B = \{2, 4, 6, \ldots\}$, find $A \cup B$.

Solution

Set A contains the odd counting numbers, and set B contains the even counting numbers. The union of A and B contains all of these elements. Therefore, $A \cup B = \{1, 2, 3, 4, 5, 6, \ldots\}$.

EXAMPLE 14

Given $A = \{1, 2, 3, 4\}$ and $B = \{2, 4, 6, 8\}$, find
a. $A \cap B$.   b. $A \cup B$.

Solution

a. The elements that are common to both A and B are 2 and 4. Therefore, $A \cap B = \{2, 4\}$.
b. The union of A and B is a set that contains all of these elements. Therefore, $A \cup B = \{1, 2, 3, 4, 6, 8\}$. Note that the elements in $A \cup B$ are elements of at least one of the two sets.

EXAMPLE 15

Given that $A = \{1, 2, 3, 4\}$, $B = \{2, 3, 4\}$ and $C = \{3, 4, 5\}$, find
a. $A \cap (B \cup C)$.   b. $(A \cap B) \cup (A \cap C)$.

Solution

a. Working inside the parentheses first, we must find $(B \cup C)$ and then find the intersection of that with A. $(B \cup C) = \{2, 3, 4, 5\}$ and now we must find $A \cap \{2, 3, 4, 5\}$, that is, $\{1, 2, 3, 4\} \cap \{2, 3, 4, 5\}$. It is $\{2, 3, 4\}$. Therefore, $A \cap (B \cup C) = \{2, 3, 4\}$.

b. Again, we must work inside the parentheses first. We have to find $(A \cap B) (A \cap C)$ and then find the union of these two sets. $(A \cap B) = \{2, 3, 4\}$ and $(A \cap C) = \{3, 4\}$. Therefore, $(A \cap B) \cup (A \cap C) = \{2, 3, 4\} \cup \{3, 4\} = \{2, 3, 4\}$.

Note that in Example 15 we obtained the same answer for $A \cap (B \cup C)$ and $(A \cap B) \cup (A \cap C)$. This suggests that $A \cap (B \cup C) = (A \cap B) \cup (A \cap C)$.

EXERCISES FOR SECTION 1.1

(Answers given to exercises marked *)

1. Determine whether each of the following is true or false.
   *a. $\{1, 3, 5\} = \{5, 3, 1\}$.
   b. $\{1, 2, 3, \ldots\} = \{1, 3, 5, \ldots\}$.
   *c. $5 \in \{1, 2, 3, \ldots\}$.
   d. $5 \notin \{1, 2, 3, \ldots\}$.
   *e. $q \in \{a, b, c, \ldots, z\}$.
   f. $0 \in \{\ \}$.
   *g. $\emptyset = \{\ \}$.
   h. $\{a, b\} = (a, b)$.
   *i. $100 \in \{x \mid x \text{ is a whole number}\}$.
   j. $9 \in \{x \mid x \text{ is a whole number}\}$.
   *k. Sunday $\in \{x \mid x \text{ is a day of the week}\}$.
   l. $\emptyset$ is a finite set.
   *m. $2 \in \{x \mid x + 0 = x\}$.
   n. $3 \in \{y \mid y \cdot 2 = 8\}$.
   *o. $\{1, 2, 3, \ldots, 10\}$ is a finite set.
   p. $\{2, 4, 6, \ldots\}$ is a finite set.
   *q. The set of odd numbers is a finite set.
   r. $\{a, b\} \subseteq \{a, b\}$.
   *s. $\emptyset \subset \{1, 2\}$.
   t. $\{2\} \subset \{1, 2, 3\}$.

The Number System

*u. $\{\emptyset\} \subset \{\ \}$.
v. $\{a\} \in \{a, b\}$.
*w. $\{8, 16\} \subset \{1, 2, 3, \ldots\}$.
x. $\{101, 102\} \subseteq \{1, 2, 3, \ldots\}$.
*y. $\{2, 4, 6, \ldots\} \subseteq \{1, 2, 3, \ldots\}$.
z. $\{1, 2, 3, \ldots\} \subset \{1, 2, 3, \ldots\}$.

2. List all possible subsets for each of the given sets.
   *a. $\{1, 2, 3\}$.
   b. $\{w, x, y, z\}$.
   *c. $\{m, a, t, h\}$.
   d. $\emptyset$

3. Use the roster method to denote:
   *a. The set of counting numbers less than 8
   b. The set of counting numbers less than or equal to 8
   *c. The set of letters in the English alphabet
   d. The set of days of the week
   *e. $\{x \mid x \text{ is a counting number less than } 7\}$
   f. $\{x \mid x \text{ is an odd counting number}\}$
   *g. $\{x \mid x \text{ is a vowel}\}$

4. Use set builder notation, $\{x \mid x \text{ has a certain property}\}$, to describe each of the following sets.
   *a. $\{1, 3, 5, 7, \ldots\}$
   b. $\{2, 4, 6, 8, \ldots\}$
   *c. $\{\text{Huron, Michigan, Erie, Superior, Ontario}\}$
   d. $\{\text{January, February}, \ldots, \text{December}\}$
   *e. $\{2, 3, 5, 7, 11, \ldots\}$

5. Given that $A = \{1, 2, 3, 4, 5\}$, $B = \{3, 4, 5, 6\}$, and $C = \{5, 6, 7, 8\}$, find:
   *a. $A \cap B$
   b. $A \cap C$
   *c. $B \cap C$
   d. $A \cup B$
   *e. $A \cup C$
   f. $B \cup C$
   *g. $A \cap \emptyset$
   h. $B \cup \emptyset$
   *i. $(A \cap B) \cup C$
   j. $(A \cup B) \cap C$
   *k. $(B \cap C) \cup (A \cap C)$
   l. $(A \cup B) \cap (B \cup C)$

1.2 NUMBERS

Regardless of which early civilization historians have studied, they have discovered that the concept of a number existed in the culture of even the most primitive tribes. Early in his history, man discovered a need for determining "how many." Hence the need for counting and numbers. The earliest type of counting was probably done by making marks or slashes in the dirt, or using pebbles to match the number of objects being counted. Later, instead of making marks in the dirt, man made marks on a stone, or on a piece of clay. Eventually, he cut notches in a stick (called a "tally stick"), and he also counted by tying knots in pieces of string or rope. Even today we sometimes use a form of this primitive type of counting. For example, when we have to count such items as ballots, we normally "tally" the votes. These tally marks (𝍩 𝍩) are used to represent numbers. The tally marks are not numbers themselves, but symbols representing numbers.

The idea of a number is an abstract concept; it exists in our minds. When we make the tally marks, 𝍩 𝍩, we are making a symbol that represents the number 10. Someone else may represent this number by using the Roman numeral that represents 10: X. Most of us would normally write it as 10. But, this too is a symbol for the number. It is a Hindu-Arabic numeral. All of these are symbols that are used to represent the number 10, and they are called numerals.

> A *numeral* is a symbol for a number; it is also a name of a number.

When you write a 5 on a piece of paper, you are writing a Hindu-Arabic numeral that represents a certain number. If we write "Julia" on a piece of paper, we are not writing a person, but we are writing a "name" or "symbol" for that person. Many people use the words "number" and "numeral" interchangeably and we understand what they mean. But to be precise, we should use the correct terminology.

Before proceeding any further, let's review some concepts previously encountered in elementary algebra. For example, in the expression $x + 5$, what names are given to the x and 5? The x is referred to as a *variable*. The expression $x + 5$ may be evaluated for any given value of x; in other words, x can vary. A variable is a symbol in an expression for which any element of a given set may be substituted. In the expression, $x + 5$, the numeral 5 is called a *constant*. The value of 5 will not change; it will remain constant. Numerals, and other symbols such as π, are examples of constants. But suppose we have an expression such as $4y + 2$ — what names are given to its various components? We can describe y as a variable and the 4 and 2 as constants (according to our previous discussion). We can better describe the 4, in the expression $4y + 2$, as a *numerical coefficient*. It is a factor of the expression 4y. Similarly, we can call the y a *literal coefficient*. Note that the expression 4y means 4 times y; 4y is the same as $4 \cdot y$, where the raised dot indicates multiplication. We can also indicate multiplication by means of parentheses. For example, $4y = 4 \cdot y = 4(y) = (4)y = (4)(y) = 4y$.

In algebra we may sometimes encounter expressions such as $x + a + y + b$. Are all of these letters called variables? Not necessarily. In mathematics it is common practice that letters from the beginning of the alphabet represent constants and letters toward the end of the alphabet represent variables. Following this practice, in the expression $x + a + y + b$, the "a" and "b" would be constants, while x and y would be variables.

The Number System 13

EXAMPLE 16

In each of the following algebraic expressions, identify the variable(s) and constant(s).

a. 4xy   b. 3x + 6   c. 4s + 7t   d. ax + by + c

Solution

a. The variables are x and y. The 4 is a numeral, therefore it is a constant. It is also a numerical coefficient.
b. The variable is x, and the constants are 3 and 6. Note that 3 is also a numerical coefficient.
c. The variables are s and t. The constants are 4 and 7.
d. The variables are x and y. The constants are a, b, and c. Recall that we let letters from the beginning of the alphabet represent constants, and those toward the end of the alphabet represent variables.

Thus far we have termed x + 5 and 4y + 2 "expressions." To be more specific, they are called *algebraic expressions* because they consist of at least a variable or a numeral. Algebraic expressions may consist of a combination of variables and numerals connected by various operation signs. All of the following are algebraic expressions.

$$x + 5 \quad 3y - 7 + a \quad x^2 - 3(x + y)$$

Algebraic expressions often occur in equations. An equation consists of algebraic expressions connected by an equal to (=) sign. An equation does not have to be a true statement. Consider 3 + 2 = 1. We would describe this as an equation (sentence) that is false. An equation is a sentence and many sentences are false. "The moon is made of green cheese" is a sentence, but it is false. An equation (sentence) such as y + 3 = 5 is called an *open sentence,* because it is neither true nor false. The variable y may assume different values. If y = 2, then the equation is a true sentence. If y = 3, then the equation is a false sentence. The equation x − 3 = 7 is an open sentence that is true for some value of x and false for some value of x. But until we substitute a numeral for the variable x, it is neither true nor false. Recall that an equation can be reversed: x − 3 = 7 and 7 = x − 3 mean exactly the same thing.

Expressions such as 3^2 and x^3 are called powers. In the expression x^3, 3 is an *exponent* and x is called the base. The expression x^3 means that there are three factors of x in that expression, that is, $x^3 = x \cdot x \cdot x$. The expression x^3 is read as "x cubed" or the "third power of x." Similarly,

x^2 means "x squared" or the "second power of x." Larger powers such as x^5 are read as "the fifth power of x." The exponent of a single variable such as x is understood to be 1, or $x = x^1$.

Algebraic expressions are said to be equivalent when they have the same numerical value after replacements are made for the corresponding variables. Therefore, the expressions $2x + 5$ and $x + 5 + x$ are equivalent expressions. Substituting some values for each, we have

$2x + 5$	$x + 5 + x$
$2 \cdot 0 + 5 = 5$	$0 + 5 + 0 = 5$
$2 \cdot 3 + 5 = 11$	$3 + 5 + 3 = 11$
$2 \cdot 7 + 5 = 19$	$7 + 5 + 7 = 19$

The algebraic expressions $2x + 5$ and $x + 5 + x$ are equivalent because they have the same value for *all* replacements of the variable.

HISTORICAL NOTE

The study of algebra began in early Egypt. The first documented occurrence of algebra-type problems appears in a papyrus copied by the Egyptian priest Ahmes (approximately 1600 B.C.). His papyrus was a copy of a book written even earlier, approximately 2200 B.C. It is significant because it illustrates the first recorded use of a "symbol" to represent an unknown quantity. The Egyptians used the word "hau" to represent any unknown quantity in a mathematical problem.

Diophantas (approximately 200 A.D.), a Greek mathematician, was one of the first mathematicians to use symbols to represent unknown quantities. He supposedly used a combination of the first two letters of the Greek word "arithmos," which means "number."

One of the most famous Arabian mathematicians was Mohammed ibn Musa al-Khowarizmi (approximately 800 A.D.), who wrote a book titled Al-Jabr wal-Mugabalah, which contributed greatly to the adoption of Hindu-Arabic numerals by the scholars of that time. It is believed that the title word "Al-Jabr" gave rise to the word algebra as we know it today.

Descartes (1596–1650), a French mathematician of great stature, urged his fellow mathematicians to use x, y, and z to represent variables and a, b, c to represent constants.

EXAMPLE 17

Evaluate each of the following algebraic expressions for the indicated numerical replacement.

a. $4z - 6$, when $z = 2$
b. $3x^2 + 7x$, when $x = 3$
c. πr^2, when $r = 4$
d. $ay^3 + by$, when $y = 2$
e. $-x^2$, when $x = 3$
f. x^2, when $x = -3$

The Number System

Solution

a. $4z - 6 = 4 \cdot 2 - 6$
$ = 8 - 6$
$ = 2$

b. $3x^2 + 7x = 3 \cdot 3^2 + 7 \cdot 3$
$ = 3 \cdot 9 + 7 \cdot 3$
$ = 27 + 21$
$ = 48$

c. $\pi r^2 = \pi \cdot 4^2$
$ = \pi \cdot 16$
or 16π

d. $ay^3 + by = a \cdot 2^3 + b \cdot 2$
$ = a \cdot 8 + b \cdot 2$
or $8a + 2b$

e. $-x^2 = -3^2$
$ = -9$

Note that the minus sign precedes the expression x^2. We are asked to square 3, not -3.

f. $x^2 = (-3)^2$
$ = 9$

Here we are asked to square -3 and $(-3)(-3) = 9$.

We identify other parts of algebraic expressions, in addition to variables and constants. If an algebraic expression consists of sums or differences (or both), then those sums or differences are called *terms* of the expression. To be more precise, *terms* are those parts of the algebraic expressions separated by plus or minus (+ or −) signs. We include the sign preceding the term as part of that term. For example, in the expression $x^2 + 7x - 6$, the terms are x^2, $+7x$, and -6. Note that the first term is not preceded by a plus sign, but it is separated from the other terms.

EXAMPLE 18

Identify the terms in each of the given algebraic expressions.
a. $5x^2 - 3x + 6$
b. $-x^2 + 7$
c. $8x^3y^2z + 2x^2y$
d. $4y^3 + 3y^2 + 2(y - 1)$

Solution

a. Three terms: $5x^2$, $-3x$, and $+6$.
b. Two terms: $-x^2$ and $+7$.
c. Two terms: $8x^3y^2z$ and $+2x^2y$.
d. Three terms: $4y^3$, $+3y^2$, and $+2(y - 1)$. (Note that $+2(y - 1)$ is that part of the expression that is separated by a plus sign. It is a term.)

An algebraic expression that consists of only one term and no division by variables is called a *monomial*. If an algebraic expression consists of two monomials, then it is called a *binomial*. A *trinomial* is an algebraic expression that consists of three monomials. A *polynomial* may be a "many termed" expression, but it may also be a monomial, binomial, or trinomial. It should be noted that some algebraic expressions are not polynomials. Those expressions that contain quotients where the variable occurs in the divisor (denominator) are *not* polynomials. For example $3x^2 + 7x$ is a polynomial, but $3x^2 + \frac{7}{x}$ is not a polynomial. Some other algebraic expressions that are not polynomials are

$$\frac{x+1}{x-1}, \quad y+\frac{2}{y^2}, \quad z^3 + 3z^2 + \frac{2}{z+1}.$$

EXAMPLE 19

Identify each of the following as a monomial, binomial, trinomial or polynomial.

a. $5x^2 + 3x$ 　　　　　b. -2

c. $3y^3 + 2y^2 + 6$ 　　d. $x^2 + \frac{1}{x}$

Solution

Note: With the exception of d., all of the expressions may be classified as polynomials. $x^2 + \frac{1}{x}$ contains a quotient where the variable occurs in the denominator of the fraction.

a. $5x^2 + 3x$, two terms, therefore a binomial
b. -2, one term, therefore a monomial
c. $3y^3 + 2y^2 + 6$, three terms, therefore a trinomial
d. Not a polynomial

Many times a polynomial will contain terms that are called *like terms*. Like terms contain the same variables and also the same exponent for each corresponding variable. For example, the expression $5x^2 + 3xy - 2x^2 + 2xy$ contains two sets of like terms. The terms $5x^2$ and $-2x^2$ are like terms, as are $+3xy$ and $+2xy$. In each case the like terms contain the same variables(s) and also the same exponents for each corresponding variable. Are $-5x^3y^2z$ and $+2x^3yz^2$ like terms? The answer is no. Granted each term contains the same variables, but note that the exponents of the variables y and z are not the same. The term $-5x^3y^2z$ has an exponent of 2 for y and 1 for z, while the term $+2x^3yz^2$ has an exponent of 1 for y and 2 for z.

EXAMPLE 20

In each of the following algebraic expressions, identify those terms that are "like."
  a. $3xy^2 + 4x^2y - 6xy^2$
  b. $r^2s^2 - 6rs + 2rs - 8r^2s^2$
  c. $2x^2y^2z^3 + xtz^2 - x^2y^2z^3 + xyz^4$
  d. $3r^4s^3t^2 - 6r^3s^4t^2 + 2r^2s^4t^2$

Solution

  a. Like terms are those terms that contain the same variables and also the same exponent for each corresponding variable. Therefore, in the expression, $3xy^2 + 4x^2y - 6xy^2$, the like terms are $3xy^2$ and $-6xy^2$.
  b. The like terms are r^2s^2, $-8r^2s^2$ and $-6rs$, $+2rs$.
  c. The like terms are $2x^2y^2z^3$ and $-x^2y^2z^3$.
  d. There are *no* like terms in the expression $3r^4s^3t^2 - 6r^3s^4t^2 + 2r^2s^4t^2$. The corresponding variables do not have the same exponent.

EXERCISES FOR SECTION 1.2
(Answers given to exercises marked *)

1. In each of the following algebraic expressions, identify the variable(s) and constant(s).
  *a. $x + 7$
  b. $3x + 2y + 6$
  *c. $3x^2 + 4y^2 + 6$
  d. $3r^2s + 2s^3$
  *e. $4s^3 + 7t^2 + 5$
  f. πr^2
  *g. $2\pi r$
  h. $4x^3y^2z - 6z^2y$
  *i. $ax^2 + bx + c$
  j. $ay^2 + by + c$

2. Evaluate each of the following algebraic expressions for the indicated numerical replacement.
  *a. $3x^3$, when $x = 3$
  b. $4y^2 + 2y + 3$, when $y = 0$
  *c. $2z^2 + 6z - 6$, when $z = 2$
  d. $2\pi r$, when $r = 4$
  *e. $x^2y - 2xy^3$, when $x = 1$, $y = -2$
  f. $2x^3y + 3xy^2$, when $x = 2$, $y = 3$
  *g. $3x^2y - 2xy + 8$, when $x = 1$, $y = 2$
  h. $3x^3y^2z^2 + 2x^3y^2z^2$, when $x = -2$, $y = 2$, $z = 2$
  *i. $ax^2 + bx + c$, when $x = 3$
  j. $ay^2 + by + c$, when $y = -2$

3. Identify the terms in each of the following algebraic expressions.
   *a. $3x^2 - 2x$
   b. $-2x + 3y - 6$
   *c. $4x^2 + 2y^2 + 3(x - y)$
   d. $ay^2 + by + c$

4. Determine whether or not the following algebraic expressions are polynomials. If the expression is a polynomial, then describe it as a monomial, binomial, or trinomial.
   *a. $3y^2 + 2y + 1$
   b. $6x^3 + 3x^2$
   *c. -2
   d. $x + \dfrac{3}{x}$
   *e. $\dfrac{1}{2}x^2 + \dfrac{1}{2}x$
   f. $\dfrac{1}{2x^2} + \dfrac{1}{2x}$
   *g. $\dfrac{x+1}{x-1}$
   h. $3z^3 + 2z^2 + z$
   *i. πr^2
   j. $\dfrac{1}{2}y^2 + \dfrac{1}{2y}$

5. Identify the like terms in each of the following algebraic expressions.
   *a. $x^2 + y^2 + 3x + 2y^2$
   b. $3x^3y + 2xy^3$
   *c. $4z^3y + 2zy^3$
   d. $s^2t^2 + 6st^2 + 3s^2t^2$
   *e. $2x^3y^2 - 3x^2y^3 - 4x^3y^2$
   f. $ax^2 + by^2 + cx^2 + dy^2$
   *g. $ax^2 + by^2 + cx + dy$
   h. $3 - x^2y + 2xy^2 - 6x^2y$

6. When are the following equations true?
   *a. $x + 4 = 6$.
   b. $2y - 4 = 6$.
   *c. $x + y = y + x$.
   d. $5 + 3 = 3 + 5$.
   *e. $x \cdot y = y \cdot x$.
   f. $4x = 12$.
   *g. $\dfrac{2}{3} + \dfrac{1}{4} = \dfrac{3}{7}$.
   h. $\dfrac{1}{2} + \dfrac{2}{3} = \dfrac{3}{5}$.
   *i. $\dfrac{2}{7} + \dfrac{3}{5} = \dfrac{31}{35}$.
   j. $x(y + z) = xy + xz$.

1.3 NATURAL NUMBERS AND WHOLE NUMBERS

Our system of numbers can be classified into different sets whose elements have certain unique characteristics. When man first discovered the concept of a number, he began with 1 and then proceeded from that point. Note that he did not begin with 0. It was not until about 700 A.D. that 0 came to be accepted as a number. As man developed his counting, he developed the set of numbers:

The Number System

$$\{1, 2, 3, 4, \ldots\}$$

This set of numbers is often referred to as the set of *counting numbers*. It is also called the set of *natural numbers*.

Any natural number can be expressed as the product of two or more natural numbers. For example, $10 = 2 \cdot 5$, $15 = 3 \cdot 5$, $6 = 3 \cdot 2$, and so on. Those numbers that are multiplied together to form the product are called *factors*. A factor of a number is a number that will divide into the given number and yield a remainder of 0. Certain natural numbers have only two factors, themselves and 1. We can also say that these certain natural numbers are divisible only by themselves and 1. Natural numbers that have this property are called *prime numbers*.

> A prime number is any natural number greater than 1 that is divisible only by itself and 1.

Note that 1 is not considered to be a prime number. Those natural numbers from 1 to 100 that are prime are:

2,3,5,7,11,13,17,19,23,29,31,37,41,43,47,53,59,61,67,71,73,79,83,89,97.

Every natural number, except 1, is either a prime number or it can be expressed as a product of prime numbers. In fact, the Fundamental Theorem of Arithmetic states that:

> Every natural number greater than 1 is either a prime number or it can be expressed as a product of prime numbers. Except for the order of the factors, this can be done in one and only one way.

For example, 6 is not prime and its prime factors are 2 and 3. Therefore, we can say that $6 = 2 \cdot 3$, since the order of the factors does not affect the value: $2 \cdot 3 = 3 \cdot 2$. This is an example of a particular property of multiplication. Do you recall what it is? It is called the *commutative property of multiplication:* the order in which two natural numbers are multiplied does not affect the product. In other words, the answer that we obtain when we multiply two natural numbers together does not depend upon the order in which we multiply the numbers.

The commutative property also holds for another operation, addition. The order in which two natural numbers are added does not affect the sum. For example, $3 + 2 = 2 + 3$. Regardless of the order, when we add 2 and 3, the sum is 5. In general terms we have

> The commutative property: *addition* $x + y = y + x$
> (If x and y are natural numbers)   *multiplication* $x \cdot y = y \cdot x$

Not all operations are commutative. If we wish to subtract two natural numbers, does the order in which we subtract two numbers affect the answer? Yes, the order does matter. For example, $5 - 3 \neq 3 - 5$; $5 - 3 = 2$ while $3 - 5 = -2$. The results are not the same. Subtraction is not commutative. Division is not commutative either. An example that illustrates this fact is $4 \div 2 \neq 2 \div 4$. If we divide 4 by 2, the quotient is 2, while 2 divided by $4 = \frac{1}{2}$ or 0.5.

EXAMPLE 21

Use the commutative property to make each of the following a true statement.
 a. $3 + 4 =$ b. $8 \cdot 6 =$
 c. $a + b =$ d. $a \cdot b =$

Solution

 a. $3 + 4 = 4 + 3$ — commutative property of addition.
 b. $8 \cdot 6 = 6 \cdot 8$ — commutative property of multiplication.
 c. $a + b = b + a$ — commutative property of addition.
 d. $a \cdot b = b \cdot a$ — commutative property of multiplication.

Suppose we want to find the sum of three numbers, for example, the sum of 2, 3 and 7, or $2 + 3 + 7$. The answer is 12, but how did we obtain the result? One person might add $2 + 3$, obtain 5 and then add that to 7 in order to obtain 12, i.e., $(2 + 3) + 7 = 12$. Someone else might add $3 + 7$, obtain 10 and then add that to 2 in order to obtain 12: $2 + (3 + 7) = 12$. For this particular example, we can say that $(2+3)+7 = 2+(3+7)$. In other words, the manner in which the numbers are grouped or associated does not affect the sum. This is an example of the *associative property of addition*.

The associative property also holds for another operation, multiplication. The manner in which three natural numbers are grouped or associated does not affect the product. For example, $(2 \cdot 3) \cdot 5 = 2 \cdot (3 \cdot 5)$, since $(2 \cdot 3) \cdot 5 = 6 \cdot 5 = 30$ and $2 \cdot (3 \cdot 5) = 2 \cdot 15 = 30$. Regardless of how we group the factors, the product is 30. In general terms, we have:

The associative property: *addition* $x + (y + z) = (x + y) + z$
(If x, y and z are   *multiplication* $x \cdot (y \cdot z) = (x \cdot y) \cdot z$
natural numbers)

The Number System 21

EXAMPLE 22

Use the associative property to make each of the following a true statement.

    a. $(3 \cdot 4)5 =$ b. $(x + 3) + 4 =$
    c. $3 + (4 + 5) =$ d. $4(3x) =$

"Daddy, 28 plus (36 plus 49) equals (28 plus 36) plus what--using the associative principle?"

(© 1969, The Register and Tribune Syndicate.)

Solution

    a. $(3 \cdot 4)5 = 3(4 \cdot 5)$ — associative property of multiplication.
    b. $(x + 3) + 4 = x + (3 + 4)$ — associative property of addition.
    c. $3 + (4 + 5) = (3 + 4) + 5$ — associative property of addition.
    d. $4(3x) = (4 \cdot 3)x$ — associative property of multiplication.

As we discovered with the commutative property, not all operations are associative. Consider the following problem, $6 - (3 - 2)$. Operating inside the parentheses first, we have $6 - (3 - 2) = 6 - 1 = 5$. Now if we regroup the numbers, we have $(6 - 3) - 2$ and $(6 - 3) - 2 = 3 - 2 = 1$. The manner in which the numbers are grouped, under the operation of subtraction, *does* affect the answer, $6 - (3 - 2) \neq (6 - 3) - 2$. This example illustrates that subtraction is not associative. One example showing that a property does not hold is sufficient to show that a property is not satisfied for all numbers in a set under a given operation.

Division is not associative. Consider the following problem: $8 \div (4 \div 2)$. Operating inside the parentheses first, we have $8 \div (4 \div 2) = 8 \div 2 = 4$. Now if we regroup the numbers, we have $(8 \div 4) \div 2$ and $(8 \div 4) \div 2 = 2 \div 2 = 1$. The manner in which the numbers are grouped, under the operation of division, *does* affect the answer, $8 \div (4 \div 2) \neq (8 \div 4) \div 2$. Therefore, division is not an associative operation.

To find the product of $13 \cdot 12$, is a calculator or pencil and paper needed? The answer could be no because we can use a technique that will enable us to obtain the answer fairly quickly. Instead of multiplying $13 \cdot 12$, we can multiply $13 \cdot 10$ and $13 \cdot 2$, and then add the two products, thus, $13 \cdot 12 = 13 \cdot 10 + 13 \cdot 2 = 130 + 26 = 156$. This technique of multiplication is an example of another property that holds for natural numbers, the *distributive property*. It is formally called the "distributive property of multiplication over addition." In general terms:

> The distributive property:
> (If x, y, and z are natural numbers) $x(y + z) = xy + xz$

To be specific, we should call this the *left* distributive property of multiplication over addition, as the x is distributed from the left. If we use the commutative property on the expression $x(y + z)$, we have $(y + z)x$, which equals $yx + zx$, and this illustrates the *right* distributive property of multiplication over addition. These distinctions are just pointed out in passing, but it is important to remember that the distributive property is illustrated by $x(y + z) = xy + xz$. It should also be noted that the distributive property holds for multiplication over subtraction, that is, $x(y - z) = xy - xz$. For example, $13 \cdot 18 = 13 \cdot 20 - 13 \cdot 2 = 260 - 26 = 234$.

EXAMPLE 23

Use the distributive property to evaluate each of the following expressions:

a. $3(5 + 2)$ b. $5 \cdot 3 + 5 \cdot 6$
c. $4(5 + 3)$ d. $2 \cdot 4 + 3 \cdot 4$
e. $(5 - 2)4$ f. $(3 - 3)2$

Solution

a. $3(5 + 2) = 3 \cdot 5 + 3 \cdot 2 = 15 + 6 = 21$.
b. $5 \cdot 3 + 5 \cdot 6 = 5(3 + 6) = 5 \cdot 9 = 45$.

The Number System

> c. $4(5 + 3) = 4 \cdot 5 + 4 \cdot 3 = 20 + 12 = 32$.
> d. $2 \cdot 4 + 3 \cdot 4 = (2 + 3)4 = 5 \cdot 4 = 20$.
> e. $(5 - 2)4 = 5 \cdot 4 - 2 \cdot 4 = 20 - 8 = 12$.
> f. $(3 - 3)2 = 3 \cdot 2 - 3 \cdot 2 = 6 - 6 = 0$.

If we add any two natural numbers, what can we say about the sum? Regardless of what two natural numbers we add, the sum will be a natural number. For example, $2 + 3 = 5$, $4 + 2 = 6$, $3 + 4 = 7$, and so on; the sum is also a natural number. This illustrates another unique property of the natural numbers, the *closure property*. A set or system of numbers is said to be *closed* with respect to an operation if, whenever we perform an operation on any two numbers in the set, the result is also a member of the set. The set of natural numbers is closed under the operation of addition, but it is not closed under subtraction. Granted that if we subtract certain natural numbers, such as $5 - 3$, or $6 - 2$, the result is a natural number. But consider $3 - 5$ and $2 - 6$; $3 - 5 = -2$ and $2 - 6 = -4$; -2 and -4 are not elements in the set of natural numbers.

The set of natural numbers is also closed under the operation of multiplication. If we multiply any two natural numbers, then the product is also a natural number. For example, $3 \cdot 4 = 12$, $6 \cdot 4 = 24$, $5 \cdot 2 = 10$, and so on. It should also be noted that the set of natural numbers is not closed under the operation of division: $5 \div 2 = 2.5$, which is not a natural number. The set of natural numbers is closed under addition and multiplication and in general terms, we have:

> The closure property:
> Given the set of natural numbers,   *addition* $x + y \in N$
> $N = \{1, 2, 3, ...\}$ and $x \in N$ and $y \in N$ *multiplication* $xy \in N$

EXAMPLE 24

Is the set $A = \{1, 3, 5\}$ closed under the operation of addition?

Solution

No, none of the sums are elements of the given set. For example, $1 + 3 = 4$ and $4 \notin A$, $3 + 5 = 8$ and $8 \notin A$.

EXAMPLE 25

Is the set $A = \{1, 3, 5, ...\}$ closed under the operation of multiplication?

Solution

Yes, set A is composed of the set of natural numbers that are odd, and whenever we multiply two odd numbers the resulting product is odd. Therefore, set A is closed under the operation of multiplication.

EXAMPLE 26

Is the set $N = \{1, 2, 3, \ldots\}$ closed under the operation of subtraction?

Solution

No, some of the differences are not elements of the given set. For example, $3 - 5 = -2$ and $-2 \notin N$; $4 - 4 = 0$ and $0 \notin N$.

If we expand the set of natural numbers to include 0, we have $\{0, 1, 2, 3, \ldots\}$ and this is called the set of *whole numbers*. It was not until about 700 A.D. that 0 came to be accepted as a number. It should be noted also that those properties that hold for the natural numbers also hold for the set of whole numbers.

What happens when we add 0 to any other whole number? For example, $3 + 0 = ?$, $4 + 0 = ?$, $5 + 0 = ?$ The result is always the same (or identical) whole number. This is true for any whole number, that is, $x + 0 = 0 + x = x$. We call this particular property the *additive identity property* of 0. Zero is the identity element for the operation of addition.

Is there an identity element for multiplication? Is there a number such that when we multiply it by a second number, the result is always the second number? The answer is yes, and the number that has this property is 1. For example, $4 \cdot 1 = 1 \cdot 4 = 4$, $3 \cdot 1 = 1 \cdot 3 = 3$. Any natural number (or whole number) that is multiplied by 1 will result in that same natural number. Therefore, 1 is the identity element for multiplication.

Identity property: *addition* Zero is the additive identity element for the set of whole numbers, $W = \{0, 1, 2, 3, \ldots\}$. For every whole number x, $x + 0 = 0 + x = x$.

multiplication One is the multiplicative identity element for the set of natural numbers, $N = \{1, 2, 3, \ldots\}$. For every natural number x, $x \cdot 1 = 1 \cdot x = x$.

Note that 1 is also the multiplicative identity element for the set of whole numbers. Since 1 is an identity element for the natural numbers, it must

The Number System

also be for the whole numbers because the set of natural numbers is a subset of the set of whole numbers. But there is no additive identity element for the set of natural numbers, since 0 is not an element of the set of natural numbers.

EXAMPLE 27

Identify which property is illustrated by each of the following:
a. $3 + 4 = 4 + 3$.
b. $4(x + 2) = 4x + 4 \cdot 2$.
c. $3 + (4 + 5) = (3 + 4) + 5$.
d. $1 \cdot y = y \cdot 1 = y$.
e. $x(yz) = (xy)z$.
f. $x + 0 = 0 + x = x$.

Solution

a. $3 + 4 = 4 + 3$ — commutative property of addition.
b. $4(x + 2) = 4x + 4 \cdot 2$ — distributive property of multiplication over addition.
c. $3 + (4 + 5) = (3 + 4) + 5$ — associative property of addition.
d. $1 \cdot y = y \cdot 1 = y$ — identity property of multiplication.
e. $x(yz) = (xy)z$ — associative property of multiplication.
f. $x + 0 = 0 + x = x$ — identity property of addition.

EXAMPLE 28

Is the set of whole numbers, $W = \{0, 1, 2, \ldots\}$, closed under the operation of
a. addition?   b. multiplication?   c. subtraction?

Solution

a. Yes, the sum of any two whole numbers is also a whole number.
b. Yes, the product of any two whole numbers is also a whole number.
c. No, the difference of any two whole numbers is not necessarily a whole number. For example, $6 - 8 = -2$, and $-2 \notin W$.

EXERCISES FOR SECTION 1.3

(Answers given to exercises marked *)

1. Identify which property is illustrated by each of the following:
   *a. $48 + 72 = 72 + 48$.
   b. $12 \cdot 13 = 13 \cdot 12$.
   *c. $4(3 + 2) = 4 \cdot 3 + 4 \cdot 2$.
   d. $1 + (2 + 3) = (1 + 2) + 3$.
   *e. $2(3 \cdot 4) = (2 \cdot 3)4$.
   f. $4(2 - 3) = 4 \cdot 2 - 4 \cdot 3$.
   *g. $8 + 0 = 0 + 8 = 8$.
   h. $17 + 13 = 13 + 17$.
   *i. $4 \cdot 1 = 1 \cdot 4 = 4$.
   j. $4 \cdot 3 + 4 \cdot 5 = 4(3 + 5)$.

2. Determine whether each of the following is true or false.
   *a. The natural numbers are commutative under the operation of addition.
   b. The whole numbers are associative under the operation of multiplication.
   *c. The natural numbers are commutative under the operation of subtraction.
   d. The natural numbers are closed under the operation of division.
   *e. The whole numbers are closed under the operation of subtraction.
   f. The set of even natural numbers, $\{2, 4, 6, \ldots\}$, is closed under the operation of addition.
   *g. The set of odd natural numbers, $\{1, 3, 5, \ldots\}$, is closed under the operation of addition.
   h. The set of odd natural numbers, $\{1, 3, 5, \ldots\}$, is closed under the operation of multiplication.
   *i. The identity element for the set of whole numbers under the operation of addition is 1.
   j. The identity element for the set of natural numbers under the operation of addition is 0.

3. Use the commutative and/or distributive properties to find equivalent expressions for each of the following:
   *a. $x \cdot 3$
   b. $4 + y$
   *c. $7 \cdot 4 + 7 \cdot 2$
   d. $3 \cdot 2 - 4 \cdot 2$
   *e. $4y + 6y$
   f. $x \cdot 3 - 2 \cdot x$
   *g. $\pi x + \pi y$
   h. $(x + y)z$
   *i. $(x + y)2 + (x + y)4$
   j. $(x + y)x + (x + y)y$
   *k. $x^3 + 3x^2$
   l. $2xy^2 + 3xy^2$
   *m. $2x^3 y^2 z - x^3 + 3x^2$
   n. $4xyz + 8x^2yz + 12x^3y^2z^2$
   *o. $(x + y)(x + z) + (x + y)(x + y)$
   p. $(x - y)(x + z) + (x + y)(x + z)$

4. Match each statement with its correct identification. Indicate the correct property by means of its Roman numeral.
   *a. $x + (y + z) = (x + y) + z$.   I. Commutative property of addition
   b. $x + y = y + x$.   II. Commutative property of multiplication
   *c. $x(y + z) = xy + xz$.   III. Associative property of addition
   d. $x + 0 = 0 + x = x$.   IV. Associative property of multiplication
   *e. $x \cdot 1 = 1x = x$.   V. The distributive property
   f. $x(yz) = (xy)z$.   VI. The closure property
   *g. $xy = yx$.   VII. The additive identity property
   h. If $x \in N$ and $y \in N$, then VIII. The multiplicative identity property
       $x + y \in N$.

1.4 THE INTEGERS

Our system of numbers began with the set of natural numbers (counting numbers), $N = \{1, 2, 3, \ldots\}$, and then we added 0 to this set to form the

The Number System

set of whole numbers, W = {0, 1, 2, 3, ...}. The set of whole numbers differs from the set of natural numbers in that it contains 0 as an element.

Before we proceed to discuss the set of integers, recall the identity element under the operation of addition; it is 0. When we add 0 to any number, the result is the same or identical number: $1 + 0 = 1$, $0 + 2 = 2$, $10 + 0 = 10$, and so on.

The *additive inverse* of a number is the number that will yield the identity element when added to the original number. In other words, the additive inverse of 5 is −5 because $5 + (-5) = 0$, which is the identity element. When we add a number "−a" (negative a) to the number "a," the result is the identity element. We say that "−a" is the *inverse* or *additive inverse* of "a."

> For each number "a," there exists its opposite, "−a" (negative a), such that $a + (-a) = 0$ and $(-a) + a = 0$. "Negative a" is the additive inverse of "a" and "a" is the additive inverse of "−a."

The additive inverse of 4 is −4 because $4 + (-4) = 0$. The additive inverse of −2 is 2 because $(-2) + 2 = 0$. We can say that 4 is the opposite of −4 and −2 is the opposite of 2. What is the opposite, or additive inverse of 0? It is 0 because $0 + 0 = 0$. It is interesting to note that 0 is its own additive inverse, that is, it is its own opposite.

The set of whole numbers, together with their additive inverses, forms the set of *integers*. So, the set of integers is:

> Integers = {..., −3, −2, −1, 0, 1, 2, 3, ...}.

Thus far we have discussed three important sets of numbers: the set of natural numbers, the set of whole numbers, and the set of integers. What are the distinguishing characteristics of each? The set of natural numbers consists of the set of counting numbers, N = {1, 2, 3, ...}. The set of whole numbers expands the natural numbers to include 0, W = {0, 1, 2, 3, ...}. The set of integers is composed of the set of whole numbers together with their additive inverses, I = {..., −3, −2, −1, 0, 1, 2, 3, ...}. It should be noted that a number such as 5 can be classified as a natural number, a whole number, or an integer. Zero is not a natural number, but it is a whole number and an integer, while a number such as −5 can only be classified as an integer (at this point of our discussion). Since −5 is an integer, we can also describe it as a negative integer, while +5 is a positive integer. The positive sign is often omitted, and +5 can be written as 5. This should cause no confusion, since the positive integers are the same as the natural numbers. Is 0 a positive or negative integer? Zero is neither positive nor negative, but still it is an integer.

The integers can be pictured on a number line, as shown in Figure 1.

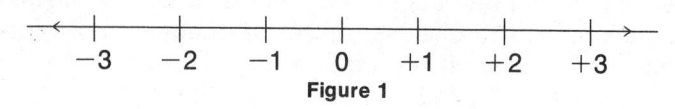

$$\leftarrow \;+\!\!-\!\!+\!\!-\!\!+\!\!-\!\!+\!\!-\!\!+\!\!-\!\!+\!\!-\!\!+\!\!-\!\!+\rightarrow$$
$$\;\;\;-3\;\;-2\;\;-1\;\;\;0\;\;+1\;\;+2\;\;+3$$

Figure 1

First, we draw a line, pick a point on the line, and label it 0. Next, we mark off equal units to the right and left of 0. The end points of the intervals to the right of 0 are labeled with the positive integers, and the end points of the intervals to the left of 0 are labeled with the negative integers. We can extend the number line indefinitely in either direction to the left or right of 0, hence the use of the "arrows" on the ends of the number line.

The place of the integers on the number line illustrates their relationship to each other, that is, their order. For example, "5 is greater than 3" (5 > 3), because 5 is to the right of 3 on the number line. For the same reason, 7 > 5, 10 > 8, and so on. Using this idea, we can also say that 0 is greater than −1 (0 > −1), for 0 is to the right of −1 on the number line. Also, −1 is greater than −9 (−1 > −9), −3 > −5, −4 > −12, and so on. As an alternate to stating that 0 > −1, we can reverse the statement: −1 < 0, which says that −1 is "less than" 0. We can say −1 < 0 because −1 is to the left of 0 on the number line. Similarly, instead of saying 5 > 3, we can say that 3 < 5.

EXAMPLE 29

Insert the correct symbol, > or <, in place of the question mark for each of the following:
a. 10 ? −2   b. 5 ? 7   c. 0 ? 2
d. 0 ? −3   e. −100 ? −101   f. −4 ? 0

Solution

a. 10 > −2 because 10 is to the right of −2 on the number line.
b. 5 < 7 because 5 is to the left of 7 on the number line.
c. 0 < 2 because 0 is to the left of 2 on the number line.
d. 0 > −3 because 0 is to the right of −3 on the number line.
e. −100 > −101 because −100 is to the right of −101 on the number line.
f. −4 < 0 because −4 is to the left of 0 on the number line.

COMBINING INTEGERS

How do we combine the integers? What are the solutions to the following problems: 2 + 1, 3 − 1, −2 − 3, (−2)(−4)? Suppose we want to

The Number System

find the sum of two positive integers, 2 + 1. We already know that 2 + 1 = 3, but let's use the number line to determine the answer. To begin, we start at 0 and proceed to the right (in a positive direction) two units to 2. To add 1, we proceed one more unit to the right, which brings us to 3. Hence 2 + 1 = 3. This procedure is illustrated in Figure 2.

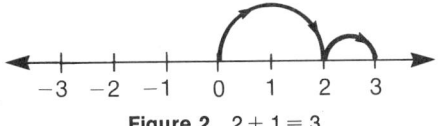

Figure 2 2 + 1 = 3.

Note that we dropped the "+" sign from the positive integers, since +1=1, +2=2, and so on. But remember that 0 also is written without a sign, and it is neither positive nor negative.

Adding the positive integers should not pose any problem since it is the same as addition of the natural numbers, which we have done most of our lives. The problem arises when we combine positive and negative integers. For example, let us evaluate −3 + 2. We can do this problem on the number line. Again, we begin at 0 and proceed three units to the left, to −3. Now we want to add 2, so we proceed two units to the *right* (+2) and this beings us to −1. Therefore, −3 + 2 = −1. This procedure is illustrated in Figure 3.

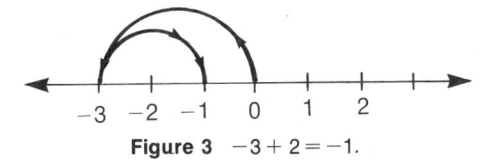

Figure 3 −3 + 2 = −1.

EXAMPLE 30

Evaluate 3 + (−5).

Solution

3 + (−5) = −2.

Analysis

We begin at 0 on the number line and proceed three units to the right to 3. Next we move five units in a negative direction (−5) to the left from 3, and this beings us to −2. This is illustrated in Figure 4.

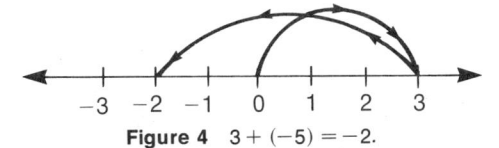

Figure 4 3 + (−5) = −2.

EXAMPLE 31

Evaluate $-3 + (-2)$.

Solution

$-3 + (-2) = -5.$

Analysis

Starting at 0 on the number line, we proceed three units to the left to -3. Next we move 2 units in a negative direction (-2) to the left from -3, and this brings us to -5. This is illustrated in Figure 5.

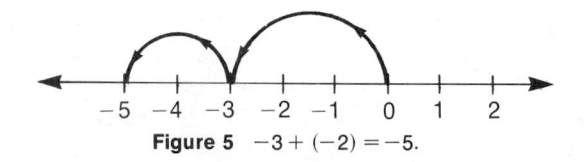

Figure 5 $-3 + (-2) = -5.$

So far we have added integers and see that the sum of two positive integers is a positive integer. If we add two negative integers, the sum is a negative integer. The sum of a positive integer and a negative integer may be a positive integer, a negative integer, or 0. Recall that the sum of an integer and its opposite is 0. Recall also that the set of integers is closed under the operation of addition.

To subtract integers, we can apply the same technique that we used for the addition of integers. The reason is that a problem such as $3 - 2$ can also be written as $3 + (-2)$. Thus, $3 - 2 = 1$ and $3 + (-2) = 1$, therefore, $3 - 2 = 3 + (-2)$. Other subtraction problems can be written in a similar manner, such as $5 - 3 = 5 + (-3)$, $10 - 7 = 10 + (-7)$. Subtracting one integer from another is the same as adding the opposite (additive inverse) of the second integer to the first integer. We have no trouble with a problem such as $5 - 3$, but consider other subtraction problems, such as $3 - 5$, $3 - (-5)$ and $-3 - (-5)$. These problems can be solved by remembering that subtraction is the same as adding the opposite of the second integer (the subtrahend). Consider the following examples:

EXAMPLE 32

Evaluate $3 - 5$.

Solution

$3 - 5 = -2.$

The Number System

Analysis

3 − 5 is the same as 3 + (−5). Using the number line, we see that 3 + (−5) = −2.

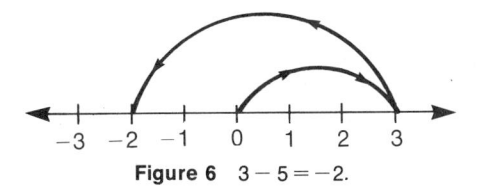

Figure 6 3 − 5 = −2.

EXAMPLE 33

Evaluate 3 − (−5).

Solution

3 − (−5) = 8.

Analysis

3 − (−5) is the same as 3 + (5) (adding the opposite of the second integer), and 3 + 5 = 8.

EXAMPLE 34

Evaluate −3 − (−5).

Solution

−3 − (−5) = 2.

Analysis

−3 − (−5) is the same as −3 + (5). Using the number line, we see that −3 + 5 = 2.

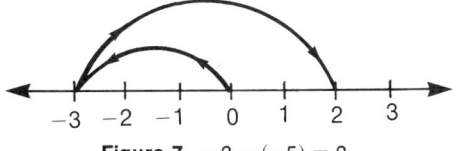

Figure 7 −3 − (−5) = 2.

Now that we have examined some examples of subtraction of integers, let's formally define subtraction of integers. Recall that in a subtraction problem such as 3 − 2 = 1, the 3 is called the "minuend," the 2 is called the "subtrahend," and the 1 is called the "difference."

Subtraction: In order to subtract two integers, add the opposite (additive inverse) of the subtrahend (second integer) to the minuend (first integer). Therefore, $x - y = x + (-y)$, where x and y represent any integers.

Many times in algebra we encounter problems that involve more than the addition or subtraction of two integers. Sometimes there are three or four integers to combine. For example, consider $3 - 5 + 4$. How do we evaluate such a problem? An efficient method is to apply the definition of subtraction first and then perform the resulting additions. Therefore, $3 - 5 + 4$ becomes $3 + (-5) + 4$. Now we can evaluate the problem $3 + (-5) + 4$ in a number of ways. For example, we could have

$$\big(3 + (-5)\big) + 4 \qquad 3 + \big((-5) + 4\big)$$
$$-2 \;+\; 4 \quad\text{OR}\quad 3 + (-1)$$
$$2 \qquad\qquad\qquad 2$$

In each instance we used the associative property of addition to evaluate the expression.

EXAMPLE 35

Evaluate $3 - (-5) + 4$.

Solution

$3 - (-5) + 4 = 12$.

Analysis

First, using the definition of subtraction, we change any subtractions to equivalent additions. Hence, $3 - (-5) + 4 = 3 + (5) + 4$ (subtraction is the same as adding the opposite of the subtrahend). $3 + (5) + 4 = 12$.

The Number System

> **EXAMPLE 36**
>
> Evaluate $-3 - (-5) - 4$.
>
> **Solution**
>
> $-3 - (-5) - 4 = -2$.
>
> **Analysis**
>
> Using the definition of subtraction, we change the subtractions to equivalent additions. (Note that -3 is not a subtraction.) Therefore, we have $-3 - (-5) - 4 = -3 + (5) + (-4) = 2 + (-4) = -2$.

Once we have mastered the operations of addition and subtraction of integers, we can proceed to an examination of the operation of multiplication of integers. Multiplication of natural numbers can be thought of as repeated additions of natural numbers. For example, $6 \cdot 2$ can also be thought of as six "2's," that is, $2 + 2 + 2 + 2 + 2 + 2$, for which the answer is 12. Using the commutative property of multiplication, $6 \cdot 2 = 2 \cdot 6$ and $2 \cdot 6 = 6 + 6 = 12$. We already know how to multiply natural numbers; since positive integers are also natural numbers, we note that a positive integer "times" a positive integer yields a positive integer. In other words, the positive integers are "closed" under the operation of multiplication.

Is the product positive or negative when we multiply a positive integer by a negative integer? For example, $2 \cdot (-6) = ?$ Since multiplication can be thought of as repeated addition, we can think of $2 \cdot (-6)$ as "two negative 6's," or, $2 \cdot (-6) = (-6) + (-6) = -12$. Instead of $2 \cdot (-6)$, suppose we had $(-2) \cdot 6$. Does this pose a problem? No, we can use the commutative property of multiplication, so $(-2) \cdot 6 = 6 \cdot (-2)$. $6 \cdot (-2)$ is the same as "six negative 2's," that is, $6 \cdot (-2) = -2 + (-2) + (-2) + (-2) + (-2) + (-2) = -12$. Therefore, $(-2) \cdot 6 = 6 \cdot (-2) = -12$. Note that a positive integer "times" a negative integer yields a negative integer. In other words, the product of a positive integer and a negative integer is negative.

Before proceeding further, it should be noted that 0 is an integer, and it is neither positive nor negative. What is the product of 0 and any integer? The product of 0 and any integer is 0:

> $x \cdot 0 = 0 \cdot x = 0$ (where x is any integer).

Next, let's consider the product of two negative integers. Is it positive or negative? We shall see that the product of two negative integers is positive. The distributive property states that $x(y + z) = xy + xz$, and we shall use this property in showing why the product of two negative integers is always positive. Consider $(-4) \cdot (0)$. We know that the product of 0 and any integer is 0. Now we shall rewrite 0 as an integer and its additive inverse: $0 = -2 + 2$, $0 = -3 + 3$, and so on. Therefore, we can say that:

$$(-4)(0) = (-4)(-3 + 3) = 0.$$

Using the distributive property, we have

$$(-4)(0) = (-4)(-3 + 3) = (-4)(-3) + (-4)(3) = 0.$$

From our previous discussion, we know that $(-4)(3) = -12$, and we have just shown that $(-4)(-3)$ and $(-4)(3)$ are additive inverses, since their sum must be 0. If $(-4)(3) = -12$, then $(-4)(-3)$ must equal 12, for the product of the two negative integers must be positive. Reviewing the steps, we have

$$(-4)(0) = 0.$$
$$(-4)(-3 + 3) = (-4)(-3) + (-4)(3) = 0.$$
$$(-4)(-3) + (-12) = 0.$$
$$12 + (-12) = 0.$$

We can summarize the various multiplication properties of integers:
1. The product of two positive integers is positive.
2. The product of a positive integer and a negative integer is negative.
3. The product of 0 and any integer is 0.
4. The product of two negative integers is positive.

EXAMPLE 37

Find the following products:
a. $3 \cdot (-2)$
b. $3 \cdot 2$
c. $(-3)(-2)$
d. $(-3)(-2 + 2)$

Solution

a. $3 \cdot (-2) = -6$. The product of a positive integer and a negative integer is negative.
b. $3 \cdot 2 = 6$. The product of two positive integers is positive.
c. $(-3) \cdot (-2) = 6$. The product of two negative integers is positive.
d. $(-3)(-2 + 2) = 0$. We can evaluate this problem in two different ways. Using the distributive property, we have $(-3)(-2+2) = (-3)(-2) + (-3)(2) = 6 + (-6) = 0$. Working inside the parentheses first, we have $(-3)(-2+2) = (-3)(0) = 0$; the product of 0 and any integer is 0.

The Number System

> **EXAMPLE 38**
>
> Find the following products:
> a. $(-2)(-3)(4)$
> b. $(-2)(3)(4)(-2)$
> c. $2(3)(-2)(4)$
>
> **Solution**
>
> Each of these examples involves finding the product of more than two integers. We can do this by using the commutative and associative properties of multiplication, that is, we can find the product of any two integers and perform repeated multiplications until we obtain the final product.
> a. $(-2)(-3)(4) = 6 \cdot 4 = 24$.
> b. $(-2)(3)(4)(-2) = (-6)(4)(-2) = (-24)(-2) = 48$ or
> $(-2)(3)(4)(-2) = (-2)(3)(-8) = (-2)(-24) = 48$.
> c. $2(3)(-2)(4) = 6(-2)(4) = 6(-8) = -48$ or
> $2(3)(-2)(4) = 2(3)(-8) = 2(-24) = -48$.

At this point, we shall briefly discuss the operation of division. For one thing, it should be noted that the set of integers is not closed under the operation of division. For example, $10 \div 3 = 3\frac{1}{3}$, which is not an integer.

> But division of integers does occur in algebra and it should be noted that:
> 1. The quotient of two positive integers is positive.
> 2. The quotient of two negative integers is positive.
> 3. The quotient of a positive and a negative integer (or a negative integer and a positive integer) is negative.

When 0 occurs in a division problem, either as the divisor or dividend (or both), it is often confusing for the student. Consider the following division problems.

$$\frac{0}{1} = ? \quad \frac{1}{0} = ? \quad \frac{0}{0} = ?$$

Again, we are using specific examples to illustrate what happens when 0 occurs in a division problem with another integer (including $0 \div 0$). What happens when we divide 0 by 1, that is, $\frac{0}{1}$ or $0 \div 1 = ?$ In dividing integers, note that

> if $\frac{x}{y} = z$, then $x = zy$ (providing $y \neq 0$).

Using this idea, then $\frac{0}{1} = z$ and $0 = z \cdot 1$. In order for this statement $(0 = z \cdot 1)$ to be true, z must equal 0. Therefore, $\frac{0}{1} = 0$.

Now consider 1 divided by 0: $1 \div 0$, $\frac{1}{0} = z$ and $1 = z \cdot 0$. What integer, z, when multiplied by 0 will result in a product of 1? There isn't any (it doesn't exist) because the product of 0 and any integer is 0. Therefore, $\frac{1}{0}$ has no solution and division by 0 is undefined, that is, $\frac{1}{0}$ is undefined.

We consider $\frac{0}{0}$ because it is a special case of division by 0. Consider $\frac{0}{0} = z$ and $0 = z \cdot 0$. What integer, z, when multiplied by 0 will result in a product of 0? We can substitute any integer for z: 0, 1, −1, 2, −2, and so on. Therefore, $\frac{0}{0}$ has no unique solution and we describe $\frac{0}{0}$ as a "meaningless" expression. It should be noted that in division, we must be sure not to divide by 0. That is why the restriction is placed on y in the statement "If $\frac{x}{y} = z$, then $x = zy$ (providing $y \neq 0$)." Division by zero is undefined.

EXAMPLE 39

Find the following quotients:

a. $\frac{0}{-2}$ b. $\frac{100}{0}$ c. $\frac{4}{3-3}$ d. $\frac{2-2}{1-1}$

Solution

a. $\frac{0}{-2} = 0$. 0 divided by any number (other than 0) will always equal 0.

b. $\frac{100}{0} =$ undefined. Division by zero is undefined.

c. $\frac{4}{3-3} = \frac{4}{0} =$ undefined.

d. $\frac{2-2}{1-1} = \frac{0}{0} =$ meaningless. ($0 \div 0$ is a meaningless expression.)

In this section we have examined the set of integers, $I = \{\ldots, -3, -2, -1, 0, 1, 2, 3, \ldots\}$ and how to combine integers under the various operations of arithmetic. It is important that we obtain the correct positive or negative sign when we add, subtract, multiply, or divide integers. It should

The Number System

be noted that the set of integers is closed under the operations of addition, subtraction, and multiplication, but not closed under the operation of division.

EXERCISES FOR SECTION 1.4

(Answers given to exercises marked *)

1. Determine the additive inverse (opposite) for each of the following integers.
   *a. 2   b. −3   *c. 100   d. −10   *e. 0   f. +10

2. The set of integers contains an additive inverse for each integer. Does it also contain a multiplicative inverse for each integer? Explain your answer.

3. Given the set of natural numbers, N = {1, 2, 3, ...}, the set of whole numbers, W = {0, 1, 2, 3, ...} and the set of integers, I = {..., −3, −2, −1, 0, 1, 2, 3, ...}, determine whether the following statements are true or false.
   *a. 2 ∈ N   b. 2 ∈ W   *c. 2 ∈ I
   d. −3 ∈ N   *e. −3 ∈ W   f. −3 ∈ I
   *g. 0 ∉ N   h. 0 ∉ W   *i. 0 ∉ I
   j. −10 ∉ N   *k. −10 ∈ W   l. −10 ∉ I
   *m. N ⊂ I   n. W ⊂ I   *o. I ⊂ N

4. Insert the correct symbol, >, <, or =, in place of the question mark in each of the following:
   *a. −2 ? 10   b. −2 ? −3   *c. −10 ? 0
   d. −2 + 5 ? 3   *e. −3 − 2 ? −6 + 2   f. (−2)(3) ? −7
   *g. (−2)(−3) ? 2·3   h. (−4)(−3) ? −12   *i. (−2)(−3)(−1) ? 6
   j. −6 ÷ 2 ? −3   *k. (−6) ÷ (−2) ? 3   l. 0 ÷ 2 ? 2

5. Evaluate each of the following:
   *a. 3 + (−4)   b. 5 + (−8)   *c. 6 + (−9)
   d. −3 + 4   *d. −5 + 8   f. −6 + 9
   *g. 3 − 4   h. 5 − 8   *i. 6 − 9
   j. 3 − (−4)   *k. 5 − (−8)   l. 6 − (−9)
   *m. −3 − (−4)   n. −5 − (−8)   *o. −6 − (−9)
   p. 3 − (−5) + 2   *q. −3 + (−2) + 4   r. −10 − (−5) + 5
   *s. −2 − 3 + 4   t. −3 + 5 − (−2)   *u. −1 + 1 − (−1)

6. Evaluate each of the following:
   *a. 3 · (−3)   b. (−3) · (−2)
   *c. (−3) · 2   d. (−4)(3)(−2)
   *e. (−1)(3)(−2)   f. (−2)(−1)(−2)(−1)
   *g. (−3)(5 − 5)   h. (−2)(−5 + 3 + 2)
   *i. (−1)(−3 + 2)   j. (−8 + 2)(−3)
   *k. (−6 + 2 − 1)(−3 + 1)   l. (−2 − 1)(−3 + 2)
   *m. −6 ÷ −3   n. (−2 + 5) ÷ (−2 + 1)
   *o. (−2 − 2) ÷ (−4)   p. (−3 + 7) ÷ (−5 + 1)
   *q. (−8 + 3) ÷ (−5 + 5)   r. (−2 − 3) ÷ (−2 + 2)

7. Determine whether each of the following statements is true or false.
   *a. The set of integers is closed under the operation of addition.
   b. The set of integers is closed under the operation of division.
   *c. Addition is a commutative operation for the set of integers.
   d. Subtraction is a commutative operation for the set of integers.
   *e. Zero is the additive inverse of zero.
   f. Zero is the identity element for the set of integers under the operation of addition.
   *g. The quotient of 0 divided by any integer is 0.
   h. Division is a commutative operation for the set of integers.
   *i. Multiplication is a commutative operation for the set of integers.
   j. Subtraction is an associative operation for the set of integers.

8. Find a replacement (integer) for x that will make each of the following open sentences a true statement.
   *a. $x + 2 = 1$.
   b. $3x = -6$
   *c. $x - 7 = -3$.
   d. $x - 1 = -1$.
   *e. $-3 + 2 = x - 1$.
   f. $-2 + 1 = 1 - x$.
   *g. $-3 - 2x = 3$.
   h. $-4 + 2 = -3 - x$.
   *i. $-5 - 3 = 4 - 4x$.

1.5 THE RATIONAL NUMBERS

In the previous section, we discussed the set of integers and used the number line to add integers. The number line pictures the relationship between natural numbers, whole numbers and integers. Are there other numbers on the number line?

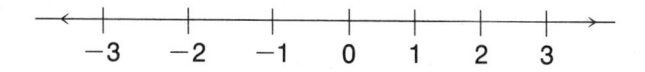

$$\xleftarrow{\quad\;\;|\quad\;\;|\quad\;\;|\quad\;\;|\quad\;\;|\quad\;\;|\quad\;\;|\quad\;\;}\rightarrow$$
$$\;\;-3\;\;-2\;\;-1\;\;\;0\;\;\;1\;\;\;2\;\;\;3$$

The answer is yes. Consider the number $\frac{1}{2}$. It is greater than 0 and less than 1, and is in the interval between 0 and 1. A number such as $\frac{1}{2}$ is not a natural number, not a whole number, and not an integer. It is a *rational number*.

> A rational number is any number that can be written in the form $\frac{x}{y}$ ($y \neq 0$), where x and y represent integers.

Any number that can be expressed as the quotient of two integers is a rational number. For example, $\frac{3}{4}, \frac{-2}{3}, \frac{5}{-2}, \frac{-3}{1}, \frac{0}{2}$, and $\frac{5}{1}$, are all rational

The Number System

numbers. The only restriction in forming rational numbers is that the divisor cannot be 0; otherwise we can use any integers. Note that numbers such as 4 and −4 are rational numbers. They can be written in the form $\frac{x}{y}$, where x and y are integers, namely, $4 = \frac{4}{1}$ and $-4 = \frac{-4}{1}$. Therefore, any natural number, any whole number, and any integer are rational numbers. But not every rational number is a natural number, or a whole number, or an integer. For example, $\frac{1}{2}$ is a rational number, but $\frac{1}{2} \notin N$, $\frac{1}{2} \notin W$, and $\frac{1}{2} \notin I$. Seeing rational numbers such as $\frac{1}{2}$ and $\frac{3}{4}$, frequently students think that a rational number has to be a fraction. But that is not the case; integers are rational numbers, too.

When we find the quotient of two integers, it will be either a terminating decimal or a repeating decimal. For example,

Terminating decimal

$$\frac{1}{4} = 0.25$$

$$\frac{1}{8} = 0.125$$

$$\frac{10}{2} = 5.0$$

Repeating decimal

$$\frac{1}{3} = 0.33\overline{3}$$

$$\frac{2}{3} = 0.66\overline{6}$$

$$\frac{23}{99} = 0.\overline{2323}$$

From these examples, we can see that a rational number can also be described as a terminating decimal or a repeating decimal. (For the repeating decimals, the bar above the 3 indicates that it repeats endlessly, and the bar above the 23 indicates that the digits 23 repeat endlessly.)

In order to express a rational number as a decimal, we must divide the numerator of the fraction by its denominator. Consider the rational number $\frac{2}{11}$. To express $\frac{2}{11}$ as a decimal, we must divide 2 by 11:

Note that we could have stopped dividing here because the remainder starts to repeat itself at this point.

```
          .1818
      11/2.0000
          1 1
            90
            88
            20
            11
            90
            88
             2
```

Hence, $\frac{2}{11} = 0.\overline{18}$

EXAMPLE 40

Express $\frac{3}{11}$ as a decimal.

Solution

In order to find the decimal equivalent for a rational number in the form $\frac{x}{y}$, we divide x by y. Therefore:

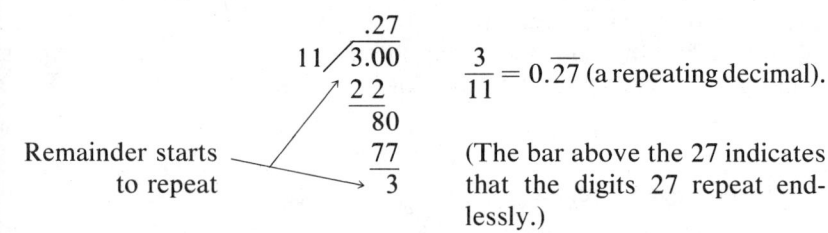

$$11\overline{)3.00}.27$$

Remainder starts to repeat

$\frac{3}{11} = 0.\overline{27}$ (a repeating decimal).

(The bar above the 27 indicates that the digits 27 repeat endlessly.)

EXAMPLE 41

Express $\frac{7}{8}$ as a decimal.

Solution

$$8\overline{)7.000}.875$$

A remainder of 0 terminates the division

$\frac{7}{8} = 0.875$ (a terminating decimal).

EXAMPLE 42

Express $\frac{2}{13}$ as a decimal.

Solution

This example illustrates that a repeating decimal may not start repeating until after a large number of divisions.

The Number System

```
        .153846
   13 /2.000000
      1 3
      ‾‾‾
       70
       65
       ‾‾
        50
        39
        ‾‾
        110
        104
        ‾‾‾
          60
          52
          ‾‾
           80
           78
           ‾‾
            20
```

Remainder starts to repeat

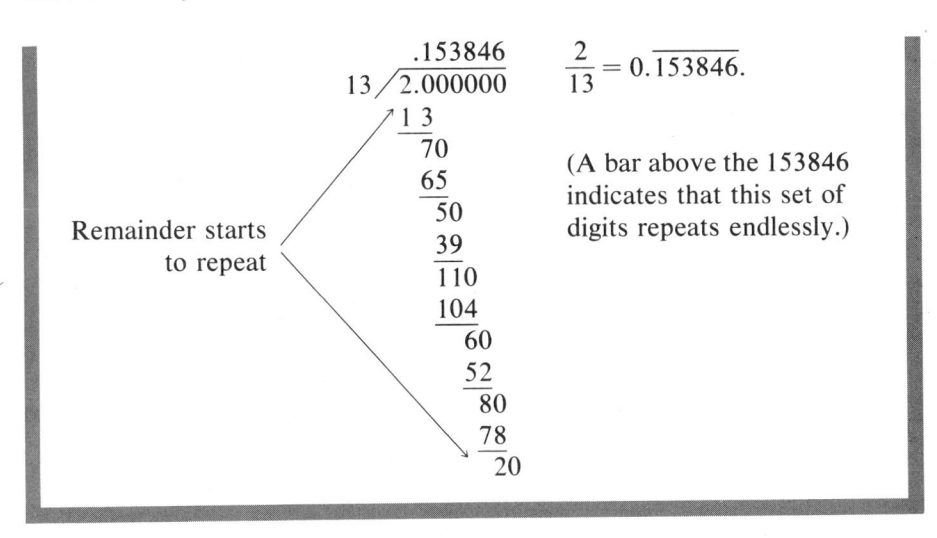

$\frac{2}{13} = 0.\overline{153846}$.

(A bar above the 153846 indicates that this set of digits repeats endlessly.)

Every rational number can be expressed as a terminating or a repeating decimal. It is also possible to express every terminating decimal and every repeating decimal as a quotient of integers, $\frac{x}{y}$ ($y \neq 0$). For example, $.5 = \frac{5}{10} = \frac{1}{2}$, $.25 = \frac{25}{100} = \frac{1}{4}$, and $.125 = \frac{125}{1000} = \frac{1}{8}$. In order to express a terminating decimal as a quotient of integers, we simply place the numerator over the denominator that is indicated by the decimal point, as we did in the illustrative examples. But how do we express a repeating decimal such as $0.\overline{33}$ as a quotient of integers? Most of us are aware that $0.\overline{33} = \frac{1}{3}$, but how do we obtain such an answer? In order to express $0.\overline{33}$ as a quotient of integers, we set up an equation:

$$x = 0.\overline{33}.$$

Now, since there is only one digit repeating, we multiply the equation by 10. If there are two digits repeating, then multiply by 100, if three digits repeating, multiply by 1000, and so on. Therefore, we now have

$10x = 3.\overline{33}.$ (Multiplying by 10, move the decimal point one place to the right.)

Note that in the equation $10x = 3.\overline{33}$, we extended the number of repeating digits in order to keep the same number of decimal places. Next, we subtract the first equation from the second (being careful to line up the decimal points):

$$\begin{array}{r} 10x = 3.\overline{33} \\ x = 0.\overline{33} \\ \hline 9x = 3 \end{array}$$

(subtracting x from 10x and $0.\overline{33}$ from $3.\overline{33}$)

$$x = \frac{3}{9} = \frac{1}{3} \quad \text{(solving for x)}$$

Therefore, $0.\overline{33} = \frac{1}{3}$.

Is $5.1\overline{23}$ a rational number? The answer is yes, since $5.1\overline{23}$ is a repeating decimal. Therefore, it can be expressed as the quotient of two integers. In order to express $5.1\overline{23}$ as a quotient of two integers, we let $x = 5.1\overline{23}$. Next we multiply this equation by 100 (two digits repeating). Therefore $100x = 512.3\overline{23}$ (extending the repeating cycle in order to keep the same number of decimal places). We now have:

$$100x = 512.3\overline{23}$$
$$x = 5.1\overline{23}$$

subtraction gives us $\quad 99x = 507.2$

solving for x, $\quad x = \dfrac{507.2}{99}$

Note, this answer is NOT CORRECT! $\dfrac{507.2}{99}$ is not the quotient of two integers: 507.2 is a decimal, not an integer. How can we eliminate the decimal point in the quotient $\dfrac{507.2}{99}$? There are two ways to do this. One way is to multiply the $\dfrac{507.2}{99}$ by 1, but in this case we let $1 = \dfrac{10}{10}$, since we have one decimal place, that is, tenths. Therefore, we would have

$$\frac{507.2}{99} \cdot \frac{10}{10} = \frac{5072}{990} = \frac{2536}{495} \quad \text{and} \quad \frac{5072}{990} = 5.1\overline{23}.$$

Another way to eliminate the decimal point is by eliminating it in the equation $99x = 507.2$. Here we simply multiply each side of the equation by 10 (one decimal place, or tenths) and the result is $990x = 5072$. Next we solve for x,

$$x = \frac{5072}{990} = 5.1\overline{23}.$$

EXAMPLE 43

Express $0.\overline{4}$ as a quotient of two integers.

Solution

Let $x = 0.\overline{4}$; multiply both sides of the equation by 10 (one digit repeating). Therefore,

The Number System

$$10x = 4.\overline{4}$$
$$x = 0.\overline{4}$$ (extending the repeating cycle in order to keep the same number of decimal places)

subtracting $\quad 9x = 4$

solving for x $\quad x = \dfrac{4}{9}$

Therefore, $\quad \dfrac{4}{9} = 0.\overline{4}.$

EXAMPLE 44

Express $1.\overline{23}$ as a quotient of two integers.

Solution

Let $x = 1.\overline{23}$; multiply both sides of the equation by 100 (two digits repeating). Therefore,

$$100x = 123.\overline{23}$$ (placing the decimal points in the same position)
$$x = 1.\overline{23}$$

subtracting $\quad 99x = 122$

solving for x $\quad x = \dfrac{122}{99}$

Therefore, $\quad \dfrac{122}{99} = 1.\overline{23}.$

EXAMPLE 45

Express $1.2\overline{345}$ as a quotient of two integers.

Solution

Let $x = 1.2\overline{345}$; multiply both sides of the equation by 100 (two digits repeating). Therefore,

$$100x = 123.45\overline{45}$$
$$x = 1.23\overline{45}$$

subtracting $\quad 99x = 122.22.$

Note that the resulting equation still contains a decimal: 122.22. It must be eliminated or we will not have the quotient of two integers. We can multiply both sides of the equation, $99x = 122.22$ by 100. Therefore, the resulting equation is $9900x = 1222$.

solving for x $\quad x = \dfrac{12222}{9900} = \dfrac{679}{550}.$

Therefore, $\quad \dfrac{679}{550} = 1.23\overline{45}.$

Fractions cause many problems for mathematics students. They are troublesome because of the different procedures of the various operations used in combining fractions. Let us review these procedures, since many rational numbers occur in the form of a fraction.

A rational number can be expressed in different ways, for there are many symbols that represent the same number. For example, $\frac{1}{2}, \frac{2}{4}, \frac{4}{8}, \frac{8}{16}$, etc., all represent the same rational number, 0.5. Two rational numbers, $\frac{a}{b}$ and $\frac{c}{d}$, are equal if and only if $a \cdot d = b \cdot c$.

$$\frac{2}{4} = \frac{8}{16} \text{ if and only if } 2 \cdot 16 = 4 \cdot 8$$

Using this definition, we can easily determine whether or not two rational numbers are equal. For example, does $\frac{4}{9} = \frac{12}{27}$? The answer is yes, since $4 \cdot 27 = 9 \cdot 12$; $108 = 108$.

How do we combine rational numbers that are in the form of fractions? For example, what is the sum of $\frac{2}{3} + \frac{4}{7}$? Instead of worrying about the "least common denominator," we shall adopt the following general rule:

For any two rational numbers, $\frac{a}{b}$ and $\frac{c}{d}$,

$$\frac{a}{b} + \frac{c}{d} = \frac{ad + bc}{bd}.$$

Since $\frac{a}{b}$ and $\frac{c}{d}$ are given as rational numbers, $b \neq 0$ and $d \neq 0$. Using this definition, we can find the sum of $\frac{2}{3}$ and $\frac{4}{7}$:

$$\frac{2}{3} + \frac{4}{7} = \frac{2 \cdot 7 + 3 \cdot 4}{3 \cdot 7} = \frac{14 + 12}{21} = \frac{26}{21}.$$

Subtraction of rational numbers can be performed in a manner similar to addition. Recall that subtraction is the same as adding the additive inverse of the subtrahend. Consider the problem $\frac{2}{3} - \frac{4}{7}$. In order to perform the indicated subtraction, we add the additive inverse of $\frac{4}{7}$:

$$\frac{2}{3} - \frac{4}{7} = \frac{2}{3} + \frac{-4}{7} \quad \left(\text{the additive inverse of } \frac{4}{7} \text{ is } \frac{-4}{7}\right)$$

The Number System

and

$$\frac{2}{3} + \frac{-4}{7} = \frac{2 \cdot 7 + 3(-4)}{3 \cdot 7} = \frac{14 - 12}{21} = \frac{2}{21}.$$

EXAMPLE 46

Add: $\frac{4}{7} + \frac{3}{5}$.

Solution

Applying the definition that $\frac{a}{b} + \frac{c}{d} = \frac{ad + bc}{bd}$,

we have $\frac{4}{7} + \frac{3}{5} = \frac{4 \cdot 5 + 7 \cdot 3}{7 \cdot 5} = \frac{20 + 21}{35} = \frac{41}{35}.$

EXAMPLE 47

Subtract: $\frac{4}{7} - \frac{3}{5}$.

Solution

The additive inverse of $\frac{3}{5}$ is $\frac{-3}{5}$, therefore

$$\frac{4}{7} - \frac{3}{5} = \frac{4}{7} + \frac{-3}{5} = \frac{4 \cdot 5 + 7(-3)}{7 \cdot 5} = \frac{20 - 21}{35} = \frac{-1}{35}.$$

EXAMPLE 48

Subtract: $\frac{2}{3} - \frac{-2}{7}$.

Solution

The additive inverse of $\frac{-2}{7}$ is $\frac{2}{7}$, therefore,

$$\frac{2}{3} - \frac{-2}{7} = \frac{2}{3} + \frac{2}{7} = \frac{2 \cdot 7 + 3 \cdot 2}{3 \cdot 7} = \frac{14 + 6}{21} = \frac{20}{21}.$$

Multiplication is probably the least confusing operation to perform with rational numbers. A rational number is a number that may be expressed as the quotient of two integers, $\frac{a}{b}$ (b ≠ 0). Therefore, in multiplication of rational numbers, the products are products of integers. In order to multiply two rational numbers (fractions), we multiply numerator times numerator and denominator times denominator.

> For any rational numbers, $\frac{a}{b}$ and $\frac{c}{d}$,
>
> $$\frac{a}{b} \cdot \frac{c}{d} = \frac{a \cdot c}{b \cdot d} \quad (b \neq 0, d \neq 0).$$

In order to find the product of $\frac{2}{3} \cdot \frac{5}{7}$, we simply multiply the numerators together to obtain the numerator of the product and then multiply the denominators together to obtain the denominator of the product. Therefore,

$$\frac{2}{3} \cdot \frac{5}{7} = \frac{2 \cdot 5}{3 \cdot 7} = \frac{10}{21}.$$

EXAMPLE 49

Multiply: $\frac{-2}{3} \cdot \frac{-5}{-7}$.

Solution

Using the definition, $\frac{a}{b} \cdot \frac{c}{d} = \frac{a \cdot c}{b \cdot d}$, we have

$$\frac{-2}{3} \cdot \frac{-5}{-7} = \frac{(-2) \cdot (-5)}{3 \cdot (-7)} = \frac{10}{-21} = -\frac{10}{21}.$$

Consider the following multiplication problem.

$$\frac{a}{b} \cdot \frac{b}{a} \text{ where } a \neq 0 \text{ and } b \neq 0.$$

Performing the indicated multiplication, we have $\frac{a \cdot b}{b \cdot a}$. Since b and a are integers, we can apply the commutative property of multiplication, $\frac{a \cdot b}{b \cdot a} = \frac{a \cdot b}{a \cdot b} = 1$. Since 1 is the identity element for multiplication, we can say that $\frac{b}{a}$ is the multiplicative inverse of $\frac{a}{b}$ ($a \neq 0$, $b \neq 0$). Two rational numbers whose product is 1 are called the multiplicative inverses of each

The Number System

other. For example, $\frac{1}{2}$ is the multiplicative inverse of 2 because $\frac{1}{2} \cdot 2 = 1$. $\frac{-3}{2}$ is the multiplicative inverse of $\frac{-2}{3}$ because $\frac{-3}{2} \cdot \frac{-2}{3} = \frac{6}{6} = 1$. The multiplicative inverse of y is $\frac{1}{y}$ (y ≠ 0) because $y \cdot \frac{1}{y} = \frac{y}{y} = 1$. It should be noted that 0 does not have a multiplicative inverse. Also note that some texts refer to multiplicative inverses as "reciprocals."

> The multiplicative inverse of a rational number $\frac{a}{b}$ is $\frac{b}{a}$ (a ≠ 0).

The multiplicative inverse of a rational number is useful in developing the definition for division of rational numbers. Consider the following fraction, $\frac{\frac{2}{3}}{\frac{5}{7}}$. This fraction is called a "complex" fraction because the numerator or denominator (or both) of the fraction is also a fraction. The fraction would not be complex if its denominator were 1. In order to convert the denominator $\left(\frac{5}{7}\right)$ to 1, we multiply it by its multiplicative inverse, that is, $\frac{7}{5}$. But if we multiply the denominator by $\frac{7}{5}$, then we must do the same to the numerator (using the property of the multiplicative identity). Therefore,

$$\frac{\frac{2}{3}}{\frac{5}{7}} = \frac{\frac{2}{3} \cdot \frac{7}{5}}{\frac{5}{7} \cdot \frac{7}{5}} = \frac{\frac{2 \cdot 7}{3 \cdot 5}}{\frac{5 \cdot 7}{7 \cdot 5}} = \frac{\frac{14}{15}}{\frac{35}{35}} = \frac{\frac{14}{15}}{1} = \frac{14}{15}.$$

$$\frac{\frac{a}{b}}{\frac{c}{d}} = \frac{\frac{a}{b} \cdot \frac{d}{c}}{\frac{c}{d} \cdot \frac{d}{c}} = \frac{\frac{a \cdot d}{b \cdot c}}{\frac{c \cdot d}{d \cdot c}} = \frac{\frac{a \cdot d}{b \cdot c}}{1} = \frac{a}{b} \cdot \frac{d}{c} \quad (b \neq 0, c \neq 0).$$

The complex fraction $\frac{\frac{2}{3}}{\frac{5}{7}}$ is another way of stating $\frac{2}{3} \div \frac{5}{7}$; the fraction indicates division. Simplifying a complex fraction is the same as dividing two rational numbers. From the illustrative example, we can see that, in order to divide two rational numbers, we multiply the first rational number (the dividend) by the multiplicative inverse of the second rational number (the divisor).

For any rational numbers, $\frac{a}{b}$ and $\frac{c}{d}$, we can state that

$$\frac{a}{b} \div \frac{c}{d} = \frac{a}{b} \cdot \frac{d}{c} \quad (b \neq 0, c \neq 0).$$

EXAMPLE 50

Divide: $\frac{-2}{3} \div \frac{-5}{-7}$.

Solution

Using the definition, $\frac{a}{b} \div \frac{c}{d} = \frac{a}{b} \cdot \frac{d}{c}$, we have

$$\frac{-2}{3} \div \frac{-5}{-7} = \frac{-2}{3} \cdot \frac{-7}{-5} = \frac{(-2) \cdot (-7)}{3 \cdot (-5)} = \frac{14}{-15} = -\frac{14}{15}.$$

EXAMPLE 51

Divide: $\frac{-4}{-5} \div 2$.

Solution

$$\frac{-4}{-5} \div 2 = \frac{-4}{-5} \cdot \frac{1}{2} = \frac{(-4) \cdot 1}{(-5) \cdot 2} = \frac{-4}{-10} = \frac{4}{10} = \frac{2}{5}.$$

The set of rational numbers satisfies the properties that we discussed earlier in this chapter. For example, the *closure* property holds for the set of rationals under all of the operations discussed—addition, subtraction, multiplication, and division. For any two rational numbers, $\frac{a}{b}$ and $\frac{c}{d}$ (where $b \neq 0$ and $d \neq 0$), the operation of:

addition, $\frac{a}{b} + \frac{c}{d}$, yields a rational number, $\frac{ad + bc}{bd}$;

subtraction, $\frac{a}{b} - \frac{c}{d}$, yields a rational number, $\frac{ad - bc}{bd}$;

multiplication, $\frac{a}{b} \cdot \frac{c}{d}$, yields a rational number, $\frac{ac}{bd}$;

division, $\frac{a}{b} \div \frac{c}{d} = \frac{a}{b} \cdot \frac{d}{c}$, yields a rational number, $\frac{ad}{bc}$ (if $c \neq 0$).

The Number System

The *commutative* property holds for the set of rational numbers for the operations of addition and multiplication, but not subtraction and division. For any two rational numbers, $\frac{a}{b}$ and $\frac{c}{d}$ (where $b \neq 0$ and $d \neq 0$), for addition we have $\frac{a}{b} + \frac{c}{d} = \frac{c}{d} + \frac{a}{b}$, and for multiplication we have $\frac{a}{b} \cdot \frac{c}{d} = \frac{c}{d} \cdot \frac{a}{b}$.

The *associative* property holds for the set of rational numbers for the operations of addition and multiplication, but not subtraction or division. For any three rational numbers, $\frac{a}{b}, \frac{c}{d}$, and $\frac{e}{f}$ (where b, d, and $f \neq 0$):

addition: $\left(\frac{a}{b} + \frac{c}{d}\right) + \frac{e}{f} = \frac{a}{b} + \left(\frac{c}{d} + \frac{e}{f}\right)$;

multiplication: $\left(\frac{a}{b} \cdot \frac{c}{d}\right) \cdot \frac{e}{f} = \frac{a}{b} \left(\frac{c}{d} \cdot \frac{e}{f}\right)$.

There exists an *identity element* in the set of rational numbers for the operations of addition and multiplication. The *additive identity* element is 0 or $\frac{0}{d}$ ($d \neq 0$) and the *multiplicative identity* element is 1 or $\frac{c}{c}$ ($c \neq 0$), therefore,

addition, $\frac{a}{b} + 0 = 0 + \frac{a}{b} = \frac{a}{b}$, and for

multiplication, $\frac{a}{b} \cdot 1 = 1 \cdot \frac{a}{b} = \frac{a}{b}$.

Every rational number, $\frac{a}{b}$ ($b \neq 0$) has an *additive inverse* and every rational number, except 0, has a *multiplicative inverse*:

addition: $\frac{a}{b} + \frac{-a}{b} = \frac{a+(-a)}{b} = \frac{0}{b} = 0$.

multiplication: $\frac{a}{b} \cdot \frac{b}{a} = \frac{ab}{ba} = 1$ ($b \neq 0, a \neq 0$).

The distributive property also is valid for the set of rational numbers. As an example, consider the left distributive property of multiplication over addition, that is, for any three rational numbers, $\frac{a}{b}, \frac{c}{d}$ and $\frac{e}{f}$ ($b \neq 0, d \neq 0, f \neq 0$),

$$\frac{a}{b} \cdot \left(\frac{c}{d} + \frac{e}{f}\right) = \frac{a}{b} \cdot \frac{c}{d} + \frac{a}{b} \cdot \frac{e}{f}.$$

One property of the set of rational numbers that natural numbers, whole numbers, and integers do not possess, is the density property. Between *any* two rational numbers, there is another rational number. Because of this density property, we can say that the rational numbers are *dense*. For example, consider the rational numbers $\frac{5}{11}$ and $\frac{6}{11}$. Can we find another rational number $\frac{a}{b}$, such that $\frac{5}{11} < \frac{a}{b} < \frac{6}{11}$. The answer is yes. We can do it in the following manner: first, find the mean (average) of the two numbers. In other words, find the sum of the two given numbers, then divide by 2 $\left(\text{or multiply by } \frac{1}{2}\right)$. Therefore,

$$x = \frac{1}{2} \cdot \left(\frac{5}{11} + \frac{6}{11}\right) = \frac{1}{2}\left(\frac{11}{11}\right) = \frac{11}{22}.$$

To verify that $\frac{5}{11} < \frac{11}{22} < \frac{6}{11}$, we can change $\frac{5}{11}$ to $\frac{10}{22}$ and $\frac{6}{11}$ to $\frac{12}{22}$. We now have $\frac{10}{22} < \frac{11}{22} < \frac{12}{22}$. There are other rational numbers between $\frac{5}{11}$ and $\frac{6}{11}$, such as $\frac{23}{44}$, but we have proved the existence of at least one rational number "between" two given rational numbers.

The density property does not hold for the natural numbers, whole numbers, or integers. For example, there is no whole number between the whole numbers 0 and 1. Given any two rational numbers $\frac{a}{b}$ and $\frac{c}{d}$, we can always find another rational number between the two given numbers by finding the arithmetic mean of the given numbers. If we let x equal the rational number we are seeking, then

$$x = \frac{1}{2} \cdot \left(\frac{a}{b} + \frac{c}{d}\right).$$

EXAMPLE 52

Find a rational number between $\frac{4}{7}$ and $\frac{5}{8}$.

Solution

$$x = \frac{1}{2} \cdot \left(\frac{4}{7} + \frac{5}{8}\right).$$

The Number System

$$x = \frac{1}{2} \cdot \frac{(4 \cdot 8 + 7 \cdot 5)}{7 \cdot 8} = \frac{1}{2} \cdot \left(\frac{32 + 35}{56}\right).$$

$$x = \frac{1}{2} \cdot \left(\frac{67}{56}\right) = \frac{67}{112}.$$

Check:

$$\frac{4}{7} = \frac{64}{112} \text{ and } \frac{5}{8} = \frac{70}{112}. \qquad \frac{64}{112} < \frac{67}{112} < \frac{70}{112}.$$

There are other rational numbers between $\frac{4}{7}$ and $\frac{5}{8}$, but by using the formula for the arithmetic mean, $x = \frac{1}{2} \cdot \left(\frac{a}{b} + \frac{c}{d}\right)$, we have found a rational number that is exactly halfway between $\frac{4}{7}$ and $\frac{5}{8}$.

EXERCISES FOR SECTION 1.5

(Answers given to exercises marked *)

1. Express each of the following as a decimal.

   *a. $\frac{1}{8}$ b. $\frac{7}{16}$ *c. $\frac{5}{6}$

   d. $\frac{2}{11}$ *e. $\frac{3}{11}$ f. $\frac{1}{33}$

   *g. $\frac{2}{33}$ h. $\frac{10}{37}$ *i. $\frac{5}{13}$

   j. $\frac{7}{13}$ *k. $\frac{1}{7}$ l. $\frac{5}{7}$

2. Express each of the following as a quotient of integers.
   *a. 0.75   b. 0.875   *c. 0.025
   d. 0.$\overline{7}$ *e. 0.$\overline{8}$ f. 0.$\overline{9}$
   *g. 0.$\overline{45}$ h. 0.$\overline{13}$ *i. 0.$\overline{123}$
   j. 2.$\overline{147}$ *k. 0.1$\overline{35}$ l. 2.1$\overline{34}$
   *m. 4.59$\overline{6}$ n. 3.1$\overline{45}$ *o. 2.4$\overline{9}$

3. Perform the indicated operation for each of the following:

   *a. $\frac{2}{3} + \frac{4}{5}$ b. $\frac{3}{7} + \frac{-4}{5}$ *c. $-\frac{1}{2} + \frac{7}{8}$

d. $\dfrac{-3}{4} - \dfrac{-2}{5}$ *e. $\dfrac{-2}{3} - \dfrac{4}{5}$ f. $\dfrac{3}{7} - \dfrac{-4}{5}$

*g. $\dfrac{2}{3} \cdot \dfrac{4}{5}$ h. $\dfrac{3}{7} \cdot \left(\dfrac{-4}{5}\right)$ *i. $\left(-\dfrac{1}{2}\right) \cdot \left(\dfrac{-2}{5}\right)$

j. $\dfrac{3}{7} \div \dfrac{-4}{5}$ *k. $-\dfrac{1}{2} \div \dfrac{-3}{5}$ l. $\dfrac{-2}{5} \div \dfrac{-5}{2}$

4. Replace the question mark with the appropriate symbol, =, >, or <, that makes each of the following a true statement. You may find it helpful to use the multiplicative identity, that is, convert the fractions under consideration so that both have the same denominator.

*a. $\dfrac{1}{2}$? $\dfrac{1}{3}$ b. $\dfrac{3}{7}$? $\dfrac{4}{9}$ *c. $\dfrac{5}{11}$? $\dfrac{6}{12}$

d. $\dfrac{5}{11}$? $\dfrac{6}{13}$ *e. $\dfrac{4}{5}$? $\dfrac{36}{45}$ f. $\dfrac{3}{7}$? $\dfrac{5}{11}$

*g. $\dfrac{6}{13}$? $\dfrac{7}{11}$ h. $\dfrac{5}{9}$? $\dfrac{7}{11}$ *i. $\dfrac{5}{3}$? $\dfrac{9}{7}$

5. Determine whether the following statements are true or false.

   *a. The set of rational numbers is closed under the operation of addition.
   b. The set of rational numbers is closed under the operation of subtraction.
   *c. The set of rational numbers is commutative under the operation of division.
   d. Zero is the multiplicative identity element for the set of rational numbers.
   *e. Every rational number has an additive inverse.
   f. Every rational number has a multiplicative inverse.
   *g. Every rational number can be expressed as either a terminating decimal or a repeating decimal.
   h. Every integer is a rational number.
   *i. Every rational number is an integer.
   j. The rational numbers are dense.

6. Find a rational number between each of the following pairs of rational numbers.

   *a. $\dfrac{1}{3}$ and $\dfrac{1}{4}$ b. $\dfrac{1}{4}$ and $\dfrac{1}{5}$ *c. $\dfrac{1}{6}$ and $\dfrac{1}{7}$

   d. $\dfrac{2}{5}$ and $\dfrac{4}{7}$ *e. $\dfrac{3}{7}$ and $\dfrac{4}{9}$ f. $\dfrac{4}{5}$ and $\dfrac{5}{7}$

   *g. $\dfrac{5}{9}$ and $\dfrac{3}{11}$ h. $\dfrac{4}{9}$ and $\dfrac{-1}{2}$ *i. $\dfrac{-3}{11}$ and $\dfrac{4}{7}$

The Number System

1.6 IRRATIONAL NUMBERS AND THE SET OF REAL NUMBERS

In the previous section, we examined the set of rational numbers. A rational number is any number that can be written in the form $\frac{x}{y}$ ($y \neq 0$) where x and y represent integers. We have seen that any rational number can be expressed as a decimal, and that this decimal will either be a terminating decimal or a repeating non-terminating decimal. For example, $\frac{1}{8}$ and $\frac{1}{7}$ are rational numbers and, expressing each as a decimal, we have

$\frac{1}{8} = 0.125$ (a terminating decimal).

$\frac{1}{7} = 0.\overline{142857}$ (a repeating non-terminating decimal).

Now consider the decimal 0.12112111222.... It is not a terminating decimal and it is not a repeating decimal (although it is a non-terminating decimal). It is not a rational number, since it is neither a terminating decimal nor a repeating non-terminating decimal. Also, it is not a member of the sets of natural numbers, whole numbers, or integers. A number whose decimal representation is "non-terminating" and "non-repeating" is called an *irrational number.*

Irrational numbers are easy to create by using patterns. For example, place any digit after the decimal point, then place another digit after the first digit, repeat the first digit twice, repeat the second digit once, repeat the first digit three times, repeat the second digit once, and so on. It looks like this: 0.32332333233332.... This is a non-terminating, non-repeating decimal, which is an irrational number.

The most famous irrational number is pi (π). Here is the value for π to the first 30 decimal places.

$\pi = 3.141592653589793238462643383279\ldots$

Square Root

Before we proceed further, let's review the definition of "square root." What is the square root of 4, written $\sqrt{4}$? You no doubt are aware that the number 4 has *two* square roots, namely +2 and −2. What about the symbol, $\sqrt{4}$? We do not want to allow both +2 and −2 to be represented by the same symbol. If we allowed this, then we would be allowing an *ambiguous* symbol into mathematics; whenever this symbol appeared in an expression we would never be sure which of the two numbers it represented. To prevent this ambiguity and uncertainty, we select one of the numbers and call it the *principal square root* of 4. It is the principal square root of 4 that is denoted by the symbol $\sqrt{4}$, and the principal square root of 4 is 2. Thus, $\sqrt{4} = 2$, and the two square roots of 4 are $\sqrt{4}$ and $-\sqrt{4}$. Similarly, every positive number n has 2 square roots; a positive and a negative one. We call the positive one the principal square root and denote it by $\sqrt{n}$. In this manner, $\sqrt{9} = 3$, and the square roots of 9 are $\sqrt{9}$ and $-\sqrt{9}$, that is, 3 and -3.

Numbers such as $\sqrt{4}$, $\sqrt{9}$, and $\sqrt{16}$ are rational numbers because $\sqrt{4} = 2$, $\sqrt{9} = 3$, $\sqrt{16} = 4$, and 2, 3, and 4 are all rational numbers. Numbers such as 4, 9, and 16 are called perfect squares because their square roots are rational numbers. A rational number such as $\frac{4}{9}$ is also a perfect square because $\sqrt{\frac{4}{9}} = \frac{2}{3}$.

The reason for reviewing square roots is that the square root of a non-negative number that is not a perfect square is an irrational number. For example, $\sqrt{2}$ is an irrational number. If we were to express $\sqrt{2}$ as a decimal, it would be a non-terminating, non-repeating decimal. Some other examples of irrational numbers are $\sqrt{3}$, $\sqrt{5}$, $\sqrt{6}$, $\sqrt{7}$, etc. These are all square roots of non-negative numbers that are not perfect squares.

Indirect Proof

At one time or another, most of us have used indirect reasoning to prove or discover something. The indirect reasoning method may have been the process of elimination or it may have involved assuming something to be true, then proving that the assumption was false. One of the best known examples of an indirect proof is the one that shows $\sqrt{2}$ cannot be rational. Consider the following proof that $\sqrt{2}$ is irrational.

We shall assume $\sqrt{2}$ to be rational, the negation of what we are trying to prove, and show that this leads to a false statement (or contradiction), indicating our assumption was false. Therefore, the original statement, that $\sqrt{2}$ cannot be rational, is true.

Assume that $\sqrt{2}$ is rational. This means $\sqrt{2} = \frac{a}{b}$ ($b \neq 0$) where $\frac{a}{b}$ is in *simplest* form, that is, a and b are relatively prime (they have no common factors). If $\sqrt{2} = \frac{a}{b}$, then $2 = \frac{a^2}{b^2}$, and $2b^2 = a^2$. Now $2b^2$ is even (it is a multiple of 2). Therefore a^2 is even and a must be even. If a is even, then we can say that $a = 2n$ where n is an integer. Thus far we have:

$\sqrt{2} = \frac{a}{b}$ (assumption)

$2 = \frac{a^2}{b^2}$ (squaring both sides)

$2b^2 = a^2$ (multiplying both sides of the equation by 2)

The Number System

$2b^2 = (2n)^2$ (substituting an even integer, 2n for a)
$2b^2 = 4n^2$ (simplifying)
$b^2 = 2n^2$ (dividing both sides of the equation by 2)

Now we see that $2n^2$ is even (it is a multiple of 2). Therefore, b^2 is even and b must be even. We have now discovered that both a and b are even. If both a and b are even, then they must have a common factor of 2, hence they are not relatively prime. Our assumption is contradicted. This means that the original statement—that $\sqrt{2}$ is irrational—is true rather than its negation, because any rational number may be expressed as a quotient of two integers in simplest form. We have proved (by contradiction) that $\sqrt{2}$ is an irrational number. This same technique can be used to show that square roots of other natural numbers that are not perfect squares are also irrational, that is, $\sqrt{3}, \sqrt{5}, \sqrt{6}, \sqrt{7}$, and so on.

HISTORICAL NOTE

Irrational numbers were studied as early as 500 B.C. by Pythagoras, a famous Greek mathematician. He discovered the fact that there is no rational number for the square root of 2, that is, there is no rational number whose square is 2.

Pythagoras and his colleagues discovered that there is no common unit of length for the length of a side of a square and its diagonal at the same time. That is, no unit of length will ever measure exactly both the diagonal of a square and its side.

These mathematicians were supposedly so upset with this discovery that they vowed not to reveal it, and threatened to punish any member of the group who did divulge this "secret." They were afraid that this information would cause people to doubt their ability as mathematicians.

EXAMPLE 53

Classify each of the following as rational or irrational numbers.
a. $0.1212\overline{12}$
b. $0.121121112\ldots$
c. $\sqrt{5}$
d. $\sqrt{36}$
e. $\sqrt{\dfrac{9}{16}}$
f. $-\sqrt{100}$

Solution

a. $0.1212\overline{12}$ is a rational number. It is a non-terminating decimal, but it is repeating.
b. $0.121121112\ldots$ is an irrational number. It is a decimal that is non-terminating and non-repeating.
c. $\sqrt{5}$ is an irrational number. The square root of a non-negative number that is not a perfect square is an irrational number.

d. $\sqrt{36}$ is a rational number; 36 is a perfect square, and $\sqrt{36} = 6$.
e. $\sqrt{\frac{9}{16}}$ is a rational number; $\frac{9}{16}$ is a perfect square, and $\sqrt{\frac{9}{16}} = \frac{3}{4}$.
f. $-\sqrt{100}$ is a rational number; 100 is a perfect square and $-\sqrt{100} = (-\sqrt{100}) = -10$.

In earlier discussions we noted that the set of natural numbers is a subset of the set of whole numbers; the set of whole numbers is a subset of the set of integers; and the set of integers is a subset of the set of rational numbers. But the set of rational numbers is *not* a subset of the set of irrational numbers. No number can be both a rational number and an irrational number. The set of rational numbers and the set of irrational numbers are disjoint sets. They have no elements in common — their intersection is empty. We have discussed the fact that there are an infinite number of rational numbers between any two integers and it can also be shown that there are an infinite number of irrational numbers between any two integers. The union of the set of rational numbers and the set of irrational numbers yields a set consisting of all possible decimals, terminating or non-terminating. A decimal must be either rational or irrational. This set of rational numbers and irrational numbers is called the set of *real numbers.* Any rational or irrational number is a real number, and any real number must be either a rational number or an irrational number. We will be dealing with the set of real numbers throughout our study of intermediate algebra. Figure 8 illustrates how the different sets of numbers fit together to form the set of real numbers.

The set of real numbers contains a separate number for each separate point on the number line. Some of the points are rational points and some are irrational points. Points that correspond to rational and irrational

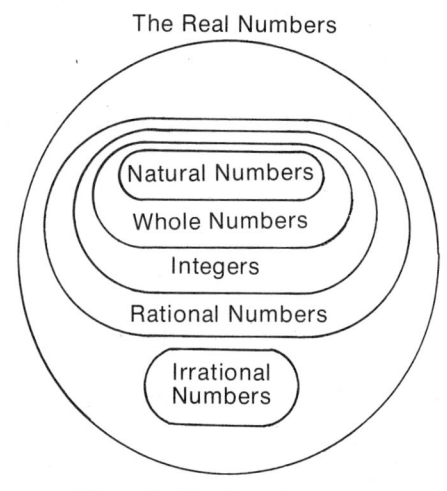

Figure 8 The Real Numbers.

The Number System

numbers can be located fairly easily on the number line. For example, suppose we want to locate the point that corresponds to $\sqrt{2}$. We can do this by laying the hypotenuse of a right triangle (whose legs are of unit length, or 1) along the number line with one vertex at 0. Recall that the square of the hypotenuse of a right triangle equals the sum of the square of the legs, $c^2 = a^2 + b^2$. If $a = 1$ and $b = 1$, then $c^2 = 1^2 + 1^2$, $c^2 = 2$ and $c = \sqrt{2}$. This length corresponds to a point on the number line. See Figure 9.

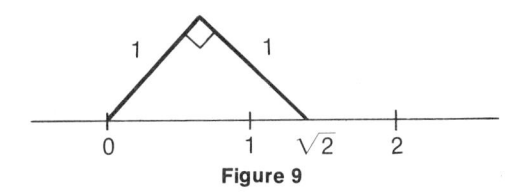

Figure 9

The set of real numbers also satisfies the properties that the set of rational numbers satisfies. We shall state them for the two basic operations, addition and multiplication, in the system of real numbers:

1. Closure. For all real numbers x and y,

$$x + y \text{ and } x \cdot y \text{ are real numbers.}$$

2. The Commutative Properties. For all real numbers x and y,

$$x + y = y + x \quad \text{and} \quad x \cdot y = y \cdot x.$$

3. The Associative Properties. For all real numbers x, y, and z,

$$(x + y) + z = x + (y + z) \quad \text{and} \quad (x \cdot y) \cdot z = x \cdot (y \cdot z).$$

4. Identity Elements. For every real number x, there exists an additive identity element, 0, such that

$$x + 0 = 0 + x = x.$$

For every real number x, there exists a multiplicative identity element, 1, such that

$$x \cdot 1 = 1 \cdot x = x.$$

5. Inverse Elements. For every real number x, there exists an additive inverse, $-x$, such that

$$x + (-x) = 0.$$

For every real number x, (x ≠ 0), there exists a multiplicative inverse, $\frac{1}{x}$, such that

$$x \cdot \frac{1}{x} = 1.$$

6. The Distributive Property. For all real numbers x, y, and z,

$$x \cdot (y + z) = x \cdot y + x \cdot z.$$

Note that the first five properties listed really amount to ten properties, as properties are both for addition and multiplication. However, the distributive property involves both addition and multiplication together. Using the distributive property, we can change a multiplication statement to a sum: x · (y + z) = x · y + x · z, or an addition statement to a product: x · y + x · z = x(y + z). So altogether, we have listed 11 different properties for the system of real numbers.

A mathematical system for which there are two operations (not necessarily addition and multiplication) and whose elements satisfy the 11 listed properties for the given operations forms a special system known as a *field*. The "field" properties can be used to prove that other rules or properties of real numbers are also true.

Axioms are statements that we accept as true without proof. Axioms are necessary in a system since not everything can be proved, and they are needed in order to derive other statements, called theorems. Theorems are proved deductively, using axioms or previously proven theorems. Here are some axioms of equality and their names.

Axioms of Equality. For all real numbers x, y, and z.
Reflexive: x = x.
Symmetric: If x = y, then y = x.
Transitive: If x = y and y = z then x = z.
Substitution: If x = y, then x may be substituted for y, or y may be substituted for x in any statement without changing the truth or falsity of the statement.

There are also axioms of inequality. (Some texts call these the order axioms for real numbers.)

Axioms of Inequality. For all real numbers x, y, and z.
Trichotomy: Exactly one of the following is true: x < y, x = y, or x > y.
Transitive: If x < y and y < z, then x < z.

The nature of this text is intuitive, but proofs in algebra do exist. Those given in this book are of an informal nature. Formal proofs are written as a sequence of deductive statements, with a reason for each statement. For example, let us prove the "multiplication law of equality," which states:

If x, y, and z are real numbers and x = y, then xz = yz, and zx = zy.

The Number System

STATEMENT	REASON
1. x, y, z are real numbers.	1. Given.
2. xz is a real number.	2. Closure property for multiplication.
3. xz = xz.	3. Reflexive axiom.
4. xz = yz.	4. Substitution axiom (substituting y for x in Statement 3, using Statement 1).
5. zx = zy.	5. Commutative property for multiplication.

Throughout this chapter we have reviewed the arithmetic operations for the different sets of numbers under consideration. Frequently we encounter expressions that are confusing for some students. For example, suppose we have to evaluate the expression:

$$6 + 8 \cdot 3 - 4 \div 2 - 8 \cdot (-2)$$

What is the answer? Unless we agree on a standard operating procedure, there will be a variety of answers. Let us agree that first, multiplications and divisions are performed as they are encountered from left to right, then additions and subtractions. Therefore, in evaluating the given expression, we have

$$6 + 8 \cdot 3 - 4 \div 2 - 8 \cdot (-2)$$
$$= 6 + 24 - 2 + 16$$
$$= 44.$$

To avoid confusion in evaluating expressions in algebra, parentheses ( ), brackets [ ], or braces { } are used. Brackets and braces have the same meaning as parentheses in algebraic expressions. All of these distinguish groupings are symbols of inclusion; they indicate which expression is to be simplified first. Always start with the innermost symbol. For example, evaluate the given expression:

$$2\{3 - 4[2 - 3(5 - 2) + 1] + 2\}.$$

Working with the innermost symbol first, we have:

$$2\{3 - 4[2 - 3(3) + 1] + 2\}$$
$$= 2\{3 - 4[2 - 9 + 1] + 2\}.$$

Next, we work inside the brackets:

$$= 2\{3 - 4[-6] + 2\}.$$

Next, we work inside the braces:

$$= 2\{3 + 24 + 2\} = 2\{29\} = 58.$$

To sum up, the operations under the system of real numbers are performed in the following order:
1. An expression contained inside a symbol of inclusion (parentheses, brackets, or braces) will be simplified first, beginning with the innermost inclusion symbol.
2. Next, we perform all multiplications and divisions as they are encountered in order from left to right.
3. Finally, additions and subtractions are performed.

EXAMPLE 54

Evaluate each of the following expressions:
a. $8 - 3 \cdot 2$ b. $4 - 2(5 - 3)$ c. $6 - 3(4 - 2) + 4$

Solution
a. According to the rules for the order of operations, multiplication is done before subtraction. Therefore, $8 - 3 \cdot 2 = 8 - 6 = 2$.
b. An expression contained inside a symbol of inclusion (parentheses) is simplified first. Therefore, $4 - 2(5 - 3) = 4 - 2(2)$. Next we perform the indicated multiplication, $4 - 2(2) = 4 - 4 = 0$.
c. The problem is similar in nature to the previous example. Therefore, $6 - 3(4 - 2) + 4 = 6 - 3(2) + 4 = 6 - 6 + 4 = 4$.

EXAMPLE 55

Evaluate each of the following expressions:
a. $(4 - 2)[5 + 3(4 - 2) + 1]$ b. $(8 \div 2)[6 - 2(5 - 2)]$
c. $-2\{4 - 3[2 + 2(3 - 5) + 1] - 2\}$

Solution

a. An expression contained inside a symbol of inclusion is simplified first, beginning with the innermost inclusion symbol. Therefore, $(4-2)[5+3(4-2)+1] = (4-2)[5+3(2)+1] = (4-2)[5+6+1] = (2)[12] = 24$.
b. We must first simplify inside the brackets. Therefore,

$(8 \div 2)[6 - 2(5 - 2)] = (8 \div 2)[6 - 2(3)] = (8 \div 2)[6 - 6]$
$= 4[0] = 0.$

c. We start with the innermost symbol of inclusion and work out. Therefore,

$$-2\{4 - 3[2 + 2(3 - 5) + 1] - 2\}$$
$$= -2\{4 - 3[2 + 2(-2) + 1] - 2\}$$
$$= -2\{4 - 3[2 - 4 + 1] - 2\}$$
$$= -2\{4 - 3[-1] - 2\}$$
$$= -2\{4 + 3 - 2\}$$
$$= -2\{5\}$$
$$= -10.$$

The Number System

Evaluating a mathematical expression where more than one operation is indicated can be tedious work, but if we follow the rules for the order of operations for the system of real numbers, we shall make our task much easier.

EXERCISES FOR SECTION 1.6
(Answers given to exercises marked *)

1. Classify each of the following as a rational or an irrational number.

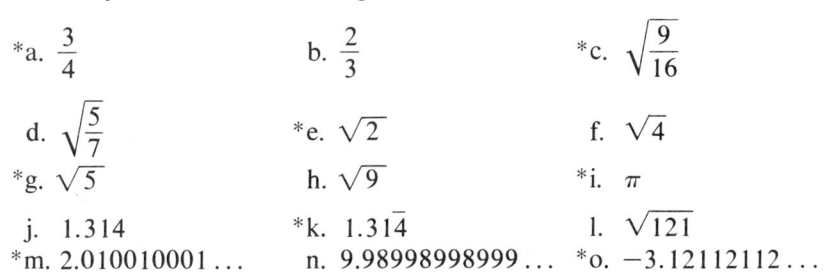

   *a. $\dfrac{3}{4}$ b. $\dfrac{2}{3}$ *c. $\sqrt{\dfrac{9}{16}}$

   d. $\sqrt{\dfrac{5}{7}}$ *e. $\sqrt{2}$ f. $\sqrt{4}$

   *g. $\sqrt{5}$ h. $\sqrt{9}$ *i. π

   j. 1.314 *k. $1.3\overline{14}$ l. $\sqrt{121}$

   *m. $2.010010001\ldots$ n. $9.98998998999\ldots$ *o. $-3.12112112\ldots$

2. Determine whether each of the following can be represented by a terminating decimal, a repeating decimal, or a non-terminating, non-repeating decimal.

   *a. $\dfrac{3}{4}$ b. $\dfrac{2}{3}$ *c. $\sqrt{\dfrac{9}{16}}$

   d. $\sqrt{2}$ *e. $\sqrt{3}$ f. $\sqrt{4}$

   *g. π h. $\dfrac{1}{7}$ *i. $\dfrac{3}{11}$

   j. $\sqrt{225}$ *k. $\sqrt{99}$ l. $\sqrt{\dfrac{4}{25}}$

*3. Arrange the following in order from smallest to largest.

   $3.141141114\ldots,\ 3.14,\ 3.\overline{14},\ 3.1\overline{4},\ 3.1441444.$

4. Arrange the following in order from smallest to largest.

   $0.21\overline{3},\ 0.2131131113\ldots,\ 0.\overline{213},\ 0.213,\ 0.2131331333\ldots$

5. Determine whether each of the following is true or false.
   *a. Every real number is a rational number.
   b. Every real number is an irrational number.
   *c. Every real number is either a rational number or an irrational number.
   d. Every rational number is a real number.
   *e. Every irrational number is a real number.
   f. If a decimal is non-terminating, then it is an irrational number.

*g. If a decimal is non-repeating, then it is an irrational number.
h. Every decimal is a real number.
*i. Every real number can be expressed as a decimal.
j. The intersection of the set of rational numbers and the set of irrational numbers is not empty.

6. Each of the following statements is an application of one of the 11 "field" properties listed in this section. Name the appropriate property exhibited.

*a. $4 + 5 = 5 + 4$.
*c. $3 \cdot 1 = 3$.
*e. $(4 + 5) + 7 = 4 + (5 + 7)$.
*g. $5 \cdot \left(\frac{1}{5}\right) = 1$.
*i. $11 \cdot 1 = 11$.
*k. $\frac{1}{3} \cdot 3 = 1$.

b. $4 \cdot 5 = 5 \cdot 4$.
d. $3 + 0 = 3$.
f. $x(y + z) = xy + xz$.
h. $-6 + 6 = 0$.
j. $(4 \cdot 5) \cdot 7 = 4 \cdot (5 \cdot 7)$.
l. $-\pi + \pi = 0$.

*m. If x and y are real numbers and $x + y = z$, then z is a real number.
n. If x and y are real numbers and $x \cdot y = z$, then z is a real number.

7.*a. The set of irrational numbers is not closed under the operation of addition. Find an example to verify this.
b. The set of irrational numbers is not closed under the operation of multiplication. Find an example to verify this.

8. State the reasons for each of the steps in the following proof of "If x, y, and z are real numbers and $x = y$, then $x + z = y + z$ and $z + x = z + y$."

STATEMENT	REASON
1. x, y and z are real numbers. $x = y$.	1.
2. $x + z$ is a real number.	2.
3. $x + z = x + z$.	3.
4. $x + z = y + z$.	4.
5. $z + x = z + y$.	5.

*9. State the reason for each of the steps in the following proof of "If x, y, and z are real numbers and $x + z = y + z$, then $x = y$."

STATEMENT	REASON
1. $x + z = y + z$	1.
2. $x + z + (-z) = y + z + (-z)$	2.
3. $z + (-z) = 0$	3.
4. $x + 0 = y + 0$	4.
5. $x = y$	5.

10. Evaluate each of the following expressions:
*a. $5 + 3 \cdot 2 - 6 \div 2$
b. $4 - 3 \cdot 2 - 6 \div 3$

The Number System

*c. $-4 \cdot 3 + 4(3-6)$
*e. $-2[3 + 4(2-1) + 2]$
*g. $\dfrac{14}{9-2} - 3$
*i. $\dfrac{2(5-3)}{2} - \dfrac{6}{-3}$
*k. $2\{2 + 2[2 + 2(2+2)]\}$
*m. $\dfrac{4(3-2) + 6(5-7)}{2(5-7) - 3(5-2)}$
*o. $\dfrac{6 - 2(3-8) + 2}{(8 \div 4) - 3(2-1)}$

d. $4 - 3(6-2) - 4$
f. $-3[2 - 1(3-2) + 1]$
h. $\dfrac{2(6-8)}{4} + 1$
j. $4 + 3[2 - 3(2+4) - 2]$
l. $4[3 + 2(2-7) + 3(5-1)]$
n. $\dfrac{2(3-5) - 2(5-3)}{3(2-1) - 3(1-2)}$

1.7 ABSOLUTE VALUE

We have already discussed the fact that numbers such as 5 and −5 are additive inverses of each other. Numbers such as these also have something else in common. If we examine their positions on the number line, we see that both 5 and −5 are five units from 0 (see Figure 10).

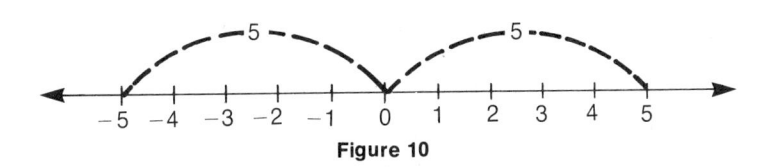

Figure 10

When we measure distance from a point we do not describe it as "negative" five units from 0. We simply say that −5 is five units from 0. This leads us to the definition of the *absolute value* of a real number. When we write $|-5|$ (read "the absolute value of −5"), we can think of this as meaning that −5 is five units from 0 on the number line. The absolute value of a real number is always non-negative. For example,

$$|-5| = 5, \quad |+5| = 5, \quad |-3| = 3, \quad |3| = 3, \quad |0| = 0.$$

Now we can state the formal definition of the absolute value operation:

If x is a real number, then $\quad |x| = \begin{cases} x, & \text{if } x > 0 \\ 0, & \text{if } x = 0 \\ -x, & \text{if } x < 0 \end{cases}$

The absolute value of a real number is never negative. This may seem confusing, considering the fact that $|x| = -x$ if $x < 0$. Does this say that the absolute value of x is negative? No, it does not. Suppose x is negative;

for example, let x = −5. According to the definition, if x < 0 then |x| = −x. Therefore, we have |−5| = −(−5) = 5. The definition is correct. The absolute value of a real number is always non-negative. The absolute value of a real number x is the distance between x and 0, and is symbolized |x|.

EXAMPLE 56

Evaluate the following expressions:
 a. $|-5|$ b. $|-5 + 3|$ c. $|-5| + |3|$

Solution

a. $|-5| = 5.$
b. $|-5 + 3| = |-2| = 2.$
c. $|-5| + |3| = 5 + 3 = 8.$

EXAMPLE 57

If x = −5 and y = 2, evaluate the following expressions.
 a. $|x + y|$ b. $|x| + |y|$ c. $|x - y|$
 d. $|x| - |y|$ e. $|xy|$ f. $\left|\dfrac{x}{y}\right|$

Solution

a. $|x + y| = |-5 + 2| = |-3| = 3.$
b. $|x| + |y| = |-5| + |2| = 5 + 2 = 7.$
c. $|x - y| = |-5 - 2| = |-7| = 7.$
d. $|x| - |y| = |-5| - |2| = 5 - 2 = 3.$
e. $|xy| = |-5 \cdot 2| = |-10| = 10.$
f. $\left|\dfrac{x}{y}\right| = \left|\dfrac{-5}{2}\right| = \dfrac{5}{2}.$

The operation of absolute value may also be included in equations; for example, |x| = 5. What is the solution to this equation? Recall that the absolute value of a real number x is the distance between x and 0 on the number line. What real numbers x are five units from zero? The answer is 5 and −5. Therefore, the solution set for the equation |x| = 5 is {5, −5}. We can use the idea of distance from 0 to solve more complicated absolute value equations. Consider the equation |x − 1| = 4. This means that the expression x − 1 is four units from 0 on the number line. See Figure 11.

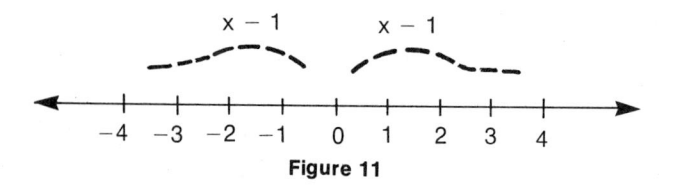

Figure 11

The Number System

Using this information we can say that x − 1 = −4 and x − 1 = 4.

In solving an absolute value equation, we are actually solving two different equations, and each equation can be solved for x.

$$x - 1 = -4 \quad \text{OR} \quad x - 1 = 4$$
$$x = -3 \quad \quad \quad \quad x = 5$$

Replacing x by −3 and 5 in the original equation, we check our solution:

Let x = −3.
|x − 1| = 4.
|−3 − 1| = 4.
|−4| = 4.
4 = 4.

Let x = 5.
|x − 1| = 4.
|5 − 1| = 4.
|4| = 4. The solutions
4 = 4. check.

EXAMPLE 58

Solve the following equations.
a. |x − 1| = 3
b. |x + 7| = 10

Solution

a. |x − 1| = 3 means that x − 1 is three units from 0 on the number line. Therefore, we can state that x − 1 = 3 or x − 1 = −3. Solving each equation we have:

$$x - 1 = 3 \quad \text{OR} \quad x - 1 = -3$$
$$x = 4 \quad \quad \quad \quad x = -2$$

b. |x + 7| = 10 means that the expression x + 7 is ten units from 0 on the number line. Therefore, we can state that x + 7 = 10 or x + 7 = −10. Solving each equation we have:

$$x + 7 = 10 \quad \text{OR} \quad x + 7 = -10$$
$$x = 3 \quad \quad \quad \quad x = -17$$

EXAMPLE 59

Solve the following equations:
a. |x| = −2
b. |x − 1| = −3

Solution

a. The solution set to the equation $|x| = -2$ is the empty set, { }, or ∅. The absolute value of a number is always non-negative. At first glance, this may seem confusing since our definition states that $|x| = -x$, if $x < 0$. But recall that when $x < 0$, or -3, we have $|-3| = -(-3) = 3$ (a non-negative value).

b. The solution is the same as in a, that is, the solution set is ∅.

EXERCISES FOR SECTION 1.7

(Answers given to exercises marked *)

1. Evaluate each of the following:
   *a. $|-3 + 2|$
   b. $|-3 - 2|$
   *c. $|3 - 2|$
   d. $|3 - (-2)|$
   *e. $|-3 - (-2)|$
   f. $|-3 + (-2)|$
   *g. $|3 + 2|$
   h. $|3 + (-2)|$
   *i. $|2 + (-3)|$
   j. $|2 - (-3)|$
   *k. $|2 - 3|$
   l. $|-2 + (-3)|$
   *m. $|-2 - (-3)|$
   n. $|-2 - 3|$
   *o. $|-2 + 3|$

2. Evaluate each of the following:
   *a. $\left|\dfrac{-8}{2}\right|$
   b. $\left|\dfrac{-8}{-2}\right|$
   *c. $\left|-\dfrac{-8}{-2}\right|$
   d. $|(-2)(-8)|$
   *e. $|2(-8)|$
   f. $|(-2)(8)|$
   *g. $|2(-2) + 4|$
   h. $\left|\dfrac{4-4}{4 \cdot 2}\right|$
   *i. $|3 \cdot 2 + 3(-2)|$

3. If $x = 4$ and $y = -3$, evaluate the following expressions:
   *a. $|x + y|$
   b. $|x| + |y|$
   *c. $|x - y|$
   d. $|x| - |y|$
   *e. $|xy|$
   f. $|x| \cdot |y|$
   *g. $\left|\dfrac{x}{y}\right|$
   h. $\dfrac{|x|}{|y|}$
   *i. $|y| - |x|$

4. Solve the following equations:
   *a. $|x| = 4$.
   b. $|x| = 5$.
   *c. $|x| = -2$.
   d. $|x - 1| = 4$.
   *e. $|x + 1| = 4$.
   f. $|x - 3| = 2$.
   *g. $|x + 2| = -1$.
   h. $|x + 2| = 3$.
   *i. $|x - (-5)| = 3$.
   j. $|-x + 3| = 2$.
   *k. $|3 - x| = 4$.
   l. $|2 - x| = 1$.
   *m. $|5 - x| = -5$.
   n. $|-x - 2| = 4$.
   *o. $|-x - 3| = 1$.

5. Replace the question mark, with the correct solution for each of the following:
   *a. $|y| = ?$, if $y > 0$.
   b. $|y| = -y$, if y is ?
   *c. $|y| = ?$, if $y = 0$.
   d. $|y| = -100$, if $y = ?$

The Number System

1.8 SUMMARY

In this chapter, we explored the topic of sets and set operations. A set may be described as a collection of objects. The objects that make up the collection are called elements or members of the set. Two sets A and B are equal or identical sets if they contain exactly the same elements. Set A is a subset of set B (A $\subseteq$ B) if every element in A is also an element in B. If A is a subset of B and there is at least one element in B not contained in A, then A is a proper subset of B, denoted by (A $\subset$ B). The intersection of two sets A and B (denoted by A $\cap$ B) is a set of elements that are members of both A and B. The union of sets A and B (denoted by A $\cup$ B) is a set of elements that are members of A, or members of B, or members of both A and B.

A numeral is a symbol for or name of a number. When we write a V or a 5, we are writing numerals that represent the number 5. The first is a Roman numeral and the second is a Hindu-Arabic numeral. Numbers are classified into different categories. The first set of numbers discussed in this chapter was the set of natural numbers, $\{1, 2, 3, \ldots\}$. By adding 0 to the natural numbers, we obtained the set of whole numbers, $\{0, 1, 2, 3, \ldots\}$. The set of whole numbers, together with their additive inverses, forms the set of integers, $\{\ldots, -3, -2, -1, 0, 1, 2, 3, \ldots\}$. The number line illustrates the order of the integers. For example, $-3 > -10$ because -3 is to the right of -10 on the number line.

The sum of two positive integers is a positive integer and the sum of two negative integers is a negative integer. But, if we add a positive integer and a negative integer, then the sum may be a positive integer, a negative integer, or 0. In order to subtract any two integers, add the additive inverse of the subtrahend to the minuend, $x - y = x + (-y)$. The product or quotient of two positive integers is positive, as is the product or quotient of two negative integers. The product or quotient of a positive integer and a negative integer is negative. The product of 0 and any integer is 0, $x \cdot 0 = 0 \cdot x = 0$. But division by 0 is undefined, $\frac{x}{0} =$ undefined.

A rational number is any number that can be written in the form $\frac{x}{y}$ ($y \neq 0$), where x and y represent integers. We can also describe rational numbers as terminating decimals or repeating non-terminating decimals. Decimals that are non-terminating and non-repeating are called irrational numbers. For example, 0.010010001 ... is an irrational number, as are $\sqrt{2}, \sqrt{3}$, and π. Square roots of other natural numbers that are not perfect squares are also irrational numbers.

The union of the set of rational numbers and the set of irrational numbers gives us the set of real numbers. The set of real numbers is the set of all decimals. The set of real numbers is composed of the set of rational numbers and the set of irrational numbers. For a diagram of how the various sets of numbers fit together to form the set of real numbers, see Figure 8 in Section 1.6. Throughout the chapter we discussed various properties for the number system. We state them here for the system of real numbers under the operations of addition and multiplication, for all real numbers x, y, and z.

1. Closure. $x + y$ and $x \cdot y$ are real numbers.
2. The Commutative Properties. $x + y = y + x$ and $x \cdot y = y \cdot x$.
3. The Associative Properties. $(x + y) + z = x + (y + z)$ and $(x \cdot y) \cdot z = x \cdot (y \cdot z)$.
4. The Identity Elements. $x + 0 = 0 + x = x$ and $x \cdot 1 = 1 \cdot x = x$.
5. Inverse Elements. $x + (-x) = 0$ and $x \cdot \frac{1}{x} = 1$ $(x \neq 0)$.
6. The Distributive Property. $x \cdot (y + z) = x \cdot y + x \cdot z$.

A mathematical system for which there are two operations and whose elements satisfy the 11 listed properties for the given operations forms a special system called a field.

In evaluating expressions under the system of real numbers, we observe the following rules for the order of operations.

1. An expression contained inside a symbol of inclusion (parentheses, brackets, or braces) is simplified first, beginning with the innermost inclusion symbol.
2. Next we perform all multiplications and divisions as they are encountered in order from left to right.
3. Finally, additions and subtractions are performed.

The absolute value of a real number is always non-negative. We can describe the absolute value of a number as the distance that number is from 0 on the number line. We define the absolute value of any real number x, $|x|$, as

$$|x| = \begin{cases} x, & \text{if } x > 0 \\ 0, & \text{if } x = 0 \\ -x, & \text{if } x < 0 \end{cases}$$

The absolute value of a real number is never negative and if $|x| = a$, then we know that $x = a$ or $x = -a$.

REVIEW EXERCISES FOR CHAPTER 1
(Answers in Appendix B)

1. Given that $A = \{-2, -1, 0, 1, 2\}$, $B = \{0, 1, 2, 3\}$ and $C = \{1, 2, 3, \ldots\}$, find:
   a. $A \cap B$
   b. $B \cap C$
   c. $A \cap C$
   d. $B \cup C$
   e. $A \cup B$
   f. $A \cap \emptyset$

The Number System

g. $B \cup \emptyset$ h. $(A \cap B) \cup C$ i. $A \cup (B \cap C)$
j. $(A \cap B) \cup (A \cap C)$

2. Use the roster method to denote:
   a. $\{x \mid x \text{ is a natural number}\}$
   b. $\{x \mid x \text{ is a whole number}\}$
   c. $\{x \mid x \text{ is an integer}\}$
   d. $\{x \mid x \text{ is a natural number that is a perfect square}\}$

3. Evaluate each of the following algebraic expressions for the indicated numerical replacement.
   a. $4x^2$, when $x = -3$
   b. $2y^2 - 3y + 2$, when $y = -2$
   c. πr^2, when $r = 5$
   d. $x^2y - 2xy^2$, when $x = 1$, $y = 2$
   e. $2x^3y - 3x^2y^2$, when $x = 2$, $y = -2$
   f. $3x^2z^2 - 2xz$, when $x = 2$, $z = -3$
   g. $ax^2 + bx + c$, when $x = -3$
   h. $ax^2 + by^2 + c$, when $x = 2$, $y = 3$

4. Determine whether each of the following is true or false.
   a. The natural numbers are commutative under the operation of multiplication.
   b. The whole numbers are closed under the operation of subtraction.
   c. The integers are commutative under the operation of addition.
   d. The identity element for the set of rational numbers under the operation of multiplication is 0.
   e. The set of irrational numbers is closed under the operation of subtraction.
   f. The set of irrational numbers is closed under the operation of multiplication.
   g. The intersection of the set of rational numbers and the set of irrational numbers is not empty.
   h. Every irrational number is a real number.
   i. The set of real numbers is formed by the union of the set of rational numbers and the set of irrational numbers.
   j. Every non-terminating decimal is an irrational number.

5. Evaluate each of the following:
   a. $5 + (-8)$
   b. $-3 + (-4)$
   c. $-2 + (-7)$
   d. $-3 - (-2)$
   e. $-2 - (-4)$
   f. $2 - 4$
   g. $4 \cdot (-2)$
   h. $(-5) \cdot (-3)$
   i. $(-2)(-3 + 2)$
   j. $-6 \div 2$
   k. $(-2 + 4) \div (-2 - 2)$
   l. $(-3 - 7) \div (-7 + 2)$

6. Express each of the following as a decimal.
   a. $\dfrac{3}{8}$
   b. $\dfrac{3}{11}$
   c. $\dfrac{1}{7}$
   d. $\dfrac{5}{13}$

7. Express each of the following as a quotient of integers.
   a. $0.\overline{4}$
   b. $0.\overline{12}$
   c. $3.4\overline{7}$
   d. $5.1\overline{23}$

8. Perform the indicated operations for each of the following.

   a. $\dfrac{3}{7} + \dfrac{4}{5}$
   b. $\dfrac{-2}{3} + \dfrac{3}{5}$
   c. $\dfrac{-2}{7} - \dfrac{-3}{5}$

   d. $\dfrac{4}{5} - \dfrac{2}{3}$
   e. $\left(\dfrac{-2}{3}\right) \cdot \left(\dfrac{5}{7}\right)$
   f. $\left(\dfrac{-3}{5}\right) \cdot \left(\dfrac{-2}{7}\right)$

   g. $\dfrac{4}{11} \div \dfrac{-1}{2}$
   h. $\dfrac{-3}{7} \div \dfrac{-5}{2}$
   i. $\dfrac{-1}{3} \div \dfrac{1}{-2}$

9. Classify each of the following as a rational or an irrational number.

   a. $0.\overline{3}$
   b. $\dfrac{1}{8}$
   c. $\sqrt{3}$

   d. $\sqrt{4}$
   e. $0.1010010001\ldots$
   f. $\sqrt{\dfrac{4}{9}}$

   g. 0.12
   h. 0.101001
   i. $\sqrt{7}$

   j. 3.14159
   k. $\sqrt{\dfrac{16}{49}}$
   l. $\sqrt{225}$

10. Arrange the following in order from smallest to largest.

    $4.5\overline{12}$, $\ 4.5121121112\ldots$, $\ 4.512$, $\ 4.\overline{512}$, $\ 4.5121221222\ldots$

11. Each of the following statements is an application of one of the 11 field properties discussed in this chapter. Name the appropriate property exhibited.

    a. $4 \cdot 1 = 4$.
    b. $4 + 5 = 5 + 4$.
    c. $5 + 0 = 5$.
    d. $-3 + 3 = 0$.
    e. $5 \cdot 7 = 7 \cdot 5$.
    f. $5 \cdot 1 = 5$.
    g. $(2 \cdot 3) \cdot 4 = 2 \cdot (3 \cdot 4)$.
    h. $4 \cdot \dfrac{1}{4} = 1$.
    i. $(2 + 3) + 4 = 2 + (3 + 4)$.
    j. $4(3 + 2) = 4 \cdot 3 + 4 \cdot 2$.
    k. $6 + (-6) = 0$.
    l. $x + y = y + x$.
    m. If x and y are real numbers and $x + y = z$, then z is a real number.
    n. If x and y are real numbers and $x \cdot y = z$, then z is a real number.

12. Evaluate each of the following expressions.

    a. $4 - 3[2 + 7(-2 + 5) - 2]$
    b. $(4 - 2)[3 - 4(5 - 7) + 1]$

    c. $\dfrac{3(5 - 2) - 6(3 - 4)}{2(5 - 7) - 3[5 - (-2)]}$
    d. $\dfrac{6 - 2(3 - 9) + 2}{(10 \div 2) - 3(5 - 8)}$

13. If $x = 5$ and $y = -2$, evaluate the following expressions.

    a. $|x + y|$
    b. $|x| + |y|$
    c. $|x - y|$
    d. $|x| - |y|$
    e. $|xy|$
    f. $|x| \cdot |y|$
    g. $|y| - |x|$
    h. $\left|\dfrac{x}{y}\right|$
    i. $\dfrac{|x|}{|y|}$

chapter 2

First Degree Equations and Inequalities in One Variable

As a result of this chapter, the student will be able to:

1. Identify an equation as a mathematical sentence that states that two quantities or expressions are equal, and identify an equation of the general form ax + b = c as a linear equation or a first degree equation in one unknown;
2. Distinguish between conditional equations and identities;
3. Solve first degree equations in one unknown using the addition property of equality and the multiplication property of equality;
4. Solve first degree equations in one unknown that contain parentheses, fractions, or decimals;
5. Solve literal equations (formulas) for the indicated variable in terms of the other symbols in the given literal equation;
6. Translate a word (verbal) problem into a linear equation and solve the equation for the following types of problems: number problems, consecutive integer problems, age problems, geometric problems, uniform motion problems, and coin problems;
7. Solve conditional inequalities in one unknown using the addition property for inequalities and the two multiplication properties for inequalities;
8. Graph the solution set of a first degree inequality in one unknown on the real number line;
9. Solve and graph the solution set of compound first degree inequalities in one unknown, e.g., $-1 < 3x + 2 < 8$;
10. Solve and graph the solution set of first degree inequalities that contain absolute value notation, e.g., $|2x - 3| \leq 5$.

2.1 INTRODUCTION

Most of the study of algebra pertains to finding the solution set for various kinds of equations and inequalities. By an *equation* we mean a mathematical sentence that states that two quantities or expressions are equal. A mathematical statement of equality is an equation. Statements which contain an equality sign (=) are called equations. When we find the solution set for the variable in an equation we are "solving" the equation.

The type of equation that we will concern ourselves with in this chapter is referred to as an "open sentence." The equation $x + 2 = 5$ is an example of an open sentence. As it is stated, it is neither true nor false. It will be true if $x = 3$ and false if x equals any other real number. The set of replacements that makes an equation true is called the "solution set." For the equation $x + 2 = 5$, $\{3\}$ is the solution set. Some equations may have one solution, others may have more than one, and still others may have no solution. If an equation has no solution, its solution set is denoted by the

First Degree Equations and Inequalities in One Variable

empty set, ∅. In this chapter, we shall be primarily concerned with equations in one variable that have one solution. They will be of the general form:

$$ax + b = c$$

(where a, b, and c are fixed real numbers, a ≠ 0).

Some specific examples are: $x + 2 = 5 (a = 1, b = 2, c = 5)$.
$2x - 3 = 7 (a = 2, b = -3, c = 7)$.
$x = 2 (a = 1, b = 0, c = 2)$.

An equation of the general form $ax + b = c$ is often called a linear equation or a first degree equation in one unknown.

If two equations have the same solution set, they are called *equivalent equations*. For example, $3x - 2x = 4$ and $x = 4$ are equivalent equations because they both have the same solution set. An equation such as $x + 4 = x + 4$ is an equation that is true for all values of x, that is, its solution set is the set of all real numbers. An equation that is true for *all* replacements of the variable is called an *identity*. A *conditional equation* is an equation that is true for only certain replacements of the variable. A conditional equation is satisfied by some value of the variable and not satisfied by others.

$2x + x = 6$ and $x = 2$ are *equivalent equations.*
$x - 5 = x - 5$ is an *identity.*
$3x + 2 = 5$ is a *conditional equation.*

We can check a solution for an equation by determining if the solution *satisfies* the equation. That is, when we substitute for the variable in the original equation, the result is a true statement for the equation. For example, we can check that $x = 2$ for the equation $2x + x = 6$ by substituting 2 for x in the equation,

$2x + x = 6$.
$2 \cdot 2 + 2 = 6$ (substituting 2 for x).
$4 + 2 = 6$.
$6 = 6$.

Therefore, $x = 2$ is a solution of the equation, $2x + x = 6$ and {2} is the solution set. Note that $x = 1$ is not a solution because $2 \cdot 1 + 1 \neq 6$.

EXERCISES FOR SECTION 2.1
(Answers given to exercises marked *)

1. Identify the following as conditional equations, identities, or neither.

   *a. $3x + 2 = 5$.
   b. $x + x = 2x$.
   *c. $y + 1 = y$.
   d. $2y - y = 5$.
   *e. $2z - z = z$.
   f. $z - 2 = -2$.
   *g. $3x + 4 = -8$.
   h. $5x - 2 = x$.
   *i. $4y - 5 = y$.
   j. $y - 10 = y$.

2. Determine which of the following pairs of equations are equivalent.

   *a. $3x + 2 = 5$ and $x = 1$.
   b. $2x + 4 = 8$ and $x + x + 4 = 8$.
   *c. $4y - 2 = 6$ and $2y + y = 8$.
   d. $y = -2$ and $y + 2 = 0$.
   *e. $z = 2$ and $|z| = 2$.
   f. $|x| = 1$ and $x = 1$.
   *g. $z = 0$ and $2z + 2 = 2$.
   h. $y = 1$ and $2y + 1 = 1$.

3. Determine if the number following each equation is a solution.

   *a. $3x + 2 = 5$, $\{1\}$.
   b. $2x - x = 5$, $\{2\}$.
   *c. $3x + 4 = -8$, $\{4\}$.
   d. $4x - 2 = 6$, $\{-2\}$.
   *e. $x + x + 4 = 8$, $\{-2\}$.
   f. $4x - 5 = x\left\{\frac{5}{3}\right\}$.
   *g. $3x + 4 = 2x - 5$, $\{-9\}$.
   h. $2y + 3 = 4y - 5$, $\{4\}$.
   *i. $z + 3z = 2z - 10$, $\{-5\}$.
   j. $5z + 7 = 2z - 5$, $\{-3\}$.

2.2 SOLVING A FIRST DEGREE EQUATION

Some conditional equations can be solved by inspection. For example, we can see that the solution set to $x + 1 = 2$ is $\{1\}$ because we can think of it as "what value (number) plus 1 equals 2?" We cannot solve most equations by inspection because of their complexity. However, we can solve conditional equations by transforming them into equivalent equations where the solution is easily determined. For example, solve the following equation for y:

$$3y + 1 = y + 5.$$
the given equation is equivalent to $3y = y + 4$.
which is equivalent to $2y = 4$.
which is equivalent to $y = 2$.

Each equation was transformed into an equivalent equation by performing various arithmetic operations on each side (or number) of the equation. Recall that an equation is a statement of equality, and we must

First Degree Equations and Inequalities in One Variable

maintain that equality or balance by performing the various operations on both sides. These transformations can be performed because of two basic equality properties (axioms) for equations.

1. Addition Property of Equality.

> If the same expression is added to both members of an equation, then the resulting equation is equivalent to the given equation.

We can also state this addition property of equality symbolically. If a, b, and c are real numbers and $a = b$, then $a = b$ is equivalent to $a + c = b + c$. We can use this addition property of equality to solve equations. Consider the following equation:

$$x - 2 = 7.$$

We can find the solution to this equation by adding 2 to each side of the equation and changing it to an equivalent equation:

$$x - 2 = 7 \quad \text{(given)}.$$
$$x - 2 + 2 = 7 + 2 \quad \text{(adding 2 to both sides)}.$$
$$x + 0 = 9.$$
$$x = 9.$$

We check by substituting 9 for x in the given equation. If the resulting statement is a true statement, then we say the "solution" checks.

Check: $9 - 2 = 7.$
 $7 = 7.$ (A true statement, therefore the solution checks.)

A subtraction property of equality does not have to be stated as such. Recall that the operation of subtraction is the same as "adding" the additive inverse of the subtrahend. For example, $5 - 3$ can be restated as $5 + (-3)$; $-2 - (-7)$ can be restated as $-2 + (+7)$ or $-2 + 7$. Therefore, we can still use the addition property of equality to solve an equation:

$$x + 3 = -2 \quad \text{(given)}.$$
$$x + 3 + (-3) = -2 + (-3) \quad \text{(adding } -3 \text{ to both sides)}.$$
$$x + 0 = -5.$$
$$x = -5.$$

Check: $-5 + 3 = -2.$
 $-2 = -2.$ (A true statement, therefore the solution checks.)

2. Multiplication Property of Equality.

> If both members of an equation are multiplied by the same expression (provided the expression does not equal 0), then the resulting equation is equivalent to the given equation.

We can also state this multiplication property of equality symbolically. If a, b, and c are real numbers ($c \neq 0$) and $a = b$, then $a = b$ is equivalent to $a \cdot c = b \cdot c$. This multiplication property of equality can be used to solve equations. Consider the following equation:

$$\frac{2}{3} y = 8.$$

We can find the solution to this equation by multiplying each side of the equation by $\frac{3}{2}$ and changing it to an equivalent equation. (Note: $\frac{3}{2} \cdot \frac{2}{3} = 1$; $\frac{3}{2}$ is the reciprocal of $\frac{2}{3}$).

$$\frac{2}{3} y = 8 \quad \text{(given)}.$$

$$\left(\frac{3}{2}\right)\left(\frac{2}{3} y\right) = \frac{3}{2} \cdot 8 \quad \left(\text{multiplying both sides by } \frac{3}{2}\right).$$

$$\left(\frac{3}{2} \cdot \frac{2}{3}\right) y = 12.$$

$$1y = 12.$$
$$y = 12.$$

Check: $\quad \frac{2}{3} \cdot 12 = 8.$

$\quad\quad\quad 8 = 8.$ (A true statement, therefore the solution checks.)

A division property of equality does not have to be stated as such. Recall that the operation of division is the same as "multiplying" the dividend by the reciprocal of the divisor. The reciprocal of a real number $x (x \neq 0)$ is $\frac{1}{x}$. We can restate a division problem, such as $10 \div 2$, as $10 \cdot \frac{1}{2}$. We can restate $\frac{2}{3} \div \frac{4}{5}$ as $\frac{2}{3} \cdot \frac{5}{4}$. Therefore, we can still use the multiplication property of equality to solve an equation:

$$5z = 30 \quad \text{(given)}.$$

$$\frac{1}{5}(5z) = \frac{1}{5} \cdot 30 \quad \left(\text{multiplying both sides by } \frac{1}{5}, \text{the reciprocal of 5}\right).$$

$$\left(\frac{1}{5} \cdot 5\right) z = 6.$$

$$1z = 6.$$
$$z = 6.$$

Check: $\quad 5 \cdot 6 = 30.$

$\quad\quad\quad 30 = 30.$ (A true statement, therefore the solution checks.)

First Degree Equations and Inequalities in One Variable

The addition property of equality and the multiplication property of equality are thus used to transform any given equation of first degree in x into an *equivalent* equation of the form x = k, therefore displaying the solution to the given equation. An equation of the form x = k has "k" as the solution, hence "k" is also the solution for the given equivalent equation.

EXAMPLE 1

Solve x + 4 = 2.

Solution

Using the addition property of equality, we add −4 (the additive inverse of 4) to both sides of the equation in order to obtain an equivalent equation.

$$x + 4 = 2.$$
$$x + 4 + (-4) = 2 + (-4) \quad \text{(adding } -4 \text{ to both sides).}$$
$$x + 0 = -2.$$
$$x = -2.$$

Check: $-2 + 4 = 2.$
$2 = 2.$

EXAMPLE 2

Solve $\frac{3}{7}z = \frac{1}{2}$.

Solution

Using the multiplication property of equality, we multiply both sides of the equation by $\frac{7}{3}$ (the reciprocal of $\frac{3}{7}$) in order to obtain an equivalent equation.

$$\frac{3}{7}z = \frac{1}{2}.$$

$$\frac{7}{3}\left(\frac{3}{7}z\right) = \frac{7}{3} \cdot \frac{1}{2}.$$

$$\left(\frac{7}{3} \cdot \frac{3}{7}\right)z = \frac{7}{6}.$$

$$1z = \frac{7}{6}.$$

$$z = \frac{7}{6}.$$

Check:
$$\frac{3}{7} \cdot \frac{7}{6} = \frac{1}{2}.$$
$$\frac{21}{42} = \frac{1}{2}.$$
$$\frac{1}{2} = \frac{1}{2}.$$

Many times it is necessary to use both the addition property of equality and the multiplication property of equality to transform a given equation to an equivalent equation. Consider the equation.

$$3x - 4 = 8.$$

In order to solve this equation, we begin by adding 4 (the additive inverse of -4) to each side of the equation:

$$3x - 4 + 4 = 8 + 4,$$

which is equivalent to $3x = 12$. Next, we multiply both sides of the equation $3x = 12$ by $\frac{1}{3}$ (the reciprocal of 3):

$$\frac{1}{3} \cdot 3x = \frac{1}{3} \cdot 12.$$
$$1x = 4.$$
$$x = 4 \quad \text{which is a simpler equivalent equation and the solution.}$$

Check: $3 \cdot 4 - 4 = 8.$
$12 - 4 = 8.$
$8 = 8.$

Note that each of the resulting equations is equivalent to the preceding equation. Therefore, the final equation is equivalent to the first equation and the solution of the final equation is also the solution of the given equation.

EXAMPLE 3

Solve $-4y + 3 = 11$.

Solution

We begin by adding (-3) to each side of the equation:

$$-4y + 3 + (-3) = 11 + (-3),$$

which is equivalent to $-4y = 8$.

Next, we multiply both sides of the equation by $\left(-\frac{1}{4}\right)$, the reciprocal of -4.

$$\left(-\frac{1}{4}\right)(-4y) = \left(-\frac{1}{4}\right)8,$$

First Degree Equations and Inequalities in One Variable

which is equivalent to $1y = -2$.
$y = -2$.
Check: $(-4)(-2) + 3 = 11$
$8 + 3 = 11$.
$11 = 11$.

EXAMPLE 4

Solve $\frac{1}{2}z + 4 = 5$.

Solution

This problem is similar to Example 3 in that we must use both the addition property of equality and the multiplication property of equality.

$$\frac{1}{2}z + 4 = 5 \quad \text{(given).}$$

$$\frac{1}{2}z + 4 + (-4) = 5 + (-4) \quad \text{(adding } (-4) \text{ to both sides).}$$

$$\frac{1}{2}z = 1.$$

Next, we multiply both sides of the equation by 2, the reciprocal of $\frac{1}{2}$.

$$2 \cdot \frac{1}{2}z = 2 \cdot 1.$$
$$1z = 2.$$
$$z = 2.$$

Check: $\frac{1}{2} \cdot 2 + 4 = 5$.
$1 + 4 = 5$.
$5 = 5$.

NOTE: Sometimes an equation will be written with a fractional coefficient in a slightly different manner. For example, instead of $\frac{1}{2}z + 4 = 5$, we can write $\frac{z}{2} + 4 = 5$. The expression $\frac{1}{2}z$ is equivalent to $\frac{z}{2}$. But in each case, the numerical coefficient is $\frac{1}{2}$.

Sometimes it may be necessary to use a property of equality more than once in order to transform a given equation to an equivalent simpler equation. For example, in order to solve the equation $4x + 2 = 2x - 6$, we have to use the addition property of equality twice. We begin by adding (-2) to both sides of the equation:

$$4x + 2 + (-2) = 2x - 6 + (-2).$$
$$4x = 2x - 8.$$

The resulting equivalent equation is still not of the general form $ax = k$ (where k is a constant). The right hand member still contains a variable term. Therefore, we must add $-2x$ (the additive inverse of $2x$) to both sides of the equation:

$$-2x + 4x = -2x + 2x - 8.$$
$$2x = -8.$$

Now the resulting equivalent equation is of the general form $ax = k$. Hence we must apply the multiplication property of equality, and multiply both sides of the equation by $\frac{1}{2}$ (the reciprocal of 2):

$$\frac{1}{2} \cdot 2x = \frac{1}{2}(-8).$$
$$x = -4.$$

Check: $\quad 4(-4) + 2 = 2(-4) - 6.$
$$-16 + 2 = -8 - 6.$$
$$-14 = -14.$$

At other times it may be necessary to simplify one or both sides of an equation before applying the addition and multiplication properties of equality. For example, in order to solve the equation $5x - 7 - 2x = 7 + x - 6$, we first simplify each side of the equation. On the left side we combine the x terms, $5x$ and $-2x$, and on the right side we combine the constant terms, 7 and -6. Therefore,

$$5x - 7 - 2x = 7 + x - 6$$

is equivalent to $\quad 3x - 7 = x + 1.$

Next, we add $-x$ to both sides of the equation,

$$-x + 3x - 7 = -x + x + 1.$$
$$2x - 7 = 1.$$

First Degree Equations and Inequalities in One Variable

Next, we add 7 to both sides of the equation,

$$2x - 7 + 7 = 1 + 7.$$
$$2x = 8.$$

Now, we multiply both sides of the equation by $\frac{1}{2}$ in order to obtain a simpler equivalent equation of the general form $x = k$,

$$\frac{1}{2} \cdot 2x = \frac{1}{2} \cdot 8,$$
$$x = 4.$$

Check: $5 \cdot 4 - 7 - (2 \cdot 4) = 7 + 4 - 6.$
$20 - 7 - 8 = 11 - 6.$
$20 - 15 = 5.$
$5 = 5.$

EXAMPLE 5

Solve $2y + 4 - y = 4y - 5$.

Solution

$2y + 4 - y = 4y - 5$ (given).
$y + 4 = 4y - 5$ (combining $2y$ and $-y$).
$-4y + y + 4 = -4y + 4y - 5$ (adding $-4y$ to both sides).
$-3y + 4 = -5.$
$-3y + 4 + (-4) = -5 + (-4)$ (adding -4 to both sides).
$-3y = -9.$

$\left(-\frac{1}{3}\right)(-3y) = \left(-\frac{1}{3}\right)(-9)$ (multiplying both sides by $-\frac{1}{3}$).

$y = 3.$

Check: $2 \cdot 3 + 4 - 3 = 4 \cdot 3 - 5.$
$6 + 4 - 3 = 12 - 5.$
$10 - 3 = 7.$
$7 = 7.$

EXAMPLE 6

Solve $z - \frac{1}{2}z + 1 = 5$.

Solution

$z - \frac{1}{2}z + 1 = 5$ (given).

$$\frac{1}{2}z + 1 = 5 \quad \left(\text{combining z and } -\frac{1}{2}z\right).$$

$$\frac{1}{2}z + 1 - 1 = 5 - 1 \quad (\text{adding } -1 \text{ to both sides}).$$

$$\frac{1}{2}z = 4.$$

$$2 \cdot \frac{1}{2}z = 2 \cdot 4 \quad (\text{multiplying both sides by 2}).$$

$$z = 8.$$

Check: $\quad 8 - \frac{1}{2} \cdot 8 + 1 = 5.$

$$8 - 4 + 1 = 5.$$
$$4 + 1 = 5.$$
$$5 = 5.$$

In Examples 5 and 6, we simplified one or both sides of the equations by combining terms before applying the addition and multiplication properties of equality. Many times we shall encounter equations that contain parentheses. For example, $2(x - 3) = 4$ and $4(y - 3) = 3(2y - 4)$ are first degree equations in one variable, and both contain parentheses. An equation containing parentheses can be solved by first applying the distributive property, $a(b + c) = ab + ac$, then proceeding in the same manner as in the previous examples. To solve $2(x - 3) = 4$, we first remove the parentheses by applying the distributive property,

$$2(x - 3) = 4 \quad (\text{given}).$$
$$2x - 6 = 4 \quad (\text{applying the distributive property}).$$

Now we can proceed in the same manner as before,

$$2x - 6 = 4.$$
$$2x - 6 + 6 = 4 + 6 \quad (\text{adding 6 to both sides}).$$
$$2x = 10.$$
$$\frac{1}{2} \cdot 2x = \frac{1}{2} \cdot 10 \quad \left(\text{multiplying both sides by } \frac{1}{2}\right).$$
$$x = 5.$$

Check: $\quad 2(5 - 3) = 4.$
$$2(2) = 4.$$
$$4 = 4.$$

First Degree Equations and Inequalities in One Variable

EXAMPLE 7

Solve $4(y-3) = 3(2y-4)$.

Solution

$4(y-3) = 3(2y-4)$ (given).
$4y - 12 = 6y - 12$ (applying the distributive property to both sides).
$4y - 12 + 12 = 6y - 12 + 12$ (adding 12 to both sides).
$4y = 6y$.
$-6y + 4y = -6y + 6y$ (adding $-6y$ to both sides).
$-2y = 0$.

$\left(-\frac{1}{2}\right)(-2y) = \left(-\frac{1}{2}\right) \cdot 0$ (multiplying both sides by $-\frac{1}{2}$).

$y = 0$.

Check: $4(0-3) = 3(2 \cdot 0 - 4)$.
$4(-3) = 3(0-4)$.
$-12 = 3(-4)$.
$-12 = -12$.

To solve a first degree equation in one unknown, we must transform the given equation into simpler equivalent equations until it is of the form $x = k$, $y = k$, and so on, (where k is a constant). The solution set of the equation $x = k$ is {k} and it is also the solution set of the given equivalent equation.

In transforming a given equation into a simpler equivalent equation, we perform a number of steps in order. First, we clear any parentheses by applying the distributive property. Next, we combine terms on each side of the equation (if possible). Then we apply the addition property of equality to transform the equation into the general form $ax = c$, where the variable term is on the left side and the numeral term is on the right side. Now we are ready to apply the multiplication property of equality to transform the equation into the general form of $x = k$, which is the solution of the given equation.

The steps required to solve a first degree equation in one unknown are:
1. Eliminate parentheses (if necessary);
2. Simplify each side of the equation by combining terms (if possible);
3. Apply the addition property of equality to obtain an equivalent equation of the form $ax = c$;
4. Apply the multiplication property of equality to obtain an equivalent equation of the form $x = k$, which is the solution of the given equation.

In the next section we shall deal with first degree equations that are a little more involved, but regardless of the complexity of a given equation, we can solve it if we transform it into a simpler equivalent equation by following the steps outlined above.

EXERCISES FOR SECTION 2.2
(Answers given to exercises marked *)

Solve the following equations and check the solutions:

*1. $x - 3 = 4$.

2. $y + 2 = 7$.

*3. $z - 4 = -6$.

4. $2y = 6$.

*5. $3x = 9$.

6. $\frac{1}{3} x = 12$.

*7. $\frac{1}{4} z = 2$.

8. $\frac{1}{4} z - 2 = 4$.

*9. $\frac{1}{3} x + 4 = -5$.

10. $\frac{1}{6} z + 3 = 2$.

*11. $\frac{x}{2} + 1 = 2$.

12. $\frac{y}{3} + 4 = 2$.

*13. $\frac{z}{2} = \frac{2}{3}$.

14. $\frac{x}{3} = \frac{3}{7}$.

*15. $\frac{y}{5} = \frac{4}{3}$.

16. $5x = 3x - 4$.

*17. $2y - 4 = 3y$.

18. $2z + 4 = 5 - z$.

*19. $3x + 4 = x - 6$.

20. $4x - 3 = x + 9$.

*21. $\frac{2}{3} y - 3 = 7$.

22. $\frac{3}{2} z - 4 = 8$.

*23. $\frac{1}{2} x - \frac{1}{8} = \frac{3}{4}$.

24. $\frac{2}{3} z + 1 = 13$.

*25. $\frac{2}{3} y - \frac{1}{2} = y + \frac{3}{4}$.

26. $\frac{3}{4} x + \frac{1}{3} = 2x - 1$.

*27. $\frac{2}{5} z + \frac{1}{2} = -z - \frac{1}{3}$.

28. $2(x - 1) = 2$.

*29. $3(y + 3) = -6$.

30. $-2(z + 3) = 14$.

*31. $3(x - 2) = 2(x + 4)$.

32. $2(2x - 4) = 3(x + 2)$.

*33. $-2(z - 1) = 2(z - 1)$.

34. $\frac{1}{2} (2z + 4) = \frac{2}{3} (3z - 6)$.

*35. $\frac{2}{3} (3y + 6) = \frac{3}{4} (4y - 8)$.

36. $\frac{1}{2} (z - 2) = \frac{3}{4} (z - 12)$.

HISTORICAL NOTE

Diophantas of Alexandria was an ancient mathematician who contributed greatly to the development of algebra. He produced three different known works in the area of mathematics. The titles are: Arithmetica, Porisms, and Polygonal Numbers. It is thought

First Degree Equations and Inequalities in One Variable 85

that all of these works dealt with the properties of rational or integral numbers. His most famous work, Arithmetica, deals with algebraic equations and certain problems that require positive rational numbers as solutions. In his presentation, Diophantas used special signs and abbreviations, such as the first letter of a word. Therefore, a great deal of his work had standardized notation that made it easier to understand. This was one of the first steps towards a formalization of algebra and the creation of mathematical notation.

Not much is known about Diophantas other than his mathematical contributions. Historians disagree as to when he lived: perhaps 75 A.D., or 250 A.D. They even disagree on the spelling of his name: Diophantas or Diophantus. Nevertheless, he was immortalized because of an epitaph that appeared in a Greek anthology. It gives the following information:

> Diophantas spent one-sixth of his life as a child, one-twelfth as a youth, one-seventh more as a bachelor. In the fifth year of his marriage he had a son. The son died four years prior to the death of Diophantas, which made the life span of the son half that of Diophantas. This information gives us the equation

$$\frac{x}{6} + \frac{x}{12} + \frac{x}{7} + 5 + \frac{x}{2} + 4 = x$$

where x equals the age of Diophantas.

2.3 MORE FIRST DEGREE EQUATIONS

Many times in the study of algebra we encounter various forms of linear or first degree equations. For example, each of the following is a first degree equation in one unknown:

$$\frac{2}{3}x - \frac{1}{2} = \frac{1}{2}x - \frac{1}{3}.$$

$$\frac{3}{4}(y - 8) = \frac{1}{2}y + 5.$$

$$0.25z + 1.3 = 0.2z - 3.7.$$

If we can transform these equations into equivalent equations of the general form $ax + b = c$ ($a \neq 0$), then we can solve them because of our experience in the previous section with the addition and multiplication properties of equality.

We have already encountered some equations similar in nature to the first example. Here is one way we can solve it:

$$\frac{2}{3}x - \frac{1}{2} = \frac{1}{2}x - \frac{1}{3} \quad \text{(given)}.$$

$$\frac{2}{3}x - \frac{1}{2} + \frac{1}{2} = \frac{1}{2}x - \frac{1}{3} + \frac{1}{2} \quad \left(\text{adding } \frac{1}{2} \text{ to both sides}\right).$$

$$\frac{2}{3}x = \frac{1}{2}x + \frac{1}{6}.$$

$$-\frac{1}{2}x + \frac{2}{3}x = -\frac{1}{2}x + \frac{1}{2}x + \frac{1}{6} \quad \left(\text{adding } -\frac{1}{2}x \text{ to both sides}\right).$$

$$\frac{1}{6}x = \frac{1}{6}.$$

$$6 \cdot \frac{1}{6}x = 6 \cdot \frac{1}{6} \quad \left(\text{multiplying both sides by 6}\right).$$

$$x = 1.$$

Equations that contain fractions can also be solved in another manner. We can simplify the process if we first multiply both sides of the equation by the least common denominator (LCD) of the fractions. We will solve the given equation again, using the LCD, after a brief review.

The Least Common Denominator

The least common multiple of two or more positive integers is the smallest number that can be divided evenly by the given integers. For example, the least common multiple of 3 and 4 is 12; the least common multiple of 2, 3, and 10 is 30. The least common denominator of two or more fractions is the least common multiple of the denominators of the fractions.

To determine the least common multiple for two or more numbers, factor the given numbers completely. If they have no common prime factors, then their least common multiple is their product. If the given numbers have prime factors in common, then their least common multiple is the product of each different prime factor, including each factor for the highest power in which it appears in any of the given numbers. For example, to find the least common multiple of 12 and 18, we have $12 = 2 \cdot 2 \cdot 3$ and $18 = 2 \cdot 3 \cdot 3$. Therefore, the least common multiple is $2 \cdot 2 \cdot 3 \cdot 3 = 36$. It is divisible by the given numbers (12 and 18) because it contains all their factors and it is the smallest such number.

EXAMPLE 8

Find the least common multiple of:
a. 10 and 12.   b. 48 and 72.

Solution

a. $10 = 2 \cdot 5$, $12 = 2 \cdot 2 \cdot 3$; the least common multiple is $2 \cdot 2 \cdot 3 \cdot 5 = 60$.
b. $48 = 2 \cdot 2 \cdot 2 \cdot 2 \cdot 3$; $72 = 2 \cdot 2 \cdot 2 \cdot 3 \cdot 3$; the least common multiple is $2 \cdot 2 \cdot 2 \cdot 2 \cdot 3 \cdot 3 = 144$.

First Degree Equations and Inequalities in One Variable 87

Let us now solve the given equation again:

$$\frac{2}{3}x - \frac{1}{2} = \frac{1}{2}x - \frac{1}{3} \quad \text{(given)}.$$

The LCD for $\frac{2}{3}, -\frac{1}{2}, \frac{1}{2},$ and $-\frac{1}{3}$ is 6. Therefore, we multiply both sides of the equation by 6.

$$6\left(\frac{2}{3}x - \frac{1}{2}\right) = 6\left(\frac{1}{2}x - \frac{1}{3}\right) \quad \text{(multiplying both sides by 6)}.$$

$$6 \cdot \frac{2}{3}x + 6 \cdot \left(-\frac{1}{2}\right) = 6 \cdot \frac{1}{2}x + 6\left(-\frac{1}{3}\right) \quad \text{(by means of the distributive property)}.$$

$$4x - 3 = 3x - 2.$$
$$4x - 3 + 3 = 3x - 2 + 3 \quad \text{(adding 3 to both sides)}.$$
$$4x = 3x + 1.$$
$$-3x + 4x = -3x + 3x + 1 \quad \text{(adding } -3x \text{ to both sides)}.$$
$$x = 1.$$

Note that the answer is the same as before, although obtained in a different way. Combining with integers is easier than combining fractions, so we see that the LCD is a valuable tool in solving equations as it enables us to eliminate fractions.

EXAMPLE 9

Solve $\frac{3}{4}y + \frac{1}{2} = \frac{2}{5}y - \frac{1}{5}.$

Solution

The LCD for $\frac{3}{4}, \frac{1}{2}, \frac{2}{5},$ and $-\frac{1}{5}$ is 20. Therefore, we multiply both sides of the equation by 20.

$$\frac{3}{4}y + \frac{1}{2} = \frac{2}{5}y - \frac{1}{5} \quad \text{(given)}.$$

$$20\left(\frac{3}{4}y + \frac{1}{2}\right) = 20\left(\frac{2}{5}y - \frac{1}{5}\right) \quad \text{(multiplying both sides by 20)}.$$

$$20 \cdot \frac{3}{4}y + 20 \cdot \frac{1}{2} = 20 \cdot \frac{2}{5}y + 20\left(-\frac{1}{5}\right) \quad \text{(by means of the distributive property)}.$$

$$15y + 10 = 8y - 4.$$
$$15y + 10 - 10 = 8y - 4 - 10 \quad \text{(adding } -10 \text{ to both sides)}.$$

$$15y = 8y - 14.$$
$$-8y + 15y = -8y + 8y - 14 \quad \text{(adding } -8y \text{ to both sides)}.$$
$$7y = -14.$$
$$\frac{1}{7} \cdot 7y = \frac{1}{7}(-14) \quad \left(\text{multiplying both sides by } \frac{1}{7}\right).$$
$$y = -2.$$

Check: $\frac{3}{4}(-2) + \frac{1}{2} = \frac{2}{5}(-2) - \frac{1}{5}.$

$$-\frac{3}{2} + \frac{1}{2} = -\frac{4}{5} - \frac{1}{5}.$$
$$-\frac{2}{2} = -\frac{5}{5}.$$
$$-1 = -1.$$

This solution shows a complete step-by-step process for solving the given equation. Many times some of these steps can be performed mentally. For example, once we have eliminated the fractions in Example 9 and obtained an equivalent equation, $15y + 10 = 8y - 4$, the addition of -10 to both sides of the equation can be done mentally, as can the addition of $-8y$. Some students may perform both of the operations simultaneously. For example,

$$\begin{array}{c} 15y - 10 = 8y - 4 \\ 15y = 8y - 14 \\ 7y = -14 \\ y = -2 \end{array} \quad \text{or} \quad \begin{array}{c} 15y - 10 = 8y - 4 \\ 7y = -14 \\ y = -2 \end{array}$$

We are not advocating one technique over another. In fact, you may feel that it is necessary to perform each of the steps separately, as shown in Example 9. That may be the best choice. The important thing to remember is that we want to be able to solve the equation *correctly*!

EXAMPLE 10

Solve $\frac{1}{3}x + \frac{3}{4} = \frac{3}{4}x - \frac{1}{2}.$

First Degree Equations and Inequalities in One Variable

Solution

The LCD for $\frac{1}{3}, \frac{3}{4}$, and $-\frac{1}{2}$ is 12. Therefore, we multiply both sides of the equation by 12.

$$\frac{1}{3}x + \frac{3}{4} = \frac{3}{4}x - \frac{1}{2} \quad \text{(given)}.$$

$$12\left(\frac{1}{3}x + \frac{3}{4}\right) = 12\left(\frac{3}{4}x - \frac{1}{2}\right) \quad \text{(multiplying both sides by 12)}.$$

$$12 \cdot \frac{1}{3}x + 12 \cdot \frac{3}{4} = 12 \cdot \frac{3}{4}x + 12\left(-\frac{1}{2}\right) \quad \text{(applying the distributive property)}.$$

$$4x + 9 = 9x - 6.$$

$$4x + 9 - 9 = 9x - 6 - 9 \quad \text{(adding } -9 \text{ to both sides)}.$$

$$4x = 9x - 15.$$

$$-9x + 4x = -9x + 9x - 15 \quad \text{(adding } -9x \text{ to both sides)}.$$

$$-5x = -15.$$

$$\left(-\frac{1}{5}\right)(-5x) = \left(-\frac{1}{5}\right)(-15) \quad \left(\text{multiplying both sides by } -\frac{1}{5}\right).$$

$$x = +3.$$

Check: $\quad \frac{1}{3} \cdot 3 + \frac{3}{4} = \frac{3}{4} \cdot 3 - \frac{1}{2}.$

$$\frac{3}{3} + \frac{3}{4} = \frac{9}{4} - \frac{1}{2}.$$

$$\frac{21}{12} = \frac{7}{4}.$$

$$\frac{7}{4} = \frac{7}{4}.$$

An equation may contain fractions in various ways. For example, $\frac{3-2x}{3} + \frac{x-2}{2} = 4$ is an equation that contains fractions. The LCD for $\frac{3-2x}{3}, \frac{x-2}{2}$, and 4 is 6. Therefore, we multiply both sides of the equation by 6:

$$6 \cdot \left(\frac{3-2x}{3} + \frac{x-2}{2}\right) = 6 \cdot 4.$$

$$6 \cdot \left(\frac{3-2x}{3}\right) + 6 \cdot \left(\frac{x-2}{2}\right) = 24.$$

$$2(3 - 2x) + 3(x - 2) = 24.$$
$$6 - 4x + 3x - 6 = 24.$$

Simplifying this equivalent equation, we have

$$-x = 24.$$

Multiplying both sides by -1, we have

$$x = -24.$$

Check: $\dfrac{3 - 2 \cdot (-24)}{3} + \dfrac{-24 - 2}{2} = 4.$

$$\dfrac{3 + 48}{3} + \dfrac{-26}{2} = 4.$$
$$\dfrac{51}{3} - 13 = 4.$$
$$17 - 13 = 4.$$
$$4 = 4.$$

Sometimes an equation will contain both fractions and parentheses. For example, recall the second equation listed at the beginning of this section, $\frac{3}{4}(y - 8) = \frac{1}{2}y + 5$. In solving this equation, we will again make use of the LCD. We use the distributive property to eliminate the parentheses:

$$\frac{3}{4}y + \frac{3}{4}(-8) = \frac{1}{2}y + 5.$$
$$\frac{3}{4}y - 6 = \frac{1}{2}y + 5.$$

The LCD for $\frac{3}{4}, -6, \frac{1}{2}$, and 5 is 4, hence we multiply both sides of the equation by 4:

$$4\left(\frac{3}{4}y - 6\right) = 4\left(\frac{1}{2}y + 5\right).$$
$$4 \cdot \frac{3}{4}y + 4(-6) = 4 \cdot \frac{1}{2}y + 4 \cdot 5.$$
$$3y - 24 = 2y + 20.$$

Now we are ready to transform the equation into an equivalent equation of the general form $ay = c$. Combining the next two steps, we add 24 to both

First Degree Equations and Inequalities in One Variable

sides of the equation and also add $-2y$ to both members, that is, add $24 - 2y$ to both sides of the equation.

$$3y - 24 + (24 - 2y) = 2y + 20 + (24 - 2y).$$

Simplifying this equation, we have

$$y = 44.$$

Check: $\frac{3}{4}(44 - 8) = \frac{1}{2} \cdot 44 + 5.$

$$\frac{3}{4}(36) = 22 + 5.$$

$$27 = 27.$$

EXAMPLE 11

Solve $\frac{1}{2}(3z - 3) = \frac{1}{3}(4z - 8)$.

Solution

The given equation contains parentheses, which we eliminate first. Applying the distributive property, we have

$$\frac{1}{2} \cdot 3z + \frac{1}{2}(-3) = \frac{1}{3} \cdot 4z + \frac{1}{3}(-8).$$

$$\frac{3}{2}z - \frac{3}{2} = \frac{4}{3}z - \frac{8}{3}.$$

The LCD for $\frac{3}{2}, -\frac{3}{2}, \frac{4}{3}$, and $-\frac{8}{3}$ is 6. Therefore, we multiply both sides of the equation by 6:

$$6\left(\frac{3}{2}z - \frac{3}{2}\right) = 6\left(\frac{4}{3}z - \frac{8}{3}\right).$$

$$6 \cdot \frac{3}{2}z + 6\left(-\frac{3}{2}\right) = 6 \cdot \frac{4}{3}z + 6\left(-\frac{8}{3}\right).$$

$$9z - 9 = 8z - 16.$$

Next, we add 9 and $-8z$ to both sides of the equation, that is, we add $9 - 8z$ to both members.

$$9z - 9 + (9 - 8z) = 8z - 16 + (9 - 8z).$$

Simplifying this equation, we have

$$z = -7.$$

Check: $\frac{1}{2}(3(-7) - 3) = \frac{1}{3}(4(-7) - 8).$

$$\frac{1}{2}(-21 - 3) = \frac{1}{3}(-28 - 8).$$

$$\frac{1}{2}(-24) = \frac{1}{3}(-36).$$

$$-12 = -12.$$

Equations can also contain decimals. See for example, the third equation listed at the beginning of this section: $0.25z + 1.3 = 0.2z - 3.7$. In solving an equation of this nature, we can again use the LCD (in a manner of speaking). Note that the given equation, $0.25z + 1.3 = 0.2z - 3.7$, can also be rewritten as an equation that contains fractions, $\frac{25}{100}z + 1\frac{3}{10} = \frac{2}{10}z - 3\frac{7}{10}$ or $\frac{25}{100}z + \frac{13}{10} = \frac{2}{10}z - \frac{37}{10}$. The LCD for this equivalent equation is 100, hence we multiply both sides of the equation by 100,

$$100\left(\frac{25}{100}z + \frac{13}{10}\right) = 100\left(\frac{2}{10}z - \frac{37}{10}\right).$$

$$25z + 130 = 20z - 370.$$

Next we add -130 and $-20z$ to both sides of the equation, that is, we add the quantity $(-130 - 20z)$ to both members.

$$25z + 130 + (-130 - 20z) = 20z - 370 + (-130 - 20z).$$

Simplifying this equation, we have

$$5z = -500.$$

Multiplying both sides by $\frac{1}{5}$ (the reciprocal of 5),

$$\frac{1}{5} \cdot 5z = \frac{1}{5}(-500).$$

therefore $z = -100.$

Check: $0.25(-100) + 1.3 = 0.2(-100) - 3.7.$
$-25 + 1.3 = -20 - 3.7.$
$-23.7 = -23.7.$

First Degree Equations and Inequalities in One Variable

It is not necessary to become involved with fractions to solve an equation containing decimals. Note that when a decimal is multiplied by 100, the decimal point is moved two places to the right. For example, 0.25 (100) = 25. Similarly,

$$0.02\ (100) = 2.$$
$$0.13\ (100) = 13.$$
$$0.045\ (100) = 4.5.$$

If a decimal is multiplied by 10, the decimal point is moved one place to the right. For example, 0.2 (10) = 2. If a decimal is multiplied by 1000, the decimal point is moved three places to the right. For example, 0.125 (1000) = 125 and 3.12 (1000) = 3120. Instead of converting each decimal in an equation to a fraction and then finding the least common denominator, we can multiply both sides of the equation by a power of 10: 10, 100, 1000, etc. The power of 10 that we multiply by depends upon the greatest number of decimal places contained in the equation. For example, in the equation $2.3x + 4.12 = 5x - 3.15$, there are at most two decimal places, therefore, we would multiply both members by 100. In the equation $4y + 3.7 = 2.50y - 4.125$, there are at most three decimal places, therefore, we would multiply both members by 1000. The original equation for this particular discussion, $0.25z + 1.3 = 0.2z - 3.7$ contains at most two decimal places in any of the decimal numbers. Therefore, we multiply both sides of the equation by 100,

$$100(0.25z + 1.3) = 100(0.2z - 3.7).$$
$$100(0.25z) + 100(1.3) = 100(0.2z) + 100(-3.7).$$
$$25z + 130 = 20z - 370.$$

Now we can transform the equation into an equivalent equation of the general form $az = c$. Hence, we add $(-130 - 20z)$ to both members.

$$25z + 130 + (-130 - 20z) = 20z - 370 + (-130 - 20z).$$

Simplifying this equation, we have

$$5z = -500.$$
$$\text{and}\quad z = -100.$$

This is exactly the same solution as we obtained before, though solved in a slightly different manner. If an equation contains decimal numbers, we can eliminate the decimals by multiplying both members by a power of ten and solve the resulting equivalent equation in the same manner as done previously.

EXAMPLE 12

Solve $2x - 5.7 = 14.3$.

Solution

The given equation contains decimal numbers that contain at most one decimal place in any of the decimals. Therefore, we multiply both sides of the equation by 10.

$$10(2x - 5.7) = 10(14.3).$$
$$10(2x) + 10(-5.7) = 143.$$
$$20x - 57 = 143.$$
$$20x - 57 + 57 = 143 + 57 \quad \text{(adding 57 to both sides).}$$
$$20x = 200.$$
$$\frac{1}{20} \cdot 20x = \frac{1}{20} \cdot 200 \quad \left(\text{multiplying both sides by } \frac{1}{20}\right).$$
$$x = 10.$$

Check:
$$2(10) - 5.7 = 14.3$$
$$20 - 5.7 = 14.3.$$
$$14.3 = 14.3.$$

EXAMPLE 13

Solve $2.2(3y - 4) = 4y + 12$.

Solution

The given equation contains a decimal number and also parentheses. The first thing we do is eliminate the parentheses. Therefore,

$$2.2(3y - 4) = 4y + 12.$$
$$6.6y - 8.8 = 4y + 12.$$

The decimal numbers contain at most one decimal place, therefore, we multiply both sides of the equation by 10.

$$10(6.6y - 8.8) = 10(4y + 12).$$
$$10(6.6y) + 10(-8.8) = 10(4y) + 10(12).$$
$$66y - 88 = 40y + 210.$$
$$66y - 88 + (88 - 40y) = 40y + 120 + (88 - 40y)$$

(adding $88 - 40y$ to both sides).

$$26y = 208.$$
$$\frac{1}{26} \cdot 26y = \frac{1}{26} \cdot 208 \quad \left(\text{multiplying both sides by } \frac{1}{26}\right).$$
$$y = 8.$$

First Degree Equations and Inequalities in One Variable

Check: $2.2(3(8) - 4) = 4(8) + 12.$
$2.2(24 - 4) = 32 + 12.$
$2.2(20) = 44.$
$44 = 44.$

In this section, we have examined first degree equations that contained fractions, parentheses and decimals. Once these equations have been transformed into equivalent equations that do not contain parentheses, fractions, or decimals, we can solve them using the addition and multiplication properties of equality. To solve a general first degree equation, we perform the following steps (if necessary):

1. If an equation contains parentheses, eliminate the parentheses by means of the distributive property.
2. If an equation contains fractions, eliminate the fractions by multiplying both members by the least common denominator (LCD).
3. If an equation contains decimal numbers, eliminate the decimals by multiplying both members by the appropriate power of 10.
4. Perform the indicated operations and simplify when possible.
5. Apply the addition property of equality to obtain an equivalent equation of the form $ax = c$.
6. Apply the multiplication property of equality to obtain an equivalent equation of the form $x = k$, which is the solution of the given equation.
7. Check the solution by substituting it for the variable in the given equation.

Note that in solving a general first degree equation, parentheses should first be eliminated, then fractions, and then decimals. A good memory device for this order is "PFD," which can represent a name or phrase, such as "Philadelphia Fire Department."

In many professions—business, engineering, aviation, navigation, medicine—errors can be costly and sometimes fatal. Mathematics makes us appreciate the importance of avoiding errors. It gives us the opportunity to recognize errors through a constant check of our work. If we strive to develop this habit of "checking," we will be able to do it quickly and efficiently, almost automatically. Most importantly, we will know whether or not our work is correct.

EXERCISES FOR SECTION 2.3
(Answers given to exercises marked *)

Solve the following equations and check the solutions.

*1. $\frac{1}{2}x + \frac{1}{3}x = 25$.

2. $\frac{2}{3}y - 2 = 8$.

*3. $\frac{2z}{3} - \frac{z}{2} = 2$.

4. $\frac{2y}{3} - \frac{3y}{5} = 1$.

*5. $\frac{4}{3}x = \frac{x}{6} + \frac{7}{2}$.

6. $\frac{1}{3}y + 5 = \frac{1}{6}y$.

*7. $\frac{z}{3} + \frac{8}{4} = 5$.

8. $\frac{x}{3} + \frac{3x}{4} = \frac{13}{2}$.

*9. $\frac{x-1}{3} = \frac{x+2}{6}$.

10. $4(x - 1) - 3(x - 2) = 10$.
*11. $4(z + 1) - 3(z - 3) = 25$.
12. $2y - 4(y + 3) = 4$.
*13. $4x - (x - 10) - 40 = 0$.
14. $3z = 2(5 - z)$.
*15. $6y - 3(y - 4) = 9$.

16. $\frac{1}{2}(2 - x) = \frac{1}{3}\left(\frac{x}{2} - 1\right)$.

*17. $\frac{2}{3}(2y - 1) - \frac{3}{4}(3y - 7) = 0$.

18. $\frac{1}{4}\left(-\frac{x}{3} - \frac{1}{2}\right) = \frac{1}{2}\left(\frac{x}{3} - \frac{1}{4}\right)$.

*19. $\frac{1}{2}(x + 1) = \frac{1}{2}\left(-\frac{x}{2} - \frac{1}{2}\right)$.

20. $\frac{1}{2}(y - 2) = \frac{1}{3}\left(1 - \frac{y}{2}\right)$.

*21. $\frac{1}{2}\left(\frac{x}{4} + \frac{1}{2}\right) = \frac{1}{2}\left(1 - \frac{x}{4}\right)$.

22. $.15x = 5.25$.
*23. $3y + 6.3 = 12.6$.
24. $0.20z - 0.16 = -0.03z + 0.30$.
*25. $2x + 3.4 = 6.8$.
26. $9x + 2.1 = 2.1 - 5x$.
*27. $2z + 5.7 = 26.3$.
28. $1.1x = 7.6 - 0.42x$.
*29. $2.9y - 20 = 1.2y - 3$.
30. $0.3t - 2.4 = 0.26t + 0.36$.
*31. $0.8x - 1.1 = 4.9 - 0.2x$.
32. $1.8z - 2.3 = 3.3z + 5.2$.
*33. $0.03x + 0.62 = -0.02x + 0.57$.

2.4 LITERAL EQUATIONS

An equation may often contain more than one variable, or it may contain other symbols representing constants, such as a, b, c, π, and so on. For example, $2x + a = b$, $ax + b = c$, and $V = lwh$ are *literal equations*. We can solve literal equations for a specific variable or for one of the other symbols in the given literal equation. The solution will be expressed in terms of the remaining symbols by using the same techniques that we developed in the previous sections. In other words, the procedure used in solving literal equations is the same as that used in solving any first degree equation. For example, to solve the equation $2x + a = b$ for x, we proceed in the following manner:

$2x + a = b$ (given).

First Degree Equations and Inequalities in One Variable

$$2x + a - a = b - a \quad \text{(adding } -a \text{ to both sides).}$$
$$2x = b - a.$$
$$\tfrac{1}{2} \cdot 2x = \tfrac{1}{2}(b - a) \quad \left(\text{multiplying both sides by } \tfrac{1}{2}\right).$$
$$x = \frac{b - a}{2}.$$

The solution is $\frac{b-a}{2}$, and we say that x has been expressed in terms of b and a. We can also check the solution in the same manner as previously:

$$\text{Check:} \quad 2\left(\frac{b-a}{2}\right) + a = b.$$
$$b - a + a = b.$$
$$b = b.$$

Many problems involve the use of formulas. The solution depends upon the use of a certain formula. Normally, these formulas contain more than one letter or variable to express a certain relationship. For example, to find the perimeter of a rectangle, we use the formula

$$\boxed{P = 2W + 2L}$$

where W represents the width and L represents the length of the rectangle. Suppose that we wish to solve this equation for L, that is, express L in terms of P and W. We proceed in the same manner as before.

$$P = 2W + 2L \quad \text{(given).}$$
$$P - 2W = 2W - 2W + 2L \quad \text{(adding } -2W \text{ to both sides).}$$
$$P - 2W = 2L.$$
$$\tfrac{1}{2}(P - 2W) = \tfrac{1}{2} \cdot 2L \quad \left(\text{multiplying both sides by } \tfrac{1}{2}\right).$$
$$\frac{P - 2W}{2} = L.$$

We have expressed L in terms of P and W. The fact that L is on the right hand side of the equation should not bother us. It does not matter whether we express the solution to an equation as x = a or a = x. Both are equally correct. Normally we express the answer as x = a because of our orientation of working from left to right, not right to left.

EXAMPLE 14

How long will it take a plane to fly 650 miles if its average speed is 520 miles per hour?

Solution

The formula needed to solve this problem is the distance formula, $d = rt$, where d represents the distance traveled; r represents the rate of speed; and t represents the elapsed time. Since we want to find the amount of time, t, to complete the trip, we should solve the formula for t.

$$d = rt \quad \text{(given)}.$$

$$\frac{1}{r} \cdot d = \frac{1}{r} \cdot rt \quad \left(\text{multiplying both sides by } \frac{1}{r}\right).$$

$$\frac{d}{r} = t.$$

Substituting the given values for d and r, we can determine the value of t.

$$\frac{d}{r} = t.$$

$$\frac{650}{520} = t.$$

$$\frac{5}{4} = t.$$

We find that the time taken for the trip is $\frac{5}{4}$ hours, or $1\frac{1}{4}$ hours, or 75 minutes.

$$\text{Check:} \quad 650 = 520 \cdot \frac{5}{4}.$$

$$650 = 650.$$

EXAMPLE 15

Given the formula $a = bc - d$, express b in terms of a, c and d.

Solution

We are asked to express b in terms of a, c and d, that is, we must solve the given formula for b.

First Degree Equations and Inequalities in One Variable

$$a = bc - d \quad \text{(given)}.$$
$$a + d = bc - d + d \quad \text{(adding d to both sides)}.$$
$$a + d = bc.$$
$$\frac{1}{c}(a + d) = \frac{1}{c} \cdot (bc) \quad \left(\text{multiplying both sides by } \frac{1}{c}\right).$$
$$\frac{a + d}{c} = b.$$

We have expressed b in terms of a, c and d.

EXAMPLE 16

Solve $y = \frac{n}{2}(x + z)$ for z.

Solution

In order to solve this equation for z, we apply the same techniques used in the previous section. First, we eliminate parentheses and fractions.

$$y = \frac{n}{2}(x + z) \quad \text{(given)}.$$

$$y = \frac{nx}{2} + \frac{nz}{2} \quad \text{(applying the distributive property)}.$$

Next, we multiply both sides of the equation by the LCD, which is 2.

$$2y = 2\left(\frac{nx}{2} + \frac{nz}{2}\right).$$

$$2y = nx + nz.$$

$$2y - nx = -nx + nx + nz \quad \text{(adding } -nx \text{ to both sides)}.$$

$$2y - nx = nz.$$

$$\frac{1}{n}(2y - nx) = \frac{1}{n} \cdot nz \quad \left(\text{multiplying both sides by } \frac{1}{n}\right).$$

$$\frac{2y - nx}{n} = z.$$

We have solved the given equation for z.

EXERCISES FOR SECTION 2.4
(Answers given to exercises marked *)

Solve the given equations for x:
* *1. $2x - b = a$.
* *3. $4ax + b = c$.
* *5. $5x + 2a = 3x + 4b$.
* 2. $3x - b - c = 0$.
* 4. $3x - b = x + c$.
* 6. $2x + 3a = 5x - 4b$.

Solve the given equations for y.
* *7. $3y - 2b = 4a$.
* *9. $a(y + b) = 3x$.
* *11. $4(y - c) = 2(y + h) + a$.
* 8. $3y + b = y - a$.
* 10. $3(y - b) = 2(y + a)$.
* 12. $\dfrac{y + b}{2} = \dfrac{y - a}{10}$.

Solve the given formulas for the specified variable:
* *13. $d = rt$. (for t)
* *15. $i = prt$. (for t)
* *17. $A = \dfrac{1}{2} bh$. (for h)
* *19. $C = 2\pi r$. (for r)
* *21. $V = \dfrac{1}{3} bh$. (for b)
* *23. $C = \dfrac{5}{9} (F - 32)$. (for F)
* *25. $A = \dfrac{h}{2} (b + B)$. (for h)
* *27. $S = 2\pi rh + 2\pi r^2$. (for h)
* *29. $y = mx + b$. (for m)
* 14. $d = rt$. (for r)
* 16. $i = prt$. (for p)
* 18. $A = \dfrac{1}{2} bh$. (for b)
* 20. $V = \dfrac{1}{3} bh$. (for h)
* 22. $F = \dfrac{9}{5} C + 32$. (for C)
* 24. $A = \dfrac{h}{2} (b + B)$. (for B)
* 26. $V = \dfrac{1}{3} \pi r^2 h$. (for h)
* 28. $y = mx + b$. (for x)
* 30. $s = vt + \dfrac{1}{2} gt^2$. (for v)

HISTORICAL NOTE

We mentioned in Chapter One that the study of algebra began in early Egypt. The first algebra-type problems occurred in a papyrus copied by Ahmes. This document also contained one of the oldest verbal problems in the world. It was stated as "Hau, its whole, its seventh, it makes nineteen," which roughly translates to "find a number such that if the whole number is added to one-seventh of it, then the result is nineteen." This problem was considered quite difficult by the Egyptians and its solution was complicated. If we translate the English phrases into mathematical phrases, we can obtain an equation and determine the solution to the problem. The desired equation is:

$$y + \frac{y}{7} = 19.$$

First Degree Equations and Inequalities in One Variable 101

Riddles and puzzles also helped to develop the study of word problems. Many ancient texts or manuscripts contained sets of various types of riddles. The following is an example of a very old children's rhyme from England:

"As I was going to St. Ives,
I met a man with seven wives,
Every wife had seven sacks,
Every sack had seven cats,
Every cat had seven kits,
Kits, cats, sacks, and wives—
How many were going to St. Ives?"

2.5 WORD PROBLEMS

First degree equations can be used to solve many word or verbal problems, by translating English sentences into mathematical sentences (equations). If the translations are done *correctly,* then the solutions to the equations are solutions to the verbal problems.

Certain procedures should be followed in solving word problems. The first step is to read the problem carefully and the second step is to reread the problem carefully. At this point we must translate the verbal statements into mathematical statements. For most students, this is the most difficult step. However, if we are able to recognize certain key phrases as mathematical phrases or mathematical operations, then our task will be simplified. It is most important to be able to recognize those key phrases that translate into the four basic operations: addition, subtraction, multiplication and division. We already know some of the basic words or phrases and their mathematical representation. For example, *plus* is represented by $+$; *minus* is represented by $-$; *times* is normally represented by $\cdot$, as in $2 \cdot a$ or "2 times a"; *divided by* is represented by $\div$, but usually in an equation it is represented by the fraction, $\frac{x}{2}$, as in "x divided by 2."

Mathematical translations of common English phrases are listed in the following table. The underlined phrase is represented in the corresponding mathematical phrase. Note that we can choose a variety of letters to represent the variable (unknown). Normally x is chosen, but it does not have to be.

English Phrase	Mathematical Phrase
Addition	
4 <u>more than</u> a number	$n + 4$
a number <u>increased by</u> 4	$x + 4$
the <u>sum</u> of x and y	$x + y$

English Phrase	Mathematical Phrase
a number <u>added to</u> 4	$4 + n$
x <u>plus</u> y	$x + y$

Subtraction

English Phrase	Mathematical Phrase
a number <u>decreased by</u> 5	$y - 5$
5 <u>less than</u> a number	$y - 5$
the <u>difference</u> between x and y	$x - y$
x <u>minus</u> y	$x - y$

Multiplication

English Phrase	Mathematical Phrase
the <u>product</u> of a and b	$a \cdot b$
x <u>multiplied by</u> a	$a \cdot x$
<u>twice</u> a number	$2 \cdot n$
$\frac{1}{2}$ <u>of</u> y	$\frac{1}{2} \cdot y$
32 <u>per cent</u> of z	$(0.32)(z)$
x <u>times</u> y	$x \cdot y$

Division

English Phrase	Mathematical Phrase
the <u>quotient</u> of x and y	$\frac{x}{y}$
the <u>quotient</u> of y and x	$\frac{y}{x}$
the <u>ratio</u> of x and y	$\frac{x}{y}$
x <u>divided by</u> y	$\frac{x}{y}$

The word "is" or some form of the verb "to be" is often used to indicate equality (=). For example, the English sentence, "The sum of a number and 10 is 32," is translated into the mathematical sentence, $x + 10 = 32$. Following are some examples of English sentences translated into mathematical sentences.

First Degree Equations and Inequalities in One Variable

English Sentence	Mathematical Sentence
A number increased by 2 is 6.	$x + 2 = 6.$
The product of a number and 3 is 18.	$3y = 18.$
The difference between a number and 3 is 10.	$n - 3 = 10.$
Two less than four times a number is 34.	$4z - 2 = 34.$
One half of a number and six more is 14.	$\frac{1}{2}n + 6 = 14.$
Two more than twice a number is 8.	$2x + 2 = 8.$
If a number is increased by 4, twice the sum is 50.	$2(y + 4) = 50.$

In order to solve a word or verbal problem, it is necessary to translate the English sentence into an equation. This equation should express the relationships stated in the problem. After reading the problem carefully, we let the unknown be represented by a variable such as x, y, z, m, n, etc., and then express the verbal relationships in a mathematical sentence. Admittedly, some of the following examples could be solved without the use of algebra, but keep in mind that we are trying to develop skill in setting up equations.

EXAMPLE 17

A tennis court is 6 feet longer than twice its width. Express the length of the tennis court in terms of its width.

Solution

Let w represent the width of the tennis court. The phrase "6 feet longer" is represented by $+6$ and "twice its width" is represented by 2W. Therefore, we have:

$$\text{length} = 2W + 6.$$

EXAMPLE 18

A number increased by 3 is 10.

Solution

Let x equal the unknown. The phrase "increased by 3" means "+3". Another word for "is" is "equals." Therefore, we have:

$$x + 3 = 10.$$

EXAMPLE 19

The sum of a number and twice that number is 21.

Solution

Let y equal the unknown. Twice that number can be expressed as 2y. Therefore, we have:

$$y + 2y = 21.$$

EXAMPLE 20

One half of a number and 10 more is 40.

Solution

Let n equal the unknown. One half of that number can be expressed as $\frac{1}{2}n$ or $\frac{n}{2}$. The phrase "and 10 more" indicates that we add 10 to $\frac{n}{2}$. Therefore, we have:

$$\frac{n}{2} + 10 = 40.$$

EXAMPLE 21

The sum of two consecutive integers is 39.

Solution

Let x equal the first integer. The next (consecutive) integer is $x + 1$. The sum of two consecutive integers is $x + (x + 1)$. Therefore, we have:

$$x + (x + 1) = 39.$$

EXAMPLE 22

Joe is two years younger than Scott. The sum of their ages is 14. Write an equation for this problem.

Solution

Let y equal Scott's age, then Joe's age is $y - 2$ (two years younger). Since the sum of their ages is 14, we have:

$$y + (y - 2) = 14.$$

Sometimes solving a word problem will require use of an unstated equality relationship. When this occurs, the relationship is implied and is

First Degree Equations and Inequalities in One Variable

assumed to be known. Examples of relationships that are assumed to be known are: distance equals rate times time, the area of a triangle is one-half the product of its base and height, the perimeter of a rectangle is the sum of twice its length and twice its width; the sum of the interior angles of a triangle is 180 degrees.

Following is a list of suggestions that will aid us in solving word or verbal problems:

1. Read the problem carefully;
2. Reread the problem carefully;
3. If possible, draw a diagram to assist in interpreting the given information;
4. Translate the English phrases into mathematical phrases and choose a variable for the unknown quantity;
5. Write the equation using all of the above information;
6. Solve the equation;
7. Check the solution to determine whether it satisfies the original problem.

EXAMPLE 23

If 11 is subtracted from four times a number, the result is 17. Find the number.

Solution

After reading the problem carefully, let x represent the number to be determined, hence four times the number is represented by 4x. Eleven subtracted from four times a number is expressed as $4x - 11$. The problem states that the result is 17, therefore, $4x - 11 = 17$. We solve the equation:

$$4x - 11 = 17.$$
$$4x = 28.$$
$$x = 7.$$

Check: $4 \cdot 7 - 11 = 17.$
$28 - 11 = 17.$
$17 = 17.$

EXAMPLE 24

The perimeter of a rectangular swimming pool is 100 feet. If the width is 2 feet more than half its length, find the length and width of the pool.

Solution

We must find both the length and width of the pool. Although both dimensions are unknown, we let y represent the length. We are given that "the width is 2 feet more than half its length" which can now be expressed as $\frac{y}{2} + 2$.

Since the perimeter of a rectangle is the sum of twice its length and twice its width, we obtain the equation $100 = 2y + 2\left(\frac{y}{2} + 2\right)$. Solving this equation, we have:

$$100 = 2y + 2\left(\frac{y}{2} + 2\right).$$
$$100 = 2y + y + 4.$$
$$100 = 3y + 4.$$
$$96 = 3y.$$
$$32 = y.$$

We find that the length of the rectangular swimming pool is 32 feet. Since the width is expressed in terms of the length, we have

$$\text{width} = \frac{y}{2} + 2.$$
$$= \frac{32}{2} + 2.$$
$$= 16 + 2.$$
$$= 18.$$

The length and width of the pool are 32 feet and 18 feet, respectively.

Check: $100 = 2 \cdot 32 + 2 \cdot 18.$
$100 = 64 + 36.$
$100 = 100.$

Uniform motion problems are those in which an object is moving at a constant rate of speed. To solve verbal problems concerning rate of speed (velocity), we must use the formula

First Degree Equations and Inequalities in One Variable

$$d = rt$$

where d represents the distance traveled, r represents the rate of speed, and t represents the time. Note that d and r must be expressed in the same unit of distance, and that r and t must be expressed in the same unit of time.

EXAMPLE 25

Two moving vans start at the same time from the same place. One van travels east at 45 miles per hour and the other van travels west at 55 miles per hour. In how many hours are they 500 miles apart?

Solution

Let t represent the time that each travels.

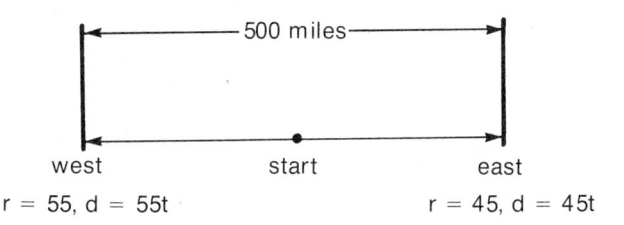

```
            ←——————— 500 miles ———————→
            |                            |
            |                            |
            ————————————•————————————
          west          start           east
       r = 55, d = 55t              r = 45, d = 45t
```

The van traveling east goes at a rate of 45 mph and since t is the time it traveled, its distance is 45t (d = rt). The van traveling west goes at a rate of 55 mph, and its distance traveled is 55t. The distance that one van traveled plus the distance that the other van traveled (in opposite directions) will equal the total distance that they are apart (500 miles). Therefore, the resulting equation is

$$55t + 45t = 500.$$
Solving, we obtain $\quad 100t = 500.$
$$t = 5 \text{ hrs.}$$
Check: $\quad 55 \cdot 5 + 45 \cdot 5 = 500.$
$$275 + 225 = 500.$$
$$500 = 500.$$

Although the next example involves coins, it can be thought of as a type of *mixture* problem. A helpful hint to remember in solving mixture problems is that in order to find the total value of something, we must multiply the unit value of the item times the number (amount) of items. For example, if we have 5 dimes, then the total value of the dimes is 50 cents:

$$10(5) = 50.$$

unit value   amount   total value

EXAMPLE 26

Scott has 95 cents in his pocket. If he has only nickels and quarters in his pocket and he has 11 coins altogether, how many nickels and quarters does he have?

Solution

Let x represent the number of quarters that Scott has in his pocket; then $(11 - x)$ represents the number of nickels. If Scott has x quarters, then the value of these quarters is $25x$. Similarly, the value of the nickels is $5(11 - x)$. The total value of the coins in his pocket is 95 cents. Therefore, the resulting equation is

$$25x + 5(11 - x) = 95.$$

value of quarters   value of nickels   total value

Solving this equation, we have

$$25x + 5(11 - x) = 95.$$
$$25x + 55 - 5x = 95.$$
$$20x = 40.$$
$$x = 2. \quad \text{quarters}$$
$$11 - x = 9. \quad \text{nickels}$$

Check: $25 \cdot 2 + 9 \cdot 5 = 95.$
        $50 + 45 = 95.$ $2 + 9 = 11.$
        $95 = 95.$ $11 = 11.$ (number of coins)

In solving integer problems, we should keep in mind that consecutive integers differ by 1. In other words, if we let x represent an integer, then the next consecutive integer would be represented by $x + 1$. Similarly, if y represents an even integer, then $y + 2$ would represent the next consecutive even integer; and if z represents an odd integer then $z + 2$ represents the next consecutive odd integer.

First Degree Equations and Inequalities in One Variable

EXAMPLE 27

The sum of three consecutive integers is 99. Find the integers.

Solution

Let z represent the first integer, then the second consecutive integer is $z + 1$. The third consecutive integer is one more than the second and therefore is represented by $z + 1 + 1$, or $z + 2$. We now have

$z =$ the first integer.
$z + 1 =$ the second consecutive integer.
$z + 2 =$ the third consecutive integer.

Since their sum is 99, the resulting equation is $z + (z + 1) + (z + 2) = 99$. Solving this equation, we have

$$z + (z + 1) + (z + 2) = 99.$$
$$3z + 3 = 99.$$
$$3z = 96.$$
$$z = 32. \quad \text{first integer}$$
$$z + 1 = 33. \quad \text{second integer}$$
$$z + 2 = 34. \quad \text{third integer}$$

Check: $32 + 33 + 34 = 99.$
$99 = 99.$

It is not uncommon for an integer problem to have no solution, that is, its solution set is the empty set. For example, consider the following problem: Find two consecutive even integers whose sum is 11. In solving this problem, we can let x represent the first even integer and then $x + 2$ represents the next consecutive even integer. The problem states that their sum is 11, therefore we obtain the equation $x + (x + 2) = 11$. Solving this equation, we have

$$x + (x + 2) = 11.$$
$$2x + 2 = 11.$$
$$2x = 9.$$
$$x = \frac{9}{2}.$$

(Field Newspapers, courtesy of Mrs. Selby Kelly.)

Since $\frac{9}{2}$ is not an integer (recall the definition of an integer), the problem has no solution. In other words, the solution set is empty set, or $\emptyset$.

EXERCISES FOR SECTION 2.5
(Answers given to exercises marked *)

Translate the following phrases into mathematical phrases.
  *1. Ten times a number.
  2. One-half the sum of 7 and a number.
  *3. Five more than three times a number.
  4. Four times the difference of a number and 5.
  *5. One more than four times a number.
  6. Seven subtracted from four times a number.
  *7. Six times the sum of a number and 8.
  8. Five increased by the product of 2 and a number.
  *9. The product of 5 and a number decreased by 3.
  10. The sum of a number and 5 subtracted from 10.

Solve each of the following and check your solution.
  *11. If 5 is added to four times a certain number, the result is 13. Find the number.
  12. If 3 is subtracted from five times a certain number, the result is 7. Find the number.
  *13. Ten less than four times a number is equal to twice that number. Find the number.
  14. One-half the sum of 5 and a number is equal to 2 less than the given number. Find the number.
  *15. Addie is 2 years younger than Bill and the sum of their ages is 70. How old is each?
  16. Ten years from now Scott will be 2 less than twice his present age. How old is Scott now?
  *17. A piece of wire 64 metres long is to be bent into the shape of a rectangle whose length is three times its width. Find the dimensions of the rectangle.

First Degree Equations and Inequalities in One Variable   111

18. The measure of the number of degrees in the largest angle of a triangle is three times the number of degrees contained in the smallest angle; the measure of the number of degrees in the second angle is twice the number of degrees contained in the smallest angle. Find the number of degrees contained in each angle of the triangle.

*19. In an isosceles triangle, the measure of each of the two equal angles is equal to one-half the measure of the largest angle. Find the number of degrees contained in each angle of the isosceles triangle.

20. The length of a rectangle is 3 centimetres less than twice its width. If the perimeter of the rectangle is 60 centimetres, find the dimensions of the rectangle.

*21. Two buses start from the same terminal at the same time. One bus travels south at 40 mph and the other bus travels north at 45 mph. In how many hours are they 510 miles apart?

22. At 12 o'clock noon, a bus leaves its terminal heading east at 40 mph. Three hours later, a second bus leaves the same terminal heading east at 60 mph. How long will it take for the second bus to overtake the first bus?

*23. Two planes took off from an airport at the same time, and flew in opposite directions. One plane traveled 120 miles per hour faster than the other. The planes were 1500 miles apart at the end of three hours. Find the rate of speed of each plane.

24. A freight train left New York City at 8:00 A.M., traveling south at 30 mph. At 11:00 A.M., a passenger train left the same station, traveling south over the same route at 55 mph. How many hours does the passenger train travel until it overtakes the freight train?

*25. Scott has $2.20 in quarters and nickels in his pocket. If he has twice as many quarters as nickels, how many of each type of coin does he have?

26. In a collection of 50 coins, the number of quarters is five more than twice the number of nickels and the number of dimes is five less than twice the number of nickels. How many of each type of coin are in the collection?

*27. Laurie emptied her bank which contained only nickels, dimes and quarters. The bank contained an equal number of nickels and dimes and 10 more quarters than dimes. If the total value of the coins was $7.30, how many of each type of coin are in the collection?

28. Joe, the newspaper carrier, has $3.25 in nickels, dimes and quarters. If he has twice as many dimes as nickels and three more quarters than nickels, how many of each type of coin does Joe have?

*29. Find three consecutive integers whose sum is 150.

30. Find three consecutive even integers whose sum is 138.

*31. Find three consecutive integers whose sum is 102.

32. Find two consecutive integers, if three times the smaller integer is 29 more than twice the larger integer.

*33. One angle of a triangle measures 12 degrees larger than a second angle. The third angle measures 20 degrees less than the sum of the other two. Find the number of degrees in each angle of the triangle.

2.6 INEQUALITIES

Expressions such as

$$3 \geq 1, -2 \leq 0, 2x - 2 \geq 6, 3x - 4 \leq 5$$

are called *inequalities*. The relationships that may exist in inequalities are: greater than (>), greater than or equal to ($\geq$), less than (<), and less than or equal to ($\leq$). For example, the inequality $5 \geq 3$ states that 5 is "greater than or equal to 3." The inequality $x < 4$ states that x is "less than 4." In other words, the variable x represents a real number that is less than 4.

Inequalities that are always true, such as

$$4 > 3, \quad 0 < 1, \quad -2 \leq 0, \quad x^2 + 1 > 0$$

are called *absolute inequalities*. These statements are always true and in the case of $x + 1 > 0$, it is true for any real number that we may substitute for x. Regardless of whether we choose a positive number, 0, or a negative number to replace x, the inequality $x^2 + 1 > 0$ will always be true. The reason that any negative number is also a solution for this inequality is that when we square a negative number, the result is positive. For example, suppose $x = -1$, then

$$x^2 + 1 > 0.$$
Let $x = -1$. $\quad (-1)^2 + 1 > 0.$
$$1 + 1 > 0.$$
$$2 > 0. \quad \text{(true)}$$

Any member of the replacement set for the variable which results in a true statement for the inequality is called a *solution* and the set of all solutions of an inequality is called the *solution set* of the inequality.

Inequalities that are true for only certain values of the variable are called *conditional inequalities*. They are not true for every replacement for the variable. The following inequalities are examples of conditional inequalities:

$$2x > 4, \quad 3y - 2 \leq 7, \quad 4z - 3 \geq z, \quad 2x - 3 < 3x + 2.$$

To solve conditional inequalities, we shall employ similar methods to those we used in solving equations. We shall perform certain operations on the inequality in order to obtain "equivalent inequalities" — inequalities that have the same solution set — until we obtain one whose solution is readily determined. As with equations, we need some fundamental properties that allow us to obtain equivalent inequalities. The first we shall consider is the addition property of inequalities. Suppose we are given the

First Degree Equations and Inequalities in One Variable

inequality $5 > 3$ and wish to add the same number to both sides of the inequality. Let's try adding 2 to both sides:

$5 > 3$, add 2 to both sides, $5 + 2 > 3 + 2$. This gives us $7 > 5$, which is also a true statement.

Next, let's try adding a negative number to the given inequality $5 > 3$. Suppose we add -10 to both sides:

$5 > 3$, add -10 to both sides, $5 + (-10) > 3 + (-10)$.
This gives us $-5 > -7$, which is also a true statement.

It is a true statement because the number farther to the right on the number line has the greater value, and -5 is to the right of -7 on the number line. These two examples illustrate the fact that adding any number (positive or negative) to both sides of an inequality produces another equivalent equality in the same direction. The inequalities $5 > 3$ and $6 > 4$ are true in the same direction or order ($>$). We can also say that these inequalities have the same *sense*.

The operation of multiplication for inequalities is one to be careful of at certain times. If we multiply both members of the inequality $5 > 3$ by 2, the result is $5 \cdot 2 > 3 \cdot 2$, which is equivalent to $10 > 6$. This inequality is true in the same order as $5 > 3$. But suppose we wish to multiply both members of the inequality $5 > 3$ by -2. Let's try it.

$$5 > 3.$$
$$5(-2) \;?\; 3(-2).$$
$$-10 \;?\; -6.$$
$$-10 < -6.$$

Note that the resulting inequality is true in the opposite direction or order; the "sense" of the inequality has been reversed. The inequality $-10 < -6$ is true because -10 is to the left of -6 on the number line. We have discovered that when we multiply an inequality by a number less than 0, we can obtain an equivalent inequality provided the sense (order) of the inequality is reversed. Here is a list of the properties that we shall need in order to obtain equivalent inequalities.

> I. If we add any number (positive or negative) to both members of an inequality, we shall obtain an equivalent inequality in the same sense (order).
> If a, b, and c are real numbers, then $a < b$ is equivalent to $a + c < b + c$.
>
> E.g., $3 < 5$ is equivalent to $3 + 2 < 5 + 2$.
> $5 < 7$.
> $3 < 5$ is equivalent to $3 + (-3) < 5 + (-3)$.
> $0 < 2$.

> **II.** If each member of an inequality is multiplied by the same *positive* number, we shall obtain an equivalent inequality in the *same* sense (order).
> If a, b, and c are real numbers and c > 0, then a < b is equivalent to ac < bc.
>
> E.g., $3 < 5$ is equivalent to $3 \cdot 2 < 5 \cdot 2$.
> $-3 < -1$ is equivalent to $(-3)(4) < (-1)(4)$.
> $-12 < -4$.
>
> **III.** If each member of an inequality is multiplied by the same *negative* number, we shall obtain an equivalent inequality in the *opposite* sense (order).
> If a, b, c are real numbers and c < 0, then a < b is equivalent to ac > bc.
>
> E.g., $3 < 5$ is equivalent to $3(-2) > 5(-2)$.
> $-6 > -10$.
> $-3 < -1$ is equivalent to $(-3)(-2) > (-1)(-2)$.
> $6 > 2$.

We have stated these inequality properties for general cases and given specific examples. It should be noted that they hold for any inequality regardless of their sense, that is, (>), (≥), (<), and (≤). Note also that we did not state the properties for subtraction and division. Subtraction is the same as adding the opposite or additive inverse of a number, and division is the same as multiplying by the reciprocal or multiplicative inverse of a number.

EXAMPLE 28

Solve the inequality $2x - 1 < 7$.

Solution

To solve a conditional inequality, we perform certain operations on it to obtain an equivalent inequality whose solution is readily determined.

$2x - 1 < 7$ (given).
$2x - 1 + 1 < 7 + 1$ (adding 1 to both members, property I).
$2x < 8$.
$\frac{1}{2}(2x) < \frac{1}{2}(8)$ (multiplying both sides by $\frac{1}{2}$, property II).
$x < 4$.

First Degree Equations and Inequalities in One Variable

Note that the sense of the inequality does not change because we multiplied both members of it by a number greater than 0. The solution set is $\{x \mid x < 4\}$, that is, the set of all x, such that x is less than 4.

EXAMPLE 29

Solve the inequality $3x - 5 < 5x + 3$.

Solution

We shall transform the given inequality into an equivalent inequality.

$3x - 5 < 5x + 3$ (given).
$3x - 5 + 5 < 5x + 3 + 5$ (adding 5 to both members, property I).
$3x < 5x + 8$.
$-5x + 3x < -5x + 5x + 8$ (adding $-5x$ to both members, property I).
$-2x < 8$.
$\left(-\frac{1}{2}\right)(-2x) > \left(-\frac{1}{2}\right)(8)$ (multiplying both sides by $-\frac{1}{2}$, property III).
$x > -4$.

Note that the sense of the inequality changed when we multiplied both members by a number less than 0. The solution set is $\{x \mid x > -4\}$.

EXAMPLE 30

Solve the inequality $-3(x + 4) \geq 3$.

Solution

As in the case of solving equations, we first eliminate the parentheses.

$-3(x + 4) \geq 3$ (given).
$-3x - 12 \geq 3$ (eliminating parentheses).
$-3x - 12 + 12 \geq 3 + 12$ (adding 12 to both members, property I).
$-3x \geq 15$.
$\left(-\frac{1}{3}\right)(-3x) \leq \left(-\frac{1}{3}\right)(15)$ (multiplying both members by $-\frac{1}{3}$, property III).
$x \leq -5$.

Note that the sense of the inequality changed (as in Example 29) because we multiplied both members by a number less than 0. The solution set is $\{x \mid x \leq -5\}$.

Besides this kind of description of the solution set for conditional inequalities, we can also draw a line graph of the solution set. In Chapter One we discussed the "real number line" and we shall use it again here to graph the solution set of conditional inequalities. Consider the inequality, x > 2. We can graph the solution set that includes all of those real numbers that are greater than 2. We indicate this by using an open circle or "hollow dot" at 2 and drawing a solid bar with an arrow, to indicate that the solution set extends to the right indefinitely. See Figure 1.

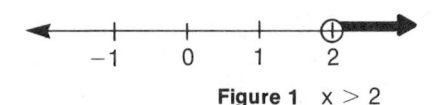

Figure 1 x > 2

If the solution set had been x ≥ 2, then the 2 would be included in the solution and the end point at 2 would also be shaded, as shown in Figure 2.

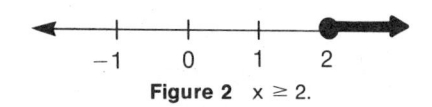

Figure 2 x ≥ 2.

If we want a graph that includes all the real numbers between −1 and 1 (excluding −1 and 1), then we must use open circles at −1 and 1 and draw a solid bar between the open circles as shown in Figure 3.

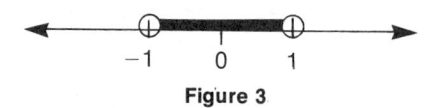

Figure 3

An algebraic statement for the graph in Figure 3 would be that −1 < x and x < 1. This expression can be shortened to read −1 < x < 1. We can interpret it as "negative 1 is less than x and x is less than 1." Another correct interpretation would be to start in the middle, "x is greater than −1 and x is less than 1." Note that if the open circles in Figure 3 were closed, or shaded in, then the graph would appear as shown in Figure 4 and the algebraic interpretation would be −1 ≤ x ≤ 1.

Figure 4 −1 ≤ x ≤ 1.

EXAMPLE 31

Solve 3x + 7 ≥ −2x − 3 and graph the solution set.

Solution

3x + 7 ≥ −2x − 3 (given).

First Degree Equations and Inequalities in One Variable

$$3x + 7 - 7 \geq -2x - 3 - 7 \quad \text{(adding } -7 \text{ to both members, property I)}.$$
$$3x \geq -2x - 10.$$
$$5x \geq -10.$$
$$\frac{1}{5}(5x) \geq \frac{1}{5}(-10) \quad \left(\text{multiplying both members by } \frac{1}{5}, \text{ property II}\right).$$
$$x \geq -2.$$

The solution set is $\{x \mid x \geq -2\}$ and the graph of the solution set looks like this:

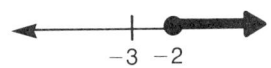

EXAMPLE 32

Solve $-3x + 2 < 2x - 3$, and graph the solution set.

Solution

In solving this inequality, we shall combine two steps, both of which involve the use of property I. We shall add $(-2x - 2)$ to both members of the inequality.

$$-3x + 2 < 2x - 3 \quad \text{(given)}.$$
$$-3x + 2 + (-2x - 2) < 2x - 3 + (-2x - 2) \quad \text{(adding } -2x - 2 \text{ to both members, property I)}.$$
$$-5x < -5.$$
$$-\frac{1}{5}(-5x) > -\frac{1}{5}(-5) \quad \left(\text{multiplying both members by } -\frac{1}{5}, \text{ property III}\right).$$
$$x > 1.$$

Note that the sense of the inequality changed because we multiplied both members of the inequality by a negative number. The solution set is $\{x \mid x > 1\}$ and the graph of the solution set is:

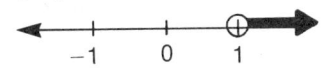

EXAMPLE 33

Graph the solution of the open sentence $-1 \leq x < 2$ where x is a real number.

Solution

The sentence $-1 \leq x < 2$ is read as "-1 is less than or equal to x and x is less than 2." The solution set for this sentence is the set of real numbers from -1 (including -1) up to 2.

The graph of the solution set is shown here.

The method used to solve conditional inequalities is similar to that used in solving equations. We perform various operations on the given inequality to obtain equivalent inequalities until we obtain one whose solution is readily seen. We must always use care in performing any operation on an inequality, but we must be particularly careful when we multiply an inequality by any number that is less than 0. Recall that if each member of an inequality is multiplied by the same negative number, then we obtain an equivalent inequality in the opposite order, that is, its sense is reversed. In addition to expressing the solution for a conditional inequality in algebraic notation, we can represent it by means of a line graph on the real number line.

EXERCISES FOR SECTION 2.6

(Answers given to exercises marked *)

Solve each of the given inequalities and graph its solution set, where x is a real number.

*1. $3x \geq 6$.
*3. $2x - 2 < 4$.
*5. $-3x \leq 6$.
*7. $2x - 2 < 3x - 4$.
*9. $2x + 4 \geq 5x - 8$.
*11. $x - 4 > 3x - 2 - x$.
*13. $3(2x + 4) < 4x - 4$.
*15. $2(3x - 4) \geq 3(x - 2)$.
*17. $2(x + 2) \geq 3(x - 2)$.
*19. $4(x - 1) - (5x - 2) \geq 1$.
*21. $5(-x + 2) - (3x + 2) > 4$.

2. $4x - 2 < 6$.
4. $3x + 2 > x - 4$.
6. $-2x - 4 > 6$.
8. $3x + 4 < 5x - 6$.
10. $3x + 10 + 2x > x - 2$.
12. $4x - 3 < 5x + 5$.
14. $3(x - 2) < 2x + 4$.
16. $2(x - 4) < -1(x + 3)$.
18. $2(3x - 4) < 3(x - 2)$.
20. $2(x - 2) + 3(2x - 1) < 1$.

Graph the solution for each of the following inequalities, where x is a real number.

22. $-3 < x \leq -1$.
*23. $-2 \leq x < 0$.
24. $0 < x < 2$.
*25. $-2 \leq x < 2$.
26. $1 \leq x \leq 3$.
*27. $1 \leq x < 3$.

2.7 MORE INEQUALITIES

In the previous section, we encountered some inequalities that were of the form $a < x < b$. These "combined" or "compound" inequalities occur frequently in different areas of mathematics, such as in the study of the domain and range of certain functions. Recall that we can interpret a statement such as $a < x < b$ as "a is less than x and x is less than b." In other words, we can take the compound inequality $a < x < b$ and restate it as two simple inequalities:

First Degree Equations and Inequalities in One Variable

$$a < x < b \quad \text{means} \quad a < x \quad \text{and} \quad x < b$$

This idea of restating a compound inequality as two separate inequalities is most useful in solving them. Consider the following inequality: $0 \leq 2x - 2 \leq 4$. In order to solve this compound inequality, we must find those values of x that satisfy both of the inequalities, that is, $0 \leq 2x - 2$ and $2x - 2 \leq 4$. We proceed in the same manner as in the previous section:

$$\begin{array}{lll} 0 \leq 2x - 2 & \text{and} & 2x - 2 \leq 4. \\ 2 \leq 2x. & & 2x \leq 6. \\ 1 \leq x & \text{and} & x \leq 3. \end{array}$$

Now we can combine these two solutions into a compound statement, $1 \leq x \leq 3$. The solution set is $\{x \mid 1 \leq x \leq 3\}$ and the graph of the solution set is shown in Figure 5.

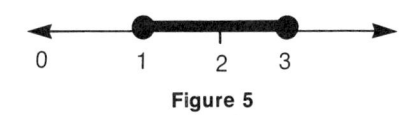

Figure 5

EXAMPLE 34

Solve and graph the solution set of the inequality $-1 < 3x + 2 \leq 8$.

Solution

In order to solve this compound inequality, we first rewrite it as two separate inequalities,

$$-1 < 3x + 2 \quad \text{and} \quad 3x + 2 \leq 8.$$

Now we solve each inequality separately.

$$\begin{array}{ll} -1 < 3x + 2. & 3x + 2 \leq 8. \\ -3 < 3x. & 3x \leq 6. \\ -1 < x. & x \leq 2. \end{array}$$

Combining the two solutions into a compound statement, we have $-1 < x \leq 2$. The solution set is $\{x \mid -1 < x \leq 2\}$ and its graph is

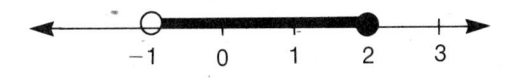

In the study of mathematics, we may also encounter inequalities that involve absolute value notation. Recall that in Chapter One (Section 1.7), we discussed the absolute value of a real number. We began with the idea that the absolute value of a number can be thought of as its distance from the origin on the number line. For example, $|-5| = 5$ because -5 is five units from zero on the number line. The absolute value of a real number x is the distance between x and 0 on the number line, and is symbolized by $|x|$. We also stated a more formal definition:

If x is a real number, then $|x| = \begin{cases} x, \text{ if } x > 0. \\ 0, \text{ if } x = 0. \\ -x, \text{ if } x < 0. \end{cases}$

It is important to understand the definition of the absolute value of a real number x because it is needed in order to solve inequalities involving absolute value. What does it mean when we say that $|x| < 1$? What values of x will satisfy this inequality? We can solve this problem in two ways. First, using the "distance idea" for absolute value, we know that the statement $|x| < 1$ means that x must be less than one unit from 0 on the number line, that is, x must be between 1 and -1. We can state this as $x < 1$ and $x > -1$ or $-1 < x < 1$. On the number line, we have the situation shown in Figure 6.

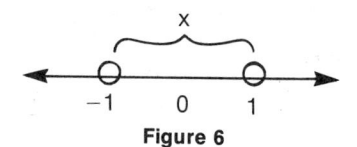

Figure 6

Another method for solving this problem, $|x| < 1$, is to apply the definition for absolute value. We have

$x < 1$, if $x > 0$. Similarly $-x < 1$, if $x < 0$.

Since we do not know whether x is positive or negative and we want to find the set of permissible replacements for x, we must consider both cases. If $x < 0$, then we have $-x < 1$. This is equivalent to $x > -1$ (multiplying both members by -1). (Note that instead of writing $x > -1$, we can also write this as $-1 < x$.) Now, for the inequality $|x| < 1$, we have the set of permissible replacements for x, that is, $-1 < x$ and $x < 1$. These two inequalities can be combined into the compound inequality $-1 < x < 1$. Therefore, the solution set for $|x| < 1$ is $\{x | -1 < x < 1\}$.

Consider the inequality $|x| > 2$. What is the set of permissible replacements for x? Applying the definition for absolute value, we have

$x > 2$, if $x > 0$. Similarly $-x > 2$, if $x < 0$.

The first inequality is already solved, and if we multiply both members of

First Degree Equations and Inequalities in One Variable

the second inequality by −1, we have x < −2. Therefore, the two solutions are x > 2, x < −2. We *cannot* say that x > 2 *and* x < −2. Why not? A real number cannot be both greater than positive 2 and less than negative 2. Therefore, the solution must be x > 2 *or* x < −2. The solution set is {x | x > 2 or x < −2}.

We can also interpret | x | > 2 as meaning that x must be more than 2 units from 0 on the number line. Now if x is more than 2 units from 0 on the number line, x must either be greater than 2, (x > 2), or else be less than −2, (x < −2). This is shown in Figure 7.

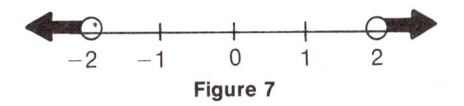

Figure 7

Regardless of the technique used, we must understand the concept and definition of absolute value. In this way we can always determine the set of permissible replacements for the variable in an inequality that contains absolute value notation.

EXAMPLE 35

Solve and graph the solution set of the inequality | x + 1 | < 2.

Solution

Applying the definition for the absolute value of a real number, we have x + 1 < 2 if (x + 1) > 0. Similarly −(x + 1) < 2, if (x + 1) < 0. Solving these inequalities, we have

$$x + 1 < 2. \qquad -(x + 1) < 2.$$
$$x < 1. \qquad -x - 1 < 2.$$
$$\qquad\qquad -x < 3.$$
$$\qquad\qquad x > -3.$$

For the set of permissible replacements for x, we have that x must be less than 1, also x must be greater than −3, that is, −3 < x < 1. The solution set is {x | −3 < x < 1} and its graph is

Alternate Solution: If we use the concept of "distance" for absolute value, then | x + 1 | < 2 means that the expression (x + 1) represents a number that is less than 2 units from 0 on the number line. If (x + 1) is less than 2 units from 0, it must be bounded by −2 and +2.

In other words, $-2 < (x + 1)$ and $(x + 1) < 2$. Solving these inequalities, we have

$$-2 < x + 1. \qquad x + 1 < 2.$$
$$-3 < x. \qquad x < 1.$$

Combining the two solutions into a compound statement we have $-3 < x < 1$. The solution set is $\{x \mid -3 < x < 1\}$ and its graph is shown in the first solution.

EXAMPLE 36

Solve and graph the solution set of the inequality $|x + 1| > 2$.

Solution

This problem is similar to the preceding example, but the order of the inequality has been reversed. Applying the definition for the absolute value of a real number, we have

$x + 1 > 2$, if $(x + 1) > 0$. Similarly, $-(x + 1) > 2$, if $(x + 1) < 0$.

Solving these inequalities, we have

$$x + 1 > 2. \qquad -(x + 1) > 2.$$
$$x > 1. \qquad -x - 1 > 2.$$
$$\qquad -x > 3.$$
$$\qquad x < -3.$$

The set of permissible replacements for x consists of all x such that $x > 1$, or $x < -3$. The solution set is $\{x \mid x < -3 \text{ or } x > 1\}$ and its graph is

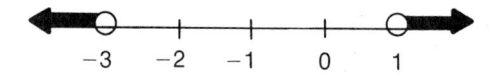

```
  ◄――○――+――+――+――○――►
     -3  -2  -1   0   1
```

Alternate Solution: Applying the distance concept for $|x+1| > 2$, we know that the expression $(x + 1)$ must be more than 2 units from 0 on the number line. In order to be more than 2 units from 0, $(x + 1)$ must be greater than 2, that is, $x + 1 > 2$, or $(x + 1)$ must be less than -2, or $x + 1 < -2$. Solving these inequalities, we have

$$x + 1 < -2. \qquad \text{or} \qquad x + 1 > 2.$$
$$x < -3. \qquad \qquad x > 1.$$

Note that x cannot be both less than -3 and greater than 1. Hence the solution is stated as $x < -3$, or $x > 1$. The solution set is $\{x \mid x < -3 \text{ or } x > 1\}$ and its graph is shown in the first solution.

First Degree Equations and Inequalities in One Variable 123

EXAMPLE 37

Solve and graph the solution set of the inequality $|2x - 3| \leq 5$.

Solution

Using the definition for the absolute value of a real number, we have $2x - 3 \leq 5$ if $(2x - 3) > 0$. Similarly $-(2x - 3) \leq 5$ if $(2x - 3) < 0$. Solving these inequalities, we have

$$
\begin{array}{lll}
2x - 3 \leq 5. & -(2x - 3) \leq 5 & \text{or} \quad 5 \geq -(2x - 3). \\
2x \leq 8. & -2x + 3 \leq 5. & \quad 5 \geq -2x + 3. \\
x \leq 4. & -2x \leq 2. & \quad 2 \geq -2x. \\
& x \geq -1. & \quad -1 \leq x.
\end{array}
$$

For the set of permissible replacements for x, we have that x must be less than or equal to 4; also x must be greater than or equal to -1, that is, $-1 \leq x$ and $x \leq 4$ or $-1 \leq x \leq 4$. The solution set is $\{x \mid -1 \leq x \leq 4\}$ and its graph is

Alternate Solution: The expression $(2x - 3)$ in the inequality $|2x - 3| \leq 5$ must be less than 5 units from 0 on the number line $(<)$ or exactly 5 units from 0 $(\leq)$. Therefore the expression $(2x - 3)$ must be bounded by -5 and 5, that is, it must have -5 and 5 as end points. In other words, $-5 \leq (2x - 3)$ and $(2x - 3) \leq 5$. Solving these inequalities, we have

$$
\begin{array}{ll}
-5 \leq 2x - 3. & 2x - 3 \leq 5. \\
-2 \leq 2x. & 2x \leq 8. \\
-1 \leq x. & x \leq 4.
\end{array}
$$

Combining the two solutions into a compound statement, we have $-1 \leq x \leq 4$. The solution set is $\{x \mid -1 \leq x \leq 4\}$ and its graph is shown in the first solution.

EXAMPLE 38

Solve and graph the solution set of the inequality $|2x - 1| \geq 3$.

Solution

According to the definition for the absolute value of a real number, we have

$2x - 1 \geq 3$ if $(2x - 1) > 0$. Similarly $-(2x - 1) \geq 3$ if $(2x - 1) < 0$.

Solving these inequalities, we have

$$2x - 1 \geq 3. \qquad -(2x - 1) \geq 3.$$
$$2x \geq 4. \qquad -2x - 1 \geq 3.$$
$$x \geq 2. \qquad -2x \geq 2.$$
$$\qquad\qquad\qquad x \leq -1.$$

The set of permissible replacements for x consists of all x such that $x \geq 2$ or $x \leq -1$. The solution set is $\{x \mid x \leq -1 \text{ or } x \geq 2\}$ and its graph is

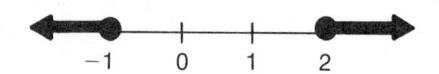

Alternate Solution: The expression $(2x - 1)$ in the inequality $|2x - 1| \geq 3$ must be at least 3 or more units from 0 on the number line ($\geq$). In order to meet these conditions, $2x - 1 \geq 3$ or $2x - 1 \leq -3$. Solving these inequalities, we have

$$2x - 1 \leq -3. \qquad 2x - 1 \geq 3.$$
$$2x \leq -2. \qquad 2x \geq 4.$$
$$x \leq -1. \qquad x \geq 2.$$

Note that x cannot be both less than (or equal to) -1 and greater than (or equal to) 2. Therefore the solution is stated as $x \leq -1$ or $x \geq 2$. The solution set is $\{x \mid x \leq -1 \text{ or } x \geq 2\}$ and its graph is shown in the first solution.

We can solve inequalities that involve absolute value notation by several means. One is to apply the definition of the absolute value of a real number. In other words, we can derive two inequalities from the given expression by assuming that the expression is greater than 0, and then assuming that the expression is less than 0. We can then solve the two inequalities, determine the solution set, and graph it on the number line.

Another way is to determine the two inequalities by using the idea of distance from the origin for the expression inside the absolute value sign. For example, in the expression $|x + a| < b$, $(x + a)$ represents a number that is less than b units from 0 on the number line. Similarly, in the expression $|x + a| > b$, $(x + a)$ represents a number that is more than b units from 0 on the number line.

Regardless of the technique used to solve inequalities that involve absolute value notation, we can check the solution by choosing values in the solution set and determining whether they satisfy the original absolute value inequality. Note that this is *not* an absolute check since it is impossible to check *all* the infinitely many members of the proposed solution set.

First Degree Equations and Inequalities in One Variable

EXERCISES FOR SECTION 2.7

(Answers given to exercises marked *)

Solve each of the given inequalities and graph its solution set, where x is a real number.

*1. $-2 < 2x < 4$.
2. $0 \leq 3x < 6$.
*3. $-2 \leq x + 1 \leq 3$.
4. $-1 < x + 4 < 5$.
*5. $-2 \leq 2x - 2 \leq 4$.
6. $0 < 3x - 3 < 3$.
*7. $-1 \leq 2x - 3 \leq 1$.
8. $0 < -2x - 2 < 2$.
*9. $-1 < -3x - 4 < 2$.
10. $|x - 1| < 1$.
*11. $|x + 1| < 2$.
12. $|x + 1| > 2$.
*13. $|x - 3| > 1$.
14. $|2x - 1| < 5$.
*15. $|2x - 3| \leq 5$.
16. $|3 - 2x| > 1$.
*17. $|2 - 3x| < 4$.
18. $|1 - 3x| > 2$.
*19. $|-1 - x| \geq 2$.
20. $|-1 - x| < 1$.
*21. $|-1 - 2x| > 1$.
22. $|-5 - 2x| < 3$.
*23. $|-5 - 2x| \geq 1$.
24. $|4x + 2| < 6$.
*25. $|2x + 1| < x - 1$.
26. $|3x - 3| < 0$.
*27. $|3x + 2| < x - 2$.

2.8 SUMMARY

Our main concern in this chapter was to develop the ability to solve first degree equations and inequalities in one variable. The equations considered were primarily of the general form $ax + b = c$ (where a, b, and c are real numbers, $a \neq 0$). Equivalent equations are equations that have the same solution set. A conditional equation is one that is true for only certain replacements of the variable, while an identity equation is one that is true for all replacements of the variable. A solution for an equation can be checked by determining if the solution satisfies the original equation, that is, results in a true statement for the equation. Conditional equations can be solved by transforming them into equivalent equations whose solution is readily seen. We can transform given conditional equations into equivalent equations by means of two properties of equality:

    The Addition Property of Equality: If the same expression is added to both members of an equation, then the resulting equation is equivalent to the given equation.

    The Multiplication Property of Equality: If both members of an equation are multiplied by the same expression (provided the expression does not equal 0), then the resulting equation is equivalent to the given equation.

In order to solve a first degree equation in one unknown, we perform the following steps (if necessary):

1. If an equation contains parentheses, eliminate them by means of the distributive property.
2. If an equation contains fractions, eliminate them by multiplying

both members of the equation by the least common denominator (LCD).
3. If an equation contains decimal numbers, eliminate them by multiplying both members of the equation by the appropriate power of 10.
4. Perform the indicated operations and simplify when possible.
5. Apply the addition property of equality to obtain an equivalent equation of the form $ax = c$.
6. Apply the multiplication property of equality to obtain an equivalent equation of the form $x = k$, which is the solution of the given equation.
7. Check the solution by substituting it for the variable in the given equation.

Equations can contain more than one variable, or other symbols representing constants. For example, $2x + a = b$ and $V = lwh$ are called literal equations. The solution to a literal equation is obtained as in solving first degree equations in one variable, but the solution is expressed in terms of the remaining symbols.

Literal equations (formulas) are useful in solving word or verbal problems. But other procedures must also be followed in solving word problems. First, we must read the problem carefully, and then probably read it a second time. If possible, we draw a diagram to assist in interpreting the given information. Next, we should translate the English phrases into mathematical phrases, choosing a variable for the unknown quantity. We then write the appropriate equation. After solving the equation, we should check the solution to determine that it satisfies the original problem.

Inequalities that are true for only certain values of the variable are called conditional inequalities. They are not true for every replacement for the variable. To solve conditional inequalities, we employ methods similar to those used in solving equations. We perform certain operations on each member of the inequality in order to obtain an equivalent inequality whose solution is readily seen. Three basic properties are used to obtain equivalent inequalities:

I. If a, b, and c are real numbers, then $a < b$ is equivalent to $a + c < b + c$.
II. If a, b, and c are real numbers and $c > 0$, then $a < b$ is equivalent to $ac < bc$.
III. If a, b, and c are real numbers and $c < 0$, then $a < b$ is equivalent to $ac > bc$.

The last property is particularly important in solving inequalities. It is important to remember that if each member of an inequality is multiplied by the same negative number, then we shall obtain an equivalent inequality in the opposite sense (order). In addition to expressing the solution set for

First Degree Equations and Inequalities in One Variable

conditional inequalities, in algebraic notation we can also draw a graph of the solution set on the real number line.

Conditional inequalities can also involve absolute value notation. In order to solve these inequalities, we can use the definition of the absolute value of a real number. We can obtain two inequalities to solve by assuming that the absolute value expression is greater than zero and then assuming that it is less than zero. Another technique used to solve these inequalities is to determine the boundaries of the expression inside the absolute value sign. For example, in the inequality $|x+a| < b$, the expression $(x + a)$ represents a number that is less than b units from 0 on the number line. Therefore, the expression $(x + a)$ must be less than b and greater than $-b$, that is, $-b < x + a$ and $x + a < b$. We can check the proposed solutions of inequalities of this nature by choosing values in the proposed solution set and determining whether they satisfy the original inequality.

REVIEW EXERCISES FOR CHAPTER 2
(Answers in Appendix B)

1. Identify the following as conditional equations, identities, or neither.
   a. $4y - 3 = 1$.
   b. $3x - x = 2x$.
   c. $z - 1 = z$.
   d. $4x - 2 = 3x + 5$.
   e. $5x - 2 + x = 2(3x - 1)$.
   f. $y - 5 = 2 + y$.

2. Solve the following equations and check the solutions.
   a. $3y - 1 = 8$.
   b. $\frac{x}{3} + 4 = -5$.
   c. $\frac{z}{2} - 1 = 3$.
   d. $5z = 2z - 6$.
   e. $2x - 4 = x - 3$.
   f. $4y - 3 = y + 9$.
   g. $\frac{3}{4}y + \frac{1}{3} = 2y - 1$.
   h. $\frac{z}{3} + 1 = 13 - \frac{z}{3}$.
   i. $\frac{2}{5}x + \frac{1}{2} = -\frac{1}{3} - x$.
   j. $2(2z - 4) = 3(z + 2)$.
   k. $\frac{2}{3}(3x + 6) = \frac{3}{4}(4x - 8)$.
   l. $4(y + 1) - 3(y - 1) = 25$.
   m. $\frac{1}{2}(2 - z) = \frac{1}{3}\left(\frac{z}{2} - 1\right)$.
   n. $1.1y = 7.6 - 1.42y$.
   o. $4.9 - 0.2y = 0.8y - 1.1$.

3. For each of the following exercises, solve the given formulas for the specified variable.
   a. $d = rt$. (for r)
   b. $i = prt$. (for r)
   c. $A = \frac{1}{2}bh$. (for h)
   d. $F = \frac{9}{5}C + 32$. (for C)
   e. $C = \frac{5}{9}(F - 32)$. (for F)
   f. $S = 2\pi rh + 2\pi r^2$. (for h)

g. $ay + b = c$. (for y)   h. $p = 2(L + W)$. (for W)
i. $ax + b = c - b$. (for x)

4. In a collection of 50 coins, the number of nickels is one-third the number of dimes and the number of quarters is ten less than twice the number of dimes. Find the value of the collection.

5. Two vans leave a certain park at the same time and travel in opposite directions along a straight road. The speed of one van is three-fourths that of the other van; in 4 hours they are 420 miles apart. What is the speed of each van?

6. A rectangular tot lot (playground) is enclosed by 1000 feet of fencing. If the length of the tot lot is 40 feet less than three times the width, find the dimensions of the tot lot.

7. Find two consecutive integers such that if five more than the first is divided by five less than the second, the quotient is 10.

8. Tickets for a particular concert cost $5 each if purchased at the advance sale and $7 each if bought at the box office on the day of the concert. For this particular concert, 1200 tickets were sold and the receipts were $6700. How many tickets were bought at the box office on the day of the concert?

9. Find three consecutive odd integers whose sum is 39.

10. Joe is 2 years younger than his brother Scott. Joe's father is 30 years older than he is. Four years ago, the father's age was 2 more than five times the sum of Joe's and Scott's ages then. Find the present ages of Joe, Scott and their father.

11. Solve each of the given inequalities and graph its solution set.
    a. $4x - 2 > 6$.   b. $3x + 2 < x - 4$.
    c. $-2x - 4 \leq 6$.   d. $3(x - 2) \geq 2x + 4$.
    e. $4(x - 1) - (5x - 2) < 1$.   f. $5(-x + 2) - (3x + 2) < 4$.

12. Solve each of the given inequalities and graph its solution set.
    a. $|x - 1| > 1$.   b. $|x - 2| < 1$.   c. $|2x - 1| \geq 5$.
    d. $|1 - 3x| < 2$.   e. $|-1 - x| \leq 2$.   f. $|2x - 2| < 0$.

13. If a long-distance telephone call from Los Angeles to New York City costs $1.25 for the first three minutes and $.30 for each additional minute or fraction of a minute, how long can a phone call be from Los Angeles to New York City such that the cost does not exceed $4.00?

chapter 3

Exponentials and Radicals

As a result of this chapter, the student will be able to:

1. Describe an exponential as an expression of the form x^n where the positive integral exponent, n, indicates the number of times the base, x, occurs as a factor;
2. Multiply and divide exponentials that have the same base where the exponents may be natural numbers, whole numbers, integers, or rational numbers;
3. Use scientific notation to evaluate expressions that contain very large or very small numbers;
4. Express a given exponential in an equivalent radical form, and express a given radical in an equivalent exponential form;
5. Simplify a given radical expression;
6. Perform the four basic operations with radical expressions, that is, add, subtract, multiply and divide radical expressions;
7. Rationalize the denominator of a fraction that has a binomial denominator where one or both terms are radicals of the second order;
8. Solve radical equations and check the solution set for extraneous roots;
9. Apply the basic laws governing operations with exponents.

$$a^m \cdot a^n = a^{m+n}. \qquad (ab)^m = a^m b^m.$$

$$(a^m)^n = a^{mn}. \qquad \frac{a^m}{a^n} = a^{m-n}.$$

$$a^{-m} = a^{1/m}. \qquad a^0 = 1.$$

$$\left(\frac{a}{b}\right)^n = \frac{a^n}{b^n}. \qquad a^{m/n} = (\sqrt[n]{a})^m = \sqrt[n]{a^m}.$$

3.1 INTRODUCTION

Expressions such as x^2, y^3, z^4, m^5, and so on, are often used in algebra. In an expression such as x^2, the 2 is called an *exponent* and x is called the *base*. The whole expression is called an *exponential* or a *power*:

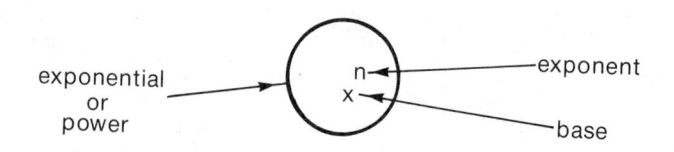

exponential or power → x^n ← exponent, base

Exponentials and Radicals

How do we interpret an expression such as x^2? What does it mean? The exponent indicates the number of times the base occurs as a factor. Therefore, x^2 means that we have two factors of x, or $x \cdot x$. Similarly, y^3 means that we have three factors of y, or $y \cdot y \cdot y$. (Note: $x = x^1$, $y = y^1$, $z = z^1$, etc.) When the exponent of a base is not indicated, it is understood to be 1, that is, $x = x^1$, and is read as "x" or "x to the first." Similarly, x^4 is read as "x to the fourth." An expression such as x^2 is "x to the second," but it is more commonly read as "x squared," just as x^3 is commonly read as "x cubed." Since the exponent indicates the number of times the base occurs as a factor, x^5 means that we have five factors of x, or $x \cdot x \cdot x \cdot x \cdot x$. An expression such as x^2y^3 means that we have two factors of x and three factors of y, $x^2y^3 = x \cdot x \cdot y \cdot y \cdot y$.

We can use this same definition to evaluate numerical expressions. For example, in the expression 2^3, 2 is the base and 3 is the exponent. Therefore, we have three factors of 2, $2^3 = 2 \cdot 2 \cdot 2 = 8$. Similarly, 3^4 means that we have four factors of 3: $3^4 = 3 \cdot 3 \cdot 3 \cdot 3 = 81$. Extreme care should be used in evaluating numerical expressions that contain a negative number in the base. Consider the following two problems: -3^2 and $(-3)^2$. In the first expression, -3^2, 3 is the base and 2 is the exponent. Therefore, $-3^2 = -(3 \cdot 3) = -9$. In the second expression, $(-3)^2$, -3 is the base and 2 is the exponent. Therefore, $(-3)^2 = (-3) \cdot (-3) = 9$. So you see, the two expressions, -3^2 and $(-3)^2$, are not equivalent.

An expression such as $x^3 \cdot x^2$ means that we have three factors of x and two more factors of x, or, $x^3 \cdot x^2 = (x \cdot x \cdot x) \cdot (x \cdot x)$. Since $(x \cdot x \cdot x) \cdot (x \cdot x) = x \cdot x \cdot x \cdot x \cdot x$, which is five factors of x, we can say that $x^3 \cdot x^2 = x^5$. Note that $3 + 2 = 5$ (the sum of the exponents). We can generalize a rule for multiplying exponentials or powers that have the same base as follows:

> If a is any real number and m and n are natural numbers, then
>
> $$a^m \cdot a^n = a^{m+n}.$$

Using this rule for multiplication of exponentials with the same base, we find that $y^4 \cdot y^5 = y^{4+5} = y^9$. To find the product of exponentials with the same base, we write that same base and add the exponents.

EXAMPLE 1

Find the indicated products:

a. $y^3 \cdot y^5$ b. $x \cdot x^2$ c. $a^2 \cdot a^3 \cdot a^5$ d. $3^2 \cdot 3^3 \cdot 3^4$.

Solution

To find the product of exponentials with the same base, write the same base and add the exponents.

a. $y^3 \cdot y^5 = y^{3+5} = y^8$.
b. $x \cdot x^2 = x^1 \cdot x^2 = x^{1+2} = x^3$.
c. $a^2 \cdot a^3 \cdot a^5 = a^{2+3+5} = a^{10}$.
d. $3^2 \cdot 3^3 \cdot 3^4 = 3^{2+3+4} = 3^9$.

EXAMPLE 2

a. $(4x^3)(2x^2)$ b. $(7y^2)(-3y^4)$ c. $(3xy^2)(2x^2y)$

Solution

a. $(4x^3)(2x^2) = (4)(2)x^3 \cdot x^2 = 8x^{3+2} = 8x^5$.
b. $(7y^2)(-3y^4) = (7)(-3)y^2 \cdot y^4 = -21y^{2+4} = -21y^6$.
c. $(3xy^2)(2x^2y) = (3)(2)x \cdot x^2 \cdot y^2 \cdot y = 6x^{1+2} \cdot y^{2+1} = 6x^3y^3$.

Division of exponentials or powers with the same base is performed in a manner similar to that of multiplication. An expression such as

$$\frac{x^5}{x^2}$$

means that we have five factors of x divided by two factors of x,

$$\frac{x^5}{x^2} = \frac{x \cdot x \cdot x \cdot x \cdot x}{x \cdot x}$$

which equals $x \cdot x \cdot x$, or x^3. Therefore, we have $\frac{x^5}{x^2} = x^3$. Similarly, $\frac{y^6}{y^4}$ means that we have six factors of y divided by four factors of y,

$$\frac{y^6}{y^4} = \frac{y \cdot y \cdot y \cdot y \cdot y \cdot y}{y \cdot y \cdot y \cdot y} = y \cdot y = y^2 \text{ or } \frac{y^6}{y^4} = y^2.$$

We can generalize a rule for dividing exponentials or powers that have the same base as follows:

> If a is any real number ($a \neq 0$) and m and n are natural numbers such that $m > n$, then
> $$\frac{a^m}{a^n} = a^{m-n}.$$

Exponentials and Radicals

To find the quotient of exponentials with the same base, we write that same base and subtract exponents.

EXAMPLE 3

Find the indicated quotients.

a. $\dfrac{y^5}{y^3}$ b. $\dfrac{z^5}{z}$ c. $\dfrac{4^3 x^5}{4^2 x^2}$

Solution

To divide exponentials with the same base, write that same base and subtract exponents.

a. $\dfrac{y^5}{y^3} = y^{5-3} = y^2.$

b. $\dfrac{z^5}{z} = \dfrac{z^5}{z^1} = z^{5-1} = z^4.$

c. $\dfrac{4^3}{4^2} \cdot \dfrac{x^5}{x^2} = 4^{3-2} \cdot x^{5-2} = 4^1 x^3$ or $4x^3.$

In the beginning of this section, we noted that an expression such as y^3 means that we have three factors of y, or, $y^3 = y \cdot y \cdot y$. Let's extend this idea a bit further. Consider the expression $(x^2)^3$. How do we interpret it? Applying our definition, $(x^2)^3$ means that we have three factors of x^2,

$$(x^2)^3 = x^2 \cdot x^2 \cdot x^2.$$

Since x^2 means that we have two factors of x, we can rewrite the expression once more,

$$(x^2)^3 = x^2 \cdot x^2 \cdot x^2 = (x \cdot x) \cdot (x \cdot x) \cdot (x \cdot x) = x^6.$$

Note that the expression $(x^2)^3$ is equivalent to the sixth power of x, $(x^2)^3 = x^6$. This is sometimes referred to as raising an exponential to a power, and in doing so we multiply the exponents.

If a is any real number (a $\neq$ 0), and m and n represent any natural numbers, then

$$(a^m)^n = a^{m \cdot n} = a^{mn}.$$

We can extend this rule to include expressions such as $(x^2y^3)^2$. Applying our definition, $(x^2y^3)^2 = x^2y^3 \cdot x^2y^3$. Now we can find the indicated product. We write the same base and add exponents, and $x^2y^3 \cdot x^2y^3 = x^4y^6$. Therefore, $(x^2y^3)^2 = x^4y^6$. Note that we could have also obtained the answer by multiplying the exponents: $(x^2y^3)^2 = x^{2 \cdot 2}y^{3 \cdot 2} = x^4y^6$. We state the general rule:

> If a and b represent any real number ($a \neq 0$), ($b \neq 0$), and m represents any natural number, then
>
> $$(ab)^m = a^m b^m.$$

EXAMPLE 4

Write an equivalent expression for each of the following:

a. $(x^4)^3$ b. $(x^2y^4)^3$ c. $(ab^2)^3$ d. $(3s^2t^3)^2$

Solution

In order to find a power of an exponential, find the product of the exponents and write the base(s) with the product(s) obtained as the exponent(s).

a. $(x^4)^3 = x^{4 \cdot 3} = x^{12}$.
b. $(x^2y^4)^3 = x^{2 \cdot 3}y^{4 \cdot 3} = x^6y^{12}$.
c. $(ab^2)^3 = a^{1 \cdot 3}b^{2 \cdot 3} = a^3b^6$.
d. $(3s^2t^3)^2 = 3^{1 \cdot 2}s^{2 \cdot 2}t^{3 \cdot 2} = 3^2s^4t^6$.

Next consider the power of a quotient. For example, $\left(\dfrac{a^2}{b^3}\right)^4$. Applying our definition, $\left(\dfrac{a^2}{b^3}\right)^4$ means that we have 4 factors of $\dfrac{a^2}{b^3}$, that is,

$$\left(\dfrac{a^2}{b^3}\right)^4 = \dfrac{a^2}{b^3} \cdot \dfrac{a^2}{b^3} \cdot \dfrac{a^2}{b^3} \cdot \dfrac{a^2}{b^3} = \dfrac{a^2 \cdot a^2 \cdot a^2 \cdot a^2}{b^3 \cdot b^3 \cdot b^3 \cdot b^3}.$$

Now, we find the indicated product of the numerator and denominator.

$$\dfrac{a^2 \cdot a^2 \cdot a^2 \cdot a^2}{b^3 \cdot b^3 \cdot b^3 \cdot b^3} = \dfrac{a^8}{b^{12}}.$$

Therefore, $\left(\dfrac{a^2}{b^3}\right)^4 = \dfrac{a^8}{b^{12}}$. We obtained this result by finding the power of the numerator and the power of the denominator. In general we have:

> If a and b represent real numbers ($b \neq 0$), then for any natural number n,
>
> $$\left(\dfrac{a}{b}\right)^n = \dfrac{a^n}{b^n}.$$

Exponentials and Radicals

EXAMPLE 5

Write an equivalent expression for each of the following:

a. $\left(\dfrac{b^3}{c^2}\right)^2$ b. $\left(\dfrac{x^2}{y}\right)^3$ c. $\left(\dfrac{a^2b}{c^3}\right)^3$ d. $\left(\dfrac{3xy^2}{2z^3}\right)^4$

Solution

In order to find a power of a fraction, find the power of the numerator and denominator and write the power of the numerator as the new numerator, and the power of the denominator as the new denominator.

a. $\left(\dfrac{b^3}{c^2}\right)^2 = \dfrac{b^{3 \cdot 2}}{c^{2 \cdot 2}} = \dfrac{b^6}{c^4}$

b. $\left(\dfrac{x^2}{y}\right)^3 = \dfrac{x^{2 \cdot 3}}{y^{1 \cdot 3}} = \dfrac{x^6}{y^3}.$

c. $\left(\dfrac{a^2b}{c^3}\right)^3 = \dfrac{a^{2 \cdot 3}b^{1 \cdot 3}}{c^{3 \cdot 3}} = \dfrac{a^6 b^3}{c^9}.$

d. $\left(\dfrac{3xy^2}{2z^3}\right)^4 = \dfrac{3^{1 \cdot 4} x^{1 \cdot 4} y^{2 \cdot 4}}{2^{1 \cdot 4} z^{3 \cdot 4}} = \dfrac{3^4 x^4 y^8}{2^4 z^{12}}.$

Thus far in our discussion of exponentials or powers, we have examined how to multiply and divide exponentials that have the same base, raise an exponential to a power, and obtain the power of a quotient. In general, if a and b represent any non-zero real numbers and m and n are natural numbers, then

1. $a^m \cdot a^n = a^{m+n}$.
2. $\dfrac{a^m}{a^n} = a^{m-n}$ (if $m > n$).
3. $(a^m)^n = a^{mn}$.
4. $(ab)^m = a^m b^m$.
5. $\left(\dfrac{a}{b}\right)^n = \dfrac{a^n}{b^n}.$

EXERCISES FOR SECTION 3.1

(Answers given to exercises marked *)

1. Write each of the following in exponential form.
  *a. $x \cdot x \cdot x$ b. $2 \cdot 2 \cdot x \cdot x \cdot x$ *c. $4 \cdot 4 \cdot 4 \cdot a \cdot a$

d. $a \cdot a \cdot a \cdot b \cdot b \cdot c \cdot c \cdot c$
f. $3a \cdot a \cdot b \cdot b \cdot b$
h. $3 \cdot 3 \cdot 3 \cdot a \cdot a \cdot a$

*e. $x \cdot x \cdot y \cdot z \cdot z \cdot z$
*g. $2 \cdot 2 \cdot 2 \cdot x \cdot x \cdot y \cdot y \cdot y$
*i. $1 \cdot 1 \cdot 1 \cdot 1 \cdot b \cdot b \cdot b \cdot c \cdot c \cdot c \cdot d$

2. Write each of the following without exponents.

*a. $x^3 y^2$
b. $3ac^2$
*c. $2^2 a^2 b^3$
d. $2^3 \cdot 3^2 x^2 y^3$
*e. $(x^2)^3$
f. $(3x^2 y)^2$

3. Evaluate each of the following:

*a. 2^3
b. 3^2
*c. 5^2
d. $(-5)^2$
*e. -5^2
f. -2^4
*g. $(-2)^4$
h. -4^3
*i. $(-2^3)^2$
j. $(-2^2)^3$
*k. $[(-2)^3]^3$
l. $[(-3)^3]^2$
*m. $\dfrac{3^2 \cdot 3^3}{3^4}$
n. $\dfrac{2^3 \cdot 2^4}{2^2 \cdot 2^5}$
*o. $\dfrac{3^4 \cdot 3^5}{3^7 \cdot 3^2}$
p. 1^{32}

4. If $x = 3$ and $y = -2$, evaluate each of the following expressions.

*a. xy
b. $2x^2 y$
*c. $3xy^2$
d. $-2x^2 y^2$
*e. $x^2 y^4$
f. xy^3
*g. $-3x^2 y^2$
h. $(xy)^2$
*i. $x^2 y^2$
j. $x^3 y^3$
*k. $(xy)^3$
l. $x^3 - y^3$

5. Find the product for each of the following. Leave your answer in exponential form (where possible) and assume that literal exponents (Exercises i–1) represent natural numbers.

*a. $x^2 \cdot x^3$
b. $y^2 \cdot y^4$
*c. $z^3 \cdot z^3$
d. $x^4 \cdot x^4 \cdot x^4$
*e. $2^2 \cdot 2^4 \cdot 2^3$
f. $(2x^2)(2^3 x^4)$
*g. $(3^2 y^3)(3y^2)$
h. $(4xy^2)(4^2 x^3 y^2)$
*i. $x^m \cdot x^n$
j. $a^{2m} \cdot a^{4m}$
*k. $b^{2x} \cdot b^{5x}$
l. $(2x^2 y^2 z)(-3x^2 z)$

6. Find the quotient for each of the following. Leave your answer in exponential form.

*a. $\dfrac{x^5}{x^2}$
b. $\dfrac{a^4}{a}$
*c. $\dfrac{y^3}{y^2}$
d. $\dfrac{c^{10}}{c^3}$
*e. $\dfrac{4^3}{4}$
f. $\dfrac{2^7}{2^5}$
*g. $\dfrac{(-4)^4}{(-4)^3}$
h. $\dfrac{2^3 x^4}{2x^3}$
*i. $\dfrac{3^2 x^3 y^4}{3x^2 y^3}$
j. $\dfrac{a^4 b^3 c^2}{a^3 bc}$
*k. $\dfrac{10^4 \cdot 10^5}{10^7}$
l. $\dfrac{2^5 \cdot 2^4}{2^3 \cdot 2^2}$

7. Simplify each of the following expressions. Leave your answer in exponential form (where possible). Assume that no denominator is equal to 0 and that literal exponents represent natural numbers.

*a. $(x^4)^2$
b. $(y^3)^4$
*c. $(z^3)^3$
d. $(a^2)^4$
*e. $(2a^2 b^3)^2$
f. $2(ab^2)^3$
*g. $(x^2 y^3)^4$
h. $(3a^2 b^3)^4$
*i. $\left(\dfrac{a^2}{b^3}\right)^2$
j. $\left(\dfrac{x^4}{y^3}\right)^2$
*k. $\left(\dfrac{x^2 y^3}{z^2}\right)^3$
l. $\left(\dfrac{a^2 b^3}{c^4}\right)^2$

Exponentials and Radicals

*m. $\left(\dfrac{a^m}{b^m}\right)^3$ n. $\left(\dfrac{x^y}{z^y}\right)^2$ *o. $\left(\dfrac{x^2}{y^3}\right)^z$ p. $\left(\dfrac{a^3}{b^4}\right)^c$

*q. $\left(\dfrac{2^2 x^3 y^2}{3^2 z^4}\right)^2$ r. $\left(\dfrac{2^2 x^4 y^3}{2x^2 y}\right)^3$ *s. $\left(\dfrac{3^3 a^2 b^4}{3b^2}\right)^3$ t. $\left(\dfrac{5^4 a^3}{5^2 a}\right)^x$

3.2 INTEGRAL EXPONENTS

In the preceding section, we defined multiplication of exponentials as $a^m \cdot a^n = a^{m+n}$ where m and n are natural numbers. We also placed restrictions on the exponents for division of exponentials, $\dfrac{a^m}{a^n} = a^{m-n}$, if m and n are natural numbers and m > n. The exponents m and n also had to be natural numbers for raising an exponential to a power and raising a quotient to a power. We shall now extend our definitions of exponents beyond the natural numbers to include the set of integers. To be consistent, our new definitions should be extensions, not changes, of our earlier definitions.

To begin with, let's examine the rule for division of exponentials with the same base. If a is any real number (a ≠ 0) and m and n are natural numbers such that m > n, then $\dfrac{a^m}{a^n} = a^{m-n}$. Now consider the case where m and n are integers. If m and n are positive integers such that m > n, then the rule still holds because the positive integers behave just like the natural numbers. Consider the case where m = n. The problem $\dfrac{a^m}{a^n}$ is then transformed to $\dfrac{a^m}{a^m}$. But this expression is equal to 1 (provided a ≠ 0), since a quantity divided by itself is equal to 1. If we could use our definition for division, we would also have

$$\dfrac{a^m}{a^m} = a^{m-m} = a^0.$$

We already have determined that $\dfrac{a^m}{a^m} = 1$. Therefore, to keep our rules for exponents consistent, we must define $a^0 = 1$.

If a is any real number (a ≠ 0) then $a^0 = 1$.

Next consider the case where m < n. Let's consider a specific example where m < n. For example, what is the quotient of $y^3 \div y^5$, that is, $\dfrac{y^3}{y^5}$? We have three factors of y divided by five factors of y,

$$\dfrac{y^3}{y^5} = \dfrac{y \cdot y \cdot y}{y \cdot y \cdot y \cdot y \cdot y} = \dfrac{1}{y \cdot y} = \dfrac{1}{y^2} \quad \text{or} \quad \dfrac{y^3}{y^5} = \dfrac{1}{y^2}.$$

What happens when we apply the rule for division of exponentials with the same base? Applying the rule for the given example, we have

$$\frac{y^3}{y^5} = y^{3-5} = y^{-2}.$$

But we have already determined that $\frac{y^3}{y^5} = \frac{1}{y^2}$ and if we want the rule $\frac{a^m}{a^n} = a^{m-n}$ to hold for all cases, then $\frac{1}{y^2}$ must be equivalent to y^{-2}. Therefore, we must define y^{-2} to be $\frac{1}{y^2}$, and the general rule is

> If a is any real number (a ≠ 0) and m represents any integer (m ≠ 0), then we define
>
> $$a^{-m} = \frac{1}{a^m}.$$

By this definition, we see that $5^{-2} = \frac{1}{5^2} = \frac{1}{25}$, and $2^{-3} = \frac{1}{2^3} = \frac{1}{8}$. Similarly, $x^{-1} = \frac{1}{x^1} = \frac{1}{x}$ and $y^{-4} = \frac{1}{y^4}$.

EXAMPLE 6

Simplify the following expressions:

a. m^0 b. $(x^2)^0$ c. $\frac{2^0}{4^0}$ d. $x^0 + y^0$

Solution

In simplifying these expressions, we shall apply the rule that $a^0 = 1$ (a ≠ 0) and we shall assume that any variable used does not equal 0.

a. $m^0 = 1$.
b. Raising an exponential to a power, we multiply exponents. Therefore $(x^2)^0 = x^{2 \cdot 0} = x^0 = 1$.
c. $\frac{2^0}{4^0} = \frac{1}{1} = 1$.
d. $x^0 + y^0 = 1 + 1 = 2$.

Exponentials and Radicals 139

EXAMPLE 7

Simplify the following expressions:

a. y^{-3} b. 4^{-2} c. $3x^{-2}$ d. $(3x)^{-2}$ e. x^{-y}

Solution

In simplifying these expressions, we shall apply the rule that $a^{-m} = \dfrac{1}{a^m}$ $(a \neq 0, m \neq 0)$, and we shall assume that any variable used does not equal 0.

a. $y^{-3} = \dfrac{1}{y^3}$.

b. $4^{-2} = \dfrac{1}{4^2} = \dfrac{1}{16}$.

c. Note that in the expression $3x^{-2}$, only the x is raised to the negative 2 power. Therefore, $3x^{-2} = 3 \cdot \dfrac{1}{x^2} = \dfrac{3}{x^2}$.

d. In the expression $(3x)^{-2}$, both the 3 and the x are raised to the negative 2 power. In raising an exponential to a power, we multiply exponents. Therefore, $(3x)^{-2} = 3^{-2}x^{-2} = \dfrac{1}{3^2} \cdot \dfrac{1}{x^2} = \dfrac{1}{9x^2}$.

e. $x^{-y} = \dfrac{1}{x^y}$ (assuming $x \neq 0$, $y \neq 0$, and that y is an integer).

EXAMPLE 8

Simplify the expression $\dfrac{a^{-2}}{b^{-3}}$.

Solution

Applying the rule that $a^{-m} = \dfrac{1}{a^m}$, we can rewrite this expression,

$$\dfrac{a^{-2}}{b^{-3}} = \dfrac{\frac{1}{a^2}}{\frac{1}{b^3}}.$$

However, this expression is a complex fraction. A complex fraction is one that contains a fraction in the numerator, denominator, or both.

Since a fraction is an indicated quotient, we can simplify the complex fraction by performing the indicated division,

$$\frac{1}{a^2} \div \frac{1}{b^3}, \text{ and } \frac{1}{a^2} \div \frac{1}{b^3} = \frac{1}{a^2} \cdot \frac{b^3}{1} = \frac{b^3}{a^2}. \text{ Therefore,}$$

$$\frac{a^{-2}}{b^{-3}} = \frac{\frac{1}{a^2}}{\frac{1}{b^3}} = \frac{1}{a^2} \cdot \frac{b^3}{1} = \frac{b^3}{a^2}.$$

Let us now check to see that our rule for multiplication of exponentials with the same base holds for integral exponents, that is, $a^m \cdot a^n = a^{m+n}$ when m and n represent any integer. This rule is true when m and n are both positive integers because the positive integers behave just like the natural numbers. Now let's consider the case where one of the exponents is a positive integer and the other is a negative integer. (It does not matter which of the exponents we consider to be positive because multiplication of real numbers is commutative.) Assume that m and n are both positive integers. We examine the product

$$a^m \cdot a^{-n} = a^m \cdot \frac{1}{a^n} \quad \text{(applying the rule for negative exponents)}$$

$$= \frac{a^m}{a^n}$$

$$= a^{m-n} \quad \text{(Dividing exponentials, we subtract exponents.)}$$

$$= a^{m+(-n)} \quad \text{(Subtraction is the same as adding the additive inverse.)}$$

Therefore $a^m \cdot a^{-n} = a^{m+(-n)}$.

Suppose that both exponents represent negative integers, would it be true that $a^{-m} \cdot a^{-n} = a^{-m+(-n)}$? Assume that m and n are both positive integers.

$$a^{-m} \cdot a^{-n} = \frac{1}{a^m} \cdot \frac{1}{a^n} \quad \text{(applying the rule for negative exponents)}$$

$$= \frac{1}{a^{m+n}}$$

$$= a^{-(m+n)} \quad \text{(applying the rule for negative exponents conversely)}$$
$$= a^{-m-n}$$
$$= a^{-m+(-n)} \quad \text{(Subtraction is the same as adding the additive inverse.)}$$

Therefore, $a^{-m} \cdot a^{-n} = a^{-m+(-n)}$.

Exponentials and Radicals

> **EXAMPLE 9**
>
> Perform the indicated operations and write equivalent expressions containing only positive exponents.
>
>   a. $x^2 \cdot x^{-5}$ b. $y^{-3} \cdot y^{-2}$ c. $\dfrac{x^3}{x^5}$ d. $\dfrac{y^{-3}}{y^{-4}}$
>
> **Solution**
>
>   a. $x^2 \cdot x^{-5} = x^{2+(-5)} = x^{-3} = \dfrac{1}{x^3}$.
>
>   b. $y^{-3} \cdot y^{-2} = y^{-3+(-2)} = y^{-5} = \dfrac{1}{y^5}$.
>
>   c. $\dfrac{x^3}{x^5} = x^{3-5} = x^{-2} = \dfrac{1}{x^2}$.
>
>   d. $\dfrac{y^{-3}}{y^{-4}} = y^{-3-(-4)} = y^{-3+4} = y^1 = y$.

If we have an exponential raised to a power that is a negative integer, we can apply the rule that $(a^m)^n = a^{m \cdot n} = a^{mn}$. Assume that both m and n are both positive integers. The problem can be then written as $(a^m)^{-n}$ and we want to show that $(a^m)^{-n} = a^{-mn}$.

$$(a^m)^{-n} = \frac{1}{(a^m)^n} \quad \text{(applying the rule for negative exponents)}$$

$$= \frac{1}{a^{mn}} \quad \text{(raising an exponential to a power)}$$

$$= a^{-mn} \quad \text{(applying the rule for negative exponents conversely)}$$

Therefore, $(a^m)^{-n} = a^{-mn}$.

Another case to consider is where both exponents are negative integers, $(a^{-m})^{-n} = a^{mn}$. This can also be verified as in the method used above.

We can also show that $\left(\dfrac{a}{b}\right)^n = \dfrac{a^n}{b^n}$ where a and b represent real numbers ($a \neq 0$, $b \neq 0$) and n represents any integer. If n is a positive integer, the rule holds because the positive integers behave just like the natural numbers. If $n = 0$, the rule still holds because $\left(\dfrac{a}{b}\right)^0 = 1$ and $\dfrac{a^0}{b^0} = \dfrac{1}{1} = 1$.

Therefore, $\left(\dfrac{a}{b}\right)^0 = \dfrac{a^0}{b^0}$. Now suppose n represents a negative integer. Let n be any positive integer. We want to show that $\left(\dfrac{a}{b}\right)^{-n} = \dfrac{a^{-n}}{b^{-n}}$.

$$\left(\frac{a}{b}\right)^{-n} = \frac{1}{\left(\frac{a}{b}\right)^n} \quad \text{(applying the rule for negative exponents)}$$

$$= \frac{1}{\frac{a^n}{b^n}} \quad \text{(raising a quotient to a power where n is positive)}$$

$$= 1 \cdot \frac{b^n}{a^n} \quad \text{(simplifying a complex fraction)}$$

$$= \frac{1}{a^n} \cdot b^n$$

$$= a^{-n} \cdot \frac{1}{b^{-n}} \quad \text{(applying the rule for negative exponents conversely)}$$

$$= \frac{a^{-n}}{b^{-n}}$$

Therefore, $\left(\frac{a}{b}\right)^{-n} = \frac{a^{-n}}{b^{-n}}.$

To find the power of a quotient (where the exponent represents any integer), find the powers of the numerator and denominator and write them as the new numerator and the new denominator.

EXAMPLE 10

Simplify each of the following expressions, and write equivalent expressions containing only positive exponents.

a. $(4^3)^{-2}$ b. $(x^3y^{-2})^{-2}$ c. $\left(\frac{2}{3}\right)^{-1}$ d. $\left(\frac{x^2}{y^3}\right)^2$

Solution

a. $(4^3)^{-2} = 4^{3(-2)} = 4^{-6} = \frac{1}{4^6}.$

b. $(x^3y^{-2})^{-2} = x^{3(-2)}y^{(-2)(-2)} = x^{-6}y^4 = \frac{1}{x^6} \cdot y^4 = \frac{y^4}{x^6}.$

c. $\left(\frac{2}{3}\right)^{-1} = \frac{2^{-1}}{3^{-1}} = \frac{\frac{1}{2^1}}{\frac{1}{3^1}} = \frac{1}{2} \cdot \frac{3}{1} = \frac{3}{2}.$

$\left(\text{Note that } \left(\frac{a}{b}\right)^{-1} = \frac{b}{a}.\right)$

d. $\left(\frac{x^2}{y^3}\right)^{-2} = \frac{x^{2(-2)}}{y^{3(-2)}} = \frac{x^{-4}}{y^{-6}} = \frac{\frac{1}{x^4}}{\frac{1}{y^6}} = \frac{1}{x^4} \cdot \frac{y^6}{1} = \frac{y^6}{x^4}.$

Exponentials and Radicals

We have now extended our definitions (rules) of exponents beyond the natural numbers to include the set of integers. In general, if a and b represent any non-zero real numbers and m and n represent any integers then

1. $a^m \cdot a^n = a^{m+n}$.
2. $\dfrac{a^m}{a^n} = a^{m-n}$.
3. $(a^m)^n = a^{mn}$.
4. $(ab)^m = a^m b^m$.
5. $\left(\dfrac{a}{b}\right)^n = \dfrac{a^n}{b^n}$.

We also developed three other rules for exponents in this section that are helpful in simplifying exponential expressions:

6. $a^0 = 1$.
7. $a^{-m} = \dfrac{1}{a^m}$.
8. $\left(\dfrac{a}{b}\right)^{-1} = \dfrac{b}{a}$.

EXERCISES FOR SECTION 3.2

(Answers given to exercises marked *)

Simplify each of the following expressions, perform the indicated operations (where possible), and write equivalent expressions containing only positive exponents.

*1. 4^0
2. y^0
*3. $4x^0$
4. $(4x)^0$

*5. $x^0 \cdot x^5$
6. $2y^2 \cdot y^0$
*7. $(a^2 \cdot b^0)^3$
8. $\dfrac{x^0}{x^3}$

*9. $\dfrac{4z^3}{z^0}$
10. $\dfrac{6x^2}{3x^0}$
*11. $\dfrac{4x^3}{(4x)^0}$
12. $\left(\dfrac{x^3}{y^2}\right)^0$

*13. m^{-3}
14. b^{-4}
*15. $3b^{-2}$
16. $3^2 b^{-2}$

*17. $(3b)^{-2}$
18. $(x^2 y)^{-3}$
*19. $(ab^2)^{-2}$
20. $(-2x^2)^{-2}$

*21. $\dfrac{x^{-2}}{x}$
22. $\dfrac{3}{a^{-3}}$
*23. $\dfrac{1}{b^{-2}}$
24. $\dfrac{x^{-1}}{y^{-2}}$

*25. $a^{-2} \cdot a^3$
26. $b^{-3} \cdot b^{-4}$
*27. $x \cdot x^{-1}$
28. $y^3 \cdot y^0 \cdot y^{-3}$

*29. $2^{-2} \cdot 2^{-4} \cdot 2^{-3}$
30. $(2x^{-2})(2^{-3} x^{-4})$
*31. $(3^2 y^{-3})(3y)$

32. $a^{-2x} \cdot a^x$ *33. $b^m \cdot b^n \cdot b^{-m}$ 34. $\dfrac{x^5}{x^{-2}}$

*35. $\dfrac{a^{-4}}{a}$ 36. $\dfrac{y^{-3}}{y^2}$ *37. $\dfrac{4^{-3}}{4^{-1}}$

38. $\dfrac{c^{-5}}{c^5}$ *39. $\dfrac{x^2}{x^{-2}}$ 40. $\dfrac{(-2)^{-5}}{(-2)^{-3}}$

*41. $\dfrac{10^{-4} \cdot 10^{-5}}{10^{-9}}$ 42. $\dfrac{2^3 \cdot 2^{-4}}{2}$ *43. $\dfrac{2^{-2} \cdot 2^5}{2^{-7}}$

44. $\dfrac{3^{-2} x^3 y^{-4}}{3 x^{-2} y^3}$ *45. $\dfrac{a^{-4} b^{-3} c^{-2}}{a^{-3} b^{-1} c^{-2}}$ 46. $(x^{-4})^2$

*47. $(y^{-3})^0$ 48. $(z^0)^{-4}$ *49. $(a^{-2})^{-4}$

50. $2(ab^{-2})^3$ *51. $(2a^2b^3)^{-2}$ 52. $(3^{-1}a^2b^{-3})^{-1}$

*53. $(2^2 x^{-3} y^{-2})^{-3}$ 54. $\left(\dfrac{a^2}{b^3}\right)^{-2}$ *55. $\left(\dfrac{x^{-4}}{y^{-3}}\right)^{-2}$

56. $\left(\dfrac{x^2 y^3}{z^{-2}}\right)^{-2}$ *57. $\left(\dfrac{a^3 b^{-2}}{a^{-2} b^3}\right)^{-3}$ 58. $\left(\dfrac{x^y}{z^y}\right)^{-z}$

*59. $\left(\dfrac{2^2 x^3 y^{-2}}{3^2 z^{-4}}\right)^{-2}$ 60. $\left(\dfrac{5^4 a^3}{5^2 a}\right)^{-x}$

3.3 SCIENTIFIC NOTATION

In scientific work, it is common to deal with numbers that are either very large or very small. For example, the Earth is approximately 93,000,000 miles from the Sun, and 1 millimicron is approximately 0.00000394 inches in length. Rather than write such numbers as they appear in the previous sentence, *scientific notation* is normally used. Scientific notation is the process of writing numbers as the product of a number between 1 and 10, and a power of 10. More specifically, scientific notation consists of writing a number in the general form $N \times 10^m$, where $1 \leq N < 10$ and m is an integer.

In order to write 93,000,000 in scientific notation, we must write it as a product of a number between 1 and 10, and a power of 10. We can rewrite 93,000,000 as $9.3 \times 10,000,000$ and $10,000,000 = 10^7$. Therefore, in scientific notation

$$93,000,000 = 9.3 \times 10^7.$$

We can do the same for 0.00000394. Since $0.00000394 = \dfrac{3.94}{1,000,000}$, we can rewrite the indicated quotient as $3.94 \times \dfrac{1}{1,000,000}$ or $3.94 \times \dfrac{1}{10^6}$ or

Exponentials and Radicals

3.94×10^{-6} (using our rule for negative exponents). Therefore, in scientific notation

$$0.00000394 = 3.94 \times 10^{-6}.$$

Note that for numbers greater than 1, the exponent of 10 is non-negative and for numbers less than 1, the exponent of 10 is negative.

Some other examples of numbers written in scientific notation are

$$35,000 = 3.5 \times 10^4.$$
$$186,000,000,000 = 1.86 \times 10^{11}.$$
$$0.23 = 2.3 \times 10^{-1}.$$
$$0.000302 = 3.02 \times 10^{-4}.$$

These are all written as the product of numbers between 1 and 10, and a power of 10: $N \times 10^m$, where $1 \leq N < 10$ and m is an integer, as stated above. Upon examining the powers of 10 for these examples, we see that m is the number of places that the decimal point has been moved. Note that m is positive if the decimal point is moved to the left, and negative if the decimal point is moved to the right. In scientific notation 3.14 is 3.14×10^0 because the decimal point does not have to be moved (3.14 is already a number between 1 and 10).

EXAMPLE 11

Express 32,000 in scientific notation.

Solution

We must express 32,000 as a product of a number between 1 and 10, and a power of 10, in other words, we must move the decimal point to obtain a number between 1 and 10. We can do so by moving the decimal point to a position between the 3 and 2, that is, 3.2. How many places was the decimal point moved? Note that it was moved four places to the left, and since it was moved to the left, the exponent of 10 is positive. Therefore,

$$32,000 = 3.2 \times 10^4.$$

EXAMPLE 12

Express 0.00245 in scientific notation.

> **Solution**
>
> To obtain a number between 1 and 10, we must move the decimal point to a position between the 2 and 4, that is, 2.45. The decimal point was moved three places to the right, so the exponent of 10 is negative. Therefore,
>
> $$0.00245 = 2.45 \times 10^{-3}.$$

Given a number in scientific notation, we can write it in normal decimal notation. For example, 3.14×10^3 can be written without the use of scientific notation. Recall that a positive power of 10 contains the number of 0's indicated in the exponent, thus $10^2 = 100$, $10^3 = 1000$, $10^4 = 10{,}000$, and so on. Therefore, $3.14 \times 10^3 = 3.14 \times 1000$ and $3.14 \times 1000 = 3140$. Note that the decimal point has been moved the same number of places that appears in the exponent of 10.

If we wish to write 3.14×10^{-2} in decimal notation, we can convert this from scientific notation by means of division of a power of 10. Using the property for negative exponents, $3.14 \times 10^{-2} = 3.14 \times \frac{1}{10^2} = 3.14 \times \frac{1}{100} = \frac{3.14}{100} = 0.0314$. Therefore, $3.14 \times 10^{-2} = 0.0314$. Note again that the decimal point has been moved the same number of places that appears in the exponent 10. We can omit the intermediate steps in converting a number from scientific notation to decimal notation by moving the decimal point the appropriate number of places that is indicated by the power of 10. If the exponent is positive, the number is greater than 1 and the decimal point must be moved to the right. If the exponent is negative, the number is less than 1 and the decimal point must be moved to the left.

> **EXAMPLE 13**
>
> Express each of the following in decimal notation.
> a. 5.21×10^5 b. 3.9×10^{-3} c. 9.3×10^7 d. 1.05×10^{-6}
>
> **Solution**
>
> a. The exponent of 10 is positive. Therefore, move the decimal point five places to the right. $5.2 \times 10^5 = 521{,}000$.
> b. The exponent of 10 is negative. Therefore, move the decimal point three places to the left. $3.9 \times 10^{-3} = 0.0039$.
> c. Move the decimal point seven places to the right. $9.3 \times 10^7 = 93{,}000{,}000$.
> d. Move the decimal point six places to the left. $1.05 \times 10^{-6} = 0.00000105$.

Exponentials and Radicals

We can use scientific notation in computing answers to problems that contain very large or very small numbers. Since scientific notation makes use of integral exponents, we can apply the rules for exponents in performing the necessary computations. Consider this multiplication problem: 12,000,000 × 130,000,000. Many of the pocket-size electronic calculators that are so popular today would not be able to handle this problem because of the number of digits involved. An electronic calculator that can handle this multiplication most likely uses scientific notation in performing the necessary computation.

To perform the indicated multiplication, we must first write the given numbers in scientific notation,

$$12{,}000{,}000 \times 130{,}000{,}000 = (1.2 \times 10^7)(1.3 \times 10^8).$$

Next we make use of the commutative and associative properties of multiplication, and write an equivalent expression,

$$(1.2 \times 10^7)(1.3 \times 10^8) = (1.2 \times 1.3)(10^7 \times 10^8).$$

Performing these multiplications, we obtain

$$(1.2 \times 1.3)(10^7 \times 10^8) = 1.56 \times 10^{15}.$$

Therefore,

$$12{,}000{,}000 \times 130{,}000{,}000 = 1.56 \times 10^{15} = 1{,}560{,}000{,}000{,}000{,}000.$$

Normally, when performing such computations, we leave the answer in scientific notation. Note that we performed the operations separately. That is, we performed the multiplication of the numbers between 1 and 10, and separately we performed the multiplication of the power of 10.

This same procedure can be used for the operation of division. Consider the following division problem, 150,000 ÷ 0.006. To begin with, we can write this as $\frac{150{,}000}{0.006}$. Now we write the given numbers in scientific notation,

$$\frac{150{,}000}{0.006} = \frac{1.5 \times 10^5}{6.0 \times 10^{-3}}.$$

Next, we separate the powers of 10 from the other numbers, and, using the properties for multiplication of fractions, we have

$$\frac{1.5 \times 10^5}{6.0 \times 10^{-3}} = \frac{1.5}{6.0} \times \frac{10^5}{10^{-3}}.$$

Performing the indicated operations separately, we obtain

$$\frac{1.5}{6.0} \times \frac{10^5}{10^{-3}} = 0.25 \times 10^8.$$

This answer appears in a form which is almost scientific notation, but it is not. In order to be in scientific notation it must be expressed as a product of a number between 1 and 10, and a power of 10. The decimal 0.25 is not a number between 1 and 10. To rewrite 0.25×10^8 in scientific notation, note that $0.25 \times 10^8 = 2.5 \times 10^7$. Therefore,

$$150{,}000 \div 0.006 = 2.5 \times 10^7.$$

Often computations will involve both multiplication and division. Consider the following problem:

$$\frac{350{,}000 \times 0.0024}{0.0003 \times 140{,}000}.$$

To evaluate this problem, we first express all numbers in scientific notation.

$$\frac{350{,}000 \times 0.0024}{0.0003 \times 140{,}000} = \frac{(3.5 \times 10^5)(2.4 \times 10^{-3})}{(3.0 \times 10^{-4})(1.4 \times 10^5)}.$$

"I won't be able to do my 'rithmetic homework tonight. Your calculator has a dead battery."

(© 1975, The Register and Tribune Syndicate.)

Exponentials and Radicals

Next we make use of the commutative and associative properties of multiplication, and write an equivalent expression

$$= \frac{(3.5 \times 2.4)(10^5 \times 10^{-3})}{(3.0 \times 1.4)(10^{-4} \times 10^5)}.$$

Separating the powers of 10 from the other numbers, and using the properties for multiplication of fractions, we have

$$= \frac{3.5 \times 2.4}{3.0 \times 1.4} \times \frac{10^5 \times 10^{-3}}{10^{-4} \times 10^5}.$$

Performing the indicated operations separately, we have

$$= \frac{8.4}{4.2} \times \frac{10^2}{10^1}.$$

$$= 2.0 \times 10^1 = 20.$$

Therefore, $\dfrac{350{,}000 \times 0.0024}{0.0003 \times 140{,}000} = 20.$

EXAMPLE 14

Use scientific notation to compute:

a. $0.0035 \times 2{,}400{,}000$
b. $0.000161 \div 2{,}300$

Solution

We first express all numbers in scientific notation and then separately perform the operations with the numbers between 1 and 10, and the powers of 10.

a. $0.0035 \times 2{,}400{,}000 = (3.5 \times 10^{-3})(2.4 \times 10^6).$
$= (3.5 \times 2.4)(10^{-3} \times 10^6).$
$= 8.4 \times 10^3.$

b. $0.000161 \div 2{,}300 = \dfrac{0.000161}{2{,}300}.$

$= \dfrac{1.61 \times 10^{-4}}{2.3 \times 10^3}.$

$= \dfrac{1.61}{2.3} \times \dfrac{10^{-4}}{10^3}.$

$= 0.7 \times 10^{-7}.$
$= 7.0 \times 10^{-8}.$

EXAMPLE 15

Use scientific notation to compute:

$$\frac{0.004 \times 18,000}{8,000 \times 0.00003}$$

Solution

First, we express all numbers in scientific notation. Next, using the commutative and associative properties of multiplication, we obtain an equivalent expression. After separating the powers of 10 from the other numbers, we perform the indicated operations separately.

$$\frac{0.004 \times 18,000}{8,000 \times 0.00003} = \frac{(4.0 \times 10^{-3})(1.8 \times 10^4)}{(8.0 \times 10^3)(3.0 \times 10^{-5})}$$

$$= \frac{(4.0 \times 1.8)(10^{-3} \times 10^4)}{(8.0 \times 3.0)(10^3 \times 10^{-5})}.$$

$$= \frac{4.0 \times 1.8}{8.0 \times 3.0} \times \frac{10^{-3} \times 10^4}{10^3 \times 10^{-5}} = \frac{7.2}{24} \times \frac{10^1}{10^{-2}}.$$

$$= 0.3 \times 10^3.$$
$$= 3.0 \times 10^2.$$

Note that in some cases it is necessary to move the decimal point in the final answer in order to write it in scientific notation. When performing computations using scientific notation, it is best to perform separately the multiplication and division of the numbers between 1 and 10, and then separately perform the multiplication and division of the powers of 10.

EXERCISES FOR SECTION 3.3

*(Answers given to exercises marked *)*

Write the given numbers in scientific notation:

*1. The Earth is approximately 4,000,000,000 years old.

2. An electron is approximately 0.0000000000001 centimetre in diameter.

*3. The weight of an electron is approximately 0.00 ... 9 grams. (27 zeros)

Exponentials and Radicals

4. The speed of light is approximately 984,000,000 feet per second.

*5. The mass of the Earth is approximately 6,000,....$\overbrace{}^{21 \text{ zeros}}$ tons.

6. An angstrom is a unit of length and it is 0.0000001 millimetre long.

Write the given expressions in decimal notation:

*7. 4.1×10^{-3}
10. 3.17×10^{-5}
8. 9.12×10^{4}
*11. 4.859×10^{7}
*9. 3.142×10^{5}
12. 5.4321×10^{-2}

Use scientific notation to evaluate each of the given expressions. Answers should be left in scientific notation.

*13. 0.000034×0.00045

14. 0.00007×0.00034

*15. $35,000,000 \times 44,000,000,000$

16. $355,000,000 \times 4,000$

*17. $\dfrac{105,000,000}{21,000}$

18. $\dfrac{255,000,000}{170,000}$

*19. $\dfrac{0.000143}{0.0013}$

20. $\dfrac{0.0204}{0.00012}$

*21. $\dfrac{10^{5} \times 10^{-6} \times 10}{10^{-3} \times 10^{-5}}$

22. $\dfrac{10 \times 10^{4} \times 10^{-5}}{10^{3} \times 10^{-2} \times 10^{-3} \times 10^{4}}$

*23. $\dfrac{0.0018 \times 12,000}{0.000027 \times 0.002}$

24. $\dfrac{24,000 \times 140,000}{210,000 \times 0.0016}$

*25. $\dfrac{320,000 \times 18,000}{0.0012 \times 0.000048}$

26. $\dfrac{0.00036 \times 0.024}{0.00012 \times 0.0015}$

3.4 RATIONAL EXPONENTS

We have developed the properties of exponents for the set of natural numbers and the set of integers. These properties are consistent, meaning they are true for the set of natural numbers and they are also true for the set of integers. For example, $a^m \cdot a^n = a^{m+n}$ where m and n represent any integers. (Note: If this property holds for the set of integers then it must hold for the set of natural numbers, since the set of natural numbers is a subset of the set of integers.) Now we shall extend these properties to rational exponents. Recall that a rational number is any number that can be written in the form $\dfrac{x}{y}$ (y ≠ 0) where x and y represent integers.

What does it mean when we write a number with a fractional exponent? For example, how should we interpret $9^{1/2}$? If our properties for exponents are to hold for rational exponents, then $9^{1/2} \cdot 9^{1/2} = 9^{1/2+1/2} = 9^1$. Therefore, $9^{1/2}$, when multiplied by itself ($9^{1/2} \cdot 9^{1/2}$), yields a result of 9. We shall define $9^{1/2}$ as the second root of 9 or, in more familiar terms, as the principal *square root* of nine. Similarly $8^{1/3}$ is the third root of 8 or the cube root of 8, and $16^{1/4}$ is the principal fourth root of 16. In general, the *nth root* of a is denoted by $a^{1/n}$. That is, $a^{1/2}$ represents the second root or square root of a, $a^{1/3}$ represents the third root or cube root of a, $a^{1/4}$ represents the fourth root of a, and so on. If n is even and "a" is positive, then "a" has a positive n^{th} root and a negative n^{th} root; $a^{1/n}$ denotes only the positive root.

The general notation for the nth root of a is $\sqrt[n]{a}$. The $\sqrt{}$ is called a *radical sign,* a is called the *radicand,* and n is called the root index or simply the *index.* The whole expression, $\sqrt[n]{a}$ is called a radical.

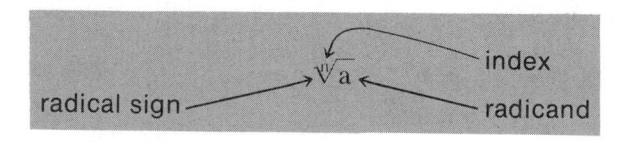

$$\sqrt[n]{a}$$
radical sign — index — radicand

Note: If $n = 2$, it is common practice not to write the 2 in place of n, that is, $\sqrt[2]{a} = \sqrt{a}$.

We now have two expressions that represent the nth root of a, $a^{1/n}$ and $\sqrt[n]{a}$.

$$a^{1/n} = \sqrt[n]{a}.$$

At this point we should elaborate on the notation $\sqrt[n]{a}$. The symbol $\sqrt[n]{a}$ represents the *principal n^{th} root of a.* If $a > 0$ and n is a positive integer, then the principal root of a is positive. If $a < 0$ and n is an odd integer, then the principal root of a is negative. If $a < 0$ and n is an even integer, then there does not exist a principal root of a in the system of real numbers. For example,

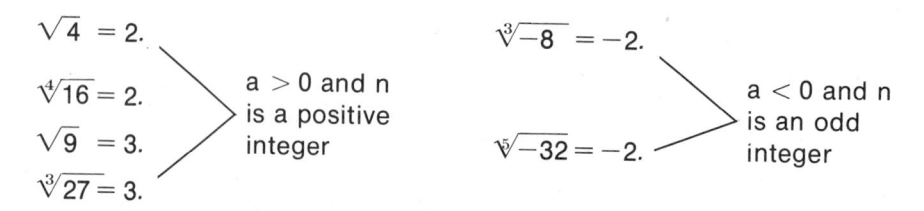

$\sqrt{4} = 2.$
$\sqrt[4]{16} = 2.$ $a > 0$ and n is a positive integer
$\sqrt{9} = 3.$
$\sqrt[3]{27} = 3.$

$\sqrt[3]{-8} = -2.$
$\sqrt[5]{-32} = -2.$ $a < 0$ and n is an odd integer

If $a < 0$ and n is an even integer, then there is no principal root of a in the system of real numbers. For example, consider the square root of a nega-

Exponentials and Radicals

tive 9, $\sqrt{-9}$. Note that $a < 0$ and n is an even integer. There is no real number that will yield a product of -9 when it is multiplied by itself. $\sqrt{-9} \neq -3$, since $(-3)(-3) = +9$, not -9.

HISTORICAL NOTE

Rather than describe a number as a square root of another number, the ancient Greeks referred to it as a side of a square number. For example, 2 is the side of the square number 4, 3 is the side of the square number 9, and so on. From this the Arabs developed the idea of a root of a number, that is, 2 is the root of 4, 3 is the root of 9, etc. Later, European mathematicians used the Latin word "radix" to indicate the root of a number. When abbreviations began to be used in algebra (about 100 A.D.), the word radix was abbreviated as R_x. Therefore, $R_x 4 = 2$, $R_x 9 = 3$. This notation was used for some time until a small r began to be used in place of R_x. $R_x 4$ became r4. Most historians believe that the radical sign $\sqrt{}$, as we know it today, is a representation of the small letter r as it was written by mathematicians at the time (since everything had to be copied by hand).

EXAMPLE 16

Determine the indicated root.

a. $\sqrt{9}$ b. $-\sqrt{9}$ c. $\sqrt[3]{-8}$ d. $-\sqrt[3]{-8}$ e. $\sqrt[4]{-81}$

Solution

In each case we are asked to find the indicated root, that is, the principal n^{th} root of the radicand.

a. The number 9 has 2 square roots, namely 3 and -3, since $(3)(3) = 9$ and $(-3)(-3) = 9$. But the principal square root of 9 is 3, $\sqrt{9} = 3$.
b. Using the result above, $-\sqrt{9} = -(3) = -3$.
c. The number -8 has one real number cube root, namely -2, since $(-2)(-2)(-2) = -8$. Therefore, the principal cube root of -8 is -2, $\sqrt[3]{-8} = -2$.
d. Using the definition from Example c, $-\sqrt[3]{-8} = -(-2) = 2$.
e. The number -81 has no real number fourth root, since $(3)(3)(3)(3) = 81$ and $(-3)(-3)(-3)(-3) = 81$. Therefore, $\sqrt[4]{-81} = \emptyset$.

EXAMPLE 17

Determine the indicated root.

a. $\sqrt{x^6}$ b. $\sqrt[3]{x^6}$ c. $\sqrt[3]{-8m^3}$ d. $\sqrt[4]{16y^{12}}$

Solution

These problems are similar in nature to those in Example 16, but the radicands now contain variables. But we still must find the principal n^{th} root for each example.

a. x^6 has two square roots, namely x^3 and $-x^3$, since $(x^3)(x^3) = x^6$ and $(-x^3)(-x^3) = x^6$. But the principal square root of x^6 is x^3, $\sqrt{x^6} = x^3$.
b. x^6 has one cube root, x^2, since $x^2 \cdot x^2 \cdot x^2 = x^6$. Therefore, the principal cube root of x^6 is x^2, $\sqrt[3]{x^6} = x^2$.
c. $-8m^3$ has one cube root, $-2m$, since $(-2m)(-2m)(-2m) = -8m^3$. Therefore, the principal cube root of $-8m^3$ is $-2m$, $\sqrt[3]{-8m^3} = -2m$.
d. $16y^{12}$ has two fourth roots, $2y^3$ and $-2y^3$, since $(2y^3)(2y^3)(2y^3)(2y^3) = 16y^{12}$ and $(-2y^3)(-2y^3)(-2y^3)(-2y^3) = 16y^{12}$. But the principal fourth root of $16y^{12}$ is $2y^3$, $\sqrt[4]{16y^{12}} = 2y^3$.

Thus far in our discussion we have only been concerned with expressions that had a fractional exponent of the form $a^{1/n}$, that is, the numerator of the fraction was 1. Suppose we have an expression such as $a^{m/n}$ where $\frac{m}{n}$ is any rational number. How shall we interpret this expression, $a^{m/n}$? We know that $a^{1/n} = \sqrt[n]{a}$, but what does $a^{m/n}$ equal? Recall one of our rules for operations with exponents, raising an exponential to a power: $(a^m)^n = a^{mn}$. If this rule is to hold for rational exponents, then

$$a^{m/n} = (a^{1/n})^m.$$

But $a^{1/n} = \sqrt[n]{a}$, therefore

$$a^{m/n} = (a^{1/n})^m = (\sqrt[n]{a})^m.$$

We can also write another equivalent expression for $a^{m/n}$. Using the rule for raising an exponential to a power, we can also state that

$$a^{m/n} = (a^m)^{1/n}.$$
$$= \sqrt[n]{a^m}.$$

Exponentials and Radicals

Therefore,

> If a represents any non-negative real number, n represents any natural number and m represents any integer.
>
> $$a^{m/n} = (a^{1/n})^m = (\sqrt[n]{a})^m$$
>
> or $a^{m/n} = (a^m)^{1/n} = \sqrt[n]{a^m}.$

Let's try an example using the above definitions. Consider the expression $9^{3/2}$. We can simplify this expression in two different ways: $9^{3/2} = (9^{1/2})^3 = (\sqrt{9})^3 = 3^3 = 27$, or $9^{3/2} = (9^3)^{1/2} = \sqrt{9^3} = \sqrt{729} = 27$. Note that we obtained the same answer using either technique. But sometimes it is more advantageous (simpler) to use one method instead of another.

EXAMPLE 18

Simplify each of the following expressions.

a. $8^{2/3}$ b. $16^{3/4}$ c. $27^{4/3}$ d. $x^{6/2}$

Solution

In simplifying each of these expressions, we shall use the fact that $a^{m/n} = (a^{1/n})^m = (\sqrt[n]{a})^m$ or $a^{m/n} = (a^m)^{1/n} = \sqrt[n]{a^m}.$

a. $8^{2/3} = (8^{1/3})^2 = (\sqrt[3]{8})^2 = 2^2 = 4.$

b. $16^{3/4} = (16^{1/4})^3 = (\sqrt[4]{16})^3 = 2^3 = 8.$

c. $27^{4/3} = (27^{1/3})^4 = (\sqrt[3]{27})^4 = 3^4 = 81.$

d. $x^{6/2} = (x^6)^{1/2} = \sqrt{x^6} = x^3.$

EXAMPLE 19

Write an exponential expression equivalent to each of the following and simplify if possible.

a. $\sqrt[3]{x^2}$ b. $\sqrt[4]{x^2}$ c. $\sqrt[6]{27}$ d. $\sqrt[4]{4}$

Solution

a. $\sqrt[3]{x^2} = (x^2)^{1/3} = x^{2/3}.$

b. $\sqrt[4]{x^2} = (x^2)^{1/4} = x^{2/4} = x^{1/2}.$

c. $\sqrt[6]{27} = (27)^{1/6} = (3^3)^{1/6} = 3^{3/6} = 3^{1/2}.$

d. $\sqrt[4]{4} = 4^{1/4} = (2^2)^{1/4} = 2^{2/4} = 2^{1/2}.$

In this section, we have defined what a rational exponent is, that is, $a^{m/n} = (a^{1/n})^m = (\sqrt[n]{a})^m$ or $a^{m/n} = (a^m)^{1/n} = \sqrt[n]{a^m}$. By using this definition, we can rewrite exponentials with fractional exponents as radicals, and we can also rewrite radical expressions as exponentials with fractional exponents.

It should also be noted that the other rules for exponents are also valid for rational exponents. For example,

$$x^{1/2} \cdot x^{1/3} = x^{1/2 + 1/3} = x^{5/6}.$$

$$\frac{m^{2/3}}{m^{1/2}} = m^{2/3 - 1/2} = m^{1/6}.$$

$$(xy)^{1/2} = x^{1/2} \cdot y^{1/2}.$$

$$\left(\frac{x}{y}\right)^{-2/3} = \frac{1}{\left(\frac{x}{y}\right)^{2/3}}.$$

EXERCISES FOR SECTION 3.4

(Answers given to exercises marked *)

1. Express each of the following in an equivalent radical form.

   a. $4^{1/2}$
   b. $2^{1/2}$
   *c. $3^{1/2}$
   d. $5^{1/3}$
   *e. $x^{1/3}$
   f. $y^{1/4}$
   *g. $x^{2/3}$
   h. $m^{3/2}$
   *i. $y^{1/4}$
   j. $n^{3/4}$
   *k. $(3x)^{1/2}$
   l. $3x^{1/2}$
   *m. $(5x^2)^{3/2}$
   n. $(4y^3)^{3/2}$
   *o. $(x+y)^{1/2}$

2. Express each of the following in an equivalent exponential form and simplify, if possible.

   a. $\sqrt{x}$
   b. $\sqrt{y}$
   *c. $\sqrt[3]{a}$
   d. $\sqrt[3]{a^4}$
   *e. $\sqrt[4]{a^3}$
   f. $\sqrt[5]{a^4}$
   *g. $\sqrt[6]{8}$
   h. $\sqrt[4]{9}$
   *i. $\sqrt[4]{16}$
   j. $\sqrt[3]{27}$
   *k. $\sqrt{2x}$
   l. $\sqrt[3]{2x^2}$
   *m. $\sqrt[4]{3x^3}$
   n. $\sqrt{x^4 y^6}$
   *o. $\sqrt[3]{x^6 y^9}$

3. Determine the indicated root for each of the following:

   *a. $\sqrt{4}$
   b. $\sqrt[3]{8}$
   *c. $\sqrt{1}$
   d. $-\sqrt{16}$
   *e. $-\sqrt[3]{-27}$
   f. $\sqrt{-1}$
   *g. $\sqrt[4]{-1}$
   h. $\sqrt{x^8}$
   *i. $\sqrt[3]{x^9}$
   j. $\sqrt[4]{y^8}$
   *k. $\sqrt{y^8}$
   l. $\sqrt[3]{-8x^6}$

Exponentials and Radicals 157

4. Simplify each of the following expressions.

*a. $4^{1/2}$
b. $16^{1/4}$
*c. $(-8)^{1/3}$
d. $-8^{2/3}$
*e. $4^{3/2}$
f. $25^{5/2}$
*g. $64^{2/3}$
h. $64^{3/2}$
*i. $8^{-1/3}$
j. $4^{-1/2}$
*k. $16^{-1/2}$
l. $(-27)^{-1/3}$
*m. $\left(\frac{1}{4}\right)^{1/2}$
n. $\left(\frac{1}{8}\right)^{1/3}$
*o. $\left(\frac{4}{9}\right)^{1/2}$
p. $\left(\frac{1}{4}\right)^{-3/2}$
*q. $\left(\frac{1}{8}\right)^{-2/3}$
r. $\left(\frac{9}{4}\right)^{-3/2}$

5. Perform the indicated operations.

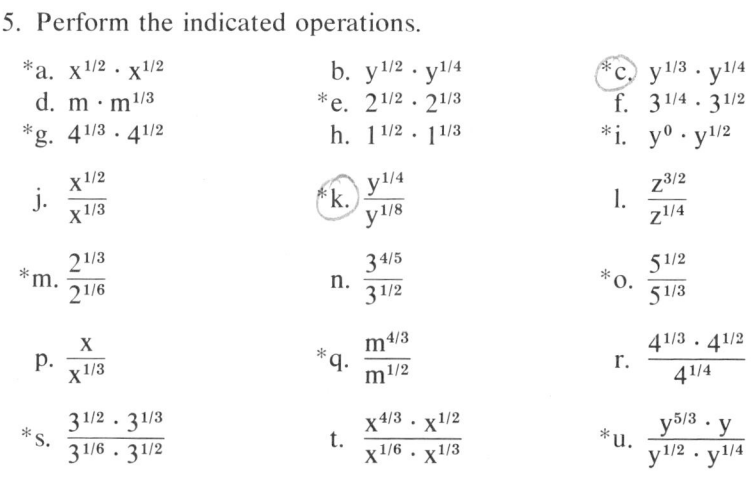

*a. $x^{1/2} \cdot x^{1/2}$
b. $y^{1/2} \cdot y^{1/4}$
*c. $y^{1/3} \cdot y^{1/4}$
d. $m \cdot m^{1/3}$
*e. $2^{1/2} \cdot 2^{1/3}$
f. $3^{1/4} \cdot 3^{1/2}$
*g. $4^{1/3} \cdot 4^{1/2}$
h. $1^{1/2} \cdot 1^{1/3}$
*i. $y^0 \cdot y^{1/2}$
j. $\dfrac{x^{1/2}}{x^{1/3}}$
*k. $\dfrac{y^{1/4}}{y^{1/8}}$
l. $\dfrac{z^{3/2}}{z^{1/4}}$
*m. $\dfrac{2^{1/3}}{2^{1/6}}$
n. $\dfrac{3^{4/5}}{3^{1/2}}$
*o. $\dfrac{5^{1/2}}{5^{1/3}}$
p. $\dfrac{x}{x^{1/3}}$
*q. $\dfrac{m^{4/3}}{m^{1/2}}$
r. $\dfrac{4^{1/3} \cdot 4^{1/2}}{4^{1/4}}$
*s. $\dfrac{3^{1/2} \cdot 3^{1/3}}{3^{1/6} \cdot 3^{1/2}}$
t. $\dfrac{x^{4/3} \cdot x^{1/2}}{x^{1/6} \cdot x^{1/3}}$
*u. $\dfrac{y^{5/3} \cdot y}{y^{1/2} \cdot y^{1/4}}$

3.5 SIMPLIFYING RADICALS

Many times expressions containing radicals (or simply radicals) can be simplified. We already can simplify expressions such as $\sqrt[3]{8}$, $\sqrt{9}$, $\sqrt{x^6}$. But we also want to be able to simplify expressions such as $\sqrt{98}$, $\sqrt[3]{16}$, $\sqrt{\dfrac{8}{9}}$, and $\sqrt[3]{\dfrac{x^2}{y^6}}$. We must again make use of our rules for exponents. Recall that $(ab)^{1/n} = a^{1/n} b^{1/n}$. In Section 3.4, we discovered that $a^{1/n} = \sqrt[n]{a}$, similarly $b^{1/n} = \sqrt[n]{b}$ and $(ab)^{1/n} = \sqrt[n]{ab}$. Therefore,

> If a and b represent non-negative real numbers, then for any n,
> $$\sqrt[n]{ab} = \sqrt[n]{a} \cdot \sqrt[n]{b}.$$

We can use this rule to simplify an expression such as $\sqrt{98}$. Note that the radicand, 98, can be factored. It can be factored in various ways, but most importantly as $98 = 49 \cdot 2$. The first factor, 49, is a perfect square, and we can determine its principal second (square) root. Therefore, when we apply the rule $\sqrt[n]{ab} = \sqrt[n]{a} \cdot \sqrt[n]{b}$, the first radical $\sqrt[n]{a}$ can be simplified completely.

$$\sqrt{98} = \sqrt{49 \cdot 2} \quad \text{(factoring)}$$
$$= \sqrt{49} \cdot \sqrt{2} \quad \text{(applying the rule } \sqrt[n]{ab} = \sqrt[n]{a} \cdot \sqrt[n]{b}\text{)}$$
$$= 7\sqrt{2}. \quad \text{(7 is the principal square root of 49.)}$$

In order to simplify the $\sqrt[3]{16}$, we will again use the rule that $\sqrt[n]{ab} = \sqrt[n]{a} \cdot \sqrt[n]{b}$. First we must factor 16 and in such a manner that one of the factors is a perfect cube. Note that $16 = 8 \cdot 2$ and 8 is a perfect cube. Thus

$$\sqrt[3]{16} = \sqrt[3]{8 \cdot 2} \quad \text{(factoring)}$$
$$= \sqrt[3]{8} \cdot \sqrt[3]{2} \quad \text{(applying the rule } \sqrt[n]{ab} = \sqrt[n]{a} \cdot \sqrt[n]{b}\text{)}$$
$$= 2\sqrt[3]{2}. \quad \text{(2 is the principal cube root of 8.)}$$

The rule $\sqrt[n]{ab} = \sqrt[n]{a} \cdot \sqrt[n]{b}$ can be used to simplify other expressions, such as $\sqrt{4x^6}$. Using this rule, we have $\sqrt{4x^6} = \sqrt{4} \cdot \sqrt{x^6} = 2x^3$.

Thus far we can simplify some radicals, but they must have radicands that can be factored such that one factor has a root of the index of the given radical.

EXAMPLE 20

Simplify the following radicals.

a. $\sqrt{18}$ b. $\sqrt[3]{54}$ c. $\sqrt{18x^7}$ d. $\sqrt[3]{16x^4y^7}$

Solution

In order to simplify a radical whose radicand can be factored such that one factor has a root of the index, we apply the rule

$$\sqrt[n]{ab} = \sqrt[n]{a} \cdot \sqrt[n]{b}$$

and then simplify the product $\sqrt[n]{a} \cdot \sqrt[n]{b}$, meaning, write the product of the principal root and the other factor.

a. $\sqrt{18} = \sqrt{9 \cdot 2} = \sqrt{9} \cdot \sqrt{2} = 3\sqrt{2}$.
b. $\sqrt[3]{54} = \sqrt[3]{27 \cdot 2} = \sqrt[3]{27} \cdot \sqrt[3]{2} = 3\sqrt[3]{2}$.
c. $\sqrt{18x^7} = \sqrt{9x^6 \cdot 2x} = \sqrt{9x^6} \cdot \sqrt{2x} = \sqrt{9} \cdot \sqrt{x^6} \cdot \sqrt{2x}$
   $= 3x^3\sqrt{2x}$.
d. $\sqrt[3]{16x^4y^7} = \sqrt[3]{8x^3y^6 \cdot 2xy} = \sqrt[3]{8x^3y^6} \cdot \sqrt[3]{2xy}$
   $= \sqrt[3]{8} \cdot \sqrt[3]{x^3} \cdot \sqrt[3]{y^6} \cdot \sqrt[3]{2xy} = 2xy^2\sqrt[3]{2xy}$.

Exponentials and Radicals

From the previous examples, we can see that a radical is not in simplified form if the radicand contains a factor that can be removed. Also, a radical is not in simplified form if the radicand contains a fraction. For example, $\sqrt{\frac{4}{9}}$ can be simplified, $\sqrt{\frac{4}{9}} = \frac{2}{3}$. But consider the expression $\sqrt{\frac{2}{3}}$. Note that $\frac{2}{3}$ is not a perfect square and yet the radicand contains a fraction, so we must still simplify $\sqrt{\frac{2}{3}}$. Note $\left(\frac{a}{b}\right)^{1/n} = \frac{a^{1/n}}{b^{1/n}}$, since $a^{1/n} = \sqrt[n]{a}$, $b^{1/n} = \sqrt[n]{b}$, and $\left(\frac{a}{b}\right)^{1/n} = \sqrt[n]{\frac{a}{b}}$. Therefore, we can state

> If a represents any non-negative real number and b represents any positive real number, then for any n
> $$\sqrt[n]{\frac{a}{b}} = \frac{\sqrt[n]{a}}{\sqrt[n]{b}} \quad (b \neq 0).$$

Using this rule, we can now write $\sqrt{\frac{2}{3}}$ as $\frac{\sqrt{2}}{\sqrt{3}}$, but this really doesn't help because a fraction that contains a radical in the denominator is not in simplified form. Recall the identity property for multiplication: if we multiply a number by 1, we do not change the identity of the given number. Therefore, we can multiply $\frac{2}{3}$ by 1. But the 1 shall assume the form of $\frac{3}{3}$. Thus

$$\sqrt{\frac{2}{3}} = \sqrt{\frac{2}{3} \cdot \frac{3}{3}} = \sqrt{\frac{6}{9}}.$$

Now we can make better use of the rule:

$$\sqrt[n]{\frac{a}{b}} = \frac{\sqrt[n]{a}}{\sqrt[n]{b}}.$$

Therefore, $\sqrt{\frac{6}{9}} = \frac{\sqrt{6}}{\sqrt{9}} = \frac{\sqrt{6}}{3}$ and we can state that $\sqrt{\frac{2}{3}} = \frac{\sqrt{6}}{3}$. Note that the radicand does not contain a fraction. If we wish to simplify $\sqrt[3]{\frac{1}{4}}$, we must again make use of the identity property for multiplication before we can apply the rule $\sqrt[n]{\frac{a}{b}} = \frac{\sqrt[n]{a}}{\sqrt[n]{b}}$. In this case, 1 should assume the form of $\frac{2}{2}$:

$$\sqrt[3]{\frac{1}{4}} = \sqrt[3]{\frac{1}{4} \cdot \frac{2}{2}} \quad \text{(multiplying by 1)}$$

$$= \sqrt[3]{\frac{2}{8}}$$

$$= \frac{\sqrt[3]{2}}{\sqrt[3]{8}} \quad \left(\text{applying the rule } \sqrt[n]{\frac{a}{b}} = \frac{\sqrt[n]{a}}{\sqrt[n]{b}}\right)$$

$$= \frac{\sqrt[3]{2}}{2}$$

Therefore, $\sqrt[3]{\frac{1}{4}} = \frac{\sqrt[3]{2}}{2}$. Note that the radicand does not contain a fraction.

We shall use this same technique when the radicand contains literal numbers. In order to simplify the radical $\sqrt{\frac{x}{y}}$, we must multiply $\frac{x}{y}$ by 1. For this example, 1 should assume the form of $\frac{y}{y}$:

$$\sqrt{\frac{x}{y}} = \sqrt{\frac{x}{y} \cdot \frac{y}{y}} \quad \text{(multiplying by 1)}$$

$$= \sqrt{\frac{xy}{y^2}}$$

$$= \frac{\sqrt{xy}}{\sqrt{y^2}} \quad \left(\text{applying the rule } \sqrt[n]{\frac{a}{b}} = \frac{\sqrt[n]{a}}{\sqrt[n]{b}}\right)$$

$$= \frac{\sqrt{xy}}{y}$$

Therefore, $\sqrt{\frac{x}{y}} = \frac{\sqrt{xy}}{y}$.

Note that in each case we chose a form of 1 that would result in a denominator that would yield a rational root of the index of the given radical. In the example $\sqrt[3]{\frac{1}{4}}$, we multiplied $\frac{1}{4}$ by $\frac{2}{2}$ because we wanted the resulting fraction to have a denominator that was a perfect cube, namely 8.

EXAMPLE 21

Simplify each of the following radicals.

a. $\sqrt[3]{\frac{5}{8}}$ b. $\sqrt{\frac{3}{7}}$ c. $\sqrt[4]{\frac{5}{8}}$ d. $\sqrt[3]{\frac{x^2}{y}}$

Exponentials and Radicals

Solution

To simplify a radical whose radicand contains a fraction, the denominator of the fraction must be a root of the index of the given radical. If this is not the case, then we apply the identity property for multiplication, that is, multiply the fraction by 1, choosing a form of 1 that will change the denominator to a perfect square, cube, and so on. Finally, we apply the rule $\sqrt[n]{\frac{a}{b}} = \frac{\sqrt[n]{a}}{\sqrt[n]{b}}$ and simplify the expression.

a. $\sqrt[3]{\frac{5}{8}} = \frac{\sqrt[3]{5}}{\sqrt[3]{8}} = \frac{\sqrt[3]{5}}{2}$.

b. $\sqrt{\frac{3}{7}} = \sqrt{\frac{3}{7} \cdot \frac{7}{7}} = \sqrt{\frac{21}{49}} = \frac{\sqrt{21}}{\sqrt{49}} = \frac{\sqrt{21}}{7}$.

c. $\sqrt[4]{\frac{5}{8}} = \sqrt[4]{\frac{5}{8} \cdot \frac{2}{2}} = \sqrt[4]{\frac{10}{16}} = \frac{\sqrt[4]{10}}{\sqrt[4]{16}} = \frac{\sqrt[4]{10}}{2}$.

d. $\sqrt[3]{\frac{x^2}{y}} = \sqrt[3]{\frac{x^2}{y} \cdot \frac{y^2}{y^2}} = \sqrt[3]{\frac{x^2 y^2}{y^3}} = \frac{\sqrt[3]{x^2 y^2}}{\sqrt[3]{y^3}} = \frac{\sqrt[3]{x^2 y^2}}{y}$.

It is not uncommon in algebra to encounter radical expressions such as $\sqrt[6]{x^2}$ or $\sqrt[4]{y^2}$. Radicals of this type can also be simplified, by the index of the given radical. For example, $\sqrt[6]{x^2}$ can be simplified because its index can be reduced. To do this, we must make use of the rule,

$$\sqrt[n]{a^m} = a^{m/n}.$$

Therefore, we can express $\sqrt[6]{x^2}$ as $x^{2/6}$. The fraction $\frac{2}{6}$ can be reduced to $\frac{1}{3}$ so $x^{2/6} = x^{1/3}$. But $x^{1/3} = \sqrt[3]{x^1}$, or simply $\sqrt[3]{x}$, and the given radical has been simplified.

$$\sqrt[6]{x^2} = x^{2/6} = x^{1/3} = \sqrt[3]{x} \quad \text{(if } x \geq 0\text{)}.$$

We can simplify $\sqrt[4]{y^2}$ in a similar manner:

$$\sqrt[4]{y^2} = y^{2/4} = y^{1/2} = \sqrt{y} \quad \text{(if } y \geq 0\text{)}.$$

To reduce the index of a given radical, we must first express the radical in an equivalent exponential form. Next, simplify the exponential expression, and then express the resulting exponential expression as a radical. Sometimes in simplifying a radical of this type we may also have to make use of the following rule: $(ab)^m = a^m b^m$. Consider this radical, $\sqrt[8]{4x^6 y^4}$. In order to simplify it, we must first express it in an equivalent exponential form:

$$\sqrt[8]{4x^6y^4} = (4x^6y^4)^{1/8} = (2^2x^6y^4)^{1/8}.$$

Next, apply the rule $(ab)^m = a^m b^m$,

$$(2^2x^6y^4) = (2^2)^{1/8} (x^6)^{1/8} (y^4)^{1/8} = 2^{1/4}x^{3/4}y^{2/4}.$$

Again, we apply the same rule (in the opposite direction),

$$2^{1/4}x^{3/4}y^{2/4} = (2x^3y^2)^{1/4}.$$

Finally, we express the resulting exponential expression as a radical.

$$(2x^3y^2)^{1/4} = \sqrt[4]{2x^3y^2}.$$

Therefore, $\sqrt[8]{4x^6y^4} = \sqrt[4]{2x^3y^2}$ (if $x \geq 0$ and $y \geq 0$).

When simplifying radical expressions of this nature and applying the rule $(ab)^m = a^m b^m$, we must be sure to simplify the exponential expression in a way such that all of the fractional exponents have the same denominator. In the previous example, all of the fractional exponents were rewritten with a denominator of 4. For other problems, the denominator will probably be some number other than 4, but all denominators must be the same.

EXAMPLE 22

Simplify the following radicals.

a. $\sqrt[6]{27}$ b. $\sqrt[4]{9}$ c. $\sqrt[4]{25x^2}$ d. $\sqrt[8]{16x^4y^6}$

Solution

a. $\sqrt[6]{27} = (27)^{1/6} = (3^3)^{1/6} = 3^{3/6} = 3^{1/2} = \sqrt{3}.$

b. $\sqrt[4]{9} = 9^{1/4} = (3^2)^{1/4} = 3^{2/4} = 3^{1/2} = \sqrt{3}.$

c. $\sqrt[4]{25x^2} = (25x^2)^{1/4} = (5^2x^2)^{1/4} = (5^2)^{1/4} (x^2)^{1/4} = 5^{1/2}x^{1/2} =$
$= (5x)^{1/2} = \sqrt{5x}.$

d. $\sqrt[8]{16x^4y^6} = (16x^4y^6)^{1/8} = (2^4x^4y^6)^{1/8} = (2^4)^{1/8} (x^4)^{1/8} (y^6)^{1/8}$
$= 2^{2/4}x^{2/4}y^{3/4} = (2^2x^2y^3)^{1/4} = \sqrt[4]{4x^2y^3}.$

Exponentials and Radicals

In this section, we have examined how to simplify various types of radicals. Basically, we have simplified three different forms of radicals. A radical is said to be simplified or in simplified form if

1. We cannot remove any factors from the radicand, that is, there is no factor in the radicand whose exponent is larger than the index.
2. The radicand does not contain any fractions, or the denominator of a fraction does not contain a radical.
3. The index of a radical cannot be reduced.

EXERCISES FOR SECTION 3.5

(Answers given to exercises marked *)

Simplify the following radicals completely.

*1. $\sqrt{4}$
2. $-\sqrt{9}$
*3. $-\sqrt{4}$
4. $\sqrt{48}$
*5. $\sqrt{50}$
6. $\sqrt{12}$
*7. $\sqrt[3]{16}$
8. $\sqrt[3]{54}$
*9. $\sqrt[3]{128}$
10. $\sqrt[3]{-8x}$
*11. $\sqrt[3]{32x^3}$
12. $\sqrt[3]{8x^5}$
*13. $\sqrt{12x^7y^5}$
14. $\sqrt[3]{16x^4y^7}$
*15. $\sqrt{8x^5y^4}$
16. $\sqrt{\dfrac{2}{9}}$
*17. $\sqrt{\dfrac{3}{4}}$
18. $\sqrt[3]{\dfrac{3}{8}}$
*19. $\sqrt{\dfrac{3}{5}}$
20. $\sqrt[3]{\dfrac{2}{3}}$
*21. $\sqrt[3]{\dfrac{4}{9}}$
22. $\sqrt{\dfrac{2}{x}}$
*23. $\sqrt[3]{\dfrac{2}{x}}$
24. $\sqrt{\dfrac{x}{y^2}}$
*25. $\sqrt[3]{\dfrac{x}{y}}$
26. $\sqrt{\dfrac{x^2}{y}}$
*27. $\sqrt{\dfrac{a}{bc}}$
28. $\sqrt[3]{\dfrac{a}{bc}}$
*29. $\sqrt{\dfrac{5}{4}}$
30. $\dfrac{\sqrt[3]{x}}{\sqrt[3]{8y^6}}$
*31. $\sqrt{\dfrac{2x}{3y}}$
32. $\sqrt[3]{\dfrac{4b}{9a^2}}$
*33. $\sqrt{\dfrac{8a^3}{3b}}$
34. $\sqrt[3]{\dfrac{3a^4b}{4c}}$
*35. $\sqrt[4]{\dfrac{3a^5b}{9a^2}}$
36. $\sqrt[6]{81}$
*37. $\sqrt[5]{32}$
38. $\sqrt[4]{4}$
*39. $\sqrt[6]{125}$
40. $\sqrt[6]{8}$
*41. $\sqrt[4]{81}$
42. $\sqrt[6]{27}$
*43. $\sqrt[8]{9}$
44. $\sqrt[8]{16}$
*45. $\sqrt[4]{4x^2}$

46. $\sqrt[6]{8y^3}$ *47. $\sqrt[4]{81x^2y^2}$ 48. $\sqrt[6]{8x^3y^3}$

*49. $\sqrt[8]{16x^4y^2}$ 50. $\sqrt[8]{9x^2y^4}$ *51. $\sqrt[6]{27x^6y^3}$

3.6 MULTIPLICATION OF RADICALS

In the previous section, we discussed how to simplify radicals or express a radical in its simplest form. We can also perform the four basic arithmetic operations (add, subtract, multiply, and divide) with radicals or radical expressions. We shall examine first the operation of multiplication because in the previous section we encountered the following rule:

> If a and b represent non-negative real numbers, then for any n,
> $$\sqrt[n]{ab} = \sqrt[n]{a} \cdot \sqrt[n]{b}.$$

Since we are quite familiar with this rule now, we shall use it to multiply radicals.

If we read the above equation from right to left, we have the necessary information for multiplying radicals, which is $\sqrt[n]{a} \cdot \sqrt[n]{b} = \sqrt[n]{ab}$. Note that in order to multiply two radicals, they must have the same index. For example,

$$\sqrt{2} \cdot \sqrt{3} = \sqrt{2 \cdot 3} = \sqrt{6}.$$
$$\sqrt[3]{4} \cdot \sqrt[3]{5} = \sqrt[3]{4 \cdot 5} = \sqrt[3]{20}.$$
$$\sqrt{x} \cdot \sqrt{y} = \sqrt{xy}.$$

When we multiply a radical by itself, such as $\sqrt{x} \cdot \sqrt{x}$, we can skip one of the intermediate steps. Note that

$$\sqrt{x} \cdot \sqrt{x} = \sqrt{x^2} = x.$$

It is not necessary (in this case) to multiply the radicands. We can simply say that $\sqrt{x} \cdot \sqrt{x} = x$. This is only true for square roots. If we have $\sqrt[3]{x} \cdot \sqrt[3]{x}$, then we must proceed in the normal manner:

$$\sqrt[3]{x} \cdot \sqrt[3]{x} = \sqrt[3]{x \cdot x} = \sqrt[3]{x^2}.$$

EXAMPLE 23

Multiply the following radicals.

a. $\sqrt{5} \cdot \sqrt{3}$ b. $\sqrt[3]{2} \cdot \sqrt[3]{3}$

c. $\sqrt[3]{x^2} \cdot \sqrt[3]{y}$ d. $\sqrt{x} \cdot \sqrt[3]{y}$

Exponentials and Radicals

Solution

In each case, we shall apply the rule $\sqrt[n]{a} \cdot \sqrt[n]{b} = \sqrt[n]{ab}$.

a. $\sqrt{5} \cdot \sqrt{3} = \sqrt{5 \cdot 3} = \sqrt{15}$.

b. $\sqrt[3]{2} \cdot \sqrt[3]{3} = \sqrt[3]{2 \cdot 3} = \sqrt[3]{6}$.

c. $\sqrt[3]{x^2} \cdot \sqrt[3]{y} = \sqrt[3]{x^2 y}$.

d. $\sqrt{x} \cdot \sqrt[3]{y}$. We have not yet discussed how to perform the indicated multiplication because the two radicals have neither the same index nor the same base.

Often when we multiply radicals, we can simplify the answer. A radical is not in simplest form if we can remove a factor from it. In other words, there can be no factor in the radical whose exponent is larger than the index. If we multiply $\sqrt{2}$ by $\sqrt{10}$, we obtain $\sqrt{20}$. But note that $\sqrt{20} = \sqrt{4 \cdot 5} = \sqrt{4} \cdot \sqrt{5} = 2\sqrt{5}$. Therefore,

$$\sqrt{2} \cdot \sqrt{10} = \sqrt{20} = \sqrt{4 \cdot 5} = \sqrt{4} \cdot \sqrt{5} = 2\sqrt{5}$$

Here is another example where the radical can be simplified after performing the multiplication:

$$\sqrt[3]{4} \cdot \sqrt[3]{6} = \sqrt[3]{24} = \sqrt[3]{8 \cdot 3} = \sqrt[3]{8} \cdot \sqrt[3]{3} = 2\sqrt[3]{3}.$$

To obtain the product of radicals *with the same index,* multiply the radicands and simplify the radical, if possible.

EXAMPLE 24

Multiply the following radicals and simplify, if possible.

a. $\sqrt{3} \cdot \sqrt{6}$ b. $\sqrt[3]{2} \cdot \sqrt[3]{4}$ c. $\sqrt[3]{x^2} \cdot \sqrt[3]{xy}$

d. $\sqrt{3x} \cdot \sqrt{6x}$

Solution

a. $\sqrt{3} \cdot \sqrt{6} = \sqrt{18} = \sqrt{9 \cdot 2} = \sqrt{9} \cdot \sqrt{2} = 3\sqrt{2}$.

b. $\sqrt[3]{2} \cdot \sqrt[3]{4} = \sqrt[3]{8} = 2$.

c. $\sqrt[3]{x^2} \cdot \sqrt[3]{xy} = \sqrt[3]{x^3 y} = \sqrt[3]{x^3} \cdot \sqrt[3]{y} = x\sqrt[3]{y}$.

d. $\sqrt{3x} \cdot \sqrt{6x} = \sqrt{18x^2} = \sqrt{9x^2 \cdot 2} = \sqrt{9x^2} \cdot \sqrt{2} = 3x\sqrt{2}$.

We can also multiply radicals that have coefficients, such as $3\sqrt{2} \cdot 4\sqrt{3}$. For this, we shall make use of the commutative and associative properties of multiplication. We can rewrite $3\sqrt{2} \cdot 4\sqrt{3}$ as $(3 \cdot 4)(\sqrt{2} \cdot \sqrt{3})$. Now we can multiply the coefficients to obtain the coefficient of the product and also multiply the radicals to obtain the radical of the product. Hence, $(3 \cdot 4)(\sqrt{2} \cdot \sqrt{3}) = 12\sqrt{6}$. We can multiply radicals with coefficients and the same index by multiplying the coefficients and the radicals (remember to simplify the radical, if possible). Therefore,

$$3\sqrt{2} \cdot 4\sqrt{3} = (3 \cdot 4)(\sqrt{2} \cdot \sqrt{3}) = 12\sqrt{6}.$$

In the next example, note that the radical can be simplified after performing the multiplication. The problem is completed by multiplying the coefficient of the product by the factor that is removed from the radical.

$$3\sqrt{2} \cdot 2\sqrt{6} = (3 \cdot 2)(\sqrt{2} \cdot \sqrt{6})$$
$$= 6\sqrt{12}$$
$$= 6\sqrt{4 \cdot 3}$$
$$= 6\sqrt{4} \cdot \sqrt{3}$$
$$= 6 \cdot 2\sqrt{3}$$
$$= 12\sqrt{3}.$$

Remember, to multiply radicals with coefficients and the same index, multiply the coefficients and multiply the radicals. Simplify the radicals, if possible.

EXAMPLE 25

Multiply the following radicals and simplify, if possible.

a. $4\sqrt{5} \cdot 3\sqrt{2}$
b. $2\sqrt[3]{4} \cdot 3\sqrt[3]{6}$
c. $2\sqrt[3]{x^2} \cdot 3\sqrt[3]{x^2 y}$
d. $3a\sqrt[3]{4a^2} \cdot 2a^2\sqrt[3]{6a^2}$

Solution

a. $4\sqrt{5} \cdot 3\sqrt{2} = (4 \cdot 3)(\sqrt{5} \cdot \sqrt{2}) = 12\sqrt{10}$.

b. $2\sqrt[3]{4} \cdot 3\sqrt[3]{6} = (2 \cdot 3)(\sqrt[3]{4} \cdot \sqrt[3]{6}) = 6\sqrt[3]{24} = 6\sqrt[3]{8 \cdot 3}$
$$= 6\sqrt[3]{8} \cdot \sqrt[3]{3} = 6 \cdot 2\sqrt[3]{3} = 12\sqrt[3]{3}.$$

c. $2\sqrt[3]{x^2} \cdot \sqrt[3]{x^2 y} = (2 \cdot 3)(\sqrt[3]{x^2} \cdot \sqrt[3]{x^2 y}) = 6\sqrt[3]{x^4 y} = 6\sqrt[3]{x^3 \cdot xy}$
$$= 6\sqrt[3]{x^3} \cdot \sqrt[3]{xy} = 6x\sqrt[3]{xy}.$$

d. $3a\sqrt[3]{4a^2} \cdot 2a^2\sqrt[3]{6a^2} = (3a \cdot 2a^2)(\sqrt[3]{4a^2} \cdot \sqrt[3]{6a^2}) = 6a^3\sqrt[3]{24a^4}$
$$= 6a^3\sqrt[3]{8a^3 \cdot 3a} = 6a^3\sqrt[3]{8a^3} \cdot \sqrt[3]{3a}$$
$$= 6a^3 \cdot 2a\sqrt[3]{3a} = 12a^4\sqrt[3]{3a}.$$

Exponentials and Radicals

Thus far in our discussion, we multiplied only those radicals that have had either the same index or the same radicand. It is possible to multiply radicals that do not have the same index or base. We can multiply radicals that have different indexes and the same radicand, and those that have different indexes and different radicands. First, let's consider radicals with different indexes and the same radicands. For example, find the product of $\sqrt{x} \cdot \sqrt[3]{x}$. To obtain the product of these two radicals, we first rewrite the radicals in exponential form,

$$\sqrt{x} \cdot \sqrt[3]{x} = x^{1/2} \cdot x^{1/3}.$$

Now, we can multiply $x^{1/2} \cdot x^{1/3}$ by applying the rule that $a^m \cdot a^n = a^{m+n}$. Therefore,

$$x^{1/2} \cdot x^{1/3} = x^{1/2 + 1/3} = x^{5/6}.$$

Since $a^{m/n} = \sqrt[n]{a^m}$, we can now write

$$x^{5/6} = \sqrt[6]{x^5}.$$

Therefore, $$\sqrt{x} \cdot \sqrt[3]{x} = \sqrt[6]{x^5}.$$

Now, let us consider the multiplication of radicals with different indexes and different radicands. For example, find the product of $\sqrt{x} \cdot \sqrt[3]{y}$. We write the radicals in exponential form,

$$\sqrt{x} \cdot \sqrt[3]{y} = x^{1/2} \cdot y^{1/3}.$$

Note that these exponentials do not have the same base. But we can rewrite each so the exponents have the same denominator,

$$x^{1/2} \cdot y^{1/3} = x^{3/6} \cdot y^{2/6}.$$

By means of the identity property for multiplication, we can rewrite any fraction. Note that $\frac{1}{2} = \frac{1}{2} \cdot \frac{3}{3} = \frac{3}{6}$ and $\frac{1}{3} = \frac{1}{3} \cdot \frac{2}{2} = \frac{2}{6}$. (It is best to use the least common denominator.) Now we have $x^{3/6} \cdot y^{2/6}$. Using the properties for multiplication of fractions, we can express the exponents as $x^{3 \cdot 1/6} \cdot y^{2 \cdot 1/6}$. Applying the rule that $(a^m b^n)^p = a^{mp} b^{np}$, we can rewrite the expression,

$$x^{3 \cdot 1/6} \cdot y^{2 \cdot 1/6} = (x^3 y^2)^{1/6}.$$

Now we can write the exponential in radical form,

$$(x^3 y^2)^{1/6} = \sqrt[6]{x^3 y^2}.$$

Therefore, $$\sqrt{x} \cdot \sqrt[3]{y} = \sqrt[6]{x^3 y^2}.$$

In general, to obtain the product of two radicals, we can use the following formula:

$$\sqrt[n]{x} \cdot \sqrt[m]{y} = \sqrt[mn]{x^m y^n}.$$

We shall apply this rule to the previous two examples that have been worked out.

$$\sqrt{x} \cdot \sqrt[3]{x}, \quad n = 2, m = 3.$$

According to this rule, $\sqrt{x} \cdot \sqrt[3]{x} = \sqrt[3 \cdot 2]{x^3 \cdot x^2} = \sqrt[6]{x^5}$

and $\sqrt{x} \cdot \sqrt[3]{y} = \sqrt[3 \cdot 2]{x^3 y^2} = \sqrt[6]{x^3 y^2}.$

Care should be taken in applying this rule because often the resulting radical can be simplified.

EXAMPLE 26

Multiply the following radicals and simplify, if possible.

a. $\sqrt{2} \cdot \sqrt[3]{2}$ b. $\sqrt{2} \cdot \sqrt[3]{3}$ c. $\sqrt[3]{x} \cdot \sqrt{3y}$

Solution

The indexes of the radicals in each example are not the same, therefore, we first rewrite the radicals in exponential form. If the bases are the same, we can add the exponents. If the bases are not the same, we rewrite each exponential so the exponents have the same denominator.

a. $\sqrt{2} \cdot \sqrt[3]{2} = 2^{1/2} \cdot 2^{1/3} = 2^{1/2 + 1/3} = 2^{5/6} = \sqrt[6]{2^5}.$

b. $\sqrt{2} \cdot \sqrt[3]{3} = 2^{1/2} \cdot 3^{1/3} = 2^{3/6} \cdot 3^{2/6} = (2^3 \cdot 3^2)^{1/6}$
$= \sqrt[6]{2^3 \cdot 3^2} = \sqrt[6]{72}.$

c. $\sqrt[3]{x} \cdot \sqrt{3y} = x^{1/3} (3y)^{1/2} = x^{2/6} (3y)^{3/6} = (x^2 (3y)^3)^{1/6}$
$= \sqrt[6]{x^2 (3y)^3} = \sqrt[6]{27 x^2 y^3}$

Alternate Solution

We shall now do the same problems using the rule that
$\sqrt[n]{x} \cdot \sqrt[m]{y} = \sqrt[mn]{x^m y^n}.$

Exponentials and Radicals

a. $\sqrt{2} \cdot \sqrt[3]{2}$, $(n = 2, m = 3) = \sqrt[3 \cdot 2]{2^3 \cdot 2^2} = \sqrt[6]{2^5}$.

b. $\sqrt{2} \cdot \sqrt[3]{3}$, $(n = 2, m = 3) = \sqrt[2 \cdot 3]{2^3 \cdot 3^2} = \sqrt[6]{8 \cdot 9} = \sqrt[6]{72}$.

c. $\sqrt[3]{x} \cdot \sqrt{3y}$, $(n = 3, m = 2) = \sqrt[2 \cdot 3]{x^2 \cdot (3y)^3} = \sqrt[6]{x^2 \cdot 27y^3}$
$= \sqrt[6]{27x^2y^3}$.

We are not advocating one technique over the other; the technique to use is the one that you understand best.

Often, in multiplying radicals, some expressions may contain more than one term with a radical. Consider the following example:

$$\sqrt{2}(\sqrt{3} + \sqrt{5})$$

In order to perform the indicated multiplication, we use the properties for multiplication of radicals and the distributive property. Recall that $a(b + c) = ab + ac$. Therefore, $\sqrt{2}(\sqrt{3} + \sqrt{5}) = \sqrt{2} \cdot \sqrt{3} + \sqrt{2} \cdot \sqrt{5} = \sqrt{6} + \sqrt{10}$. This particular example illustrates multiplication of a monomial times a binomial. In a later section, we shall examine multiplication of a binomial times a binomial. But regardless of the number of terms in the expression, we apply the distributive property.

EXAMPLE 27

Perform the indicated multiplication and simplify, if possible.

a. $\sqrt{2}(1 + \sqrt{2})$

b. $\sqrt[3]{x}(\sqrt[3]{2} - \sqrt[3]{y^2})$

c. $\sqrt{x}(3\sqrt{x} + 2\sqrt{xy})$

Solution

a. $\sqrt{2}(1 + \sqrt{2}) = \sqrt{2} \cdot 1 + \sqrt{2} \cdot \sqrt{2} = \sqrt{2} + 2$.

b. $\sqrt[3]{x}(\sqrt[3]{2} - \sqrt[3]{y^2}) = \sqrt[3]{x} \cdot \sqrt[3]{2} - \sqrt[3]{x} \cdot \sqrt[3]{y^2} = \sqrt[3]{2x} - \sqrt[3]{xy^2}$.

c. $\sqrt{x}(3\sqrt{x} + 2\sqrt{xy}) = \sqrt{x} \cdot 3\sqrt{x} + \sqrt{x} \cdot 2\sqrt{xy}$
$= 3\sqrt{x^2} + 2\sqrt{x^2y} = 3x + 2x\sqrt{y}$.

Using the properties of exponents, we can multiply radicals, whether or not they have the same index or radicands.

EXERCISES FOR SECTION 3.6

(Answers given to exercises marked *)

Perform the indicated multiplication and simplify, if possible.

*1. $\sqrt{3} \cdot \sqrt{5}$
2. $\sqrt{3} \cdot \sqrt{7}$
*3. $\sqrt[3]{2} \cdot \sqrt[3]{2}$
4. $\sqrt[3]{2} \cdot \sqrt[3]{4}$
*5. $3\sqrt[3]{5} \cdot 2\sqrt[3]{25}$
6. $3\sqrt{x} \cdot 2\sqrt{3}$
*7. $4\sqrt{2x} \cdot 2\sqrt{3y}$
8. $3\sqrt[3]{2a} \cdot 2\sqrt[3]{66}$
*9. $3\sqrt{2x} \cdot 4\sqrt{3x}$
10. $2\sqrt[3]{2xy} \cdot \sqrt[3]{3x^2y}$
*11. $3\sqrt[3]{4x^2} \cdot 5\sqrt[3]{2x}$
12. $(\sqrt{3x})^2$
*13. $(\sqrt[3]{3x})^2$
14. $(\sqrt{3x})^3$
*15. $\sqrt{3} \cdot \sqrt[3]{4}$
16. $\sqrt{a} \cdot \sqrt[3]{b}$
*17. $\sqrt[3]{x^2} \cdot \sqrt{y}$
18. $\sqrt[3]{2x} \cdot \sqrt{2x}$
*19. $2\sqrt[3]{4x^2} \cdot 3\sqrt{2y}$
20. $4\sqrt{5y} \cdot 2\sqrt[4]{x^2}$
*21. $3\sqrt[3]{2xy} \cdot \sqrt[4]{x^2y}$
22. $\sqrt[3]{3a^2} \cdot 2\sqrt{4b}$
*23. $\sqrt{3ab} \cdot \sqrt[3]{2a^2b^2}$
24. $3\sqrt{2xy} \cdot \sqrt[3]{x^2y}$
*25. $\sqrt[3]{4y} \cdot 2\sqrt{3xy}$
26. $\sqrt[3]{4x^2y} \cdot 2\sqrt{2y}$
*27. $2\sqrt[3]{2a^2b^2} \cdot 3\sqrt{ab}$
28. $\sqrt{2}(3-\sqrt{3})$
*29. $\sqrt{2}(\sqrt{3}-\sqrt{5})$
30. $\sqrt{x}(\sqrt{x}-\sqrt{y})$
*31. $\sqrt{y}(3\sqrt{y}+2\sqrt{xy})$
32. $\sqrt[3]{y}(\sqrt[3]{x}-\sqrt[3]{2})$
*33. $\sqrt[3]{y}(\sqrt[3]{y^2}+\sqrt[3]{x})$
34. $2\sqrt{x}(3\sqrt{y}+2\sqrt{z})$
*35. $3\sqrt{2y}(2\sqrt{2x}-\sqrt{z})$
36. $2\sqrt{xy}(2\sqrt{xy}+1)$
*37. $(\sqrt[3]{3x^2y}+2\sqrt[3]{2y^2})\sqrt[3]{4xy}$
38. $(\sqrt{3xy}-\sqrt{5x})(\sqrt{2xy})$
*39. $\sqrt{2}(\sqrt[3]{3}-\sqrt[3]{2})$
40. $\sqrt[3]{2}(\sqrt{3}+\sqrt[3]{5})$
*41. $\sqrt[3]{x}(\sqrt{x}-\sqrt[3]{y})$
42. $\sqrt{xy}(\sqrt[3]{x}-\sqrt{y})$

3.7 ADDITION AND SUBTRACTION OF RADICALS

When two or more monomials with radicals have the same index and the same radicand, they are called *like radicals.* Some texts also refer to like radicals as similar radicals. The following pairs of monomials are like radicals because they have the same index *and* the same radicand—the different coefficients do not affect their likeness.

$$3\sqrt{2} \text{ and } 2\sqrt{2}$$

$$\sqrt[3]{x^2y} \text{ and } 5\sqrt[3]{x^2y}$$

Note that radicals such as $\sqrt{2}$ and $\sqrt{3}$ are not like radicals because they

Exponentials and Radicals

do not have the same radicand. Similarly, $\sqrt[3]{2}$ and $\sqrt{2}$ are not like radicals because they do not have the same index.

We can combine like radicals by means of the distributive property. Recall that the distributive property states that for all real numbers, x, y and z

$$x(y + z) = xy + xz.$$

Using the commutative property for multiplication and the symmetric property of equality, we can rewrite the distributive property as

$$yx + zx = (y + z)x.$$

It is this form of the distributive property that is used for combining other like terms. To combine $5x^2$ and $3x^2$, we can rewrite this expression (using the distributive property) as $5x^2 + 3x^2 = (5 + 3)x^2 = 8x^2$. This same technique can be used for combining like radicals. For example,

$$3\sqrt{2} + 2\sqrt{2} = (3 + 2)\sqrt{2} = 5\sqrt{2}.$$

Similarly,

$$\sqrt[3]{x^2y} + 5\sqrt[3]{x^2y} = (1 + 5)\sqrt[3]{x^2y} = 6\sqrt[3]{x^2y}.$$

If the radicals are not like, then we can only indicate their sum or difference. For example, $2\sqrt{2} + 5\sqrt{3} = 2\sqrt{2} + 5\sqrt{3}$, and $3\sqrt[3]{x} + 2\sqrt[3]{y} = 3\sqrt[3]{x} + 2\sqrt[3]{y}$.

Sometimes expressions will contain radicals that are not like, but one more of the radicals can be simplified. Note that $\sqrt{8} + 3\sqrt{2}$ do not have the same radicand, but $\sqrt{8}$ can be simplified: $\sqrt{8} = \sqrt{4 \cdot 2} = \sqrt{4} \cdot \sqrt{2} = 2\sqrt{2}$. After simplifying the radicals in a given expression, we may be able to combine the radicals. Therefore, we can perform the indicated addition for the given example,

$$\sqrt{8} + 3\sqrt{2} = \sqrt{4 \cdot 2} + 3\sqrt{2}$$
$$= \sqrt{4} \cdot \sqrt{2} + 3\sqrt{2}$$
$$= 2\sqrt{2} + 3\sqrt{2}$$
$$= (2 + 3)\sqrt{2}$$
$$= 5\sqrt{2}.$$

Let's examine another example—combine $\sqrt{18} - \sqrt{8}$. At first glance it appears that we cannot combine these two radicals because they are not like. But both of these radicals can be simplified, thus,

$$\sqrt{18} = \sqrt{9 \cdot 2} = \sqrt{9} \cdot \sqrt{2} = 3\sqrt{2}$$

and $\sqrt{8} = \sqrt{4 \cdot 2} = \sqrt{4} \cdot \sqrt{2} = 2\sqrt{2}.$

Therefore, $\sqrt{18} - \sqrt{8} = 3\sqrt{2} - 2\sqrt{2} = (3-2)\sqrt{2} = 1\sqrt{2}$ or $\sqrt{2}.$

Radical expressions that do not contain like radicals cannot be combined unless the radicals can be transformed into equivalent like radicals. In simplifying expressions involving radicals and addition or subtraction, we make use of the distributive property and the simplification techniques used in previous sections.

EXAMPLE 28

Perform the indicated operations.

a. $4\sqrt{2} + 5\sqrt{2}$ b. $7\sqrt{x} - 2\sqrt{x}$ c. $3\sqrt{2} + 2\sqrt{3}$

Solution

a. $4\sqrt{2} + 5\sqrt{2} = (4+5)\sqrt{2} = 9\sqrt{2}.$

b. $7\sqrt{x} - 2\sqrt{x} = (7-2)\sqrt{x} = 5\sqrt{x}.$

c. $3\sqrt{2} + 2\sqrt{3} = 3\sqrt{2} + 2\sqrt{3}.$ This sum can only be indicated because each radical is in simplest form and they are not like radicals.

EXAMPLE 29

Perform the indicated operations.

a. $5\sqrt{2} - \sqrt{18}$ b. $\sqrt{18} - \sqrt{50}$ c. $3x\sqrt[3]{2x} + 2\sqrt[3]{16x^4}$
d. $3\sqrt{x^3 y} + 2x\sqrt{xy} - x\sqrt{xy}$

Solution

To add or subtract radicals, we simplify the radicals in each term and then combine like radicals.

a. $5\sqrt{2} - \sqrt{18} = 5\sqrt{2} - \sqrt{9 \cdot 2}$
$= 5\sqrt{2} - \sqrt{9} \cdot \sqrt{2}$
$= 5\sqrt{2} - 3\sqrt{2}$
$= (5-3)\sqrt{2}$
$= 2\sqrt{2}.$

Exponentials and Radicals

b. $\sqrt{18} - \sqrt{50} = \sqrt{9 \cdot 2} - \sqrt{25 \cdot 2}$
$= 3\sqrt{2} - 5\sqrt{2}$
$= (3 - 5)\sqrt{2}$
$= -2\sqrt{2}.$

c. $3x\sqrt[3]{2x} + 2\sqrt[3]{16x^4} = 3x\sqrt[3]{2x} + 2\sqrt[3]{8x^3 \cdot 2x}$
$= 3x\sqrt[3]{2x} + 2\sqrt[3]{8x^3}\sqrt[3]{2x}$
$= 3x\sqrt[3]{2x} + 2 \cdot 2x\sqrt[3]{2x}$
$= 3x\sqrt[3]{2x} + 4x\sqrt[3]{2x}$
$= (3x + 4x)\sqrt[3]{2x}$
$= 7x\sqrt[3]{2x}.$

d. $3\sqrt{x^3 y} + 2x\sqrt{xy} - x\sqrt{xy} = 3\sqrt{x^2 \cdot xy} + 2x\sqrt{xy} - x\sqrt{xy}$
$= 3\sqrt{x^2}\sqrt{xy} + 2x\sqrt{xy} - x\sqrt{xy}$
$= 3x\sqrt{xy} + 2x\sqrt{xy} - x\sqrt{xy}$
$= (3x + 2x - x)\sqrt{xy}$
$= 4x\sqrt{xy}.$

In order to add or subtract radicals, the radicals must be like radicals, that is, they must have the same index and the same radicand. We have already discovered that it is sometimes possible to simplify one or more radicals in an expression and then combine the radicals. It is also sometimes possible to change the indexes of certain radicals in a given expression in order to combine like terms. Can we combine $3\sqrt{x} + \sqrt[6]{x^3}$? These radicals are not like radicals, but note that $\sqrt[6]{x^3}$ can be simplified. We can rewrite $\sqrt[6]{x^3}$ as $(x^3)^{1/6}$, which is equivalent to $x^{1/2}$ and can be rewritten as $\sqrt{x}$. Therefore,

$$3\sqrt{x} + \sqrt[6]{x^3} = 3\sqrt{x} + x^{3/6}$$
$$= 3\sqrt{x} + x^{1/2}$$
$$= 3\sqrt{x} + \sqrt{x}$$
$$= (3 + 1)\sqrt{x}$$
$$= 4\sqrt{x}.$$

Note that in order to use this technique, the index of a given radical must be reducible.

EXAMPLE 30

Perform the indicated operations:

a. $\sqrt{3} + \sqrt[4]{9}$ b. $2\sqrt[6]{4} - 3\sqrt[3]{16}$

Solution

It appears that these radicals cannot be combined, but note that some of the radicals can be simplified by reducing the index.

a. $\sqrt{3} + \sqrt[4]{9} = \sqrt{3} + (9)^{1/4} = \sqrt{3} + (3^2)^{1/4}$
$= \sqrt{3} + 3^{2/4} = \sqrt{3} + 3^{1/2}$
$= \sqrt{3} + \sqrt{3} = (1 + 1)\sqrt{3}$
$= 2\sqrt{3}.$

b. $2\sqrt[6]{4} - 3\sqrt[3]{16} = 2(4)^{1/6} - 3\sqrt[3]{8 \cdot 2}$
$= 2(2^2)^{1/6} - 3\sqrt[3]{8}\sqrt[3]{2}$
$= 2(2^{2/6}) - 3 \cdot 2\sqrt[3]{2}$
$= 2(2^{1/3}) - 6\sqrt[3]{2}$
$= 2\sqrt[3]{2} - 6\sqrt[3]{2}$
$= (2 - 6)\sqrt[3]{2}$
$= -4\sqrt[3]{2}.$

In simplifying expressions involving radicals and addition or subtraction, we make use of the distributive property and the simplification technique used in previous sections.

EXERCISES FOR SECTION 3.7

(Answers given to exercises marked *)

Perform the indicated operations.

*1. $3\sqrt{2} + 2\sqrt{2}$ 2. $3\sqrt{2} - 2\sqrt{2}$

*3. $\sqrt[3]{3} + 2\sqrt[3]{3}$ 4. $\sqrt[3]{3} - 2\sqrt[3]{3}$

*5. $\sqrt{2} + \sqrt{3}$ 6. $\sqrt{2} - \sqrt{3}$

*7. $3\sqrt{2} + 4\sqrt{2} - 5\sqrt{2}$ 8. $2\sqrt[3]{3} - 3\sqrt[3]{3} - 4\sqrt[3]{3}$

*9. $\sqrt{x} + 2\sqrt{x} - 3\sqrt{x}$ 10. $5\sqrt{2} - \sqrt{8}$

*11. $4\sqrt{2} - 3\sqrt{8}$ 12. $\sqrt{18} + \sqrt{50}$

Exponentials and Radicals

*13. $\sqrt{27} + \sqrt{48}$
*15. $\sqrt[3]{16} + \sqrt[3]{54}$
*17. $2\sqrt{32} - \sqrt{98} + 3\sqrt{50}$
*19. $3\sqrt{2x} - 2\sqrt{8x} + 3\sqrt{8x}$
*21. $x\sqrt{.25} + \sqrt{4x^2} + 2\sqrt{0.01x^2}$
*23. $3\sqrt[3]{x^2} - \sqrt[3]{x^5} + \sqrt[6]{x^4}$
*25. $\sqrt{x} + \sqrt[3]{x}$
*27. $\sqrt{x^3} + \sqrt[3]{x^4}$
*29. $2\sqrt{3} + 3\sqrt[4]{9}$
*31. $2\sqrt{2} + 3\sqrt[4]{4} + \sqrt[6]{8}$
*33. $3\sqrt{3} - 2\sqrt{12} + 2\sqrt[4]{9}$

14. $\sqrt{48} + \sqrt{75}$
16. $\sqrt{18} - 2\sqrt{50} + \sqrt{98}$
18. $\sqrt{27} - 2\sqrt{48} + 2\sqrt{75}$
20. $3\sqrt{3y} - 2\sqrt{2y} + 4\sqrt{27y}$
22. $2\sqrt{x} + 3\sqrt{x^3} - \sqrt{4x}$
24. $2\sqrt{y} + \sqrt[4]{y^2} + \sqrt[6]{y^3}$
26. $2\sqrt{y} - \sqrt[3]{y}$
28. $\sqrt{2} + \sqrt[4]{4}$
30. $2\sqrt[6]{4} + \sqrt[3]{54}$
32. $3\sqrt{2} - 2\sqrt[4]{4} + \sqrt[6]{8}$

3.8 MULTIPLICATION OF RADICAL EXPRESSIONS

In Section 3.6, we examined multiplication of two radicals. We multiplied radicals with and without the same index. At the end of the section, we considered problems such as $\sqrt{2}(\sqrt{3}+\sqrt{5})$, or, a monomial times a binomial. In this section, we will again consider multiplication of a monomial times a binomial (in general) and also multiplication of a binomial times a binomial. Recall that in order to perform the indicated multiplication for $\sqrt{2}(\sqrt{3}+\sqrt{5})$, we apply the distributive property, $\sqrt{2}(\sqrt{3}+\sqrt{5}) = \sqrt{2}\cdot\sqrt{3}+\sqrt{2}\cdot\sqrt{5} = \sqrt{6}+\sqrt{10}$. Often, when multiplying radical expressions, we can simplify the answer by combining like terms. Consider the example: $\sqrt{2}(\sqrt{6}+\sqrt{24}) = \sqrt{2}\cdot\sqrt{6}+\sqrt{2}\cdot\sqrt{24} = \sqrt{12}+\sqrt{48}$. Both radicals $\sqrt{12}$ and $\sqrt{48}$ can be simplified. Note that $\sqrt{12} = \sqrt{4\cdot 3} = \sqrt{4}\cdot\sqrt{3} = 2\sqrt{3}$ and $\sqrt{48} = \sqrt{16\cdot 3} = \sqrt{16}\cdot\sqrt{3} = 4\sqrt{3}$. Therefore, $\sqrt{12}+\sqrt{48} = 2\sqrt{3}+4\sqrt{3}$. Combining like terms, we obtain $6\sqrt{3}$. Hence, $\sqrt{2}(\sqrt{6}+\sqrt{24}) = 6\sqrt{3}$.

Let's consider another example, $\sqrt{3}(\sqrt{6}+\sqrt{24})$. Applying the distributive property, we obtain

$$\sqrt{3}(\sqrt{6}+\sqrt{24}) = \sqrt{3}\cdot\sqrt{6}+\sqrt{3}\cdot\sqrt{24}$$
$$= \sqrt{18}+\sqrt{72}$$

Both radicals can be simplified
$$= \sqrt{9\cdot 2}+\sqrt{36\cdot 2}$$
$$= \sqrt{9}\cdot\sqrt{2}+\sqrt{36}\cdot\sqrt{2}$$
$$= 3\sqrt{2}+6\sqrt{2}$$

Combining like terms
$$= 9\sqrt{2}.$$

EXAMPLE 31

Perform the indicated operations and simplify, if possible.

a. $2\sqrt{2}\,(2\sqrt{6}+3\sqrt{2})$ b. $3\sqrt{2}\,(\sqrt{24}+3\sqrt{6})$

c. $2\sqrt{3}\,(\sqrt{6}+3\sqrt{24})$

Solutions

We multiply radicals with coefficients and the same index by multiplying the coefficients and multiplying the radicals.

a. $2\sqrt{2}\,(2\sqrt{6}+3\sqrt{2}) = (2\sqrt{2})(2\sqrt{6}) + (2\sqrt{2})(3\sqrt{2})$

$\qquad = 4\sqrt{12} + 6\sqrt{4}$

$\qquad = 4\sqrt{4\cdot 3} + 6\cdot 2$ (simplifying radicals)

$\qquad = 4\cdot\sqrt{4}\cdot\sqrt{3} + 12$

$\qquad = 4\cdot 2\sqrt{3} + 12$

$\qquad = 8\sqrt{3} + 12.$ (Note: we cannot combine these two terms.)

b. $3\sqrt{2}\,(\sqrt{24}+3\sqrt{6}) = (3\sqrt{2})(\sqrt{24}) + (3\sqrt{2})(3\sqrt{6})$

$\qquad = 3\sqrt{48} + 9\sqrt{12}$

$\qquad = 3\sqrt{16\cdot 3} + 9\sqrt{4\cdot 3}$ (simplifying radicals)

$\qquad = 3\sqrt{16}\,\sqrt{3} + 9\sqrt{4}\,\sqrt{3}$

$\qquad = 3\cdot 4\sqrt{3} + 9\cdot 2\sqrt{3}$

$\qquad = 12\sqrt{3} + 18\sqrt{3}$

$\qquad = 30\sqrt{3}.$ (combining like terms)

c. $2\sqrt{3}\,(\sqrt{6}+3\sqrt{24}) = (2\sqrt{3})(\sqrt{6}) + (2\sqrt{3})(3\sqrt{24})$

$\qquad = 2\sqrt{18} + 6\sqrt{72}$

$\qquad = 2\sqrt{9\cdot 2} + 6\sqrt{36\cdot 2}$ (simplifying radicals)

$\qquad = 2\sqrt{9}\,\sqrt{2} + 6\sqrt{36}\,\sqrt{2}$

$\qquad = 2\cdot 3\sqrt{2} + 6\cdot 6\sqrt{2}$

$\qquad = 6\sqrt{2} + 36\sqrt{2}$

$\qquad = 42\sqrt{2}.$ (combining like terms)

Exponentials and Radicals

When multiplying a binomial by a binomial, we can obtain the product in one of two ways. Consider the example: $(\sqrt{2}+\sqrt{3})(\sqrt{3}+\sqrt{2})$. To obtain the product of these two binomials, we can write one binomial below the other. One binomial is called the multiplicand and the other binomial is the multiplier.

$$\sqrt{2}+\sqrt{3} \quad \text{(multiplicand)}$$
$$\sqrt{3}+\sqrt{2} \quad \text{(multiplier)}$$

Now we multiply each term (monomial) in the multiplier by each term (monomial) in the multiplicand. These products are called partial products and we list like or similar partial products in the same column (if any occur).

$$\begin{array}{l} \sqrt{2}+\sqrt{3} \\ \sqrt{3}+\sqrt{2} \\ \sqrt{6}+\sqrt{9} \quad \text{(multiplying by } \sqrt{3}) \\ \sqrt{6} \quad\quad +\sqrt{4} \quad \text{(multiplying by } \sqrt{2}) \end{array}$$

Next, add the partial products $\quad 2\sqrt{6}+\sqrt{9}+\sqrt{4}$
Simplifying $\quad\quad\quad\quad\quad\quad\quad\quad 2\sqrt{6}+3+2$
$\quad\quad\quad\quad\quad\quad\quad\quad\quad\quad\quad 2\sqrt{6}+5$
Therefore $\quad\quad\quad\quad\quad (\sqrt{2}+\sqrt{3})(\sqrt{3}+\sqrt{2}) = 2\sqrt{6}+5.$

We can also obtain the product of a binomial times a binomial by means of the distributive property. Recall that $a(b+c) = ab + ac$. We can expand this property by substituting a binomial in place of a. We could let $a = (d+f)$. Therefore, $a(b+c)$ can be rewritten as $(d+f)(b+c)$. Instead of having

$$a(b+c) = ab + ac,$$

we have $\quad (d+f)(b+c) = (d+f)b + (d+f)c.$

Using the distributive property again, $\quad = db + fb + dc + fc.$

Note that the product of two binomials is the sum of the products obtained by multiplying each term of one binomial by each term of the other binomial. Applying this technique to the original example, we have

$$(\sqrt{2}+\sqrt{3})(\sqrt{3}+\sqrt{2}) = (\sqrt{2}+\sqrt{3})\sqrt{3} + (\sqrt{2}+\sqrt{3})\sqrt{2}$$
$$= \sqrt{6}+\sqrt{9} \quad\quad +\sqrt{4}+\sqrt{6}$$
$$= \sqrt{6}+3 \quad\quad\quad + 2+\sqrt{6}$$
$$= 2\sqrt{6}+5.$$

178   Chapter 3

The technique to use is the one that you understand the best. Regardless of the technique used, we must multiply each term of one binomial by each term in the other binomial, and add the results.

EXAMPLE 32

Perform the indicated operation and simplify, if possible.

$$(\sqrt{2}+\sqrt{5})(\sqrt{3}+\sqrt{2})$$

Solution

Applying the distributive property, we have

$$(\sqrt{2}+\sqrt{5})(\sqrt{3}+\sqrt{2}) = (\sqrt{2}+\sqrt{5})\sqrt{3} + (\sqrt{2}+\sqrt{5})\sqrt{2}$$
$$= \sqrt{6}+\sqrt{15} \quad\quad +\sqrt{4}+\sqrt{10}$$
$$= \sqrt{6}+\sqrt{15} \quad\quad + 2 + \sqrt{10}.$$

Alternate Solution

We write one binomial below the other, multiply each term of the multiplier by each term in the multiplicand, and add the results.

$(\sqrt{2}+\sqrt{5})(\sqrt{3}+\sqrt{2}) = \sqrt{2}+\sqrt{5}$ (multiplicand)
$\phantom{(\sqrt{2}+\sqrt{5})(\sqrt{3}+\sqrt{2}) = }\sqrt{3}+\sqrt{2}$ (multiplier)
$\phantom{(\sqrt{2}+\sqrt{5})(\sqrt{3}+\sqrt{2}) = }\sqrt{6}+\sqrt{15}$ (multiplying by $\sqrt{3}$)
$\phantom{(\sqrt{2}+\sqrt{5})(\sqrt{3}+\sqrt{2}) = xxxxxxxx}+\sqrt{4}+\sqrt{10}$ (multiplying by $\sqrt{2}$)
$\phantom{(\sqrt{2}+\sqrt{5})(\sqrt{3}+\sqrt{2})} = \sqrt{6}+\sqrt{15}+\sqrt{4}+\sqrt{10}$ (adding the partial products)
$\phantom{(\sqrt{2}+\sqrt{5})(\sqrt{3}+\sqrt{2})} = \sqrt{6}+\sqrt{15}+2+\sqrt{10}.$

EXAMPLE 33

Perform the indicated operation and simplify, if possible:

$$(2\sqrt{2}-3\sqrt{3})(\sqrt{3}+3\sqrt{2})$$

Exponentials and Radicals

Solution

$$(2\sqrt{2} - 3\sqrt{3})(\sqrt{3} + 3\sqrt{2}) = (2\sqrt{2} - 3\sqrt{3})\sqrt{3} + (2\sqrt{2} - 3\sqrt{3})3\sqrt{2}$$
$$= 2\sqrt{6} - 3\sqrt{9} + 6\sqrt{4} - 9\sqrt{6}$$
$$= 2\sqrt{6} - 3 \cdot 3 + 6 \cdot 2 - 9\sqrt{6}$$
$$= 2\sqrt{6} - 9 + 12 - 9\sqrt{6}$$
$$= 3 - 7\sqrt{6}.$$

Alternate Solution

$$(2\sqrt{2} - 3\sqrt{3})(\sqrt{3} + 3\sqrt{2}) = \begin{array}{r} 2\sqrt{2} - 3\sqrt{3} \\ \sqrt{3} + 3\sqrt{2} \\ \hline 2\sqrt{6} - 3\sqrt{9} \\ -9\sqrt{6} \qquad + 6\sqrt{4} \\ \hline \end{array}$$

(Note that we listed like partial products in the same column.)

$$= -7\sqrt{6} - 3\sqrt{9} + 6\sqrt{4}$$
$$= -7\sqrt{6} - 9 + 12$$
$$= -7\sqrt{6} + 3.$$

EXAMPLE 34

Perform the indicated operation and simplify, if possible.

$$(x + \sqrt{2})(x - \sqrt{2})$$

Solution

$$(x + \sqrt{2})(x - \sqrt{2}) = (x + \sqrt{2})x + (x + \sqrt{2})(-\sqrt{2})$$
$$= x^2 + x\sqrt{2} - x\sqrt{2} - \sqrt{4}$$
$$= x^2 - 2.$$

Alternate Solution

$$(x + \sqrt{2})(x - \sqrt{2}) = \begin{array}{r} x + \sqrt{2} \\ x - \sqrt{2} \\ \hline x^2 + x\sqrt{2} \\ -x\sqrt{2} - \sqrt{4} \\ \hline \end{array}$$
$$= x^2 \qquad - \sqrt{4}$$
$$= x^2 - 2.$$

EXAMPLE 35

Perform the indicated operation and simplify, if possible.

$$(2\sqrt{3} + 3\sqrt{5})^2$$

Solution

We are asked to square the binomial, that is, multiply $(2\sqrt{3} + 3\sqrt{5})$ by $(2\sqrt{3} + 3\sqrt{5})$. Therefore,

$$\begin{aligned}
(2\sqrt{3} + 3\sqrt{5})^2 &= (2\sqrt{3} + 3\sqrt{5})(2\sqrt{3} + 3\sqrt{5}) \\
&= (2\sqrt{3} + 3\sqrt{5})2\sqrt{3} + (2\sqrt{3} + 3\sqrt{5})3\sqrt{5} \\
&= 4\sqrt{9} + 6\sqrt{15} + 6\sqrt{15} + 9\sqrt{25} \\
&= 4 \cdot 3 + 6\sqrt{15} + 6\sqrt{15} + 9 \cdot 5 \\
&= 12 + 12\sqrt{15} + 45 \\
&= 57 + 12\sqrt{15}.
\end{aligned}$$

Alternate Solution

$$\begin{aligned}
(2\sqrt{3} + 3\sqrt{5})^2 &= (2\sqrt{3} + 3\sqrt{5})(2\sqrt{3} + 3\sqrt{5}) \\
&= 2\sqrt{3} + 3\sqrt{5} \\
& \underline{2\sqrt{3} + 3\sqrt{5}} \\
& 4\sqrt{9} + 6\sqrt{15} \\
& \underline{ + 6\sqrt{15} + 9\sqrt{25}} \\
&= 4\sqrt{9} + 12\sqrt{15} + 9\sqrt{25} \\
&= 12 + 12\sqrt{15} + 45 \\
&= 57 + 12\sqrt{15}.
\end{aligned}$$

In order to multiply expressions that contain radicals, where some of the expressions may contain more than one term, we can use the distribution property or we can obtain the product by writing one polynomial below the other and finding the partial products. Regardless of the procedure used, after obtaining the product we should simplify the radical expressions when possible.

Exponentials and Radicals

EXERCISES FOR SECTION 3.8

(Answers given to exercises marked *)

Perform the indicated operations and simplify, if possible.

*1. $\sqrt{2}(\sqrt{3}-1)$
2. $\sqrt{2}(2\sqrt{3}+3)$
*3. $2\sqrt{3}(3\sqrt{2}-2)$
4. $2\sqrt{3}(3\sqrt{2}-\sqrt{3})$
*5. $2\sqrt{2}(4\sqrt{2}-2\sqrt{3})$
6. $2\sqrt[3]{4}(\sqrt[3]{5}-\sqrt[3]{2})$
*7. $2\sqrt{x}(3\sqrt{x}-5\sqrt{xy})$
8. $3\sqrt{y}(2\sqrt{xy}-2\sqrt{2})$
*9. $5\sqrt{x}(3\sqrt{x}+2\sqrt{y})$
10. $(\sqrt{3}+2)(\sqrt{3}-4)$
*11. $(2\sqrt{3}+2)(3\sqrt{3}+5)$
12. $(5\sqrt{3}-\sqrt{2})(2\sqrt{3}+2\sqrt{2})$
*13. $(\sqrt{5}-\sqrt{3})(\sqrt{2}+\sqrt{3})$
14. $(\sqrt{2}+\sqrt{5})(\sqrt{3}+\sqrt{1})$
*15. $(2\sqrt{2}-\sqrt{5})(3\sqrt{3}+2\sqrt{5})$
16. $(\sqrt{x}+\sqrt{y})(3\sqrt{x}-\sqrt{y})$
*17. $(2\sqrt{x}+3\sqrt{y})(3\sqrt{x}-\sqrt{y})$
18. $(\sqrt{x}+\sqrt{7})(\sqrt{y}-\sqrt{2})$
*19. $(2\sqrt{5}+3\sqrt{2})(2\sqrt{3}-3\sqrt{2})$
20. $(3\sqrt{5}+2\sqrt{3})(3\sqrt{2}-\sqrt{3})$
*21. $(2\sqrt{5}+\sqrt{3})(2\sqrt{5}-\sqrt{3})$
22. $(\sqrt{x}+\sqrt{y})(\sqrt{x}-\sqrt{y})$
*23. $(3\sqrt{2}+\sqrt{3})(3\sqrt{2}-\sqrt{3})$
24. $(5\sqrt{x}+\sqrt{2})(3\sqrt{2}+2\sqrt{x})$
*25. $(\sqrt{x}+\sqrt{3})^2$
26. $(2\sqrt{2}+\sqrt{5})^2$
*27. $(3\sqrt{2}-2\sqrt{5})^2$
28. $(2\sqrt{x}+3\sqrt{y})^2$
*29. $(3\sqrt{2x}+4\sqrt{2y})^2$
30. $(2\sqrt{3x}+5\sqrt{2y})^2$
*31. $(\sqrt{2}+\sqrt{3}+\sqrt{5})^2$
32. $(\sqrt{2}+\sqrt{3}+1)(\sqrt{3}+\sqrt{2}+2)$
*33. $(2+\sqrt{x}+\sqrt{y})(3+2\sqrt{x}-\sqrt{y})$

3.9 DIVISION OF RADICAL EXPRESSIONS (RATIONALIZING A RADICAL EXPRESSION)

In Section 3.5, we saw that a radical is not in simplest form if the radicand contains a fraction or if a radical occurs in the denominator of a fraction. For example, $\sqrt{\frac{2}{3}}$ and $\frac{5}{\sqrt{2}}$ are two expressions that are not in simplest form. By using the rule that $\sqrt[n]{\frac{a}{b}} = \frac{\sqrt[n]{a}}{\sqrt[n]{b}}$ ($b \neq 0$) we can simplify

$\sqrt{\frac{2}{3}}$. Recall that in order to do this, we multiply $\frac{2}{3}$ by a form of 1, $\frac{3}{3}$. Therefore, we have

$$\sqrt{\frac{2}{3}} = \sqrt{\frac{2}{3} \cdot \frac{3}{3}} = \sqrt{\frac{6}{9}} = \frac{\sqrt{6}}{\sqrt{9}} = \frac{\sqrt{6}}{3}.$$

We can use this same identity property (multiplying by 1) to simplify $\frac{5}{\sqrt{2}}$. We shall multiply the fraction by 1, and if we use $\frac{\sqrt{2}}{\sqrt{2}}$ as a form of 1, the fraction will be simplified thus,

$$\frac{5}{\sqrt{2}} = \frac{5}{\sqrt{2}} \cdot \frac{\sqrt{2}}{\sqrt{2}} = \frac{5\sqrt{2}}{2}.$$

Note that the denominator of the given fraction ($\sqrt{2}$) is an irrational number, and by multiplying the fraction by $\frac{\sqrt{2}}{\sqrt{2}}$ we obtained an equivalent fraction whose denominator is a rational number. This technique is called *rationalizing the denominator*.

The expression $\frac{5}{\sqrt{2}}$ is another way of stating $5 \div \sqrt{2}$. Hence, we are also dividing radical expressions. If we wanted a decimal approximation for this expression, we would probably use a pocket calculator, a slide rule, or paper and pencil with a table of values.

To simplify a radicand that contains a fraction, or to simplify a fraction that contains a radical in the denominator, we multiply the fraction by an appropriate form of the number 1. Therefore, we can simplify $\sqrt{\frac{2}{3}}$ in a slightly different manner than shown in the beginning of this section. Since $\sqrt[n]{\frac{a}{b}} = \frac{\sqrt[n]{a}}{\sqrt[n]{b}}$ ($b \neq 0$), we can write $\sqrt{\frac{2}{3}}$ as an equivalent fraction, that is, $\sqrt{\frac{2}{3}} = \frac{\sqrt{2}}{\sqrt{3}}$. Now, we can multiply the fraction by an appropriate form of 1, namely $\frac{\sqrt{3}}{\sqrt{3}}$. Therefore,

$$\sqrt{\frac{2}{3}} = \frac{\sqrt{2}}{\sqrt{3}} = \frac{\sqrt{2}}{\sqrt{3}} \cdot \frac{\sqrt{3}}{\sqrt{3}} = \frac{\sqrt{6}}{3}.$$

The only difference between this solution and the other shown earlier is that here we multiplied by 1 after stating that $\sqrt{\frac{2}{3}} = \frac{\sqrt{2}}{\sqrt{3}}$. In the solution

Exponentials and Radicals

shown earlier, we multiplied the fraction $\frac{2}{3}$ by 1 and then applied the rule $\sqrt[n]{\frac{a}{b}} = \frac{\sqrt[n]{a}}{\sqrt[n]{b}}$. We shall continue to use the technique that involves multiplying the numerator and denominator by a radical so that the denominator becomes a rational number.

EXAMPLE 36

Simplify the following:

a. $\sqrt{\frac{1}{2}}$ b. $\sqrt{\frac{2}{5}}$ c. $\frac{4}{\sqrt{3}}$ d. $\frac{5}{\sqrt[3]{2}}$

Solution

If necessary, we shall apply the rule, $\sqrt[n]{\frac{a}{b}} = \frac{\sqrt[n]{a}}{\sqrt[n]{b}}$ and then multiply the fraction by 1, choosing the appropriate form of 1 that will allow us to rationalize the denominator.

a. $\sqrt{\frac{1}{2}} = \frac{\sqrt{1}}{\sqrt{2}} = \frac{1}{\sqrt{2}} = \frac{1}{\sqrt{2}} \cdot \frac{\sqrt{2}}{\sqrt{2}} = \frac{\sqrt{2}}{2}$.

b. $\sqrt{\frac{2}{5}} = \frac{\sqrt{2}}{\sqrt{5}} = \frac{\sqrt{2}}{\sqrt{5}} \cdot \frac{\sqrt{5}}{\sqrt{5}} = \frac{\sqrt{10}}{5}$.

c. $\frac{4}{\sqrt{3}} = \frac{4}{\sqrt{3}} \cdot \frac{\sqrt{3}}{\sqrt{3}} = \frac{4\sqrt{3}}{3}$.

d. $\frac{5}{\sqrt[3]{2}} = \frac{5}{\sqrt[3]{2}} \cdot \frac{\sqrt[3]{4}}{\sqrt[3]{4}} = \frac{5\sqrt[3]{4}}{\sqrt[3]{8}} = \frac{5\sqrt[3]{4}}{2}$. (Note: in this example we had to use $\sqrt[3]{4} \div \sqrt[3]{4}$ as a form of 1 in order to obtain a rational number in the denominator.)

Sometimes it is necessary to rationalize a fraction that contains a binomial in the denominator. If we are asked to find the quotient of $1 \div (2 + \sqrt{3})$, we can express this as a fraction, $\frac{1}{2 + \sqrt{3}}$. To rationalize the denominator of this fraction, it is not enough to multiply the fraction by $\frac{\sqrt{3}}{\sqrt{3}}$, for we would still have a radical in the denominator:

$$\frac{1}{2 + \sqrt{3}} \cdot \frac{\sqrt{3}}{\sqrt{3}} = \frac{1 \cdot \sqrt{3}}{(2 + \sqrt{3})\sqrt{3}} = \frac{\sqrt{3}}{2\sqrt{3} + 3}.$$

But if we choose the numerator and denominator of our expression for 1 to be $2-\sqrt{3}$, we will obtain a new denominator, which, when simplified, is free of radicals.

$$= \frac{1}{(2+\sqrt{3})} \cdot \frac{(2-\sqrt{3})}{(2-\sqrt{3})}$$

$$= \frac{2-\sqrt{3}}{(2+\sqrt{3})2 + (2+\sqrt{3})(-\sqrt{3})}$$

$$= \frac{2-\sqrt{3}}{4+2\sqrt{3}-2\sqrt{3}-3} = \frac{2-\sqrt{3}}{1} = 2-\sqrt{3}.$$

Note that the binomial used in the numerator and denominator of the expression for 1 contains the same terms as the binomial in the denominator of the given expression, except that the terms are connected by the opposite sign. In general, if the denominator of the fraction is $a-b$, then we should use $\frac{a+b}{a+b}$, and if the denominator of the fraction is $a+b$, then we should use $\frac{a-b}{a-b}$. Therefore, the rationalization of the denominator will always involve the product $(a+b)(a-b)$, and

$$(a+b)(a-b) = (a+b)a + (a+b)(-b)$$
$$= a^2 + ab - ab - b^2$$
$$= a^2 - b^2.$$

The two factors, $(a+b)$ and $(a-b)$ are called *conjugates* of each other. Two binomials that differ only in the sign of the second term are conjugates of each other. Note that the product of a binomial and its conjugate is the difference of the squares of the two terms. Therefore, if one or both terms of the denominator of a fraction are radicals of the second order (index of two), we can multiply the numerator and denominator of the given fraction by the conjugate of the denominator and obtain an equivalent fraction without radicals in the denominator.

Using the fact that $(a+b)(a-b) = a^2 - b^2$, we can eliminate some of the steps in rationalizing the expression $\frac{1}{2+\sqrt{3}}$. We choose the numerator and denominator of our expression for 1 to be the conjugate of $2+\sqrt{3}$, or $2-\sqrt{3}$. Therefore, we have

$$\frac{1}{2+\sqrt{3}} = \frac{1}{2+\sqrt{3}} \cdot \frac{2-\sqrt{3}}{2-\sqrt{3}} = \frac{1(2-\sqrt{3})}{(2+\sqrt{3})(2-\sqrt{3})} = \frac{2-\sqrt{3}}{(2)^2 - (\sqrt{3})^2}$$

$$= \frac{2-\sqrt{3}}{4-3} = \frac{2-\sqrt{3}}{1} = 2-\sqrt{3}.$$

$\begin{pmatrix}\text{The denominator is of the}\\ \text{form } (a+b)(a-b).\end{pmatrix}$ $(a+b)(a-b) = a^2 - b^2.$

Exponentials and Radicals 185

In order to rationalize the binomial denominator of a fraction where one or both terms are radicals of the second order, multiply both numerator and denominator by the conjugate of the denominator.

EXAMPLE 37

Simplify $\dfrac{5}{\sqrt{3}-1}$.

Solution

The conjugate of $\sqrt{3}-1$ is $\sqrt{3}+1$. Therefore,

$$\frac{5}{\sqrt{3}-1} = \frac{5}{\sqrt{3}-1} \cdot \frac{\sqrt{3}+1}{\sqrt{3}+1} = \frac{5(\sqrt{3}+1)}{(\sqrt{3}-1)(\sqrt{3}+1)} = \frac{5\sqrt{3}+5}{(\sqrt{3})^2-(1)^2}$$

$$= \frac{5\sqrt{3}+5}{3-1} = \frac{5\sqrt{3}+5}{2}.$$

EXAMPLE 38

Simplify $\dfrac{\sqrt{2}+\sqrt{3}}{\sqrt{5}+\sqrt{2}}$.

Solution

The conjugate of $\sqrt{5}+\sqrt{2}$ is $\sqrt{5}-\sqrt{2}$. Therefore,

$$\frac{\sqrt{2}+\sqrt{3}}{\sqrt{5}+\sqrt{2}} = \frac{\sqrt{2}+\sqrt{3}}{\sqrt{5}+\sqrt{2}} \cdot \frac{\sqrt{5}-\sqrt{2}}{\sqrt{5}-\sqrt{2}} = \frac{(\sqrt{2}+\sqrt{3})(\sqrt{5}-\sqrt{2})}{(\sqrt{5}+\sqrt{2})(\sqrt{5}-\sqrt{2})}$$

$$= \frac{(\sqrt{2}+\sqrt{3})\sqrt{5}+(\sqrt{2}+\sqrt{3})(-\sqrt{2})}{(\sqrt{5})^2-(\sqrt{2})^2}$$

$$= \frac{\sqrt{10}+\sqrt{15}-2-\sqrt{6}}{5-2} = \frac{\sqrt{10}+\sqrt{15}-2-\sqrt{6}}{3}.$$

EXAMPLE 39

Simplify $\dfrac{\sqrt{3}-\sqrt{5}}{-\sqrt{2}+\sqrt{3}}$.

Solution

Note that the conjugate of $-\sqrt{2}+\sqrt{3}$ is $-\sqrt{2}-\sqrt{3}$. Two binomials that differ only in the sign of the second term are conjugates of each other.

$$\frac{\sqrt{3}-\sqrt{5}}{-\sqrt{2}+\sqrt{3}} = \frac{\sqrt{3}-\sqrt{5}}{-\sqrt{2}+\sqrt{3}} \cdot \frac{-\sqrt{2}-\sqrt{3}}{-\sqrt{2}-\sqrt{3}} = \frac{(\sqrt{3}-\sqrt{5})(-\sqrt{2}-\sqrt{3})}{(-\sqrt{2}+\sqrt{3})(-\sqrt{2}-\sqrt{3})}$$

$$= \frac{(\sqrt{3}-\sqrt{5})(-\sqrt{2}) + (\sqrt{3}-\sqrt{5})(-\sqrt{3})}{(-\sqrt{2})^2 - (\sqrt{3})^2}$$

$$= \frac{\sqrt{6}+\sqrt{10}-3+\sqrt{15}}{2-3} = \frac{\sqrt{6}+\sqrt{10}-3+\sqrt{15}}{-1}.$$

When one or both terms of the denominator of a fraction are radicals of the second order (index of two), we can multiply the numerator and denominator of the given fraction by the conjugate of the denominator (a form of the number 1) and obtain an equivalent fraction without radicals in the denominator.

EXERCISES FOR SECTION 3.9

*(Answers given to exercises marked *)*

Simplify the following by rationalizing the denominator.

*1. $\sqrt{\dfrac{1}{3}}$
2. $\sqrt{\dfrac{2}{3}}$
*3. $\sqrt{\dfrac{3}{2}}$

4. $\sqrt{\dfrac{4}{5}}$
*5. $\sqrt{\dfrac{x}{y}}$
6. $\sqrt{\dfrac{3}{2\pi}}$

*7. $\sqrt[3]{\dfrac{1}{2}}$
8. $\sqrt[4]{\dfrac{1}{8}}$
*9. $\sqrt[3]{\dfrac{2}{9}}$

10. $\dfrac{5}{\sqrt{2}}$
*11. $\dfrac{x}{\sqrt{y}}$
12. $\dfrac{\sqrt{2x}}{\sqrt{3}}$

*13. $\dfrac{\sqrt[3]{2}}{\sqrt[3]{4}}$
14. $\dfrac{\sqrt[3]{x}}{\sqrt[3]{y}}$
*15. $\dfrac{\sqrt[3]{y}}{\sqrt[3]{x}}$

16. $\dfrac{1}{2+\sqrt{3}}$
*17. $\dfrac{2}{3-\sqrt{2}}$
18. $\dfrac{3}{\sqrt{2}+3}$

*19. $\dfrac{x}{\sqrt{2x}+1}$
20. $\dfrac{y}{\sqrt{3}-y}$
*21. $\dfrac{3}{\sqrt{2}+\sqrt{3}}$

22. $\dfrac{2}{\sqrt{2}-1}$
*23. $\dfrac{4}{-\sqrt{5}+\sqrt{3}}$
24. $\dfrac{2}{-\sqrt{2}-\sqrt{3}}$

Exponentials and Radicals 187

*25. $\dfrac{\sqrt{2}+1}{\sqrt{2}-1}$ 26. $\dfrac{2+\sqrt{3}}{2-\sqrt{3}}$ *27. $\dfrac{\sqrt{5}-\sqrt{3}}{\sqrt{5}+\sqrt{3}}$

28. $\dfrac{\sqrt{7}+\sqrt{5}}{\sqrt{3}-\sqrt{2}}$ *29. $\dfrac{\sqrt{3}+\sqrt{2}}{\sqrt{5}-\sqrt{2}}$ 30. $\dfrac{\sqrt{5}-\sqrt{2}}{\sqrt{3}-\sqrt{7}}$

*31. $\dfrac{2\sqrt{3}+3\sqrt{2}}{2\sqrt{5}-3\sqrt{3}}$ 32. $\dfrac{4\sqrt{3}-2\sqrt{2}}{5\sqrt{5}+2\sqrt{3}}$ 33. $\dfrac{3\sqrt{5}+2\sqrt{7}}{-2\sqrt{3}-3\sqrt{2}}$

*34. $\dfrac{\sqrt{x}+\sqrt{y}}{\sqrt{x}-\sqrt{y}}$ *35. $\dfrac{3\sqrt{x}+\sqrt{y}}{2\sqrt{y}-3\sqrt{x}}$ 36. $\dfrac{4\sqrt{x}+2\sqrt{y}}{3\sqrt{5}-3\sqrt{x}}$

37. $\dfrac{4\sqrt{y}-\sqrt{3}}{5\sqrt{2}+\sqrt{x}}$ *38. $\dfrac{3\sqrt{2}+2\sqrt{x}}{3\sqrt{y}-3\sqrt{5}}$ 39. $\dfrac{5\sqrt{2x}-2\sqrt{3y}}{2\sqrt{2y}+3\sqrt{5x}}$

3.10 EQUATIONS WITH RADICALS

Equations sometimes contain variables in a radicand, raised to a fractional power. Equations of this type are called radical, or irrational, equations. For example, $\sqrt{x}=2$ is a radical equation. Note that we can also express this equation as $x^{1/2}=2$. To solve a radical equation, we first eliminate the radical(s), by raising both members of the equation to the appropriate power. For this particular example, $\sqrt{x}=2$, we can square each member,

$$\sqrt{x}=2. \qquad \text{Check:} \quad \sqrt{4}=2.$$
$$(\sqrt{x})^2=(2)^2. \qquad\qquad\qquad 2=2.$$
$$x=4.$$

Whenever we are trying to eliminate a radical in a radical equation, we isolate that radical on one side of the equation. In order to solve the equation $\sqrt[3]{x}+3=0$, we must first rewrite the equation so that the radical is contained in one member of the equation by itself, then proceed,

$$\sqrt[3]{x}+3=0. \qquad \text{Check:} \quad \sqrt[3]{-27}+3=0.$$
$$\sqrt[3]{x}=-3. \qquad\qquad\qquad -3+3=0.$$
$$(\sqrt[3]{x})^3=(-3)^3. \qquad\qquad\qquad 0=0.$$
$$x=-27.$$

It should be noted that the solution set of the original equation is a subset of the solution set of the new equation, obtained by raising both members of the given equation to the appropriate power. Consider the equation $\sqrt{2x-4}+2=0$. To solve this equation, we first rewrite so that the radical is contained in one member of the equation by itself, and then proceed,

$$\sqrt{2x - 4} + 2 = 0.$$
$$\sqrt{2x + 4} = -2.$$
$$(\sqrt{2x - 4})^2 = (-2)^2.$$
$$2x - 4 = 4.$$
$$2x = 8.$$
$$x = 4.$$

Check:
$$\sqrt{2 \cdot 4 - 4} + 2 = 0.$$
$$\sqrt{8 - 4} + 2 = 0.$$
$$\sqrt{4} + 2 = 0.$$
$$2 + 2 \neq 0.$$

The answer does not check because no principal square root produces a negative number. Therefore, the solution set of the given equation, $\sqrt{2x - 4} + 2 = 0$, is the empty set, ∅. Note that we stated that the solution set of the original equation is a subset of the solution set of the new equation, obtained by raising both members of the given equation to the appropriate power, and the empty set is a subset of every set. The answer, $x = 4$, is known as an extraneous root. It is not a root of the original equation, but it is a root of some other equation that occurs in the derivation. Extraneous roots occur because the process of raising both members of an equation to a power does not always produce an equivalent equation. For example, the solution to the equation, $2x = 6$, is $x = 3$. However, if we square both members of the given equation, $2x = 6$, we obtain $4x^2 = 36$, and from this, $x^2 = 9$. There are two solutions to this new equation, $x^2 = 9$, namely 3 and −3, and −3 is not a member of the solution set of the original equation. As we stated previously, always check the solution in the original equation. This is an absolute must whenever we square both sides of an equation.

EXAMPLE 40

Find the solution set for the given equations.

a. $\sqrt{x + 2} = 2.$
b. $\sqrt{3x - 6} = 3.$

Solution

To solve these radical equations, we eliminate the radicals by squaring both members, since in each case one member contains a radical as its only term.

a. $\sqrt{x + 2} = 2.$
$(\sqrt{x + 2})^2 = (2)^2.$
$x + 2 = 4.$
$x = 2, \{2\}.$

Check:
$\sqrt{2 + 2} = 2.$
$\sqrt{4} = 2.$
$2 = 2.$

Exponentials and Radicals

b. $\sqrt{3x-6}=3$. Check: $\sqrt{3(5)-6}=3$.
$(\sqrt{3x-6})^2 = (3)^2$. $\sqrt{15-6}=3$.
$3x-6=9$. $\sqrt{9}=3$.
$3x=15$. $3=3$.
$x=5, \{5\}$.

EXAMPLE 41

Find the solution set for the given equations.

a. $\sqrt{x+2}-1=0$. b. $\sqrt[3]{x+2}-3=0$. c. $\sqrt{5x-10}+5=0$.

Solution

Before raising each member of the equation to the appropriate power, we first rewrite the equation so that one member contains a radical as its only term.

a. $\sqrt{x+2}-1=0$. Check: $\sqrt{-1+2}-1=0$.
$\sqrt{x+2}=1$. $\sqrt{1}-1=0$.
$(\sqrt{x+2})^2 = (1)^2$. $1-1=0$.
$x+2=1$. $0=0$.
$x=-1, \{-1\}$.

b. $\sqrt[3]{x+2}-3=0$. Check: $\sqrt[3]{25+2}-3=0$.
$\sqrt[3]{x+2}=3$. $\sqrt[3]{27}-3=0$.
$(\sqrt[3]{x+2})^3 = (3)^3$. $3-3=0$.
$x+2=27$. $0=0$.
$x=25, \{25\}$.

c. $\sqrt{5x-10}+5=0$. Check: $\sqrt{5\cdot 7-10}+5=0$.
$\sqrt{5x-10}=-5$. $\sqrt{35-10}+5=0$.
$(\sqrt{5x-10})^2 = (-5)^2$. $\sqrt{25}+5=0$.
$5x-10=25$. $5+5 \neq 0$.
$5x=35$. (7 is an extraneous root.) $10 \neq 0$.
$x=7, \{\quad\}$.

The solution does not check, therefore the solution set is the empty set, $\emptyset$.

> To solve a radical equation we must first have a radical by itself in one member of the equation. Next, we raise both members to the appropriate power. (If necessary, repeat this process.) Solve the resulting equation. Check to determine if the solution(s) is a solution(s) of the original equation, noting any extraneous roots that may occur.

EXERCISES FOR SECTION 3.10
(Answers given to exercises marked *)

Find the solution set for the given equations.

*1. $\sqrt{x} = 3$.
2. $\sqrt{x} = 4$.
*3. $\sqrt{x} - 2 = 0$.
4. $\sqrt{x} - 3 = 0$.
*5. $\sqrt{y} + 2 = 0$.
6. $\sqrt{x+1} = 1$.
*7. $\sqrt{y+2} - 2 = 0$.
8. $\sqrt{x+3} + 2 = 0$.
*9. $\sqrt{3x+1} - 5 = 0$.
10. $\sqrt{3x-2} - 5 = 0$.
*11. $\sqrt{2y+3} + 1 = 0$.
12. $\sqrt{x+1} + 2 = 0$.
*13. $\sqrt[3]{x-1} = 2$.
14. $\sqrt[3]{z+1} - 3 = 0$.
*15. $\sqrt[3]{2x+1} + 3 = 0$.
16. $\sqrt[4]{x+1} - 2 = 0$.
*17. $\sqrt[4]{2x-1} - 3 = 0$.
18. $\sqrt[3]{x+1} + 2 = 0$.
*19. $(y+2)^{1/2} = 2$.
20. $(3x+1)^{1/2} = 5$.
*21. $(y+3)^{1/2} = -2$.
22. $(2y+3)^{1/2} + 1 = 0$.
*23. $(x+1)^{1/3} - 3 = 0$.
24. $(2y+1)^{1/3} - 3 = 0$.
*25. $\sqrt{x-6} = \sqrt{3}$.
26. $\sqrt{2x} - \sqrt{4x-8} = 0$.
*27. $\sqrt{3} - \sqrt{3x+6} = 0$.
28. $\sqrt{3x-4} - \sqrt{x} = 0$.
*29. $\sqrt{2y-4} + \sqrt{y-5} = 0$.
30. $\sqrt{2x-3} - \sqrt{3x+5} = 0$.

3.11 SUMMARY

In an expression such as x^3, 3 is called the exponent and x is called the base. The whole expression, x^3, is called an exponential or a power. In this chapter, we examined the properties of exponentials whose exponents were natural numbers, integers, and rational numbers. Listed below are the basic laws governing operations with exponents. In general, if a and b represent positive real numbers and m and n represent any integers, then

1. $a^m \cdot a^n = a^{m+n}$.

2. $(ab)^m = a^m b^m$.

3. $(a^m)^n = a^{mn}$.

4. $\dfrac{a^m}{a^n} = a^{m-n}$.

Exponentials and Radicals

5. $a^{-m} = \dfrac{1}{a^m}$.

6. $\left(\dfrac{a}{b}\right)^n = \dfrac{a^n}{b^n}$.

7. $a^{m/n} = (\sqrt[n]{a})^m = \sqrt[n]{a^m}$.

8. $a^0 = 1$.

Scientific notation is a direct application of exponentials whose exponents are integers. Scientific notation is the process of writing a number that is greater than or equal to 1 and less than 10, times an integral power of 10. Scientific notation, therefore, consists of $N \times 10^m$, where $1 \leq N < 10$ and m is an integer.

An application of the rule $a^{m/n} = (\sqrt[n]{a})^m = \sqrt[n]{a^m}$ is to find the indicated root of a number. More importantly, we expanded this rule to obtain the rule that $\sqrt[n]{ab} = \sqrt[n]{a} \cdot \sqrt[n]{b}$, which we used in simplifying radicals. A radical is in simplified form if

    a. We cannot remove any factors from the radicand, that is, there is no factor in the radicand whose exponent is larger than the index;

    b. The radicand does not contain any fractions, or the denominator of a fraction does not contain a radical;

    c. The index of a radical cannot be reduced.

After simplifying radicals, we examined the four basic operations—addition, subtraction, multiplication, and division—with radicals. To add or subtract radicals, they must be like radicals, meaning, they must have the same index and the same radicand. We can then combine the radicals by means of the distributive property. To obtain the product of two radicals, we can use the formula $\sqrt[n]{x} \cdot \sqrt[m]{y} = \sqrt[mn]{x^m y^n}$. We can multiply radicals, whether or not they have the same index or radicands. Division of radical expressions is also referred to as rationalizing a radical expression. When the denominator of a fraction contains a radical (an irrational number), we can rationalize it by means of the multiplicative identity property. For example, to rationalize the fraction $\dfrac{\sqrt{a}}{\sqrt{b}}$, we would multiply it by a form of the number 1, $\dfrac{\sqrt{b}}{\sqrt{b}}$. In order to rationalize the denominator of a fraction that has a binomial denominator, where one or both terms are radicals of the same order, multiply both numerator and denominator by the conjugate of the denominator. Two binomials that differ only in the sign of the second term are conjugates of each other. For example, $(a + b)$ is the conjugate of $(a - b)$. To rationalize the fraction $\dfrac{1}{\sqrt{2}+\sqrt{3}}$, we multiply the fraction by $\dfrac{\sqrt{2}-\sqrt{3}}{\sqrt{2}-\sqrt{3}}$, a form of the number 1.

Equations that contain variables in a radicand are called radical equations. To solve radical equations we eliminate the radical, by raising both members of the equation to the appropriate power. In performing this operation, we must be sure that one member of the equation contains only a radical. Note that the solution set of the original radical equation is a subset of the solution set of the new equation, obtained by raising both members of the given equation to the appropriate power. Therefore, we must always check the solution in the original equation, checking for extraneous roots. Sometimes the solution set of the original equation will be the empty set.

REVIEW EXERCISES FOR CHAPTER 3
(Answers in Appendix B)

1. Simplify each of the following expressions, perform the indicated operations, and write equivalent expressions containing only positive exponents.

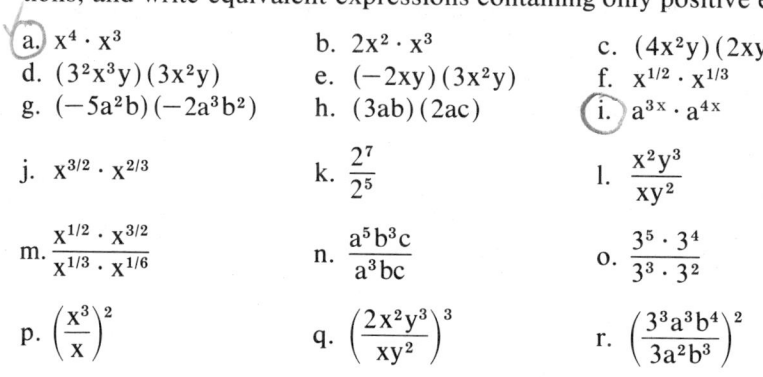

    a. $x^4 \cdot x^3$
    b. $2x^2 \cdot x^3$
    c. $(4x^2y)(2xy^3)$
    d. $(3^2x^3y)(3x^2y)$
    e. $(-2xy)(3x^2y)$
    f. $x^{1/2} \cdot x^{1/3}$
    g. $(-5a^2b)(-2a^3b^2)$
    h. $(3ab)(2ac)$
    i. $a^{3x} \cdot a^{4x}$
    j. $x^{3/2} \cdot x^{2/3}$
    k. $\dfrac{2^7}{2^5}$
    l. $\dfrac{x^2y^3}{xy^2}$
    m. $\dfrac{x^{1/2} \cdot x^{3/2}}{x^{1/3} \cdot x^{1/6}}$
    n. $\dfrac{a^5b^3c}{a^3bc}$
    o. $\dfrac{3^5 \cdot 3^4}{3^3 \cdot 3^2}$
    p. $\left(\dfrac{x^3}{x}\right)^2$
    q. $\left(\dfrac{2x^2y^3}{xy^2}\right)^3$
    r. $\left(\dfrac{3^3a^3b^4}{3a^2b^3}\right)^2$

2. Simplify each of the following expressions, perform the indicated operations, and write equivalent expressions containing only positive exponents.

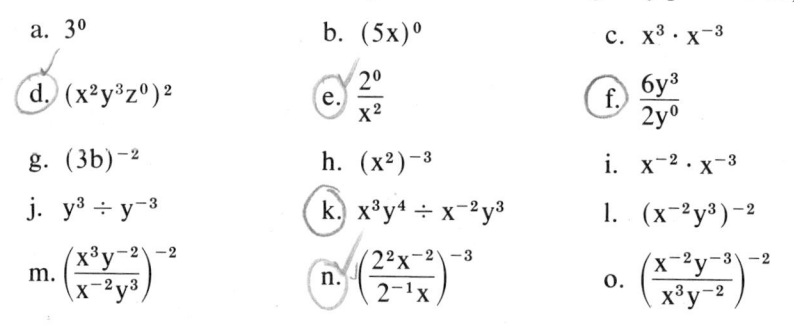

    a. 3^0
    b. $(5x)^0$
    c. $x^3 \cdot x^{-3}$
    d. $(x^2y^3z^0)^2$
    e. $\dfrac{2^0}{x^2}$
    f. $\dfrac{6y^3}{2y^0}$
    g. $(3b)^{-2}$
    h. $(x^2)^{-3}$
    i. $x^{-2} \cdot x^{-3}$
    j. $y^3 \div y^{-3}$
    k. $x^3y^4 \div x^{-2}y^3$
    l. $(x^{-2}y^3)^{-2}$
    m. $\left(\dfrac{x^3y^{-2}}{x^{-2}y^3}\right)^{-2}$
    n. $\left(\dfrac{2^2x^{-2}}{2^{-1}x}\right)^{-3}$
    o. $\left(\dfrac{x^{-2}y^{-3}}{x^3y^{-2}}\right)^{-2}$

3. Use scientific notation to evaluate each of the given expressions. Answers should be left in scientific notation.

    a. $355{,}000 \times 8{,}000$
    b. 0.000065×0.00084
    c. $\dfrac{24{,}000 \times 140{,}000}{210{,}000 \times 0.0016}$
    d. $\dfrac{0.024 \times 0.00036}{0.0015 \times 0.00012}$

4. Express each of the following in an equivalent radical form; simplify, if possible.

    a. $x^{2/3}$
    b. $x^{2/3}y^{1/3}$
    c. $(x+y)^{1/2}$

Exponentials and Radicals

d. $\left(\dfrac{x}{y}\right)^{1/2}$ e. $\dfrac{3y^{1/2}}{x^{3/2}}$ f. $\left(\dfrac{x^2 y}{z^2}\right)^{1/2}$

5. Express each of the following in an equivalent exponential form and simplify, if possible.

   a. $\sqrt[3]{x^2}$ b. $\sqrt[4]{2x^3}$ c. $\sqrt[3]{x^6 y^9}$

   d. $\sqrt{x+y}$ e. $\sqrt[3]{x^2 y}$ f. $\sqrt{x^2 + y^2}$

6. Simplify the following radicals completely.

   a. $\sqrt{12}$ b. $\sqrt[3]{54}$ c. $\sqrt[3]{-8x}$ d. $\sqrt{12x^5 y^7}$

   e. $\sqrt{\dfrac{2}{9}}$ f. $\sqrt{\dfrac{2}{3}}$ g. $\sqrt[3]{\dfrac{3}{2}}$ h. $\sqrt{\dfrac{8x^3}{3y}}$

   i. $\sqrt[4]{4}$ j. $\sqrt[6]{8}$ k. $\sqrt[6]{8x^3}$ l. $\sqrt[8]{9x^2 y^4}$

7. Perform the indicated operations and simplify, if possible.

   a. $3\sqrt{2} + 4\sqrt{2}$ b. $2\sqrt[3]{3} - 3\sqrt[3]{3}$

   c. $\sqrt{48} + \sqrt{75}$ d. $\sqrt{18} - 2\sqrt{50} + \sqrt{98}$

   e. $3\sqrt{3y} - 2\sqrt{12y} + 4\sqrt{27y}$ f. $2\sqrt{2} + \sqrt[4]{4} + 3\sqrt[6]{8}$

8. Perform the indicated operations and simplify, if possible.

   a. $\sqrt{3} \cdot \sqrt{7}$ b. $\sqrt{x} \cdot \sqrt{y}$

   c. $3\sqrt[3]{2a} \cdot 2\sqrt[3]{6b}$ d. $(\sqrt{3y})^3$

   e. $\sqrt{2x}(\sqrt{y} - 3\sqrt{xy})$ f. $3\sqrt{y}(2\sqrt{xy} + 2\sqrt{2})$

   g. $(\sqrt{3} + \sqrt{2})(\sqrt{3} - \sqrt{3})$ h. $(\sqrt{2} + \sqrt{5})(\sqrt{3} - \sqrt{2})$

   i. $(\sqrt{x} + \sqrt{y})(\sqrt{x} - \sqrt{y})$ j. $(5\sqrt{x} + \sqrt{2})(3\sqrt{2} + 2\sqrt{x})$

   k. $\dfrac{\sqrt{x}}{\sqrt{y}}$ l. $\dfrac{3}{\sqrt{3}}$

   m. $\dfrac{1}{\sqrt{2}}$ n. $\dfrac{\sqrt[3]{3}}{\sqrt[3]{2}}$ o. $\dfrac{2}{1 + \sqrt{3}}$ p. $\dfrac{3}{\sqrt{5} - 2}$

   q. $\dfrac{2 + \sqrt{3}}{\sqrt{2} - 3}$ r. $\dfrac{\sqrt{3} + \sqrt{2}}{\sqrt{3} - \sqrt{2}}$ s. $\dfrac{2\sqrt{x} + \sqrt{y}}{3\sqrt{x} - 2\sqrt{y}}$ t. $\dfrac{3\sqrt{2} + \sqrt{x}}{2\sqrt{y} - 2\sqrt{3}}$

9. Find the solution set for the given equations and check.

   a. $\sqrt{y} = 2$.   b. $\sqrt{x+2} = 0$.   c. $\sqrt{x+2} - 1 = 0$.

   d. $\sqrt{3y-2} = 5$.   e. $\sqrt{2x+3} = -1$.   f. $\sqrt{y+1} + 1 = 0$.

   g. $\sqrt{2x+1} - 3 = 0$.   h. $\sqrt[3]{x+2} - 2 = 0$.   i. $\sqrt{2x+2} - \sqrt{4x-8} = 0$.

chapter 4

Polynomials

As a result of this chapter, the student will be able to:

1. Describe a polynomial as an algebraic expression where the exponents of the variables are natural numbers, and the variables do not occur in the denominator of a fraction;
2. Describe a polynomial consisting of only one term as a monomial, a polynomial containing two terms as a binomial, and a polynomial containing three terms as a trinomial;
3. Add and subtract polynomials;
4. Multiply polynomials, that is, multiply a monomial by a monomial, a monomial by a polynomial, a binomial by a binomial, a binomial by a polynomial, and a polynomial by a polynomial;
5. Determine certain products of polynomials by inspection, that is, a binomial by a binomial, the square of a binomial, the product of two conjugates, and the cube of a binomial;
6. Divide polynomials: a polynomial by a monomial and a polynomial by a polynomial;
7. Use the Remainder Theorem to determine the remainder when a polynomial P(x) is divided by a divisor of the form $x - b$;
8. Use the technique of synthetic division to divide a polynomial P(x) by a divisor of the form $x - b$.

4.1 INTRODUCTION

We briefly discussed polynomials in Chapter One. In this chapter, we shall examine polynomials in much greater detail, and also their addition, subtraction, multiplication, and division. Recall that the sum or differences contained in algrebaic expressions are called *terms* of the expression. A term may consist of one or more factors multiplied together. These factors may consist of numbers, and one or more variables raised to powers. For example, in the algebraic expression $x^3 + 2x^2y + 3xy^4 - 2y^5 - 7$, x^3, $+2x^2y$, $+3xy^4$, $-2y^5$, and -7 are all terms. A factor or group of factors in a given term are called *coefficients* of the remaining factors. We can say that $2x^2$ is the coefficient of y in the term $2x^2y$. Similarly, we can also say that 2y is the coefficient of x^2 in the same term. Normally, when we speak of coefficients we are referring to numerical coefficients (unless otherwise indicated). In this sense, the coefficient of $2x^2y$ is 2, the coefficient of $-2y^5$ is -2. If a term does not appear to have a numerical coefficient as a factor, it is understood to be 1. For example, the coefficient of x^3 is 1, therefore, $x^3 = 1x^3$.

An algebraic expression that consists of one term is called a *monomial*. A *binomial* is an algebraic expression that consists of two terms, and a *trinomial* is an algebraic expression that consists of three terms. A *polynomial* is any algebraic expression with one or more terms. The word

Polynomials

polynomial can also be used to describe a monomial, binomial, or trinomial, but we give special names to these algebraic expressions because they occur so frequently in algebra. A polynomial can consist of any number of terms, provided the variable does not occur in the denominator of a fraction, and the exponents of the variables are natural numbers. For example, the following algebraic expressions are all polynomials.

> monomials: x^2, $4x^2$, $-2xy^3$, 3
> binomials: $x^2 + y^2$, $3 - x^2$, $2x + 3y$
> trinomials: $2x + 3y + 4$, $5x^2 - 2y^2 - 6$. $x^2 + 3x - 2$

However, none of the following expressions are polynomials:

$$x + \frac{1}{x}, \quad \frac{x+y}{y}, \quad y^2 + 3y - \frac{1}{y}, \quad x^3 + \frac{2}{x^2 + 1}$$

If a monomial contains only one variable, then the *degree* of that monomial is indicated by the exponent of the variable. The monomial $3x^2$ is of the second degree, and $-4y^3$ is the third degree. If a term (monomial) contains more than one variable, then the degree of that term is determined by the sum of the exponents of the variables. For example, $4x^2y$ is of the third degree, $-3xy^2z^3$ is of the sixth degree, and x^2y^2 is of the fourth degree. The degree of a polynomial is determined by the term of the polynomial that has the highest degree. The polynomial $x^2 + 2x + 1$ is of the second degree, $x^4 - 1$ is of the fourth degree, while $x^4 + x^3y + 2x^2y^3$ is of the fifth degree. Regardless of how the terms are arranged in a polynomial, the degree remains the same. Normally, polynomials are written so that the powers of one of the variables are arranged in descending or ascending order. For example, $3x^4 + 5x^2 - 2x + 1$ is written in descending powers of x; $2 + 3y + 2xy^2 - 3y^3$ is written in ascending powers of y.

EXAMPLE 1

Arrange $3x^2y^2z + 2xy^3z^3 - 5x^5y^4 + yz^2$ in:

a. ascending powers of y;
b. ascending powers of x;
c. descending powers of z.

Solution

a. $yz^2 + 3x^2y^2z + 2xy^3z^3 - 5x^5y^4$
b. $yz^2 + 2xy^3z^3 + 3x^2y^2z - 5x^5y^4$
c. $2xy^3z^3 + yz^2 + 3x^2y^2z - 5x^5y^4$

Note that in Example 1–b, the first term is yz^2 because we can interpret this as x^0yz^2 and $x^0 = 1$. Therefore, $yz^2 = 1yz^2 = x^0yz^2$. Similarly, in Example 1-c, $-5x^5y^4 = -5x^5y^4 \cdot 1 = -5x^5y^4z^0$.

EXERCISES FOR SECTION 4.1

(Answers given to exercises marked*)

1. Identify the terms in each of the following algebraic expressions.
   *a. $x - 2$
   b. $3x^2 - 2y + 1$
   *c. $4xy - 1$
   d. $-2 + 3xy$
   *e. $4y^2 - 3y + 2$
   f. $4y^2 - 3y + 1$
   *g. $ax^2 + bx$
   h. $8x^2 - 2x - 3$

2. Identify each expression in Exercise 1 as a monomial, binomial, or trinomial.

3. Identify the numerical coefficient of each term in the following polynomials.
   *a. $3x^2 + 2x$
   b. $x^2 - 2x$
   *c. $x^3 - 2xy$
   d. $-x^3 + xy$
   *e. $3x^3 + 2x^2 - 2xy$
   f. $x^3y^2 + x^2y + x - y$

4. Determine the degree of each of the following polynomials.
   *a. $2x^2 + 3x$
   b. $x^2 + xy^2 - 2$
   *c. $x^3 + x^2y + xy^2$
   d. $x^2y^2 + x^3$
   *e. $3x^4 + x^3y - x^2y^4$
   f. $4y^2 - 3y + 2x^2y$
   *g. $xyz - 3x^2$
   h. $x^2yz - 2y^3$
   *i. $x^3y^2z + x^4$

5. Arrange $x^3y^2z - 2xyz^2 + 3x^2y^3z^3 + 2$ in:
   *a. ascending powers of x;
   b. ascending powers of y;
   *c. descending powers of z;
   d. descending powers of x.

6. Determine which of the following expressions are not polynomials.
   *a. $x^2 + 3x - 4$
   b. $\frac{x^2 + 4x}{2}$
   *c. $\frac{x+3}{x-1}$
   d. $x^2 + \frac{1}{x} - \frac{1}{2}$
   *e. $\frac{1}{3}x^2 + \frac{1}{2}x - \frac{1}{2}$
   f. $\frac{x^2}{2} + \frac{3}{x} - \frac{1}{2}$

7. Evaluate each of the following polynomials for the indicated numerical replacement.
   *a. $4x^2 - 2x - 3$ when $x = 2$
   b. $x^2 - 3x - 4$, when $x = -1$
   *c. $x^3 + 2x^2 - 1$, when $x = -2$
   d. $3x^2 - 2xy + y^2$, when $x = 2, y = 3$
   *e. $2x^3y + 3xy^2$, when $x = 2, y = 3$
   f. $x^3 - y^3$, when $x = 1, y = -1$

4.2 SUMS AND DIFFERENCES OF POLYNOMIALS

We can sometimes simplify polynomials by means of the distributive property. For example, $3x^2 + 4x^2$ can be rewritten as $7x^2$: $3x^2 + 4x^2 =$

Polynomials

$(3 + 4) x^2 = 7x^2$. Similarly, $5x^2y - 2x^2y = (5 - 2) x^2y = 3x^2y$. This illustrates that terms of a polynomial containing exactly the same variables, as well as the same exponents on each variable, can be combined (added or subtracted). Terms that contain exactly the same variables, and also the same exponent on each corresponding variable, are called *like terms*. For example, $3x^2y$ and $-2x^2y$ are like terms, as are $4x^2yz^3$ and $2x^2yz^3$, but $2xy^2$ and $3x^2y$ are not like terms, because the corresponding variables do not have the same exponents. Note that constant terms, such as 7 and -2, are also considered to be like terms. We can simplify polynomials by combining like terms, but unlike terms cannot be combined. The sum or difference of unlike terms can only be indicated.

Let us simplify $3x^2 + 2y + 2x^2 + 4y$ by combining like terms:

$3x^2 + 2y + 2x^2 + 4y = 3x^2 + 2x^2 + 2y + 4y$ (applying the commutative property of addition)

$= (3 + 2)x^2 + (2 + 4)y$ (applying the distributive property)

$= 5x^2 + 6y.$

The expression obtained by combining like terms is equivalent to the original polynomial because the distributive law guarantees that it represents the same number for all real number replacements of the variable(s). For example, if $x = 2$ and $y = 1$, then $3x^2 + 2y + 2x^2 + 4y = 3(2)^2 + 2(1) + 2(2)^2 + 4(1) = 3 \cdot 4 + 2 + 2 \cdot 4 + 4 = 12 + 2 + 8 + 4 = 26$, and $5x^2 + 6y = 5(2)^2 + 6(1) = 5 \cdot 4 + 6 = 20 + 6 = 26$. The two expressions represent the same number for the indicated numerical replacements, and the distributive law guarantees that the two expressions represent the same number for all numerical replacements.

EXAMPLE 2

Simplify the following polynomials.

a. $3xy^2 + 4x^2y + 6xy^2$
b. $3x^2 + 2x + 5x^2 - 5x$
c. $8r^2s^2 + 6rs + 2rs - r^2s^2$

Solution

To simplify a polynomial, we combine like terms. By means of the commutative and associative properties of addition, we can regroup the terms and then apply the distributive property.

a. $3xy^2 + 4x^2y + 6xy^2$
$= 3xy^2 + 6xy^2 + 4x^2y$ (applying the commutative property)
$= (3xy^2 + 6xy^2) + 4x^2y$ (applying the associative property)
$= (3 + 6)xy^2 + 4x^2y$ (applying the distributive property)
$= 9xy^2 + 4x^2y.$

b. $3x^2 + 2x + 5x^2 - 5x$
$= 3x^2 + 5x^2 + 2x - 5x$
$= (3x^2 + 5x^2) + (2x - 5x)$
$= (3 + 5)x^2 + (2 - 5)x$
$= 8x^2 - 3x.$

c. $8r^2s^2 + 6rs + 2rs - r^2s^2$
$= 8r^2s^2 - r^2s^2 + 6rs + 2rs$
$= (8r^2s^2 - r^2s^2) + (6rs + 2rs)$
$= (8 - 1)r^2s^2 + (6 + 2)rs$ (Note: $-r^2s^2 = -1r^2s^2$.)
$= 7r^2s^2 + 8rs.$

Much of the work shown in Example 2 can be performed mentally, but it is important to understand the procedures involved in combining like terms. The "why" of algebra is just as important as the "how to" of algebra.

If a polynomial does not contain any like terms, then it is in simplest form and cannot be simplified. But we can apply the same procedures used in Example 2 to find the sum of two or more polynomials. For example, $3x^2 + 2x - 3$ and $4x^2 + 6x + 7$ are two polynomials that are in simplest form and we can find the sum, $(3x^2 + 2x - 3) + (4x^2 + 6x + 7)$, by combining like terms. To perform the indicated addition, we apply the commutative and associative properties of addition:

$(3x^2 + 2x - 3) + (4x^2 + 6x + 7)$
$= 3x^2 + 2x - 3 + 4x^2 + 6x + 7$
$= 3x^2 + 4x^2 + 2x + 6x - 3 + 7$ (applying the commutative property)
$= (3x^2 + 4x^2) + (2x + 6x) + (-3 + 7)$ (applying the associative property)
$= (3 + 4)x^2 + (2 + 6)x + (-3 + 7)$ (applying the distributive property)
$= 7x^2 + 8x + 4.$

Many algebra students prefer to add two or more polynomials by adding like terms in a vertical column, that is, arranging all like terms in the same column. In this way, the sum of $(3x^2 + 2x - 3) + (4x^2 + 6x + 7)$ is found by rewriting thus:

$$\begin{array}{r} 3x^2 + 2x - 3 \\ + 4x^2 + 6x + 7 \\ \hline 7x^2 + 8x + 4 \end{array}$$

It should be noted that this technique also makes use of the commutative and associative properties of addition. For some students, the mechanics of adding two polynomials is easier to follow with this procedure.

Polynomials

EXAMPLE 3

Add the following polynomials.
a. $3x^2 + 2x + 2$ and $2x^2 - 4x + 3$
b. $4y^3 + 3y^2 + y - 2$ and $2y^3 + y^2 + 2y + 4$
c. $x^3 + 3x^2 + 2$ and $2x^4 + 3x$

Solution

To find the sum of two or more polynomials, combine like terms.
a. $(3x^2 + 2x + 2) + (2x^2 - 4x + 3)$
 $= 3x^2 + 2x + 2 + 2x^2 - 4x + 3$
 $= (3x^2 + 2x^2) + (2x - 4x) + (2 + 3)$
 $= (3 + 2)x^2 + (2 - 4)x + (2 + 3) = 5x^2 - 2x + 5.$
b. $(4y^3 + 3y^2 + y - 2) + (2y^3 + y^2 + 2y + 4)$
 $= 4y^3 + 3y^2 + y - 2 + 2y^3 + y^2 + 2y + 4$
 $= (4y^3 + 2y^3) + (3y^2 + y^2) + (y + 2y) + (-2 + 4)$
 $= (4 + 2)y^3 + (3 + 1)y^2 + (1 + 2)y + (-2 + 4)$
 $= 6y^3 + 4y^2 + 3y + 2.$
c. $(x^3 + 3x^2 + 2) + (2x^4 + 3x)$
 $= x^3 + 3x^2 + 2 + 2x^4 + 3x.$ Note that there are no like terms in the two polynomials. Therefore, we indicate the sum as $2x^4 + x^3 + 3x^2 + 3x + 2.$

Alternate Solution

Arranging the polynomials so that like terms are in the same column:

a. $3x^2 + 2x + 2$
 $\underline{2x^2 - 4x + 3}$
 $5x^2 - 2x + 5$

b. $4y^3 + 3y^2 + y - 2$
 $\underline{2y^3 + y^2 + 2y + 4}$
 $6y^3 + 4y^2 + 3y + 2$

c. $ x^3 + 3x^2 + 2$
 $\underline{2x^4 + 3x }$
 $2x^4 + x^3 + 3x^2 + 3x + 2$

The operation of subtraction can also be thought of as addition, since $x - y = x + (-y)$. In Chapter One we used this technique to subtract integers. For example, $-5 - (-2) = -5 + (2) = -3$. We can use this same technique to simplify polynomials. The polynomial $x^2 - (3x + 2)$ can be rewritten by using the fact that subtraction is the same as adding the additive inverse (opposite) of the subtrahend, or $x^2 - (3x + 2) = x^2 + -(3x + 2) = x^2 + -3x - 2 = x^2 - 3x - 2$. This example illustrates that to form the opposite of a polynomial $(3x + 2)$, we change the sign of each term $(-3x - 2)$. In other words, $-(a + b) = -1(a + b) = -a - b$.

We can apply this property when we have to find the difference of two polynomials. Consider the following subtraction problem: $(3x^2 + 4x - 5) - (x^2 + 2x + 3)$. We can find the difference (or simplify) by applying the rule that $-(a + b) = -1(a + b) = -a - b$. Therefore,

$$(3x^2 + 4x - 5) - (x^2 + 2x + 3) = 3x^2 + 4x - 5 - x^2 - 2x - 3$$
$$= 2x^2 + 2x - 8.$$

We can also subtract polynomials by using the column technique. The given example $(3x^2 + 4x - 5) - (x^2 + 2x + 3)$ would be written as:

$$\begin{array}{r} 3x^2 + 4x - 5 \\ -(x^2 + 2x + 3) \end{array} \quad \text{or} \quad \begin{array}{r} 3x^2 + 4x - 5 \\ -\ x^2 - 2x - 3 \\ \hline 2x^2 + 2x - 8 \end{array}$$

Note that in order to subtract two polynomials using the column technique, we must change the sign of each term in the subtrahend. Instead of the problem being given as $(3x^2 + 4x - 5) - (x^2 + 2x + 3)$, it might be given as "subtract $x^2 + 2x + 3$ from $3x^2 + 4x - 5$." For example, if we want to subtract $5x^2 - 2x + 3$ from $8x^2 + 4x - 5$ with the column method, then we would have

$$\begin{array}{r} 8x^2 + 4x - 5 \\ -(5x^2 - 2x + 3) \end{array} \quad \begin{array}{c} \text{Changing the signs of} \\ \text{each term in the} \\ \text{subtrahend:} \end{array} \quad \begin{array}{r} 8x^2 + 4x - 5 \\ \ominus 5x^2 \oplus 2x \ominus 3 \\ \hline 3x^2 + 6x - 8 \end{array}$$

The changed signs are circled to indicate the correct sign of each term in the subtrahend in combining terms. In essence, we are replacing each term in the subtrahend with its additive inverse. Using the horizontal method, the given problem would be written as $8x^2 + 4x - 5 - (5x^2 - 2x + 3) = 8x^2 + 4x - 5 - 5x^2 + 2x - 3 = 3x^2 + 6x - 8$.

EXAMPLE 4

Simplify:
a. $(4x^2 - x + 3) - (5x^2 + 2x - 4)$
b. $(3x^2 - 4x - 7) - (8x^2 + 2x - 9)$

Solution

To simplify the given expressions, we must perform the indicated subtraction. To do this, we must make use of the rule that $-(a + b) = -1(a + b) = -a - b$.

a. $(4x^2 - x + 3) - (5x^2 + 2x - 4) = 4x^2 - x + 3 - 5x^2 - 2x + 4$
$= -x^2 - 3x + 7$.

b. $(3x^2 - 4x - 7) - (8x^2 + 2x - 9) = 3x^2 - 4x - 7 - 8x^2 - 2x + 9$
$= -5x^2 - 6x + 2$.

Polynomials

EXAMPLE 5

Subtract the first polynomial from the second polynomial.
a. $4x^2 + 5x - 7$ from $3x^2 + 2x - 5$
b. $3x + 2y - 1$ from $4x + y - 3$

Solution

We shall use the column technique, changing the signs of each term in the subtrahend, as indicated by the circles.

a. $3x^2 + 2x - 5$
    $\underline{\ominus 4x^2 \ominus 5x \oplus 7}$
    $x^2 - 3x + 2$

b. $4x + y - 3$
    $\underline{\ominus 3x \ominus 2y \oplus 1}$
    $x - y - 2$

Regardless of the technique used in subtracting polynomials, we are adding the additive inverse of the subtrahend to the minuend, or $x - y = x + (-y)$.

EXERCISES FOR SECTION 4.2

(Answers given to exercises marked*)

Simplify the given polynomials:
*1. $4x^2 + 3x + 6 + 4x + 8x^2$
2. $3x - 2 + 4x + 5 + x^2$
*3. $3y - 2y^2 + y + 4 + y^2$
4. $4xy + 8x + 2y + 3x + 2xy$
*5. $3x + 2xy + 2y$
6. $x^2 - 2x + 3 + 4x + x^2 - 2x$
*7. $8y + y^2 - 3 - y^2 + 2y + 4$
8. $3xy - 2y + 4 + 2x - 8xy$
*9. $4 - x + 3 + x^2 - 2x + 3x^2$
10. $3y^3 + 4y + 2y^2 - 6y^3 - 8y$
*11. $x^3 + 3x^2 + 2x - 4x - 7x^2$
12. $x^3 + 2x + 3 + 4y + y^3$

Add the given polynomials:
*13. $4x - 7$ and $3x + 5$
14. $3y + 4x + 7$ and $2x - 3y - 2$
*15. $x^2 + 3x + 2$ and $2x^2 - 3x - 1$
16. $3xy + 4x - 2$ and $7 - 2x - 4xy$
*17. $x^3 - 2x^2 + 3x + 1$ and $4x^2 - 3x - 4$
18. $2xy + 3x + 4y$ and $4x - 2y - 3xy$
*19. $-3x^2 - 4x - 5$ and $-7x^2 - 5x - 3$
20. $4y^3 + y - 1$ and $y^2 + 3y$

Subtract the first polynomial from the second polynomial:
*21. $4x + 3y - 5$ from $2x + 5y - 7$
22. $x + y + 1$ from $-x - y - 1$
*23. $x^2 - 3x - 2$ from $x^2 + 2x + 4$
24. $y^2 - 2y - 1$ from $3y^2 + 2y + 4$
*25. $y^3 + 2y^2 + 1$ from 0
26. $8x^2 - 3x + 1$ from $x^2 + x - 1$

*27. $3x^2 - 11x + 5$
  from $-2x^2 - 7x - 4$

28. $y^2 - 2y + 1$
  from $3y + 5$

*29. $-x^2 - x + 7$ from $2x^4 + 4x^2$

Perform the indicated operations and simplify by combining like terms:
30. $(4x^2 + 2x) + (3x^2 - 7)$
*31. $-(x^2 + 2x) + (x^2 + 2x)$
32. $(x^2 + 3x^2 + 2) - (x + 3)$
*33. $(4y^2 - 2y + 1) - (-3y^2 + 2y - 1)$
34. $(5x - 3y + 2) - (-4x - 2y - 3)$
*35. $(x^2 - 2x - 3) - (-2x^2 - 7x + 4)$
36. $(8y^2 - 2y + 3) - (4y^2 + 3y - 2)$
*37. $(4x^2 - 2x - 1) - (-3x^2 + 2x - 1)$
38. $(4x^3 + 2x + x + 1) - (x^3 - 2x^2 + x + 1)$
*39. $3x^2 - 2x - [3 - (x^2 + x)]$
40. $4x + 2 - [3 - (x + 1)]$
*41. $y^2 + 3y - 2[3 - (y + 2)]$
42. $x - 3[2x - 2(x - 2) + 1]$
*43. $-\{3x - 2[x - 2(x + 2) + 1] - 1\}$
44. $x - 3\{2 + 3[x - 2(x + 2) - 1] - 1\}$

4.3 MULTIPLICATION OF POLYNOMIALS

Multiplication of polynomials can take on many different forms. The simplest is when we multiply a monomial by a monomial. But there are many other forms, such as a monomial times a binomial, a binomial times a binomial, a binomial times a trinomial, and so on. To multiply a monomial by a monomial, recall the rule for multiplying exponentials (Chapter 3),

$$x^m \cdot x^n = x^{m+n}$$

Granted, we do not have to use this rule when multiplying two monomials that do not have the same literal factors, for example, $x \cdot y = xy$ and $3x^2 \cdot 2y^3 = 6x^2y^3$. But to find the product of x^2 and x^4, we have $x^2 \cdot x^4 = x^{2+4} = x^6$. Similarly, $(3x^2y^3)(2xy^2) = (3 \cdot 2)(x^2 \cdot x)(y^3 \cdot y^2) = 6x^3y^5$.

To find the product of a monomial and a binomial (polynomial), we use the rule $x^m \cdot x^n = x^{m+n}$, and also the distributive property. Recall that $a(b + c) = ab + ac$. Therefore,

$$3x^2(4x^3 - 2x) = (3x^2)(4x^3) + (3x^2)(-2x)$$
$$= 12x^5 - 6x^3.$$

Polynomials

This same procedure can be used when multiplying a monomial times any polynomial. For example,

$$-2x^2(3x^3 - 2x^2 - x - 1) = (-2x^2)(3x^3) + (-2x^2)(-2x^2) + (-2x^2)(-x) \\ + (-2x^2)(-1)$$
$$= -6x^5 + 4x^4 + 2x^3 + 2x^2.$$

We can also use the distributive property to find the product of two binomials. To find the product of $(a + b)(c + d)$, we apply the distributive property (as we did in Chapter 3 when multiplying radicals),

$$(a + b)(c + d) = (a + b)c + (a + b)d$$
$$= ac + bc + ad + bd.$$

For example, to find the product of $2x - 5$ and $3x + 2$, we have $(2x - 5)(3x + 2)$ and

$$(2x - 5)(3x + 2) = (2x - 5)(3x) + (2x - 5)(2).$$

Applying the distributive property again,

$$= (2x)(3x) + (-5)(3x) + (2x)(2) + (-5)(2)$$
$$= 6x^2 - 15x + 4x - 10.$$

Combining like terms,

$$= 6x^2 - 11x - 10.$$

Note that the product of two binomials may contain as many as four terms $(ac + bc + ad + bd)$. We shall see later that the product of two binomials may contain as few as two terms. The product of two polynomials can be obtained by finding the product of each term of one polynomial by each term of the other polynomial. This is illustrated when we perform the indicated multiplication vertically,

$$\begin{array}{ll} 2x - 5 & \\ 3x + 2 & \\ \hline 6x^2 - 15x & \text{(multiplying by } 3x\text{)} \\ + 4x - 10 & \text{(multiplying by } 2\text{)} \\ \hline 6x^2 - 11x - 10 & \text{(combining like terms)} \end{array}$$

Using this technique, we have written one polynomial beneath the other and found the product of each term of one polynomial by each term of the other. Note that like partial products ($-15x$ and $+4x$) are written in the same column. Regardless of the technique used, to multiply two polynomials, we must multiply each term of one polynomial by each term of the other polynomial, and then combine like terms (if any).

EXAMPLE 6

Perform the indicated operations.

a. $4x^2(2x^2 + 3x)$ b. $3y(2y^2 + y - 3)$

Solution

To find the product of a monomial and a polynomial, we apply the distributive property.

a. $4x^2(2x^2 + 3x) = (4x^2)(2x^2) + (4x^2)(3x)$
$= 8x^4 + 12x^3.$
b. $3y(2y^2 + y - 3) = (3y)(2y^2) + (3y)(y) + (3y)(-3)$
$= 6y^3 + 3y^2 - 9y.$

EXAMPLE 7

Perform the indicated operations, and simplify.

a. $(3x + 2)(2x - 5)$ b. $(y - 2)(2y + 3)$ c. $(x + y)(x - y)$

Solution

As in Example 6, we apply the distributive property.

a. $(3x + 2)(2x - 5) = (3x + 2)(2x) + (3x + 2)(-5)$
$= 6x^2 + 4x - 15x - 10$
$= 6x^2 - 11x - 10.$
b. $(y - 2)(2y + 3) = (y - 2)(2y) + (y - 2)(3)$
$= 2y^2 - 4y + 3y - 6$
$= 2y^2 - y - 6.$
c. $(x + y)(x - y) = (x + y)(x) + (x + y)(-y)$
$= x^2 + xy - xy - y^2$
$= x^2 - y^2.$

Alternate Solution

Performing the computations using the vertical (column) technique, we have

a. $3x + 2$
$\underline{2x - 5}$
$6x^2 + 4x$
$\underline{ - 15x - 10}$
$6x^2 - 11x - 10$

b. $y - 2$
$\underline{2y + 3}$
$2y^2 - 4y$
$\underline{ + 3y - 6}$
$2y^2 - y - 6$

c. $x + y$
$\underline{x - y}$
$x^2 + xy$
$\underline{ - xy - y^2}$
$x^2 - y^2$

Polynomials

The product of a binomial and a trinomial, or the product of any two polynomials, can be found in the same manner as illustrated in Example 7 — we can apply the distributive property, or we can use the vertical technique. Regardless of the technique used, extreme care should be used because of the number of opportunities for making an error. Consider the following example: $(2x - 3)(x^2 + 2x + 1)$. Applying the distributive property, we have

$$(2x - 3)(x^2 + 2x + 1) = (2x - 3)(x^2) + (2x - 3)(2x) + (2x - 3)(1)$$
$$= 2x^3 - 3x^2 + 4x^2 - 6x + 2x - 3$$
$$= 2x^3 + x^2 - 4x - 3.$$

Note that we multiplied each term of one polynomial by each term of the other polynomial, and then combined like terms. This is also the case when we use the vertical technique.

$$
\begin{array}{ll}
x^2 + 2x + 1 & \\
\underline{2x - 3} & \\
2x^3 + 4x^2 + 2x & \text{(multiplying by 2x)} \\
\underline{-3x^2 - 6x - 3} & \text{(multiplying by } -3\text{)} \\
2x^3 + x^2 - 4x - 3 & \text{(combining like terms)}
\end{array}
$$

Note that the partial products are arranged so that like terms are in the same column, which facilitates the final step of addition.

EXAMPLE 8

Perform the indicated operations and simplify.
a. $(3x - 2)(2x^2 - 3x + 4)$ b. $(2y - 3x^2)(4y^2 + 6x^2y + 9x^4)$

Solution

Applying the distributive property, we have
a. $(3x - 2)(2x^2 - 3x + 4) = (3x - 2)(2x^2) + (3x - 2)(-3x)$
$$+ (3x - 2)(4)$$
$$= 6x^3 - 4x^2 - 9x^2 + 6x + 12x - 8$$
$$= 6x^3 - 13x^2 + 18x - 8.$$
b. $(2y - 3x^2)(4y^2 + 6x^2y + 9x^4) = (2y - 3x^2)(4y^2)$
$$+ (2y - 3x^2)(6x^2y)$$
$$+ (2y - 3x^2)(9x^4)$$
$$= 8y^3 - 12x^2y^2 + 12x^2y^2$$
$$- 18x^4y + 18x^4y - 27x^6$$
$$= 8y^3 - 27x^6.$$

Alternate Solution

Performing the computations with the vertical (column) technique, we have

a. $2x^2 - 3x + 4$
$3x - 2$
$\overline{6x^3 - 9x^2 + 12x}$
$-4x^2 + 6x - 8$
$\overline{6x^3 - 13x^2 + 18x - 8}$

b. $4y^2 + 6x^2y + 9x^4$
$2y - 3x^2$
$\overline{8y^3 + 12x^2y^2 + 18x^4y}$
$-12x^2y^2 - 18x^4y - 27x^6$
$\overline{8y^3 - 27x^6}$

To find the product of any two polynomials, multiply each term of one polynomial by each term of the other, and then combine like terms (if possible).

EXERCISES FOR SECTION 4.3

(Answers given to exercises marked*)

Perform the indicated operations and simplify, if possible.

*1. $3x^2 \cdot 2x^3$
2. $x \cdot y$
*3. $-2y^3 \cdot 3y^4$
4. $3(x - y)$
*5. $3x(x + y)$
6. $-2x(3x^2 + 4x)$
*7. $-2x^2(3xy + 4x^2)$
8. $3xy(2xy + 4y)$
*9. $2x^2y(3xy^2 + 4xy)$
10. $2xy(3x^2 + 4xy + y^2)$
*11. $-3x^2y(2y^2 + 3xy + 4x^2)$
12. $-2x^2y^2z(3xy + 2y^2z)$
*13. $(x + 1)(x - 2)$
14. $(y + 3)(y + 2)$
*15. $(2y + 1)(3y - 2)$
16. $(3x - 2)(2x - 3)$
*17. $(3y + 2)(2y - 3)$
18. $(4x - 1)(2x + 3)$
*19. $(x^2 + 3)(2x - 1)$
20. $(3x^2 + 1)(2x + 2)$
*21. $(3y^2 - 1)(2y^2 + 3)$
22. $(3x^2 + 2x)(x^2 - 2)$
*23. $(3y^2 + y)(y - 2)$
24. $(2y + 3)(y^2 + 3y)$
*25. $(3x + 5y)(2x - y)$
26. $(5y - 3x)(y - x)$
*27. $(3x + 2y)(2y - x)$
28. $(r + s)(2r^2 + 3s)$
*29. $(3r^2 - 2s^3)(r^3 + s^2)$
30. $(x^2 + 2y)(x^3 - y^2)$
*31. $(2t^2 + u^3)(3u^2 - 2t)$
32. $(3v + 4u^2)(2v^3 - u)$
*33. $(x + 2)(x^2 + 3x - 1)$
34. $(2y - 3)(y^2 + 3y - 2)$
*35. $(3y^2 - 2y)(2y^2 + 3y + 4)$
36. $(2r^2 - s)(3r^2 + 2s^2 - 1)$
*37. $(3r^2 + 2s)(4r^3 + 2rs + s^2)$
38. $(2x - y)(4x^2 + 2xy + y^2)$
*39. $(x + 2y)(x^2 - 2xy + 4y^2)$
40. $(x + y + 2)(x + y - 3)$
*41. $(2x + 3y + 1)(x - 2y + 3)$
42. $(r + s + t)(r - s - t)$
*43. $(x^2 + 2y + 3)(2x^2 + 3y^2 - 2)$
44. $(x^2 + x + 2)(x^2 + 3x - 1)$
*45. $(y^2 + 2y + 1)(y^2 - 4y + 3)$
46. $(x + 2)(x - 1)(x + 3)$
*47. $(r + 2)(r - 1)(r + 1)$
48. $(x + 2)(x - 2)(x^2 + 4)$

Polynomials

4.4 SPECIAL PRODUCTS

Many products of polynomials occur so frequently in algebra (and in other branches of mathematics) that they deserve special consideration. This is also the case in arithmetic. Many routine computations can be done rather quickly because we can almost automatically determine the product of two integers. For example, we automatically write that $3 \cdot 2 = 6$ without recalling that multiplication is repeated addition. We use the multiplication tables that we learned in the study of arithmetic. Similarly, we want to be able to determine the answer to certain multiplication problems in algebra with a minimum amount of written work. In fact, we will be able to determine some products by inspection.

In Section 4.3, to find the product of two binomials we used the distributive property. Recall that $(a + b)(c + d) = (c + d)(a + b)$, by reason of the commutative property of multiplication. Therefore,

$$(a + b)(c + d) = (c + d)(a + b) = (c + d)a + (c + d)b = ac + ad + bc + bd.$$

Note that ac is the product of the *first* terms of each binomial, ad is the product of the *outer* terms of each binomial, bc is the product of the *inner* terms of each binomial, and bd is the product of the *last* terms of each binomial. Many students use a memory device with great success to remember this particular type of multiplication. Taking the first letter from each of the words: first, outer, inner, last, we get FOIL. With the FOIL technique, we obtain the products of the first terms, outer terms, inner terms, and last terms, and then find the sum of these products to obtain the product of two binomials.

To find the product of $(x - 2)(x + 3)$ using the FOIL technique, we have

$$(x - 2)(x + 3) = \underset{\substack{\uparrow \\ \left(\begin{array}{c}\text{product of} \\ \text{first terms}\end{array}\right)}}{x^2} \quad \underset{\substack{\uparrow \\ \left(\begin{array}{c}\text{product of} \\ \text{outer terms}\end{array}\right)}}{+ 3x} \quad \underset{\substack{\uparrow \\ \left(\begin{array}{c}\text{product of} \\ \text{inner terms}\end{array}\right)}}{- 2x} \quad \underset{\substack{\uparrow \\ \left(\begin{array}{c}\text{product of} \\ \text{last terms}\end{array}\right)}}{- 6.}$$

Combining like terms $(x - 2)(x + 3) = x^2 + x - 6$. Many times the products of the outer and inner terms will yield like terms. When this happens, we can eliminate another step by combining these terms mentally,

$$(x - 2)(x + 3) = \underset{\substack{\uparrow \\ \left(\begin{array}{c}\text{product of} \\ \text{first terms}\end{array}\right)}}{x^2} \quad \underbrace{+ x}_{\left(\begin{array}{c}\text{product of} \\ \text{outer terms}\end{array}\right) + \left(\begin{array}{c}\text{product of} \\ \text{inner terms}\end{array}\right)} \quad \underset{\substack{\uparrow \\ \left(\begin{array}{c}\text{product of} \\ \text{last terms}\end{array}\right)}}{- 6.}$$

The FOIL technique for finding the product of two binomials is a method that enables us to quickly determine the product with a minimum amount of written work. If the two binomials contain like terms, then the sum of the "outer" and "inner" products will give us one term of the product.

EXAMPLE 9

Perform the indicated operations, and simplify.
a. $(x + 3)(x - 5)$
b. $(y - 1)(y + 2)$
c. $(3x + 2y)(2x - y)$
d. $(x + y)(x + y)$

Solution

Using the FOIL method, we shall determine these products mentally so that the answer can be written by means of inspection. The sum of the outer and inner terms is indicated by the curved lines.

a. $(x + 3)(x - 5) = x^2 - 2x - 15$.
b. $(y - 1)(y + 2) = y^2 + y - 2$.
c. $(3x + 2y)(2x - y) = 6x^2 + xy - 2y^2$.
d. $(x + y)(x + y) = x^2 + 2xy + y^2$.

The FOIL technique can be used to find the product of any two binomials, as we did in Example 9. Note that Example 9–d $(x + y)(x + y)$ can also be written as $(x + y)^2$. Therefore, we can also use the FOIL technique to *square a binomial.* Regardless of the nature of the terms contained in the binomial, when we square a binomial, it will always be of the general form

$$(a + b)^2 = (a + b)(a + b) = a^2 + 2ab + b^2.$$

Since the "square of a binomial" will always have this same general form, we have found a short cut for determining this type of product. The square of a binomial is the sum of the squares of each term in the binomial, plus twice the product of the two terms, in other words, the square of the first term, plus twice the product of the two terms, plus the square of the last term. For example,

$$(x + y)^2 = x^2 + 2xy + y^2$$

$$\begin{pmatrix} \text{square of} \\ \text{the first term} \end{pmatrix} \begin{pmatrix} \text{twice the product} \\ \text{of the two terms} \end{pmatrix} \begin{pmatrix} \text{square of the} \\ \text{last term} \end{pmatrix}$$

EXAMPLE 10

Perform the indicated operations, and simplify.
a. $(3x + 4)^2$
b. $(x^3 + 2y)^2$
c. $(2x - 3y)^2$

Solution

In each case, we are squaring a binomial.
a. Square of the first term = $(3x)^2 = 9x^2$.
   Twice the product of the two terms = $2(3x)(4) = 24x$.
   Square of the last term = $(4)^2 = 16$.
   $(3x + 4)^2 = 9x^2 + 24x + 16$.
b. Square of the first term = $(x^3)^2 = x^6$.
   Twice the product of the two terms = $2(x^3)(2y) = 4x^3y$.
   Square of the last term = $(2y)^2 = 4y^2$.
   $(x^3 + 2y)^2 = x^6 + 4x^3y + 4y^2$.
c. Square of the first term = $(2x)^2 = 4x^2$.
   Twice the product of the two terms = $2(2x)(-3y) = -12xy$.
   Square of the last term = $(-3y)^2 = 9y^2$.
   $(2x - 3y)^2 = 4x^2 - 12xy + 9y^2$.

Note that in each solution we have listed each step in detail, but as in the FOIL technique, these steps should be done mentally so that the answer can be written by inspection (as illustrated in Example 11).

EXAMPLE 11

Perform the indicated operations.
a. $(2x + 3y)^2$ b. $(3x^2 + 2y^3)^2$ c. $(x^2 - 4y)^2$

Solution

a. $(2x + 3y)^2 = 4x^2 + 12xy + 9y^2$.
b. $(3x^2 + 2y^3)^2 = 9x^4 + 12x^2y^3 + 4y^6$.
c. $(x^2 - 4y)^2 = x^4 - 8x^2y + 16y^2$.

It should be noted at this point that the products obtained in Examples 10 and 11 are of a special nature. In these examples, we squared a binomial, and the resulting trinomials are called *perfect squares*. A trinomial is a perfect square if it can be expressed as the square of a binomial. For example, $x^2 + 2xy + y^2$ is a perfect square because

$$x^2 + 2xy + y^2 = (x + y)^2.$$

Note that x^2 is the square of x, y^2 is the square of y, and $+2xy$ is twice the product of x and y. Similarly, $25x^6 - 20x^3y^2 + 4y^4$ is a perfect square, since $25x^6 - 20x^3y^2 + 4y^4 = (5x^3 - 2y^2)^2$.

As we stated earlier, the FOIL technique can be used to find the product of any two binomials. Consider the indicated product, $(a + b)(a - b)$. If we perform the indicated multiplication (using the FOIL technique), we have

$$(a + b)(a - b) = a^2 - ab + ab - b^2 = a^2 - b^2.$$

The result, $a^2 - b^2$ is called the *difference of two squares.* Whenever we find the product of two binomials of the form $(a + b)(a - b)$, the result will always be of the form $a^2 - b^2$. Recall that in Chapter Three we discussed expressions of the form $a + b$ and $a - b$. They are called "conjugates" of each other, and the product of two conjugates will be the difference of two squares. We can also describe this as the product of the sum and difference of two monomials. The product is the square of the first monomial minus the square of the second monomial. For example, the product of $(3x + 2y)(3x - 2y)$ is $9x^2 - 4y^2$. Note that $(3x + 2y)(3x - 2y)$ is the product of the sum and difference of two monomials, and the resulting product is the square of the first monomial minus the square of the second monomial. Any indicated product is of the general form $(a + b)(a - b) = a^2 - b^2$, and to find the product of a pair of conjugates, we should perform the work mentally and write the answer directly (by inspection).

EXAMPLE 12

Perform the indicated operations, and simplify.
a. $(x + 3y)(x - 3y)$ b. $(2x^2 + 3y^3)(2x^2 - 3y^3)$

Solution

In each case, we have the indicated product of conjugates, that is, the product of the sum and difference of two monomials. The resulting product is the square of the first monomial minus the square of the second monomial.

a. $(x + 3y)(x - 3y) = (x)^2 - (3y)^2 = x^2 - 9y^2.$
b. $(2x^2 + 3y^3)(2x^2 - 3y^3) = (2x^2)^2 - (3y^3)^2 = 4x^4 - 9y^6.$

We can also use this type of special product to find the product of integers. For example, most of us cannot multiply $(31)(29)$ mentally, because of the way we have learned to multiply. But if we rewrite the given expression as $(30 + 1)(30 - 1)$, we still have the same indicated product and we can determine the product by inspection,

$$\begin{aligned}(31)(29) &= (30 + 1)(30 - 1) \\ &= (30)^2 - (1)^2 \\ &= 900 - 1 \\ &= 899.\end{aligned}$$

Polynomials

Similarly, to find the product of (54)(46), we can rewrite this as (50 + 4)(50 − 4), which equals $(50)^2 − (4)^2 = 2500 − 16 = 2484$.

EXAMPLE 13

Perform the indicated operations.
a. (63)(57)   b. (46)(34)

Solution

In each case, we can rewrite the indicated product in the form of $(a + b)(a − b)$ and make use of the fact that $(a + b)(a − b) = a^2 − b^2$.
a. $(63)(57) = (60 + 3)(60 − 3) = (60)^2 − (3)^2 = 3600 − 9$
  $= 3591$.
b. $(46)(34) = (46 + 6)(40 − 6) = (40)^2 − (6)^2 = 1600 − 36$
  $= 1564$.

Note that in Examples 12 and 13, we listed the detailed steps in the solution, but as in the other special products, these steps should be performed mentally and the answer written directly (by means of inspection).

EXERCISES FOR SECTION 4.4

(Answers given to exercises marked *)

Perform the indicated operations, and simplify. All products should be determined by inspection.

*1. $2x(3x + 2y)$
*3. $4x^2y(3xy^2 − 2y)$
*5. $(y + 5)(y − 3)$
*7. $(x − 1)(x − 7)$
*9. $(y − 10)(y + 2)$
*11. $(4x + 3)(2x − 1)$
*13. $(5y − 3)(2y + 1)$
*15. $(2x + y)(3x − 2y)$
*17. $(3y − 2z)(2y + 5z)$
*19. $(2x^2 + 3y)(3x^2 + 4y)$
*21. $(3y^2 − 4x)(2y^2 + 5x)$
*23. $(5y^3 + 2z^2)(y^3 − 3z^2)$
*25. $(x + y)^2$
*27. $(x + 2y)^2$
*29. $(z + 5)^2$
*31. $(3z + 2x)^2$

2. $5xy(2x − 3y)$
4. $(x + 3)(x − 4)$
6. $(z + 2)(z − 3)$
8. $(y − 2)(y − 3)$
10. $(2x + 3)(3x − 2)$
12. $(3y + 2)(y − 1)$
14. $(2x + 3y)(x − y)$
16. $(2x + 4y)(5x − 3y)$
18. $(x^2 + 2y)(x^2 − 3y)$
20. $(5x^3 + 2y^2)(2x^3 − y^2)$
22. $(3y^2 + 2x^2)(4y^2 − 2x^2)$
24. $(x^4 + 2y^3)(3x^4 − y^3)$
26. $(x − y)^2$
28. $(3x + 2y)^2$
30. $(3x − 2z)^2$
32. $(−3x − 2)^2$

*33. $(-4y - 5)^2$
34. $(-5x - 2y)^2$
*35. $(4y + 5)^2$
36. $(-2x + 3y)^2$
*37. $(-3x^2 + 4y)^2$
38. $(5x^2 + 3y^3)^2$
*39. $(4y^3 - 2z^2)^2$
40. $(5x^3 + 3y)^2$
*41. $(-2x^3 - 3y^2)^2$
42. $(-5z^2 - 3y^3)^2$
*43. $(x + y)(x - y)$
44. $(2x + 3z)(2x - 3z)$
*45. $(3x + 2y)(3x - 2y)$
46. $(8z + 2y)(8z - 2y)$
*47. $(x^2 + 4y)(x^2 - 4y)$
48. $(y^3 - 3x^2)(y^3 + 3x^2)$
*49. $(5x^3 - 3y)(5x^3 + 3y)$
50. $(4y^2 - 6x)(4y^2 + 6x)$
*51. $(3z^4 - y^3)(3z^4 + y^3)$
52. $(4xy + 3z)(4xy - 3z)$
*53. $(5x^2y + 3z^2)(5x^2y - 3z^2)$
54. $(4xyz + 1)(4xyz - 1)$
*55. $(42)(38)$
56. $(33)(27)$
*57. $(23)(17)$
58. $(25)(15)$
*59. $(24)(16)$
60. $(32)(28)$
*61. $(18)(22)$
62. $(19)(21)$

Replace the question mark by a term that will make the trinomial a perfect square.

*63. $x^2 + ? + y^2$
64. $4x^2 - 12xy + ?$
*65. $? + 4xy + y^2$
66. $x^4 + 6x^2y + ?$
*67. $9x^2 + 12xy^2 + ?$
68. $16x^4 + ? + 9y^6$

4.5 MORE SPECIAL PRODUCTS

The special products covered in Section 4.4 are frequently found in algebra and other branches of mathematics. In this section, we shall discuss some other special products that also occur rather frequently, but because of their nature should be considered separately from those covered earlier. Previously, we discussed the square of a binomial, $(a + b)^2$. Now, let us consider the cube of a binomial, $(a + b)^3$. By means of the properties for exponents, we can express $(a + b)^3$ as $(a + b)^2(a + b)$. Next, we expand $(a + b)^2$, perform the indicated multiplication, and combine like terms:

$(a + b)^3 = (a + b)^2(a + b)$

$\qquad = (a^2 + 2ab + b^2)(a + b)$
$\qquad = (a^2 + 2ab + b^2)(a) + (a^2 + 2ab + b^2)(b)$ (by means of the distributive property)

$\qquad = a^3 + 2a^2b + ab^2 + a^2b + 2ab^2 + b^3$
$\qquad = a^3 + 3a^2b + 3ab^2 + b^3$.

Therefore, to cube a binomial, we have

$$(a + b)^3 = a^3 + 3a^2b + 3ab^2 + b^3.$$

This is the general form for the cube of any binomial. For example, to

Polynomials

determine $(x + 2y)^3$, let $a = x$ and $b = 2y$; substitute in the general expression for a and b and we have

$$(x + 2y)^3 = x^3 + 3x^2(2y) + 3x(2y)^2 + (2y)^3$$
$$= x^3 + 6x^2y + 3x(4y^2) + 8y^3$$
$$= x^3 + 6x^2y + 12xy^2 + 8y^3.$$

EXAMPLE 14

Expand the following:
a. $(2x + 3y)^3$ b. $(3x^2 - 1)^3$

Solution

To cube a binomial, we can use the general form $(a + b)^3 = a^3 + 3a^2b + 3ab^2 + b^3$, where a is the first term and b is the second term.

a. Let $a = x$, $b = 3y$,
$(a + b)^3 = a^3 + 3a^2b + 3ab^2 + b^3$.
$(2x + 3y)^3 = (2x)^3 + 3(2x)^2(3y) + 3(2x)(3y)^2 + (3y)^3$
$= 8x^3 + 3(4x^2)(3y) + 6x(9y^2) + 27y^3$
$= 8x^3 + 36x^2y + 54xy^2 + 27y^3$.

b. Let $a = 3x^2$, $b = -1$,
$(a + b)^3 = a^3 + 3a^2b + 3ab^2 + b^3$.
$(3x^2 - 1)^3 = (3x^2)^3 + 3(3x^2)^2(-1) + 3(3x^2)(-1)^2 + (-1)^3$
$= 27x^6 + 3(9x^4)(-1) + 9x^2(1) - 1$
$= 27x^6 - 27x^4 + 9x^2 - 1.$

SPECIAL NOTE

We have already considered the special products $(a + b)^2$ and $(a + b)^3$. There exists a special formula for the expansion of $(a + b)^n$ where n is any positive integer. We can readily verify, by repeated multiplication, the following:

$(a + b)^1 = a + b$
$(a + b)^2 = a^2 + 2ab + b^2$
$(a + b)^3 = a^3 + 3a^2b + 3ab^2 + b^3$
$(a + b)^4 = a^4 + 4a^3b + 6a^2b^2 + 4ab^3 + b^4$
$(a + b)^5 = a^5 + 5a^4b + 10a^3b^2 + 10a^2b^3 + 5ab^4 + b^5$

Observe that for these expansions of $(a + b)^n$, the exponent of a decreases by 1 and the exponent of b increases by 1 for each term. The sum of the exponents of a and b is n for each term. Note also that in each expansion, there are n + 1 terms. The first term is a^n and the last term is b^n. If we write down only the numerical coefficients for each of the expansions, we have

$(a + b)^1$			1 1			These numbers are called the binomial coefficients for the indicated expansions, and this array of numbers is known as *Pascal's Triangle*.
$(a + b)^2$		1	2	1		
$(a + b)^3$		1 3		3 1		
$(a + b)^4$	1	4	6	4	1	
$(a + b)^5$	1 5	10		10	5 1	

We can construct such an array by the following method. In each row, the first and last numbers are always 1. Any other number in the array is the sum of the two numbers on the line above that it lies between. Therefore, to obtain the coefficients for $(a + b)^6$, begin with 1. The next number 6, is obtained by adding 1 and 5 from the fifth row. Next, we obtain 15 by adding 5 and 10, and so on.

$$\begin{array}{ccccccc} (a+b)^5 & 1 & 5 & 10 & 10 & 5 & 1 \\ & & \vee & \vee & \vee & \vee & \vee \\ (a+b)^6 \; 1 & 6 & 15 & 20 & 15 & 6 & 1 \end{array}$$

This corresponds to the fact that $(a + b)^6 = a^6 + 6a^5b + 15a^4b^2 + 20a^3b^3 + 15a^2b^4 + 6ab^5 + b^6$.

Blaise Pascal (1623–1662), famous French mathematician and author, contributed greatly to the development of geometry and probability. Pascal did not invent the triangle which bears his name, since some Chinese works of the early 1300's contain such an array of numbers. But because he discovered and applied many of its properties, this triangular array has become known as Pascal's Triangle.

The following special products will be most useful when we discuss factoring polynomials. They are presented here to acquaint you with their form. Consider $(a + b)(a^2 - ab + b^2)$. To determine the product, we apply the distributive property and combine like terms:

$$\begin{aligned} (a + b)(a^2 - ab + b^2) &= (a + b)(a^2) + (a + b)(-ab) + (a + b)(b^2) \\ &= a^3 + a^2b - a^2b - ab^2 + ab^2 + b^3 \\ &= a^3 + b^3. \end{aligned}$$

This special product results in the sum of two cubes. Note that one factor is a binomial that is the sum of the cube roots, and the other factor is a trinomial containing two terms that are squares of the cube roots. The third term is the additive inverse of the product of the cube roots.

If we perform the indicated multiplications for $(a - b)(a^2 + ab + b^2)$, we obtain the difference of two cubes,

$$\begin{aligned} (a - b)(a^2 + ab + b^2) &= (a - b)(a^2) + (a - b)(ab) + (a - b)(b^2) \\ &= a^3 - a^2b + a^2b - ab^2 + ab^2 - b^3 \\ &= a^3 - b^3. \end{aligned}$$

Polynomials

This special product is the difference of two cubes, where one factor is a binomial that is the difference of the cube roots, and the other factor is a trinomial containing two terms that are squares of the cube roots. The third term is the additive inverse of the product of the cube roots.

Ideally, you should be able to recognize the indicated product for the sum and difference of two cubes. If you do not, then simply apply the distributive property (or use the vertical technique) to obtain the product. Note that by means of the symmetric property, we can state that

$$a^3 + b^3 = (a + b)(a^2 - ab + b^2)$$

and

$$a^3 - b^3 = (a - b)(a^2 + ab + b^2).$$

That is, we can also use the above information to find the *factors* of the sum and difference of two cubes.

EXAMPLE 15

Perform the indicated operations,
a. $(x + y)(x^2 - xy + y^2)$ b. $(y - 2)(y^2 + 2y + 4)$
c. $(x + 2y)(x^2 - 2xy + 4y^2)$

Solution

In each case, one factor is the sum or difference of two terms, and the other factor is a trinomial where two terms are squares of the terms in the binomial. The third term of the trinomial is the additive inverse of the product of the terms in the binomial. Therefore, the product is the sum or difference of two cubes, depending upon the sign between the two terms in the binomial.
a. $(x + y)(x^2 - xy + y^2) = x^3 + y^3.$
b. $(y - 2)(y^2 + 2y + 4) = y^3 - (2)^3 = y^3 - 8.$
c. $(x + 2y)(x^2 - 2xy + 4y^2) = x^3 + (2y)^3 = x^3 + 8y^3.$

Rather than concern ourselves with special formulas for products such as $(x + y + z)^2$ and $(x + y + 4)(x + y - 2)$, it is recommended that we use the distributive property or the vertical (column) technique that was discussed in Section 4.3. For example, $(x + y + z)^2 = (x + y + z)(x + y + z)$ and

$$(x + y + z)(x + y + z) = (x + y + z)x + (x + y + z)y + (x + y + z)z$$
$$= x^2 + xy + xz + xy + y^2 + yz + xz + yz + z^2$$
$$= x^2 + y^2 + z^2 + 2xy + 2xz + 2yz.$$

To find the product of any two polynomials, we multiply each term of one polynomial by each term of the other, and then combine like terms. Reviewing the vertical technique for multiplication of polynomials, for (x + y + 4)(x + y − 2) we have

$$\begin{array}{l} x + y + 4 \\ x + y - 2 \\ \hline x^2 + xy + 4x \\ + xy + y^2 + 4y \\ - 2x - 2y - 8 \\ \hline x^2 + 2xy + 2x + y^2 + 2y - 8. \end{array}$$

Recognizing special products enables us to save time by performing most of the work mentally and then writing the answer directly. But in order to do this, we must first recognize the correct form of the indicated product, and also use care in determining the product.

EXERCISES FOR SECTION 4.5

(Answers given to exercises marked *)

Perform the indicated operations, and simplify. For Exercises 1 through 33, the products should be determined by inspection.

*1. $(x + y)^3$
2. $(x + 1)^3$
*3. $(x - y)^3$
4. $(x - 2)^3$
*5. $(x + 3)^3$
6. $(x + 2)^3$
*7. $(x - 4)^3$
8. $(y + 4)^3$
*9. $(2x + 1)^3$
10. $(x + 2y)^3$
*11. $(3x - y)^3$
12. $(3x + 2y)^3$
*13. $(x + 3y)^3$
14. $(2x - 3y)^3$
*15. $(3x - 4y)^3$
16. $(4y - 2x)^3$
*17. $(2y + 4x)^3$
18. $(x^2 + y)^3$
*19. $(y^2 - x)^3$
20. $(2x^2 + 3y)^3$
*21. $(3x^2 - 2y)^3$
22. $(x + y)(x^2 - xy + y^2)$
*23. $(x - y)(x^2 + xy + y^2)$
24. $(x - 2)(x^2 + 2x + 4)$
*25. $(x + 3)(x^2 - 3x + 9)$
26. $(2x + 1)(4x^2 - 2x + 1)$
*27. $(3x - 2)(9x^2 + 6x + 4)$
28. $(x + 2y)(x^2 - 2xy + 4y^2)$
*29. $(2x - 3y)(4x^2 + 6xy + 9y^2)$
30. $(x^2 + y)(x^4 - x^2y + y^2)$
*31. $(2x^2 - 3y)(4x^4 + 6x^2y + 9y^2)$
32. $(x^3 - 2y^2)(x^6 + 2x^3y^2 + 4y^4)$
*33. $(2x^3 + 3y^2)(4x^6 - 6x^3y^2 + 9y^4)$
34. $(x + y - 3)(x + y + 2)$
*35. $(2x + 3y + 1)(x - 2y + 3)$
36. $(x^2 - 2x + 1)(x^2 + 3x + 2)$
*37. $(y^2 - 3y + 4)(y^2 - 2y - 5)$
38. $(x^2 + 2xy + 1)(x^2 + 3xy + 4)$
*39. $(x^2 + 2xy + y^2)(x^2 + 2xy - y^2)$

Polynomials

4.6 DIVISION OF POLYNOMIALS

In order to perform division of polynomials, we should first examine division of monomials. Recall that in Chapter Three, we performed the operation of division for exponentials, $x^m \div x^n = x^{m-n}$. Therefore, to obtain the quotient of two monomials, we should obtain the quotient of the numerical coefficients (if any), and then obtain the quotient of the exponentials. For example, $x^5 \div x^2 = x^{5-2} = x^3$, $10y^7 \div 5y^3 = 2y^4$, and $18x^4y^3 \div 3x^2y = 6x^2y^2$.

In the simplest type of division of polynomials that we shall consider, the divisor is a monomial. For example, $(24x^3 + 18x^2 + 12x) \div 6x$. Recall that in order to add two or more fractions with the same denominator, we write the same denominator in the sum and add the numerators of the given fractions:

$$\frac{a}{d} + \frac{b}{d} + \frac{c}{d} = \frac{a+b+c}{d}.$$

By means of the symmetric property, we can also state that

$$\frac{a+b+c}{d} = \frac{a}{d} + \frac{b}{d} + \frac{c}{d}.$$

This is the concept that we use in division of polynomials where the divisor is a monomial. The problem $(24x^3 + 18x^2 + 12x) \div 6x$ can be rewritten as

$$\frac{24x^3 + 18x^2 + 12x}{6x} = \frac{24x^3}{6x} + \frac{18x^2}{6x} + \frac{12x}{6x} = 4x^2 + 3x + 2, \ x \neq 0.$$

To divide a polynomial by a monomial, we must divide each term of the polynomial by that monomial. The algebraic sum of the resulting quotients is the quotient of the polynomial divided by the monomial.

EXAMPLE 16

Perform the indicated operations.
a. $(18x^5 + 12x^3 + 24x^2) \div 3x^2$
b. $(10x^2y^3 + 5xy^2 - 15xy) \div 5xy$
c. $(y^4 + 3y^2 - y + 2) \div y^2$

Solution

To divide a polynomial by a monomial, divide each term of the polynomial by the monomial.

a. $(18x^5 + 12x^3 + 24x^2) \div 3x^2 = \dfrac{18x^5 + 12x^3 + 24x^2}{3x^2}$

$= \dfrac{18x^5}{3x^2} + \dfrac{12x^3}{3x^2} + \dfrac{24x^2}{3x^2}$

$= 6x^3 + 4x + 8.$

b. $\dfrac{10x^2y^3 + 5xy^2 - 15xy}{5xy} = \dfrac{10x^2y^3}{5xy} + \dfrac{5xy^2}{5xy} - \dfrac{15xy}{5xy}$

$= 2xy^2 + y - 3.$

c. $\dfrac{y^4 + 3y^2 - y + 2}{y^2} = \dfrac{y^4}{y^2} + \dfrac{3y^2}{y^2} - \dfrac{y}{y^2} + \dfrac{2}{y^2}$

$= y^2 + 3 - \dfrac{1}{y} + \dfrac{2}{y^2}.$

The general type of division of polynomials to consider is that in which the divisor is also a polynomial (not a monomial). To divide one polynomial by another, we use a method similar to the method for long division in arithmetic. For example,

$$\begin{array}{r} 12 \\ 24\overline{)288} \\ \underline{24} \\ 48 \\ \underline{48} \\ 0 \end{array}$$ Therefore, $\dfrac{288}{24} = 12$

$$\begin{array}{r} 12 \\ 35\overline{)428} \\ \underline{35} \\ 78 \\ \underline{70} \\ 8 \end{array}$$ Therefore, $\dfrac{428}{35} = 12 + \dfrac{8}{35}$

We shall now expand upon this technique to see how division of polynomials is performed. In the division example above, recall that 24 is called the divisor, 288 is called the dividend, 12 is the quotient and 0 is the remainder. The operation of division is terminated as soon as the remainder is less than the divisor (or zero).

To divide two polynomials such as $(6x^2 - 5x - 6)$ and $(2x - 3)$, we first make sure that the polynomials are arranged in descending powers of the variable. If they are not, we apply the commutative property of addition and arrange them in descending order. Next we set the problem up as we would a division problem in arithmetic. $(6x^2 - 5x - 6) \div (2x - 3)$ appears as

$$2x - 3\overline{)6x^2 - 5x - 6}$$

To obtain the quotient, we begin by dividing the first term of the divisor into the first term of the dividend, that is, divide $2x$ into $6x^2$. The quotient is $3x$ which we place over the x term in the dividend. Thus far we have

$$2x - 3\overline{)\stackrel{\displaystyle 3x}{6x^2 - 5x - 6}}$$

Polynomials

Recall that in arithmetic we would now multiply the partial quotient by the divisor, and we do the same thing here,

$$\begin{array}{r} 3x \\ 2x-3 \overline{\smash{)}\, 6x^2 - 5x - 6} \\ 6x^2 - 9x \end{array}$$

We are now ready to subtract. To subtract two polynomials, we must change the signs of the terms in the subtrahend. Performing the subtraction, we obtain

$$\begin{array}{r} 3x \\ 2x-3 \overline{\smash{)}\, 6x^2 - 5x - 6} \\ 6x^2 - 9x \\ \hline 4x \end{array}$$

Now, we "bring down" the next term and since $4x - 6$ is not less than the divisor (the degree of the remainder is not less than the degree of the divisor), we repeat the process. We now have

$$\begin{array}{r} 3x \\ 2x-3 \overline{\smash{)}\, 6x^2 - 5x - 6} \\ 6x^2 - 9x \\ \hline 4x - 6 \end{array}$$

and we divide $2x$ into $4x$ (the first term of the new dividend). The quotient is $+2$ and we place it over the constant term in the dividend.

$$\begin{array}{r} 3x + 2 \\ 2x-3 \overline{\smash{)}\, 6x^2 - 5x - 6} \\ 6x^2 - 9x \\ \hline 4x - 6 \end{array}$$

We now multiply $2x - 3$ by $+2$ and subtract the product from $4x - 6$.

$$\begin{array}{r} 3x + 2 \\ 2x-3 \overline{\smash{)}\, 6x^2 - 5x - 6} \\ 6x^2 - 9x \\ \hline 4x - 6 \\ 4x - 6 \\ \hline 0 \end{array}$$

(Remember to change the signs of the subtrahend when subtracting.)

Since the remainder is 0, the division is completed. Therefore $(6x^2 - 5x - 6) \div (2x - 3) = 3x + 2$. We can check our work by multiplying the divisor by the quotient; the resulting product should be the dividend. Checking, we have $(2x - 3)(3x + 2) = 6x^2 - 5x - 6$, which is the desired result. The solution is correct.

To divide two polynomials, we first write the polynomials in descending powers of the variable. Therefore, to divide $(8x^3 + 6x + 4)$ by $(2x - 3)$ we should rewrite $8x^3 + 6x + 4$. Note that the x^2 term is missing. Hence, $8x^3 + 6x + 4 = 8x^3 + 0x^2 + 6x + 4$. The $0x^2$ term acts as a place holder; it does not change the value of the dividend, and allows us to perform the necessary operations more easily. Setting the problem up, we have

$$2x - 3 \overline{\smash{)}8x^3 + 0x^2 + 6x + 4}$$

We begin by dividing $2x$ into $8x^3$. The quotient is $4x^2$ and we place it over the x^2 term in the dividend.

$$\begin{array}{r} 4x^2 \\ 2x - 3 \overline{\smash{)}8x^3 + 0x^2 + 6x + 4} \end{array}$$

Next, multiply $2x - 3$ by $4x^2$, subtract the product $8x^3 - 12x^2$ from $8x^3 + 0x^2$, and "bring down" the next term, $6x$.

$$\begin{array}{r} 4x^2 \\ 2x - 3 \overline{\smash{)}8x^3 + 0x^2 + 6x + 4} \\ \underline{8x^3 - 12x^2 } \\ 12x^2 + 6x \end{array}$$

(Changing the signs of the subtrahend, and bringing down $6x$)

Divide $2x$ into $12x^2$, obtaining $+ 6x$. Place it over the x term in the dividend.

$$\begin{array}{r} 4x^2 + 6x \\ 2x - 3 \overline{\smash{)}8x^3 + 0x^2 + 6x + 4} \\ \underline{8x^3 - 12x^2 } \\ 12x^2 + 6x \end{array}$$

Multiply $2x - 3$ by $6x$. Subtract the product, $12x^2 - 18x$, from $12x^2 + 6x$, and "bring down" the next term, 4.

$$\begin{array}{r} 4x^2 + 6x \\ 2x - 3 \overline{\smash{)}8x^3 + 0x^2 + 6x + 4} \\ \underline{8x^3 - 12x^2 } \\ 12x^2 + 6x \\ \underline{12x^2 - 18x } \\ 24x + 4 \end{array}$$

Divide $2x$ into $24x$, obtaining $+ 12$. Place it over the constant term in the dividend. Multiply $2x - 3$ by 12. Subtract the product, $24x - 36$, from $24x + 4$.

Polynomials

$$\begin{array}{r} 4x^2 + 6x + 12 \\ 2x - 3 \overline{\smash{\big)}\, 8x^3 + 0x^2 + 6x + 4} \\ \underline{8x^3 - 12x^2 } \\ 12x^2 + 6x \\ \underline{12x^2 - 18x } \\ 24x + 4 \\ \underline{24x - 36} \\ 40 \end{array}$$

The remainder, 40, is of degree 0 while the divisor, $2x - 3$, is of degree 1. The division terminates when the degree of the remainder is less than the degree of the divisor, or the remainder is 0. Therefore,

$$(8x^3 + 6x + 4) \div (2x - 3) = 4x^2 + 6x + 12 + \frac{40}{2x - 3}.$$

EXAMPLE 17

Divide $(21x^2 + 29x - 10)$ by $(3x + 5)$.

Solution

The polynomials are written in descending powers of the variable, and we can proceed as follows:

$$\begin{array}{r} 7x \\ 3x + 5 \overline{\smash{\big)}\, 21x^2 + 29x - 10} \\ \underline{21x^2 + 35x } \\ -6x - 10 \end{array}$$
Divide $3x$ into $21x^2$, obtaining $7x$. Multiply $(3x + 5)$ by $7x$ and subtract the product from $21x^2 + 29x$. Bring down next term, -10.

$$\begin{array}{r} 7x - 2 \\ 3x + 5 \overline{\smash{\big)}\, 21x^2 + 29x - 10} \\ \underline{21x^2 + 35x } \\ -6x - 10 \\ \underline{-6x - 10} \\ 0 \end{array}$$
Divide $3x$ into $-6x$, obtaining -2. Multiply $3x + 5$ by -2 and subtract the product from $-6x - 10$. The remainder is 0, the division is complete.

Therefore $(21x^2 + 29x - 10) \div (3x + 5) = 7x - 2$. The problem is separated into parts so that you can follow the steps involved.

EXAMPLE 18

Divide $(18y^5 + 12y^3 + 24y)$ by $(3y^2 + 4)$.

Solution

We must first write the polynomials in descending powers of the variable. Therefore, $18y^5 + 12y^3 + 24y = 18y^5 + 0y^4 + 12y^3 + 0y^2 + 24y + 0$ and $3y^2 + 4 = 3y^2 + 0y + 4$. Now we can proceed in the usual fashion.

$$\begin{array}{r} 6y^3 - 4y \\ 3y^2 + 0y + 4 \overline{) 18y^5 + 0y^4 + 12y^3 + 0y^2 + 24y + 0 } \\ \underline{18y^5 + 0y^4 + 24y^3} \\ -12y^3 + 0y^2 + 24y \\ \underline{-12y^3 + 0y^2 - 16y} \\ 40y \end{array}$$

Here we had to bring down two terms. The division is complete, since the degree of the remainder is less than the degree of the divisor.

Therefore,

$$(18y^5 + 12y^3 + 24y) \div (3y^2 + 4) = 6y^3 - 4y + \frac{40y}{3y^2 + 4}.$$

Many times when discussing polynomials such $x^3 + 3x^2 + 4x + 2$ and $y^2 + 3y + 2$, we can represent them by symbols such as P(x) and Q(y), where the letter in the parentheses designates the variable. For example, we could let

$$P(x) = x^3 + 3x^2 + 4x + 2$$

and

$$Q(y) = y^2 + 3y + 2.$$

This notation can also be used to represent values of the polynomial for specific values of the variable. The notation P(1) represents the value of the polynomial P(x) when x = 1. If $P(x) = x^3 + 3x^2 + 4x + 2$, then we determine P(1) by substituting 1 for x in the given polynomial. For example,

if $P(x) = x^3 + 3x^2 + 4x + 2$
then $P(1) = 1^3 + 3(1)^2 + 4(1) + 2$
$= 1 + 3 \cdot 1 + 4 + 2$
$= 1 + 3 + 4 + 2$
$= 10.$

Similarly, if $Q(y) = y^2 + 3y + 2$,
then $Q(-2) = (-2)^2 + 3(-2) + 2$
$= 4 - 6 + 2$
$= 0.$

Polynomials

> **EXAMPLE 19**
>
> If $P(x) = x^2 - 3x + 2$, find: a. $P(1)$; b. $P(-1)$; c. $P(-3)$.
>
> **Solution**
>
> Substituting the indicated value for the variable in the given polynomial, we have:
> a. $P(1) = 1^2 - 3(1) + 2 = 1 - 3 + 2 = 0$.
> b. $P(-1) = (-1)^2 - 3(-1) + 2 = 1 + 3 + 2 = 6$.
> c. $P(-3) = (-3)^2 - 3(-3) + 2 = 9 + 9 + 2 = 20$.

If we divide $x^2 - 3x + 2$ by $x + 3$, we obtain a quotient of $x - 6$ and a remainder of 20. $(x^2 - 3x + 2) \div (x + 3) = x - 6 + \dfrac{20}{x + 3}$. The work is shown below.

$$\begin{array}{r} x - 6 \\ x+3 \overline{\smash{)}\, x^2 - 3x + 2} \\ \underline{x^2 + 3x} \\ -6x + 2 \\ \underline{-6x - 18} \\ 20 \end{array}$$

In Example 19-c, we had $P(x) = x^2 - 3x + 2$ and $P(-3) = (-3)^2 - 3(-3) + 2 = 9 + 9 + 2 = 20$. Note that $P(-3)$ has the same value as the remainder in the division problem when we divided $(x^2 - 3x + 2)$ by $(x + 3)$. Similarly, if we divide $(x^2 - 3x + 2)$ by $(x + 1)$, we obtain a quotient of $x - 4$ and a remainder of 6: $(x^2 - 3x + 2) \div (x + 1) = x - 4 + \dfrac{6}{x + 1}$. In Example 19-b, we had $P(x) = x^2 - 3x + 2$ and $P(-1) = (-1)^2 - 3(-1) + 2 = 1 + 3 + 2 = 6$. Again, we note that $P(-1)$ has the same value as the remainder when $x^2 - 3x + 2$ is divided by $x + 1$. In general, we can say:

> If $P(x)$ is a polynomial and $P(x)$ is divided by a divisor in the form $x - b$, then the remainder is equal to $P(b)$.

This is known as the *Remainder Theorem*. The proof is beyond the scope of this text, but we have shown some examples that would lead to the discovery of this rule. Note that $P(x)$ must be a polynomial and the divisor must be of the form $x - b$. The Remainder Theorem will also apply when the divisor is of the form $x + b$. The reason is that we can rewrite $x + b$ as $x - (-b)$ and this fulfills the requirements for the theorem. In the preceding discussion, we determined the remainder when $x^2 - 3x + 2$ was divided by $x + 1$ by finding $P(-1)$ because $x + 1 = x - (-1)$, which is of the general form $x - b$.

What is the remainder when $x^2 - 3x + 2$ is divided by $x - 1$? Since we are only concerned with the remainder, we can apply the Remainder Theorem. To find the remainder when P(x) is divided by $x - b$, we find P(b). The divisor is $x - 1$, hence we find P(1), and $P(1) = 1^2 - 3(1) + 2 = 1 - 3 + 2 = 0$. The fact that the remainder is 0 indicates that the divisor is exact, that is, $x - 1$ divides $x^2 - 3x + 2$ exactly, with "0" remainder. Therefore, $x - 1$ must be a factor of $x^2 - 3x + 2$.

EXAMPLE 20

Find the remainder when $3x^3 - 2x^2 + 4x - 3$ is divided by $x - 2$.

Solution

By the Remainder Theorem, the remainder is equal to P(2).

$$P(2) = 3(2)^3 - 2(2)^2 + 4(2) - 3 = 3 \cdot 8 - 2 \cdot 4 + 8 - 3$$
$$= 24 - 8 + 8 - 3 = 21.$$

Therefore, the remainder is 21.

EXAMPLE 21

Find the remainder when $4x^2 - 3x + 5$ is divided by $x + 3$.

Solution

Recall that the theorem states that the divisor must be of the form $x - b$. The divisor is $x + 3$. The value of b is -3 because $x + 3 = x - (-3)$, which is the correct form of the divisor. Therefore, the remainder is equal to P(−3).

$$P(-3) = 4(-3)^2 - 3(-3) + 5 = 4(9) + 9 + 5 = 36 + 9 + 5 = 50.$$

Therefore, the remainder is 50.

In order to apply the Remainder Theorem, the divisor must be of the form $x - b$. Recall that if $P(b) = 0$, then the division is exact and $x - b$ is a factor of P(x).

Polynomials

EXERCISES FOR SECTION 4.6

(Answers given to exercises marked *)

For Exercises 1 through 45, perform the indicated divisions, that is, find the quotients.

*1. $10x^3 \div 5x$

2. $(-18x^2y^3) \div 6xy^2$

*3. $24x^3y^2z \div (-12xy^2)$

4. $34xy^3 \div 17y^2$

*5. $48x^5y^3z^2 \div 16x^3y^2z$

6. $16a^3b^2c \div 32ab^4$

*7. $\dfrac{-32x^3y}{8x^2y}$

8. $\dfrac{51xy^2z}{3x^2yz^2}$

*9. $\dfrac{14x^3y^5z^2}{7x^4y^3z}$

10. $\dfrac{4x^4y^3z^3 + 8x^4y^2z^2}{8x^3yz^2}$

*11. $\dfrac{21a^3b^4c^2 - 14a^2b^3c}{7a^2b^2c}$

12. $\dfrac{-18t^6u^2 + 12t^5u^3}{-6t^4u}$

*13. $(10x^3 + 5x^2 - 10) \div 5$

14. $(2x^3 + 4x^2 + 8x) \div 2x$

*15. $(16y^3 + 8y^2 + 4y) \div 4y$

16. $(36x^3y^2 - 18xy^2 - 27xy) \div 9xy$

*17. $(48x^3y^2 - 32x^2y^4 - 80x^2y^2) \div 16x^2y^2$

18. $(8a^3b^2c - 4a^4b^3c^2) \div 4abc$

*19. $(15t^3u^4 - 10t^2u^3 + 5t^2u^5) \div (-5t^2u^3)$

20. $(12a^3b^4 - 16a^4b^5) \div 4a^6b^8$

*21. $(20x^2y^3 - 16x^3y^2) \div 4x^5y^6$

22. $(x^2 - 2x - 3) \div (x + 1)$

*23. $(x^2 - 2x - 3) \div (x - 3)$

24. $(x^2 - x - 20) \div (x + 4)$

*25. $(x^2 - x - 20) \div (x - 5)$

26. $(6x^2 + x - 2) \div (3x + 2)$

*27. $(8y^2 + 10y - 3) \div (2y + 3)$

28. $(15y^2 - y - 6) \div (3y - 2)$

*29. $(6x^2 + x - 2) \div (2x - 1)$

30. $(15x^2 - x - 6) \div (5x + 3)$

*31. $(8x^2 + 10x - 3) \div (4x - 1)$

32. $(12x^2 - 5x + 3) \div (3x + 4)$

*33. $(18y^2 + 12y + 3) \div (6y - 2)$

34. $(4x^3 + 4x^2 + 5x - 4) \div (2x - 1)$

*35. $(3y^3 + 8y^2 - 5y - 6) \div (3y + 2)$

36. $(6y^3 + 5y^2 + 6y - 8) \div (3y - 2)$

*37. $(12x^3 + 13x^2 - 39x + 5) \div (4x - 5)$

38. $(y^4 + 4y^2 - 6) \div (y - 2)$

*39. $(x^5 + 3x^3 + 2x) \div (x + 1)$

40. $(y^3 + 1) \div (y + 1)$

*41. $(x^3 - 8) \div (x - 2)$ 42. $(x^4 - 16) \div (x^2 - 4)$

*43. $(4y^5 + 3y^3 + 5y) \div (2y - 3)$ 44. $(8x^4 + 2x^2 + 4) \div (4x^2 + 1)$

*45. $(8x^6 + 4x^4 + 2x^3 - x + 3) \div (x^2 + 4)$

46. If $P(x) = 3x^2 - 5x + 4$, find: a. $P(1)$; b. $P(2)$; c. $P(-3)$; d. $P(0)$.

*47. If $Q(y) = y^3 - 3y^2 + 2y + 1$, find: a. $Q(-1)$; b. $Q(2)$; c. $Q(-3)$; d. $Q(0)$.

48. Find the remainder when $x^3 + 2x^2 - 3x + 4$ is divided by $x - 1$.

*49. Find the remainder when $3x^4 + 2x^3 - x^2 + 4x - 5$ is divided by $x - 2$.

50. Find the remainder when $4y^3 - 2y^2 + 2y - 5$ is divided by $y - 3$.

*51. Find the remainder when $5x^5 + 16x^4 + 12x^2 + 4x - 8$ is divided by $x + 1$.

52. Find the remainder when $y^{16} - 8y^{12} + 4y^6 + 2$ is divided by $y + 1$.

*53. Find the remainder when $x^6 + 2x^3 + 4x + 2$ is divided by x.

54. What is the remainder when $x^4 - 3x^3 + 3x^2 - 4x + 3$ is divided by $x - 1$? What does this prove?

4.7 SYNTHETIC DIVISION

In the previous section, we examined the process of division. For example, to divide $(4x^3 + 2x^2 - 3x + 2)$ by $(x + 2)$, we can proceed as follows:

$$\begin{array}{r} 4x^2 - 6x + 9 \\ x + 2 \overline{\smash{\big)}\, 4x^3 + 2x^2 - 3x + 2} \\ \underline{4x^3 + 8x^2} \\ -6x^2 - 3x \\ \underline{-6x^2 - 12x} \\ 9x + 2 \\ \underline{9x + 18} \\ -16 \quad \text{(remainder)} \end{array}$$

Often it is necessary to divide a polynomial $P(x)$ by a divisor of the form $x - b$. There exists a shorter (and easier) method that we can use to obtain the quotient and remainder. The technique is called *Synthetic Division*.

To divide two polynomials, they must be arranged in descending powers of the variable. This is also the case for synthetic division. Since the divisor will always be of the form $x - b$, we need only concern ourselves with arranging the terms of the dividend. But the terms of the dividend are written down in order, using only the coefficients of the terms. For example, $4x^3 + 2x^2 - 3x + 2$ is written as: $4 + 2 - 3 + 2$. In other words, we detach the

Polynomials

coefficients from the terms (some texts refer to this operation as division by detached coefficients). If a term is missing in the dividend, then the detached coefficient for that term is 0. For example $4x^3 + 3x - 1$ can be expressed as $4x^3 + 0x^2 + 3x - 1$ and the detached coefficients would be $4 + 0 + 3 - 1$. Thus far, we have the coefficients for the dividend, $4 + 2 - 3 + 2$. The detached coefficients for the divisor $x + 2$ are $1 + 2$. Now we shall repeat the division process shown at the beginning of this section, using only the detached coefficients.

$$
\begin{array}{r}
4 - 6 + 9 \\
1 + 2 \overline{\smash{)}\,4 + 2 - 3 + 2} \\
(4) + 8 \\
\overline{- 6 \,\,(-3) } \\
(-6) - 12 \\
\overline{9 + (2) } \\
(9) + 18 \\
\overline{-16}
\end{array}
$$

The numbers written in parentheses are repeated, that is, they are repetitions of the numbers directly above them. Note that it will not change the problem if they are not written twice:

$$
\begin{array}{r}
4 - 6 + 9 \\
1 + 2 \overline{\smash{)}\,4 + 2 - 3 + 2} \\
+ 8 \\
\overline{- 6 } \\
- 12 \\
\overline{9 } \\
+ 18 \\
\overline{-16}
\end{array}
$$

Condensing all of this into two lines, we have

$$
\begin{array}{r}
4 - 6 + 9 \\
1 + 2 \overline{\smash{)}\,4 + 2 - 3 + 2} \\
+ 8 - 12 + 18 \\
\overline{4 - 6 + 9 - 16}
\end{array}
$$

Since whenever we subtract two numbers we are adding the additive inverse of the subtrahend, we can change the signs on the second line and use the operation of addition instead of subtraction:

$$
\begin{array}{r}
4 - 6 + 9 \\
1 + 2 \overline{\smash{)}\,4 + 2 - 3 + 2} \\
- 8 + 12 - 18 \\
\overline{4 - 6 + 9 - 16}
\end{array}
$$

Note that the numbers in the bottom line are the coefficients of the terms in the quotient and the last number (−16) is the remainder, that is, 4 − 6 + 9 − 16 represents $4x^2 - 6x + 9$, $r = -16$. We can obtain these numbers if we delete the 1 in the divisor and use the additive inverse of + 2, namely, (−2). Rearranging the problem, we have

$$\begin{array}{r} 4 + 2 - 3 + 2 \underline{|-2} \\ -8 + 12 - 18 \\ \hline 4 - 6 + 9 - 16 \end{array}$$

(The quotient is $4x^3 - 6x + 9$ with a remainder of −16.)

This final form is what we have been leading up to and it is the process of synthetic division that we must master at this point. The purpose of the preceding discussion is to illustrate how this final form came about.

This final form is called Horner's Method of Synthetic Division. In order to use this technique we must first write down the terms of the dividend (in order), using only the detached coefficients, 4 + 2 − 3 + 2. The x term of the divisor is deleted and the additive inverse of the second term is used, −2. Next, draw a line under the dividend (leaving room for a row of integers). Bring down the first term (4), and multiply it by the divisor (−2). The product, −8, is written under the +2 of the dividend. Add the integers, obtaining −6. Now repeat the process: multiply −6 by −2, obtaining +12. Write +12 under the next term of the dividend, −3. Add the integers, obtaining +9. Repeat the process, $(+9)(-2) = -18$; write −18 under the next term of the dividend, +2. Add the integers, obtaining −16. We are at the end of the line, therefore, the process terminates. The bottom line is 4 − 6 + 9 − 16, which are the coefficients of the quotient (and the remainder). Since the degree of the quotient must be one less than the degree of the dividend, we have $4x^2 - 6x + 9$, $r = -16$.

Now, let's try another example. Consider the following division problem: $(3x^3 + 4x^2 - 3x - 5) \div (x + 3)$. Using the technique of synthetic division, we would use only the coefficients of the dividend and the additive inverse of the second term in the divisor. Setting the problem up, we have

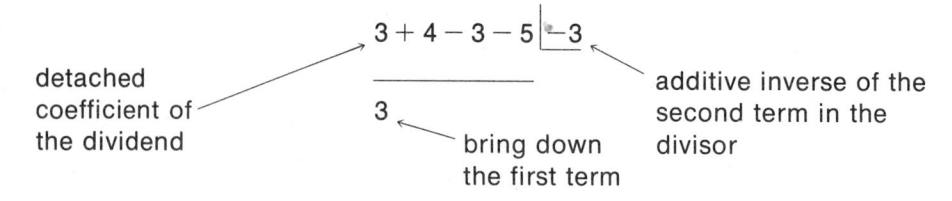

$$3 + 4 - 3 - 5 \,\underline{|-3}$$
$$\overline{3}$$

detached coefficient of the dividend · · · · bring down the first term · · · · additive inverse of the second term in the divisor

Performing the necessary computation, we have

$$\begin{array}{rl} A & 3 + 4 - 3 - 5 \,\underline{|-3} \\ B & -9 + 15 - 36 \\ C & 3 - 5 + 12 - 41 \end{array}$$

(The quotient is $3x^2 - 5x + 12$ and the remainder is −41.)

Polynomials

The three lines have been labeled A, B, and C to enable us to describe the steps involved in obtaining the quotient.
1. 3 is "brought down" from line A to line C.
2. −9, the product of 3 and −3, is written under +4 on line B.
3. −5, the sum of +4 and −9, is written on line C.
4. +15, the product of −5 and −3, is written under −3 on line B.
5. +12, the sum of −3 and +15, is written on line C.
6. −36, the product of +12 and −3, is written under −5 on line B.
7. −41, the sum of −5 and −36, is written on line C.
8. The process is terminated since we are at the end of the line.
9. Line C represents the coefficients of the quotient and the remainder. The degree of the quotient must be one less than the degree of the dividend.
10. Therefore, the quotient is $3x^2 - 5x + 12$ and the remainder is -41, or

$$(3x^3 + 4x^2 - 3x - 5) \div (x + 3) = 3x^2 - 5x + 12 + \frac{-41}{x+3}$$

To divide a polynomial P(x) by a divisor of the form x − b, using the technique of synthetic division, we must first write the detached coefficients of P(x) in order (representing a missing term by means of a 0 coefficient). Bring down the first coefficient, and multiply it by b. Add the product to the second coefficient, and multiply the resulting sum by b. Add the product of the next coefficient, etc. Repeat the process until all terms of the dividend have been exhausted. The last sum obtained is the remainder and the preceding terms represent the coefficients of the quotient (in order).

EXAMPLE 22

By synthetic division, find the quotient and remainder when $4x^3 + 6x^2 - 3$ is divided by $x - 3$.

Solution

The dividend $4x^3 + 6x^2 - 3$ has the x term missing and we can rewrite it as $4x^3 + 6x^2 + 0x - 3$. The detached coefficients are (in order) $4 + 6 + 0 - 3$ and so we have

$$\begin{array}{r} 4 + 6 + 0 - 3 \underline{|3} \\ +12 + 54 + 162 \\ \hline 4 + 18 + 54 + 159 \end{array} \leftarrow \text{indicates division by } x - 3$$

The quotient is $4x^2 + 18x + 54$ and the remainder is $+159$.

EXAMPLE 23

By synthetic division, find the quotient and remainder when $5x^4 - 6x^2 - 3$ is divided by $x + 2$.

Solution

As in Example 22, the dividend does not contain all the terms (in order). The x^3 term and the x term are missing. Therefore, $5x^4 - 6x^2 - 3$ is rewritten as $5x^4 + 0x^3 - 6x^2 + 0x - 3$. The divisor is $x + 2$ and it should be of the form $x - b$. Therefore, $x + 2$ is rewritten as $x - (-2)$. We can now proceed as follows:

$$\begin{array}{r} 5 + 0 - 6 + 0 - 3 \underline{|-2} \\ \underline{-10 + 20 - 28 + 56} \\ 5 - 10 + 14 - 28 + 53 \end{array}$$ ←indicates division by $x - (-2)$, or, $x + 2$.

The quotient is $5x^3 - 10x^2 + 14x - 28$ and the remainder is $+53$.

We can always check our work in a division problem because the dividend must be equal to the product of the divisor and quotient plus the remainder. We can also make a quick check of our work by means of the Remainder Theorem. In other words, if the remainders check, we can be reasonably sure of our answer. This does not guarantee that the quotient is correct — it only indicates that it is probably correct. For instance, in the second example, when $5x^4 - 6x^2 - 3$ was divided by $x + 2$, we obtained a quotient of $5x^3 - 10x^2 + 14x - 28$ with a remainder of $+53$. By the Remainder Theorem, the remainder is equal to $P(-2)$. $P(x) = 5x^4 - 6x^2 - 3$ and $P(-2) = 5(-2)^4 - 6(-2)^2 - 3 = 5(16) - 6(4) - 3 = 80 - 24 - 3 = 80 - 27 = 53$. The remainders check, hence we can be reasonably sure of the accuracy of the answer. To be absolutely sure of the correctness of the solution, we should multiply the divisor by the quotient, add the remainder, and determine whether the result is equal to the dividend.

EXERCISES FOR SECTION 4.7
(Answers given to exercises marked *)

By means of synthetic division, find the quotient and remainder for each of the following:

*1. $(x^2 - 2x - 3) \div (x - 3)$
2. $(x^2 - 2x - 3) \div (x + 1)$
*3. $(6x^3 + 2x^2 + 3x - 2) \div (x - 1)$
4. $(5x^3 + 4x^2 + 2x + 3) \div (x + 2)$
*5. $(x^3 - 3x^2 + 5x - 4) \div (x + 1)$
6. $(4x^3 + 3x^2 - 2x + 1) \div (x - 1)$

Polynomials   233

*7. $(x^3 - 8) \div (x - 2)$ 8. $(x^4 - 16) \div (x + 2)$

*9. $(3x^3 + x - 5) \div (x - 1)$ 10. $(4x^4 + 2x^2 + 5) \div (x - 2)$

*11. $(6x^4 + 4x^2 + 3x) \div (x + 1)$ 12. $(x^5 - x^3 - x) \div (x - 2)$

*13. $(x^5 + x^3 + x) \div (x + 2)$ 14. $(5x^6 - 4x^3 - 3x - 1) \div (x + 3)$

*15. $(x^6 - x^4 + x^2) \div (x + 1)$ 16. $(x^6 - x^3 - x) \div (x - 1)$

*17. $(x^6 + 1) \div (x + 1)$ 18. $(x^6 - 8) \div (x - 2)$

*19. $(x^6 - 27) \div (x - 3)$ 20. $(x^6 + 3x^3 + x) \div (x - 2)$

*21. If $P(x) = x^3 - x^2 - 4x + 4$, find $P(1)$, $P(2)$, $P(-2)$. What are the factors of $P(x)$?

22. Which, if any, of the following are factors of $x^4 - 5x^2 + 4$? $(x+2)$, $(x-2)$, $(x+1)$, $(x-1)$.

4.8 SUMMARY

In this chapter, we examined the topic of polynomials and the four basic arithmetic operations involving polynomials, addition, subtraction, multiplication, and division. A polynomial can consist of any number of terms, provided the variable does not occur in the denominator of a fraction. The expression

$$a_0x^n + a_1x^{n-1} + \ldots + a_{n-1}x + a_n$$

is the general form of a polynomial in one variable where x can represent any variable, n is a non-negative integer, and $a_0, a_1, \ldots, a_n$ represent numerical coefficients. The word polynomial can be used to describe a monomial, binomial, or trinomial. The degree of a polynomial is the same as the degree of that term in the polynomial that has the highest degree. For example, $x^3 + 2x^2 + 1$ is of the third degree, while $x^4 + x^3y + x^2y^3$ is of the fifth degree.

We can simplify polynomials by combining like terms, that is, finding the sum or the difference of like terms. Unlike terms cannot be combined—their sum or difference must be indicated. If a polynomial does not contain any like terms, then it is in simplest form and cannot be simplified. We can find the sum of two or more polynomials by combining like terms. To subtract two polynomials, we apply the rule that $x - y = x + (-y)$. That is, subtraction is the same as adding the additive inverse (opposite) of the subtrahend.

Multiplication of polynomials can occur in different forms. To multiply a monomial by a monomial, apply the rule for multiplying exponentials, $x^m \cdot x^n = x^{m+n}$. For example, $(3x^2y^3)(2xy^2) = 6x^3y^5$. To find the product of a monomial and a polynomial, we use the above rule and the distributive property. For example, $3x^2(4x^3 - 2x) = (3x^2)(4x^3) + (3x^2)(-2x) = 12x^5 - 6x^3$.

We can also use the distributive property to find the product of two binomials, $(a + b)(c + d) = (a + b)c + (a + b)d = ac + bc + ad + bd$. The vertical technique of multiplication can also be used to find the product of any two polynomials. Using this technique, we write one polynomial beneath the other and multiply each term of one polynomial by each term of the other polynomial (listing like partial products in the same column), and then combine like terms.

In Section 4.4, we considered some special products because they occur so frequently. To find the product of two binomials, $(a + b)(c + d)$, we used the FOIL technique, the product of the *first* terms, *outer* terms, *inner* terms, *last* terms, and determined the sum of these products to obtain the final result. Another special product is the square of a binomial. The square of a binomial is the sum of the squares of each term in the binomial, plus twice the product of the two terms, $(a + b)^2 = a^2 + 2ab + b^2$. Whenever we find the product of two binomials of the form $(a + b)(a - b)$, the result will always be of the form $a^2 - b^2$, the difference of two squares. The product of a pair of conjugates is the square of the first monomial minus the square of the second monomial.

One type of division of polynomials that we considered is where the divisor is a monomial, for example $(24x^3 + 18x^2 + 12x)$ and $6x$. To divide a polynomial by a monomial, we must divide each term of the polynomial by the monomial. The algebraic sum of the resulting quotients is the quotient of the polynomial divided by the monomial. To divide two polynomials, such as $(6x^2 - 5x - 6)$ and $(2x - 3)$, we must first have the polynomials arranged in descending powers of the variable. We can then set the problem up as we would set up a division problem in arithmetic and perform the long division (see Section 4.6).

If we are concerned only with the remainder in a division problem (to determine whether it is 0 or not), we can use the Remainder Theorem. It states: If $P(x)$ is a polynomial and $P(x)$ is divided by a divisor of the form $x - b$, then the remainder is equal to $P(b)$.

It may be necessary to divide a polynomial $P(x)$ by a divisor of the form $x - b$. The technique of synthetic division can be used to obtain the quotient and remainder. This technique is formally known as Horner's Method of Synthetic Division. To use this technique to find $P(x) \div (x - b)$, we must first write down the terms of the dividend (in order) using only the detached coefficients, representing a missing term by means of a 0 coefficient. Bring down the first coefficient, and multiply it by b. Add the product to the second coefficient. Multiply the resulting sum by b, and add the product to the next coefficient. Repeat the process until all terms of the dividend have been exhausted. The last sum obtained is the remainder and the preceding terms represent the coefficients of the quotient (in order). The degree of the quotient must be one less than the degree of the dividend. Regardless of the division technique used, if $P(x)$ is divided by $(x - b)$ and the remainder is 0, then $x - b$ is a factor of $P(x)$. We can also determine if $x - b$ is a factor of $P(x)$ in another manner. Using the Remainder Theorem, we can say that $x - b$ divides $P(x)$ exactly if $P(b) = 0$.

Polynomials

REVIEW EXERCISES FOR CHAPTER 4
(Answers in Appendix B)

1. Simplify the given polynomials, that is, combine like terms.

   a. $5x^2 + 3x - 4x^3 - 8x + 2x^3 + 5x^2$ b. $8xy + 2x - 4x^2 + 2xy - 8x$
   c. $3y^3 + 2y^2 - y^3 - y^2 + y$

2. Add the given polynomials.

   a. $3x^2 + 2x - 7$ and $4x^2 + 5x - 6$ b. $8x^2 + 3x - 2$ and $x^2 - 4x + 3$
   c. $4x^2 + x + 2$ and $y^2 - y + 3$

3. Subtract the first polynomial from the second polynomial.

   a. $2x + 3y + 5$ from $4x - y - 6$ b. $x^2 + 3x + 4$ from $2x^2 + 2x + 5$
   c. $x^2 + 2x + 1$ from 0

4. Perform the indicated operations and simplify by combining like terms.

   a. $(4x^2 + 3x - 2) - (x^2 - 3x - 1)$ b. $y^2 + 3y - 3[2 - (y + 1)]$
   c. $x - 3[2x - 2(x - 1) + 1]$

5. Perform the indicated operations and simplify, if possible.

   a. $-2x(4x^2 - 3x)$ b. $2x^3y^2(3xy^2 - 4x^2y)$ c. $(x + 3)(x - 2)$
   d. $(4x^2 + 3)(2y + 3)$ e. $(2y - 3)(y^2 + 3y - 2)$
   f. $(2x + 3y + 4)(3x - 2y - 5)$

6. Perform the indicated operations, and simplify. All products should be determined by inspection.

   a. $(x - 4)(x + 3)$ b. $(y - 10)(y + 5)$ c. $(2x + 3)(3x - 3)$
   d. $(2x + 3y)(x - 2y)$ e. $(x^2 + 2y)(x^2 - 3y)$ f. $(2y^2 + 3x)(3y^2 - 2x)$
   g. $(x + y)^2$ h. $(x - 3y)^2$ i. $(2x + 3y)^2$
   j. $(x^2 + 2y)^2$ k. $(5x^2 - 3y^3)^2$ l. $(x + y)(x - y)$
   m. $(x^2 + 2y)(x^2 - 2y)$ n. $(3x^2y + 1)(3x^2y - 1)$ o. $(x + y)^3$
   p. $(x - y)^3$ q. $(2x + 3y)^3$ r. $(x - 2)(x^2 + 2x + 4)$
   s. $(x + 2y)(x^2 - 2xy + 4y^2)$ t. $(x^2 + y)(x^4 - x^2y + y^2)$
   u. $(3x - 1)(9x^2 + 3x - 1)$

7. Replace the question mark by a term that will make the trinomial a perfect square.

   a. $y^2 + ? + x^2$ b. $? - 12xy + 9y^2$ c. $16x^4 + 24x^2y + ?$

8. Perform the indicated divisions, that is, find the quotients.

   a. $(16x^4 + 12x^3 - 4x^2) \div 2x^2$ b. $(4x^4y^3z^3 - 8x^3y^2z^2) \div 4x^2y^2z$
   c. $(36x^3y^2 + 18xy^2 - 27xy) \div 9xy$ d. $(x^2 - 2x - 3) \div (x + 1)$

e. $(6x^2 + x - 2) \div (3x + 2)$ f. $(15x^2 - x - 6) \div (5x + 3)$
g. $(8x^4 + 2x^2 + 4) \div (4x^2 + 1)$ h. $(x^4 - 16) \div (x^2 - 4)$
i. $(y^3 + 1) \div (y + 1)$

9. Find the remainder when $4y^3 - 2y^2 + 2y - 5$ is divided by:

   a. $y - 3$ b. $y + 1$

10. By means of synthetic division, find the quotient and remainder for each of the following:

    a. $(x^2 - 2x - 3) \div (x - 1)$ b. $(2x^2 + 3x + 4) \div (x + 1)$
    c. $(4x^3 + 3x^2 + 4x + 1) \div (x - 2)$ d. $(4x^4 + 4x^2 + 4) \div (x + 2)$
    e. $(5x^5 + 3x^3 + 1) \div (x - 3)$ f. $(x^6 - 8) \div (x + 2)$

11. Which, if any, of the following: $(x - 2)$, $(x + 2)$, $(x + 3)$, $(x + 5)$, are factors of $x^4 - 2x^3 - 19x^2 + 8x - 60$?

CUMULATIVE REVIEW

(Answers in Appendix B)

1. Given that $A = \{0,1,2,3,4,5\}$, $B = \{4,5,6,7,8,9\}$, and $C = \{3,4,5,6,7\}$, find:

   a. $A \cap B$.   b. $B \cap C$.   c. $A \cup C$.
   d. $(A \cap B) \cup C$.   e. $(B \cup C) \cap A$.   f. $(A \cup B) \cap (B \cup C)$.
   g. $(A \cup B) \cup (A \cap C)$.   h. $(B \cup C) \cap \emptyset$.   i. $(A \cup \emptyset) \cap (C \cap \emptyset)$.

2. Determine whether each of the following is true or false.

   a. The whole numbers are closed under the operation of subtraction.
   b. The integers are commutative under the operation of division.
   c. Every non-terminating decimal is an irrational number.
   d. The set of irrational numbers is closed under the operation of multiplication.
   e. The identity element for the set of rational numbers under the operation of multiplication is zero.
   f. The intersection of the set of rational numbers and the set of irrational numbers is not empty.
   g. The set of real numbers is formed by the union of set of rational numbers and the set of irrational numbers.

3. Arrange the following in order from smallest to largest.

   $2.3\overline{12}$, $2.3121121112\ldots$, 2.312, $2.3122122212222\ldots$, $2.\overline{312}$

Polynomials 237

4. Solve the following equations and check the solutions.

   a. $2(3x - 4) = 3(x + 2)$.

   b. $\frac{2}{3}(3x + 6) = \frac{3}{4}(4x - 8)$.

   c. $\frac{2}{5}y + \frac{1}{2} = -\frac{1}{3} - y$.

   d. $4(x + 1) - 3(x - 1) = 25$.

   e. $\frac{1}{2}(2 - x) = \frac{1}{3}\left(\frac{x}{2} - 1\right)$.

   f. $4.9 - 0.2x = 0.8x - 1.1$.

5. A piece of wire, whose length is 100 centimetres, is bent to form an isosceles triangle (a triangle with two sides of equal length). The third side is five centimetres shorter than one of the two equal sides. Find the lengths of the sides of the triangle.

6. The sum of three consecutive odd integers is 39. Find the three integers.

7. At 12 o'clock noon, a Trailways bus leaves its terminal heading west at 40 mph. Two hours later, a second bus leaves the same terminal heading west at 60 mph. At what time did the second bus overtake the first bus?

8. Solve each of the given inequalities and graph its solution set.

   a. $4(x - 1) - 5(x - 2) < 1$.

   b. $3(x - 2) \geq 2x + 4$.

   c. $|-1 - x| \leq 2$.

9. Simplify each of the following expressions, perform the indicated operations, and write equivalent expressions containing only positive exponents.

   a. $x^{1/2} \cdot x^{1/3}$

   b. $x^{3/2} \div x^{2/3}$

   c. $(4y^{2/3})^{1/2}$

   d. $(x^4y^3)(x^{-5}y^{-4})$

   e. $\dfrac{x^{-2}y^{-3}}{x^3y^{-2}}$

   f. $\dfrac{x^4y^{-2}z}{x^3y^5z^4}$

10. Use scientific notation to evaluate each of the given expressions. Answers should be left in scientific notation.

    a. $\dfrac{24{,}000 \times 140{,}000}{0.0016 \times 210{,}000}$

    b. $\dfrac{0.024 \times 0.00036}{0.0015 \times 0.00012}$

11. Perform the indicated operations and simplify, if possible.

    a. $3\sqrt{20} - 6\sqrt{45}$

    b. $\sqrt[4]{4} - 3\sqrt[6]{8} + 2\sqrt{2}$

    c. $(5\sqrt{x} + 3\sqrt{2})(4\sqrt{2} - 2\sqrt{x})$

    d. $\dfrac{2}{1 + \sqrt{3}}$

    e. $\dfrac{2}{\sqrt[3]{2}}$

    f. $\dfrac{\sqrt{2} + \sqrt{3}}{\sqrt{3} - \sqrt{2}}$

12. Find the solution set for the given equations and check.

    a. $\sqrt{y+2} = 1$.   b. $\sqrt{2x+1} - 5 = 0$.   c. $\sqrt{x+1} + 1 = 0$.

13. Perform the indicated operations and simplify, if possible.

    a. $x^2 + 2x - 3[2 - (x+1) + x]$
    b. $(4x^2 + 3x - 2) - (5x^2 - 2x + 3)$
    c. $(x + 3y)(2x + 5y)$
    d. $(2x - 3)(x^2 + 5x - 3)$
    e. $(x - 2y)^2$
    f. $(x + y)^3$
    g. $(x + y - 2)(x + y + 3)$
    h. $(2x^3 + 3x - 4) \div (x - 2)$
    i. $(5x^5 + 3x^3 + 2x^2 + 1) \div (x + 1)$

chapter 5

Factoring

As a result of this chapter, the student will be able to:

1. Express a given polynomial as a product of its greatest common monomial factor and a simpler polynomial, e.g., $5x^2 + 20x = 5x(x + 4)$;
2. Factor a polynomial containing four (or more) terms by the grouping method, e.g., $ax + ay + bx + by = a(x + y) + b(x + y) = (a + b)(x + y)$;
3. Express the difference of two squares as the product of the sum and difference of their square roots, e.g., $x^2 - y^2 = (x + y)(x - y)$;
4. Recognize a perfect square trinomial, and express it as the square of a binomial, e.g., $x^2 + 2xy + y^2 = (x + y)^2$;
5. Express a trinomial of the general form $x^2 + bx + c$ as the product of two binomial factors (if possible), using the grouping technique, e.g., $x^2 + 5x + 6 = x^2 + 2x + 3x + 6 = x(x + 2) + 3(x + 2) = (x + 3)(x + 2)$;
6. Express a trinomial of the general form $ax^2 + bx + c$ as the product of two binomial factors (if possible), using the grouping technique, e.g., $6x^2 - x - 5 = 6x^2 - 6x + 5x - 5 = 6x(x - 1) + 5(x - 1) = (6x + 5)(x - 1)$;
7. Express the sum of two cubes as a product of factors, e.g., $x^3 + y^3 = (x + y)(x^2 - xy + y^2)$;
8. Express the difference of two cubes as a product of factors, e.g., $x^3 - y^3 = (x - y)(x^2 + xy + y^2)$.

5.1 INTRODUCTION

In the previous chapter, we examined the four basic operations of addition, subtraction, multiplication, and division with polynomials. In this chapter, we will further examine the operation of multiplication. We have discussed methods for finding the product of any two polynomials and now we will learn to reverse the process. We want to be able to take a given polynomial and express it as a product of two or more simpler polynomials. In other words, we want to be able to *factor* a given polynomial. To factor a polynomial means to find two or more polynomials whose product is the given polynomial. We say that a polynomial is factored completely if its factored form contains polynomials that cannot be factored further. An exception to this statement is the factored form of a monomial. Normally we would not factor an expression such as $6x^2y^3$, since its prime factors, $2 \cdot 3 \cdot x \cdot x \cdot y \cdot y \cdot y$ are fairly obvious and are not usually written out. Most of the time we will consider polynomials with integral coefficients. A polynomial with integral coefficients is completely factored if it is written as a product of polynomials with integral coefficients that cannot be factored further (except into monomial factors).

Recall that to multiply a monomial by a polynomial of more than one

Factoring

term, we used the distributive property of multiplication over addition, $a(b + c) = ab + ac$. For example,

$$2(x + 5) = 2x + 10.$$
$$5x(x + y) = 5x^2 + 5xy.$$
$$3x^2y(2x^2y - 3) = 6x^4y^2 - 9x^2y.$$

The distributive property also shows us how to factor a polynomial when the terms have a common factor. It can be used to transform products to sums, and also to transform sums to products. By means of the symmetric property, we can rewrite each of the given examples,

$2(x + 5) = 2x + 10$ as $2x + 10 = 2(x + 5).$

$5x(x + y) = 5x^2 + 5xy$ as $5x^2 + 5xy = 5x(x + y).$

$3x^2y(2x^2y - 3) = 6x^4y^2 - 9x^2y$ as $6x^4y^2 - 9x^2y = 3x^2y(2x^2y - 3).$

Factoring a polynomial is the reverse of multiplying the factors to obtain the given polynomial. When we factor a polynomial (product), we are separating it into its factors.

We can apply the distributive property directly when a polynomial contains a factor that is common to each term, that is, when each term contains the same monomial factor. For example,

$$2x + 10 = 2x + 2 \cdot 5 = 2(x + 5).$$

Note that we can check our factoring by reversing the process, and performing the indicated multiplication: $2(x + 5) = 2x + 10$.

Consider the polynomial, $14x^2 + 21x$. If we cannot ascertain the common factor, we should write the factors of each term and then determine by inspection the common monomial factor.

$$14x^2 = 2 \cdot 7 \cdot x \cdot x \quad \text{and} \quad 21x = 3 \cdot 7 \cdot x.$$

The factors which are common to each term are 7 and x, and the common monomial factor is $7 \cdot x = 7x$. Now, to determine the other factor, we can divide the given polynomial by the factor 7x, and the quotient is the other factor:

$$\frac{14x^2 + 21x}{7x} = 2x + 3 \quad \text{and} \quad 14x^2 + 21x = 7x(2x + 3).$$

We can also remove a common monomial factor in a slightly different manner. After determining the common monomial factor, we can rewrite the polynomial with the indicated common factor. For example,

$$14x^2 + 21x = 7x \cdot 2x + 7x \cdot 3.$$

Now we can apply the distributive property.

$$14x^2 + 21x = 7x \cdot 2x + 7x \cdot 3 = 7x(2x + 3).$$

common factor → 7x

sum of the remaining factors → (2x + 3)

To factor $30x^3y^2 + 75x^2y^3$, we should first write the factors of each term and then determine by inspection the common monomial factor.

$$30x^3y^2 = 2 \cdot 3 \cdot 5 \cdot x \cdot x \cdot x \cdot y \cdot y \quad \text{and}$$
$$75x^2y^3 = 3 \cdot 5 \cdot 5 \cdot x \cdot x \cdot y \cdot y \cdot y.$$

The factors that are common to each term are 3, 5, x, x, y, y, and the common monomial factor is $3 \cdot 5 \cdot x \cdot x \cdot y \cdot y = 15x^2y^2$. To determine the terms of the other factor, divide the given polynomial by $15x^2y^2$. The quotient is the other factor.

$$\frac{30x^3y^2 + 75x^2y^3}{15x^2y^2} = 2x + 5y \quad \text{and} \quad 30x^3y^2 + 75x^2y^3 = 15x^2y^2(2x + 5y).$$

We also can factor the given polynomial by rewriting the given polynomial with the indicated common factor. The other factor is the sum of the remaining factors, after removing the common factor:

$$30x^3y^2 + 75x^2y^3 = 15x^2y^2 \cdot 2x + 15x^2y^2 \cdot 5y = 15x^2y^2(2x + 5y).$$

Regardless of which technique we use, we can always check our work by reversing the process and performing the indicated multiplication: $15x^2y^2(2x + 5y) = 30x^3y^2 + 75x^2y^3$.

EXAMPLE 1

Factor each of the following completely:

a. $6x^2 + 9x$ b. $16x^3y + 8x^2y$ c. $3x^2 + 9xy - 6x$

Solution

To factor the given polynomial, we first write the factors of each term and then determine by inspection the common monomial factor. Next, divide the given polynomial by the common factor. The quotient is the other factor and the indicated product of the two factors is the desired result.

a. $6x^2 + 9x = 2 \cdot 3 \cdot x \cdot x + 3 \cdot 3 \cdot x$. The common monomial

Factoring

factor is 3x. We find $\frac{6x^2 + 9x}{3x} = 2x + 3$ and therefore $6x^2 + 9x = 3x(2x + 3)$.

b. $16x^3y + 8x^2y = 2 \cdot 2 \cdot 2 \cdot 2 \cdot x \cdot x \cdot x \cdot y + 2 \cdot 2 \cdot 2 \cdot x \cdot x \cdot y$. The common monomial factor is $8x^2y$. We find $\frac{16x^3y + 8x^2y}{8x^2y} = 2x + 1$, and therefore $16x^3y + 8x^2y = 8x^2y(2x + 1)$.

c. $3x^2 + 9xy - 6x = 3 \cdot x \cdot x + 3 \cdot 3 \cdot x \cdot y - 2 \cdot 3 \cdot x$. The common monomial factor is 3x. We find $\frac{3x^2 + 9xy - 6x}{3x} = x + 3y - 2$, and therefore $3x^2 + 9xy - 6x = 3x(x + 3y - 2)$.

Alternate Solution

After determining the common monomial factor, we can factor the given polynomial by rewriting it with the indicated common factor. The other factor is the sum of the remaining factors, after removing the common factor.

a. $6x^2 + 9x = 3x \cdot 2x + 3x \cdot 3 = 3x(2x + 3)$.
b. $16x^3y + 8x^2y = 8x^2y \cdot 2x + 8x^2y \cdot 1 = 8x^2y(2x + 1)$.
c. $3x^2 + 9xy - 6x = 3x \cdot x + 3x \cdot 3y - 3x \cdot 2 = 3x(x + 3y - 2)$.

Consider the polynomial $-3x^2 + 6x$. If we write the factors of each term, we have $-3x^2 + 6x = -1 \cdot 3 \cdot x \cdot x + 2 \cdot 3 \cdot x$, and the common factor is 3x. Therefore, $-3x^2 + 6x = 3x(-x + 2)$. But note that we can also write the factors of each term in the following manner: $-3x^2 + 6x = -1 \cdot 3 \cdot x \cdot x + (-1)(-1)2 \cdot 3 \cdot x$. Now the common monomial factor is $-3x$ and $-3x^2 + 6x = -3x(x - 2)$. Since $3x(-x + 2) = -3x^2 + 6x$ and $-3x(x - 2) = -3x^2 + 6x$, both factored forms are equally correct. From this example, we can see that -1 is a common factor of any polynomial. Normally, -1 is factored from a polynomial only when it is desirous to have the first term of the other factor positive.

EXAMPLE 2

Factor completely: $-5x^3 + 10x^2 - 5x$.

Solution

Writing the factors of each term, we have $(-1)5 \cdot x \cdot x \cdot x + 2 \cdot 5 \cdot x \cdot x + (-1)5 \cdot x$. Using this form, the common monomial factor is 5x. Therefore, $-5x^3 + 10x^2 - 5x = 5x(-x^2) + 5x \cdot 2x + 5x(-1) = 5x(-x^2 + 2x - 1)$.

Alternate Solution

Since -1 is a common factor of any polynomial, we can also write the factors of each term as $-1 \cdot 5 \cdot x \cdot x \cdot x + (-1)(-1) \cdot 2 \cdot 5 \cdot x \cdot x + (-1) \cdot 5 \cdot x$. Now the common monomial factor is $-5x$. Therefore, $-5x^3 + 10x^2 - x = (-5x)x^2 + (-5x)(-2x) + (-5x)1 = -5x(x^2 - 2x + 1)$.

It should be noted that both forms of the solutions in Example 2 are equally correct. Usually, we will use the form shown in the first solution, where the common factor removed is positive. But sometimes it may be advantageous to use the other form, where the common factor removed is negative.

Many times an expression will contain a common factor that is a binomial. For example, in the expression

$$(x + y)x + (x + y)5,$$

the binomial $(x + y)$ is a common factor. By means of the distributive property, $ab + ac = a(b + c)$, we can express $(x + y)x + (x + y)5$ in factored form,

$$(x + y)x + (x + y)5 = (x + y)(x + 5).$$

Using the distributive property, $(x + y)$ corresponds to a, x corresponds to b and c corresponds to 5. We can also factor the given expression by removing the indicated common factor, and the other factor is the sum of the remaining factors. Note that by means of the commutative property we can restate the distributive property. Instead of $ab + ac = a(b + c)$, we can also have $ba + ca = (b + c)a$. Therefore, we can factor $x(x + y) + 5(x + y)$ into the form $(x + 5)(x + y)$. Whether the common factor occurs first or last in the set of factors, we can still remove it, and the other factor is the sum of the remaining factors.

EXAMPLE 3

Factor each of the following completely:

a. $(x + 2)y + (x + 2)3$ b. $x(y - 1) + 2(y - 1)$
c. $(x - 1)y - (x - 1)z$ d. $(x + y)x + (x + y)y + (x + y)z$

Solution

In each case, the common factor is a binomial. We can remove the

Factoring

> common factor and the other factor is the sum of the remaining factors.
> a. $(x+2)y + (x+2)3 = (x+2)(y+3)$.
> b. $x(y-1) + 2(y-1) = (x+2)(y-1)$.
> c. $(x-1)y - (x-1)z = (x-1)(y-z)$.
> d. $(x+y)x + (x+y)y + (x+y)z = (x+y)(x+y+z)$.

Many times it appears that an expression cannot be factored, particularly those expressions that contain four terms. For example, it appears that the expression $ax + ay + bx + by$ does not contain a common factor. But, if we examine the first and second terms separately, and the third and fourth terms separately, we see that these pairs of terms contain a common factor. If we remove the common factor from each pair of terms, we have

$$ax + ay + bx + by = a(x+y) + b(x+y).$$

Now we can use the procedure for removing a common binomial factor from an expression (see Example 3). Treating it as a single quantity, we have

$$ax + ay + bx + by = a(x+y) + b(x+y) = (a+b)(x+y).$$

In order to factor an expression containing four (or more) terms, we first group those terms that have common monomial factors in order to remove them. Next, we remove the common binomial factor. Sometimes we can group the terms of a given polynomial in more than one way. For example, instead of having $ax + ay + bx + by$, we can apply the commutative property and rewrite the given polynomial as $ax + bx + ay + by$. Removing the common monomial factor from the two groups of terms, we have $(a+b)x + (a+b)y$ and this can be expressed as $(a+b)(x+y)$. Reviewing these steps, we have

$$\begin{aligned}ax + ay + bx + by &= ax + bx + ay + by \\ &= (a+b)x + (a+b)y \\ &= (a+b)(x+y).\end{aligned}$$

Sometimes it is not so obvious which terms should be grouped so that each term has a common factor. But remember that we want a common factor for a group of terms, not necessarily a common factor for all the terms.

EXAMPLE 4

Factor each of the following completely:

a. $y^2 + ay + by + ab$
b. $x^2 + 4x + xy + 4y$
c. $x^3 + 2x^2 - 2x - 4$
d. $ax + bz + ay + bx + az + by$

Solution

In each case, there is no common factor for all the terms. Therefore, we look for a common factor for a group of terms. If necessary, group those terms that have a common factor.

a. $y^2 + ay + by + ab = y(y + a) + b(y + a) = (y + b)(y + a)$.
b. $x^2 + 4x + xy + 4y = x(x + 4) + y(x + 4) = (x + y)(x + 4)$.
c. $x^3 + 2x^2 - 2x - 4 = x^2(x + 2) + (-2)(x + 2) = (x^2 - 2)(x + 2)$.
d. For this problem, we must first group those terms that have a common factor. Therefore,

$$ax + bz + ay + bx + az + by = ax + bx + ay + by + az + bz$$
$$= (a + b)x + (a + b)y + (a + b)z$$
$$= (a + b)(x + y + z).$$

Regardless of which technique we use to factor a polynomial, we can always check our work by reversing the process and performing the indicated multiplication. The expression "factor completely" means to factor so that the coefficients of the factorization will be integers.

EXERCISES FOR SECTION 5.1

(Answers given to exercises marked *)

Factor each of the following completely:

*1. $2x + 10$
2. $3x + 6$
*3. $5x + 10$
4. $8y + 4$
*5. $6y + 3$
6. $4z + 2$
*7. $2x - 6$
8. $3y - 3$
*9. $8y - 4$
10. $x^2 + 5x$
*11. $y^2 + 3y$
12. $2x^2 + 3x$
*13. $3y^2 + 5y$
14. $4z^2 - 2z$
*15. $2x^2 + 6x$
16. $3y^3 - 6y^2$
*17. $xy^2 + xy$
18. $-4xy + 8x^2y$
*19. $-10xy + 5x^2y$
20. $2x + 4y + 6z$
*21. $4y^3 + 8y^2 + 2y$
22. $2xy + 4x^2y + 8xy^2$
*23. $4xyz + 8xz^2 - 2yz$
24. $x^2y^2 + 4x^3y^2 - 2x^2y^3$
*25. $-3x^2y^2 + 6x^4y^2 - 9x^2y^4$
26. $(x + 2)z + (x + 2)4$
*27. $x(y + 2) + 5(y + 2)$
28. $(x - y)x + (x - y)z$
*29. $(x + 3)y - (x + 3)3$

Factoring

30. $x(2x + 1) - 2(2x + 1)$
32. $ax + by + ay + bx$
34. $2ax + by + 4ax + 2by$
36. $st + uv + sv + ut$
38. $x^2 + ax + bx + ab$
40. $y^3 + 2y^2 - 4y - 8$
42. $4x^2 + 16x + 12xy + 48y$
44. $ax + bx - ay - by + a + b$
*45. $3ax - 6ay + 2bx - 4by - 2cx + 4cy$

*31. $(3x - 2)y - (3x - 2)2$
*33. $ax - by + ay - bx$
*35. $4ay + 2by + 2bx + 4ax$
*37. $uv + sv - st - ut$
*39. $y^2 + 4y + ay + 4a$
*41. $y^2 + 8y + xy + 8x$
*43. $2x^3 + 4x^2 - 4x - 8$

5.2 THE DIFFERENCE OF TWO SQUARES

In the previous chapter, we considered certain special products. One that we considered occurred when we multiplied two binomials together so that one binomial was the conjugate of the other, in other words, the product of the sum and difference of two monomials: $(a - b)(a + b)$. We can use different techniques to obtain this product, but here we shall apply the distributive property. We have

$$(a - b)(a + b) = (a - b)a + (a - b)b$$
$$= a^2 - ba + ab - b^2$$
$$= a^2 - b^2.$$

This special product is called the *difference of two squares,* and we recognize the special product as $(a - b)(a + b) = a^2 - b^2$. By means of the symmetric property, we can rewrite this as

$$\boxed{a^2 - b^2 = (a - b)(a + b) \text{ or } a^2 - b^2 = (a + b)(a - b).}$$

Therefore, we can factor the difference of two squares into the product of the *sum and difference of their square roots.* For example, to factor $x^2 - 4$, we find the principal square root of each square, or, $\sqrt{x^2} = x$ and $\sqrt{4} = 2$. The binomial factors are the sum and difference of the two square roots, $(x + 2)$ and $(x - 2)$. Therefore,

$$x^2 - 4 = (x + 2)(x - 2).$$

sum and difference of the square roots

To use this technique of factoring, we must start with the difference of two squares. The following expressions are not factorable because they are not the difference of two perfect squares: $x^3 - 4, y^5 - 9, 4x^2 - 5, x^2 + 4$.

Sometimes the difference of two squares is not easily recognized. We should also look for common monomial factors. For example, $8y^2 - 2$ does not appear to be the difference of two squares. But each term contains a common factor of 2. Therefore,

$$8y^2 - 2 = 2(4y^2 - 1).$$

The second factor is the difference of two squares. To factor the difference of two squares, find the square root of each square, and the binomial factors are the sum and difference of the two square roots. Hence, we have $\sqrt{4y^2} = 2y$ and $\sqrt{1} = 1$, and the binomial factors are $(2y + 1)$ and $(2y - 1)$. Summarizing the steps, we have

$$8y^2 - 2 = 2(4y^2 - 1) = 2(2y + 1)(2y - 1).$$

EXAMPLE 5

Factor $9x^2 - 4y^2$ completely.

Solution

Find the square root of each square. $\sqrt{9x^2} = 3x$, $\sqrt{4y^2} = 2y$, therefore, $9x^2 - 4y^2 = (3x + 2y)(3x - 2y)$.

Carefully examine monomials that contain variables raised to some power. For example, x^6 is a perfect square because $x^6 = x^3 \cdot x^3$. Similarly, x^4 and y^{10} are perfect squares, since $x^4 = x^2 \cdot x^2$ and $y^{10} = y^5 \cdot y^5$.

EXAMPLE 6

Factor $3x^6 - 27$ completely.

Solution

$3x^6 - 27$ does not appear to be the difference of two squares. But the terms have a common factor of 3. Hence, $3x^6 - 27 = 3(x^6 - 9)$. The second factor is the difference of two squares. $\sqrt{x^6} = x^3, \sqrt{9} = 3$. Therefore, $3x^6 - 27 = 3(x^3 + 3)(x^3 - 3)$.

EXAMPLE 7

Factor $x^4 - 1$ completely.

Factoring

Solution

Find the square root of each square. $\sqrt{x^4} = x^2$, $\sqrt{1} = 1$. Therefore, $x^4 - 1 = (x^2 + 1)(x^2 - 1)$. But note that the second factor, $(x^2 - 1)$, is also the difference of two squares, and we can express $x^2 - 1$ as $(x + 1)(x - 1)$. Hence,

$$x^4 - 1 = (x^2 + 1)(x^2 - 1) = (x^2 + 1)(x + 1)(x - 1).$$

As in the previous section, we may sometimes encounter expressions that do not appear factorable. For example, $x^2 - y^2 + 2x - 2y$ does not resemble any factorable polynomial that we have encountered in this section. However, recall that when this occurs we should look for a common factor for a group of terms. The first two terms, $x^2 - y^2$, form the difference of two squares and we can express $x^2 - y^2$ as $(x + y)(x - y)$. The second group of terms, $+2x - 2y$, contains a common factor of 2. Therefore, $+2x - 2y = 2(x - y)$. Now we have

$$x^2 - y^2 + 2x - 2y = (x + y)(x - y) + 2(x - y).$$

We can now use the procedure for removing a common binomial factor from an expression. The other factor is the sum of the remaining factors,

$$(x + y)(x - y) + 2(x - y) = (x + y + 2)(x - y).$$

Summarizing these steps, we have

$$x^2 - y^2 + 2x - 2y = (x + y)(x - y) + 2(x - y)$$
$$= (x + y + 2)(x - y).$$

EXAMPLE 8

Factor $x^2 + 4y + 2x - 4y^2$ completely.

Solution

This problem is similar in nature to the one that we just discussed. First, we must group the terms: $x^2 + 4y + 2x - 4y^2 = x^2 - 4y^2 + 2x + 4y$. We can now factor the two groups of terms.

$$x^2 - 4y^2 + 2x + 4y = (x + 2y)(x - 2y) + 2(x + 2y).$$

Next, remove the common binomial factor, $(x + 2y)(x - 2y) + 2(x + 2y) = (x + 2y)(x - 2y + 2)$. Therefore, $x^2 - 4y^2 + 2x + 4y = (x + 2y)(x - 2y + 2)$.

Given the difference of two squares, we can factor this difference by finding the square root of each square. The binomial factors are the sum and difference of the two square roots.

EXERCISES FOR SECTION 5.2

(Answers given to exercises marked *)

Factor each of the following completely:

*1. $x^2 - 9$ 2. $y^2 - 4$ *3. $y^2 - 16$

4. $x^2 - 25$ *5. $9 - x^2$ 6. $1 - x^2$

*7. $1 - y^2$ 8. $x^6 - 9$ *9. $y^{10} - 4$

10. $4y^2 - 9$ *11. $9x^2 - 16$ 12. $25 - 4x^2$

*13. $49 - 16y^2$ 14. $9z^2 - 25$ *15. $16z^2 - 49$

16. $1 - z^4$ *17. $16x^2 - 1$ 18. $x^4 - y^4$

*19. $4x^3y^2 - 4xy^4$ 20. $8x^2y - 2y^3$ *21. $27x^7 - 3xy^4$

22. $32x^6 - 50x^2y^2$ *23. $x^2 - y^2 - 2x - 2y$ 24. $4x - 4y + x^2 - y^2$

*25. $x^2 + 2x + 2y - y^2$ 26. $4x^2 + 6x - 9y - 9y^2$

*27. $9x^2 + 4y - y^2 + 12x$ 28. $(x+y)^2 - (a-b)^2$

*29. $(2x+3)^2 - y^2$ 30. $(x-y)^2 - (z-w)^2$

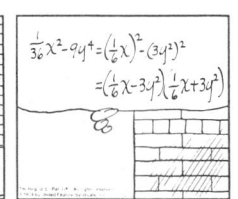

(© 1974 United Features Syndicate, Inc.)

NOTE OF INTEREST

We are given the equation, $x = y$, and shall perform some operations on it (in particular, some factoring takes place). Recall that whatever we do to one side of an equation we must do to the other side. We can think of an equation as a balance scale, and in performing the operations we try to keep the equation in balance.

$x = y$.   given

$x^2 = xy$.   multiplying both sides by x

$x^2 - y^2 = xy - y^2$. subtracting y^2 from both sides

Factoring

$(x + y)(x - y) = y(x - y).$ factoring both sides

$\dfrac{(x + y)(x - y)}{(x - y)} = \dfrac{y(x - y)}{(x - y)}.$ dividing both sides by $(x - y)$

$x + y = y.$ This is the result.

$y + y = y.$ Since $x = y$, we may substitute y for x.

$2y = y.$ combining $y + y$

$\dfrac{2y}{y} = \dfrac{y}{y}.$ dividing both sides by y

Therefore, $2 = 1.$ This is the conclusion.

Can you find the error?

5.3 TRINOMIALS THAT ARE PERFECT SQUARES

One of the special products that we considered in the previous chapter was the "square of a binomial", $(a + b)^2$ or $(a - b)^2$. If we square a binomial, the product is a trinomial, and this trinomial is called a perfect square.

$$\left.\begin{array}{l}(a + b)^2 = a^2 + 2ab + b^2 \\ (a - b)^2 = a^2 - 2ab + b^2\end{array}\right\} \text{perfect squares}$$

Note that a trinomial that is a perfect square has the following characteristics: 1. Two of the terms are squares of monomials; 2. The "middle" term has the same absolute value as twice the product of the two monomials. For example, $x^2 + 6x + 9$ is a perfect square. Two of the terms, x^2 and 9, are squares of monomials, $x^2 = (x)^2$ and $9 = 3^2$. The middle term, $+6x$, has the same absolute value as twice the product of the two monomials, or, $|6x| = |2 \cdot 3 \cdot x|$. Since $x^2 + 6x + 9$ is a perfect square, it is the square of a binomial. That binomial is of the form $(a + b)$ or $(a - b)$, where a is the square root of one of the squared terms, and b is the square root of the other squared term. The plus or minus sign is the sign of the middle term of the perfect square. Therefore, $x^2 + 6x + 9 = (x + 3)^2$ or $x^2 + 6x + 9 = (3 + x)^2$. Both forms are correct because either $(x + 3)^2$ or $(3 + x)^2$ will yield the trinomial $x^2 + 6x + 9$.

Can we factor $4x^2 - 12xy + 9y^2$? We can find the factors of this polynomial if it is a perfect square. First, we shall determine whether it is a perfect square. Two of the terms are squares of monomials, $4x^2 = (2x)^2$

and $9y^2 = (3y)^2$. The middle term, $-12xy$, has the same absolute value as twice the product of the two monomials, $|-12xy| = |2 \cdot 2x \cdot 3y|$. Hence, we can factor $4x^2 - 12xy + 9y^2$. The two monomials are square roots of the squared terms, $\sqrt{4x^2} = 2x$, $\sqrt{9y^2} = 3y$. The second monomial is preceded by the sign of the middle term in the given trinomial. Therefore,

$$4x^2 - 12xy + 9y^2 = (2x - 3y)^2.$$

Note that we can also have $4x^2 - 12xy + 9y^2 = (3y - 2x)^2$ because both factored forms will yield the given trinomial, $4x^2 - 12xy + 9y^2$. Therefore, both factored forms are equally correct.

EXAMPLE 9

Factor $25x^2 - 10xy + y^2$ completely.

Solution

We can factor the given trinomial because it is a perfect square. It satisfies the criteria, that is, $25x^2 = (5x)^2$, $y^2 = (y)^2$ and $|-10xy| = |2 \cdot 5x \cdot y|$. The terms of the binomial factor are $5x$ and y. The sign preceding the second term of the binomial factor is minus, the same as the sign of the middle term of the given trinomial. Therefore, $25x^2 - 10xy + y^2 = (5x - y)^2$. Note that $25x^2 - 10xy + y^2 = (y - 5x)^2$ is also correct.

EXAMPLE 10

Factor $49y^2 + 42x^2y + 9x^4$ completely.

Solution

The given trinomial is a perfect square; it satisfies the criteria: $49y^2 = (7y)^2$, $9x^4 = (3x^2)^2$ and $|+42x^2y| = |2 \cdot 3x^2 \cdot 7y|$. The terms of the binomial factor are $7y$ and $3x^2$. The sign preceding the second term of the factor is plus, the sign of the middle term of the trinomial. Therefore, $49y^2 + 42x^2y + 9x^4 = (7y + 3x^2)^2$.

EXAMPLE 11

Factor $12x^3 - 36x^2y + 27xy^2$ completely.

Factoring   253

> **Solution**
>
> The given trinomial is not a perfect square. But the terms have a common factor of 3x. (The first step in factoring any expression is to determine whether the terms have a common factor.) Hence, $12x^3 - 36x^2y + 27xy^2 = 3x(4x^2 - 12xy + 9y^2)$. The second factor is a perfect square, $4x^2 = (2x)^2$, $9y^2 = (3y)^2$ and $|-12xy| = |2 \cdot 2x \cdot 3y|$ and $4x^2 - 12xy + 9y^2 = (2x - 3y)^2$. Therefore, $12x^3 - 36x^2y + 27xy^2 = 3x(2x - 3y)^2$. Note that $3x(3y - 2x)^2$ is also a correct factored form.

Some trinomials can be recognized as a perfect square and can be factored by inspection. A trinomial that is a perfect square contains two terms that are squares of monomials, and the third term has the same absolute value as twice the product of the two monomials. The sign preceding the second term of the binomial factors is the same as the sign of the term that is not a square of a monomial.

EXERCISES FOR SECTION 5.3

*(Answers given to exercises marked *)*

Factor each of the following completely:

*1. $x^2 + 4x + 4$ 2. $y^2 - 6y + 9$

*3. $y^2 + 2y + 1$ 4. $z^2 - 10z + 25$

*5. $z^2 - 8z + 16$ 6. $1 + 2x + x^2$

*7. $1 - 2x + x^2$ 8. $x^2 + 2xy + y^2$

*9. $x^2 - 2xy + y^2$ 10. $4x^2 + 20xy + 25y^2$

*11. $9x^2 - 30xy + 25y^2$ 12. $16y^2 - 24xy + 9x^2$

*13. $25y^2 - 20xy + 4x^2$ 14. $a^2x^2 + 4axy + 4y^2$

*15. $a^2x^2 - 10axy + 25y^2$ 16. $25x^4 + 10x^2 + 1$

*17. $4x^4 + 12x^2y^2 + 9y^4$ 18. $x^6 + 2x^3y^4 + y^8$

*19. $-4x^2 + 12xy - 9y^2$ 20. $xy^2 - 6xy + 9x$

*21. $3y^3 + 6y^2 + 3y$ 22. $4x^4 + 8x^3 + 4x^2$

*23. $8x^2z + 40xyz + 50y^2z$ 24. $x^2 + 2x + 1 - y^2$

*25. $y^2 - 6y + 9 - x^2$ 26. $z^2 - 8z + 16 - 9y^2$

*27. $x^2 - 2xy + y^2 - 4z^2$

5.4 TRINOMIALS OF THE GENERAL FORM $x^2 + bx + c$

Thus far in our discussion of factoring, we have factored a common monomial factor from a polynomial, factored the difference of two squares, and factored trinomials that were perfect squares. Now we want to examine trinomials that may not be perfect squares. In this section, we shall consider only those trinomials of the form $x^2 + bx + c$. Note that the coefficient of the x^2 term is 1, and that b and c are integers. We shall examine cases where the coefficient of the x^2 term is some integer other than 1 in Section 5.5. Here we want to be able to find binomial factors of $x^2 + bx + c$—we want to have the following:

$$x^2 + bx + c = (x + p)(x + q) \quad \text{where p and q are integers.}$$

If we find the product of the two binomial factors $(x+p)$ and $(x+q)$, we have

$$\begin{aligned}(x + p)(x + q) &= (x + p)x + (x + p)q \\ &= x^2 + px + qx + pq \\ &= x^2 + (p + q)x + pq.\end{aligned}$$

Therefore, we now have

$$x^2 + bx + c = x^2 + (p + q)x + pq.$$

Since these two trinomials are identical, the corresponding parts must be equal and we see that the numbers p and q must satisfy the following two requirements:

$$1. \quad pq = c \quad \text{and} \quad 2. \quad p + q = b.$$

Therefore, to factor a trinomial of the form $x^2 + bx + c$, we have to find (if possible) integers p and q such that $pq = c$ and $p + q = b$. To do this, we first find all the pairs of integers whose product is c, and then choose the pair whose sum is b, $p + q = b$.

For example, to factor $x^2 + 7x + 12$, we must find two integers whose product is 12 and whose sum is 7. The possible pairs of factors of 12 are: 1 and 12, 2 and 6, 3 and 4. (Note that we shall not consider negative factors for this example, since $p + q = +7$.) By trial and error, we see that 3 and 4 are the desired integers, $3 \cdot 4 = 12$ and $3 + 4 = 7$. Recall that earlier in this section we discovered that

$$x^2 + bx + c = x^2 + px + qx + pq.$$

Factoring

Therefore, we rewrite the middle term

$$x^2 + 7x + 12 = x^2 + 3x + 4x + 12$$
$$= x(x + 3) + 4(x + 3). \quad \text{factoring the two groups of terms}$$
$$x^2 + 7x + 12 = (x + 4)(x + 3). \quad \text{removing the common binomial factor}$$

Note that the order in which 3x and 4x are placed in the rewritten equation does not affect the result (except for the order of the factors):

$$x^2 + 7x + 12 = x^2 + 4x + 3x + 12$$
$$= x(x + 4) + 3(x + 4). \quad \text{factoring the two groups of terms}$$
$$x^2 + 7x + 12 = (x + 3)(x + 4). \quad \text{removing the common binomial factor}$$

To factor $x^2 - 3x - 10$, we must find two integers whose product is -10 and whose sum is -3. The factors of -10 must be of opposite sign, since pq equals a negative quantity. The possible pairs of factors of -10 are 1 and -10, -1 and 10, 2 and -5, -2 and 5. By trial and error, we see that 2 and -5 are the desired integers, since $(2)(-5) = -10$ and $2 + (-5) = -3$. Therefore, we can rewrite the middle term as $-5x + 2x$.

$$x^2 - 3x - 10 = x^2 - 5x + 2x - 10$$
$$= x(x - 5) + 2(x - 5).$$
$$x^2 - 3x - 10 = (x + 2)(x - 5).$$

EXAMPLE 12

Factor $x^2 + 9x + 20$.

Solution

We must find two integers whose product is 20, and whose sum is 9. The possible pairs of factors of 20 are 1 and 20, 2 and 10, 4 and 5. We see that 4 and 5 are the desired integers: $4 \cdot 5 = 20$ and $4 + 5 = 9$. Therefore, we can rewrite the middle terms as $4x + 5x$,

$$x^2 + 9x + 20 = x^2 + 4x + 5x + 20 = x(x + 4) + 5(x + 4)$$
$$= (x + 5)(x + 4).$$

EXAMPLE 13

Factor $y^2 - 6y + 5$.

Solution

Here we look for two integers whose product is +5 and whose sum is −6. The factors of +5 that satisfy these conditions are −5 and −1, since $(-1)(-5) = +5$ and $-1 + (-5) = -6$. Therefore, we can rewrite the middle term as $-y - 5y$, or, $y^2 - 6y + 5 = y^2 - 1y - 5y + 5 = y(y - 1) - 5(y - 1) = (y - 5)(y - 1)$.

EXAMPLE 14

Factor $y^2 + 3y + 5$.

Solution

We must find two integers whose product is 5, and whose sum is 3. No such integers exist. The only possible factors of 5 are 5 and 1 and −1 and −5. Since the sum of these factors is not 3, we say that $y^2 + 3y + 5$ is prime. It cannot be factored into binomials with integral coefficients.

From Example 14, we see that not all polynomials of the general form $x^2 + bx + c$ (or $y^2 + by + c$, etc.) can be expressed as the product of two binomials with integral coefficients. In general, any polynomial that does not contain a common monomial factor for each term, or cannot be factored into two other polynomials with integer coefficients, is said to be prime. When we say that a polynomial is prime, it is understood that the polynomial cannot be factored over the set of integers.

EXAMPLE 15

Factor $3y^3 - 6y^2 - 45y$ completely.

Solution

To begin with, note that the terms have a common factor of $3y$. (The first step in factoring any expression is to determine whether the terms have a common factor.) Hence $3y^3 - 6y^2 - 45y = 3y(y^2 - 2y - 15)$. The second factor is of the general form $y^2 + by + c$. We need two integers whose product is −15 and whose sum is −2. The factors of −15 that satisfy these conditions are −5 and 3: $(-5)(3) = -15$ and $-5 + 3 = -2$. We can now rewrite the middle term. There-

Factoring

fore, $y^2 - 2y - 15 = y^2 - 5y + 3y - 15 = y(y-5) + 3(y-5) = (y+3)(y-5)$. We can now express the given polynomial in completely factored form, $3y^3 - 6y^2 - 45y = 3y(y-5)(y+3)$.

EXAMPLE 16

Factor $x^4 + 5x^2 + 6$.

Solution

This polynomial is not of the form $x^2 + bx + c$. But if we substitute a variable for x^2 in the expression, let $y = x^2$, then we have $y^2 + 5y + 6$ and this is in the general form that we can factor. We need two factors of 6 whose product is $+6$ and whose sum is 5; they are 3 and 2: $3 \cdot 2 = 6$ and $3 + 2 = 5$. Therefore, $y^2 + 5y + 6 = y^2 + 3y + 2y + 6 = y(y+3) + 2(y+3) = (y+2)(y+3)$. But, since we let $y = x^2$, we can write $x^4 + 5x^2 + 6 = (x^2 + 2)(x^2 + 3)$.

EXAMPLE 17

Factor $y^4 - 2y^2 - 8$ completely.

Solution

This problem is similar in nature to Example 16. We can express $y^4 - 2y^2 - 8$ as $(y^2)^2 - 2(y^2) - 8$; it does fit the general form. Let $x = y^2$ and we have $x^2 - 2x - 8$. We must find two integers whose product is -8 and whose sum is -2. The factors of -8 that satisfy these conditions are -4 and 2, $(-4)(2) = -8$, and $-4 + 2 = -2$. Therefore, $x^2 - 2x - 8 = x^2 - 4x + 2x - 8 = x(x-4) + 2(x-4) = (x+2)(x-4)$. Substituting for x, we have $(y^2 + 2)(y^2 - 4)$. Since the second factor is the difference of two squares, we factor it again, obtaining

$$y^4 - 2y^2 - 8 = (y^2 + 2)(y + 2)(y - 2).$$

To factor a trinomial of the general form $x^2 + bx + c$ into the product of two binomials, $x^2 + bx + c = (x + p)(x + q)$, we must find two integers p and q such that $pq = c$ and $p + q = b$. Then we can rewrite the middle term and remove a common factor from the two groups of terms. Finally, we factor out the common binomial factor such that $x^2 + bx + c = (x + p)(x + q)$. Recall that not all polynomials of this form can be expressed as the product of two binomials—the polynomial $x^2 + bx + c$ may be prime.

EXERCISES FOR SECTION 5.4
(Answers given to exercises marked *)

Factor each of the following completely:

*1. $x^2 + 2x + 1$
2. $x^2 - 2x + 1$
*3. $x^2 - 8x + 12$
4. $x^2 - 4x - 12$
*5. $x^2 + 11x - 12$
6. $x^2 - 13x + 12$
*7. $x^2 - 7x + 12$
8. $x^2 - x - 12$
*9. $x^2 + x - 12$
10. $y^2 - 9y + 18$
*11. $y^2 - 3y - 18$
12. $y^2 - 11y + 18$
*13. $y^2 - 17y - 18$
14. $z^2 - 6z - 55$
*15. $z^2 + 3z - 28$
16. $x^2 + 5x + 4$
*17. $y^2 + 3y - 4$
18. $2x^3 + 4x^2 - 70x$
*19. $3y^3 - 6y^2 - 105y$
20. $2x^2y - 24xy + 70y$
*21. $3x^2y^2z^2 + 6xy^2z^2 - 24y^2z^2$
22. $48 + 14x + x^2$
*23. $-4 - 3y + y^2$
24. $-30 - 4y + 2y^2$
*25. $30 - 5x - 5x^2$
26. $x^4 + 2x^2 + 1$
*27. $x^4 - 2x^2 + 1$
28. $y^4 - 4y^2 - 12$
*29. $y^4 - 7y^2 + 12$
30. $x^{2n} - 2x^n + 1$
*31. $y^{2n} - 7y^n + 12$
32. $a^2 - ab - 6b^2$
*33. $a^2 + 2ab - 24b^2$

5.5 TRINOMIALS OF THE GENERAL FORM $ax^2 + bx + c$

In Section 5.4, we discussed a method for factoring trinomials of the general form $x^2 + bx + c$. We shall now expand upon this technique so that we can factor trinomials of the more general form $ax^2 + bx + c$. Previously, the coefficient of the second-degree term was 1 and now we will consider those polynomials, $ax^2 + bx + c$, where the coefficient of the second-degree term (a) is not necessarily 1. In this section, we want to be able to find binomial factors of $ax^2 + bx + c$, that is, we want to have

$$ax^2 + bx + c = (mx + p)(nx + q).$$

If we find the product of the two binomial factors $(mx + p)$ and $(nx + q)$, we have

$$(mx + p)(nx + q) = (mx + p)nx + (mx + p)q$$
$$= mnx^2 + pnx + mqx + pq$$
$$= mnx^2 + (pn + mq)x + pq.$$

Factoring

We then have
$$ax^2 + bx + c = mnx^2 + (pn + mq)x + pq.$$

Since these two expressions are identical, we see that a corresponds to mn, pq corresponds to c, and the sum (pn + mq) corresponds to b. At first glance, it appears that it is impossible to find integers that satisfy all of these conditions. But that is not the case! Notice that the sum (pn + mq) is a combination of the factors of c and a: a = mn, c = pq, and b = pn + mq. If we multiply a times c, we have ac = mn · pq = mnpq. The sum (pn + mq) is a combination of the factors mnpq. In other words, to factor an expression of the form $ax^2 + bx + c$, we first find the product of a and c, ac, and then find two integers whose product is ac and whose sum is b. We can then use these two integers to rewrite the middle term and proceed as we did in Section 5.4.

Let's examine this technique through an example. Consider the polynomial $6x^2 + 13x + 6$. It is of the general form $ax^2 + bx + c$. Note that a = 6, b = 13 and c = 6. First, we find ac, 6 · 6 = 36. Now we want two integers whose product is 36 and whose sum is 13 (b). Thirty-six has many different pairs of factors, but the integers whose product is 36 and whose sum is 13 are 4 and 9. Therefore, we can now rewrite the middle term of the given polynomial.

$$6x^2 + 13x + 6 = 6x^2 + 9x + 4x + 6.$$

Now, we can factor the two groups of terms, $= 3x(2x + 3) + 2(2x + 3)$.
Next, we remove the common binomial factor, $= (3x + 2)(2x + 3)$.
Therefore, $6x^2 + 13x + 6 = (3x + 2)(2x + 3)$.

Note that the order in which 9x and 4x are placed in the rewritten equation does not affect the result (except for the order of the factors),

$$\begin{aligned}6x^2 + 13x + 6 &= 6x^2 + 4x + 9x + 6\\ &= 2x(3x + 2) + 3(3x + 2).\end{aligned}$$ factoring the two groups of terms

$$6x^2 + 13x + 6 = (2x + 3)(3x + 2).$$ removing the common binomial factor

To factor an expression of the form $ax^2 + bx + c$, we first find the product of a and c. Next, we find two integers whose product is ac and whose sum is b. To factor the polynomial $8x^2 + 2x - 15$, we first find the product of a and c, 8(-15) = -120. Now we must find two integers whose product is -120 and whose sum is 2. The two integers that satisfy these conditions are 12 and -10, 12(-10) = -120 and 12 + (-10) = 2. We can now rewrite the middle term of the polynomial,

$$\begin{aligned}8x^2 + 2x - 15 &= 8x^2 + 12x - 10x - 15\\ &= 4x(2x + 3) - 5(2x + 3).\end{aligned}$$ factoring the two groups of terms

$8x^2 + 2x - 15 = (4x - 5)(2x + 3)$.   removing the common binomial factor

We should always check our work, and we can quickly determine if the polynomial is factored correctly by finding the product of the factors. In this case, we see that $(4x-5)(2x+3) = 8x^2 + 2x - 15$, therefore, the work is correct.

EXAMPLE 18

Factor $8x^2 + 14x + 3$

Solution

First, we find the product of a and c, a = 8, c = 3 and ac = 8 · 3 = 24. Now we must find two integers whose product is 24 and whose sum is b, 14. The factors of 24 that satisfy these conditions are 12 and 2; 12 · 2 = 24 and 12 + 2 = 14. Rewriting the middle term of the polynomial (using 12 and 2) we have $8x^2 + 14x + 3 = 8x^2 + 2x + 12x + 3$. Now we can factor the two groups of terms and then remove the common binomial factor,

$8x^2 + 2x + 12x + 3 = 2x(4x + 1) + 3(4x + 1) = (2x + 3)(4x + 1)$.

EXAMPLE 19

Factor $15x^2 + 14x - 8$.

Solution

a = 15, b = 14, c = −8. Find the product of a and c; 15(−8) = −120. Next, find two integers whose product is −120 and whose sum is b (14). The factors of −120 that satisfy these conditions are 20 and −6, 20(−6) = −120 and 20 + (−6) = 14. Rewriting the middle term of the polynomial (using 20 and −6):

$15x^2 + 14x - 8 = 15x^2 + 20x - 6x - 8$
$= 5x(3x + 4) - 2(3x + 4)$.   factoring the two groups of terms

$15x^2 + 14x - 8 = (5x - 2)(3x + 4)$.   removing the common binomial factor

Admittedly there exist a number of possible pairs of factors for −120. But keep in mind that we are looking for a pair whose product is −120 and whose sum is +14. Therefore, we immediately exclude some pairs of factors, such as 60 and −2, 40 and −3, 120 and −1, etc. The factors we are looking for must satisfy both conditions — their product must equal ac and their sum must equal b.

Factoring

EXAMPLE 20

Factor $3y^2 + y + 1$.

Solution

$a = 3$, $b = 1$, $c = 1$. The product of a and c is $3 \cdot 1 = 3$. We need to find two integers whose product is 3 and whose sum is 1. The factors of 3 are 3, 1 and $-3, -1$. Neither pair of factors has a sum of 1. There are no two integers whose product is 3 and whose sum is 1. Therefore, we say that $3y^2 + y + 1$ is prime. It cannot be factored using integral coefficients.

EXAMPLE 21

Factor $6y^4 + 3y^2 + 5$.

Solution

This polynomial is not of the form $ay^2 + by + c$. But if we substitute a variable for y^2 in the given expression, say $x = y^2$, then we have $6x^2 + 13x + 5$, and this is in the general form that we can factor. $a = 6$, $b = 13$, $c = 5$. The product of a and c is $6 \cdot 5 = 30$. We need to find two integers whose product is 30 and whose sum is 13. They are 10 and 3; $10 \cdot 3 = 30$ and $10 + 3 = 13$. Next, we rewrite the middle term of the polynomial (using 10 and 3):

$$6x^2 + 13x + 5 = 6x^2 + 3x + 10x + 5$$
$$= 3x(2x + 1) + 5(2x + 1)$$
$$= (3x + 5)(2x + 1).$$

Since $x = y^2$, we write:

$$6y^4 + 13y^2 + 5 = (3y^2 + 5)(2y^2 + 1).$$

EXAMPLE 22

Factor $18x^3 - 3x^2 - 45x$ completely.

Solution

The first step in factoring any expression is to determine whether the terms have a common factor. Note that the terms have a common factor of $3x$. Hence, $18x^3 - 3x^2 - 45x = 3x(6x^2 - x - 15)$. The second factor is of the general form ax^2+bx+c where $a=6$, $b=-1$ and $c=-15$. The product of a and c is $6(-15)=-90$. We must find two integers whose product is -90 and whose sum is -1. They are -10 and 9, $(-10)9=-90$ and $-10+9=-1$. Rewriting the middle term of the second factor, we have $6x^2-x-15=6x^2+9x-10x-15=3x(2x+3)-5(2x+3)=(3x-5)(2x+3)$. We can now express the given polynomial in completely factored form: $18x^3-3x^2-45x=3x(3x-5)(2x+3)$.

EXAMPLE 23

Factor $3x^2+7mx-6m^2$.

Solution

$a=3$, $b=7m$, $c=-6m^2$. The product of a and c is $3(-6m^2)=-18m^2$. We need to find two factors whose product is $-18m^2$ and whose sum is $7m$. They are $-2m$ and $+9m$. Rewriting the middle term of the polynomial,

$3x^2+7mx-6m^2 = 3x^2+9mx-2mx-6m^2$
$\qquad\qquad\qquad\quad = 3x(x+3m)-2m(x+3m)$. factoring the two groups of terms

$3x^2+7mx-6m^2 = (3x-2m)(x+3m)$. removing the common binomial factor

EXAMPLE 24

Factor $6x^2+xy-2y^2+9x+6y$.

Solution

Since no term contains a common factor and the given polynomial is not of the general form ax^2+bx+c, we group terms (the second degree terms in one group and the first degree terms in another group). Hence, we have $(6x^2+xy-2y^2)+(9x+6y)$. Now we can factor the first group, which is of the form ax^2+bx+c, $a=6$, $b=y$ and $c=-2y^2$. We can also factor the second group, as each term contains a common factor of 3: $9x+6y=3(3x+2y)$. For the first group of terms, $ac=6(-2y^2)=-12y^2$. We must find two factors whose product is $-12y^2$ and whose sum is y. They are $4y$ and $-3y$. Therefore, $6x^2+xy-2y^2=6x^2+4xy-3xy-2y^2=2x(3x+2y)-y(3x+2y)=(2x-y)(3x+2y)$. The given polynomial $6x^2+xy-2y^2+9x+6y$ now appears as $(2x-y)(3x+2y)+3(3x+2y)$ and we can remove the common binomial factor $(3x+2y)$. Hence, $6x^2+xy-2y^2+9x+6y=(2x-y)(3x+2y)+3(3x+2y)=(2x-y+3)(3x+2y)$.

Factoring

To factor a trinomial of the general form $ax^2 + bx + c$ into the product of two binomials, $(mx + p)(nx + q)$, we first find the product of a and c, ac. Next, we find two factors of ac such that their product is ac and their sum is b. Then we can rewrite the middle term and remove a common factor from the two groups of terms. Finally, we factor out the common binomial factor such that $ax^2 + bx + c = (mx + p)(nx + q)$.

EXERCISES FOR SECTION 5.5

(Answers given to exercises marked *)

Factor each of the following completely.

*1. $2x^2 + 5x + 2$ 2. $6x^2 - x - 1$ *3. $3y^2 + 5y + 2$
 4. $6x^2 - 11x + 3$ *5. $5y^2 - 6y + 1$ 6. $3x^2 + 13x + 12$
*7. $9x^2 + 30x + 9$ 8. $12z^2 + 14z + 2$ *9. $5z^2 + z - 6$
10. $3x^2 + 7x + 2$ *11. $6y^2 + 13y + 6$ 12. $6y^2 + 11y + 5$
*13. $15z^2 + 22z + 8$ 14. $15z^2 - 22z + 8$ *15. $15y^2 - 2y - 8$
16. $15y^2 + 2y - 8$ *17. $2 - 17x + 21x^2$ 18. $-2 - 7x + 15x^2$
*19. $-3 + 5x + 2x^2$ 20. $15 + 13z + 2z^2$ *21. $6y^4 + 5y^2 + 1$
22. $3x^4 + 13x^2 + 12$ *23. $6y^4 - 11y^2 + 3$ 24. $5x^4 - 6x^2 + 1$
*25. $6x^3 + 26x^2 + 24x$ 26. $15y^3 - 18y^2 + 3y$ *27. $12x^2y - 2xy - 2y$
28. $12x^2 + 8ax - 15a^2$ *29. $2x^2 + 5xy + 3y^2$ 30. $2x^2 - 5ax - 12a^2$
*31. $12x^2 - 29xy + 15y^2$ 32. $3y^2 + xy - 2x^2$ *33. $y^2 + xy - 2x^2$
34. $x^2 + 3xy + 2y^2 + 3x + 6y$
36. $m^2 + 2mn + n^2 + 2m + 2n$
38. $m^2 - 2n^2 + 6m - 6n + mn$
*35. $x^2 + xy - 2y^2 - 3x + 3y$
*37. $x^2 + 2xy - 2x + 2y - 3y^2$
*39. $10b + 6ab + 2a + a^2 + 5b^2$

5.6 THE SUM OR DIFFERENCE OF TWO CUBES

In the previous chapter, we considered certain special products. One was obtained when we multiplied a binomial by a trinomial. We specifically studied $(a + b)(a^2 - ab + b^2)$. We can use various techniques to obtain this product, but here we shall apply the distributive property.

We have:

$$(a + b)(a^2 - ab + b^2) = (a + b)a^2 + (a + b)(-ab) + (a + b)(b^2)$$
$$= a^3 + a^2b - a^2b - ab^2 + ab^2 + b^3$$
$$= a^3 + b^3.$$

This special product is called the *sum of two cubes,* and we recognize it as $(a + b)(a^2 - ab + b^2) = a^3 + b^3$. By means of the symmetric property, we can rewrite this as

$$\boxed{a^3 + b^3 = (a + b)(a^2 - ab + b^2)}$$

Therefore, we can express the sum of two cubes as the indicated product of two factors. One factor is a binomial composed of the sum of the cube roots, and the other factor is a trinomial—two terms are squares of the cube roots while the third term is the additive inverse of the product of the cube roots. (Note that the trinomial cannot be factored.) For example, to factor $x^3 + 8$, we find the cube roots of each cube, $\sqrt[3]{x^3} = x$ and $\sqrt[3]{8} = 2$. The binomial factor is the sum of the cube roots, $(x + 2)$. The trinomial factor is composed of the squares of the cube roots and the additive inverse of their product, $(x^2 - 2x + 4)$. Therefore,

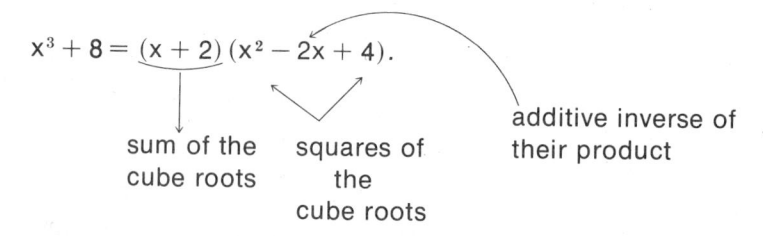

$$x^3 + 8 = \underbrace{(x + 2)}_{\text{sum of the cube roots}} \underbrace{(x^2 - 2x + 4)}_{\substack{\text{squares of} \\ \text{the} \\ \text{cube roots}}}.$$

additive inverse of their product

Thus far we have only discussed the sum of two cubes, but whether we wish to factor the sum or difference of two cubes, we must be sure to recognize a perfect cube. A monomial containing variables is a perfect cube if the exponents of the variables are multiples of 3, that is, 3, 6, 9, 12, 15, and so on. The following monomials are examples of perfect cubes:

$$x^3 = (x^1)^3,\ y^6 = (y^2)^3,\ z^9 = (z^3)^3,\ x^{12} = (x^4)^3,\ x^6y^9 = (x^2y^3)^3.$$

Integers that are perfect cubes are sometimes difficult to recognize, but with practice you will develop the ability to recognize them. The following integers are examples of perfect cubes:

$$1 = (1)^3,\ 8 = (2)^3,\ 27 = (3)^3,\ 64 = (4)^3,\ 125 = (5)^3,\ 216 = (6)^3.$$

Similarly, $-1 = (-1)^3$, $-8 = (-2)^3$, $-27 = (-3)^3$, $-64 = (-4)^3$, and so on.

Factoring 265

A perfect cube monomial that contains variables can also have an integer coefficient. The integer coefficient is also a perfect cube. For example,

$$8x^3 = (2x)^3, \; 64y^6 = (4y^2)^3, \; 125z^{12} = (5z^4)^3.$$

EXAMPLE 25

Factor $27y^3 + x^3$.

Solution

To factor the sum of two cubes, first find the cube root of each cube, $\sqrt[3]{27y^3} = 3y$, $\sqrt[3]{x^3} = x$. One factor is the sum of the cube roots, $(3y + x)$. The other factor is composed of the squares of the cube roots $(9y^2$ and $x^2)$, and the additive inverse of their product $(-3yx)$, $(9y^2 - 3yx + x^2)$. Therefore, $27y^3 + x^3 = (3y + x)(9y^2 - 3yx + x^2)$.

EXAMPLE 26

Factor $64y^6 + 1$.

Solution

Find the cube root of each cube, $\sqrt[3]{64y^6} = 4y^2$, $\sqrt[3]{1} = 1$. One factor is the sum of the cube roots, $(4y^2 + 1)$. The other factor is composed of the squares of the cube roots $(16y^4$ and $1)$, and the additive inverse of their product $(-1 \cdot 4y^2 = -4y^2)$, or, $(16y^4 - 4y^2 + 1)$. Therefore, $64y^6 + 1 = (4y^2 + 1)(16y^4 - 4y^2 + 1)$. (Note that the trinomial cannot be factored.)

Another special product that we considered in the previous chapter was $(a - b)(a^2 + ab + b^2)$. If we perform the indicated multiplication, we have

$$(a - b)(a^2 + ab + b^2) = (a - b)a^2 + (a - b)ab + (a - b)b^2$$
$$= a^3 - a^2b + a^2b - ab^2 + ab^2 - b^3$$
$$= a^3 - b^3.$$

This special product is called the *difference of two cubes*, and we recognize the special product as $(a - b)(a^2 + ab + b^2) = a^3 - b^3$. By means of the symmetric property, we have

$$\boxed{a^3 - b^3 = (a - b)(a^2 + ab + b^2).}$$

We can express the difference of two cubes as the indicated product of two factors. Note that this is similar in nature to the sum of two cubes. One factor is a binomial composed of the difference of the cube roots, and the other factor is a trinomial—two terms are squares of the cube roots while the third term is the additive inverse of the product of the terms in the binomial factor. For example, $x^3 - 27$ is the difference of two cubes, and we can express it in factored form as

$$x^3 - 27 = (x - 3)(x^2 + 3x + 9).$$

- $(x - 3)$: difference of the cube roots
- $(x^2 + 9)$: squares of the cube roots
- $3x$: product of the cube roots

EXAMPLE 27

Factor $8x^3 - 27y^6$.

Solution

To factor the difference of two cubes, first find the cube root of each cube, $\sqrt[3]{8x^3} = 2x$, $\sqrt[3]{27y^6} = 3y^2$. One factor is the difference of the cube roots, $(2x - 3y^2)$. The other factor is composed of the squares of the cube roots ($4x^2$ and $9y^4$) and their product ($6xy^2$), so the second factor is $(4x^2 + 6xy^2 + 9y^4)$. Therefore, $8x^3 - 27y^6 = (2x - 3y^2)(9y^2 + 6xy^2 + 9y^4)$.

EXAMPLE 28

Factor $x^6 - 1$ completely.

Solution

We can factor this expression in two different ways. $x^6 - 1$ can be considered to be the difference of two squares, $x^6 - 1 = (x^3)^2 - (1)^2$, or it can also be considered as the difference of two cubes, $x^6 - 1 = (x^2)^3 - (1)^3$. First, we shall consider it as the difference of two squares. Hence, $x^6 - 1 = (x^3)^2 - (1)^2$ and we find the square root of each square,

$$\sqrt{x^6} = x^3, \sqrt{1} = 1. \text{ Therefore,}$$
$$x^6 - 1 = (x^3 + 1)(x^3 - 1).$$

Factoring

Now we have the sum $(x^3 + 1)$ and the difference $(x^3 - 1)$ of two cubes and they can both be factored.

$(x^3 + 1) = (x + 1)(x^2 - x + 1)$ and $(x^3 - 1) = (x - 1)(x^2 + x + 1)$.

Hence,

$$\begin{aligned} x^6 - 1 &= (x^3 + 1)(x^3 - 1) \\ &= (x + 1)(x^2 - x + 1)(x - 1)(x^2 + x + 1). \end{aligned}$$

If we treat $x^6 - 1$ as the difference of two cubes, $x^6 - 1 = (x^2)^3 - (1)^3$, we have

$$x^6 - 1 = (x^2 - 1)(x^4 + x^2 + 1).$$

The factor $(x^2 - 1)$ is the difference of two squares and we can factor it, $(x^2 - 1) = (x + 1)(x - 1)$. But can we factor $x^4 + x^2 + 1$? Normally, we cannot factor the trinomial factor in the difference of two cubes. Note that we can rewrite this new expression as $(x^2 + 1)^2 - x^2$. We again have the difference of two squares and $(x^2 + 1)^2 - x^2 = (x^2 + 1 + x)(x^2 + 1 - x)$. Therefore,

$$\begin{aligned} x^6 - 1 &= (x^2 - 1)(x^4 + x^2 + 1) \\ &= (x + 1)(x - 1)(x^2 + 1 + x)(x^2 + 1 - x). \end{aligned}$$

NOTE OF INTEREST

Many times we encounter expressions that do not appear factorable, that is, they appear to be prime. But often we can factor a polynomial by using the identity property for addition, $x + 0 = x$. In Example 28 we had the polynomial $x^4 + x^2 + 1$, which is not factorable in its present form. The polynomial $x^4 + x^2 + 1$ closely approximates a perfect square trinomial. If it were a perfect square, then we could factor it. In order to be a perfect square, it would have to be of the form $x^4 + 2x^2 + 1$. We can change the given polynomial to this form by adding x^2 to it. But if we add x^2, then we must subtract x^2, that is, we are adding 0. Hence,

$$\begin{aligned} x^4 + x^2 + 1 &= x^4 + x^2 + 1 + x^2 - x^2 \\ &= x^4 + x^2 + x^2 + 1 - x^2 \quad \text{by means of the commutative property} \\ &= x^4 + 2x^2 + 1 - x^2 \\ &= (x^2 + 1)^2 - x^2 \\ &= (x^2 + 1 + x)(x^2 + 1 - x). \quad \text{factoring the difference of two squares} \end{aligned}$$

You should be aware that sometimes an expression can be factored even though it appears to be prime. As an exercise, show that $4x^4 + 1$ can be expressed as $(2x^2 + 1 + 2x)(2x^2 + 1 - 2x)$.

An expression that can be considered either the difference of two squares or the difference of two cubes can be factored by either technique. But when we factored the expression $x^6 - 1$ as the difference of two cubes, we encountered a lengthy and cumbersome technique to arrive at the final factored form. It is generally a good idea to factor the expression (when you have a choice) as the difference of two squares. In this way, you will avoid the situation that arose in the second solution to Example 28.

EXAMPLE 29

Factor $x^9 - y^9$ completely.

Solution

The expression $x^9 - y^9$ can only be considered as the difference of two cubes. Finding the cube roots of each cube, we have $\sqrt[3]{x^9} = x^3$, $\sqrt[3]{y^9} = y^3$. One factor is the difference of the cube roots, $(x^3 - y^3)$. The trinomial factor is the sum of the squares of the cube roots plus the product of the cube roots, $(x^6 + x^3y^3 + y^6)$. Hence, $x^9 - y^9 = (x^3 - y^3)(x^6 + x^3y^3 + y^6)$. But note that the binomial factor $(x^3 - y^3)$ is also the difference of two cubes and $(x^3 - y^3) = (x - y)(x^2 + xy + y^2)$. Therefore,

$$x^9 - y^9 = (x^3 - y^3)(x^6 + x^3y^3 + y^6)$$
$$= (x - y)(x^2 + xy + y^2)(x^6 + x^3y^3 + y^6).$$

EXAMPLE 30

Factor $x^4 - x^3 + 27x - 27$.

Solution

This expression is not the sum or the difference of two cubes, and the terms do not have a common factor. Hence, we look for a common factor for a group of terms, and $x^4 - x^3 + 27x - 27 = x^3(x - 1) + 27(x - 1)$. Removing the common binomial factor, we have $(x^3 + 27)(x - 1)$. Note that $(x^3 + 27)$ is the sum of two cubes and $(x^3 + 27) = (x + 3)(x^2 - 3x + 9)$. Summarizing the steps, we have,

$$x^4 - x^3 + 27x - 27 = x^3(x - 1) + 27(x - 1)$$
$$= (x^3 + 27)(x - 1)$$
$$= (x + 3)(x^2 - 3x + 9)(x - 1).$$

Factoring

Examining the special products found in Chapter Four, we found that

$$(a + b)(a^2 - ab + b^2) = a^3 + b^3$$

and

$$(a - b)(a^2 + ab + b^2) = a^3 - b^3.$$

We can use these equations to factor expressions that are of these general forms, that is, we can factor the sum or difference of two cubes.

EXERCISES FOR SECTION 5.6

(Answers given to exercises marked *)

Factor each of the following completely:

*1. $y^3 + 64$ 2. $x^3 + 27$ *3. $8z^3 + 125$
4. $8x^6 + 27$ *5. $27y^6 + 1$ 6. $x^3 + y^3$
*7. $8x^3 + 27y^3$ 8. $64x^6 + 27y^9$ *9. $125y^9 + 64x^{12}$
10. $x^3 - 64$ *11. $y^3 - 1$ 12. $8z^3 - 27$
*13. $27x^6 - y^3$ 14. $8y^6 - 27x^9$ *15. $64z^3 - 8x^9$
16. $x^3 - 8z^{12}$ *17. $125z^9 - 64x^6$ 18. $343x^9 - 64y^6$
*19. $x^3y^6 + 8z^9$ 20. $27x^3y^6z^9 - 1$ *21. $8x^6y^3 - 27z^3$
22. $z^6 - 1$ *23. $y^6 - 64$ 24. $x^6 + 2x^3 - 3$
*25. $y^6 - y^3 - 2$ 26. $y^4 - y^3 + 8y - 8$ *27. $z^4 + 2z^3 - z - 2$
28. $z^6 + 3z^3 - 4$ *29. $x^4 + 3x^3 - x - 3$ 30. $x^5 + 5x^3 + x^2 + 5$

5.7 SUMMARY

When we factor an expression, we find the numbers or algebraic expressions whose product is the given expression. In this chapter, we have discussed those techniques of factoring that are used most often in solving problems.

To factor any expression, we should first determine whether or not the terms have a common factor; if so, then we should immediately factor out the common factor. For example, each of the terms in $5x^3 - 10x^2 + 5x$ has a common factor of $5x$, and $5x^3 - 10x^2 + 5x = 5x(x^2 - 2x + 1)$. Next, we determine whether the remaining expression can be factored. In order to determine this, we must be able to recognize certain factorable forms, of which there are a variety. For example, the remaining expression could be one of the different types of trinomials. Some trinomials are "perfect squares," that is, they are of the form $a^2 + 2ab + b^2 = (a + b)^2$ or $a^2 - 2ab + b^2 = (a - b)^2$. They could also be of the form $x^2 + bx + c = (x + p)(x + q)$ or $ax^2 + bx + c = (mx + p)(nx + q)$. If the remaining expression is a binomial, then it could be one of three different factorable forms: the difference of two squares, $a^2 - b^2 = (a + b)(a - b)$; the sum of two cubes, $a^3 + b^3 = (a + b)(a^2 - ab + b^2)$; or the difference of two cubes, $a^3 - b^3 = (a - b)(a^2 + ab + b^2)$. If the remaining expression contains more than three terms, it may still be factorable by grouping those terms that have a common monomial factor and then removing the common factor. Whatever technique we use to factor an expression, we can quickly determine whether the polynomial is factored correctly by finding the product of the factors.

Examples of the factoring techniques that we have discussed in this chapter follow:

Expression	Explanation	Factors
$5x^2 + 10x$	a common factor	$5x(x + 2)$
$x^2 - 4$	difference of two squares	$(x + 2)(x - 2)$
$x^3 + 8$	sum of two cubes	$(x + 2)(x^2 - 2x + 4)$
$x^6 - y^3$	difference of two cubes	$(x^2 - y)(x^4 + x^2y + y^2)$
$x^2 + 2xy + y^2$ $a^2 - 2ab + b^2$	perfect squares	$(x + y)^2$ $(a - b)^2$
$x^2 + 4x + 3$	trinomial of the form $x^2 + bx + c$	$(x + 3)(x + 1)$
$6x^2 + 11x - 10$	trinomial of the form $ax^2 + bx + c$	$(3x - 2)(2x + 5)$
$y^2 + ay + by + ab$	factoring by grouping	$(y + b)(y + a)$

REVIEW EXERCISES FOR CHAPTER 5

(Answers in Appendix B)

Factor each of the following completely:

1. $x^2 + 3x$
2. $4y^2 - 2y$
3. $xy^2 - xy$
4. $-10xy + 5x^2y$
5. $2x^2y^2 + 4x^3y + 6x^2y^3$
6. $x(y + 3) + 5(y + 3)$
7. $(x + y)x + (x + y)z$
8. $(x + y)x - (x + y)3$
9. $(2x + 3)y - (2x + 3)4$

Factoring

10. $ax + by + ay + bx$
11. $2ax + by + 4ax + 2by$
12. $st + uv + sv + ut$
13. $y^2 - 9$
14. $4x^2 - 9$
15. $16 - y^2$
16. $9x^2 - 16$
17. $25 - 4x^2$
18. $y^{10} - 4$
19. $x^4 - 1$
20. $y^4 - 81$
21. $16 - x^4$
22. $27x^7 - 3xu^4$
23. $4x^3y^2 - 4xy^4$
24. $32x^6 - 50x^2y^2$
25. $x^2 - y^2 - 2x - 2y$
26. $x^2 + 2x + 2y - y^2$
27. $9x^2 + 4y - y^2 + 12x$
28. $(2x+1)^2 - y^2$
29. $(x+y)^2 - (a+b)^2$
30. $x^2 - 8x + 16$
31. $16y^2 - 24xy + 9x^2$
32. $4x^4 + 12x^2y^2 + 9y^4$
33. $4y^4 + 8y^3 + 4y^2$
34. $x^6 + 2x^3y^4 + y^8$
35. $x^2 + 11x - 12$
36. $x^2 - 11x + 18$
37. $x^2 - 6x - 55$
38. $48 + 14y + y^2$
39. $2x^3 + 4x^2 - 70x$
40. $x^4 - 4x^2 - 12$
41. $y^4 - 7y^2 + 12$
42. $6y^2 - y - 1$
43. $3z^2 + 13z + 12$
44. $6x^2 + 11x + 5$
45. $15z^2 - 22z + 8$
46. $6y^3 + 26y^2 + 24y$
47. $15x^3 - 18x^2 + 3x$
48. $y^2 + xy - 2x^2$
49. $3x^2 + xy - 2y^2$
50. $x^2 + xy - 2y^2 - 3x + 3y$
51. $x^2 + 3xy + 2y^2 + 3x + 6y$
52. $x^3 + y^3$
53. $8z^3 + 125$
54. $125y^9 + 64x^{12}$
55. $8z^3 - 27$
56. $8y^6 - 27x^9$
57. $x^6 - 1$
58. $z^6 - 64$
59. $x^4 - x^3 + 8x - 8$
60. $z^4 + 2z^3 - z - 2$

chapter 6

**Rational Expressions
(Algebraic Fractions)**

As a result of this chapter, the student will be able to:

1. Describe a rational expression as an algebraic expression that is the quotient of two polynomials where the denominator is not equal to 0;
2. Determine whether or not two rational expressions are equal;
3. Simplify rational expressions, that is, reduce algebraic fractions by means of the Fundamental Principle of Fractions;
4. Multiply and divide rational expressions and simplify the results, if possible;
5. Find the sum or difference of rational expressions that have the same denominator;
6. Combine two rational expressions with unlike denominators by means of the "cross-product rule," $\frac{a}{b} + \frac{c}{d} = \frac{ad + bc}{bd}$;
7. Combine two or more rational expressions with unlike denominators by means of the least common denominator;
8. Describe a complex fraction as a fraction that has a fraction in its numerator or denominator, or both;
9. Simplify a complex fraction by combining the fractions in the numerator and denominator (if necessary) and performing the indicated division, or by multiplying the complex fraction by 1 in the form $\frac{m}{m}$ where m is the least common denominator of the fractions contained in the complex fraction;
10. Describe a fractional equation as an equation containing fractions where the variable occurs in the denominator;
11. Solve fractional equations by applying the techniques used in adding, subtracting, and multiplying rational expressions;
12. Realize that the solution set of a fractional equation must be checked, since solving the given fractional equation may produce an extraneous solution;
13. Translate word problems into fractional equations and solve the resulting equation, for example, in uniform motion problems and work problems.

6.1 INTRODUCTION

Thus far in our study of algebra, we have limited our discussion to polynomials. We have discussed monomials, binomials, trinomials, and polynomials in general. In our study of polynomials, we have examined the operations of addition, subtraction, multiplication, and division. Recall that in the development of the number system, we began with the set of natural numbers, proceeded to the set of whole numbers, and then

Rational Expressions (Algebraic Fractions)

to the set of integers. Next we discussed the set of rational numbers, and determined that a rational number is any number that can be expressed as the quotient of two integers, that is, $\frac{a}{b}$, $b \neq 0$. A rational expression is an algebraic expression of the form $\frac{P}{Q}$, where P and Q are polynomials, $Q \neq 0$. In other words, a rational expression is the quotient of two polynomials where the denominator is not equal to 0. Some examples of rational expressions are:

$$5, \frac{3}{4}, \frac{2}{x}, \frac{x}{y}, \frac{x+y}{x-y}, \frac{x^2+2x+3}{x^2+5x}, \frac{3x}{(2x-1)(2x+3)}.$$

Any possible value for the variables that will produce a denominator equal to 0 is excluded since division by 0 is undefined. We shall assume, therefore, that all denominators of rational expressions are not equal to 0.

In this chapter, we shall examine the various properties of rational expressions and examine the operations of addition, subtraction, multiplication, and division of rational expressions. We shall generalize the techniques for combining rational numbers to include rational expressions, since the techniques involved are essentially the same. Also, we shall sometimes refer to a rational expression as an *algebraic fraction* or simply, a *fraction*.

6.2 SIMPLIFYING FRACTIONS

Recall that we can say that two fractions such as $\frac{4}{8}$ and $\frac{3}{6}$ are equal, $\frac{4}{8} = \frac{3}{6}$, because when we cross multiply, the resulting products are equal.

$$\frac{4}{8} = \frac{3}{6}, \quad 4 \cdot 6 = 8 \cdot 3, \quad 24 = 24.$$

In general, we have $\frac{a}{b} = \frac{c}{d}$ whenever $ad = bc$. This is also the case for any two rational expressions. If $\frac{P}{Q}$ and $\frac{R}{S}$ are rational expressions, then $\frac{P}{Q} = \frac{R}{S}$ whenever $PS = QR$. For example, $\frac{x}{2y} = \frac{4x}{8y}$, since $x \cdot 8y = 2y \cdot 4x$. Similarly, $\frac{x+y}{x-y} = \frac{x^2+2xy+y^2}{x^2-y^2}$, because when we cross multiply, we have $(x+y)(x^2-y^2) = (x-y)(x^2+2xy+y^2)$ or $x^3 - xy^2 + yx^2 - y^3 = x^3 + x^2y - xy^2 - y^3$. We can use this property of equal fractions to verify that a rational expression has been reduced. We can verify that $\frac{3}{6} = \frac{1}{2}$, since $3 \cdot 2 = 6 \cdot 1$, or, 6 = 6. Similarly, we can verify that $\dfrac{x^2-y^2}{x^2+2xy+y^2} = \dfrac{x-y}{x+y}$ by examining the products $(x^2-y^2)(x+y)$ and $(x^2+2xy+y^2)(x-y)$. If the products are the same, then the fractions are equal. Hence, we have

$$(x^2 - y^2)(x+y) = x^3 + x^2y - xy^2 - y^3$$

and

$$(x^2 + 2xy + y^2)(x-y) = x^3 - x^2y + 2x^2y - 2xy^2 + xy^2 - y^3$$
$$= x^3 + x^2y - xy^2 - y^3.$$

We have verified that

$$\dfrac{x^2-y^2}{x^2+2xy+y^2} = \dfrac{x-y}{x+y}.$$

In the previous two chapters, we discussed special products and factoring. We can make use of these concepts to simplify rational expressions, that is, to reduce fractions. To write a simpler expression for $\dfrac{10xy}{15y^2}$, we can factor both the numerator and denominator, and then rewrite both the numerator and denominator of the given fraction, $\dfrac{10xy}{15y^2} = \dfrac{5y \cdot 2x}{5y \cdot 3y}$ or $\dfrac{5y}{5y} \cdot \dfrac{2x}{3y}$.

Now, since $\dfrac{5y}{5y} = 1$ and $1 \cdot \dfrac{2x}{3y} = \dfrac{2x}{3y}$, the original fraction $\dfrac{10xy}{15y^2}$ equals $\dfrac{2x}{3y}$. Since $\dfrac{P}{Q} = \dfrac{R}{S}$ whenever $PS = QR$, we can verify that $\dfrac{10xy}{15y^2} = \dfrac{2x}{3y}$ by checking the products, $10xy \cdot 3y$ and $15y^2 \cdot 2x$. Performing the indicated multiplication, we have $10xy \cdot 3y = 30xy^2$ and $15y^2 \cdot 2x = 30xy^2$. The products are equal, so we have verified that $\dfrac{10xy}{15y^2} = \dfrac{2x}{3y}$.

If we wish to simplify $\dfrac{5x+10y}{x^2-4y^2}$, we must first factor the numerator and denominator and then rewrite them.

$$\dfrac{5x+10y}{x^2-4y^2} = \dfrac{5(x+2y)}{(x+2y)(x-2y)} = \dfrac{(x+2y)5}{(x+2y)(x-2y)} = \dfrac{(x+2y)}{(x+2y)} \cdot \dfrac{5}{(x-2y)} = \dfrac{5}{(x-2y)}.$$

Therefore, the original fraction, $\dfrac{5x+10y}{x^2-4y^2}$, is equivalent to $\dfrac{5}{x-2y}$. We verify this by checking the products $(5x+10y)(x-2y)$ and $(x^2-4y^2)5$. Performing the indicated multiplication, we have $(5x+10y)(x-2y) = 5x^2 - 20y^2$ and $(x^2-4y^2)5 = 5x^2 - 20y^2$. We have verified that the two rational expressions are equivalent.

Recall that in Chapter 1 (Section 1.5), it was stated that to multiply

Rational Expressions (Algebraic Fractions)

two rational numbers (fractions), we multiply numerator times numerator and denominator times denominator, that is:

For any rational numbers, $\frac{a}{b}$ and $\frac{c}{d}$,

$$\frac{a}{b} \cdot \frac{c}{d} = \frac{a \cdot c}{b \cdot d} \quad (b \neq 0, d \neq 0).$$

We have been using a form of this rule to simplify the fractions that we have discussed thus far. Actually, we have been using the idea that $\frac{a}{b} = \frac{ca}{cb}$ ($c \neq 0$, $b \neq 0$). This is referred to as the Fundamental Principle of Fractions. The formal name is the Fundamental Law of Rational Expressions. It states:

If a, b and c are real numbers ($b \neq 0$, $c \neq 0$), then

$$\frac{a}{b} = \frac{ca}{cb}.$$

We can interpret this statement to mean that the value of a fraction is not changed when we multiply its numerator and denominator by the same non-zero expression.

The student may be more familiar with the idea of canceling common factors to simplify fractions. But it is not necessary to cancel common factors if we apply the Fundamental Principle of Fractions. Since $\frac{a}{b} = \frac{ca}{cb}$, we can also state that $\frac{ca}{cb} = \frac{a}{b}$ (using the symmetric property). The idea behind this is that $\frac{ca}{cb} = \frac{c}{c} \cdot \frac{a}{b}$. Since $\frac{c}{c} = 1$, we have $1 \cdot \frac{a}{b}$, and $1 \cdot \frac{a}{b} = \frac{a}{b}$ (using the identity property for the operation of multiplication in the system of real numbers). For example, to simplify the fraction $\frac{x^2 - 9}{x^2 + 6x + 9}$, we factor the numerator and denominator,

$$\frac{x^2 - 9}{x^2 + 6x + 9} = \frac{(x + 3)(x - 3)}{(x + 3)(x + 3)}.$$

By means of the Fundamental Principle of Fractions, $\left(\frac{a}{b} = \frac{ca}{cb}\right)$, we can rewrite the given fraction in a simpler form:

$$\frac{x^2 - 9}{x^2 + 6x + 9} = \frac{x - 3}{x + 3}.$$

We can say that the rational expression is written in *lowest terms* or *simplified* completely when neither the numerator nor the denominator contains a common factor other than ±1. Therefore, $\frac{x-3}{x+3}$ is written in lowest terms.

EXAMPLE 1

Simplify the following fractions.

a. $\frac{30}{42}$ b. $\frac{140}{330}$ c. $\frac{4}{9}$

Solution

To simplify (or reduce) a fraction, we factor both the numerator and denominator of the given fraction, and then apply the Fundamental Principle, $\frac{a}{b} = \frac{ca}{cb}$.

a. $\frac{30}{42} = \frac{2 \cdot 3 \cdot 5}{2 \cdot 3 \cdot 7} = \frac{2 \cdot 3}{2 \cdot 3} \cdot \frac{5}{7} = \frac{5}{7}$.

b. $\frac{140}{330} = \frac{2 \cdot 2 \cdot 5 \cdot 7}{2 \cdot 3 \cdot 5 \cdot 11} = \frac{2 \cdot 5 \cdot 2 \cdot 7}{2 \cdot 5 \cdot 3 \cdot 11} = \frac{2 \cdot 7}{3 \cdot 11} = \frac{14}{33}$.

c. $\frac{4}{9} = \frac{2 \cdot 2}{3 \cdot 3} = \frac{4}{9}$. (There are no common factors in the numerator and denominator other than ±1.)

EXAMPLE 2

Simplify the following:

a. $\frac{x^2 - 5x + 6}{x^2 - 7x + 10}$ b. $\frac{x^2 - y^2}{x^2 + 2xy + y^2}$

Solution

a. $\frac{x^2 - 5x + 6}{x^2 - 7x + 10} = \frac{(x-2)(x-3)}{(x-2)(x-5)} = \frac{(x-2)}{(x-2)} \cdot \frac{(x-3)}{(x-5)} = \frac{x-3}{x-5}$.

Note that in this step, we are applying the Fundamental Principle of Fractions.

b. $\frac{x^2 - y^2}{x^2 + 2xy + y^2} = \frac{(x+y)(x-y)}{(x+y)(x+y)} = \frac{(x+y)}{(x+y)} \cdot \frac{(x-y)}{(x+y)} = \frac{x-y}{x+y}$.

In solution b, we cannot simplify the fraction $\frac{x-y}{x+y}$ any further.

Rational Expressions (Algebraic Fractions)

The numerator and denominator do not contain any common factors, although they do contain common terms. Hence, it is in simplest form, or, it is written in lowest terms.

EXAMPLE 3

Simplify the following:

a. $\dfrac{6x^2 - 7x - 3}{15x^2 + 2x - 1}$ b. $\dfrac{x^3 - y^3}{x^2 - y^2}$

Solution

a. $\dfrac{6x^2 - 7x - 3}{15x^2 + 2x - 1} = \dfrac{(3x + 1)(2x - 3)}{(5x - 1)(3x + 1)}$

$= \dfrac{(3x + 1)}{(3x + 1)} \cdot \dfrac{(2x - 3)}{(5x - 1)} = \dfrac{2x - 3}{5x - 1}.$

b. We have the difference of two cubes in the numerator, and the difference of two squares in the denominator. Factoring both numerator and denominator, we have

$\dfrac{x^3 - y^3}{x^2 - y^2} = \dfrac{(x - y)(x^2 + xy + y^2)}{(x + y)(x - y)} =$

$= \dfrac{(x - y)}{(x - y)} \cdot \dfrac{(x^2 + xy + y^2)}{(x + y)} = \dfrac{x^2 + xy + y^2}{x + y}.$

Many times students try to reduce fractions of the form $\dfrac{x + y}{x - y}$ or $\dfrac{x - y}{x + y}$. But these two fractions are in simplest form, since the numerator and denominator do not contain a common factor. Note that in each case only one of the terms in the numerator and denominator differs in sign, that is, $x + y$ and $x - y$. Recall that binomials of this form are called *conjugates* of each other. Their sum is not 0. Expressions such as $x + y$ and $-x - y$ are *opposites* of each other and their sum is 0, $(x + y) + (-x - y) = 0$. If we have a fraction in which the numerator and denominator are opposites of each other, then the fractions can be simplified. Note that $x + y = 1(x + y)$ and $-x - y = -1(x + y)$. We can use this information to simplify a fraction in which the numerator and denominator are opposites of each other. For example,

$\dfrac{-x - y}{x + y} = \dfrac{-1(x + y)}{1(x + y)} = \dfrac{-1}{1} \cdot \dfrac{(x + y)}{(x + y)} = \dfrac{-1}{1} = -1.$

Polynomials that are opposites of each other must have opposite signs for each corresponding term. The binomials $x - y$ and $y - x$ are opposites of each other. The corresponding x terms have opposite signs, as do the corresponding y terms. Therefore, the fraction $\frac{x-y}{y-x}$ can be simplified.

$$\frac{x-y}{y-x} = \frac{1(x-y)}{-1(-y+x)} = \frac{1}{-1} \cdot \frac{(x-y)}{(x-y)} = \frac{1}{-1} = -1.$$

by means of the commutative property

In the two examples involving opposites that we just discussed, we encountered expressions of the form $\frac{-1}{1}$ and $\frac{1}{-1}$. Both expressions are equivalent to -1. It is sometimes helpful to recognize the different equivalent forms of fractions. For example, each of the following fractions is equal, $-\frac{3}{4} = \frac{-3}{4} = \frac{3}{-4}$. All of them are equivalent to -0.75. In general, we have

$$\boxed{-\frac{a}{b} = \frac{-a}{b} = \frac{a}{-b}.}$$

There are also equivalent expressions for fractions whose sign is positive. By means of the Fundamental Principle, we have $\frac{a}{b} = \frac{ca}{cb}$. Consider the case where $c = -1$; then we have $\frac{a}{b} = \frac{-1}{-1} \cdot \frac{a}{b} = \frac{-a}{-b}$. Using this property, we can also have the following cases: $\frac{a}{b} = \frac{-1}{-1} \cdot \frac{a}{b} = -\frac{-a}{b}$ and $\frac{a}{b} = \frac{-1}{-1} \cdot \frac{a}{b} = -\frac{a}{-b}$. Therefore, we can state that

$$\boxed{\frac{a}{b} = \frac{-a}{-b} = -\frac{-a}{b} = -\frac{a}{-b}.}$$

It is sometimes advantageous to write a fraction equivalent to a given fraction.

Rational Expressions (Algebraic Fractions)

EXAMPLE 4

Simplify the following:

a. $\dfrac{y - 3x}{3x - y}$ b. $\dfrac{x^2 - y^2}{2y^2 + xy - 3x^2}$ c. $\dfrac{2x + 3y}{2x - 3y}$

Solution

a. The binomials $y - 3x$ and $3x - y$ are opposites of each other. Their sum is 0. Therefore, the given fraction can be simplified.

$$\frac{y - 3x}{3x - y} = \frac{-1(-y + 3x)}{1(3x - y)} = \frac{-1}{1} \cdot \frac{(3x - y)}{(3x - y)} = \frac{-1}{1} = -1.$$

b. Factor both the numerator and denominator of the given fraction, and then apply the Fundamental Principle.

$$\frac{x^2 - y^2}{2y^2 + xy - 3x^2} = \frac{(x - y)(x + y)}{(y - x)(2y + 3x)}.$$

The numerator and denominator do not appear to contain any common factors, but note that $(x - y)$ and $(y - x)$ are opposites of each other. Therefore,

$$\frac{x^2 - y^2}{2y^2 + xy - 3x^2} = \frac{(x - y)(x + y)}{(y - x)(2y + 3x)} = \frac{-1(-x + y)(x + y)}{(y - x)(2y + 3x)}$$

$$= \frac{(y - x)}{(y - x)} \cdot \frac{(-1)(x + y)}{(2y + 3x)} = -\frac{x + y}{2y + 3x}.$$

There are also other equivalent answers, such as $\dfrac{-x - y}{2y + 3x}$ and $\dfrac{x + y}{-2y - 3x}$.

c. The fraction $\dfrac{2x + 3y}{2x - 3y}$ cannot be simplified. The numerator and denominator do not contain any common factors, and the binomials $2x + 3y$ and $2x - 3y$ are not opposites of each other. Their sum is not 0.

The simplification of fractions makes use of the special products and factoring that we have discussed previously. In order to simplify a rational expression, we factor both the numerator and denominator of the given fraction, and then apply the Fundamental Principle of Fractions,

$\frac{a}{b} = \frac{ca}{cb}$ (b ≠ 0, c ≠ 0). It should be noted that the variables contained in any fraction cannot be assigned values that will involve division by 0. In this text, it will be assumed that all examples are defined for all permissible replacements of the variable.

EXERCISES FOR SECTION 6.2

(Answers given to exercises marked *)

Determine whether the pairs of rational expressions are equal or not.

*1. $\frac{15}{20}, \frac{39}{52}$

2. $\frac{14}{16}, \frac{101}{169}$

*3. $\frac{x}{y^2}, \frac{x^3}{x^2y^2}$

4. $\frac{2xy}{z^3}, \frac{xy}{2z^3}$

*5. $\frac{x+y}{x-y}, \frac{y-x}{y+x}$

6. $\frac{a^2+b^2}{a^2+b^2}, \frac{a^2-b^2}{a^2+b^2}$

*7. $\frac{x^3-y^3}{x-y}, \frac{x^2+xy+y^2}{1}$

8. $\frac{x^2-y^2}{x+y}, \frac{y-x}{-1}$

*9. $\frac{-2x-3y}{x-y}, \frac{2x+3y}{y-x}$

10. $\frac{x^2+2xy+y^2}{x+y}, \frac{x^2+y^2}{x+y}$

*11. $\frac{-1}{x+y}, \frac{x-y}{x^2-y^2}$

12. $-\frac{x+y}{x-y}, \frac{-x-y}{y-x}$

Simplify each of the following, if possible.

*13. $\frac{210}{660}$

14. $\frac{221}{323}$

*15. $\frac{a^2b^3c}{a^2b^2c^2}$

16. $\frac{14x^3y^2}{7x^2y^3}$

*17. $\frac{12x^2y^3z}{18x^3y^4}$

18. $\frac{24xy^2}{36x^2y^3}$

*19. $\frac{-18s^2t^3}{27s^3t^4}$

20. $\frac{21x^4y^5}{-35x^5y^4}$

*21. $-\frac{12a^3b^4c^2}{36a^2b^5c^3}$

22. $\frac{8xy}{4x-12y}$

*23. $\frac{3x-6y}{12xy}$

24. $\frac{16x^2}{4x^3-8x^2}$

25. $\frac{6y^2-12y^3}{36y^3}$

26. $\frac{2y^2-4y}{4xy+8y}$

*27. $\frac{6xy+12y^2}{2y^2-4y}$

28. $\frac{5x-10y}{x^2-4y^2}$

*29. $\frac{x^2-y^2}{x^2+2xy+y^2}$

30. $\frac{x^3-y^3}{x^2-y^2}$

*31. $\frac{2x^2+9x+4}{2x^2-5x-3}$

32. $\frac{2y^2-y-3}{2y^2-5y+3}$

*33. $\frac{x^2+y^2}{x^2-y^2}$

34. $\frac{x^2-x-6}{x^2-8x+15}$

35. $\frac{y^2-5y+6}{-2y^2+10y-12}$

36. $\frac{x^2-4x-5}{-3x^2+12x+15}$

*37. $\frac{6x^2-13xy+6y^2}{2x^2-xy-3y^2}$

38. $\frac{2x^2-xy-15y^2}{6x^2+17xy+5y^2}$

*39. $\frac{2x^2-xy-y^2}{2x^2+3xy+y^2}$

Rational Expressions (Algebraic Fractions)

40. $\dfrac{2x^2+xy-y^2}{-2x^2-3xy-y^2}$ *41. $\dfrac{3x^2+2xy-y^2}{-6x^2-xy+y^2}$ 42. $\dfrac{x^2-y^2}{y^2+xy-2x^2}$

*43. $\dfrac{x^2-y^2}{2y^2-xy-x^2}$ 44. $\dfrac{2x+4y-6z}{-3x-6y+9z}$ *45. $\dfrac{x^2-y^2}{y^4-x^4}$

6.3 MULTIPLICATION AND DIVISION OF RATIONAL EXPRESSIONS

To multiply two rational numbers, we multiply the two numerators and the two denominators:

$$\dfrac{a}{b} \cdot \dfrac{c}{d} = \dfrac{a \cdot c}{b \cdot d} \quad (b \neq 0, d \neq 0).$$

Multiplication of rational expressions is performed in the same manner. To multiply two algebraic fractions, we obtain the numerator of the product by multiplying the numerators. To obtain the denominator of the product, we multiply the denominators. For example, $\dfrac{3x}{2y} \cdot \dfrac{5x}{7y^2} = \dfrac{3x \cdot 5x}{2y \cdot 7y^2} = \dfrac{15x^2}{14y^3}$. Many times we can rearrange common factors of the numerator and denominator (by means of the commutative and associative properties) so that the Fundamental Principle can be applied to show factors of 1; the resulting product is consequently simplified. For example,

$$\dfrac{5a^2}{7c^3} \cdot \dfrac{21c^4}{10a^3} = \dfrac{5a^2 \cdot 21c^4}{7c^3 \cdot 10a^3} = \dfrac{5 \cdot a \cdot a \cdot 3 \cdot 7 \cdot c \cdot c \cdot c \cdot c}{7 \cdot c \cdot c \cdot c \cdot 2 \cdot 5 \cdot a \cdot a \cdot a}$$

$$= \dfrac{7 \cdot 5 \cdot a^2 c^3}{7 \cdot 5 \cdot a^2 c^3} \cdot \dfrac{3c}{2a} = 1 \cdot \dfrac{3c}{2a} = \dfrac{3c}{2a}.$$

Note that the numerators and denominators of the given fractions are factored, and those factors common to the numerator and denominator are grouped to show factors of 1. We can use this procedure whenever we multiply two rational expressions. Considering another example, we have

$$\dfrac{x^2-4}{x^4} \cdot \dfrac{x^2}{x+2}.$$

First, we indicate the multiplication, and then factor both the numerator and denominator:

$$\dfrac{x^2-4}{x^4} \cdot \dfrac{x^2}{x+2} = \dfrac{(x^2-4)x^2}{x^4(x+2)} = \dfrac{(x+2)(x-2) \cdot x \cdot x}{x \cdot x \cdot x \cdot x(x+2)}.$$

Since multiplication is both commutative and associative, we arrange those factors common to both the numerator and denominator to illustrate factors of 1, and apply the Fundamental Principle. We then have

$$\frac{(x+2)(x-2) \cdot x \cdot x}{x \cdot x \cdot x \cdot x(x+2)} = \underbrace{\frac{x^2(x+2)}{x^2(x+2)}}_{\text{an expression equal to 1}} \cdot \frac{(x-2)}{x^2} = 1 \cdot \frac{(x-2)}{x^2} = \frac{x-2}{x^2}.$$

Normally, some of the steps shown here are omitted, but we have illustrated the complete process. The solution can be briefly shown as

$$\frac{x^2-4}{x^4} \cdot \frac{x^2}{x+2} = \frac{x^2(x+2)}{x^2(x+2)} \cdot \frac{(x-2)}{x^2} = \frac{x-2}{x^2}.$$

Let's try another example. Consider $\frac{3x^2-3x}{x^2-1} \cdot \frac{(x+1)^2}{6x^2}$. Factoring both the numerators and denominators of the indicated product, we have $\frac{3x(x-1)}{(x+1)(x-1)} \cdot \frac{(x+1)(x+1)}{3 \cdot 2 \cdot x \cdot x}$. Next, we group those factors common to the numerator and denominator of the product to illustrate factors of 1, and then apply the Fundamental Principle. The complete solution is:

$$\frac{3x^2-3x}{x^2-1} \cdot \frac{(x+1)^2}{6x^2} = \frac{3x(x-1)}{(x+1)(x-1)} \cdot \frac{(x+1)(x+1)}{3 \cdot 2 \cdot x \cdot x}$$

$$= \frac{3x(x-1)(x+1)}{3x(x-1)(x+1)} \cdot \frac{(x+1)}{2x} = 1 \cdot \frac{(x+1)}{2x} = \frac{x+1}{2x}.$$

EXAMPLE 5

Multiply $\frac{7ab^2}{11cx^3} \cdot \frac{22c^2x^2}{14abd}$.

Solution

First, we factor both the numerators and denominators of the indicated product, then group those factors common to the numerator and denominator to illustrate factors of 1, and then apply the Fundamental Principle.

$$\frac{7ab^2}{11cx^3} \cdot \frac{22c^2x^2}{14abd} = \frac{7 \cdot a \cdot b \cdot b}{11 \cdot c \cdot x \cdot x \cdot x} \cdot \frac{2 \cdot 11 \cdot c \cdot c \cdot x \cdot x}{2 \cdot 7 \cdot a \cdot b \cdot d}$$

$$= \frac{2 \cdot 7 \cdot 11 \cdot abcx^2}{2 \cdot 7 \cdot 11 \cdot abcx^2} \cdot \frac{bc}{xd} = 1 \cdot \frac{bc}{xd} = \frac{bc}{xd}.$$

After you gain experience, you will probably eliminate some of these steps.

Rational Expressions (Algebraic Fractions)

EXAMPLE 6

Multiply $\dfrac{9x^2 - 4}{2x^2 - x - 1} \cdot \dfrac{6x^2 + x - 1}{3x^2 + x - 2}$.

Solution

$$\dfrac{9x^2 - 4}{2x^2 - x - 1} \cdot \dfrac{6x^2 + x - 1}{3x^2 + x - 2} = \dfrac{(3x + 2)(3x - 2)}{(2x + 1)(x - 1)} \cdot \dfrac{(2x + 1)(3x - 1)}{(3x - 2)(x + 1)}$$

$$= \dfrac{(3x - 2)(2x + 1)}{(3x - 2)(2x + 1)} \cdot \dfrac{(3x + 2)(3x - 1)}{(x - 1)(x + 1)}$$

$$= \dfrac{(3x + 2)(3x - 1)}{(x - 1)(x + 1)}.$$

Note: The answers are usually left in factored form. A rational expression is simplified completely when it is in lowest terms and the parentheses have been eliminated. But, as we stated, it is common practice to express an answer in factored form.

EXAMPLE 7

Multiply $\dfrac{2y^2 - 2y - 4}{2y^2 - 5y + 2} \cdot \dfrac{2y^2 + 3y - 2}{3y^2 + 6y}$.

Solution

$$\dfrac{2(y + 1)(y - 2)}{(2y - 1)(y - 2)} \cdot \dfrac{(2y - 1)(y + 2)}{3y(y + 2)} = \dfrac{(y - 2)(2y - 1)(y + 2)}{(y - 2)(2y - 1)(y + 2)} \cdot \dfrac{2(y + 1)}{3y}$$

$$= \dfrac{2(y + 1)}{3y}.$$

EXAMPLE 8

Multiply $\dfrac{z^2 - 10z + 25}{5xz - xz^2} \cdot \dfrac{2z + 4}{z^2 - 3z - 10}$.

Solution

$$\dfrac{(z - 5)(z - 5)}{xz(5 - z)} \cdot \dfrac{2(z + 2)}{(z + 2)(z - 5)} = \dfrac{(z - 5)(z + 2)}{(z - 5)(z + 2)} \cdot \dfrac{(z - 5)}{(5 - z)} \cdot \dfrac{2}{xz}$$

the numerator and denominator are "opposites"

$$= 1 \cdot \frac{-1(-z+5)}{(5-z)} \cdot \frac{2}{xz}$$

$$= 1(-1) \cdot \frac{2}{xz} = \frac{-2}{xz}.$$

Note: The answer can also be written as $\frac{2}{-xz}$ or $-\frac{2}{xz}$.

EXAMPLE 9

Multiply $\dfrac{x^3-y^3}{x^2+xy+y^2} \cdot \dfrac{x+y}{x^2-y^2}$.

Solution

$$\frac{x^3-y^3}{x^2+xy+y^2} \cdot \frac{x+y}{x^2-y^2} = \frac{(x-y)(x^2+xy+y^2)}{(x^2+xy+y^2)} \cdot \frac{(x+y)}{(x+y)(x-y)}$$

$$= \frac{(x-y)(x+y)(x^2+xy+y^2)}{(x-y)(x+y)(x^2+xy+y^2)} = 1.$$

In Chapter One (Section 1.5), we discussed the set of rational numbers and the different arithmetic operations for the rational numbers. To divide two rational numbers, we multiply the first rational number (the dividend) by the multiplicative inverse of the second rational number (the divisor):

For any rational numbers, $\dfrac{a}{b}$ and $\dfrac{c}{d}$,

$$\frac{a}{b} \div \frac{c}{d} = \frac{a}{b} \cdot \frac{d}{c} \quad (b \neq 0, c \neq 0).$$

To divide $\frac{3}{4}$ by $\frac{5}{7}$, we have $\frac{3}{4} \div \frac{5}{7}$ and we can rewrite this as $\frac{3}{4} \cdot \frac{7}{5}$, which equals $\frac{21}{20}$. Summarizing the steps, we have $\frac{3}{4} \div \frac{5}{7} = \frac{3}{4} \cdot \frac{7}{5} = \frac{3 \cdot 7}{4 \cdot 5} = \frac{21}{20}$. To divide $\frac{3x}{4y^2}$ by $\frac{27x^2}{8y^3}$, we can rewrite the division problem as the multiplication problem (using the above definition), and then proceed as we normally would for any multiplication problem. Hence,

$$\frac{3x}{4y^2} \div \frac{27x^2}{8y^3} = \frac{3x}{4y^2} \cdot \frac{8y^3}{27x^2} = \frac{3 \cdot 4 \cdot x \cdot y^2}{3 \cdot 4 \cdot x \cdot y^2} \cdot \frac{2y}{9x} = 1 \cdot \frac{2y}{9x} = \frac{2y}{9x}.$$

When simplifying a problem involving division of rational expressions,

Rational Expressions (Algebraic Fractions)

first rewrite the indicated quotient as a multiplication problem (using the multiplicative inverse of the divisor), then factor the numerators and denominators of the indicated product. Finally, apply the Fundamental Principle by regrouping those factors common to the numerator and denominator of the indicated product.

EXAMPLE 10

Divide $\dfrac{x^2 - 4}{x^2 - 9}$ by $\dfrac{x - 2}{x + 3}$.

Solution

$$\dfrac{x^2 - 4}{x^2 - 9} \div \dfrac{x - 2}{x + 3} = \dfrac{x^2 - 4}{x^2 - 9} \cdot \dfrac{x + 3}{x - 2} = \dfrac{(x + 2)(x - 2)}{(x + 3)(x - 3)} \cdot \dfrac{(x + 3)}{(x - 2)}$$

$$= \dfrac{(x - 2)(x + 3)}{(x - 2)(x + 3)} \cdot \dfrac{(x + 2)}{(x - 3)} = \dfrac{x + 2}{x - 3}.$$

EXAMPLE 11

Divide and simplify: $\dfrac{x^2 - y^2}{x^2 + 2xy + y^2} \div \dfrac{x - y}{x + y}$

Solution

$$\dfrac{x^2 - y^2}{x^2 + 2xy + y^2} \div \dfrac{x - y}{x + y} = \dfrac{x^2 - y^2}{x^2 + 2xy + y^2} \cdot \dfrac{x + y}{x - y}$$

$$= \dfrac{(x + y)(x - y)}{(x + y)(x + y)} \cdot \dfrac{(x + y)}{(x - y)}$$

$$= \dfrac{(x + y)(x + y)(x - y)}{(x + y)(x + y)(x - y)} = 1.$$

EXAMPLE 12

Divide and simplify: $\dfrac{4y^2 + 8y + 3}{2y^2 - 5y - 3} \div \dfrac{4y^2 - 9}{3y^2 - 9y}$

Solution

$$\dfrac{4y^2 + 8y + 3}{2y^2 - 5y - 3} \div \dfrac{4y^2 - 9}{3y^2 - 9y} = \dfrac{4y^2 + 8y + 3}{2y^2 - 5y - 3} \cdot \dfrac{3y^2 - 9y}{4y^2 - 9}$$

$$= \dfrac{(2y + 3)(2y + 1)}{(2y + 1)(y - 3)} \cdot \dfrac{3y(y - 3)}{(2y + 3)(2y - 3)}$$

$$= \dfrac{(2y + 3)(2y + 1)(y - 3)}{(2y + 3)(2y + 1)(y - 3)} \cdot \dfrac{3y}{(2y - 3)}$$

$$= \dfrac{3y}{2y - 3}.$$

EXAMPLE 13

Divide and simplify: $\dfrac{x-y}{3x+6y} \div \dfrac{x^2 - 3xy + 2y^2}{x^2 + 7xy + 10y^2}$

Solution

$$\dfrac{x-y}{3x+6y} \div \dfrac{x^2 - 3xy + 2y^2}{x^2 + 7xy + 10y^2} = \dfrac{x-y}{3x+6y} \cdot \dfrac{x^2 + 7xy + 10y^2}{x^2 - 3xy + 2y^2}$$

$$= \dfrac{(x-y)}{3(x+2y)} \cdot \dfrac{(x+2y)(x+5y)}{(x-y)(x-2y)}$$

$$= \dfrac{(x-y)(x+2y)}{(x-y)(x+2y)} \cdot \dfrac{(x+5y)}{3(x-2y)}$$

$$= \dfrac{x+5y}{3(x-2y)}.$$

To multiply rational expressions, we first factor the numerators and denominators of the indicated product, next group those factors common to the numerator and denominator (by means of the commutative and associative properties) to show factors of 1, and then apply the Fundamental Principle. When simplifying a problem involving division of rational expressions, we must first rewrite the indicated quotient as a multiplication problem (using the multiplicative inverse of the divisor). We can then proceed in the same manner as we normally would for a multiplication problem.

EXERCISES FOR SECTION 6.3

(Answers given to exercises marked *)

Multiply and simplify:

*1. $\left(-\dfrac{4}{9}\right) \cdot \dfrac{27}{64}$ 2. $\left(-\dfrac{3}{8}\right)\left(-\dfrac{16}{12}\right)$ *3. $\dfrac{22}{21} \cdot \dfrac{35}{66}$

4. $\dfrac{35}{99} \cdot \dfrac{143}{105}$ *5. $\dfrac{3}{8} \cdot \dfrac{9}{11} \cdot \dfrac{16}{27}$ 6. $\left(-\dfrac{3}{8}\right)\left(-\dfrac{24}{36}\right)\left(-\dfrac{12}{27}\right)$

*7. $\dfrac{6xy}{4z} \cdot \dfrac{8xz}{18y^2}$ 8. $\dfrac{4xy}{3z} \cdot \dfrac{15xz}{25x^2}$ *9. $\dfrac{a^2b}{c^3d} \cdot \dfrac{c^2d^2}{a^4}$

10. $\dfrac{3c^2d^3}{4ab^2} \cdot \dfrac{12a^3b}{9c^3d}$ *11. $\dfrac{7u^3v^2}{11s^2} \cdot \dfrac{55s^4v}{35u^2}$ 12. $\dfrac{18x^2y}{25y^3} \cdot \dfrac{5xy}{9xy^2}$

Rational Expressions (Algebraic Fractions)   289

*13. $\dfrac{x+2}{x-3} \cdot \dfrac{x^2-5x-6}{x^2-4}$ 14. $\dfrac{y+2}{y+5} \cdot \dfrac{y^2-25}{2y^2+3y-2}$ *15. $\dfrac{z^2-6z+9}{z^2-2z+1} \cdot \dfrac{z-1}{z+3}$

16. $\dfrac{x^2-y^2}{x+y} \cdot \dfrac{3x^2+3xy}{x^2-2xy+y^2}$ *17. $\dfrac{x^3+y^3}{x+y} \cdot \dfrac{3x+2y}{x^2-xy+y^2}$

18. $\dfrac{x-2}{x+3} \cdot \dfrac{x^2+2x+4}{x^3-8}$ *19. $\dfrac{9y^2-25}{y^2-1} \cdot \dfrac{2y-2}{6y-10}$

20. $\dfrac{60xy}{10-5z} \cdot \dfrac{4z-8}{12x^2y^2}$ *21. $\dfrac{x^3-8}{x^2+2x+4} \cdot \dfrac{4-x^2}{x^2+x-2}$

22. $\dfrac{6x^2+7x-3}{3x^2+4x+1} \cdot \dfrac{x^2-1}{2x^2+x-3}$ *23. $\dfrac{3y^2-y-2}{4y^2+7y+3} \cdot \dfrac{8y^2+6y+1}{2-y-y^2}$

24. $\dfrac{x^6-y^6}{x^3+y^3} \cdot \dfrac{3x^2}{x^3-y^3}$ *25. $\dfrac{xs+xt+ys+yt}{ac+ad+bc+bd} \cdot \dfrac{ax-ay+bx-by}{sc-sd+tc-td}$

26. $\dfrac{xm+xn-zm-zn}{za+zb+xa+xb} \cdot \dfrac{bx-by+ax-ay}{ny-nx+my-mx}$

*27. $\dfrac{x^2+2x-15}{x^3-2x^2+x-2} \cdot \dfrac{6-3x}{x^2-25}$

Divide and simplify:

28. $\dfrac{6x^2}{x^2+2x-15} \div \dfrac{3x}{2x-6}$ *29. $\dfrac{3x+6}{4y-12} \div \dfrac{-2x+4}{y^2-3y}$

30. $\dfrac{yx+y^2}{x^2-xy} \div \dfrac{3x^2+3xy}{x^2-xy}$ *31. $\dfrac{4x^2-12x}{2y^2+4y} \div \dfrac{6y-2yx}{3xy+6x}$

32. $\dfrac{y^2-5y+6}{y^2+5y+4} \div \dfrac{2y-4}{3y+12}$ *33. $\dfrac{y^2+2y+1}{y^2+y-6} \div \dfrac{y^2+6y+5}{y^2+3y-10}$

34. $\dfrac{3x^2-x-2}{2x^2-x-1} \div \dfrac{3x^2+5x+2}{2x^2+3x+1}$ *35. $\dfrac{y^2-4y+4}{3y^2-5y-2} \div \dfrac{4-y^2}{3y^2+7y+2}$

36. $\dfrac{x^3-y^3}{2x^2-5xy+3y^2} \div \dfrac{x^2+xy+y^2}{2x^2-xy-3y^2}$

*37. $\dfrac{y^2-2y+4}{6y^2+8y+1} \div \dfrac{y^3+8}{2y^2+5y+2}$

38. $\dfrac{6x^2+5xy-6y^2}{6x^2+11xy+3y^2} \div \dfrac{2y^2-5xy+3x^2}{3x^2+4xy+y^2}$

*39. $\dfrac{4y^2+8xy+3x^2}{2y^2-xy-x^2} \div \dfrac{6y^2+11xy+3x^2}{x^2+2xy-3y^2}$

40. $\dfrac{ac+ad+bc+bd}{xs+xt+ys+yt} \div \dfrac{3c+3d}{2s+2t}$

6.4 ADDITION AND SUBTRACTION OF RATIONAL EXPRESSIONS

To find the sum of two rational expressions with the same denominator, we write down the common denominator for the sum and add the numerators. For example, $\frac{2}{x}+\frac{3}{x}=\frac{2+3}{x}=\frac{5}{x}$. Similarly, $\frac{2x}{z}+\frac{3y}{z}=\frac{2x+3y}{z}$ and $\frac{x+3}{x+1}+\frac{2x+4}{x+1}=\frac{x+3+2x+4}{x+1}=\frac{3x+7}{x+1}$. Recall that the operation of subtraction can be performed by adding the additive inverse of the subtrahend, i.e., $a-b=a+(-b)$. Therefore, to find the difference of two rational expressions with the same denominator, we can rewrite the problem as an addition problem and proceed as we normally would for an addition problem. For example, $\frac{3}{x}-\frac{2}{x}=\frac{3}{x}+\frac{-2}{x}=\frac{3+(-2)}{x}=\frac{1}{x}$. Similarly $\frac{2x}{z}-\frac{3y}{z}=\frac{2x}{z}+\frac{-3y}{z}=\frac{2x-3y}{z}$, and $\frac{x+3}{x+1}-\frac{2x+4}{x+1}=\frac{x+3}{x+1}+\frac{-(2x+4)}{x+1}=\frac{x+3-(2x+4)}{x+1}=\frac{x+3-2x-4}{x+1}=\frac{-x-1}{x+1}$. Note that the numerator and denominator of the last example are "opposites," so their sum is 0. Hence, we can simplify the rational expression $\frac{-x-1}{x+1}$, that is, $\frac{-x-1}{x+1}=\frac{-1(x+1)}{1(x+1)}=\frac{-1}{1}=-1$. We have shown that $\frac{x+3}{x+1}-\frac{2x+4}{x+1}=-1$. This example also illustrates the point that we should examine the resulting sums and differences for common factors and simplify (reduce) the rational expression, if possible.

Consider the following addition problem:

$$\frac{x}{x^2-y^2}+\frac{y}{x^2-y^2}$$

We can perform the indicated operation (addition) since both rational expressions (fractions) have the same denominator. Therefore, we write down the common denominator for the sum and add the numerators. Hence, $\frac{x}{x^2-y^2}+\frac{y}{x^2-y^2}=\frac{x+y}{x^2-y^2}$. But, the denominator, x^2-y^2, is the difference of two squares and can be factored into $(x+y)(x-y)$. We can simplify the resulting sum $\frac{x+y}{x^2-y^2}$, that is, $\frac{x+y}{x^2-y^2}=\frac{1(x+y)}{(x-y)(x+y)}=\frac{(x+y)}{(x+y)}\cdot\frac{1}{(x-y)}=\frac{1}{x-y}$. We should always examine the sums and differences for common factors and simplify, if possible.

Rational Expressions (Algebraic Fractions)

EXAMPLE 14

Perform the indicated operations and simplify, if possible.

a. $\dfrac{3x+1}{y} + \dfrac{2x-5}{y}$
b. $\dfrac{2x+3}{x+2} - \dfrac{x-1}{x+2}$

c. $\dfrac{-2x+1}{x^2-9} + \dfrac{3x+2}{x^2-9}$
d. $\dfrac{x+2}{x^2+2x+1} + \dfrac{2x+1}{x^2+2x+1}$

Solution

Note that in each exercise, the rational expressions have the same denominator. Therefore, we can write down the common denominator and combine the numerators.

a. $\dfrac{3x+1}{y} + \dfrac{2x-5}{y} = \dfrac{3x+1+2x-5}{y} = \dfrac{5x-4}{y}.$

b. To find the difference of two rational expressions with the same denominator, we can rewrite the problem as an addition problem (recall $a-b = a+(-b)$) and proceed in a normal manner. Hence,

$$\dfrac{2x+3}{x+2} - \dfrac{x-1}{x+2} = \dfrac{2x+3}{x+2} + \dfrac{-(x-1)}{x+2} = \dfrac{2x+3-(x-1)}{x+2}$$

$$= \dfrac{2x+3-x+1}{x+2} = \dfrac{x+4}{x+2}.$$

c. $\dfrac{-2x+1}{x^2-9} + \dfrac{3x+2}{x^2-9} = \dfrac{-2x+1+3x+2}{x^2-9} = \dfrac{x+3}{x^2-9}$

$$= \dfrac{x+3}{(x+3)(x-3)} = \dfrac{1(x+3)}{(x+3)(x-3)}$$

$$= \dfrac{(x+3)}{(x+3)} \cdot \dfrac{1}{(x-3)} = \dfrac{1}{x-3}.$$

d. $\dfrac{x+2}{x^2+2x+1} + \dfrac{2x+1}{x^2+2x+1} = \dfrac{x+2+2x+1}{x^2+2x+1} = \dfrac{3x+3}{x^2+2x+1}$

$$= \dfrac{3(x+1)}{(x+1)(x+1)} = \dfrac{(x+1)}{(x+1)} \cdot \dfrac{3}{(x+1)} = \dfrac{3}{x+1}.$$

The Cross-Product Rule

We can combine fractions that have the same denominator by writing down the common denominator and combining the numerators (as illustrated in Example 14). But we must be careful (and use a different procedure) when we combine fractions with unlike denominators. To combine two fractions with unlike denominators, we have two options. One is to use the rule that we used in Chapter 1 (Section 1.5), which states:

> For any two rational numbers, $\dfrac{a}{b}$ and $\dfrac{c}{d}$ ($b \neq 0$, $d \neq 0$),
>
> $$\dfrac{a}{b} + \dfrac{c}{d} = \dfrac{ad + bc}{bd}.$$

For example, $\dfrac{2}{3} + \dfrac{4}{5} = \dfrac{2 \cdot 5 + 3 \cdot 4}{3 \cdot 5} = \dfrac{10 + 12}{15} = \dfrac{22}{15}$. This is sometimes referred to as the *cross-product rule*. Considering other examples using this rule, we have $\dfrac{x+2}{5} + \dfrac{2x-1}{7} = \dfrac{(x+2)7 + 5(2x-1)}{5 \cdot 7} = \dfrac{7x + 14 + 10x - 5}{35} = \dfrac{17x + 9}{35}$; also, $\dfrac{5}{12y} + \dfrac{3}{4y} = \dfrac{5(4y) + (12y)3}{(12y)(4y)} = \dfrac{20y + 36y}{48y^2} = \dfrac{56y}{48y^2} = \dfrac{8y \cdot 7}{8y \cdot 6y} = \dfrac{7}{6y}$.

The LCD Option

The other option in combining fractions with unlike denominators is to use the *least common denominator*, which we abbreviate LCD. The least common denominator of two or more rational expressions is the *least common multiple* of the denominators. To determine the least common multiple for two or more numbers, factor the given numbers completely. If they have no common factors, then their least common multiple is their product. If the given numbers have prime factors in common, then their least common multiple is the product of each different prime factor, including each factor to the highest power in which it appears in any of the given numbers. For example, to find the least common multiple of 12 and 18, we have $12 = 2 \cdot 2 \cdot 3$ and $18 = 2 \cdot 3 \cdot 3$. Therefore, the least common multiple is $2 \cdot 2 \cdot 3 \cdot 3 = 36$. It is the smallest multiple of 12 and 18, the smallest number that contains all of their factors. Hence, to combine $\dfrac{5}{12} + \dfrac{7}{18}$ (using the LCD), we must first rewrite each fraction so their denominators are the LCD (36). Using the Fundamental Principle of Fractions, $\dfrac{a}{b} = \dfrac{ca}{cb}$, we have

$$\dfrac{5}{12} + \dfrac{7}{18} = \dfrac{5}{12} \cdot \dfrac{3}{3} + \dfrac{7}{18} \cdot \dfrac{2}{2} = \dfrac{15}{36} + \dfrac{14}{36} = \dfrac{15 + 14}{36} = \dfrac{29}{36}.$$

To determine the least common multiple of $x^3 - 4x$ and $x^2 + 4x + 4$ we first factor the given expressions: $x^3 - 4x = x(x^2 - 4) = x(x+2)(x-2)$ and $x^2 + 4x + 4 = (x+2)(x+2)$. Therefore, the least common multiple

Rational Expressions (Algebraic Fractions)

of $x^3 - 4x$ and $x^2 + 4x + 4$ is $x(x-2)(x+2)(x+2)$ or $x(x-2)(x+2)^2$. Note that the factor $(x + 2)$ occurs twice in the expression $x^2 + 4x + 4$, hence it must occur twice in the least common multiple. Also, the least common multiple is given in factored form.

To combine $\dfrac{7}{10y} + \dfrac{11}{15y}$ using the LCD, we must first find the least common multiple of the denominators. Hence, we factor the denominators of each fraction completely:

$$10y = 2 \cdot 5 \cdot y \quad \text{and} \quad 15y = 3 \cdot 5 \cdot y.$$

The least common multiple of 10y and 15y is $2 \cdot 3 \cdot 5 \cdot y = 30y$. Therefore, the LCD of the given rational expressions is 30y. Now, we rewrite each fraction so that the denominator of each is the LCD (30y).

$$\frac{7}{10y} + \frac{11}{15y} = \frac{7}{10y} \cdot \frac{3}{3} + \frac{11}{15y} \cdot \frac{2}{2} = \frac{21}{30y} + \frac{22}{30y} = \frac{21 + 22}{30y} = \frac{43}{30y}.$$

Let's consider another example. To combine $\dfrac{5}{4x - 12} - \dfrac{7}{4x + 12}$ using the LCD, we must factor the denominators of each fraction:

$$4x - 12 = 4(x - 3) \quad \text{and} \quad 4x + 12 = 4(x + 3).$$

The LCD is $4(x - 3)(x + 3)$. Rewriting each fraction with an LCD of $4(x - 3)(x + 3)$, we have

$$\frac{5}{4x - 12} - \frac{7}{4x + 12} = \frac{5}{4(x - 3)} \cdot \frac{(x + 3)}{(x + 3)} - \frac{7}{4(x + 3)} \cdot \frac{(x - 3)}{(x - 3)}$$

Note: In algebra the LCD is usually given in factored form.

$$= \frac{5(x + 3)}{4(x - 3)(x + 3)} - \frac{7(x - 3)}{4(x + 3)(x - 3)}$$

$$= \frac{5(x + 3) - 7(x - 3)}{4(x - 3)(x + 3)} = \frac{5x + 15 - 7x + 21}{4(x - 3)(x + 3)}$$

$$= \frac{-2x + 33}{4(x - 3)(x + 3)}.$$

The numerator and denominator do not contain any common factors. Hence, the result is in simplest form.

We have discussed two different techniques for combining algebraic fractions with unlike denominators. One method is to use the cross-product rule, $\dfrac{a}{b} + \dfrac{c}{d} = \dfrac{ad + bc}{bd}$. The cross-product method is suitable when the LCD of the given fractions is the product of the denominators. It is also suitable when the work is performed mentally. For example, to add

$\frac{y}{3} + \frac{2y}{7}$, we have $\frac{y}{3} + \frac{2y}{7} = \frac{13y}{21}$ (mentally computing $\frac{y \cdot 7 + 3 \cdot 2y}{3 \cdot 7}$). But to combine two *or more* rational expressions when the denominators are unlike and may also contain some common factors, we will generally find it easier to use the technique involving the least common denominator (LCD).

EXAMPLE 15

Perform the indicated operations.

a. $\frac{4x}{7} + \frac{3x}{5}$ b. $\frac{x+3}{2} + \frac{2x+1}{5}$ c. $\frac{3y-2}{3} - \frac{2y-1}{5}$

Solution

Since in each case the product of the denominators is the LCD, we shall use the cross-product rule, $\frac{a}{b} + \frac{c}{d} = \frac{ad + bc}{bd}$.

a. $\frac{4x}{7} + \frac{3x}{5} = \frac{4x \cdot 5 + 7 \cdot 3x}{7 \cdot 5} = \frac{20x + 21x}{35} = \frac{41x}{35}$.

b. $\frac{x+3}{2} + \frac{2x+1}{5} = \frac{5(x+3) + 2(2x+1)}{2 \cdot 5} = \frac{5x + 15 + 4x + 2}{10}$

$= \frac{9x + 17}{10}$.

c. $\frac{3y-2}{3} - \frac{2y-1}{5} = \frac{5(3y-2) - 3(2y-1)}{3 \cdot 5}$

$= \frac{15y - 10 - 6y + 3}{15} = \frac{9y - 7}{15}$.

Note: care must be exercised to obtain the correct signs when simplifying an expression such as $-3(2y - 1)$.

EXAMPLE 16

Perform the indicated operations and simplify, if possible.

a. $\frac{2x+7}{60} + \frac{x-5}{90}$ b. $\frac{2x+3}{24} - \frac{x-4}{36}$ c. $\frac{y}{2} + \frac{2y}{5} - \frac{3y}{10}$

Solution

Since in each case the product of the denominators is *not* the LCD

Rational Expressions (Algebraic Fractions)

and the denominators do contain some common factors, we shall use the technique involving the LCD.

a. We first factor completely the denominators of each fraction. $60 = 2 \cdot 2 \cdot 3 \cdot 5, 90 = 2 \cdot 3 \cdot 3 \cdot 5$. The LCD is the least common multiple of 60 and 90, $2 \cdot 2 \cdot 3 \cdot 3 \cdot 5 = 180$. Next, we rewrite each fraction so its denominator is the LCD (180).

$$\frac{2x+7}{60} + \frac{x-5}{90} = \frac{2x+7}{60} \cdot \frac{3}{3} + \frac{x-5}{90} \cdot \frac{2}{2}$$

$$= \frac{3(2x+7)}{180} + \frac{2(x-5)}{180}$$

$$= \frac{3(2x+7) + 2(x-5)}{180} = \frac{6x + 21 + 2x - 10}{180}$$

$$= \frac{8x + 11}{180}.$$

b. We must use care when subtracting in order to obtain all of the correct signs. $24 = 2 \cdot 2 \cdot 2 \cdot 3, 36 = 2 \cdot 2 \cdot 3 \cdot 3$. Therefore, the LCD $= 2 \cdot 2 \cdot 2 \cdot 3 \cdot 3 = 72$.

$$\frac{2x+3}{24} - \frac{x-4}{36} = \frac{2x+3}{24} \cdot \frac{3}{3} - \frac{x-4}{36} \cdot \frac{2}{2}$$

$$= \frac{3(2x+3)}{72} - \frac{2(x-4)}{72}$$

$$= \frac{3(2x+3) - 2(x-4)}{72} = \frac{6x + 9 - 2x + 8}{72}$$

$$= \frac{4x + 17}{72}.$$

c. The LCD technique is particularly useful when combining more than two terms. The least common multiple for 2, 5, and 10 is 10. Two and five are prime and $10 = 2 \cdot 5$. Hence, the LCD is 10.

$$\frac{y}{2} + \frac{2y}{5} - \frac{3y}{10} = \frac{y}{2} \cdot \frac{5}{5} + \frac{2y}{5} \cdot \frac{2}{2} - \frac{3y}{10} = \frac{5y}{10} + \frac{4y}{10} - \frac{3y}{10}$$

$$= \frac{5y + 4y - 3y}{10} = \frac{6y}{10} = \frac{2 \cdot 3y}{2 \cdot 5} = \frac{3y}{5}.$$

EXAMPLE 17

Perform the indicated operations and simplify, if possible.

a. $\dfrac{3}{2x + 10} + \dfrac{4}{3x + 15}$ b. $\dfrac{7y}{y^2 - 4} + \dfrac{3y}{y^2 + 2y}$

Solution

a. $2x + 10 = 2(x + 5)$, $3x + 15 = 3(x + 5)$. The LCD is $2 \cdot 3(x + 5)$. Therefore,

$$\frac{3}{2x + 10} + \frac{4}{3x + 15} = \frac{3}{2(x + 5)} \cdot \frac{3}{3} + \frac{4}{3(x + 5)} \cdot \frac{2}{2}$$

$$= \frac{9}{6(x + 5)} + \frac{8}{6(x + 5)} = \frac{9 + 8}{6(x + 5)}$$

$$= \frac{17}{6(x + 5)}.$$

b. $y^2 - 4 = (y + 2)(y - 2)$, $y^2 + 2y = y(y + 2)$. The LCD is $y(y + 2)(y - 2)$. Therefore,

$$\frac{7y}{y^2 - 4} + \frac{3y}{y^2 + 2y} = \frac{7y}{(y + 2)(y - 2)} \cdot \frac{y}{y} + \frac{3y}{y(y + 2)} \cdot \frac{(y - 2)}{(y - 2)}$$

$$= \frac{7y^2}{y(y + 2)(y - 2)} + \frac{3y(y - 2)}{y(y + 2)(y - 2)}$$

$$= \frac{7y^2 + 3y(y - 2)}{y(y + 2)(y - 2)} = \frac{7y^2 + 3y^2 - 6y}{y(y + 2)(y - 2)}$$

$$= \frac{10y^2 - 6y}{y(y + 2)(y - 2)} = \frac{2y(5y - 3)}{y(y + 2)(y - 2)}$$

$$= \frac{y \cdot 2(5y - 3)}{y \cdot (y + 2)(y - 2)} = \frac{2(5y - 3)}{(y + 2)(y - 2)}.$$

The numerator and denominator do not contain any *other* common factors. Hence, the result is in simplest form and we leave the final result in factored form.

Sometimes we may be asked to combine fractions where it appears that we must use either the cross-product technique or the LCD, such as $\frac{2}{x - 1} + \frac{3}{1 - x}$. Note that the denominators are unlike, but they are opposites of each other. Their sum is 0, $x - 1 + 1 - x = 0$. When two fractions have denominators that are opposites of each other, we can rewrite one of them so that their denominators will be like. Recall that the Fundamental Principle of Fractions states that $\frac{a}{b} = \frac{ca}{cb}$ ($b \neq 0$, $c \neq 0$). Therefore, to combine $\frac{2}{x - 1} + \frac{3}{1 - x}$, we can multiply one of the fractions by 1. But we shall write 1 in the form of $\frac{-1}{-1}$. Hence, we have

$$\frac{2}{x - 1} + \frac{3}{1 - x} = \frac{2}{x - 1} + \frac{-1}{-1} \cdot \frac{3}{(1 - x)} = \frac{2}{x - 1} + \frac{-3}{-1 + x}.$$

Rational Expressions (Algebraic Fractions)

Now the denominators are like and we write down the common denominators for the sum and combine the numerators.

$$\frac{2}{x-1} + \frac{-3}{-1+x} = \frac{2-3}{x-1} = \frac{-1}{x-1}.$$

Note that other possible correct solutions are $\frac{1}{1-x}$ and $-\frac{1}{x-1}$. It should also be noted that this technique should be used only when the denominators of the given fractions are opposites.

EXAMPLE 18

Perform the indicated operations and simplify, if possible.

a. $\dfrac{2x}{x-y} + \dfrac{3y}{y-x}$ b. $\dfrac{x-2}{1-x} - \dfrac{x-3}{x-1}$

Solution

In each case the denominators are opposites of each other. Therefore, we can rewrite one of the fractions by multiplying it by 1, that is, $\dfrac{-1}{-1}$.

a. $\dfrac{2y}{x-y} + \dfrac{3y}{y-x} = \dfrac{2y}{x-y} + \dfrac{-1}{-1} \cdot \dfrac{3y}{(y-x)} = \dfrac{2x}{x-y} + \dfrac{-3y}{-y+x}$

$= \dfrac{2x-3y}{x-y}.$

b. $\dfrac{x-2}{1-x} - \dfrac{x-3}{x-1} = \dfrac{-1}{-1} \cdot \dfrac{(x-2)}{(1-x)} - \dfrac{x-3}{1-x} = \dfrac{-x+2}{-1+x} - \dfrac{x-3}{x-1}$

$= \dfrac{-x+2-(x-3)}{x-1} = \dfrac{-x+2-x+3}{x-1} = \dfrac{-2x+5}{x-1}.$

When the LCD for a set of algebraic fractions contains more than two factors, we must exercise extreme care in performing the work. It is easy to make an error in simplifying problems of this nature.

EXAMPLE 19

Perform the indicated operation and simplify, if possible.

$$\frac{y+3}{y^2+y-6} + \frac{y+7}{y^2-3y+2}$$

Solution

First, factor completely the denominators of each fraction. $y^2 + y - 6 = (y + 3)(y - 2)$; $y^2 - 3y + 2 = (y - 2)(y - 1)$. The LCD is $(y + 3)(y - 2)(y - 1)$. Next, rewrite each fraction so that its denominator is the LCD.

$$\frac{y+3}{(y+3)(y-2)} + \frac{y+7}{(y-2)(y-1)}$$

$$= \frac{y+3}{(y+3)(y-2)} \cdot \frac{(y-1)}{(y-1)} + \frac{y+7}{(y-2)(y-1)} \cdot \frac{(y+3)}{(y+3)}$$

$$= \frac{(y+3)(y-1)}{(y+3)(y-2)(y-1)} + \frac{(y+7)(y+3)}{(y+3)(y-2)(y-1)}$$

$$= \frac{(y+3)(y-1) + (y+7)(y+3)}{(y+3)(y-2)(y-1)} = \frac{y^2 + 2y - 3 + y^2 + 10y + 21}{(y+3)(y-2)(y-1)}$$

$$= \frac{2y^2 + 12y + 18}{(y+3)(y-2)(y-1)} = \frac{2(y+3)(y+3)}{(y+3)(y-2)(y-1)}$$

$$= \frac{2(y+3)}{(y-2)(y-1)}.$$

We can leave the answer in factored form to indicate that it is in simplest form.

EXAMPLE 20

Perform the indicated operation and simplify, if possible.

$$\frac{2}{x+1} + \frac{1}{x+2} - \frac{3}{x+3}$$

Solution

The LCD is the product of the denominators, $(x+1)(x+2)(x+3)$. Therefore, we have

$$\frac{2}{x+1} + \frac{1}{x+2} - \frac{3}{x+3}$$

$$= \frac{2}{x+1} \cdot \frac{(x+2)(x+3)}{(x+2)(x+3)} + \frac{1}{x+2} \cdot \frac{(x+1)(x+3)}{(x+1)(x+3)}$$

$$- \frac{3}{x+3} \cdot \frac{(x+1)(x+2)}{(x+1)(x+2)}$$

$$= \frac{2(x+2)(x+3) + 1(x+1)(x+3) - 3(x+1)(x+2)}{(x+1)(x+2)(x+3)}$$

$$= \frac{2x^2 + 10x + 12 + x^2 + 4x + 3 - 3x^2 - 9x - 6}{(x+1)(x+2)(x+3)}$$

$$= \frac{5x + 9}{(x+1)(x+2)(x+3)}.$$

Rational Expressions (Algebraic Fractions)

EXERCISES FOR SECTION 6.4

(Answers given to exercises marked *)

Perform the indicated operations and simplify, if possible.

*1. $\dfrac{4}{y} + \dfrac{x}{y}$

2. $\dfrac{x}{y} - \dfrac{3}{y}$

*3. $\dfrac{4}{x} + \dfrac{y-2}{x}$

4. $\dfrac{x-3}{x+1} + \dfrac{2x+1}{x+1}$

*5. $\dfrac{3x-2}{x-2} - \dfrac{x+1}{x-2}$

6. $\dfrac{y+1}{y-1} - \dfrac{y-1}{y-1}$

*7. $\dfrac{x}{x^2-y^2} - \dfrac{y}{x^2-y^2}$

8. $\dfrac{y}{y^2-4} + \dfrac{2}{y^2-4}$

*9. $\dfrac{x}{x^2-9} - \dfrac{3}{x^2-9}$

10. $\dfrac{3x+5}{x^2+2x+1} - \dfrac{2x+4}{x^2+2x+1}$

*11. $\dfrac{-3x-6}{x^2+3x+1} + \dfrac{4x+8}{x^2+3x+1}$

12. $\dfrac{x+1}{2x^2-x-6} + \dfrac{x+2}{2x^2-x-6}$

*13. $\dfrac{2}{x} + \dfrac{3}{2x}$

14. $\dfrac{4}{y} - \dfrac{3}{2y}$

*15. $\dfrac{1}{x-1} - \dfrac{1}{x}$

16. $\dfrac{3}{x+1} + \dfrac{2}{x+2}$

*17. $\dfrac{4}{y-2} + \dfrac{3}{y+2}$

18. $\dfrac{x}{x+y} + \dfrac{y}{x+y}$

*19. $\dfrac{x}{x+y} - \dfrac{y}{x+y}$

20. $\dfrac{2x}{x+3} + \dfrac{3x}{x-1}$

*21. $\dfrac{3y}{y+2} - \dfrac{3y}{y-1}$

22. $\dfrac{2}{x} + \dfrac{3}{y} + \dfrac{4}{z}$

*23. $\dfrac{3}{x} - \dfrac{2}{y} - \dfrac{5}{z}$

24. $\dfrac{5}{4x} + \dfrac{3}{2y} - \dfrac{2}{8z}$

*25. $\dfrac{2}{4x} + \dfrac{3}{2x} + \dfrac{5}{8x}$

26. $\dfrac{4}{15y} + \dfrac{2}{5y} + \dfrac{1}{12y}$

*27. $\dfrac{3}{2y} - \dfrac{1}{9y} + \dfrac{2}{6y}$

28. $\dfrac{2x+3}{20} + \dfrac{x+1}{10}$

*29. $\dfrac{3y+2}{30} - \dfrac{y+1}{20}$

30. $\dfrac{y+1}{33} + \dfrac{2y-1}{44}$

*31. $\dfrac{2x+1}{3x+2} + \dfrac{x-2}{x+1}$

32. $\dfrac{y+1}{y+2} + \dfrac{3y+2}{2y+1}$

*33. $\dfrac{2y-1}{3y+1} - \dfrac{y+1}{y-1}$

34. $\dfrac{x+2}{3x+6} + \dfrac{x+1}{4x+8}$

*35. $\dfrac{x+1}{4x-12} + \dfrac{x+2}{2x-6}$ 36. $\dfrac{3y+1}{y+2} - \dfrac{y+1}{3y+6}$

*37. $\dfrac{x+1}{x^2-1} + \dfrac{x-1}{x+1}$ 38. $\dfrac{y+2}{y-2} - \dfrac{y+1}{y^2-4}$

39. $\dfrac{y-3}{y^2-9} - \dfrac{y+3}{y-3}$ 40. $\dfrac{x+2}{x-2} + \dfrac{x+3}{2-x}$

*41. $\dfrac{y+1}{y-1} + \dfrac{y-2}{1-y}$ 42. $\dfrac{x+3}{x-3} + \dfrac{x+3}{3-x}$

*43. $\dfrac{x+y}{x-y} - \dfrac{y-x}{y-x}$ 44. $\dfrac{x-1}{x^2+3x+2} + \dfrac{x+1}{x^2+2x}$

45. $\dfrac{y+1}{y^2-4y+3} - \dfrac{y+2}{y^2-3y}$ 46. $\dfrac{x-3}{x^2+6x+9} + \dfrac{x+1}{x^2-9}$

*47. $\dfrac{2x+1}{x^2-4} + \dfrac{3x+2}{x^2-4x+4}$ 48. $\dfrac{2y+1}{y^2-10y+25} + \dfrac{y-1}{y^2-25}$

49. $\dfrac{3}{x^2-2x+1} + \dfrac{2}{x^2+x-2}$ 50. $\dfrac{4}{y^2-6y+9} + \dfrac{3}{y^2-4y+3}$

*51. $\dfrac{y+2}{y^2-4y+4} - \dfrac{y-1}{y^2-y-2}$ 52. $\dfrac{x+1}{x^2+x-2} + \dfrac{x-1}{x^2+5x+6}$

*53. $\dfrac{x+2}{x^2-1} + \dfrac{x+3}{x^2-4x+3}$ 54. $\dfrac{y-1}{y^2+2y+1} + \dfrac{y+2}{y^2-1}$

*55. $\dfrac{1}{x+1} + \dfrac{2}{x-1} + \dfrac{3}{x+2}$ 56. $\dfrac{2}{y-1} + \dfrac{3}{y+1} - \dfrac{1}{y-2}$

*57. $\dfrac{3}{x-1} - \dfrac{2}{x+2} - \dfrac{3}{x-3}$

6.5 COMPLEX FRACTIONS

In Chapter 1 (Section 1.5), we briefly discussed complex fractions. A *complex fraction* contains a fraction in its numerator or denominator, or both.

For example, $\dfrac{\tfrac{2}{3}}{\tfrac{5}{7}}$ is a complex fraction, as are $\dfrac{4}{\tfrac{2}{3}}$ and $\dfrac{3}{\tfrac{1}{2}}$.

A rational expression is an algebraic fraction and the quotient of two rational expressions is also a complex fraction, so the numerator or denominator (or both) contains a fraction. The fraction $\dfrac{\tfrac{x}{x+y}}{\tfrac{y}{x-y}}$ is another

Rational Expressions (Algebraic Fractions)

example of a complex fraction. Usually the major fraction line (division line) is made longer to avoid confusion about which is the numerator and which is the denominator of the complex fraction.

We can simplify a complex fraction in two ways. Since a fraction bar indicates division, we can perform the indicated division and simplify. For example, $\dfrac{\frac{2}{3}}{\frac{5}{7}}$ is a complex fraction and we have $\dfrac{2}{3} \div \dfrac{5}{7}$. To divide two fractions, we multiply the first fraction (the dividend) by the multiplicative inverse of the second fraction (the divisor). Therefore,

$$\frac{\frac{2}{3}}{\frac{5}{7}} = \frac{2}{3} \cdot \frac{7}{5} = \frac{14}{15}.$$

An alternate technique for simplifying complex fractions is to multiply the fraction by 1. But the form of 1 should also be a fraction, and its numerator and denominator should be the LCD of the fractions contained in the complex fraction. The LCD of the fractions in $\dfrac{\frac{2}{3}}{\frac{5}{7}}$ is 21. Hence, we multiply the complex fraction by 1, or $\dfrac{21}{21}$. We have

$$\frac{\frac{2}{3}}{\frac{5}{7}} \cdot 1 = \frac{\frac{2}{3}}{\frac{5}{7}} \cdot \frac{21}{21} = \frac{\frac{2}{3} \cdot 21}{\frac{5}{7} \cdot 21} = \frac{14}{15}.$$

We are not advocating one technique over the other. Both are useful for simplifying complex fractions. The one you understand is the best technique to use.

Sometimes a complex fraction may contain more than one fraction in the numerator or denominator (or both). For example, $\dfrac{\frac{1}{2}+\frac{1}{3}}{\frac{2}{3}+\frac{1}{5}}$ is also a complex fraction. In order to simplify it, we can first combine those fractions contained in the numerator and denominator, and then simplify the resulting complex fraction. Combining the fractions in the numerator and denominator of the example, we have $\frac{1}{2}+\frac{1}{3}=\frac{3+2}{6}=\frac{5}{6}$ and $\frac{2}{3}+\frac{1}{5}=\frac{10+3}{15}=\frac{13}{15}$. Therefore,

$$\frac{\frac{1}{2}+\frac{1}{3}}{\frac{2}{3}+\frac{1}{5}} = \frac{\frac{5}{6}}{\frac{13}{15}} = \frac{5}{6} \cdot \frac{15}{13} = \frac{75}{78} = \frac{25}{26}.$$

Since the LCD of $\frac{1}{2}, \frac{1}{3}, \frac{2}{3}$ and $\frac{5}{6}$ is 30, we can also simplify the complex fraction by multiplying it by $\frac{30}{30}$:

$$\frac{\frac{1}{2}+\frac{1}{3}}{\frac{2}{3}+\frac{1}{5}} = \frac{30}{30} \cdot \frac{\frac{1}{2}+\frac{1}{3}}{\frac{2}{3}+\frac{1}{5}} = \frac{30\left(\frac{1}{2}+\frac{1}{3}\right)}{30\left(\frac{2}{3}+\frac{1}{5}\right)} = \frac{30 \cdot \frac{1}{2} + 30 \cdot \frac{1}{3}}{30 \cdot \frac{2}{3} + 30 \cdot \frac{1}{5}} = \frac{15+10}{20+6} = \frac{25}{26}.$$

> To simplify a complex fraction, we can first combine the fractions in the numerator and those in the denominator (if necessary) and then simplify the resulting complex fraction by performing the indicated division, that is, multiply the fraction in the numerator by the multiplicative inverse of the fraction in the denominator. We can also simplify a complex fraction by multiplying it by 1, and the 1 should be of the form $\frac{m}{m}$, where m is the LCD of the fractions contained in the complex fraction.

EXAMPLE 21

Simplify $\dfrac{2-\frac{1}{3}}{3+\frac{1}{2}}$.

Solution

$$\frac{2-\frac{1}{3}}{3+\frac{1}{2}} = \frac{\frac{6}{3}-\frac{1}{3}}{\frac{6}{2}+\frac{1}{2}} = \frac{\frac{5}{3}}{\frac{7}{2}} = \frac{5}{3} \cdot \frac{2}{7} = \frac{10}{21}.$$

Rational Expressions (Algebraic Fractions) 303

Alternate Solution

The LCD of $2, \frac{1}{3}, 3$ and $\frac{1}{2}$ is 6. Hence,

$$\frac{2-\frac{1}{3}}{3+\frac{1}{2}} = \frac{6}{6} \cdot \frac{2-\frac{1}{3}}{3+\frac{1}{2}} = \frac{6\left(2-\frac{1}{3}\right)}{6\left(3+\frac{1}{2}\right)} = \frac{6\cdot 2 - 6\cdot\frac{1}{3}}{6\cdot 3 + 6\cdot\frac{1}{2}} = \frac{12-2}{18+3} = \frac{10}{21}.$$

EXAMPLE 22

Simplify $\dfrac{\frac{1}{x}+\frac{1}{y}}{\frac{1}{x}-\frac{1}{y}}$.

Solution

$$\frac{\frac{1}{x}+\frac{1}{y}}{\frac{1}{x}-\frac{1}{y}} = \frac{\frac{y+x}{xy}}{\frac{y-x}{xy}} = \frac{y+x}{xy} \cdot \frac{xy}{y-x} = \frac{y+x}{y-x}.$$

Alternate Solution

The LCD of $\frac{1}{x}$ and $\frac{1}{y}$ is xy. Therefore,

$$\frac{\frac{1}{x}+\frac{1}{y}}{\frac{1}{x}-\frac{1}{y}} = \frac{xy}{xy} \cdot \frac{\frac{1}{x}+\frac{1}{y}}{\frac{1}{x}-\frac{1}{y}} = \frac{xy\left(\frac{1}{x}+\frac{1}{y}\right)}{xy\left(\frac{1}{x}-\frac{1}{y}\right)} = \frac{xy\cdot\frac{1}{x} + xy\cdot\frac{1}{y}}{xy\cdot\frac{1}{x} - xy\cdot\frac{1}{y}} = \frac{y+x}{y-x}.$$

EXAMPLE 23

Simplify $\dfrac{\frac{x^2-4}{x-5}}{\frac{x-2}{x^2-25}}$.

Solution

Since we do not need to combine any fractions, we perform the indicated division.

$$\frac{\dfrac{x^2-4}{x-5}}{\dfrac{x-2}{x^2-25}} = \frac{x^2-4}{x-5} \cdot \frac{x^2-25}{x-2} = \frac{(x+2)(x-2)}{(x-5)} \cdot \frac{(x+5)(x-5)}{(x-2)}$$

$$= (x+2)(x+5).$$ We can leave the answer in factored form.

Alternate Solution

The LCD of the fractions is $x^2 - 25$. Hence,

$$\frac{\dfrac{x^2-4}{x-5}}{\dfrac{x-2}{x^2-25}} = \frac{x^2-25}{x^2-25} \cdot \frac{\left(\dfrac{x^2-4}{x-5}\right)}{\left(\dfrac{x-2}{x^2-25}\right)} = \frac{(x+5)(x-5)}{(x^2-25)} \cdot \frac{\dfrac{(x+2)(x-2)}{(x-5)}}{\dfrac{x-2}{x^2-25}}$$

$$= \frac{(x+5)(x+2)(x-2)}{(x-2)} = (x+5)(x+2).$$

Note: In this example, the first solution is less confusing, but the choice of techniques is a matter of personal preference.

EXAMPLE 24

Simplify $\dfrac{\dfrac{1}{x+y} + \dfrac{1}{x-y}}{\dfrac{1}{x+y} - \dfrac{1}{x-y}}$.

Solution

$$\frac{\dfrac{1}{x+y} + \dfrac{1}{x-y}}{\dfrac{1}{x+y} - \dfrac{1}{x-y}} = \frac{\dfrac{x-y+x+y}{(x+y)(x-y)}}{\dfrac{x-y-x-y}{(x+y)(x-y)}} = \frac{\dfrac{2x}{(x+y)(x-y)}}{\dfrac{-2y}{(x+y)(x-y)}}$$

$$= \frac{2x}{(x+y)(x-y)} \cdot \frac{(x+y)(x-y)}{-2y} = \frac{2x}{-2y} = -\frac{x}{y}.$$

Alternate Solution

The LCD is $(x+y)(x-y)$. Therefore,

$$\frac{\dfrac{1}{x+y} + \dfrac{1}{x-y}}{\dfrac{1}{x+y} - \dfrac{1}{x-y}} = \frac{(x+y)(x-y)}{(x+y)(x-y)} \cdot \frac{\left(\dfrac{1}{x+y} + \dfrac{1}{x-y}\right)}{\left(\dfrac{1}{x+y} - \dfrac{1}{x-y}\right)}$$

Rational Expressions (Algebraic Fractions)

$$= \frac{(x+y)(x-y) \cdot \frac{1}{(x+y)} + (x+y)(x-y) \cdot \frac{1}{(x-y)}}{(x+y)(x-y) \cdot \frac{1}{(x+y)} - (x+y)(x-y) \cdot \frac{1}{(x-y)}}$$

$$= \frac{x-y+x+y}{x-y-(x+y)} = \frac{2x}{-2y} = -\frac{x}{y}.$$

EXAMPLE 25

Simplify $1 + \dfrac{1}{1 + \dfrac{1}{x}}$.

Solution

To simplify this indicated sum, we must first simplify the complex fraction.

$$\frac{1}{1+\frac{1}{x}} = \frac{1}{\frac{x+1}{x}} = 1 \cdot \frac{x}{x+1} = \frac{x}{x+1}. \text{ Therefore,}$$

$$1 + \frac{1}{1+\frac{1}{x}} = 1 + \frac{x}{x+1} = \frac{x+1+x}{x+1} = \frac{2x+1}{x+1}.$$

EXERCISES FOR SECTION 6.5

(Answers given to exercises marked*)

Simplify each of the following:

*1. $\dfrac{\frac{3}{11}}{\frac{4}{9}}$

2. $\dfrac{\frac{2}{13}}{\frac{5}{7}}$

*3. $\dfrac{1+\frac{1}{2}}{2-\frac{1}{3}}$

4. $\dfrac{3+\frac{1}{7}}{2-\frac{1}{5}}$

*5. $\dfrac{\frac{3}{7}+\frac{4}{5}}{\frac{2}{3}-\frac{1}{2}}$

6. $\dfrac{\frac{3}{5}-\frac{1}{3}}{\frac{4}{7}-\frac{3}{11}}$

*7. $\dfrac{2+\frac{1}{x}}{3-\frac{1}{x}}$

8. $\dfrac{5-\frac{1}{y}}{3+\frac{2}{y}}$

*9. $\dfrac{x+\frac{1}{y}}{y+\frac{1}{x}}$

10. $\dfrac{\dfrac{2}{x}+\dfrac{1}{y}}{\dfrac{1}{x}+\dfrac{2}{y}}$
*11. $\dfrac{\dfrac{3}{x}+\dfrac{2}{y}}{\dfrac{2}{x}+\dfrac{3}{y}}$
12. $\dfrac{\dfrac{x}{y}-\dfrac{y}{x}}{\dfrac{x}{y}+\dfrac{y}{x}}$

*13. $\dfrac{\dfrac{y}{x}-\dfrac{x}{y}}{\dfrac{x}{y}-\dfrac{y}{x}}$
14. $\dfrac{\dfrac{1}{x}+\dfrac{1}{y}}{\dfrac{1}{x^2}+\dfrac{1}{y^2}}$
*15. $\dfrac{\dfrac{1}{x}-\dfrac{1}{y}}{\dfrac{1}{x^2}-\dfrac{1}{y^2}}$

16. $\dfrac{1+\dfrac{x}{y}}{1-\dfrac{x^2}{y^2}}$
*17. $\dfrac{2+\dfrac{y}{x}}{4-\dfrac{y^2}{x^2}}$
18. $\dfrac{3x+\dfrac{1}{y}}{\dfrac{1}{y^2}-9x^2}$

*19. $\dfrac{\dfrac{x}{x+1}+\dfrac{1}{x}}{\dfrac{1}{x}-\dfrac{1}{x+1}}$
20. $\dfrac{\dfrac{x}{x+y}+\dfrac{1}{y}}{\dfrac{x}{y}+\dfrac{1}{x+y}}$
*21. $\dfrac{\dfrac{x-y}{x}+\dfrac{x}{x+y}}{\dfrac{x-y}{x+y}+\dfrac{1}{x}}$

22. $\dfrac{\dfrac{3}{x+y}+\dfrac{2}{x-y}}{\dfrac{2}{x+y}-\dfrac{3}{x-y}}$
*23. $\dfrac{\dfrac{1}{x+y}+\dfrac{1}{x-y}}{\dfrac{2x}{x^2-y^2}}$
24. $\dfrac{\dfrac{x+y}{x-y}+\dfrac{x-y}{x+y}}{\dfrac{x-y}{x+y}-\dfrac{x+y}{x-y}}$

*25. $\dfrac{\dfrac{x+2}{x+3}+\dfrac{x+1}{x-2}}{\dfrac{x-1}{x+5}-\dfrac{x-1}{x+2}}$
26. $\dfrac{\dfrac{y-1}{y+1}+\dfrac{y+1}{y-1}}{\dfrac{y-1}{y+1}-\dfrac{y+1}{y-1}}$
*27. $\dfrac{\dfrac{y+2}{y-3}-\dfrac{y-2}{y-1}}{\dfrac{y-2}{y-1}-\dfrac{y+1}{y-2}}$

28. $\dfrac{\dfrac{x-2}{x-1}-\dfrac{x+1}{x+2}}{\dfrac{x-2}{x+1}-\dfrac{x-1}{x+2}}$
*29. $\dfrac{x+\dfrac{1}{x}+1}{x^2-\dfrac{1}{x}}$
30. $\dfrac{y+2+\dfrac{4}{y}}{y^2-\dfrac{8}{y}}$

*31. $\dfrac{x+4}{x-\dfrac{4}{x+3}}$
32. $\dfrac{x+1}{x}+\dfrac{x}{x-\dfrac{1}{x}}$
*33. $y+\dfrac{1}{1+\dfrac{1}{y}}$

34. $1+\dfrac{1}{1+\dfrac{1}{x}}$
*35. $1+\dfrac{1}{1+\dfrac{1}{1+\dfrac{1}{1+1}}}$

36. $1-\dfrac{1}{1+\dfrac{1}{1-\dfrac{1}{1+\dfrac{1}{1-\dfrac{1}{1+1}}}}}$

Rational Expressions (Algebraic Fractions)

HISTORICAL NOTE

Many of the symbols that we now use in algebra were adopted through repeated use by different people and their exact origins are not known. For example, the sign for plus (+) is believed to have been adapted from the symbol for "and," the ampersand, &. After repeatedly writing the symbol for "and," it began to appear as + instead of &. The plus sign did not originate with mathematicians—it was first used by merchants in the 1400s to indicate excess. For example, if a bolt of cloth was supposed to contain 10 yards of cloth but it actually contained 13 yards, then a merchant had an excess of 3 yards and would mark "10 + 3" on the package.

The exact origin of the symbol for equality (=) is known. It first appeared in an English algebra book, Whetstone of Witte (1557), written by Robert Recorde. The author stated that he chose a pair of horizontal line segments to represent equality because "noe 2 thynges can be moare equalle." Recorde's symbol for equality was greeted cautiously and not universally accepted until the early 1700s.

6.6 SOLVING FRACTIONAL EQUATIONS

In Chapter 2 (Section 2.3), we discussed the process for solving equations that contained fractions. But those equations were polynomial equations, which means the variable did *not* occur in the denominator of the fraction. In this section, we shall examine the process for solving equations containing rational expressions, where the variable does occur in the denominator of a fraction. These are called *fractional equations*. For example, $\frac{3}{x} = \frac{2}{x+1}$ is a fractional equation. Similarly, $\frac{2}{y-1} - \frac{3}{y} = 0$ and $\frac{1}{2y} + \frac{2}{y-1} = \frac{3}{y}$ are fractional equations. We shall see that many of the techniques used in adding, subtracting, and multiplying rational expressions will be used to solve fractional equations. The LCD for rational expressions is of particular importance. As an example, let us consider the fractional equation

$$\frac{3}{x} = \frac{2}{x+1}.$$

We already learned to solve an equation if it is of the general form $ax + b = c$. To transform the given equation into a similar general form, we must eliminate the denominators. How can we do this? We can use the multiplication property of equality: multiply both sides of the equation by some expression that will eliminate the denominators of both fractions. The expression that we should use is the LCD of the fractions contained in the equation. The LCD for the fractions in the given equation is $x(x+1)$. Therefore, we multiply both members by $x(x+1)$.

$$\frac{3}{x} = \frac{2}{x+1}.$$

$$x(x+1)\left(\frac{3}{x}\right) = x(x+1)\left(\frac{2}{x+1}\right).$$

By means of the associative and commutative properties we can simplify this equation:

$$(x+1)3 = x \cdot 2 \text{ or } 3(x+1) = 2x.$$

Simplifying again, we have

$$3x + 3 = 2x.$$
$$x + 3 = 0.$$
$$x = -3.$$

Recall that the solution set of the original equation is a subset of the solution set of the new equation. Therefore, we must check the solution, $x = -3$, in the original equation. The reason for this is that we have multiplied both sides of the equation by an expression involving the unknown. Since the variable occurs in the denominator of a fraction, the solution of the derived equation may make the denominator of one of the fractions in the original equation equal to 0, and division by 0 is impossible. We check the proposed solution.

Check: $\quad \dfrac{3}{x} = \dfrac{2}{x+1},\quad$ if $x = -3$.

$$\frac{3}{-3} = \frac{2}{-3+1}.$$

$$-1 = \frac{2}{-2}.$$

$-1 = -1.\quad$ The solution set is $\{-3\}$.

Consider the fractional equation:

$$\frac{4}{x-2} - \frac{5}{x+2} = \frac{20}{x^2 - 4}.$$

We can solve this equation if we can eliminate the denominators of the fractions, by transforming it into an equivalent equation where the variable does not occur in the denominator of a fraction. We first determine

Rational Expressions (Algebraic Fractions)

the LCD. Note that $x^2 - 4 = (x + 2)(x - 2)$. The LCD is $(x + 2)(x - 2)$ and we multiply both members of the equation by it.

$$(x + 2)(x - 2)\left(\frac{4}{x - 2} - \frac{5}{x + 2}\right) = (x + 2)(x - 2) \cdot \frac{20}{x^2 - 4}$$

by means of the distributive property

$$(x + 2)(x - 2) \cdot \frac{4}{(x - 2)} - (x + 2)(x - 2) \cdot \frac{5}{(x + 2)}$$

$$= (x + 2)(x - 2) \cdot \frac{20}{(x + 2)(x - 2)}.$$

$$4(x + 2) - 5(x - 2) = 20.$$

$$4x + 8 - 5x + 10 = 20.$$

$$-x + 18 = 20.$$

$$-x = 2.$$

$$x = -2.$$

We check the solution of the derived equation.

$$\frac{4}{-2 - 2} - \frac{5}{-2 + 2} = \frac{20}{4 - 4}$$

$$\frac{4}{-4} - \frac{5}{0} = \frac{20}{0}.$$

The solution does not check. Division by 0 is not possible and, therefore, we must reject the solution of the derived equation. It is an extraneous root. The solution set of the original equation is the empty set, ∅. It is a subset of the solution set of the new equation. Since the solution set of the given equation is ∅, we say that the given equation has no solution.

EXAMPLE 26

Find the solution set for

$$\frac{3}{2x} + \frac{1}{6} = \frac{2}{3x}.$$

Solution

To solve a fractional equation, we first determine the LCD of the fractions in the equation. Next, we multiply both sides of the equation by the LCD, simplify the new equation, and solve it. The LCD is $2 \cdot 3 \cdot x = 6x$.

$$6x\left(\frac{3}{2x}+\frac{1}{6}\right) = 6x\left(\frac{2}{3x}\right).$$

Check: $\frac{3}{2(-5)} + \frac{1}{6} \stackrel{?}{=} \frac{2}{3(-5)}.$

$$6x \cdot \frac{3}{2x} + 6x \cdot \frac{1}{6} = 6x \cdot \frac{2}{3x}.$$

$$-\frac{3}{10} + \frac{1}{6} \stackrel{?}{=} \frac{-2}{15}.$$

$$3 \cdot 3 + x \cdot 1 = 2 \cdot 2.$$

$$-\frac{9}{30} + \frac{5}{30} = \frac{-2}{15}.$$

$$9 + x = 4.$$

$$\frac{-4}{30} = \frac{-2}{15}.$$

$$x = -5.$$

$$\frac{-2}{15} = \frac{-2}{15}.$$

The solution set is $\{-5\}$.

EXAMPLE 27

Find the solution set for

$$\frac{1}{x^2-1} + \frac{1}{x} = \frac{1}{x+1}.$$

Solution

$x^2 - 1 = (x+1)(x-1)$ and the LCD is $x(x+1)(x-1)$.

$$x(x+1)(x-1)\left(\frac{1}{x^2-1} + \frac{1}{x}\right) = x(x+1)(x-1) \cdot \left(\frac{1}{x+1}\right).$$

$$x(x+1)(x-1) \cdot \frac{1}{(x+1)(x-1)} + x(x+1)(x-1) \cdot \frac{1}{x}$$

$$= x(x+1)(x-1) \cdot \frac{1}{x+1}.$$

$$x \cdot 1 + (x+1)(x-1) \cdot 1 = x(x-1) \cdot 1.$$

$$x + x^2 - 1 = x^2 - x.$$

$$2x = 1.$$

$$x = \frac{1}{2}.$$

Check: $\frac{1}{\left(\frac{1}{2}\right)^2 - 1} + \frac{1}{\frac{1}{2}} \stackrel{?}{=} \frac{1}{\frac{1}{2}+1}.$

$$\frac{1}{\frac{1}{4}-1} + 2 \stackrel{?}{=} \frac{1}{\frac{3}{2}}.$$

Rational Expressions (Algebraic Fractions) 311

$$\frac{1}{\frac{-3}{4}} + 2 = \frac{2}{3}.$$

$$\frac{-4}{3} + 2 = \frac{2}{3}.$$

$$\frac{-4}{3} + \frac{6}{3} = \frac{2}{3}.$$

$$\frac{2}{3} = \frac{2}{3}. \quad \text{The solution set is } \left\{\frac{1}{2}\right\}.$$

EXAMPLE 28

Find the solution set for

$$\frac{x+2}{x-3} = \frac{x+1}{x-1}.$$

Solution

The LCD is $(x-3)(x-1)$.

$$(x-3)(x-1) \cdot \frac{x+2}{x-3} = (x-3)(x-1) \cdot \frac{x+1}{x-1}.$$

$$(x-1)(x+2) = (x-3)(x+1).$$
$$x^2 + x - 2 = x^2 - 2x - 3.$$
$$3x = -1.$$

$$x = -\frac{1}{3}.$$

Check: $$\frac{-\frac{1}{3} + 2}{-\frac{1}{3} - 3} \stackrel{?}{=} \frac{-\frac{1}{3} + 1}{-\frac{1}{3} - 1}.$$

$$\frac{\frac{5}{3}}{\frac{-10}{3}} = \frac{\frac{2}{3}}{-\frac{4}{3}}.$$

$$-\frac{5}{10} = -\frac{2}{4}.$$

$$-\frac{1}{2} = -\frac{1}{2}. \quad \text{The solution set is } \left\{-\frac{1}{3}\right\}.$$

EXAMPLE 29

Find the solution set for

$$\frac{3}{x+2} + \frac{4}{x-3} = \frac{20}{x^2 - x - 6}.$$

Solution

$x^2 - x - 6 = (x-3)(x+2)$ and the LCD is $(x-3)(x+2)$.

$$(x-3)(x+2)\left(\frac{3}{x+2} + \frac{4}{x-3}\right) = (x-3)(x+2) \cdot \frac{20}{x^2 - x - 6}.$$

$$(x-3)(x+2) \cdot \frac{3}{x+2} + (x-3)(x+2) \cdot \frac{4}{x-3}$$

$$= (x-3)(x+2) \cdot \frac{20}{(x-3)(x+2)}.$$

$$3(x-3) + 4(x+2) = 20.$$
$$3x - 9 + 4x + 8 = 20.$$
$$7x - 1 = 20.$$
$$7x = 21.$$
$$x = 3.$$

Check: $\quad \dfrac{3}{3+2} + \dfrac{4}{3-3} \stackrel{?}{=} \dfrac{20}{9-3-6}.$

$$\frac{3}{5} + \frac{4}{0} = \frac{20}{0}.$$

The solution does not check. Division by zero is not possible; $x = 3$ is an extraneous root. Therefore, the solution set is ∅.

To solve fractional equations, meaning equations that contain rational expressions, we use the multiplication property of equality. We can transform the given equation into an equation that does not contain any fractions by multiplying both sides of the equation by the LCD in the denominator. After obtaining a solution to the new transformed equation, we check it to make sure that it satisfies the original equation. If the solution of the transformed equation does not satisfy the given equation, then we say that the given equation has no solution and its solution set is the empty set.

Rational Expressions (Algebraic Fractions)

EXERCISES FOR SECTION 6.6

(Answers given to exercises marked *)

Find the solution set for each of the following equations:

*1. $\dfrac{y}{2} + \dfrac{y}{3} = 5$.

2. $\dfrac{2x}{3} + \dfrac{3x}{4} = x + 5$.

*3. $\dfrac{3x}{5} - \dfrac{2x}{3} = 1$.

4. $\dfrac{2y-5}{3} + \dfrac{3y+2}{2} = 1$.

*5. $\dfrac{y-1}{3} - \dfrac{2y+1}{5} = 1$.

6. $\dfrac{x-2}{2} - \dfrac{x+1}{3} = \dfrac{1}{6}$.

*7. $\dfrac{1}{x} + \dfrac{1}{2x} = \dfrac{1}{4}$.

8. $\dfrac{2}{y} - \dfrac{1}{3y} = \dfrac{5}{6}$.

*9. $\dfrac{3}{5y} - \dfrac{1}{2y} = \dfrac{1}{10}$.

10. $\dfrac{2}{x+5} + \dfrac{3}{x-5} = \dfrac{1}{x-5}$.

*11. $\dfrac{2}{x-2} + \dfrac{3}{x+2} = \dfrac{8}{x^2-4}$.

12. $\dfrac{1}{x} + \dfrac{2}{x-2} = \dfrac{3}{x^2-2x}$.

*13. $\dfrac{5}{x-3} - \dfrac{2}{x+3} = \dfrac{13}{x^2-9}$.

14. $\dfrac{4}{y+1} = \dfrac{3}{y-1}$.

*15. $\dfrac{5}{y+3} = \dfrac{4}{y+2}$.

16. $\dfrac{1}{y^2-1} + \dfrac{1}{y} - \dfrac{1}{y+1} = 0$.

*17. $\dfrac{3}{x^2-9} + \dfrac{2}{x+3} - \dfrac{2}{x} = 0$.

18. $\dfrac{y+2}{y-3} = \dfrac{y+1}{y-1}$.

*19. $\dfrac{y-1}{y+2} = \dfrac{y+3}{y+5}$.

20. $\dfrac{x+1}{x+2} = \dfrac{x-1}{x-2}$.

*21. $\dfrac{y+3}{y-1} = \dfrac{y-2}{y+2}$.

22. $\dfrac{3}{2x+2} + \dfrac{2}{3x+3} = \dfrac{1}{6}$.

*23. $\dfrac{4}{5x-10} - \dfrac{2}{3x-6} = \dfrac{1}{15}$.

24. $\dfrac{2}{x-1} - \dfrac{1}{1-x} = \dfrac{1}{2}$.

*25. $\dfrac{4}{y+1} + \dfrac{2}{y-2} = \dfrac{6}{y^2-y-2}$.

26. $\dfrac{5}{x-1} - \dfrac{2}{x+3} = \dfrac{-2}{x^2+2x-3}$.

*27. $\dfrac{3}{x+1} + \dfrac{2}{x+2} = \dfrac{7}{x^2+3x+2}$.

28. $\dfrac{3}{x+2} - \dfrac{2}{x-1} = \dfrac{2x+1}{x^2+x-2}$.

*29. $\dfrac{x-2}{x^2+4x+3} - \dfrac{2}{x+3} = \dfrac{-2}{x+1}$.

30. $\dfrac{3x+2}{x^2+x} + \dfrac{1}{x} = \dfrac{2}{x+1}$.

*31. $\dfrac{x-3}{x-2} - \dfrac{x+4}{x+2} = \dfrac{x+6}{x^2-4}$.

32. $\dfrac{x+1}{x+2} - \dfrac{x-2}{x-3} = \dfrac{2-x}{x^2-x-6}$.

*33. $\dfrac{y-4}{y+2} - \dfrac{y-3}{y+3} = \dfrac{y-4}{y^2+5y+6}$.

(© 1968 United Features Syndicate, Inc.)

6.7 APPLICATIONS

In Chapter 2 (Section 2.5), we discussed the general procedures for solving word problems. Here we shall again examine various types of word problems, but we shall deal with those applications whose solution is obtained from fractional equations. That is, when we set up the equation from the stated problem, the equation will contain rational expressions. Recall the list of suggestions to assist us in solving word or verbal problems:

1. Read the problem carefully;
2. Reread the problem carefully;
3. If possible, draw a diagram to assist in interpreting the given information;
4. Translate the English phrases into mathematical phrases, and choose a variable for the unknown quantity;
5. Write the equation using all of the above information;
6. Solve the equation;
7. Check the solution to determine whether it satisfies the original problem.

Now let us examine a few examples that will result in fractional equations.

EXAMPLE 30

Benny's super-charged van can travel a distance of 200 miles in the same time that Larry's van can travel a distance of 150 miles. Benny's van goes 20 miles per hour faster than Larry's. What is the speed of each van?

Solution

To solve problems concerning rate of speed, we must use the formula $d = rt$, where d represents the distance traveled, r represents the rate of speed and t represents the time.

Rational Expressions (Algebraic Fractions)

Read the problem carefully. Let x represent the speed of Larry's van. Then x + 20 represents the speed of Benny's van. Since both vans traveled the same amount of time, we can use this fact in determining an equation. But instead of using d = rt, we solve the formula for t and $t = \frac{d}{r}$.

Larry traveled 150 miles in his van at a rate of x mph. Hence, Larry's time is

$$t = \frac{150}{x}.$$

Benny traveled 200 miles in his van at a rate of x + 20. Benny's time is

$$t = \frac{200}{x + 20}.$$

Since they both traveled the same amount of time, we have

$$\frac{200}{x + 20} = \frac{150}{x} \quad \text{(a fractional equation)}.$$

The LCD is x(x + 20).

$$x(x + 20) \cdot \frac{200}{x + 20} = x(x + 20) \cdot \frac{150}{x}.$$
$$x \cdot 200 = (x + 20) 150.$$
$$200x = 150x + 3000.$$
$$50x = 3000.$$
$$x = 60.$$
$$x + 20 = 80.$$

Therefore, Larry's van traveled at a rate of 60 mph, and Benny's van traveled at a rate of 80 mph.

Check: Are the times equal?

$$\frac{200}{60 + 20} \stackrel{?}{=} \frac{150}{60}.$$

$$\frac{200}{80} \stackrel{?}{=} 2.5.$$

$$2.5 = 2.5. \quad \text{The solution checks.}$$

Another type of application that sometimes results in a fractional equation is the solving of "work" problems. If a machine can produce 30 items (such as bolts) in one minute, then it can produce 150 items in five minutes. In other words, the amount of work done by the machine is equal to the rate of work multiplied by the amount of time of work, or work = rate × time, or w = rt. We can also use this formula in two other ways: $r = \frac{w}{t}$, the rate of work is equal to the amount of work divided by the time of work, and $t = \frac{w}{r}$, the time of work is equal to the amount of work divided by the rate of work.

If Scott mows half the lawn and Joe mows half the lawn, then together they have completed the job and the sum of their work is 1, $\frac{1}{2} + \frac{1}{2} = 1$. The total work accomplished by machines or people working together is determined by adding the parts. If the job is completed, then the sum of the parts must be 1. For example, if Scott can split a cord of wood alone in five hours and together Scott and Joe can split a cord of wood in three hours, how long does it take Joe to split a cord of wood alone? Let x equal the time it takes Joe to split a cord of wood alone. It takes Scott five hours to do the job alone; working with Joe, he worked three hours, hence, he has completed $\frac{3}{5}$ of the job. Similarly, Joe has completed $\frac{3}{x}$ of the job and working together, they complete the job:

$$\frac{3}{5} + \frac{3}{x} = 1 \quad \text{(a fractional equation)}.$$

The LCD is 5x. $5x \left(\frac{3}{5} + \frac{3}{x} \right) = 5x \cdot 1.$

$$5x \cdot \frac{3}{5} + 5x \cdot \frac{3}{x} = 5x.$$
$$3x + 15 = 5x.$$
$$15 = 2x.$$
$$7.5 = x.$$

Therefore, Joe can split the cord of wood alone in seven and a half hours. Checking our solution, we must determine whether the job can be completed in three hours using the respective times. Scott does $\frac{3}{5}$ of the job and Joe does $\frac{3}{7.5}$ of the job—is the sum equal to 1?

Check: $\quad \frac{3}{5} + \frac{3}{7\frac{1}{2}} \stackrel{?}{=} 1.$

Rational Expressions (Algebraic Fractions)

$$\frac{3}{5} + \frac{\frac{3}{15}}{2} \stackrel{?}{=} 1.$$

$$\frac{3}{5} + \frac{6}{15} \stackrel{?}{=} 1.$$

$$\frac{3}{5} + \frac{2}{5} \stackrel{?}{=} 1.$$

$$\frac{5}{5} = 1.$$

$$1 = 1. \qquad \text{The solution checks.}$$

EXAMPLE 31

The total time available to service a car at Sparky's Garage is three hours. After Sparky had been working on the car for one hour, another mechanic, Axelrod, who would have taken half as many hours to do the job alone, started working on the car together with Sparky. Sparky and Axelrod finished the job exactly on time. How many hours would it have taken Sparky to service the car alone?

Solution

Since Axelrod would have taken half as many hours as Sparky to complete the job alone, we have

$x =$ the number of hours it would have taken Axelrod to do the job.

$2x =$ the number of hours it would have taken Sparky to do the job.

Sparky worked on the job three hours, hence he completed $\frac{3}{2x}$ of the job. Axelrod worked on the job two hours, and he completed $\frac{2}{x}$ of the job. Since they completed the job, the resulting equation is

$$\frac{3}{2x} + \frac{2}{x} = 1.$$

The LCD is 2x. $2x\left(\frac{3}{2x} + \frac{2}{x}\right) = 2x \cdot 1.$

$$2x \cdot \frac{3}{2x} + 2x \cdot \frac{2}{x} = 2x.$$

$$3 + 4 = 2x.$$

$$7 = 2x.$$

This is the solution, since 2x equals the number of hours it would have taken Sparky to do the job alone and that is what we are asked to find. 2x = 7 hours, and x = $3\frac{1}{2}$ hours.

Check: $\frac{3}{7} + \frac{2}{3\frac{1}{2}} \stackrel{?}{=} 1.$

$\frac{3}{7} + \frac{2}{\frac{7}{2}} \stackrel{?}{=} 1.$

$\frac{3}{7} + \frac{4}{7} \stackrel{?}{=} 1.$

$\frac{7}{7} = 1.$

$1 = 1.$ The solution checks.

EXAMPLE 32

If a barrel of wine can be filled by an inlet pipe in three hours, and emptied by an outlet pipe in five hours, then how long will it take to fill the barrel of wine if both pipes are open?

Solution

We let x represent the amount of time required to fill the barrel of wine when both pipes are open. Therefore, $\frac{1}{x}$ is the amount of the job that will be completed (how much of the barrel that will be filled) after one hour. Similarly, the inlet pipe will fill $\frac{1}{3}$ of the barrel at the end of one hour and the outlet pipe will empty $\frac{1}{5}$ of the barrel in the same amount of time. Therefore, the total work accomplished by both pipes working together is

$$\frac{1}{3} - \frac{1}{5} = \frac{1}{x}.$$

We subtract, since one pipe is emptying the barrel.

The LCD is 15x. $15x\left(\frac{1}{3} - \frac{1}{5}\right) = 15x \cdot \frac{1}{x}.$

$15x \cdot \frac{1}{3} - 15x \cdot \frac{1}{5} = 15.$

Rational Expressions (Algebraic Fractions)

$$5x - 3x = 15.$$
$$2x = 15.$$
$$x = 7.5 \quad \text{hours.}$$

With both pipes open, 7.5 hours is required to fill the barrel of wine.

Check: $\dfrac{1}{3} - \dfrac{1}{5} \stackrel{?}{=} \dfrac{1}{7\frac{1}{2}}.$ $\quad\quad \dfrac{7.5}{3} - \dfrac{7.5}{5} \stackrel{?}{=} 1.$

$\dfrac{5-3}{15} \stackrel{?}{=} \dfrac{1}{\frac{15}{2}}.$ OR $2.5 - 1.5 \stackrel{?}{=} 1.$
$\quad\quad\quad\quad\quad\quad\quad\quad\quad\quad 1 = 1.$

$\dfrac{2}{15} = \dfrac{2}{15}.$

EXAMPLE 33

One number is three times larger than another number, and the sum of their reciprocals (multiplicative inverses) is $6\frac{2}{3}$. Find the numbers.

Solution

Let x equal the smaller number, then 3x equals the larger number. The reciprocal of x is $\dfrac{1}{x}$ and the reciprocal of 3x is $\dfrac{1}{3x}$. The sum of the reciprocals is $6\frac{2}{3}$. Therefore, we have

$$\frac{1}{3x} + \frac{1}{x} = 6\frac{2}{3} \quad \text{or} \quad \frac{1}{3x} + \frac{1}{x} = \frac{20}{3}.$$

The LCD is 3x. $3x\left(\dfrac{1}{3x} + \dfrac{1}{x}\right) = 3x \cdot \dfrac{20}{3}.$

$$3x \cdot \frac{1}{3x} + 3x \cdot \frac{1}{x} = x \cdot 20.$$
$$1 + 3 = 20x.$$
$$4 = 20x.$$
$$\frac{4}{20} = x.$$
$$\frac{1}{5} = x \quad \text{and} \quad 3x = 3 \cdot \frac{1}{5} = \frac{3}{5}.$$

The numbers are $\dfrac{1}{5}$ and $\dfrac{3}{5}$.

Check: $\frac{3}{5}$ is three times larger than $\frac{1}{5}$ and

$$\frac{1}{\frac{3}{5}} + \frac{1}{\frac{1}{5}} \stackrel{?}{=} \frac{20}{3}.$$

$$\frac{5}{3} + \frac{5}{1} \stackrel{?}{=} \frac{20}{3}.$$

$$\frac{5 + 15}{3} = \frac{20}{3}.$$

$$\frac{20}{3} = \frac{20}{3}. \quad \text{The solution checks.}$$

EXAMPLE 34

What number must be added to the denominator of $\frac{1}{2}$ to make the resulting fraction equal to $\frac{4}{5}$?

Solution

Let x equal the number that must be added to the denominator of $\frac{1}{2}$. Since the resulting fraction is $\frac{4}{5}$, we have

$$\frac{1}{2 + x} = \frac{4}{5}.$$

The LCD is $5(2 + x)$. $5(2 + x) \cdot \frac{1}{2 + x} = 5(2 + x) \cdot \frac{4}{5}.$

$$5 \cdot 1 = (2 + x)4.$$
$$5 = 8 + 4x.$$
$$-3 = 4x.$$
$$-\frac{3}{4} = x.$$

Check: $\quad \dfrac{1}{2 - \frac{3}{4}} \stackrel{?}{=} \dfrac{4}{5}.$

$$\dfrac{1}{\frac{8}{4} - \frac{3}{4}} \stackrel{?}{=} \dfrac{4}{5}.$$

$$\dfrac{1}{\frac{5}{4}} \stackrel{?}{=} \dfrac{4}{5}.$$

$$\dfrac{4}{5} = \dfrac{4}{5}. \quad \text{The solution checks.}$$

Rational Expressions (Algebraic Fractions)

To solve any word problem, we must read the problem carefully, translate the English phrases into mathematical phrases (choosing a variable for the unknown quantity), and then write the equation using the given information. After solving, we must check the solution to determine whether it satisfies the original problem.

EXERCISES FOR SECTION 6.7

*(Answers given to exercises marked *)*

*1. An express passenger train can travel a distance of 300 miles in the same amount of time that a freight train can travel a distance of 150 miles. If the express train travels at a rate of 30 miles per hour faster than the freight train, what is the speed of each train?

2. A truck went from Rochester to New York City, a distance of 300 miles, and then traveled from New York City to Washington, D.C., a distance of 200 miles. If the truck traveled at the same rate of speed for both trips, and it took two hours longer to go from Rochester to New York than from New York to Washington, find the time for each trip.

*3. One morning at 8 A.M., Jim left his house and jogged 16 miles to his friend Bill's house. Jim rested for an hour and then, since he had developed a blister, Bill drove him home at a rate eight times faster than Jim jogged. If they arrived at Jim's house promptly at 1 P.M., what was Jim's rate of jogging?

4. Jim walks or jogs ten miles every day. He can jog three times faster than he can walk. On Saturday, he jogged six miles and walked four miles in three hours. At what rate did Jim jog the six miles?

*5. Irene is taking flying lessons. The plane she uses flies 150 miles per hour in still air. One day it took Irene the same amount of time to fly 250 miles with the wind as it did to return a distance of 200 miles against the same wind. Find the rate of the wind. (Express your answer to the nearest tenth.)

6. Bill can seal his driveway in four hours. If his son, Scott, helps him, then working together they can seal the driveway in three hours. How long would it take Scott, working alone, to seal the driveway?

*7. Craig can paint the living room alone in four hours and Tom can paint the living room alone in six hours. Craig began painting the living room and at the end of two hours Tom joined him and together they completed the job. How long did Tom paint?

8. One pipe can fill an oil tank in two hours. A second pipe can fill the same oil tank in three hours. If both pipes are open, how long will it take to fill the oil tank?

*9. Axelrod can install a new set of brakes on a truck in five hours. If Sparky helps him, then working together they can complete the job in three hours. How long would it take Sparky working alone to install the set of brakes?

10. When two snow plows are in operation together, they can plow a certain parking lot in three hours. The first snow plow alone takes twice as long as the second snow plow alone to plow the parking lot. How long does it take the first snow plow to do the job alone?

*11. Jim had to paint a building in three days. After painting for two days, he saw that he could not finish on time unless he had another helper. He hired Larry, who would have taken twice as long to do the job alone. Jim and Larry finished painting the building exactly on time. How long would it have taken Larry to paint the building alone?

12. If an oil tank can be filled by an inlet pipe in three hours, and emptied by an outlet pipe in six hours, then how long will it take to fill the oil tank if both pipes are open?

*13. A large wine barrel can be filled by an inlet pipe in four hours, and emptied by an outlet pipe in six hours. Both pipes are on for two hours and then the outlet pipe is turned off. How much more time will be required to fill the barrel?

14. If a gasoline tank can be filled by an inlet pipe in 12 hours, and emptied by an outlet pipe in 16 hours, then how long will it take to fill the gasoline tank if both pipes are open?

*15. A large barrel of apple cider can be filled by an inlet pipe in 10 hours, and emptied by an outlet pipe in 15 hours. The inlet pipe is open for five hours and then (by mistake) the outlet pipe is also opened. How much more time will be required to fill the barrel of cider?

16. The denominator of a certain fraction exceeds its numerator by 4. If the numerator and the denominator are each increased by 1, then the value of the resulting fraction is $\frac{1}{2}$. Find the original fraction.

*17. The numerator of a certain fraction is 4 less than its denominator. If the numerator and the denominator are each decreased by 1, then the value of the resulting fraction is $\frac{1}{2}$. Find the original fraction.

Rational Expressions (Algebraic Fractions)

18. The denominator of a certain fraction is 3 less than twice its numerator. If the numerator and denominator are each decreased by 1, then the value of the fraction is $\frac{2}{3}$. Find the original fraction.

*19. The denominator of a certain fraction is 2 more than twice its numerator. If the numerator is decreased by 1, and the denominator is increased by 1, then the value of the resulting fraction is $\frac{2}{9}$. Find the original fraction.

20. One number is five times larger than another number, and the sum of their reciprocals is $\frac{2}{5}$. Find the numbers.

*21. Bonnie, a stock clerk, is assigned to do a job that she can complete in eight hours. After Bonnie has been working for two hours, another stock clerk, Clyde, who can do the job alone in ten hours, is assigned to help her. In how many hours will the two clerks, Bonnie and Clyde, working together complete the job?

22. A small single engine plane flies at 100 miles per hour in still air. On a certain day it took the same amount of time to fly 300 miles with the wind as it did to return a distance of 250 miles against the same wind. Find the rate of the wind. Express your answer to the nearest tenth.

*23. It takes 50 minutes for a boat to travel upstream a distance of 5 miles, and it takes 30 minutes for the boat to travel the same distance downstream. Find the rate of the boat in still water, and the rate of the current.

24. Joe, working alone, takes twice as long as Scott to seal a driveway. Scott starts sealing the driveway alone, and after half the job is done, he is joined by Joe. They work together for six hours and complete the job. How long would it have taken each to seal the driveway, working alone?

*25. The denominator of a certain fraction is 3 more than twice its numerator. If the numerator is decreased by 3 and the denominator is increased by 5, then the value of the resulting fraction is $\frac{1}{9}$. Find the original fraction.

6.8 SUMMARY

A rational expression is an algebraic expression of the form $\frac{P}{Q}$, where P and Q are polynomials, Q $\neq$ 0. A rational expression is the quotient of two polynomials where the denominator is not equal to 0. Any possible values for the variables that will produce a denominator equal to 0 are excluded. If $\frac{P}{Q}$ and $\frac{R}{S}$ are rational expressions, then $\frac{P}{Q} = \frac{R}{S}$ whenever $PS = QR$, that is, in a proportion, the product of the means equals the product of the extremes. For example, $\frac{5x}{3y} = \frac{10x}{6y}$, since $5x \cdot 6y = 3y \cdot 10x$, $30xy = 30xy$.

To simplify rational expressions (algebraic fractions), we can use the Fundamental Principle of Fractions. It states:

If a, b, c are real numbers ($b \neq 0$, $c \neq 0$), then

$$\frac{a}{b} = \frac{ca}{cb}.$$

To simplify the fraction $\frac{x^2 - 1}{x^2 + 2x + 1}$, we factor the numerator and denominator, and rewrite the given fraction in a simpler form (using the Fundamental Principle):

$$\frac{x^2 - 1}{x^2 + 2x + 1} = \frac{(x+1)(x-1)}{(x+1)(x+1)} = \frac{x-1}{x+1}.$$

A rational expression is written in lowest terms when both the numerator and denominator do not contain a common factor.

To multiply two algebraic fractions, we obtain the numerator of the product by multiplying the numerators, and obtain the denominator of the product by multiplying the denominators. Since multiplication is both commutative and associative, we can arrange those factors common to both the numerator and denominator to illustrate factors of 1, and apply the Fundamental Principle. For example,

$$\frac{y^2 - 1}{y^3} \cdot \frac{y^2}{y - 1} = \frac{(y^2 - 1)y^2}{y^3(y - 1)} = \frac{y^2 \cdot (y - 1)(y + 1)}{y^2 \cdot (y - 1)y} = \frac{y + 1}{y}.$$

When simplifying a problem involving division of rational expressions, first rewrite the indicated quotient as a multiplication problem (using the multiplicative inverse of the divisor), and then proceed in the same manner that is used to obtain the product of two rational expressions.

To combine two rational expressions that have the same denominator, write down the common denominator for the indicated sum or difference and combine the numerators. To combine two rational expressions that do not have the same denominator, we can use the cross-product rule, $\frac{a}{b} + \frac{c}{d} = \frac{ad + bc}{bd}$. But to combine two or more rational expressions where the denominators are unlike and may also contain some common factors, we use the technique involving the least common denominator (LCD).

Rational Expressions (Algebraic Fractions)

The LCD of two or more rational expressions is the least common multiple of the denominators. The least common multiple of two or more numbers is the product of each different prime factor, including each factor to the highest power in which it appears in any of the given numbers.

Many of the techniques used in adding, subtracting, and multiplying rational expressions are also used to solve fractional equations. If the variable occurs in the denominator of a fraction (in an equation) then the equation is called a fractional equation. We can solve a fractional equation by eliminating the denominators of the fractions, that is, we transform the given equation into an equivalent equation such that the variable does not occur in the denominator of a fraction. After determining the LCD of the fractions in the given equation, we multiply both members of the equation by the LCD. Next, by means of the associative and commutative properties, we can transform the given equation and solve the derived equation. The solution set of the original equation is a subset of the solution set of the new equation. Therefore, we must check the solution in the original equation, because we may have produced an extraneous solution if the solution set of the derived equation makes the denominator of one of the fractions in the original equation equal to 0. If the solution of the transformed equation does not satisfy the given equation, then we say that the given equation has no solution and its solution set is the empty set.

A complex fraction is a fraction that contains a fraction in its numerator or denominator or both. To simplify a complex fraction, we must combine the fractions in the numerator and those in the denominator (if necessary), then simplify the resulting complex fraction by performing the indicated division: multiplying the fraction in the numerator by the multiplicative inverse of the fraction in the denominator. We can also simplify a complex fraction by multiplying it by 1, and the 1 should be of the form $\frac{m}{m}$ where m is the LCD of the fractions contained in the complex fraction.

In this chapter, we again examined some applications (word problems) and the solutions were obtained from fractional equations. When we set up the equation from the stated problem, the equation contained rational expressions. Following is a list of suggestions for solving word problems:

1. Read the problem carefully;
2. Reread the problem carefully;
3. If possible, draw a diagram to assist in interpreting the given information;
4. Translate the English phrases into mathematical phrases, and choose a variable for the unknown quantity;
5. Write an equation using all of the above information;
6. Solve the equation;
7. Check the solution to determine that it satisfies the original problem.

REVIEW EXERCISES FOR CHAPTER 6
(Answers in Appendix B)

Determine whether the pairs of rational expressions are equal.

1. $\dfrac{14}{26}, \dfrac{101}{169}$

2. $\dfrac{2xy}{z^3}, \dfrac{xy}{2z^3}$

3. $\dfrac{x^2 - y^2}{x + y}, \dfrac{y - x}{-1}$

4. $-\dfrac{x + y}{x - y}, \dfrac{-x - y}{y - x}$

Simplify each expression, if possible.

5. $\dfrac{221}{323}$

6. $\dfrac{14x^3y^2}{7x^2y^3}$

7. $\dfrac{24xy^2}{36x^2y^3}$

8. $\dfrac{8xy}{4x - 12y}$

9. $\dfrac{2y^2 - 4y}{4xy + 8y}$

10. $\dfrac{5x + 10y}{x^2 - 4y^2}$

11. $\dfrac{x^3 - y^3}{x^2 - y^2}$

12. $\dfrac{2x^2 - x - 3}{2x^2 + x - 1}$

13. $\dfrac{y^2 - y - 6}{y^2 - 8y + 15}$

14. $\dfrac{y^2 - 4y - 5}{-3y^2 + 12y + 15}$

15. $\dfrac{2x^2 - xy - 15y^2}{6x^2 + 17xy + 5y^2}$

16. $\dfrac{2x^2 + xy - y^2}{-2x^2 - 3xy - y^2}$

Multiply and simplify, if possible.

17. $\dfrac{4xy}{3z} \cdot \dfrac{15xz}{25x^2}$

18. $\dfrac{3c^2d^3}{4ab^2} \cdot \dfrac{12a^3b}{9c^3d}$

19. $\dfrac{x + 2}{x + 5} \cdot \dfrac{x^2 - 25}{2x^2 + 3x - 2}$

20. $\dfrac{x^2 - y^2}{x + y} \cdot \dfrac{3x^2 + 3xy}{x^2 - 2xy + y^2}$

21. $\dfrac{y - 2}{y + 3} \cdot \dfrac{y^2 + 2y + 4}{y^3 - 8}$

22. $\dfrac{6y^2 + 7y - 3}{3y^2 + 5y + 1} \cdot \dfrac{y^2 - 1}{2y^2 + y - 3}$

Divide and simplify, if possible.

23. $\dfrac{6y^2}{y^2 + 2y - 15} \div \dfrac{3y}{2y - 6}$

24. $\dfrac{yx + y^2}{x^2 - xy} \div \dfrac{3x^2 + 3xy}{x^2 - xy}$

25. $\dfrac{x^2 - 5x + 6}{x^2 + 5x + 4} \div \dfrac{2x - 4}{3x + 12}$

26. $\dfrac{3y^2 - y - 2}{2y^2 - y - 1} \div \dfrac{3y^2 + 5y + 2}{2y^2 + 3y + 1}$

27. $\dfrac{x^3 - y^3}{2x^2 - 5xy - 3y^2} \div \dfrac{x^2 + xy + y^2}{2x^2 - xy - 3y^2}$

28. $\dfrac{6x^2 + 5xy - 6y^2}{6x^2 + 11xy + 3y^2} \div \dfrac{2y^2 - 5xy - 3x^2}{3x^2 + 4xy + y^2}$

Rational Expressions (Algebraic Fractions)

Perform the indicated operations and simplify, if possible.

29. $\dfrac{x}{x^2-4}+\dfrac{2}{x^2-4}$

30. $\dfrac{y+1}{2y^2-y-6}+\dfrac{y+2}{2y^2-y-6}$

31. $\dfrac{2x}{x+3}+\dfrac{3x}{x-1}$

32. $\dfrac{x+1}{33}+\dfrac{2x-1}{44}$

33. $\dfrac{2x-1}{3x+1}-\dfrac{x+1}{x-1}$

34. $\dfrac{3x+1}{x+2}-\dfrac{x+1}{3x+6}$

35. $\dfrac{x+2}{x-2}-\dfrac{x+1}{x^2-4}$

36. $\dfrac{y-1}{y^2+3y+2}+\dfrac{y+1}{y^2+2y}$

37. $\dfrac{2y+1}{y^2-10y+25}+\dfrac{y-1}{y^2-25}$

38. $\dfrac{y+1}{y^2+y-2}-\dfrac{y-1}{y^2+5y+6}$

39. $\dfrac{1}{y+1}-\dfrac{2}{y-1}+\dfrac{3}{y+2}$

40. $\dfrac{2}{x-1}+\dfrac{3}{x+1}-\dfrac{1}{x-2}$

Simplify each of the complex fractions.

41. $\dfrac{\dfrac{3}{5}-\dfrac{1}{3}}{\dfrac{4}{7}-\dfrac{3}{11}}$

42. $\dfrac{\dfrac{1}{x}-\dfrac{1}{y}}{\dfrac{1}{x^2}-\dfrac{1}{y^2}}$

43. $\dfrac{\dfrac{x-1}{x+1}+\dfrac{x+1}{x-1}}{\dfrac{x-1}{x+1}-\dfrac{x+1}{x-1}}$

44. $\dfrac{y+1}{y}+\dfrac{y}{y-\dfrac{1}{y}}$

Find the solution set for each of the following equations.

45. $\dfrac{2y}{3}+\dfrac{3y}{4}=y+5$.

46. $\dfrac{2x-5}{3}+\dfrac{3x+2}{2}=1$.

47. $\dfrac{4}{x+1}=\dfrac{3}{x-1}$.

48. $\dfrac{x+2}{x-3}=\dfrac{x+1}{x-1}$.

49. $\dfrac{4}{5y-10}-\dfrac{2}{3y-6}=\dfrac{1}{15}$.

50. $\dfrac{y+1}{y+2}-\dfrac{y-2}{y-3}=\dfrac{2-y}{y^2-y-6}$.

51. A comparison of two Indianapolis racing cars found that car A can travel a distance of 400 miles in the same time that car B can travel a distance of 350 miles. If car A goes 25 miles per hour faster than car B, what is the speed of each racing car?

52. Scott has to open a bushel of clams and he has three hours to complete the job. After Scott had been opening the clams for two hours, his brother Joe (who would have taken twice as long to do the job alone) started opening the clams together with Scott. They finished the job

exactly on time. How many hours would it have taken Scott to open the bushel of clams alone?

53. At a certain refinery, a large oil tank can be filled by an inlet pipe in six hours, and emptied by an outlet pipe in eight hours. How long will it take to fill the oil tank if both pipes are open?

54. One number is ten times larger than another number, and the sum of their reciprocals is $2\frac{1}{5}$. Find the numbers.

55. It takes three hours for a boat to travel upstream a distance of six miles, and it takes two hours for the boat to travel the same distance downstream. Find the rate of the boat in still water, and the rate of the current.

chapter 7

Complex Numbers and Quadratic Equations

As a result of this chapter, the student will be able to:

1. Identify the square root of a negative number as an imaginary number and express it in terms of i (e.g., $\sqrt{-1} = i$, $\sqrt{-4} = 2i$);
2. Add, subtract, multiply, and divide imaginary numbers, and also simplify powers of i;
3. Identify an expression of the form $a + bi$ (where a is a real number and bi is an imaginary number) as a complex number;
4. Add, subtract, multiply, and divide complex numbers, and simplify complex number expressions;
5. Identify a quadratic equation as a second degree equation in one variable that contains the second power of that variable, and no higher power of that variable (e.g., $x^2 + 3x + 2 = 0$);
6. Write the standard form of a quadratic equation as $ax^2 + bx + c = 0$;
7. Solve incomplete quadratic equations of the form: $ax^2 = 0$, $ax^2 + c = 0$, and $ax^2 + bx = 0$;
8. Solve certain complete quadratic equations by means of factoring;
9. Solve any quadratic equation by *completing the square;*
10. Solve any quadratic equation by means of the *quadratic formula:* If $ax^2 + bx + c = 0$, then

$$x = \frac{-b \pm \sqrt{b^2 - 4ac}}{2a};$$

11. Translate word problems into equations whose solution requires the solution of quadratic equations (geometric problems, uniform motion problems, and work problems);
12. Solve quadratic inequalities and graph their solution set on the real number line.

7.1 INTRODUCTION

Our system of real numbers began with the set of natural numbers, $\{1, 2, 3, \ldots\}$, and then as the need arose, the set of numbers was expanded to include the set of whole numbers, the set of integers, the set of rational numbers, and the set of irrational numbers. For example, the equation $x + 1 = 0$ has no solution in the set of natural numbers, but if we expand the universal set to include the set of integers, then $x = -1$. Similarly, $2x - 1 = 0$ has no solution in the set of natural numbers or the set of integers, but if we expand the universal set to include the set of rational numbers, then $x = \frac{1}{2}$. There also exist equations that have no solution in the system of real

Complex Numbers and Quadratic Equations

numbers. For example, consider the equation $x^2 + 1 = 0$. What value of x will satisfy this equation: If we let $x = 1$, then $(1)^2 + 1 = 1 + 1 = 2 \neq 0$. Similarly, if we let $x = -1$, then $(-1)^2 + 1 = 1 + 1 = 2 \neq 0$. There does not exist any real number that when squared and added to 1 will result in a sum of 0. If $x = \sqrt{-1}$, then $(\sqrt{-1})^2 + 1 = -1 + 1 = 0$. This solution does satisfy the given equation. But recall that the symbol $\sqrt[n]{a}$ represents the principal n^{th} root of a. If $a < 0$ and n is an even integer, then there does not exist a principal n^{th} root of a in the system of real numbers (see Section 3.4). Therefore, $\sqrt{-1}$ is not a real number.

However, mathematicians developed a system of numbers that gave meaning to $\sqrt{-1}$. The square root of -1 is called an *imaginary number*, usually designated by the letter i. (Note: in some technical areas, such as electronics, the letter j is used to represent $\sqrt{-1}$, since i usually represents current, but throughout this text we shall use i to represent $\sqrt{-1}$.) An imaginary number i whose square root is negative 1 must also have the property that its square is equal to negative 1, that is,

$$i^2 = -1 \text{ when } i = \sqrt{-1}.$$

Recall that $\sqrt{ab} = \sqrt{a}\sqrt{b}$ if $a > 0$, $b > 0$. We can use this property to express the square roots of a negative number. For example,

$$\sqrt{-2} = \sqrt{-1 \cdot 2} = \sqrt{-1} \cdot \sqrt{2} = i\sqrt{2}.$$

$$\sqrt{-4} = \sqrt{-1 \cdot 4} = \sqrt{-1} \cdot \sqrt{4} = i \cdot 2 \text{ or } 2i.$$

$$\sqrt{-8} = \sqrt{-1 \cdot 8} = \sqrt{-1} \cdot \sqrt{8} = i\sqrt{4} \cdot \sqrt{2} = i2\sqrt{2} \text{ or } 2i\sqrt{2}.$$

Using the fact that $\sqrt{-1} = i$, we can express the square root of a negative number as the indicated product of a real number and an imaginary number.

EXAMPLE 1

Simplify the following and express in terms of i.

    a. $\sqrt{-9}$ b. $\sqrt{-7}$ c. $-\sqrt{-5}$ d. $\sqrt{-27}$

Solution

    a. $\sqrt{-9} = \sqrt{-1 \cdot 9} = \sqrt{-1} \cdot \sqrt{9} = i \cdot 3 = 3i.$

    b. $\sqrt{-7} = \sqrt{-1 \cdot 7} = \sqrt{-1} \cdot \sqrt{7} = i\sqrt{7}.$

c. $-\sqrt{-5} = -\sqrt{-1 \cdot 5} = -\sqrt{-1} \cdot \sqrt{5} = -i\sqrt{5}$.
d. $\sqrt{-27} = \sqrt{-1 \cdot 27} = \sqrt{-1} \cdot \sqrt{27} = i\sqrt{9} \cdot \sqrt{3} = i \cdot 3\sqrt{3}$
$= 3i\sqrt{3}$.

Note: it is normal procedure to write the i before the radical sign.

Earlier in our discussion, i^2 was defined to be equal to -1, $i^2 = -1$. Using this fact, we can simplify other powers of i. For example, i^3 is equivalent to $-i$, because $i^3 = i^2 \cdot i = -1 \cdot i = -i$. Similarly, it can be shown that $i^4 = 1$, since $i^4 = i^2 \cdot i^2 = (-1)(-1) = 1$. What is an equivalent expression for i^5? We can express i^5 as $i^4 \cdot i$ and since $i^4 = 1$, $i^5 = i^4 \cdot i = 1 \cdot i = i$. Note that the first four powers of i are all different, but after that they begin to repeat. For example, $i^{12} = (i^4)^3 = (1)^3 = 1$ and $i^{14} = i^{12} \cdot i^2 = 1 \cdot (-1) = -1$. We can simplify i^{14} in another way, $i^{14} = (i^2)^7 = (-1)^7 = -1$. Recall that a negative number raised to an odd integral power is negative, and a negative number raised to an even integral power is positive. Some simplified powers of i are listed:

$i = \sqrt{-1}$.

$i^2 = -1$.

$i^3 = i^2 \cdot i = -1 \cdot i = -i$.

$i^4 = i^2 \cdot i^2 = (-1)(-1) = 1$.

$i^6 = i^4 \cdot i^2 = 1 \cdot (-1) = -1$.

$i^8 = i^4 \cdot i^4 = 1 \cdot 1 = 1$.

$i^{13} = i^4 \cdot i^4 \cdot i^4 \cdot i = 1 \cdot 1 \cdot 1 \cdot i = i$.

$i^{20} = i^4 \cdot i^4 \cdot i^4 \cdot i^4 \cdot i^4 = 1 \cdot 1 \cdot 1 \cdot 1 \cdot 1 = 1$.

EXAMPLE 2

Simplify and evaluate each of the following:
a. i^6
b. i^{17}
c. $(-3i)^3$
d. $(-3i)^4$

Solution

a. $i^6 = i^4 \cdot i^2 = 1(-1) = -1$.
b. $i^{17} = i^4 \cdot i^4 \cdot i^4 \cdot i^4 \cdot i = 1 \cdot 1 \cdot 1 \cdot 1 \cdot i = i$.
c. $(-3i)^3 = (-3)^3 \cdot i^3 = -27 \cdot i^2 \cdot i = -27 \, (-1)i = 27i$.
d. $(-3i)^4 = (-3)^4 \cdot i^4 = 81 \cdot 1 = 81$.

Complex Numbers and Quadratic Equations   333

Alternate Solution for a and b

a. $i^6 = (i^2)^3 = (-1)^3 = -1.$
b. $i^{17} = i^{16} \cdot i = (i^2)^8 \cdot i = (-1)^8 \cdot i = 1 \cdot i = i.$

We can add, subtract, multiply, and divide imaginary numbers in a manner similar to that used for real numbers. For example, we can find the sum of 3i and 5i by means of the distributive property, $3i + 5i = (3 + 5)i = 8i$. Similarly, $3i - 5i = (3 - 5)i = -2i$. The product of 2i and 3i is obtained by multiplying 2 times 3 and i times i and simplifying the result, $(2i)(3i) = 6i^2 = 6(-1) = -6$. Care must be exercised when multiplying imaginary numbers that have not been expressed in terms of i. For example, consider $\sqrt{-4} \cdot \sqrt{-1}$. Both numbers are imaginary and we must first express them in terms of i and then perform the multiplication, $\sqrt{-4} \cdot \sqrt{-1} = 2i \cdot i = 2i^2 = 2(-1) = -2$. If we had multiplied the radicands directly (instead of expressing them in terms of i first) we would have obtained $\sqrt{-4} \cdot \sqrt{-1} = \sqrt{4} = 2$, which is *not* correct! To obtain the quotient of two imaginary numbers, we can use the property that $\sqrt[n]{\frac{a}{b}} = \frac{\sqrt[n]{a}}{\sqrt[n]{b}}$ ($b \neq 0$). For example, to obtain the quotient of $\sqrt{-4} \div \sqrt{-2}$, we first express the imaginary numbers in terms of i and then perform the indicated division.

$$\frac{\sqrt{-4}}{\sqrt{-2}} = \frac{i\sqrt{4}}{i\sqrt{2}} = \frac{\sqrt{4}}{\sqrt{2}} = \sqrt{\frac{4}{2}} = \sqrt{2}.$$

We can also have the case where only one of the quantities in the indicated quotient is imaginary. For example,

$$\frac{\sqrt{-6}}{\sqrt{2}} = \frac{i\sqrt{6}}{\sqrt{2}} = i\sqrt{\frac{6}{2}} = i\sqrt{3}.$$

Next, consider the case where the divisor is an imaginary number and the dividend is a real number, for example, $\sqrt{6} \div \sqrt{-2}$. Performing the indicated division, we have

$$\frac{\sqrt{6}}{\sqrt{-2}} = \frac{\sqrt{6}}{i\sqrt{2}} = \frac{1}{i} \cdot \frac{\sqrt{6}}{\sqrt{2}} = \frac{1}{i}\sqrt{\frac{6}{2}} = \frac{1}{i}\sqrt{3} = \frac{\sqrt{3}}{i}.$$

But since i occurs in the denominator of the fraction, we can simplify the expression further. Using the identity property of 1, we have

$$\frac{\sqrt{3}}{i} = \frac{\sqrt{3}}{i} \cdot \frac{i}{i} = \frac{i\sqrt{3}}{i^2} = \frac{i\sqrt{3}}{-1} = -i\sqrt{3}.$$

Note that we can apply the identity property at any time in our simplification process. Therefore, the division can also be performed in the following manner:

$$\frac{\sqrt{6}}{\sqrt{-2}} = \frac{\sqrt{6}}{i\sqrt{2}} = \frac{\sqrt{6}}{i\sqrt{2}} \cdot \frac{i}{i} = \frac{i}{i^2} \cdot \frac{\sqrt{6}}{\sqrt{2}} = \frac{i}{-1} \cdot \sqrt{\frac{6}{2}} = -i\sqrt{3}.$$

EXAMPLE 3

Perform the indicated operations.

a. $6i - 2i$
b. $\sqrt{-4} + \sqrt{-9}$
c. $\sqrt{25} - \sqrt{-4}$
d. $\sqrt{9} - \sqrt{-9}$

Solution

a. $6i - 2i = (6-2)i = 4i.$

b. Before performing the indicated operations, we must first express the imaginary numbers in terms of i.

$\sqrt{-4} + \sqrt{-9} = i\sqrt{4} + i\sqrt{9} = 2i + 3i = 5i.$

c. $\sqrt{25} - \sqrt{-4} = 5 - i\sqrt{4} = 5 - 2i.$ Since we cannot combine real and imaginary numbers, the result is left in the form $a + bi$.

d. $\sqrt{9} - \sqrt{-9} = 3 - i\sqrt{9} = 3 - 3i.$

EXAMPLE 4

Perform the indicated operations and simplify, if possible.

a. $5\sqrt{-4}$
b. $2i\sqrt{-9}$
c. $\sqrt{-2} \cdot \sqrt{3}$
d. $\sqrt{-2} \cdot \sqrt{-3}$

Solution

a. $5\sqrt{-4} = 5 \cdot i\sqrt{4} = 5 \cdot 2i = 10i.$

b. $2i\sqrt{-9} = 2i \cdot i\sqrt{9} = 2i \cdot 3i = 6i^2 = 6(-1) = -6.$

c. $\sqrt{-2} \cdot \sqrt{3} = i\sqrt{2} \cdot \sqrt{3} = i\sqrt{6}.$

d. $\sqrt{-2} \cdot \sqrt{-3} = i\sqrt{2} \cdot i\sqrt{3} = i^2\sqrt{6} = -1 \cdot \sqrt{6} = -\sqrt{6}.$

Complex Numbers and Quadratic Equations

> **EXAMPLE 5**
> Perform the indicated operations and simplify, if possible.
>
> a. $\dfrac{\sqrt{-18}}{\sqrt{-6}}$ b. $\dfrac{\sqrt{-18}}{\sqrt{6}}$ c. $\dfrac{\sqrt{18}}{\sqrt{-6}}$ d. $\dfrac{\sqrt{3}}{\sqrt{-2}}$
>
> **Solution**
>
> a. $\dfrac{\sqrt{-18}}{\sqrt{-6}} = \dfrac{i\sqrt{18}}{i\sqrt{6}} = \dfrac{\sqrt{18}}{\sqrt{6}} = \sqrt{\dfrac{18}{6}} = \sqrt{3}.$
>
> b. $\dfrac{\sqrt{-18}}{\sqrt{6}} = \dfrac{i\sqrt{18}}{\sqrt{6}} = i\sqrt{\dfrac{18}{6}} = i\sqrt{3}.$
>
> c. $\dfrac{\sqrt{18}}{\sqrt{-6}} = \dfrac{\sqrt{18}}{i\sqrt{6}} = \dfrac{\sqrt{18}}{i\sqrt{6}} \cdot \dfrac{i}{i} = \dfrac{i\sqrt{18}}{i^2\sqrt{6}} = \dfrac{i}{-1}\sqrt{\dfrac{18}{6}} = -i\sqrt{3}.$
>
> d. $\dfrac{\sqrt{3}}{\sqrt{-2}} = \dfrac{\sqrt{3}}{i\sqrt{2}} = \dfrac{\sqrt{3}}{i\sqrt{2}} \cdot \dfrac{i\sqrt{2}}{i\sqrt{2}} = \dfrac{i\sqrt{6}}{i^2 \cdot 2} = \dfrac{i\sqrt{6}}{-2} = \dfrac{-i\sqrt{6}}{2}.$
>
> Note the form of 1 chosen in order to rationalize the denominator.

Mathematicians decided to call numbers such as $\sqrt{-1}$ imaginary because for any number of the form $\sqrt[n]{a}$ where n is an even integer and a < 0, there does not exist a principal n^{th} root of a in the system of real numbers. Also, they thought that imaginary numbers had no practical value. But today these imaginary numbers occur in many practical applications, particularly in various phases of engineering.

EXERCISES FOR SECTION 7.1

(Answers given to exercises marked *)

Simplify each of the following and express it in terms of i.

*1. $\sqrt{-1}$ 2. $\sqrt{-4}$ *3. $\sqrt{-16}$

4. $\sqrt{-25}$ *5. $\sqrt{-49}$ 6. $\sqrt{-8}$

*7. $-\sqrt{-27}$ 8. $-\sqrt{-12}$ *9. $2\sqrt{-18}$

10. $2\sqrt{-20}$ *11. $3\sqrt{-12}$ 12. $5\sqrt{-45}$

Simplify and evaluate each of the following.

*13. i^5
14. i^9
*15. i^{21}
16. $-i^7$
*17. $-i^2$
18. $-i^{27}$
*19. $(2i)^3$
20. $(-i)^4$
*21. $-2i^4$
22. $(-2i)^5$
*23. $(4i)^2$
24. $-2i^5$
*25. $2i^6$
26. $(2i)^6$
*27. $(-2i)^6$
28. $(2i)^{-6}$
*29. $2i^{89}$
30. $4i^{101}$

Perform the indicated operations and simplify, if possible.

*31. $7i + 3i$
32. $2i - 5i$
*33. $-3i - 8i$
34. $\sqrt{-1} + \sqrt{-4}$
*35. $\sqrt{-4} + \sqrt{-25}$
36. $\sqrt{-36} - \sqrt{-9}$
*37. $\sqrt{25} + \sqrt{-9}$
38. $\sqrt{36} - \sqrt{-25}$
*39. $\sqrt{8} - \sqrt{-6}$
40. $3\sqrt{-9}$
*41. $5\sqrt{-16}$
42. $2\sqrt{-4}$
*43. $3i\sqrt{-4}$
44. $-2i\sqrt{-9}$
*45. $-6i\sqrt{-25}$
46. $2i\sqrt{-27}$
*47. $3i\sqrt{-8}$
48. $-7i\sqrt{-27}$
*49. $\sqrt{-2} \cdot \sqrt{-5}$
50. $\sqrt{3} \cdot \sqrt{-3}$
*51. $\sqrt{5} \cdot \sqrt{-5}$
52. $2i\sqrt{-3} \cdot 3\sqrt{-2}$
*53. $-2i\sqrt{-2} \cdot 3i\sqrt{-3}$
54. $-3i^3\sqrt{-2} \cdot 2i\sqrt{-3}$
*55. $\sqrt[3]{-8} \cdot \sqrt{-4}$
56. $\sqrt{-8} \cdot \sqrt[3]{-27}$
*57. $\sqrt[3]{-1} \cdot \sqrt[3]{-8}$
58. $\dfrac{\sqrt{-12}}{\sqrt{-2}}$
*59. $\dfrac{\sqrt{-18}}{\sqrt{-6}}$
60. $\dfrac{\sqrt{-12}}{\sqrt{2}}$
*61. $\dfrac{\sqrt{-18}}{\sqrt{6}}$
62. $\dfrac{\sqrt{12}}{\sqrt{-2}}$
*63. $\dfrac{\sqrt{18}}{\sqrt{-6}}$
64. $\dfrac{\sqrt{-5}}{\sqrt{-2}}$
*65. $\dfrac{\sqrt{6}}{\sqrt{-5}}$
66. $\dfrac{\sqrt{-8}}{\sqrt{-12}}$
*67. $\dfrac{-2\sqrt{18}}{\sqrt{-8}}$
68. $\dfrac{3\sqrt{-20}}{\sqrt{8}}$
*69. $\dfrac{5\sqrt{-27}}{\sqrt{-50}}$

7.2 COMPLEX NUMBERS

Whenever we combine a real number and an imaginary number, we obtain a sum (or difference) of the form a + bi. For example, $\sqrt{25} + \sqrt{-4} = 5 + 2i$ and $\sqrt{4} - \sqrt{-9} = 2 - 3i$. In the first example, 5 + 2i, a corresponds to 5 and b corresponds to 2, while in the second example, 2 − 3i, a corresponds to 2 and b corresponds to −3. Both expressions are of the form a + bi, that is, we have the sum of a real number and an imaginary number. Any expression of the general form a + bi is called a *complex number*.

Complex Numbers and Quadratic Equations

Note that $2 - 3i$ is of the form $a + bi$ since we can express $2 - 3i$ as $2 + (-3)i$, that is, $a = 2$ and $b = -3$. Two complex numbers, $a + bi$ and $c + di$, are equal whenever their corresponding parts are equal, or, $a = c$ and $b = d$. In general, we have

$$a + bi = c + di, \text{ if and only if } a = c \text{ and } b = d.$$

Therefore, $3 + 2i = (2 + 1) + (1 + 1)i$, but $3 + 2i \neq 2 + 3i$.

The standard form of a complex number is $a + bi$, the sum of a real number and an imaginary number. The real number a is referred to as the *real part* of the complex number $a + bi$. Since bi is an imaginary number, we refer to b as the *imaginary part* of the complex number. If we have the complex number $a + bi$ where $b = 0$ and $a \neq 0$, we can write it as $a + 0i$ or simply as a. Note that the complex number $a + 0i$ is also a real number. In fact, all real numbers can be expressed as a complex number by letting $b = 0$. For example, $3 = 3 + 0i$ and $-2 = -2 + 0i$. In the case where $a = 0$ and $b \neq 0$, we can express $a + bi$ as $0 + bi$ or simply as bi. The complex number $0 + bi$ is called a *pure imaginary* number.

Complex numbers have the same general form as binomials, and we add two complex numbers $a + bi$ and $c + di$ by adding their real parts and adding their imaginary parts:

$$(a + bi) + (c + di) = (a + c) + (b + d)i.$$

For example, $(2 + 3i) + (5 + 2i) = (2 + 5) + (3 + 2)i = 7 + 5i$, and $(-2 - 4i) + (-5 + 2i) = (-2 - 5) + (-4 + 2)i = -7 - 2i$.

It can be shown that the system of complex numbers satisfies the same properties as the system of real numbers under the operations of addition and multiplication—the closure property, the commutative property, the associative property, the identity property, and the inverse property. For example, the identity element for the operation of addition is $0 + 0i$. Whenever we add it to a complex number $a + bi$, we obtain the original element: $(a + bi) + (0 + 0i) = (a + 0) + (b + 0)i = a + bi$.

To subtract two complex numbers, we use the same technique that we use for real numbers, which is to add the additive inverse of the subtrahend to the minuend: $x - y = x + (-y)$, where x and y represent any real numbers. It should be noted that the additive inverse of $c + di$ is $-c - di$, since $(c + di) + (-c - di) = (c - c) + (d - d)i = 0 + 0i$. Therefore, to subtract $c + di$ from $a + bi$, we have

$$(a + bi) - (c + di) = (a + bi) + (-c - di) = (a - c) + (b - d)i.$$

For example, $(2 + 3i) - (5 + 2i) = (2 + 3i) + (-5 - 2i) = (2 - 5) + (3 - 2)i = -3 + 1i$ or $-3 + i$, and $(-3 + 2i) - (-2 - 3i) = (-3 + 2i) + (2 + 3i) = (-3 + 2) + (2 + 3)i = -1 + 5i$.

EXAMPLE 6

Perform the indicated operations and simplify, if possible.
a. $(2 + 3i) + (3 + 4i)$ b. $(-3 - 2i) + (5 - 2i)$
c. $(2 + 3i) - (3 + 4i)$ d. $(-3 - 2i) - (5 - 2i)$

Solution

a. $(2 + 3i) + (3 + 4i) = (2 + 3) + (3 + 4)i = 5 + 7i$.
b. $(-3 - 2i) + (5 - 2i) = (-3 + 5) + (-2 - 2)i = 2 - 4i$.
c. $(2 + 3i) - (3 + 4i) = (2 + 3i) + (-3 - 4i)$
$\qquad = (2 - 3) + (3 - 4)i = -1 - i$.
d. $(-3 - 2i) - (5 - 2i) = (-3 - 2i) + (-5 + 2i)$
$\qquad = (-3 - 5) + (-2 + 2)i$
$\qquad = -8 + 0i = -8$.

The product of two complex numbers can be obtained in the same manner as the product of two binomials (either the distributive property or the FOIL method), since we can consider a + bi and c + di to be binomials. Therefore,

$(a + bi)(c + di) = (a + bi)c + (a + bi)di$
$\qquad = ac + bci + adi + bdi^2$ ⟵ *Note:* $i^2 = -1$.
$\qquad = ac + bci + adi + bd(-1)$
$\qquad = (ac - bd) + (ad + bc)i.$ (collecting the real and imaginary parts)

It is imperative to remember that $i^2 = -1$, and to replace i^2 by -1, which allows us to combine the real and imaginary parts and express the final result in the form of a + bi. For example, we can determine the product of $(4 - 3i)(2 + 5i)$ by means of the distributive property or by the FOIL technique. Both methods are illustrated.

$(4 - 3i)(2 + 5i) = (4 - 3i)2 + (4 - 3i)5i$
$\qquad = 8 - 6i + 20i - 15i^2$
$\qquad \qquad \qquad \qquad \qquad i^2 = -1$
$\qquad = 8 + 14i - 15(-1)$
$\qquad = 8 + 14i + 15$
$\qquad = 23 + 14i.$

By means of the FOIL technique, we have

$(4 - 3i)(2 + 5i) = \quad 8 \qquad\qquad + 20i \qquad\qquad -6i \qquad\qquad -15i^2$
                          ↑   ↑   ↑   ↑
                product of   product of   product of   product of
                first terms   outer terms   inner terms   last terms

Complex Numbers and Quadratic Equations

Replacing i^2 by -1 and combining the real parts and imaginary parts,

$$(4 - 3i)(2 + 5i) = 23 + 14i.$$

The FOIL technique for finding the product of two complex numbers enables us to quickly determine the product with a minimum amount of written work. Note that the sum of the "outer" and "inner" products will give us one term of the product.

EXAMPLE 7

Perform the indicated operations and simplify, if possible.
a. $(2 + 3i)(4 + 5i)$ b. $(-3 - 2i)(5 - 2i)$
c. $(-5 + 3i)(2 - 4i)$ d. $(2 + 3i)(2 + 3i)$

Solution

For Example 7-a, we shall find the product by means of the distributive property and the FOIL technique, but since we will also have to multiply complex numbers when we obtain the quotient of two complex numbers, we shall advocate the FOIL technique for the remainder.

a. $(2 + 3i)(4 + 5i) = (2 + 3i)4 + (2 + 3i)5i$
$= 8 + 12i + 10i + 15i^2$ → $i^2 = -1$
$= 8 + 22i - 15$
$= -7 + 22i.$

Alternate Solution

a. $(2 + 3i)(4 + 5i) = 8 + 10i + 12i + 15i^2$ → $i^2 = -1$
$= 8 + 22i - 15$
$= -7 + 22i.$

b. $(-3 - 2i)(5 - 2i) = -15 + 6i - 10i + 4i^2$ → $i^2 = -1$
$= -15 - 4i - 4$
$= -19 - 4i.$

c. $(-5 + 3i)(2 - 4i) = -10 + 20i + 6i - 12i^2$ → $i^2 = -1$
$= -10 + 26i + 12$
$= 2 + 26i.$

d. $(2 + 3i)(2 + 3i) = 4 + 6i + 6i + 9i^2$ → $i^2 = -1$
$= 4 + 12i - 9$
$= -5 + 12i.$

The FOIL technique can be used to find the product of any two complex numbers, as we did in Example 7. Note that the last example $(2+3i)(2+3i)$ can also be written as $(2+3i)^2$. It is of the form of *squaring a binomial,* or, $(x+y)^2$.

> To square a binomial, we must perform three steps: 1. square the first term, 2. find twice the product of the two terms, 3. square the last term. The square of the binomial is the sum of these three products. $(x+y)^2 = x^2 + 2xy + y^2$. We can use this technique to evaluate $(2+3i)^2$,
>
> $(2+3i)^2 = 4 + 2 \cdot 6i + 9i^2 = 4 + 12i - 9 = -5 + 12i$.
>
> square of the first term   twice the product of the two terms   square of the last term

Recall that in Chapters 3, 4, and 5, we encountered expressions of the form $a+b$ and $a-b$. They are called *conjugates* of each other, and the product of two conjugates will be the difference of two squares. The product of the sum and difference of two monomials is the difference of two squares, $(x+y)(x-y) = x^2 - y^2$. The product is the square of the first monomial minus the square of the second monomial. For example, $(3+2i)(3-2i) = (3)^2 - (2i)^2 = 9 - 4i^2 = 9 - 4(-1) = 9 + 4 = 13$. Note that this product is a real number. It can be shown that the product of any complex number and its conjugate is a real number, and that $(a+bi)(a-bi) = a^2 + b^2$.

EXAMPLE 8

Perform the indicated operation and simplify, if possible.
a. $(4+3i)^2$ b. $(1-i)^2$
c. $(3+4i)(3-4i)$ d. $(1+i)(1-i)$

Solution

To square a complex number, perform the following steps: 1. square the real part, 2. find twice the product of the two terms, 3. square the imaginary part. The sum of these three products is the square of the complex number.
a. $(4+3i)^2 = (4)^2 + 2 \cdot (4)(3i) + (3i)^2 = 16 + 24i + 9i^2$
$= 16 + 24i - 9 = 7 + 24i$.
b. $(1-i)^2 = (1)^2 + 2(1)(-i) + (-i)^2 = 1 - 2i + i^2$
$= 1 - 2i - 1 = -2i$.
The product of two conjugates is the difference of two squares.

Complex Numbers and Quadratic Equations

The product is the square of the real part minus the square of the imaginary part. Note that the product is a real number, and $(a + bi)(a - bi) = a^2 + b^2$.

c. $(3 + 4i)(3 - 4i) = (3)^2 - (4i)^2 = 9 - 16i^2 = 9 - 16(-1)$
$= 9 + 16 = 25$.

d. $(1 + i)(1 - i) = (1)^2 - (i)^2 = 1 - i^2 = 1 - (-1)$
$= 1 + 1 = 2$.

Recall that the identity element under the operation of multiplication is 1, that is, we can multiply an expression by 1 and not change its value. We have used this property to rationalize radical expressions and in combining rational expressions (refer to Chapters 3 and 6). We shall again make use of this property when dividing complex numbers. For example, to divide $3 + 5i$ by $2 + 3i$, we can first express it as a fraction, $\frac{3 + 5i}{2 + 3i}$. Next, we multiply the fraction by 1, but the form of 1 that is chosen is important. It should be of a form such that its numerator and denominator are the conjugate of the divisor. For this particular example, the conjugate of $2 + 3i$ is $2 - 3i$, and the form of 1 is $\frac{2 - 3i}{2 - 3i}$. Therefore, we now have

$$\frac{3 + 5i}{2 + 3i} = \frac{3 + 5i}{2 + 3i} \cdot \frac{2 - 3i}{2 - 3i} = \frac{6 - 9i + 10i - 15i^2}{(2)^2 - (3i)^2} = \frac{6 + i + 15}{4 - (-9)} = \frac{21 + i}{13}.$$

The quotient of $(3 + 5i) \div (2 + 3i)$ is $\frac{21 + i}{13}$. We can also express the result as $\frac{21}{13} + \frac{i}{13}$; either form is correct.

To find the quotient of two complex numbers, multiply the numerator and the denominator of the fraction by the conjugate of the denominator. In essence, we are using the concept of multiplication by 1. In general, we have

$$\frac{a + bi}{c + di} = \frac{a + bi}{c + di} \cdot \frac{c - di}{c - di} = \frac{ac - adi + bci - bdi^2}{c^2 - (di)^2}$$

$$= \frac{ac - adi + bci - bd(-1)}{c^2 - (d^2i^2)}$$

$$= \frac{(ac + bd) + (bc - ad)i}{c^2 + d^2}.$$

The general case is illustrated to show the procedure involved in dividing any two complex numbers. Consider the case where $(3 - 2i) \div 4i$. Again, we must find the quotient of two complex numbers, and we have $\frac{3 - 2i}{4i}$. We must multiply both the numerator and denominator by the conjugate of the denominator, 4i. What is the conjugate of 4i? It is a pure imaginary number and we can rewrite it as $0 + 4i$. The conjugate of $0 + 4i$ is $0 - 4i$ or simply $-4i$. Therefore, we have

$$\frac{3-2i}{4i} = \frac{3-2i}{4i} \cdot \frac{-4i}{-4i} = \frac{-12i + 8i^2}{-16i^2} = \frac{-12i + 8(-1)}{-16(-1)} = \frac{-8-12i}{16}.$$

Note that we can reduce the result, since both the numerator and denominator contain 4 as a common factor:

$$\frac{-8-12i}{16} = \frac{4(-2-3i)}{4 \cdot 4} = \frac{-2-3i}{4}.$$

We can check the result by multiplying the divisor by the quotient to see if we obtain the dividend. Checking this particular example, we have

$$\underbrace{\frac{-2-3i}{4}}_{\text{the quotient}} \cdot \underbrace{4i}_{\text{the divisor}} = \frac{-8i - 12i^2}{4} = \frac{-8i - 12(-1)}{4} = \frac{12 - 8i}{4} = \underbrace{3 - 2i}_{\text{the dividend}}.$$

To obtain the quotient of two complex numbers, multiply the numerator and denominator of the fraction by the conjugate of the denominator.

EXAMPLE 9

Perform the indicated operations and simplify, if possible.

a. $\dfrac{-5 + 2i}{3i}$
b. $\dfrac{3 - 2i}{2 + 5i}$

c. $\dfrac{1 + 2i}{-3 - 5i}$
d. $\dfrac{1 + i\sqrt{2}}{1 - i\sqrt{2}}$

Solution

a. $\dfrac{-5 + 2i}{3i} = \dfrac{-5 + 2i}{3i} \cdot \dfrac{-3i}{-3i} = \dfrac{15i - 6i^2}{-9i^2} = \dfrac{15i - 6(-1)}{-9(-1)}$

$= \dfrac{6 + 15i}{9} = \dfrac{2 + 5i}{3}.$

b. $\dfrac{3 - 2i}{2 + 5i} = \dfrac{3 - 2i}{2 + 5i} \cdot \dfrac{2 - 5i}{2 - 5i} = \dfrac{6 - 15i - 4i + 10i^2}{(2)^2 - (5i)^2}$

$= \dfrac{6 - 19i - 10}{4 - (-25)} = \dfrac{-4 - 19i}{29}.$

Complex Numbers and Quadratic Equations

c. $\dfrac{1+2i}{-3-5i} = \dfrac{1+2i}{-3-5i} \cdot \dfrac{-3+5i}{-3+5i} = \dfrac{-3+5i-6i+10i^2}{(-3)^2-(5i)^2}$

$= \dfrac{-3-i-10}{9-(-25)}$

$= \dfrac{-13-i}{34}.$

Note: conjugates differ only in sign preceding the second term.

d. $\dfrac{1+i\sqrt{2}}{1-i\sqrt{2}} = \dfrac{1+i\sqrt{2}}{1-i\sqrt{2}} \cdot \dfrac{1+i\sqrt{2}}{1+i\sqrt{2}} = \dfrac{1+i\sqrt{2}+i\sqrt{2}+2i^2}{(1)^2-2i^2}$

$= \dfrac{1+2i\sqrt{2}-2}{1-(-2)} = \dfrac{-1+2i\sqrt{2}}{3}.$

Real numbers are also complex numbers, since any real number "a" can be expressed as a +0i. The operations with complex numbers have been defined in such a manner that they are consistent with those for real numbers. But it should be noted that unlike the real numbers, the set of complex numbers has no order. We cannot say that one complex number is greater than another. In fact, we cannot describe a given complex number as positive or negative. For example, it is not possible to determine whether $-3+2i$ is greater than or less than $3-2i$. The properties of inequality do not hold for the set of complex numbers.

HISTORICAL NOTE

It was Leonard Euler (1707–1783), a great Swiss mathematician, who first introduced the letter i to represent the square root of negative 1, $i = \sqrt{-1}$. This symbol began to appear about 1748. Approximately 100 years earlier, Descartes had decided to call *imaginary* a number such as $\sqrt{-1}$. Until that time, such numbers were called fictitious. Much later (1832), it was Gauss who thought it necessary to distinguish between imaginary numbers such as 4i and 3 + 4i. Consequently, he invented the term *complex number* to describe an expression that contained both a real number and an imaginary number of the form a + bi.

EXERCISES FOR SECTION 7.2

(Answers given to exercises marked *)

Perform the indicated operation for each of the following and simplify, if possible.

*1. $(2+3i) + (3+5i)$ 　　　　 2. $(4+2i) + (2+3i)$

*3. $(2 + 5i) + (1 + i)$
4. $(3 + i) + (2 + 3i)$
*5. $(1 + i) + (1 + i)$
6. $(3 - 2i) + (-2 + 3i)$
*7. $(-2 - 3i) + (-4 - i)$
8. $(-2 - 3i) + (2 - 3i)$
*9. $(1 - i) + (-1 + i)$
10. $(3 + 2i) - (2 + i)$
*11. $(4 - 3i) - (2 - i)$
12. $(5 - 3i) - (2 - 5i)$
*13. $(-3 + 2i) - (-4 - 5i)$
14. $(-11 + 2i) - (9 - 3i)$
*15. $(-1 + i) - (1 - i)$
16. $(-8 + 3i) - (-8 - 3i)$
*17. $(-2 + 3i) - (2 + 3i)$
18. $(2 - 3i) - (2 - 3i)$
*19. $3(2 - 5i)$
20. $3i(2 - 5i)$
*21. $5i(-3 + 2i)$
22. $(1 + i)(3 + 2i)$
*23. $(5 + 2i)(1 + 3i)$
24. $(-2 + i)(3 - 2i)$
*25. $(3 - 2i)(-2 + i)$
26. $(-2 - i)(-3 + 2i)$
*27. $(-2 + 5i)(-3 - 2i)$
28. $(2 - 3i)(-2 + 3i)$
*29. $(3 - i)(4 + 3i)$
30. $(1 + i)(2 + 2i)$
*31. $(2 + 3i)^2$
32. $(3 - 5i)^2$
*33. $(-2 - 3i)^2$
34. $(-2 - i)^2$
*35. $(1 + i)^2$
36. $(2 + 3i)(2 - 3i)$
*37. $(1 + i)(1 - i)$
38. $(-2 - 3i)(-2 + 3i)$
*39. $(-3 - i)(-3 + i)$
40. $(5 - 2i)(5 + 2i)$
*41. $(-3 + 2i)(-3 - 2i)$
42. $(-1 - i)(-1 + i)$
*43. $(1 + 2i)(1 - 2i)(3 + 2i)$
44. $(3 - 2i)(5 + 4i)(3 + 2i)$
*45. $(-1 + i)(1 - i)(-1 - i)$
46. $\dfrac{-2 + 3i}{2i}$
*47. $\dfrac{3 + 2i}{i}$
48. $\dfrac{1 + i}{2i}$
*49. $\dfrac{3}{1 + i}$
50. $\dfrac{-5}{2 - i}$
*51. $\dfrac{1}{1 - i}$
52. $\dfrac{1 + i}{1 - i}$
*53. $\dfrac{3 + 2i}{2 - 3i}$
54. $\dfrac{-1 + 2i}{1 - 3i}$
*55. $\dfrac{-3 - 5i}{-3 + 5i}$
56. $\dfrac{2 - 3i}{-2 + 3i}$

Complex Numbers and Quadratic Equations 345

*57. $\dfrac{3-i}{1-i}$

58. $\dfrac{-2+7i}{-1+2i}$

*59. $-\dfrac{1+i}{3+2i}$

60. $-\dfrac{-3-2i}{-1-i}$

*61. $(2+i)^{-1}$

62. $(1+2i)^{-2}$

*63. $(-2-3i)^{-2}$

64. $\dfrac{-1+i}{1-i} \cdot \dfrac{1-i}{-1-i}$

*65. $\dfrac{2+i}{3+i} \cdot \dfrac{3+i}{1-i}$

66. $\dfrac{-1+2i}{3-i} \cdot \dfrac{3-i}{-1+2i}$

*67. $(1+i)^3$

68. $(1-i)^3$

*69. $(-1+i)^3$

7.3 QUADRATIC EQUATIONS

Almost all of our work with equations thus far has been with equations of the first degree, that is, those in which the highest degree of the variable has been one. Now, we shall examine second degree equations in one variable; these contain the second but no higher power of that variable. An equation of this type is formally called a quadratic equation. For example, $2x^2 + 3x - 2 = 0$ is a quadratic (second degree) equation in one variable. A quadratic equation in x can be written in the *standard form,*

$$ax^2 + bx + c = 0$$

where a, b, and c are real numbers and $a \neq 0$. Note that a is the coefficient of the x^2 term, b is the coefficient of the x term, and c is the constant. For example, $2x^2 + 3x - 2 = 0$ is a quadratic equation written in the standard form where $a = 2$, $b = 3$, and $c = -2$. The equation $4x^2 = 2x + 6$ is also a quadratic equation, but it is not written in standard form. To be in standard form, it must be of the form $ax^2 + bx + c = 0$. Therefore, to write $4x^2 = 2x + 6$ in standard form we must add $-2x$ and -6 to both sides of the equation. Hence, $4x^2 = 2x + 6$ can be written as $4x^2 - 2x - 6 = 0$, where $a = 4$, $b = -2$, and $c = -6$. It should be noted that $x^2 - 25 = 0$ and $2x^2 + 4x = 0$ are also quadratic equations. To consider $x^2 - 25 = 0$ in standard form, we rewrite it as $1x^2 + 0x - 25 = 0$ where $a = 1$, $b = 0$, and $c = -25$. Similarly, we can rewrite $2x^2 + 4x = 0$ in standard form as $2x^2 + 4x + 0 = 0$ where $a = 2$, $b = 4$, and $c = 0$. The last two equations are examples of quadratic equations that are said to be *incomplete.* They are called incomplete quadratic equations because $b = 0$ or $c = 0$, or we can also have the case where b and c are both equal to 0. For example, $x^2 = 0$ is an incomplete quadratic equation where both b and $c = 0$. Other examples of incomplete quadratic equations are: $2x^2 - 8 = 0$ ($b = 0$), $3x^2 + 6x = 0$ ($c = 0$), and $5x^2 = 0$ ($b = 0$, $c = 0$). The general incomplete quadratic equations are $ax^2 + c = 0$ ($b = 0$), $ax^2 + bx = 0$ ($c = 0$), and $ax^2 = 0$ ($b = 0$, $c = 0$). The solution of an incomplete quadratic equation is easier to obtain than that of a complete quadratic equation. Hence, we shall examine the solution of incomplete quadratic equations first.

Suppose the product of two numbers is 0, as in $ab = 0$. What can we conclude? If the product of two numbers equals 0, then at least one of the factors equals 0: If $a = 0$ and $b \neq 0$, then $0 \cdot b = 0$. Similarly, if $a \neq 0$ and $b = 0$, then $a \cdot 0 = 0$. We can also have the case where $a = 0$ and $b = 0$, then $0 \cdot 0 = 0$. Therefore, if one or the other or both factors equal 0, then the product of the two factors equals 0. We can state this formally as:

$$\boxed{ab = 0, \text{ if and only if } a = 0 \text{ or } b = 0.}$$

The word "or" is used in the inclusive sense: one, or the other, or both, that is, $a = 0$, or $b = 0$, or both $a = 0$, $b = 0$.

Note that a quadratic equation is of the second degree, and it has two roots. If we consider the incomplete quadratic equation $ax^2 = 0$, it should have two roots.

$$ax^2 = 0. \quad \text{given}$$
$$x^2 = 0. \quad \text{multiplying both sides by } \frac{1}{a}$$
$$x \cdot x = 0. \quad \text{definition of exponent}$$

Now we have the situation where the product of two factors equals 0. From our previous discussion, we conclude that

$$x = 0 \text{ or } x = 0.$$

Note that 0 is a double root of the given equation; we have two solutions, but they are not distinct. If we have an incomplete quadratic equation where $b = 0$ and $c = 0$, that is, $ax^2 = 0$, then both roots of the given quadratic equation are 0. For example, the following equations are incomplete quadratic equations where $b = 0$, $c = 0$, and both roots are 0: $2x^2 = 0$, $4y^2 = 0$, $5z^2 = 0$.

Next, we shall consider the case where only $b = 0$ in an incomplete quadratic equation. For example, $x^2 - 1 = 0$ is an incomplete quadratic equation where $a = 1$, $b = 0$ and $c = -1$. Perhaps you can solve this particular equation by trial and error. What numbers will satisfy the given equation? After some searching, we see that $\{1, -1\}$ is the solution set. Checking, we have $(1)^2 - 1 = 1 - 1 = 0$ and $(-1)^2 - 1 = 1 - 1 = 0$. The members of the solution set satisfy the original equation. We could have also

Complex Numbers and Quadratic Equations 347

solved this equation by solving it for x^2 and then taking the square root of both sides:

$x^2 - 1 = 0.$
$x^2 = 1.$
$x = \pm\sqrt{1}.$ (The symbol $\pm$ is read "positive or negative.")
$x = \pm 1.$

Therefore, the solution set is $\{\pm 1\}$.

If we have an incomplete quadratic equation of the form $ax^2 + c = 0$ where $a \neq 0$, $c \neq 0$, and $b = 0$, we can always solve it for x^2 and take the square root of both sides,

$$ax^2 + c = 0.$$

$$ax^2 = -c.$$

$$x^2 = -\frac{c}{a}.$$

$$x = \pm\sqrt{-\frac{c}{a}}.$$

Therefore, whenever $b = 0$ and $c \neq 0$, $x = \sqrt{-\frac{c}{a}}$ or $x = -\sqrt{-\frac{c}{a}}$. Note that the two roots are almost identical — they differ only in sign. They are additive inverses of each other. It should also be noted that the expression $-\frac{c}{a}$ is not necessarily negative. If c is negative and a is positive, then $-\frac{c}{a}$ is positive. Similarly, if a is negative and c is positive, then $-\frac{c}{a}$ is positive. The expression $-\frac{c}{a}$ is negative only when a and c have like signs, both positive or both negative. If $ax^2 + c = 0$ is to have real roots, then $-\frac{c}{a}$ cannot be negative.

The equation $4x^2 - 36 = 0$ is an incomplete quadratic where $b = 0$ and $a \neq 0$, $c \neq 0$. We can find its solution set by solving it for x^2 and taking the square root of both sides.

$$4x^2 - 36 = 0.$$
$$4x^2 = 36.$$
$$x^2 = \frac{36}{4} = 9.$$
$$x = \pm\sqrt{9} = \pm 3.$$

Check:

If x = 3, $4(3)^2 - 36 \stackrel{?}{=} 0$ Or   if x = −3, $4(−3)^2 − 36 \stackrel{?}{=} 0$.
$4 \cdot 9 - 36 \stackrel{?}{=} 0$. $4(9) − 36 \stackrel{?}{=} 0$.
$36 - 36 = 0$. $36 − 36 = 0$.
$0 = 0$. $0 = 0$.

The notation x = ±3 is an abbreviation for the statement that x = 3 or x = −3.

EXAMPLE 10

Solve $3x^2 - 9 = 0$.

Solution

$$3x^2 - 9 = 0.$$
$$3x^2 = 9.$$
$$x^2 = 3.$$
$$x = \pm\sqrt{3}.$$

Check:

If x = $\sqrt{3}$, $3(\sqrt{3})^2 - 9 \stackrel{?}{=} 0$.   If x = $-\sqrt{3}$, $3(-\sqrt{3})^2 - 9 \stackrel{?}{=} 0$.
$3 \cdot 3 - 9 = 0$. $3(3) - 9 = 0$.
$9 - 9 = 0$. $9 - 9 = 0$.
$0 = 0$. $0 = 0$.

The solution set is $\{\pm\sqrt{3}\}$

EXAMPLE 11

Solve $4x^2 - 3 = 0$.

Solution

$$4x^2 - 3 = 0.$$
$$4x^2 = 3.$$
$$x^2 = \frac{3}{4}.$$
$$x = \pm\sqrt{\frac{3}{4}} = \pm\frac{\sqrt{3}}{2}.$$

Complex Numbers and Quadratic Equations 349

Check:

If $x = \frac{\sqrt{3}}{2}$, $4\left(\frac{\sqrt{3}}{2}\right)^2 - 3 \stackrel{?}{=} 0.$ If $x = -\frac{\sqrt{3}}{2}$, $4\left(-\frac{\sqrt{3}}{2}\right)^2 - 3 \stackrel{?}{=} 0.$

$\phantom{If x = \frac{\sqrt{3}}{2}, }4\left(\frac{3}{4}\right) - 3 = 0.$ $\phantom{If x = -\frac{\sqrt{3}}{2}, }4\left(\frac{3}{4}\right) - 3 = 0.$

$\phantom{If x = \frac{\sqrt{3}}{2}, 4}3 - 3 = 0.$ $\phantom{If x = -\frac{\sqrt{3}}{2}, 4}3 - 3 = 0.$
$\phantom{If x = \frac{\sqrt{3}}{2}, 4}0 = 0.$ $\phantom{If x = -\frac{\sqrt{3}}{2}, 4}0 = 0.$

The solution set is $\left\{\pm\frac{\sqrt{3}}{2}\right\}$.

EXAMPLE 12

Solve $2x^2 + 4 = 0$.

Solution

$2x^2 + 4 = 0.$
$2x^2 = -4.$
$x^2 = -2.$

$x = \pm\sqrt{-2}.$ We have the square root of a negative
$x = \pm i\sqrt{2}.$ number, hence the roots are imaginary.

Check:

If $x = i\sqrt{2}$, $2(i\sqrt{2})^2 + 4 \stackrel{?}{=} 0.$ If $x = -i\sqrt{2}$, $2(-i\sqrt{2})^2 + 4 \stackrel{?}{=} 0.$
$\phantom{If x = i\sqrt{2}, }2(i^2 2) + 4 = 0.$ $\phantom{If x = -i\sqrt{2}, }2(i^2 2) + 4 = 0.$
$\phantom{If x = i\sqrt{2}, 2}4i^2 + 4 = 0.$ $\phantom{If x = -i\sqrt{2}, 2}4i^2 + 4 = 0.$
$\phantom{If x = i\sqrt{2}, 2}-4 + 4 = 0.$ $\phantom{If x = -i\sqrt{2}, 2}-4 + 4 = 0.$
$\phantom{If x = i\sqrt{2}, 2}0 = 0.$ $\phantom{If x = -i\sqrt{2}, 2}0 = 0.$

The solution set is $\{\pm i\sqrt{2}\}$.

EXAMPLE 13

Solve $3x^2 + 4 = 0$.

Solution

$3x^2 + 4 = 0.$
$3x^2 = -4.$
$x^2 = -\frac{4}{3}.$

$$x = \pm\sqrt{-\frac{4}{3}}.$$

$$x = \pm i\sqrt{\frac{4}{3}}.$$

$$x = \pm i\frac{2}{\sqrt{3}}.$$

$$x = \pm i\frac{2\sqrt{3}}{3} = \pm\frac{2i\sqrt{3}}{3}.$$

Check:

If $x = \frac{2i\sqrt{3}}{3}$, $3\left(\frac{2i\sqrt{3}}{3}\right)^2 + 4 \stackrel{?}{=} 0$.

$3\left(\frac{4i^2 \cdot 3}{9}\right) + 4 = 0.$

$4i^2 + 4 = 0.$

$0 = 0.$

The check for $x = \frac{-2i\sqrt{3}}{3}$ is left for the student.

Recall our previous discussion regarding the case where the product of two numbers is equal to 0. That is, ab = 0 if and only if a = 0 or b = 0. Expanding upon this idea, consider the case where (x − 1)(x + 2) = 0. Again, we have the product of two numbers equal to 0 and (x − 1)(x + 2) = 0 if and only if (x − 1) = 0 or (x + 2) = 0. If x − 1 = 0, then x = 1, and if x + 2 = 0, then x = −2. Similarly, if x(x + 1) = 0, then x = 0 or x + 1 = 0 and solving x + 1 = 0 for x, x = −1. Therefore, the expression x(x + 1) = 0 only when x = 0 or x = −1. We shall use this information to solve the quadratic equation $ax^2 + bx + c = 0$ when $a \neq 0$, $b \neq 0$ and c = 0. We can solve the equation $ax^2 + bx = 0$ for x by removing the common factor x, and determining those values that will make the two factors equal to 0. Proceeding, we have

$$ax^2 + bx = 0.$$
$$x(ax + b) = 0.$$

The product equals zero, if and only if

$$x = 0 \text{ or } ax + b = 0.$$
$$ax = -b.$$
$$x = -\frac{b}{a}.$$

Therefore, $x = 0$ or $x = -\frac{b}{a}$.

Complex Numbers and Quadratic Equations   351

To solve $3x^2 + 2x = 0$, we remove the common factor x, $x(3x + 2) = 0$. The indicated product equals 0 if and only if $x = 0$ or $3x + 2 = 0$. Solving the equation $3x + 2 = 0$, we obtain $x = -\frac{2}{3}$. Therefore, the solution set for the equation $3x^2 + 2x = 0$ is $\{0, -\frac{2}{3}\}$.

EXAMPLE 14

Solve $4x^2 - 3x = 0$.

Solution

$$4x^2 - 3x = 0.$$
$$x(4x - 3) = 0.$$
$$x = 0 \text{ or } 4x - 3 = 0.$$
$$4x = 3.$$
$$x = \frac{3}{4}.$$

Check:

If $x = 0$, $4(0)^2 - 3(0) \stackrel{?}{=} 0$.   If $x = \frac{3}{4}$, $4\left(\frac{3}{4}\right)^2 - 3\left(\frac{3}{4}\right) \stackrel{?}{=} 0$.

$$4 \cdot 0 - 0 = 0.$$ $$4 \cdot \frac{9}{16} - \frac{9}{4} = 0.$$

$$0 - 0 = 0.$$ $$\frac{9}{4} - \frac{9}{4} = 0.$$

$$0 = 0.$$ $$0 = 0.$$

The solution set is $\{0, \frac{3}{4}\}$.

EXAMPLE 15

Solve $2x^2 + 6x = 0$.

Solution

$2x^2 + 6x = 0.$
$2x(x + 3) = 0.$ removing the common factor 2x
$2x = 0$ or $x + 3 = 0.$
$x = 0$ or $x = -3.$

Check:

If x = 0, $2(0)^2 + 6(0) \stackrel{?}{=} 0$.　　If x = −3, $2(−3)^2 + 6(−3) \stackrel{?}{=} 0$.
$2 \cdot 0 + 0 = 0$.　　　　　　　　　$2(9) − 18 = 0$.
$0 + 0 = 0$.　　　　　　　　　　$18 − 18 = 0$.
$0 = 0$.　　　　　　　　　　　$0 = 0$.

The solution set is $\{0, −3\}$.

EXAMPLE 16

Solve $x^2 + x = 0$.

Solution

Note that a = 1 and b = 1, hence
$$x^2 + x = 0.$$
$$x(x + 1) = 0.$$
$$x = 0 \text{ or } x + 1 = 0.$$
$$x = −1.$$

Check:

If x = 0, $0^2 + 0 = 0$.　　If x = −1, $(−1)^2 + (−1) \stackrel{?}{=} 0$.
$0 = 0$.　　　　　　　　$1 − 1 = 0$.
　　　　　　　　　　　　$0 = 0$.

The solution set is $\{0, −1\}$.

Thus far we have discovered methods for solving the different types of incomplete quadratic equations. The standard form of a quadratic equation is $ax^2 + bx + c = 0$. If b = 0 or c = 0, then the quadratic equation is incomplete. If both b = 0 and c = 0, then $ax^2 = 0$ and it has 0 as a double root. If b = 0, then $ax^2 + c = 0$ can be solved for x by solving the given equation for x^2 and taking the square root of both sides, and $x = \pm\sqrt{-\frac{c}{a}}$. If c = 0, then $ax^2 + bx = 0$ can be solved for x by removing the common factor x, and determining those values that will make the two factors equal to 0 and $x = 0$ or $x = -\frac{b}{a}$.

Complex Numbers and Quadratic Equations 353

EXERCISES FOR SECTION 7.3
(Answers given to exercises marked *)

Solve each of the following incomplete quadratic equations.

*1. $4x^2 = 0$.
2. $3y^2 + 2 = 2$.
*3. $5x^2 + 1 = 1$.
4. $3x^2 - 3 = 0$.
*5. $4y^2 - 4 = 0$.
6. $2y^2 - 8 = 0$.
*7. $3x^2 - 6 = 0$.
8. $4y^2 - 12 = 0$.
*9. $3y^2 - 27 = 0$.
10. $4x^2 - 9 = 0$.
*11. $3x^2 - 15 = 0$.
12. $4x^2 - 7 = 0$.
*13. $3y^2 + 6 = 0$.
14. $2y^2 + 10 = 0$.
*15. $2y^2 + 8 = 0$.
16. $2x^2 - 7 = 0$.
*17. $3y^2 - 5 = 0$.
18. $5x^2 - 16 = 0$.
*19. $3y^2 + 8 = 0$.
20. $4x^2 + 5 = 0$.
*21. $5y^2 + 2 = 0$.
22. $2x^2 + 3 = 4 - x^2$.
*23. $y^2 - 1 = -2 - y^2$.
24. $x^2 - 5x = 0$.
*25. $3y^2 - 2y = 0$.
26. $4y^2 + 2y = 0$.
*27. $3y^2 = 5y$.
28. $5x^2 = 5x$.
*29. $3y^2 + 2y = y^2 - 2y$.
30. $x^2 - 2x = 2x^2 + x$.
*31. $4y^2 - 3y = 3y^2 + 3y$.
32. $5x^2 + 6x = 4x^2 + 7x$.
*33. $2y^2 - 5y = y^2 + y$.
34. $ax^2 + ax = 0$.
*35. $ay^2 - by = 0$.
36. $mx^2 = nx$.

7.4 SOLVING COMPLETE QUADRATIC EQUATIONS BY FACTORING

In the beginning of our discussion of quadratic equations, we said that the standard form of a quadratic equation is $ax^2 + bx + c = 0$. In the previous section, we learned methods for solving *incomplete* quadratic equations. We can solve a quadratic equation, $ax^2 + bx + c = 0$, when $b = 0$ or $c = 0$, or when both b and c equal 0. Now let us consider the case where $a \neq 0$, $b \neq 0$ and $c \neq 0$. We shall solve the *complete* quadratic equation, $ax^2 + bx + c = 0$. In Chapter 5, we discussed techniques for factoring polynomials of the form $x^2 + bx + c$ and $ax^2 + bx + c$. We shall make use of these techniques, since often a complete quadratic equation can be solved by factoring $ax^2 + bx + c$ into two binomials. We have already discussed the case where the product of two numbers is 0. That is, $ab = 0$, if and only if $a = 0$ or $b = 0$. Therefore, if $(x-2)(x+1) = 0$, then $(x-2) = 0$ or $(x+1) = 0$, and $x = 2$ or $x = -1$. We can use all of this information to solve the complete equation $x^2 - x - 2 = 0$. Note that it is the standard form $ax^2 + bx + c = 0$.

$$\begin{aligned} x^2 - x - 2 &= 0. \quad &\text{given} \\ (x-2)(x+1) &= 0. \quad &\text{factoring the left side} \\ x - 2 = 0 \text{ or } x + 1 &= 0. \quad &ab = 0, \text{ if and only if } a = 0 \text{ or } b = 0 \\ x = 2 \text{ or } x &= -1. \quad &\text{solving each equation} \end{aligned}$$

Check:

If $x = 2$, $2^2 - 2 - 2 \stackrel{?}{=} 0$. If $x = -1$, $(-1)^2 - (-1) - 2 \stackrel{?}{=} 0$.
$\qquad 4 - 4 = 0$. $\qquad\qquad\qquad 1 + 1 - 2 = 0$.
$\qquad\quad 0 = 0$. $\qquad\qquad\qquad\qquad 0 = 0$.

The solution set is $\{2, -1\}$.

To solve a quadratic equation by factoring, we must have it in standard form, that is, one side of the equation must be 0. Next, factor the quadratic expression $ax^2 + bx + c$. Since $ab = 0$ if and only if $a = 0$ or $b = 0$, set each factor equal to 0 and solve the resulting equations. It should be noted that *not all* quadratic equations of the form $ax^2 + bx + c = 0$ can be solved by means of factoring. We shall explore this in detail later, and for now, consider only those that can.

The first step in solving the equation, $6x^2 = 2 - x$, is to write it in standard form, and then proceed in the manner outlined previously.

$$\begin{aligned} 6x^2 &= 2 - x. \\ 6x^2 + x - 2 &= 0. \quad &\text{writing the equation in standard form} \\ (3x + 2)(2x - 1) &= 0. \quad &\text{factoring the left side} \end{aligned}$$

Therefore,

$$\begin{aligned} 3x + 2 = 0 \text{ or } 2x - 1 &= 0. \quad &ab = 0, \text{ if and only if } a = 0 \text{ or } b = 0. \\ 3x = -2 \quad \text{ or } \quad 2x &= 1. \quad &\text{solving each equation} \\ x = -\frac{2}{3}. \qquad x &= \frac{1}{2}. \end{aligned}$$

Checking the solutions in the original equation:

If $x = -\frac{2}{3}$, $6\left(-\frac{2}{3}\right)^2 \stackrel{?}{=} 2 - \left(-\frac{2}{3}\right)$. If $x = \frac{1}{2}$, $6\left(\frac{1}{2}\right)^2 \stackrel{?}{=} 2 - \frac{1}{2}$.

Complex Numbers and Quadratic Equations 355

$$6 \cdot \frac{4}{9} = 2 + \frac{2}{3}.$$

$$\frac{24}{9} = 2\frac{2}{3}.$$

$$\frac{8}{3} = \frac{8}{3}.$$

$$6 \cdot \frac{1}{4} = \frac{3}{2}.$$

$$\frac{6}{4} = \frac{3}{2}.$$

$$\frac{3}{2} = \frac{3}{2}.$$

The solution set is $\left\{-\frac{2}{3}, \frac{1}{2}\right\}$.

EXAMPLE 17

Solve $x^2 - 4x - 5 = 0$.

Solution

$x^2 - 4x - 5 = 0.$	the equation is in standard form
$(x - 5)(x + 1) = 0.$	factoring the left side
$x - 5 = 0$ or $x + 1 = 0.$	$ab = 0$ if and only if $a = 0$ or $b = 0.$
$x = 5$ or $x = -1.$	solving each equation

Check:

If $x = 5$, $(5)^2 - 4(5) - 5 \stackrel{?}{=} 0.$ If $x = -1$, $(-1)^2 - 4(-1) - 5 \stackrel{?}{=} 0.$
$25 - 20 - 5 = 0.$ $1 + 4 - 5 = 0.$
$0 = 0.$ $0 = 0.$

The solution set is $\{5, -1\}$.

EXAMPLE 18

Solve $12x^2 = 11x - 2$.

Solution

To solve the given quadratic equation by means of factoring, we first write in the standard form and then proceed in the normal manner.

$$12x^2 = 11x - 2.$$
$$12x^2 - 11x + 2 = 0.$$
$$(4x - 1)(3x - 2) = 0.$$
$$4x - 1 = 0 \text{ or } 3x - 2 = 0.$$
$$4x = 1. \qquad 3x = 2.$$
$$x = \frac{1}{4}. \qquad x = \frac{2}{3}.$$

Check:

$$\text{If } x = \frac{1}{4}, \ 12\left(\frac{1}{4}\right)^2 \stackrel{?}{=} 11\left(\frac{1}{4}\right) - 2. \qquad \text{If } x = \frac{2}{3}, \ 12\left(\frac{2}{3}\right)^2 \stackrel{?}{=} 11\left(\frac{2}{3}\right) - 2.$$

$$12 \cdot \frac{1}{16} = \frac{11}{4} - 2. \qquad\qquad 12 \cdot \frac{4}{9} = \frac{22}{3} - 2.$$

$$\frac{3}{4} = \frac{3}{4}. \qquad\qquad\qquad \frac{16}{3} = \frac{16}{3}.$$

The solution set is $\left\{\frac{1}{4}, \frac{2}{3}\right\}$.

In the previous section, we solved incomplete quadratic equations. One of the forms considered was where $b = 0$, and $a \neq 0$, $c \neq 0$. For example, $x^2 - 36 = 0$. We found its solution by solving the equation for x^2 and then taking the square root of both sides,

$$x^2 - 36 = 0.$$
$$x^2 = 36.$$
$$x = \pm\sqrt{36}.$$
$$x = \pm 6.$$

Note that we can also solve this equation by means of factoring. The expression $x^2 - 36$ is the difference of two squares and we can factor it, $x^2 - 36 = (x + 6)(x - 6)$.

$$x^2 - 36 = 0.$$
$$(x + 6)(x - 6) = 0.$$
$$x + 6 = 0 \text{ or } x - 6 = 0.$$
$$x = -6. \qquad x = 6.$$

Check:

$$\text{If } x = -6, \ (-6)^2 - 36 \stackrel{?}{=} 0. \qquad \text{If } x = 6, \ (6)^2 - 36 \stackrel{?}{=} 0.$$
$$36 - 36 = 0. \qquad\qquad\qquad 36 - 36 = 0.$$
$$0 = 0. \qquad\qquad\qquad\qquad 0 = 0.$$

The solution set is $\{\pm 6\}$.

Normally, we should solve the quadratic equation $ax^2 + bx + c = 0$ where $b = 0$ and $a \neq 0$, $c \neq 0$, by solving the equation for x^2 and taking the square root of both sides. But if the incomplete quadratic equation $ax^2 + c = 0$ is the difference of two squares, then we can also solve it by factoring.

In the previous chapter, we solved equations containing rational

Complex Numbers and Quadratic Equations

expressions. Recall that we solved them by first multiplying both sides of the equation by the least common denominators of the fractions contained in the equation. For example, to solve the equation $\frac{1}{x} + \frac{1}{2} = 1$, we multiply both sides by 2x (the LCD of $\frac{1}{x}$ and $\frac{1}{2}$), and we have

$$\frac{1}{x} + \frac{1}{2} = 1.$$

$$2x\left(\frac{1}{x} + \frac{1}{2}\right) = 2x \cdot 1.$$

$$2x \cdot \frac{1}{x} + 2x \cdot \frac{1}{2} = 2x.$$

$$2 + x = 2x.$$
$$2 = x.$$

Many times the transforming of such equations will result in quadratic equations, and to solve the resulting quadratic equation, we can employ the methods just discussed. For example, consider the fractional equation $1 = \frac{1}{x} + \frac{2}{x^2}$.

To eliminate the fractions, multiply both sides of the equation by x^2 (the LCD of $\frac{1}{x}$ and $\frac{1}{x^2}$),

$$1 = \frac{1}{x} + \frac{2}{x^2}.$$

$$x^2 \cdot 1 = x^2\left(\frac{1}{x} + \frac{2}{x^2}\right).$$

$$x^2 = x^2 \cdot \frac{1}{x} + x^2 \cdot \frac{2}{x^2}.$$

$$x^2 = x + 2.$$
$$x^2 - x - 2 = 0. \quad \text{(a quadratic equation in standard form)}$$
$$(x - 2)(x + 1) = 0. \quad \text{(factoring } x^2 - x - 2\text{)}$$
$$x - 2 = 0 \text{ or } x + 1 = 0.$$
$$x = 2. \qquad x = -1.$$

Check:

if $x = 2$, $1 \stackrel{?}{=} \frac{1}{2} + \frac{2}{2^2}.$ If $x = -1$, $1 \stackrel{?}{=} \frac{1}{-1} + \frac{2}{(-1)^2}.$

$\qquad\qquad 1 = \frac{1}{2} + \frac{2}{4}.$ $\qquad\qquad\qquad 1 = -1 + 2.$

$\qquad\qquad 1 = 1.$ $\qquad\qquad\qquad\qquad 1 = 1.$

The solution set is {2, −1}.

It is important to check the solutions of the resulting transformed equations in the original equation, since the roots may be extraneous.

EXAMPLE 19

Solve $\dfrac{2}{x+2} + \dfrac{3}{x-2} = 1$.

Solution

The LCD of the fractions contained in the equation is $(x + 2)(x − 2)$. Hence, we multiply both sides of the equation by it.

$$\frac{2}{x+2} + \frac{3}{x-2} = 1.$$

$$(x+2)(x-2) \cdot \left(\frac{2}{x+2} + \frac{3}{x-2}\right) = (x+2)(x-2) \cdot 1.$$

$$(x+2)(x-2) \cdot \frac{2}{(x+2)} + (x+2)(x-2) \cdot \frac{3}{(x-2)} = x^2 - 4.$$

$$(x-2)2 + (x+2)3 = x^2 - 4.$$
$$2x - 4 + 3x + 6 = x^2 - 4.$$
$$5x + 2 = x^2 - 4.$$
$$0 = x^2 - 5x - 6.$$
$$0 = (x-6)(x+1).$$
$$x - 6 = 0 \text{ or } x + 1 = 0.$$
$$x = 6. \qquad x = -1.$$

Check:

If $x = 6$, $\dfrac{2}{6+2} + \dfrac{3}{6-2} \stackrel{?}{=} 1$.   If $x = -1$, $\dfrac{2}{-1+2} + \dfrac{3}{-1-2} \stackrel{?}{=} 1$.

$\qquad\qquad \dfrac{2}{8} + \dfrac{3}{4} = 1$. $\qquad\qquad \dfrac{2}{1} + \dfrac{3}{-3} = 1$.

$\qquad\qquad\qquad 1 = 1$. $\qquad\qquad\qquad 2 - 1 = 1$.
$\qquad\qquad\qquad\qquad\qquad\qquad\qquad\qquad\quad 1 = 1$.

The solution set is {6, −1}.

Complex Numbers and Quadratic Equations 359

EXAMPLE 20

Solve $\dfrac{x}{x+1} + \dfrac{2}{x-1} = \dfrac{4}{x^2-1}$.

Solution

The LCD is $(x+1)(x-1)$ and we multiply both sides of the equation by it.

$$\dfrac{x}{x+1} + \dfrac{2}{x-1} = \dfrac{4}{x^2-1}.$$

$$(x+1)(x-1) \cdot \left(\dfrac{x}{x+1} + \dfrac{2}{x-1}\right) = (x+1)(x-1) \cdot \dfrac{4}{x^2-1}.$$

$$(x+1)(x-1) \cdot \dfrac{x}{(x+1)} + (x+1)(x-1) \cdot \dfrac{2}{(x-1)} = 4.$$

$$(x-1)x + (x+1)2 = 4.$$
$$x^2 - x + 2x + 2 = 4.$$
$$x^2 + x - 2 = 0.$$
$$(x+2)(x-1) = 0.$$
$$x + 2 = 0 \text{ or } x - 1 = 0.$$
$$x = -2, \qquad x = 1.$$

Check:

If $x = -2$, $\dfrac{-2}{-2+1} + \dfrac{2}{-2-1} \stackrel{?}{=} \dfrac{4}{(-2)^2-1}.$

$$\dfrac{-2}{-1} + \dfrac{2}{-3} = \dfrac{4}{4-1}.$$

$$2 - \dfrac{2}{3} = \dfrac{4}{3}.$$

$$\dfrac{4}{3} = \dfrac{4}{3}.$$

If $x = 1$, $\dfrac{1}{1+1} + \dfrac{2}{1-1} \stackrel{?}{=} \dfrac{4}{1^2-1}.$

$$\dfrac{1}{2} + \dfrac{2}{0} \neq \dfrac{4}{0}.$$

Division by zero is impossible; $x = 1$ is not a solution of the given equation.

The solution set is $\{-2\}$.

Thus far we have examined how to solve a quadratic equation by means of factoring. It should be noted that we can also reverse the process. If we are given the roots of a quadratic equation, then we can construct a quadratic equation that has the given roots. We can solve the equation $x^2 + 3x + 2 = 0$ by means of factoring.

$$x^2 + 3x + 2 = 0.$$
$$(x + 2)(x + 1) = 0.$$
$$x + 2 = 0 \text{ or } x + 1 = 0.$$
$$x = -2, \quad x = -1.$$

To construct a quadratic equation that has the given roots, we reverse the process. If we are given that -2 and -1 are roots of a quadratic equation, then we can state that

$$x = -2 \quad \text{or} \quad x = -1.$$
$$x + 2 = 0 \text{ or } x + 1 = 0.$$

Since $ab = 0$ if and only if $a = 0$ or $b = 0$, we can state that

$$(x + 2)(x + 1) = 0.$$

Performing the indicated multiplication, we have the quadratic equation

$$x^2 + 3x + 2 = 0.$$

EXAMPLE 21

Construct a quadratic equation which has -1 and 3 as roots.

Solution

$$x = -1 \quad \text{or} \quad x = 3.$$
$$x + 1 = 0, \text{ or } x - 3 = 0.$$
$$(x + 1)(x - 3) = 0.$$
$$x^2 - 2x - 3 = 0.$$

EXAMPLE 22

Construct a quadratic equation which has $\dfrac{3}{2}$ and $-\dfrac{3}{5}$ as roots.

Solution

$$x = \frac{3}{2} \text{ or } x = -\frac{3}{5}.$$

Complex Numbers and Quadratic Equations

It is convenient to eliminate the fractions at this point. Using the equations above, we have

$$2x = 3 \quad \text{or} \quad 5x = -3.$$
$$2x - 3 = 0 \text{ or } 5x + 3 = 0.$$
$$(2x - 3)(5x + 3) = 0.$$
$$10x^2 - 9x - 9 = 0.$$

In this section, we have examined how to solve various quadratic equations by means of factoring. It should be noted that *not all* quadratic equations can be solved this way. In the next two sections, we shall develop a method for solving any quadratic equation. All of the quadratic equations in the following exercises, however, can be solved by factoring.

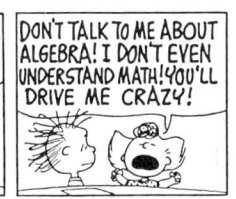

© 1965 United Features Syndicate, Inc.

EXERCISES FOR SECTION 7.4

(Answers given to exercises marked *)

Solve each of the following by factoring:

*1. $x^2 + 3x + 2 = 0$.
2. $y^2 + 2y + 1 = 0$.
*3. $y^2 + 5y + 6 = 0$.
4. $x^2 - 7x + 10 = 0$.
*5. $x^2 - 4x + 3 = 0$.
6. $y^2 + y - 2 = 0$.
*7. $y^2 + 2y - 3 = 0$.
8. $x^2 - 4x - 5 = 0$.
*9. $x^2 - 4x - 21 = 0$.
10. $2x^2 - x - 1 = 0$.
*11. $3y^2 - 2y - 1 = 0$.
12. $4x^2 + 5x + 1 = 0$.
*13. $2x^2 + 6 = 7x$.
14. $3y^2 = 8y + 3$.
*15. $2y^2 = 7 - 5y$.
16. $5x^2 = 6x - 1$.
*17. $6y^2 = 1 - y$.
18. $4y^2 = 4y - 1$.

*19. $9y^2 = -4 - 12y$.

20. $4x^2 = 20x - 25$.

*21. $9x^2 + 16 = 24x$.

22. $x^2 - 1 = 0$.

*23. $y^2 - 4 = 0$.

24. $x^2 - 25 = 0$.

*25. $4y^2 - 9 = 0$.

26. $9x^2 - 16 = 0$.

*27. $25x^2 - 1 = 0$.

28. $x(x - 3) = -2$.

*29. $y(2y - 1) = 1$.

30. $x(6x - 5) = -1$.

*31. $4x(x - 1) = -1$.

32. $3(y^2 + 2) = 11y$.

*33. $2(2x^2 + 1) = 6x$.

34. $x^2 = \dfrac{x}{2} + \dfrac{1}{2}$.

*35. $y^2 + 3 = \dfrac{12y}{3}$.

36. $\dfrac{1}{x} = \dfrac{1}{x^2}$.

*37. $\dfrac{3}{y} = \dfrac{4}{y^2}$.

38. $2 - \dfrac{7}{x} + \dfrac{6}{x^2} = 0$.

*39. $4 = \dfrac{20}{y} - \dfrac{25}{y^2}$.

40. $3 = \dfrac{8}{x} + \dfrac{3}{x^2}$.

*41. $5 = \dfrac{6}{y} - \dfrac{1}{y^2}$.

42. $\dfrac{2}{x + 2} + \dfrac{3}{x - 2} = 1$.

*43. $\dfrac{3}{x + 1} + \dfrac{2}{x - 1} = 3$.

44. $\dfrac{5}{x + 4} - \dfrac{3}{x - 2} = 4$.

*45. $\dfrac{x}{x^2 - 1} + \dfrac{1}{x + 1} = 1$.

46. $\dfrac{x}{x^2 - x - 2} + \dfrac{1}{x + 1} = 1$.

*47. $\dfrac{1}{x^2 - 2x + 1} + \dfrac{2}{x - 1} - 3 = 0$.

48. $\dfrac{3y}{y - 1} = 4 + \dfrac{3y}{y + 1}$.

*49. $\dfrac{2}{x^2 + 4x + 3} = \dfrac{1}{x^2 + 2x + 1}$.

50. $\dfrac{1}{x} + \dfrac{1}{x + 1} = \dfrac{1}{x + 1}$.

*51. $\dfrac{y - 4}{y + 2} = \dfrac{3y}{y^2 - 4}$.

Construct a quadratic equation in standard form with integral coefficients, given the indicated roots.

52. 1 and 2

*53. -1 and 3

54. 5 and -1

*55. 1 and -5

56. 2 and -2

*57. $\dfrac{1}{2}$ and $\dfrac{2}{3}$

58. $\dfrac{3}{4}$ and $-\dfrac{2}{3}$

*59. $\dfrac{5}{7}$ and $-\dfrac{2}{5}$

Complex Numbers and Quadratic Equations 363

60. $\sqrt{3}$ and $-\sqrt{3}$
*61. $1+\sqrt{2}$ and $1-\sqrt{2}$
62. $1+2\sqrt{3}$ and $1-2\sqrt{3}$
*63. $2-\sqrt{3}$ and $2+\sqrt{3}$
64. i and $-i$
*65. $1+2i$ and $1-2i$
66. $1+i$ and $1-i$

7.5 SOLVING QUADRATIC EQUATIONS BY COMPLETING THE SQUARE

In the previous section, we noted that not all quadratic equations can be solved by factoring. In this section, we shall develop a method that allows us to solve any quadratic equation whether it is factorable or not. One of the first techniques we used to solve quadratic equations such as $x^2 = 1$ was to take the square root of both sides, $x^2 = 1$ and $x = \pm\sqrt{1} = \pm 1$. We can use this same technique to solve the equation $(x+1)^2 = 25$.

$(x+1)^2 = 25.$ given

$x + 1 = \pm\sqrt{25}.$ taking the square root of both sides

$x + 1 = \pm 5.$

Therefore,

$x + 1 = 5$ or $x + 1 = -5.$
$x = 4$ or $x = -6.$

Check:

If $x = 4$, $(4+1)^2 \stackrel{?}{=} 25.$ If $x = -6$, $(-6+1)^2 \stackrel{?}{=} 25.$
$5^2 = 25.$ $(-5)^2 = 25.$
$25 = 25.$ $25 = 25.$

The solution set is $\{4, -6\}$.

EXAMPLE 23

Solve $(2x - 1)^2 = 9$.

Solution

$(2x - 1)^2 = 9.$

$$2x - 1 = \pm\sqrt{9}.$$
$$2x - 1 = \pm 3.$$

$$2x - 1 = 3 \quad \text{or} \quad 2x - 1 = -3.$$
$$2x = 4. \qquad\qquad 2x = -2.$$
$$x = 2. \qquad\qquad x = -1.$$

It can be verified by checking that the solution set is $\{2, -1\}$.

EXAMPLE 24

Solve $(x + 2)^2 = -1$

Solution

$$(x + 2)^2 = -1.$$
$$x + 2 = \pm\sqrt{-1}.$$
$$x + 2 = \pm i.$$

$$x + 2 = i \quad \text{or} \quad x + 2 = -i.$$
$$x = -2 + i \qquad\qquad x = -2 - i.$$

Check:

If $x = -2 + i$, $(-2 + i + 2)^2 \stackrel{?}{=} -1$.
$$i^2 = -1.$$
$$-1 = -1.$$

If $x = -2 - i$, $(-2 - i + 2)^2 \stackrel{?}{=} -1$.
$$(-i)^2 = -1.$$
$$i^2 = -1.$$
$$-1 = -1.$$

Note that the roots are complex numbers and the solution set is $\{-2 + i, -2 - i\}$ or $\{-2 \pm i\}$.

It is important to understand the procedure demonstrated in the examples, since this is an integral part of the process that we shall develop for solving any quadratic equation.

Now let's try to use the previous technique to solve the equation $x^2 + 2x - 24 = 0$. We cannot proceed directly since the given equation is not the same form as those in Examples 23 and 24. They were of the form

Complex Numbers and Quadratic Equations

$(x + a)^2 = n$, that is, the left member of the equation is a *perfect square*. If we expand the left side, we have $x^2 + 2ax + a^2 = n$. Therefore, to solve the equation $x^2 + 2x - 24 = 0$, we must get a perfect square on the left and it should be of the form $x^2 + 2ax + a^2$. As a first step, we add 24 to both sides of the equation.

$$x^2 + 2x - 24 = 0.$$
$$x^2 + 2x = 24.$$

Now we must make the left side a perfect square. Recall that a trinomial that is a perfect square has the following characteristics: 1. Two of the terms are squares of monomials; 2. The "middle" term has the same absolute value as twice the product of the two monomials. For example, $x^2 + 6x + 9$ is a perfect square. Two of the terms, x^2 and 9, are squares of monomials, $x^2 = (x)^2$ and $9 = (3)^2$. The middle term, $+ 6x$, has the same absolute value as twice the product of the two monomials, $|6x| = |2 \cdot 3 \cdot x|$. Therefore, $x^2 + 6x + 9 = (x + 3)^2$.

The left side of the equation $x^2 + 2x = 24$ will be a perfect square if we add 1 to it, $x^2 + 2x + 1$. But if we add 1 to the left side of the equation, then we must add 1 to the right side of the equation. The process now appears as

$$x^2 + 2x - 24 = 0.$$
$$x^2 + 2x = 24.$$
$$x^2 + 2x + 1 = 24 + 1.$$

Since $x^2 + 2x + 1$ is a perfect square, we can rewrite the equation as

$$(x + 1)^2 = 25.$$
$$x + 1 = \pm\sqrt{25}.$$
$$x + 1 = \pm 5.$$

$x + 1 = 5$ or $x + 1 = -5.$
$x = 4.$ $x = -6.$

It can be verified by checking that the solution set is $\{4, -6\}$.

Admittedly, the given equation $x^2 + 2x - 24 = 0$ can be solved by factoring. But it was used to illustrate the technique of forming a perfect square on one side of an equation. This method of solution is formally called *completing the square*. Let us next solve a quadratic equation that cannot be factored. Consider $x^2 + 4x + 1 = 0$. First, we add -1 to both sides of the equation.

$$x^2 + 4x + 1 = 0.$$
$$x^2 + 4x = -1.$$

Now we must make the left side a perfect square. If $x^2 + 4x$ is to be part of a perfect square, then 4x must be twice the product of the two monomials, $4x = 2 \cdot a \cdot x$. We see that $a = 2$ and since $(x + a)^2 = x^2 + 2ax + a^2$, we can make $x^2 + 4x$ into a perfect square by adding 4, $(2)^2$, to it. But if we add 4 to the left side of the equation, then we must add 4 to the right side. The process now appears as

$$x^2 + 4x + 1 = 0.$$
$$x^2 + 4x = -1.$$
$$x^2 + 4x + 4 = -1 + 4.$$

Since $x^2 + 4x + 4$ is a perfect square, we can rewrite the equation as

$$(x + 2)^2 = 3.$$
$$x + 2 = \pm\sqrt{3}.$$

$$x + 2 = \sqrt{3} \quad \text{or} \quad x + 2 = -\sqrt{3}.$$

$$x = -2 + \sqrt{3}. \qquad x = -2 - \sqrt{3}.$$

Check:

If $x = -2 + \sqrt{3}$, $(-2 + \sqrt{3})^2 + 4(-2 + \sqrt{3}) + 1 \stackrel{?}{=} 0.$ It can also be

$$4 - 4\sqrt{3} + 3 - 8 + 4\sqrt{3} + 1 = 0. \quad \text{verified that}$$
$$7 - 8 + 1 = 0. \quad -2 - \sqrt{3} \text{ is a}$$
$$0 = 0. \quad \text{root of the}$$
$$\text{equation.}$$

The solution set is $\{-2 + \sqrt{3}, -2 - \sqrt{3}\}$ or $\{-2 \pm \sqrt{3}\}$.

To solve the quadratic equation by *completing the square,* we must perform the following steps:
1. Write the equation in standard form, if necessary.
2. If $a \neq 1$, then multiply both sides of the equation by $\frac{1}{a}$, making the x^2 coefficient equal to 1.
3. Rewrite the equation so that the constant is on one side of the equals sign and the terms containing variables are on the other side.
4. Take one-half of the x coefficient and square it.
5. Complete the square by adding the square obtained in step 4 to both sides of the equation.
6. Factor the perfect square trinomial.

Complex Numbers and Quadratic Equations   367

> 7. Take the square root of both sides and solve the resulting equations.
>
> Many times an equation will be of such a form that steps 1 and 2 are not necessary, but the remaining steps are important and must be done carefully, particularly steps 3, 4, and 5.

EXAMPLE 25

Solve for x by completing the square.

$$x^2 + 6x + 3 = 0.$$

Solution

Since the given equation is in standard form and the coefficient of the x^2 term is 1, our first step is to rewrite the equation so that the constant is on one side of the equals sign.

$x^2 + 6x + 3 = 0.$	given
$x^2 + 6x = -3.$	adding -3 to both sides
$\frac{1}{2}$ of $6 = 3$ and $3^2 = 9.$	taking $\frac{1}{2}$ of the x coefficient and squaring it
$x^2 + 6x + 9 = -3 + 9.$	completing the square by adding 9 to both sides
$x^2 + 6x + 9 = 6.$	
$(x + 3)^2 = 6.$	factoring the perfect square trinomial
$x + 3 = \pm\sqrt{6}.$	taking the square root of both sides

$x + 3 = \sqrt{6}$ or $x + 3 = -\sqrt{6}.$

$x = -3 + \sqrt{6}.$ $x = -3 - \sqrt{6}.$

Check:

If $x = -3 + \sqrt{6},$ $(-3 + \sqrt{6})^2 + 6(-3 + \sqrt{6}) + 3 \stackrel{?}{=} 0.$

$$9 - 6\sqrt{6} + 6 - 18 + 6\sqrt{6} + 3 = 0.$$

$$9 + 6 - 18 + 3 = 0.$$

$$0 = 0.$$

It can also be verified that $-3 -\sqrt{6}$ is a root of the equation.

The solution set is $\{-3 \pm \sqrt{6}\}$.

EXAMPLE 26

Solve for x by completing the square.

$$2 = 6x - 2x^2.$$

Solution

$2 = 6x - 2x^2.$	given
$2x^2 - 6x + 2 = 0.$	writing the equation in standard form
$\frac{1}{2}(2x^2 - 6x + 2) = \frac{1}{2} \cdot 0.$	multiplying both sides by $\frac{1}{2}$
$x^2 - 3x + 1 = 0.$	making the x^2 coefficient equal to 1
$x^2 - 3x = -1.$	adding -1 to both sides
$\frac{1}{2}$ of $(-3) = -\frac{3}{2}$ and $\left(-\frac{3}{2}\right)^2 = \frac{9}{4}.$	taking $\frac{1}{2}$ of the x coefficient and squaring it
$x^2 - 3x + \frac{9}{4} = -1 + \frac{9}{4}.$	completing the square by adding $\frac{9}{4}$ to both sides
$x^2 - 3x + \frac{9}{4} = \frac{5}{4}.$	
$\left(x - \frac{3}{2}\right)^2 = \frac{5}{4}.$	factoring the perfect square trinomial
$x - \frac{3}{2} = \pm\sqrt{\frac{5}{4}}.$	taking the square root of both sides
$x - \frac{3}{2} = \pm\frac{\sqrt{5}}{2}.$	
$x - \frac{3}{2} = \frac{\sqrt{5}}{2} \quad \text{or} \quad x - \frac{3}{2} = -\frac{\sqrt{5}}{2}.$	

Complex Numbers and Quadratic Equations

$$x = \frac{3}{2} + \frac{\sqrt{5}}{2}. \qquad x = \frac{3}{2} - \frac{\sqrt{5}}{2}.$$

$$x = \frac{3 + \sqrt{5}}{2}. \qquad x = \frac{3 - \sqrt{5}}{2}.$$

Check:

If $x = \frac{3 + \sqrt{5}}{2}$, $2 \stackrel{?}{=} 6\left(\frac{3 + \sqrt{5}}{2}\right) - 2\left(\frac{3 + \sqrt{5}}{2}\right)^2.$

$$2 = \frac{18 + 6\sqrt{5}}{2} - 2\left(\frac{9 + 6\sqrt{5} + 5}{4}\right).$$

$$2 = 9 + 3\sqrt{5} - 2\left(\frac{14 + 6\sqrt{5}}{4}\right).$$

$$2 = 9 + 3\sqrt{5} + \frac{-28 - 12\sqrt{5}}{4}.$$

$$2 = 9 + 3\sqrt{5} - 7 - 3\sqrt{5}.$$

$$2 = 2.$$

It can also be verified that $\frac{3 - \sqrt{5}}{2}$ is a root of the equation.

Therefore, the solution set is $\left\{\frac{3 \pm \sqrt{5}}{2}\right\}$.

EXAMPLE 27

Solve for x by completing the square.

$$3x^2 - x + 2 = 0.$$

Solution

$3x^2 - x + 2 = 0.$	standard form
$\frac{1}{3}(3x^2 - x + 2) = \frac{1}{3} \cdot 0.$	multiplying both sides by $\frac{1}{3}$, making the x^2 coefficient equal to 1
$x^2 - \frac{1}{3}x + \frac{2}{3} = 0.$	
$x^2 - \frac{1}{3}x \phantom{+ \frac{2}{3}} = -\frac{2}{3}$	adding $-\frac{2}{3}$ to both sides

$$\tfrac{1}{2} \text{ of } -\tfrac{1}{3} = -\tfrac{1}{6} \text{ and } \left(-\tfrac{1}{6}\right)^2 = \tfrac{1}{36}.$$ squaring $\tfrac{1}{2}$ of the x coefficient

$$x^2 - \tfrac{1}{3}x + \tfrac{1}{36} = -\tfrac{2}{3} + \tfrac{1}{36}.$$ completing the square by adding $\tfrac{1}{36}$ to both sides

$$x^2 - \tfrac{1}{3}x + \tfrac{1}{36} = -\tfrac{23}{36}.$$

$$\left(x - \tfrac{1}{6}\right)^2 = -\tfrac{23}{36}$$ factoring the perfect square trinomial

$$x - \tfrac{1}{6} = \pm\sqrt{-\tfrac{23}{36}}$$

$$x - \tfrac{1}{6} = \pm\tfrac{\sqrt{-23}}{6}.$$

$$x - \tfrac{1}{6} = \tfrac{\pm i\sqrt{23}}{6}.$$

$$x - \tfrac{1}{6} = \tfrac{i\sqrt{23}}{6} \quad \text{or} \quad x - \tfrac{1}{6} = \tfrac{-i\sqrt{23}}{6}.$$

$$x = \tfrac{1}{6} + \tfrac{i\sqrt{23}}{6}. \qquad x = \tfrac{1}{6} - \tfrac{i\sqrt{23}}{6}.$$

$$x = \tfrac{1 + i\sqrt{23}}{6}. \qquad x = \tfrac{1 - i\sqrt{23}}{6}.$$

It can be verified that $\tfrac{1 + i\sqrt{23}}{6}$ and $\tfrac{1 - i\sqrt{23}}{6}$ are roots of the given equation.

Therefore, the solution set is $\left\{\tfrac{1 \pm i\sqrt{23}}{6}\right\}$.

EXERCISES FOR SECTION 7.5

(Answers given to exercises marked *)

Solve each of the following equations by completing the square.

*1. $x^2 + 6x + 8 = 0$.
2. $x^2 - 2x - 8 = 0$.
*3. $y^2 + 2y - 3 = 0$.
4. $y^2 + 4y + 3 = 0$.
*5. $x^2 + 8x + 15 = 0$.
6. $x^2 - 2x - 15 = 0$.
*7. $y^2 - 8y + 12 = 0$.
8. $y^2 + 4y - 12 = 0$.

Complex Numbers and Quadratic Equations   371

*9. $x^2 - 3x - 10 = 0$.   10. $y^2 + 3y + 2 = 0$.
*11. $y^2 = 7y - 10$.   12. $y^2 = 5y + 14$.
*13. $x^2 = 3x + 4$.   14. $2x^2 + 6x - 20 = 0$.
*15. $2x^2 + 4x - 6 = 0$.   16. $3x^2 + 3x - 6 = 0$.
*17. $4y^2 + 12y + 8 = 0$.   18. $3x^2 = 2 - x$.
*19. $2x^2 = 3 + 5x$.   20. $5x^2 = 2 - 9x$.
*21. $x^2 + x + 1 = 0$.   22. $y^2 - y - 1 = 0$.
*23. $x^2 + x + 2 = 0$.   24. $y^2 + 3y + 3 = 0$.
*25. $y^2 + 2y + 2 = 0$.   26. $2y^2 + y + 2 = 0$.
*27. $3x^2 - x + 6 = 0$.   28. $2y^2 - 3y + 9 = 0$.
*29. $2y^2 + 3y = 1$.   30. $3x^2 + 1 = x$.
*31. $5x^2 = 2x - 5$.   32. $4y^2 - 3y + 8 = 0$.
*33. $3x^2 + 5x + 7 = 0$.

7.6 THE QUADRATIC FORMULA

In the previous section, we developed a technique called completing the square for solving any quadratic equation of the form $ax^2 + bx + c = 0$. Now let us apply this method to derive the general solution to the quadratic equation in standard form, $ax^2 + bx + c = 0$. Since the equation is already in standard form, the next step is to multiply both sides by $\frac{1}{a}$, making the x^2 coefficient equal to 1.

$$ax^2 + bx + c = 0. \quad \text{given}$$

$$\frac{1}{a}(ax^2 + bx + c) = \frac{1}{a} \cdot 0. \quad \text{multiplying both sides by } \frac{1}{a}$$

$$\frac{1}{a} \cdot ax^2 + \frac{1}{a} \cdot bx + \frac{1}{a} \cdot c = 0.$$

$$x^2 + \frac{b}{a}x + \frac{c}{a} = 0.$$

$$x^2 + \frac{b}{a}x = -\frac{c}{a}. \quad \text{adding } -\frac{c}{a} \text{ to both sides}$$

$$\frac{1}{2} \text{ of } \frac{b}{a} = \frac{b}{2a} \text{ and } \left(\frac{b}{2a}\right)^2 = \frac{b^2}{4a^2}.$$

Take $\frac{1}{2}$ of the x coefficient and square it.

$$x^2 + \frac{b}{a}x + \frac{b^2}{4a^2} = -\frac{c}{a} + \frac{b^2}{4a^2}. \quad \text{completing the square by adding } \frac{b^2}{4a^2} \text{ to both sides}$$

$$x^2 + \frac{b}{a}x + \frac{b^2}{4a^2} = -\frac{c}{a} \cdot \frac{4a}{4a} + \frac{b^2}{4a^2} = -\frac{4ac}{4a^2} + \frac{b^2}{4a^2}.$$

$$x^2 + \frac{b}{a}x + \frac{b^2}{4a^2} = \frac{b^2 - 4ac}{4a^2} \quad \longleftarrow \quad \text{combining } -\frac{c}{a} + \frac{b^2}{4a^2}$$

$$(x + \frac{b}{2a})^2 = \frac{b^2 - 4ac}{4a^2}. \quad \text{factoring the perfect square trinomial}$$

$$x + \frac{b}{2a} = \pm\sqrt{\frac{b^2 - 4ac}{4a^2}}. \quad \text{taking the square root of both sides}$$

$$x + \frac{b}{2a} = \pm\frac{\sqrt{b^2 - 4ac}}{2a}.$$

$$x + \frac{b}{2a} = \frac{\sqrt{b^2 - 4ac}}{2a} \quad \text{or} \quad x + \frac{b}{2a} = -\frac{\sqrt{b^2 - 4ac}}{2a}.$$

$$x = -\frac{b}{2a} + \frac{\sqrt{b^2 - 4ac}}{2a}. \quad x = -\frac{b}{2a} - \frac{\sqrt{b^2 - 4ac}}{2a}.$$

$$x = \frac{-b + \sqrt{b^2 - 4ac}}{2a} \quad \text{or} \quad x = \frac{-b - \sqrt{b^2 - 4ac}}{2a}.$$

It can be verified that $\frac{-b + \sqrt{b^2 - 4ac}}{2a}$ and $\frac{-b - \sqrt{b^2 - 4ac}}{2a}$ are roots of the equation. Therefore, the solution set is $\left\{\frac{-b \pm \sqrt{b^2 - 4ac}}{2a}\right\}$.

We can also state that the two solutions of the equation, $ax^2 + bx + c = 0$, are

$$\boxed{x = \frac{-b \pm \sqrt{b^2 - 4ac}}{2a}.}$$

This form of the solution is commonly known as the *quadratic formula* and since it is the solution for the general quadratic equation $ax^2 + bx + c = 0$, we can use it to solve any quadratic equation. The method of com-

Complex Numbers and Quadratic Equations

pleting the square is actually seldom used but we discussed the process in detail so that the derivation of the quadratic formula could be better understood, as the formula was obtained by completing the square.

Now let's examine a couple of examples to see how we can apply the formula. Consider again the equation $x^2 + 2x - 24 = 0$. We shall use the quadratic formula to find the roots of the equation; $x = \dfrac{-b \pm \sqrt{b^2 - 4ac}}{2a}$.

Since the equation is already in standard form, our next task is to determine the values for a, b, and c. Note that $x^2 + 2x - 24 = 0$ is the same as $1x^2 + 2x - 24 = 0$, and the standard form is $ax^2 + bx + c = 0$. Therefore, we have

$a = 1$, $b = 2$, and $c = -24$. Note that the value of c is -24, not 24.

Next, we substitute these values in the quadratic formula,

$$x = \dfrac{-b \pm \sqrt{b^2 - 4ac}}{2a}.$$

$$x = \dfrac{-2 \pm \sqrt{2^2 - 4(1)(-24)}}{2(1)}.$$

Simplifying this expression, we obtain

$$x = \dfrac{-2 \pm \sqrt{4 + 96}}{2}.$$

$$x = \dfrac{-2 \pm \sqrt{100}}{2}.$$

$$x = \dfrac{-2 \pm 10}{2}.$$

$$x = \dfrac{-2 + 10}{2} \quad \text{or} \quad x = \dfrac{-2 - 10}{2}.$$

$$x = \dfrac{8}{2}. \qquad\qquad x = \dfrac{-12}{2}.$$

$$x = 4. \qquad\qquad x = -6.$$

Check:

If $x = 4$, $4^2 + 2(4) - 24 \stackrel{?}{=} 0$. If $x = -6$, $(-6)^2 + 2(-6) - 24 \stackrel{?}{=} 0$.
$\qquad\qquad 16 + 8 - 24 = 0$. $\qquad\qquad 36 - 12 - 24 = 0$.
$\qquad\qquad 24 - 24 = 0$. $\qquad\qquad\quad 24 - 24 = 0$.
$\qquad\qquad\qquad 0 = 0$. $\qquad\qquad\qquad\quad 0 = 0$.

The solution set is $\{4, -6\}$.

To solve a quadratic equation by means of the quadratic formula, we must be sure that the equation is in standard form before we determine the values for a, b, and c. Also, in determining these values we must be sure to include the correct signs for a, b, and c.

EXAMPLE 28

Solve for x by means of the quadratic formula.

$$2x^2 - 3x + 1 = 0.$$

Solution

The equation is in standard form. Therefore, $a = 2, b = -3, c = 1$.

$$x = \frac{-b \pm \sqrt{b^2 - 4ac}}{2a}.$$

$$x = \frac{-(-3) \pm \sqrt{(-3)^2 - 4(2)(1)}}{2(1)}.$$

$$x = \frac{3 \pm \sqrt{9 - 8}}{2}.$$

$$x = \frac{3 \pm \sqrt{1}}{2}.$$

$$x = \frac{3 \pm 1}{2}.$$

$$x = \frac{3 + 1}{2} \quad \text{or} \quad x = \frac{3 - 1}{2}.$$

$$x = \frac{4}{2}. \qquad\qquad x = \frac{2}{2}.$$

$$x = 2. \qquad\qquad x = 1.$$

It can be verified by checking that the solution set is {2, 1}.

EXAMPLE 29

Solve for x by means of the quadratic formula.
$$2x^2 = -2x - 1.$$

Complex Numbers and Quadratic Equations

Solution

First, write the equation in standard form. $2x^2 = -2x - 1$ can be rewritten as

$$2x^2 + 2x + 1 = 0.$$

Now that the equation is in standard form and $a = 2$, $b = 2$, and $c = 1$:

$$x = \frac{-b \pm \sqrt{b^2 - 4ac}}{2a}.$$

$$x = \frac{-2 \pm \sqrt{2^2 - 4(2)(1)}}{2(2)}.$$

$$x = \frac{-2 \pm \sqrt{4 - 8}}{4}.$$

$$x = \frac{-2 \pm \sqrt{-4}}{4} = \frac{-2 \pm i\sqrt{4}}{4}.$$

$$x = \frac{-2 \pm 2i}{4}.$$

$$x = \frac{-2 + 2i}{4} \quad \text{or} \quad x = \frac{-2 - 2i}{4}.$$

$$x = \frac{-1 + i}{2}. \qquad\qquad x = \frac{-1 - i}{2}.$$

Check:

If $x = \frac{-1+i}{2}$, $2\left(\frac{-1+i}{2}\right)^2 \stackrel{?}{=} -2\left(\frac{-1+i}{2}\right) - 1$. It can also be verified that $\frac{-1-i}{2}$ is a root of the equation.

$$2\left(\frac{1 - 2i + i^2}{4}\right) = \frac{2 - 2i}{2} - 1.$$

$$2\left(\frac{1 - 2i - 1}{4}\right) = 1 - i - 1.$$

$$\frac{-4i}{4} = -i.$$

$$-i = -i.$$

The solution set is $\left\{\frac{-1 \pm i}{2}\right\}$.

EXAMPLE 30

Solve for x by means of the quadratic formula.

$$3x^2 + 1 = 2x.$$

Solution

First, write the equation in standard form,

$$3x^2 - 2x + 1 = 0.$$

Now the equation is in standard form and $a = 3$, $b = -2$ and $c = 1$.

$$x = \frac{-b \pm \sqrt{b^2 - 4ac}}{2a}.$$

$$x = \frac{-(-2) \pm \sqrt{(-2)^2 - 4(3)(1)}}{2(3)}.$$

$$x = \frac{2 \pm \sqrt{4 - 12}}{6}.$$

$$x = \frac{2 \pm \sqrt{-8}}{6} = \frac{2 \pm i\sqrt{8}}{6}.$$

$$x = \frac{2 \pm 2i\sqrt{2}}{6}.$$

$$x = \frac{2 + 2i\sqrt{2}}{6} \text{ or } x = \frac{2 - 2i\sqrt{2}}{6}.$$

$$x = \frac{1 + i\sqrt{2}}{3}. \qquad x = \frac{1 - i\sqrt{2}}{3}.$$

It can be verified by checking that the solution set is $\left\{\frac{1 \pm i\sqrt{2}}{3}\right\}$.

In Section 3.10, we solved some radical equations, such as $\sqrt{x + 2} + 2 = 0$. Recall that to solve a radical equation, we first get a radical by itself in one member of the equation, then we raise both members to the appropriate power to eliminate the radical, and, if necessary, repeat this process. We solve the resulting equation and check to determine whether the

Complex Numbers and Quadratic Equations

solution(s) solves the original equation, noting any extraneous roots that may occur. For example, to solve the equation $\sqrt{x+2}+2=0$:

$$\sqrt{x+2}+2=0. \qquad \text{check:} \quad \sqrt{2+2}+2 \stackrel{?}{=} 0.$$

$$\sqrt{x+2} = -2. \qquad\qquad\qquad \sqrt{4}+2=0.$$

$$(\sqrt{x+2})^2 = (-2)^2. \qquad\qquad 2+2 \neq 0.$$

$$x+2 = 4. \qquad\qquad\qquad 4 \neq 0.$$

$$x = 2.$$

Note that 2 is an extraneous root. The solution set is the empty set, { }.

We can apply this same technique to solve radical equations whose solution involves quadratic equations. For example, consider the following equation:

$$x = \sqrt{3x-2}.$$

The radical expression is by itself in one member of the equation. Therefore, we can square both sides of the equation and solve the resulting equation.

$$x = \sqrt{3x-2}. \qquad \text{check:} \quad \text{If } x=2, \ 2 \stackrel{?}{=} \sqrt{3 \cdot 2 - 2}.$$

$$(x)^2 = (\sqrt{3x-2})^2. \qquad\qquad\qquad 2 = \sqrt{6-2}.$$

$$x^2 = 3x - 2. \qquad\qquad\qquad\qquad 2 = \sqrt{4}.$$

$$x^2 - 3x + 2 = 0. \qquad\qquad\qquad\qquad 2 = 2.$$

$$(x-2)(x-1) = 0. \qquad\qquad \text{If } x=1, \ 1 \stackrel{?}{=} \sqrt{3 \cdot 1 - 2}.$$

$$x-2=0 \quad \text{or} \quad x-1=0. \qquad\qquad 1 = \sqrt{3-2}.$$

$$x = 2, \qquad\qquad x = 1. \qquad\qquad\qquad 1 = 1.$$

The solution set is {2, 1}.

EXAMPLE 31

Solve for x: $x = \sqrt{2x+3}$.

Solution

$$x = \sqrt{2x+3}.$$

$$(x)^2 = (\sqrt{2x+3})^2.$$
$$x^2 = 2x + 3.$$
$$x^2 - 2x - 3 = 0 \quad \text{where } a = 1, b = 2, c = -3,$$
$$x = \frac{-b \pm \sqrt{b^2 - 4ac}}{2a}.$$
$$x = \frac{-(-2) \pm \sqrt{(-2)^2 - 4(1)(-3)}}{2(1)}.$$
$$x = \frac{2 \pm \sqrt{4+12}}{2} = \frac{2 \pm \sqrt{16}}{2} = \frac{2 \pm 4}{2}.$$
$$x = \frac{2+4}{2} \quad \text{or} \quad x = \frac{2-4}{2}.$$
$$x = \frac{6}{2}. \qquad\qquad x = \frac{-2}{2}.$$
$$x = 3. \qquad\qquad x = -1.$$

Check:

If $x = 3$, $3 \stackrel{?}{=} \sqrt{2 \cdot 3 + 3}.$ If $x = -1$, $-1 \stackrel{?}{=} \sqrt{2(-1) + 3}.$

$\qquad\qquad 3 = \sqrt{6+3}.$ $-1 = \sqrt{-2+3}.$

$\qquad\qquad 3 = 3.$ $-1 \neq \sqrt{1}.$

$\qquad\qquad\qquad\qquad\qquad\qquad\qquad -1 \neq 1.$

One root does not check in the original equation; it is extraneous. Therefore, the solution set is {3}.

EXAMPLE 32

Solve for x: $x + 2 - \sqrt{x+2} = 0.$

Solution

$$x + 2 - \sqrt{x+2} = 0.$$
$$x + 2 = \sqrt{x+2}.$$

Complex Numbers and Quadratic Equations

Note that here $(x + 2)^2 = (\sqrt{x + 2})^2$.
we are squaring $x^2 + 4x + 4 = x + 2$.
a binomial. $x^2 + 3x + 2 = 0$, where $a = 1, b = 3, c = 2$.

$$x = \frac{-b \pm \sqrt{b^2 - 4ac}}{2a}. \qquad x = \frac{-3 \pm \sqrt{3^2 - 4(1)(2)}}{2(1)}.$$

$$x = \frac{-3 \pm \sqrt{9 - 8}}{2} = \frac{-3 \pm \sqrt{1}}{2} = \frac{-3 \pm 1}{2}.$$

$$x = \frac{-3 + 1}{2}. \qquad x = \frac{-3 - 1}{2}.$$

$$x = \frac{-2}{2}. \qquad x = \frac{-4}{2}.$$

$$x = -1. \qquad x = -2.$$

Check: If $x = -1$, $-1 + 2 - \sqrt{-1 + 2} \stackrel{?}{=} 0$.

$$1 - \sqrt{1} = 0.$$
$$1 - 1 = 0.$$
$$0 = 0.$$

If $x = -2$, $-2 + 2 - \sqrt{-2 + 2} \stackrel{?}{=} 0$.

$$0 - \sqrt{0} = 0.$$
$$0 = 0.$$

The solution set is $\{-1, -2\}$.

In this section we have developed a formula that can be used to solve any quadratic equation. The solution set of most quadratic equations is obtained by means of this formula. Therefore, for the sake of convenience, this formula should be memorized just as you would memorize your student identification number.

$$\boxed{x = \frac{-b \pm \sqrt{b^2 - 4ac}}{2a}.}$$

SPECIAL INTEREST NOTE

It is interesting to note that complex numbers occur in the Fundamental Theorem of Algebra. This states formally: "Every polynomial equation whose coefficients are real or complex numbers has at least one root, which is either real or complex." As a result of the Fundamental Theorem, it can be shown that an integral rational equation of degree n has n roots. For example, $2x = 6$ has 1 root, $x^2 = 4$ has 2 roots, and $x^3 = 1$ has 3 roots. The equation $x^3 = 1$ has one real root and two complex roots, which is shown in the solution below.

$$x^3 = 1. \quad \text{given}$$
$$x^3 - 1 = 0.$$
$$(x - 1)(x^2 + x + 1) = 0. \quad \text{factoring the difference of two cubes}$$
$$x - 1 = 0 \text{ or } x^2 + x + 1 = 0.$$
$$x = 1.$$
$$x = \frac{-1 \pm \sqrt{(-1)^2 - 4(1)(1)}}{2(1)}. \quad \text{solving by means of the quadratic formula, where } a = 1, b = 1, c = 1$$
$$x = \frac{-1 \pm \sqrt{1 - 4}}{2}.$$
$$x = \frac{-1 \pm i\sqrt{3}}{2}.$$

The solution set is $\left\{ 1, \dfrac{-1 \pm i\sqrt{3}}{2} \right\}$.

EXERCISES FOR SECTION 7.6
(Answers given to exercises marked *)

Solve each of the following equations by means of the quadratic formula.

*1. $2x^2 + 5x + 3 = 0$.
2. $3x^2 + 4x + 1 = 0$.
*3. $6x^2 + 7x + 2 = 0$.
4. $x^2 + 3x + 2 = 0$.
*5. $y^2 - 5y + 6 = 0$.
6. $x^2 - 2x - 3 = 0$.
*7. $y^2 - 3y - 4 = 0$.
8. $y^2 - 8y + 15 = 0$.
*9. $6x^2 + 6x - 36 = 0$.
10. $15x^2 + 8x + 1 = 0$.
*11. $6y^2 - 7y - 5 = 0$.
12. $15y^2 - y - 2 = 0$.
*13. $6x^2 = 1 - x$.
14. $3y^2 = 4y + 4$.
*15. $4y^2 = 2 - 3y$.
16. $6x^2 + 1 = 7x$.
*17. $5x^2 + 7x = 6$.
18. $8x^2 = 6x - 1$.

Complex Numbers and Quadratic Equations   381

*19. $y^2 + y + 1 = 0$.   20. $y^2 = y + 1$.
*21. $2y^2 - y - 1 = 0$.   22. $x^2 = 6x + 3$.
*23. $x^2 + 3x = 1$.   24. $x^2 + 2x = 2$.
*25. $x^2 - 2 = 2x$.   26. $y^2 + y = 4$.
*27. $y^2 - 2y = 5$.   28. $y^2 = 1 + y$.
*29. $x^2 + x + 2 = 0$.   30. $y^2 + 3y + 3 = 0$.
*31. $y^2 + 2y + 2 = 0$.   32. $2y^2 + y + 2 = 0$.
*33. $3x^2 - x + 6 = 0$.   34. $2y^2 - 3y + 9 = 0$.
*35. $2y^2 + 3y = 1$.   36. $3x^2 + 1 = x$.
*37. $5x^2 = 2x - 5$.   38. $4y^2 - 3y + 8 = 0$.
*39. $3x^2 + 5x + 7 = 0$.   40. $x = \sqrt{2x + 3}$.
*41. $\sqrt{x - 2} = x$.   42. $y = \sqrt{2 - y}$.
*43. $\sqrt{3y - 2} = y$.   44. $\sqrt{6 - 5y} = y$.
*45. $2x = \sqrt{x + 1}$.   46. $x + 1 = \sqrt{x + 1}$.
*47. $x - \sqrt{x} = 1$.   48. $y - 3 = \sqrt{y - 3}$.
*49. $\sqrt{x + 1} + x + 1 = 0$.   50. $x + 1 = \sqrt{3x + 1}$.
*51. $2x - 1 + \sqrt{x + 1} = 0$.

7.7 APPLICATIONS

In the previous chapter, we discussed fractional equations and those applications (word problems) whose solution involved fractional equations. We are now ready to examine some word problems whose solution involves quadratic equations. For example, consider this problem: Find two consecutive integers whose product is 42. We begin the solution by letting x equal the first integer and x + 1 equals the next consecutive integer. Since their product is 42, we have the equation $x(x + 1) = 42$. Solving this equation, we have

$$x(x + 1) = 42.$$
$$x^2 + x = 42.$$
$$x^2 + x - 42 = 0.$$
$$(x + 7)(x - 6) = 0.$$
$$x + 7 = 0 \quad \text{or} \quad x - 6 = 0.$$
$$x = -7 \quad \text{or} \quad x = 6.$$

If $x = -7$, $x + 1 = -7 + 1 = -6$.   If $x = 6$, $x + 1 = 6 + 1 = 7$.

Check:
$(-7)(-6) = 42.$ $6 \cdot 7 = 42.$
$ 42 = 42.$ $ 42 = 42.$

Therefore, the consecutive integers whose product is 42 are 6 and 7, or -7 and -6.

EXAMPLE 33

The length of a rectangle is 1 metre more than its width. If the area of the rectangle is 72 square metres, find its dimensions.

Solution

Let $x =$ width of the rectangle, then $x + 1 =$ the length of the rectangle. We have:

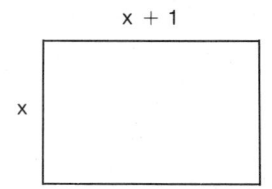

Figure 1 $A = l \cdot w.$

The resulting equation is

$$x(x + 1) = 72.$$
$$x^2 + x = 72.$$
$$x^2 + x - 72 = 0.$$
$$(x + 9)(x - 8) = 0.$$
$$x + 9 = 0 \quad \text{or} \quad x - 8 = 0.$$
$$x = -9. \quad \text{or} \quad x = 8.$$

Since the width cannot be -9 metres, $x = 8$ metres is the desired width and the length is $8 + 1 = 9$ metres. Check: $9 \cdot 8 = 72$. It should be noted that -9 is a root of the original equation, but it is not considered since it does not satisfy the physical conditions stated in the problem.

EXAMPLE 34

The sum of the squares of two consecutive integers is 85. Find the numbers.

Complex Numbers and Quadratic Equations

Solution

Let x = first integer, x + 1 = next integer. Since the sum of their squares is 85, we have the equation

$$x^2 + (x + 1)^2 = 85.$$
$$x^2 + x^2 + 2x + 1 = 85.$$
$$2x^2 + 2x + 1 = 85.$$
$$2x^2 + 2x - 84 = 0.$$
$$x^2 + x - 42 = 0.$$
$$(x + 7)(x - 6) = 0.$$
$$x + 7 = 0 \quad \text{or} \quad x - 6 = 0.$$
$$x = -7. \qquad\qquad x = 6.$$

If x = −7, x + 1 = −7 + 1 = −6   If x = 6, x + 1 = 6 + 1 = 7.

Check:
$$(-7)^2 + (-6)^2 = 85. \qquad 6^2 + 7^2 = 85.$$
$$49 + 36 = 85. \qquad 36 + 49 = 85.$$
$$85 = 85. \qquad 85 = 85.$$

Therefore, the consecutive integers whose sum of their squares is 85 are 6 and 7, or −7 and −6.

EXAMPLE 35

The length of a rectangular piece of sheet metal is 2 more than its width. An open box is formed by cutting 2-inch squares from the corners and bending up the sides. If the volume of the box is 48 cubic inches, find the dimensions of the original piece of metal.

Solution

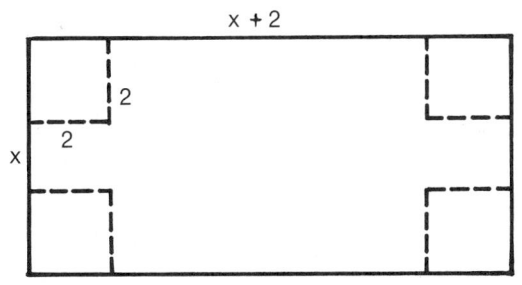

Figure 2 $V = l \cdot w \cdot h.$

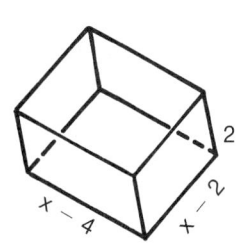

Figure 3

Let x = width, then x + 2 = length. But since a 2-inch square is cut from each corner, the width of the resulting box is x − 4, the length is (x + 2) − 4 = x − 2, and the box has a height of 2 inches.

The resulting equation is

$$2(x-4)(x-2) = 48.$$
$$2(x^2 - 6x + 8) = 48.$$
$$2x^2 - 12x + 16 = 48.$$
$$2x^2 - 12x - 32 = 0.$$
$$x^2 - 6x - 16 = 0.$$
$$(x-8)(x+2) = 0.$$
$$x - 8 = 0 \quad \text{or} \quad x + 2 = 0.$$
$$x = 8. \qquad\qquad x = -2.$$

We exclude x = −2, since a piece of metal cannot have a negative measurement. If x = 8, then x + 2 = 10.

Check: The length, 10, is 2 more than the width, 8.

After 2-inch squares are cut from the corners, the box will have the following dimensions: l = 6, w = 4, h = 2. V = l · w · h = 6 · 4 · 2 = 48 cubic inches.

EXAMPLE 36

On a 75-mile trip to Buffalo, Ike's average speed in his moving van for the first 15 miles was 10 mph less than his average speed for the rest of the trip. If it took Ike two hours to complete the trip, find his average speed for the second part of the trip.

Solution

Let x = Ike's speed for the second part of the trip (60 miles), x − 10 = Ike's speed for the first 15 miles. Using the formula d = rt, we have $t = \frac{d}{r}$. Therefore, the time required for the first 15 miles is $\frac{15}{x-10}$ and the time required for the remainder of the trip is $\frac{60}{x}$. The total time for the trip was two hours. The resulting equation is: time for 15 miles + time for 60 miles = total time. $\frac{15}{x-10} + \frac{60}{x} = 2$.

The LCD is x(x − 10) and we multiply both sides of the equation by it.

$$x(x-10) \cdot \left(\frac{15}{x-10} + \frac{60}{x}\right) = x(x-10)2.$$

Complex Numbers and Quadratic Equations   385

$$x(x-10) \cdot \frac{15}{x-10} + x(x-10) \cdot \frac{60}{x} = 2x^2 - 20x.$$

$$15x + (x-10)60 = 2x^2 - 20x.$$
$$15x + 60x - 600 = 2x^2 - 20x.$$
$$75x - 600 = 2x^2 - 20x.$$
$$0 = 2x^2 - 95x - 600.$$
$$0 = (2x - 15)(x - 40).$$
$$2x - 15 = 0 \quad \text{or} \quad x - 40 = 0.$$
$$2x = 15. \qquad\qquad x = 40 \text{ mph and } x - 10 = 30 \text{ mph.}$$
$$x = 7.5$$

Reject, since $x - 10$ is a negative quantity.

Check:

If $x = 40$, $\dfrac{15}{40-10} + \dfrac{60}{40} \stackrel{?}{=} 2$.   The answer to the problem is 40 mph.

$$\frac{15}{30} + \frac{3}{2} = 2.$$

$$\frac{1}{2} + \frac{3}{2} = 2.$$

$$\frac{4}{2} = 2.$$

$$2 = 2.$$

It should be noted that $x = 7.5$ is a root of the quadratic equation and it does satisfy the original equation, but it must be rejected since it is not actually meaningful. If $x = 7.5$ then $x - 10 = -2.5$, and the speed of the moving van cannot be negative.

EXERCISES FOR SECTION 7.7

(Answers given to exercises marked *)

*1. Find two consecutive integers whose product is 110.

2. Find two consecutive odd integers whose product is 143.

*3. Find two consecutive odd integers whose product is 195.

4. The sum of the squares of two consecutive integers is 181. Find the numbers.

*5. The sum of the squares of two consecutive odd integers is 202. Find the numbers.

6. The sum of the squares of two consecutive even integers is 340. Find the numbers.

*7. The length of a rectangle is three more than twice its width. If the area of the rectangle is 44 square metres, find its dimensions.

8. The width of a rectangle is two less than one-half its length. If the area of the rectangle is 160 square inches, find its dimensions.

*9. An open box is formed by cutting 2-inch squares from a square piece of sheet metal and bending up the sides. If the volume of the box is 72 cubic inches, find the dimensions of the box.

10. The length of a rectangular piece of cardboard is three more than twice its width. An open box is formed by cutting 3-inch squares from the corners and bending up the sides. If the volume of the box is 135 cubic inches, find the dimensions of the original piece of cardboard.

*11. A rectangular piece of sheet metal is twice as long as it is wide. From each of its four corners a 2-inch square is cut out. The flaps are then turned up to form an open box. If the volume of the box is 320 cubic inches, find the dimensions of the box.

12. A rectangular field is 50 metres wide and 60 metres long. If both the width and length are increased by the same amount, the area is increased by 1200 square metres. How much are the length and width increased?

*13. The area of a rectangle is 30 square centimetres. If the length of the rectangle is increased by 4 centimetres and the width is decreased by 1 centimetre, then the area is increased by 3 square centimetres. Find the dimensions of the original rectangle.

14. One leg of a right triangle is one inch longer than the other leg and the hypotenuse is 5 inches in length. Find the lengths of the two legs of the right triangle.

*15. One leg of a right triangle is one centimetre less than twice the length of the other leg. If the hypotenuse is 17 centimetres in length, find the length of the two legs of the right triangle.

16. A Greyhound bus is scheduled to make a trip from New York City to Riverhead, a distance of 90 miles, in a certain period of time. If the bus is 45 minutes late in leaving New York City, it can arrive in Riverhead on time by increasing its usual average speed by 10 miles per hour. Find the usual average speed of the bus.

Complex Numbers and Quadratic Equations   387

*17. Larry traveled a distance of 96 miles on his motorcycle while Benny traveled a distance of 100 miles in his van. Larry's average speed was 8 miles an hour less than Benny's and Larry's trip took one-half hour more than Benny's. Find the average speed of each.

18. If Barbara increases her usual average driving speed by 6 miles an hour, she can cover the 168 miles from her college to her home in a half hour less time than she usually does. Find Barbara's usual average driving speed.

*19. Nelson drove his motorcycle from his home to his summer cottage one weekend, in exactly 4 hours. Going home, Nelson followed a different route that was 18 miles longer, but by increasing his average speed by 12 miles per hour, he was able to make the return trip in $\frac{3}{5}$ of an hour less time. Find Nelson's average speed for the trip from his home to his summer cottage.

20. Hugh's motorboat can go 10 miles per hour in still water. Hugh traveled 12 miles upstream and then immediately returned to his starting point. If the round trip required $2\frac{1}{2}$ hours, find the rate of the current.

*21. It takes Julia seven hours longer to hook a rug than it does Addie, but together, Julia and Addie can hook the rug in 12 hours. How long would it take Addie to hook the rug alone?

22. It takes Scott three hours longer to seal a certain driveway than it takes his brother Joe. The brothers worked together for three hours, then Scott left and Joe finished sealing the driveway alone in one hour. How many hours would it have taken Scott to do the job alone?

*23. To produce a certain type of mercury switch alone, it takes one machine 10 minutes longer than another machine. But if the two machines work together, they can produce the mercury switch in 12 minutes. How long does it take the slower machine to produce a mercury switch alone?

24. It takes Stan, the mason, four hours longer than Steve, the mason, to build a wall. Steve began to build the wall alone and worked on it for three hours; he was then called away. Stan finished the wall in five more hours. How long would it have taken Steve alone to do the whole job?

*25. Pam can type a certain report in 10 minutes less time than Angie. Pam worked on the report alone for 15 minutes and then Angie finished the report alone in 20 minutes. How long would it take Pam and Angie working alone to type the report?

7.8 QUADRATIC INEQUALITIES

In Chapter 2, we solved conditional inequalities such as $-2x - 4 > 6$. The discussion of these inequalities evolved from the solution of first degree equations. Now that we have examined the solution of quadratic equations, it follows that we should examine the solution of quadratic inequalities. Consider the following quadratic inequality:

$$x^2 + x - 6 > 0.$$

We wish to find its solution set and one of the first things we can do is factor the quadratic expressions of the inequality,

$$x^2 + x - 6 > 0.$$
$$(x + 3)(x - 2) > 0.$$

Note that the product of the two binomial factors is greater than 0. This means that the product of the two factors is positive. When the product of two numbers is positive, they both must be positive or they both must be negative—they both must have the same sign. Therefore, using the factored form of the inequality, $(x + 3)(x - 2) > 0$, we have

$$(x + 3 > 0 \quad \text{and} \quad x - 2 > 0) \quad \text{or} \quad (x + 3 < 0 \quad \text{and} \quad x - 2 < 0).$$

Rewriting each of the inequalities gives us the following statement:

$$(x > -3 \quad \text{and} \quad x > 2) \quad \text{or} \quad (x < -3 \quad \text{and} \quad x < 2).$$

Recall that the word "and" corresponds to the operation of "intersection" and we can solve the two compound inequalities. The statement $(x > -3$ and $x > 2)$ means that we want the set of all x such that x is greater than -3 and x is greater than 2 (at the same time). The points on the number line that satisfy both of these conditions are those that are greater than 2, that is,

$$\{x \mid x > -3\} \cap \{x \mid x > 2\} = \{x \mid x > 2\}.$$

Similarly, the statement $(x < -3$ and $x < 2)$ is equivalent to $\{x \mid x < -3\} \cap \{x \mid x < 2\}$, and this intersection is $\{x \mid x < -3\}$. From this discussion, we can now rewrite the statement $(x > -3$ and $x > 2)$ or $(x < -3$ and $x < 2)$, as

$$x > 2 \quad \text{or} \quad x < -3.$$

The set of permissible replacements for x for the inequality $x^2 + x - 6 > 0$

Complex Numbers and Quadratic Equations

consists of $x > 2$, or $x < -3$. The solution set is $\{x \mid x < -3 \text{ or } x > 2\}$ and its graph is:

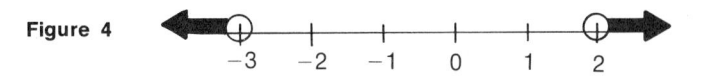

Figure 4

We can also solve the inequality $x^2 + x - 6 > 0$ in another manner. Again, we factor the quadratic expression to obtain $(x + 3)(x - 2) > 0$. From this factored form of the inequality, we see that the values $x = -3$ and $x = 2$ are values that make the expression equal to 0. If we plot these points on the number line, they divide the number line into three intervals, as shown in Figure 5.

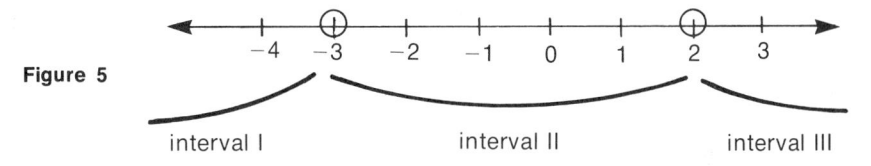

Figure 5

interval I   interval II   interval III

Now, if one point in a given interval satisfies the inequality, then all of the points in that interval will also satisfy it. (It should also be noted that if one point in an interval does not satisfy the inequality, then none of the points will.) We can choose any point in an interval to determine whether or not that interval is part of the solution set. Note that we cannot use -3 and 2 in this case as they are not contained in the intervals, but are boundaries of the intervals. Now let us check the intervals.

Interval I: Choosing $x = -4$, substituting it in the inequality $(x+3)(x-2) \overset{?}{>} 0$, we have $(-4 + 3)(-4 - 2) \overset{?}{>} 0$.

$$(-1)(-6) > 0.$$
$$6 > 0. \quad \text{(true)}$$

Therefore, the points in interval I are members of the solution set, that is, $x < -3$.

Interval II: Let $x = 0$, $(0 + 3)(0 - 2) \overset{?}{>} 0$.
$$(3)(-2) > 0.$$
$$-6 > 0. \quad \text{(false)}$$

Therefore, the points in interval II are not members of the solution set.

Interval III: Let $x = 3$, $(3 + 3)(3 - 2) \overset{?}{>} 0$.
$$(6)(1) > 0.$$
$$6 > 0. \quad \text{(true)}$$

Therefore, the points in interval III are members of the solution set, $x > 2$.

The set of permissible replacements for x in the inequality $x^2 + x - 6 > 0$ consists of all x such that $x < -3$, or $x > 2$. The solution set is $\{x \mid x < -3 \text{ or } x > 2\}$ and its graph is shown in Figure 4.

Let's examine and solve another inequality, using both techniques just discussed. Consider the inequality $x^2 - x - 2 < 0$. Factoring the quadratic expression, we have

$$x^2 - x - 2 < 0.$$
$$(x - 2)(x + 1) < 0.$$

Note that the product of the two binomial factors is less than 0, that is, their product is negative. When two numbers have a negative product, they must be opposite in sign. Therefore, one factor must be positive and the other must be negative. Using the factored form of the inequality, we have

$$(x - 2 > 0 \text{ and } x + 1 < 0) \text{ or } (x - 2 < 0 \text{ and } x + 1 > 0).$$

Rewriting each of the inequalities gives us the following statement:

$$(x > 2 \text{ and } x < -1) \text{ or } (x < 2 \text{ and } x > -1).$$

For the statement ($x > 2$ and $x < -1$), we want the set of all x such that x is greater than 2 and x is less than -1 (at the same time). This is not possible! No number can be greater than a positive number and less than a negative number at the same time. Therefore, from the first statement ($x > 2$ and $x < -1$), we have $\{x \mid x > 2\} \cap \{x \mid x < -1\} = \emptyset$. Their intersection is empty.

For the second statement ($x < 2$ and $x > -1$), we have $\{x \mid x < 2\} \cap \{x \mid x > -1\}$. The intersection of these two sets contains all of the points between -1 and $+2$. If $-1 < x$ and $x < 2$, then x will satisfy the given statement, $\{x \mid x < 2\} \cap \{x \mid x > -1\} = \{x \mid -1 < x < 2\}$. From this discussion we can now rewrite the statement ($x > 2$ and $x < -1$), or ($x < 2$ and $x > -1$), as

$$\emptyset \text{ or } -1 < x < 2.$$

The set of permissible replacements for x for the inequality $x^2 - x - 2 < 0$ consists of $-1 < x < 2$. The solution set is $\{x \mid -1 < x < 2\}$, and its graph is shown in Figure 6.

Figure 6

Complex Numbers and Quadratic Equations

Solving this same inequality by the interval technique, we again factor the quadratic expression and note those values that will make the expression equal to 0. Hence,

$$x^2 - x - 2 < 0.$$
$$(x - 2)(x + 1) < 0.$$

If $x = 2$ or $x = -1$, then $x^2 - x - 2 = 0$.

Next we plot these points on the number line and divide it into three intervals as shown in Figure 7.

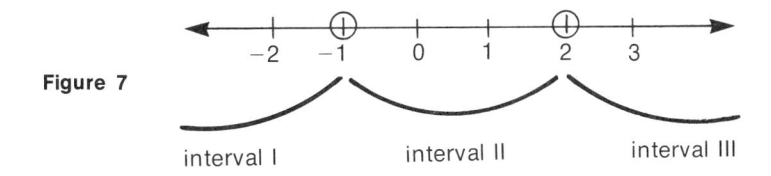

Figure 7

interval I   interval II   interval III

If one point in a given interval satisfies the inequality, then all of the points in that interval satisfy the inequality, and if one point in an interval does not satisfy the inequality, then none of the points in that interval will satisfy it. Checking the intervals, we have:

Interval I: Checking $x = -2$, substituting it in the inequality $(x-2)(x+1) \overset{?}{<} 0$, we have $(-2-2)(-2+1) < 0.$
$$(-4)(-1) < 0.$$
$$4 < 0. \quad \text{(false)}$$

Therefore, the points in interval I are not members of the solution set.

Interval II: Let $x = 0$, $(0-2)(0+1) \overset{?}{<} 0.$
$$(-2)(1) < 0.$$
$$-2 < 0. \quad \text{(true)}$$

Therefore, the points in interval II are members of the solution set, $-1 < x < 2$.

Interval III: Let $x = 3$, $(3-2)(3+1) \overset{?}{<} 0.$
$$(1)(4) < 0.$$
$$4 < 0. \quad \text{(false)}$$

Therefore, the points in interval III are not members of the solution set.

The set of permissible replacements for x in the inequality $x^2 - x - 2 < 0$ consists of $-1 < x < 2$. The solution set is $\{x \mid -1 < x < 2\}$ and its graph is shown in Figure 6.

> We have solved quadratic inequalities using two different techniques. One method is to use the idea of intersection and union of sets of points, while the other method is to use the idea of dividing the number line into intervals and testing points in the intervals to determine whether they satisfy the given inequality. The latter technique is less formal than the former, but both are equally effective.

Following are two more examples, each solved in a different manner.

EXAMPLE 37

Solve the given inequality and graph its solution set.

$$x^2 + 5x + 4 > 0.$$

Solution

Factoring the inequality, we have

$$(x + 4)(x + 1) > 0.$$

Since the product of the two binomial factors is positive, they both must have the same sign, both positive or both negative. Therefore, $(x + 4)(x + 1) > 0$ means that

$$(x + 4 > 0 \quad and \quad x + 1 > 0) \quad or \quad (x + 4 < 0 \quad and \quad x + 1 < 0).$$

Rewriting each of the inequalities,

$$(x > -4 \quad and \quad x > -1) \quad or \quad (x < -4 \quad and \quad x < -1).$$

The points on the number line that are greater than -4 and -1 (at the same time) are those that are greater than -1,

$$\{x \mid x > -4\} \cap \{x \mid x > -1\} = \{x \mid x > -1\}.$$

Similarly, the points on the number line that are less than -4 and -1 (at the same time) are those that are less than -4,

$$\{x \mid x < -4\} \cap \{x \mid x < -1\} = \{x \mid x < -4\}.$$

Complex Numbers and Quadratic Equations 393

Those points that satisfy the inequality are $\{x \mid x < -4\} \cup \{x \mid x > -1\}$ and the solution set is $\{x \mid x < -4 \text{ or } x > -1\}$ and its graph is:

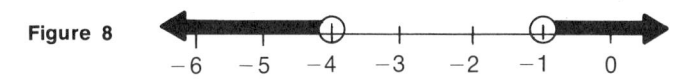

Figure 8

EXAMPLE 38

Solve the given inequality and graph its solution set:
$$x^2 - 2x - 15 \leq 0.$$

Solution

Factoring the inequality, we have

$$(x - 5)(x + 3) \leq 0.$$

From this factored form, we see that the values $x = 5$ and $x = -3$ are values that make the expression equal to 0. We plot these points on the number line to divide it into three intervals as shown in Figure 9.

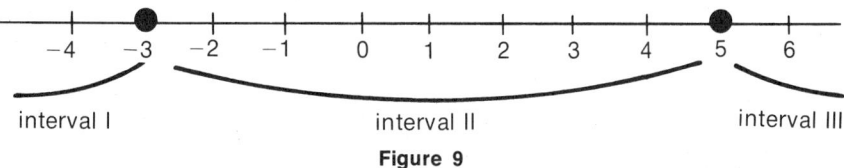

interval I   interval II   interval III
Figure 9

Checking the intervals, we have

Interval I: Checking $x = -4$, substituting it in the inequality $(x - 5)(x + 3) \leq 0$, we have

$$(-4 - 5)(-4 + 3) \stackrel{?}{\leq} 0.$$
$$(-9)(-1) \leq 0.$$
$$9 \leq 0. \quad \text{(false)}$$

Therefore, the points in interval I are not members of the solution set.

Interval II: Let $x = 0$, $(0 - 5)(0 + 3) \stackrel{?}{\leq} 0.$
$$(-5)(3) \leq 0.$$
$$-15 \leq 0. \quad \text{(true)}$$

Therefore, the points in interval II are members of the solution set, that is, $-3 \leq x \leq 5$. Note that the end points of the interval are included since the inequality ($\leq$) contains the equal sign, "less than or equal to."

Interval III: Let x = 6, $(6-5)(6+3) \stackrel{?}{\leq} 0$.
$$(1)(9) \leq 0.$$
$$9 \leq 0. \quad \text{(false)}$$

Therefore, the points in interval III are not members of the solution set. The set of permissible replacements for x in the inequality $x^2 - 2x - 15 \leq 0$ consists of $-3 \leq x \leq 5$. The solution set is $\{x \mid -3 \leq x \leq 5\}$ and its graph is:

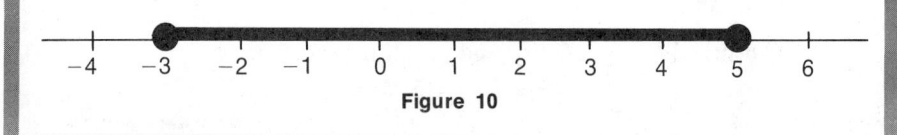

Figure 10

Recall that complex numbers do not have any order. For example, it is not possible to determine whether $-3 + 2i$ is greater than or less than $3 - 2i$ since it is not possible to describe a complex number as positive or negative. The properties of inequality do not hold for the set of complex numbers. Therefore, all quadratic inequalities that we consider here are to be solved for real number replacements.

It should be noted that some quadratic inequalities have no solution under the set of real numbers while other quadratic inequalities will have a solution set consisting of all real numbers. For example, consider the inequality $x^2 + 1 < 0$. It has no solution; its solution set is the empty set, $\emptyset$. Rewriting the inequality $x^2 + 1 < 0$ as $x^2 < -1$, we see that it is impossible to replace x by any real number such that its square will be less than -1. Regardless of the value we choose for x, all real numbers are non-negative squares. Therefore, $x^2 + 1 < 0$ has no solution. Next, consider the inequality $x^2 + 1 > 0$. Its solution set consists of the set of real numbers, that is, all of the points on the number line. Regardless of what value we choose for x, it will satisfy the inequality $x^2 + 1 > 0$. Since x^2 will always be non-negative and we are adding 1 to it, the sum will always be positive and greater than 0. For example, if $x = -2$, $(-2)^2 + 1 > 0$, $4 + 1 > 0$, $5 > 0$. Similarly, if $x = 0$, $0^2 + 1 > 0$, $0 + 1 > 0$, $1 > 0$ and all positive values of x will also satisfy the given inequality.

EXERCISES FOR SECTION 7.8

(Answers given to exercises marked *)

Find the solution set for each of the following inequalities and graph its solution set on the real number line.

*1. $x^2 - 4 < 0$.
2. $x^2 - 4 > 0$.
*3. $x^2 - 1 < 0$.
4. $x^2 - 1 > 0$.

Complex Numbers and Quadratic Equations   395

*5. $x^2 - 9 < 0$.   6. $x^2 - 9 > 0$.
*7. $x^2 - 2x + 1 < 0$.   8. $x^2 - 2x + 1 \geq 0$.
*9. $x^2 - x + 1 < 0$.   10. $x^2 + 3x + 2 \geq 0$.
*11. $x^2 + 4x - 5 \geq 0$.   12. $x^2 - 2x - 3 < 0$.
*13. $x^2 - x + 5 > 0$.   14. $x^2 - 2x + 4 \geq 0$.
*15. $x^2 - 2x + 7 > 0$.   16. $x^2 + x + 2 < 0$.
*17. $x^2 + 2x + 8 < 0$.   18. $x^2 + 3x + 9 < 0$.
*19. $x^2 + 3x + 2 \leq 0$.   20. $x^2 - 3x + 2 > 0$.
*21. $x^2 + x - 2 \geq 0$.   22. $x^2 - x - 2 < 0$.
*23. $x^2 - 3x - 4 < 0$.   24. $x^2 - 5x + 4 \geq 0$.
*25. $x^2 + 3x - 4 > 0$.   26. $x^2 + 5x + 4 < 0$.
*27. $x^2 + 7x + 10 \geq 0$.   28. $x^2 - 7x + 10 < 0$.
*29. $x^2 + 3x - 10 \leq 0$.   30. $x^2 - 3x - 10 > 0$.
*31. $2x^2 - 3x - 2 \leq 0$.   32. $2x^2 - x - 3 > 0$.
*33. $6x^2 - x - 1 < 0$.   34. $6x^2 + 5x - 6 < 0$.
*35. $5x^2 - x - 4 \leq 0$.   36. $6x^2 + 29x + 35 \geq 0$.

7.9 SUMMARY

The symbol $\sqrt[n]{a}$ represents the principal n^{th} root of a. If $a < 0$ and n is an even integer, then there does not exist a principal n^{th} root of a in the system of real numbers. Therefore, $\sqrt{-1}$ is not a real number. The square root of negative one is called an imaginary number and it is usually designated by the letter i. That is, $i = \sqrt{-1}$ and from this we see that $i^2 = -1$, $i^3 = -i$ and $i^4 = +1$. Using these four powers of i we can simplify any other power of i. For example, $i^{25} = i$, $i^{12} = 1$ and $i^{14} = -1$. Since $\sqrt{-1} = i$, we can use this fact to express other imaginary numbers in terms of i, that is, $\sqrt{-4} = \sqrt{-1 \cdot 4} = \sqrt{-1} \cdot \sqrt{4} = i \cdot 2$ or 2i. Similarly, $\sqrt{-8} = \sqrt{-1 \cdot 8} = \sqrt{-1} \cdot \sqrt{8} = i\sqrt{4} \cdot \sqrt{2} = i \cdot 2\sqrt{2}$ or $2i\sqrt{2}$. We add, subtract, multiply, and divide imaginary numbers in a manner similar to that which is used for real numbers.

Whenever we combine a real number and an imaginary number, we obtain a sum (or difference) of the form $a + bi$. An expression of the form $a + bi$ is called a complex number. Two complex numbers, $a + bi$ and $c + di$, are equal whenever their corresponding parts are equal, or, $a = c$ and $b = d$. In general, we have

$$a + bi = c + di, \text{ if and only if } a = c \text{ and } b = d.$$

Complex numbers have the same general form as binomials, and we add two complex numbers, $a+bi$ and $c+di$, by adding their real and imaginary parts, $(a+bi)+(c+di)=(a+c)+(b+d)i$. The product of two complex numbers can be obtained in the same manner as the product of two binomials. Therefore,

$$\begin{aligned}(a+bi)(c+di) &= (a+bi)c + (a+bi)di \\ &= ac + bci + adi + bdi^2 \\ &= ac + bci + adi - bd \\ &= (ac - bd) + (ad + bc)i.\end{aligned}$$

We can use the distributive property or the FOIL method to obtain the product of two complex numbers. Remember that i^2 should be replaced by -1 so that the real and imaginary parts can be combined, allowing us to express the final result in the general form of $a+bi$.

To obtain the quotient of two complex numbers, multiply the numerator and denominator of the fraction by the conjugate of the denominator, that is, multiply the fraction by the appropriate form of 1. In general, we have

$$\begin{aligned}\frac{a+bi}{c+di} &= \frac{a+bi}{c+di} \cdot \frac{c-di}{c-di} = \frac{ac - adi + bci - bdi^2}{c^2 - (di)^2} \\ &= \frac{ac - adi + bci + bd}{c^2 + d^2} \\ &= \frac{(ac+bd) + (bc-ad)i}{c^2 + d^2}\end{aligned}$$

We can always check the result by multiplying the divisor by the quotient to determine whether we obtain that dividend. A complex number cannot be described as positive or negative and, therefore, we cannot say that one complex number is greater than another. The properties of inequality do not hold for the set of complex numbers.

A second degree equation in one variable contains the second power of that variable and no higher power of that variable. An equation of this type is formally called a quadratic equation. For example, $3x^2 + 4x + 2 = 0$ is a quadratic equation in one variable. A general quadratic equation in x can be written in standard form $ax^2 + bx + c = 0$, where a, b, and c are real numbers and $x \neq 0$. A quadratic equation that has b or c equal to 0 is called an incomplete quadratic equation. For example, $x^2 + 1 = 0$ (b=0) and $x^2 + x = 0$ (c = 0) are incomplete quadratic equations. General incomplete quadratic equations are: $ax^2 + c = 0$ (b=0), $ax^2 + bx = 0$ (c=0), and $ax^2 = 0$ (b=0, c=0). If both b=0 and c=0, then $ax^2 = 0$ has 0 as a double root. If b=0, then $ax^2 + c = 0$ can be solved for x by solving the given equation for x^2 and taking the square root of both sides, and $x = \pm\sqrt{-\frac{c}{a}}$. If c = 0, then $ax^2 + bx = 0$ can be solved for x by removing

Complex Numbers and Quadratic Equations

the common factor x, and determining those values that will make the two factors equal to 0, and $x = 0$ or $x = -\frac{b}{a}$.

Often a complete quadratic equation can be solved by factoring $ax^2 + bx + c$ into two binomials. To solve a quadratic equation by factoring, we must have it in standard form, meaning one side of the equation must be 0. Next, factor the quadratic expression $ax^2 + bx + c = 0$, and since $ab = 0$, if and only if $a = 0$ or $b = 0$, set each factor equal to 0 and solve the resulting equations. Not all quadratic equations of the form $ax^2 + bx + c = 0$ can be solved by factoring. One method that we can use to solve any quadratic equation is *completing the square*. To solve a quadratic equation by completing the square, we must perform the following steps:

1. If necessary, write the equation in standard form;
2. If $a \neq 1$, then multiply both sides of the equation by $\frac{1}{a}$, making the x^2 coefficient equal to 1;
3. Rewrite the equation so that the constant is on one side of the equals sign and the terms containing variables are on the other side;
4. Take one-half of the x coefficient and square it;
5. Complete the square by adding the square (obtained in step 4) to both sides of the equation;
6. Factor the perfect square trinomial;
7. Take the square root of both sides and solve the resulting equations.

The method of *completing the square* was used to solve the general quadratic equation $ax^2 + bx + c = 0$. The two solutions of this equation are

$$x = \frac{-b \pm \sqrt{b^2 - 4ac}}{2a}.$$

This form of the solution is commonly known as the quadratic formula and since it is the solution of the general quadratic equation, we can use it to solve any quadratic equation. In order to solve a quadratic equation by means of the formula, we must make sure that the equation is in standard form before we determine the values for a, b, and c. Also, when determining these values we must be sure to include the correct sign for a, b, and c.

Many times the solution of a radical equation will involve the solution of a quadratic equation. To solve a radical equation, we first get a radical by itself in one member of the equation, then we raise both members of the equation to the appropriate power to eliminate the radical, and if necessary, repeat this process. Next, solve the resulting equation and check to determine whether the solution(s) is a solution(s) of the original equation, noting any extraneous roots that may occur.

In this chapter we developed two methods for solving quadratic inequalities. Regardless of the technique used, we must have 0 on one side of the inequality and then factor the other side. In the first technique, we must note whether the inequality is greater or less than 0. If the inequality is greater than 0, then the product of the two factors is positive and both factors have the same sign, both positive or both negative. If the inequality is less than 0, then the product of the two factors is negative and one factor must be positive while the other is negative. Using this information, we rewrite the factored form of the inequality as a compound statement, using the key words "and" and "or." For example, $(x+3)(x-2) > 0$ means that ($x > -3$ *and* $x > 2$) *or* ($x < -3$ *and* $x < 2$). Since "and" corresponds to intersection, ($x > -3$ and $x > 2$) is equivalent to $\{x \mid x > -3\} \cap \{x \mid x > 2\}$, which equals $\{x \mid x > 2\}$. Similarly, ($x < -3$ and $x < 2$) is equivalent to $\{x \mid x < -3\} \cap \{x \mid x < 2\}$ equals $\{x \mid x < -3\}$, and the solution set is $\{x \mid x < -3 \text{ or } x > 2\}$.

The second technique requires finding those values that will make the factored form equal to 0. For example, using $(x+3)(x-2) > 0$, we note that the expression will equal 0 when $x = -3$ or $x = 2$. Next, we plot these points on the number line and they divide the line into three intervals. If one point in a given interval satisfies the inequality, then all the points in that interval satisfy the inequality, and if one point in a given interval does not satisfy the inequality, then none of the points will. We can choose any point in an interval to determine whether that interval is part of the solution set, except that we cannot use the end points of the interval. We must be sure to test all three intervals.

All quadratic inequalities are solved under the system of real numbers since complex numbers do not have any order. Some quadratic inequalities have no solution under the set of real numbers, while other quadratic inequalities may have a solution set consisting of all real numbers.

REVIEW EXERCISES FOR CHAPTER 7
(Answers in Appendix B)

Simplify each of the following and express it in terms of i.

1. $\sqrt{-4}$
2. $\sqrt{-25}$
3. $\sqrt{-8}$
4. $-\sqrt{-18}$
5. $3\sqrt{-20}$
6. $5\sqrt{-45}$

Simplify and evaluate each of the following.

7. i^9
8. $-i^{11}$
9. $-i^{27}$
10. $(-i)^8$
11. $(-2i)^5$
12. $2i^{-6}$

Complex Numbers and Quadratic Equations

Perform the indicated operations and simplify, if possible.

13. $\sqrt{-1} + \sqrt{-4}$
14. $2\sqrt{-4}$
15. $2i\sqrt{-27}$
16. $\sqrt{3} \cdot \sqrt{-3}$
17. $2i\sqrt{-3} \cdot 3\sqrt{-2}$
18. $\sqrt{-8} \cdot \sqrt[3]{-27}$
19. $\dfrac{\sqrt{12}}{\sqrt{-2}}$
20. $\dfrac{\sqrt{-5}}{\sqrt{-2}}$
21. $\dfrac{3\sqrt{-20}}{\sqrt{8}}$

Perform the indicated operations and simplify, if possible.

22. $(-2 - 3i) + (2 - 3i)$
23. $(-11 + 2i) - (9 - 3i)$
24. $(2 - 3i) - (2 - 3i)$
25. $3i(2 - 5i)$
26. $(-2 - i)(-3 + 2i)$
27. $(3 - 5i)^2$
28. $\dfrac{1 + i}{1 - i}$
29. $\dfrac{2 - 3i}{-2 + 3i}$
30. $-\dfrac{-3 - 2i}{-1 - i}$

Solve the given incomplete quadratic equations.

31. $3x^2 + 2 = 3$.
32. $4y^2 - 4 = 0$.
33. $x^2 - 4x = 0$.
34. $2x^2 + 3 = 4 - x^2$.
35. $x^2 - 2x = 2x^2 + x$.
36. $ax^2 + bx = 0$.

Solve the given equations by factoring.

37. $2y^2 - y - 1 = 0$.
38. $4x^2 + 5x + 1 = 0$.
39. $5y^2 = 6y - 1$.
40. $3(x^2 + 2) = 11x$.
41. $2 - \dfrac{7}{x} + \dfrac{6}{x^2} = 0$.
42. $\dfrac{5}{x + 4} + \dfrac{3}{x - 2} = 4$.

Solve the given equations by completing the square.

43. $x^2 - 2x - 15 = 0$.
44. $2y^2 + 6y - 20 = 0$.
45. $3x^2 = 2 - x$.
46. $2x^2 + x + 2 = 0$.
47. $3y^2 + 1 = y$.
48. $4x^2 - 3x + 8 = 0$.

Solve the given equations by means of the quadratic formula.

49. $3y^2 = 4y + 4$.
50. $8x^2 = 6x - 1$.
51. $y^2 - y = 1$.
52. $x^2 + 2x = 2$.
53. $x^2 + 3x + 3 = 0$.
54. $4x^2 - 3x + 8 = 0$.

55. $2x^2 - 3x + 9 = 0$.

56. $\sqrt{6 - 5x} = x$.

57. $y + 1 = \sqrt{3y + 1}$.

Find the solution set for each of the following inequalities and graph its solution set on the real number line.

58. $x^2 - 9 < 0$.

59. $x^2 - 2x + 1 \geq 0$.

60. $x^2 + x - 2 < 0$.

61. $x^2 - x - 2 < 0$.

62. $6x^2 + 5x - 6 < 0$.

63. $6x^2 + 29x + 35 \geq 0$.

64. Find two consecutive odd integers whose product is 63.

65. The sum of the squares of two consecutive integers is 145. Find the numbers.

66. The length of a rectangular piece of sheet metal is three more than twice its width. An open box is formed by cutting 3-centimetre squares from the corners and bending up the sides. If the volume of the box is 135 cubic centimetres, find the dimensions of the original piece of sheet metal.

67. A certain bus makes a daily trip of 168 miles. Last Tuesday, the bus was one-half hour late in leaving its starting point. In order to arrive at its destination on time, the bus driver had to increase his usual average speed by six miles an hour. What is the usual average speed of the bus?

68. It takes Mary seven hours longer to wallpaper a room than it does Barbara. Together, Mary and Barbara can wallpaper the room in 12 hours. How long would it take Barbara to wallpaper the room alone?

69. One leg of a right triangle is two centimetres longer than twice the length of the other leg. If the hypotenuse is 13 centimetres in length, find the lengths of the two legs of the right triangle.

chapter 8

Relations and Functions

As a result of this chapter, the student will be able to:

1. Define a relation as a set of ordered pairs;
2. Determine the domain and range of a relation, R, and also determine the inverse of R;
3. Define a function as a relation in which no two ordered pairs have the same first coordinate and a different second coordinate;
4. Determine whether a given relation is a function or not;
5. Plot a point on the Cartesian plane, given the coordinates of the point, and also determine the coordinates of a point on the plane;
6. Find the slope of a line, given the coordinates of two different points on the line, by means of the formula

$$m = \frac{y_2 - y_1}{x_2 - x_1};$$

7. Determine the equation of a line, given the coordinates of two different points on the line, by means of the point-slope formula: $y - y_1 = m(x - x_1)$ and the slope-intercept formula: $y = mx + b$;
8. Find the distance between two points on the Cartesian plane by means of the distance formula

$$d = \sqrt{(x_2 - x_1)^2 + (y_2 - y_1)^2}$$

and also determine the coordinates of the midpoint of the line segment connecting two points on the plane:

$$\text{midpoint} = \left(\frac{x_1 + x_2}{2}, \frac{y_1 + y_2}{2}\right);$$

9. Identify a variation problem as direct variation, inverse variation, or joint variation and solve it accordingly;
10. Identify and graph the equations of:

$ax + by = c$	straight line	$ax^2 + by^2 = c$	ellipse
$y = ax^2 + bx + c$	parabola	(if $a \neq b$ and they have the same sign)	
$(x - h)^2 + (y - k)^2 = r^2$			
$x^2 + y^2 = r^2$	circle	$ax^2 + by^2 = c$	hyperbola
$ax^2 + by^2 = c$		(if a and b are opposite signs)	
(if $a = b$ and they have the same sign)		$xy = c$	hyperbola

Relations and Functions

8.1 INTRODUCTION

A relation is formally described as the way in which one thing is connected with another. There are many different ways that two items can be connected (or compared) with each other. For example, in comparing two cars, car A is "older than" car B, car A is "faster than" car B, car A is "smaller than" car B, car A is "lighter than" car B, and so on. If car A is older than car B, we can say that car A is in that relation with car B.

If we consider a finite number of cars in a certain parking lot, we can compare all the cars. We probably could find all pairs of cars such that the first car is older than the second car. For example, car A is older than car B, car A is older than car C, car A is older than car D, car B is older than car D, and so on. We are forming all possible *ordered pairs* such that the first car is older than the second. Since we know the relation between the two cars, we can write the ordered pairs as (car A, car B), (car A, car C), (car A, car D), (car B, car D). An ordered pair is a pair of items such that there is a *first* element and a *second* element. The order of the elements is important. Consider the relation "put on" and the two items: sock and shoe. It is important which item is listed first and which item is listed second. Using the given relation "put on," the correct ordered pair is (sock, shoe). If we have (shoe, sock), we still have an ordered pair, but try putting on your shoe and then your sock! It should be noted that relations can exist between more than two items, such as the ordered triple (a, b, c) where b is "between" a and c. But here we shall consider only those relations that involve pairs of items.

Consider the set of numbers {1, 2, 3, 4, 5} and the relation "greater than." The set of ordered pairs that represent this relation for the given set of numbers is:

{(2, 1), (3, 1), (4, 1), (5, 1), (3, 2), (4, 2), (5, 2), (4, 3), (5, 3), (5, 4)}

Using the same set of numbers {1, 2, 3, 4, 5} and the relation "less than," we have the following set of ordered pairs:

{(1, 2), (1, 3), (1, 4), (1, 5), (2, 3), (2, 4), (2, 5), (3, 4), (3, 5), (4, 5)}

A set of ordered pairs can be used to represent a relation. This means that the set of ordered pairs {(4, 2), (2, 1), (6, 3), (9, 3)} is a relation even though we may not know what the relation is between the first and second element. Formally defined, a relation is a set of ordered pairs.

EXAMPLE 1

Given the set of vowels, {a, e, i, o, u} and the relation "occurs before in the alphabet," write the set of ordered pairs representing this relation.

Solution

{(a, e), (a, i), (a, o), (a, u), (e, i), (e, o), (e, u), (i, o), (i, u), (o, u)}.

EXAMPLE 2

Given the set of presidents, {Kennedy, Johnson, Nixon, Ford, Carter} and the relation "served before," write the set of ordered pairs representing this relation.

Solution

{(Kennedy, Johnson), (Kennedy, Nixon), (Kennedy, Ford), (Kennedy, Carter), (Johnson, Nixon), (Johnson, Ford), (Johnson, Carter), (Nixon, Ford), (Nixon, Carter), (Ford, Carter)}.

A relation is a set of ordered pairs and an ordered pair means that we have a pair of elements where one element is first and the other is second. That is, an ordered pair contains a first element and a second element. The first element is often called the first coordinate and the second element is called the second coordinate.

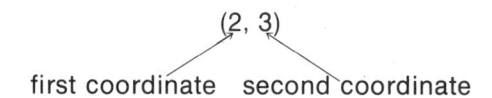

(2, 3)
first coordinate   second coordinate

The *domain* of a relation is the set of all first coordinates of the given relation and the *range* of a relation is the set of all second coordinates of the given relation. For example, in the relation {(2, 1), (3, 2), (5, 3), (7, 5)}, the domain is the set of all first coordinates, {2, 3, 5, 7} and the range is the set of all second coordinates, {1, 2, 3, 5}. Note that the domain and range do not necessarily contain the same elements. In fact, they do not even have to contain the same number of different elements. Consider the relation

{(1, 2), (1, 3), (1, 4), (2, 3), (2, 4), (2, 5)}.

The domain is the set of all first coordinates, {1, 2} and the range is the set of all second coordinates, {2, 3, 4, 5}.

Relations and Functions

> **EXAMPLE 3**
>
> Given A = {1, 2, 3, 4}, list the ordered pairs that can be formed from A such that the first coordinate is less than the second coordinate. What is the domain and range of the relation?
>
> **Solution**
>
> a. {(1, 2), (1, 3), (1, 4), (2, 3), (2, 4), (3, 4)}
> b. Domain: {1, 2, 3}
> c. Range: {2, 3, 4}

Consider the following relation: {(1, 2), (3, 2), (4, 3), (5, 3), (6, 4), (7, 4)}. The domain of a relation is the set of all first coordinates, and the domain for this relation is {1, 3, 4, 5, 6, 7}. The range of a relation is the set of all second coordinates, and the range for this relation is {2, 3, 4}. Since a relation is defined to be a set of ordered pairs, we can form a new relation from the given relation by interchanging the first and second coordinates, {(2, 1), (2, 3), (3, 4), (3, 5), (4, 6), (4, 7)}. Note that the domain of the new relation is {2, 3, 4}, and the range is {1, 3, 4, 5, 6, 7}. The second relation is called the *inverse* of the first relation. Note that the domain of the original relation became the range of the inverse relation and the range of the original relation became the domain of the inverse relation.

The "inverse" of a given relation can be formed by interchanging the first and second coordinates in all ordered pairs of the given relation.

> **EXAMPLE 4**
>
> Given the relation A = {(a, b), (a, c), (a, d), (a, e)}, determine its inverse relation.
>
> **Solution**
>
> The inverse of A is {(b, a), (c, a), (d, a), (e, a)}.

> A relation is a set of ordered pairs. An ordered pair consists of a first element and a second element. The first element in an ordered pair is called the first coordinate and the second element is called the second coordinate. The domain of a relation is the set of all first coordinates of the given relation, and the range is the set of all second coordinates. The inverse of a relation is formed by interchanging the first and the second coordinates in all ordered pairs of the given relation.

EXERCISES FOR SECTION 8.1

(Answers given to exercises marked *)

*1. Given A = {1, 2, 3} and B = {4, 5, 6}, determine the relation R such that the first coordinate is obtained from A and the second coordinate is obtained from B.

2. Given V = {a, e, i, o, u} and X = {x, y, z}, determine the relation R such that the first coordinate is obtained from V and the second coordinate is obtained from X.

*3. Given A = {0, 2, 4, 6}, determine the relation stated for each of the following:

   a. The first coordinate is less than the second.
   b. The second coordinate is less than the first.
   c. The first and second coordinates are identical.
   d. The first coordinate is divisible by the second.
   e. The second coordinate is divisible by the first.

4. Given B = {1, 2, 3, 4, 5}, determine the relation stated for each of the following:

   a. The first coordinate is a multiple of the second.
   b. The first coordinate is a prime number and the second is not.
   c. The first coordinate is divisible by the second.
   d. The second coordinate is not divisible by the first.
   e. The first coordinate is less than the second.

*5. Given A = {1, 3, 5, 7}, determine the relation R such that the first coordinate is less than the second coordinate. Then determine the following:

   a. What is the domain of R?
   b. What is the range of R?
   c. Determine the inverse of R.

6. Given V = {a, e, i, o, u} and X = {x, y, z}, determine the relation R such that the first coordinate is obtained from X and the second coordinate is obtained from V.

   a. What is the domain of R?
   b. What is the range of R?
   c. Determine the inverse of R.

*7. Given A = {−2, −1, 0, 1, 2}, determine the relation R such that the first coordinate is less than the second coordinate.

   a. What is the domain of R?

Relations and Functions

b. What is the range of R?
c. Determine the inverse of R.
d. What is the domain of the inverse relation?
e. What is the range of the inverse relation?

8. Given C = {−6, −3, 0, 3, 6}, determine the relation R such that the first coordinate is greater than or equal to the second coordinate.

   a. What is the domain of R?
   b. What is the range of R?
   c. Determine the inverse of R.
   d. What is the domain of the inverse relation?
   e. What is the range of the inverse relation?

*9. Given I = {−2, −1, 0, 1, 2}, determine the relation R such that the first coordinate is divisible by the second coordinate.

   a. What is the domain of R?
   b. What is the range of R?
   c. Determine the inverse of R.
   d. What is the domain of the inverse relation?
   e. What is the range of the inverse relation?

10. Given A = {a, b, c, d, e, f, g, h, i}, determine the relation R such that the first coordinate is a vowel in the English alphabet and the second coordinate is not.

    a. What is the domain of R?
    b. What is the range of R?
    c. Determine the inverse of R.
    d. What is the domain of the inverse relation?
    e. What is the range of the inverse relation?

8.2 RELATIONS AND FUNCTIONS

In the previous section, we considered various types of relations. Most of them contained ordered pairs whose elements were real numbers, but some contained ordered pairs whose elements were letters or words. We shall now restrict our discussion to only those relations whose sets of ordered pairs contain real numbers. These relations can be pictured or graphed on a coordinate system. The coordinate system is a plane and since a plane contains sets of points, the set of all points on the plane will correspond to the set of all ordered pairs in the real number system. The coordinate plane consists of two perpendicular real number lines, one horizontal and the other vertical. The two perpendicular lines intersect at their origins (zero values). The horizontal line is called the *x axis*, and the vertical line is called the *y axis*. The x axis is positive to the right

Figure 1

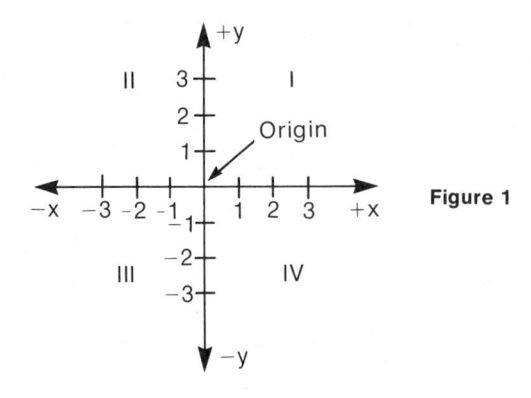

of the origin and negative to the left of the origin, while the y axis is positive above the origin and negative below the origin. Note that the axes or coordinate axes divide the plane into four quadrants. They are numbered I to IV in a counterclockwise direction. See Figure 1. The coordinate system contains the x and y axes and they intersect at a common point, the origin. We can describe this point by means of an ordered pair, (0, 0), that is, $x = 0$ and $y = 0$. Whenever we are given an ordered pair of real numbers such as (3, 2) the first coordinate is always the x value and the second coordinate is always the y value. Therefore, to plot or graph the ordered pair (3, 2) we start at the origin and move three units in a positive x direction, and then two units in a positive y direction. Similarly, the point represented by (2, −3) is obtained by starting at the origin and moving two units in a positive x direction, and then three units in a negative y direction. The point represented by (−2, 3) is obtained by starting at the origin and moving two units in a negative x direction, and then three units in a positive y direction. The point represented by (−3, −2) is obtained by starting at the origin and moving three units in a negative x direction, and then moving two units in a negative y direction. These points are graphed and labeled in Figure 2.

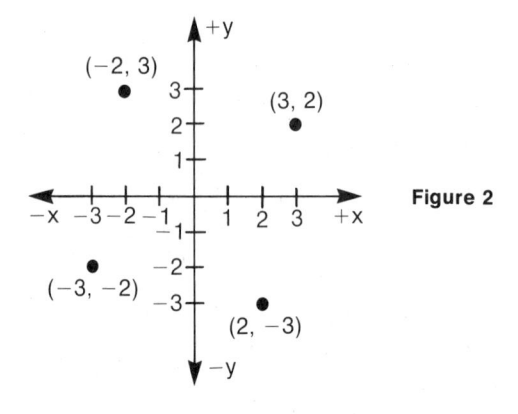

Figure 2

Relations and Functions

The coordinate system illustrated is often called the *rectangular coordinate* system. The plane itself is frequently called the *Cartesian plane* and the ordered pairs of real numbers representing points on the plane are called *Cartesian coordinates* (in honor of the famous French mathematician René Descartes). An ordered pair of real numbers is always given with the x value first and the y value second. Formally, we call the x value the *abscissa,* and the y value the *ordinate*; together they form a coordinate point. Every point of a Cartesian plane can be represented by an ordered pair of real numbers. Also, every ordered pair of real numbers can be used to locate a point of the plane. Formally, we can say that there is a one-to-one correspondence between the set of points on the plane and the set of ordered pairs of real numbers.

EXAMPLE 5

Graph each of the given points on the coordinate plane: A(1, 4), B(−2, 3), C(2, −3), D(−1, −4), E(5, 0), F(0, 1).

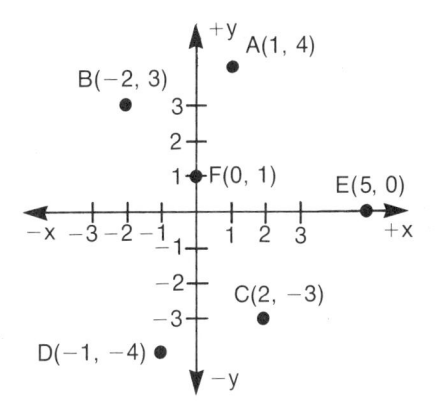

Solution

Whenever we are given an ordered pair of real numbers (a, b), the first coordinate is the abscissa (the x-value) and the second coordinate is the ordinate (y-value). To graph an ordered pair (a, b), we start at the origin and move a units in the appropriate x direction and then b units in the appropriate y direction.

EXAMPLE 6

Write the coordinates of each point whose graph is shown on the following page.

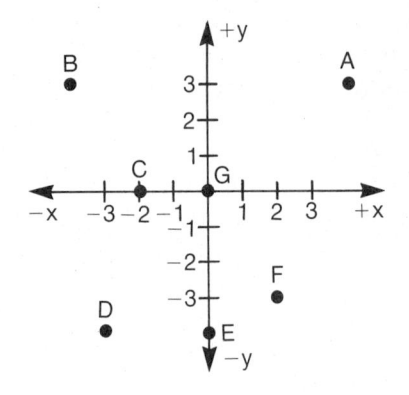

Solution

A(4, 3), B(−4, 3), C(−2, 0), D(−3, −4), E(0, −4), F(2, −3), G(0, 0). Note that point G is at the origin and the coordinates of the origin are (0, 0).

A relation is defined to be a set of ordered pairs and therefore we can graph a relation on the Cartesian plane. For example, consider the relation A = {(2, 2), (−2, 2), (−2, −2), (2, −2)}. By graphing each of the ordered pairs on the plane, we can graph the relation A. See Figure 3.

Consider the following relations:

$$A = \{(1, 2), (1, 3), (2, 4), (2, 5)\}$$

$$B = \{(1, 2), (2, 3), (4, 5), (6, 7)\}$$

Both relations are composed of four sets of pairs of real numbers. But they are different in nature. Note that *some* of the ordered pairs in relation A have the same first coordinate while *none* of the ordered pairs in relation B have the same first coordinate. This is significant. Relation B is

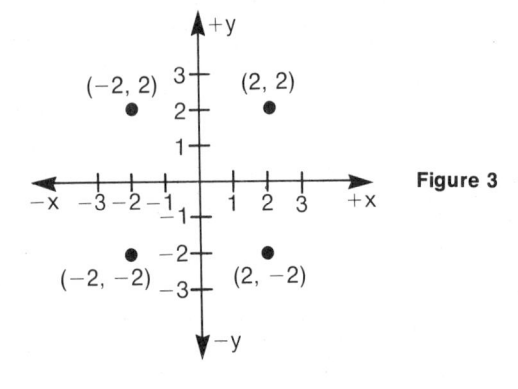

Figure 3

Relations and Functions

a special type of relation because no two ordered pairs have the same first coordinate. Relations such as B that have this unique property are given a special name. They are called *functions.* We formally define a function:

> A function is a relation in which no two ordered pairs have the same first coordinate and a different second coordinate.

Therefore relation A is not a function since it has ordered pairs that have the same first coordinate and a different second coordinate.

EXAMPLE 7

Determine whether the following relations are functions or not.

a. $\{(1, -1), (-2, 3), (3, 1), (4, 2)\}$
b. $\{(2, 0), (3, 1), (4, -2), (3, -2)\}$
c. $\{(1, 1), (2, 1), (3, 1), (4, 1)\}$

Solution

a. yes   b. no   c. yes

Note that the relation given in Example 3-c is a function. A function is a relation in which no two ordered pairs have the same first coordinate and a different second coordinate. Therefore, a function can be a relation where two or more ordered pairs have the same second coordinates, but different first coordinates.

EXAMPLE 8

Determine whether the following relations are functions or not.

a. $\{(0, 2), (-1, -2), (3, 5), (2, 0)\}$
b. $\{(-1, 1), (-2, 3), (-1, 2), (-2, 4)\}$
c. $\{(1, 1), (2, 2), (3, 1), (4, 2)\}$

Solution

a and c are functions; no two ordered pairs have the same first coordinate.

Figure 4-a

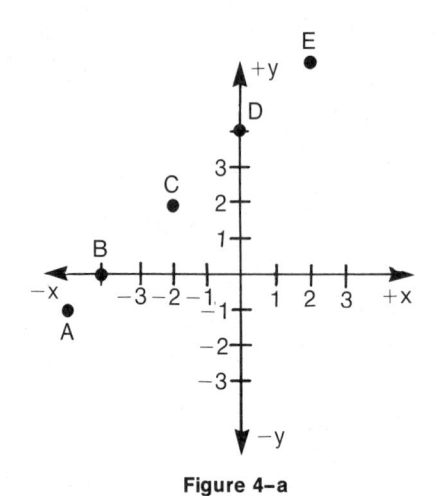

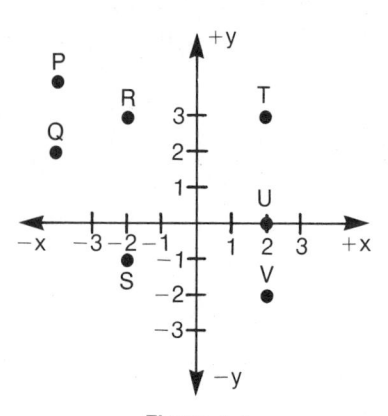

Figure 4-b

Sometimes we may have to determine from the graph of a relation whether it is a function or not. Consider the two relations pictured in Figures 4-a and 4-b.

Note that the coordinates of the points on the graph in Figure 4-a are: A(−5, −1), B(−4, 0), C(−2, 2), D(0, 4), E(2, 6). The relation graphed in Figure 4-a is a function since no two ordered pairs have the same first coordinate. But the coordinates of the points on the graph in Figure 4-b are: P(−4, 4), Q(−4, 2), R(−2, 3), S(−2, −1), T(2, 3), U(2, 0), V(2, −2). The relation graphed in Figure 4-b is not a function since at least two ordered pairs do have the same first coordinates; the points are such that one is directly below the other. If we connect two such points, then we can form a vertical line segment. Therefore, any time the graph of a relation has more than one point on the same *vertical* line, it is *not* a function. Also, if no vertical line can contain two points of the graph of a relation, then the relation is a function. Figures 5-a and 5-b are graphs of relations that are *not* functions.

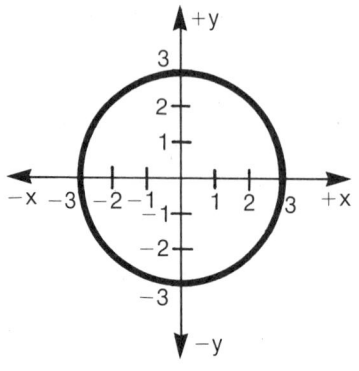

Figure 5-a

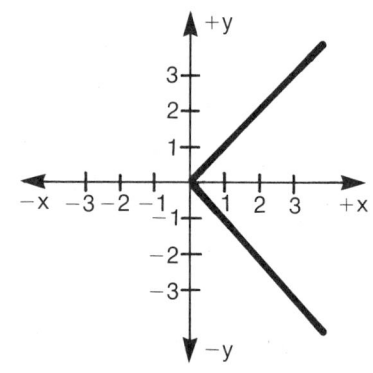

Figure 5-b

Relations and Functions

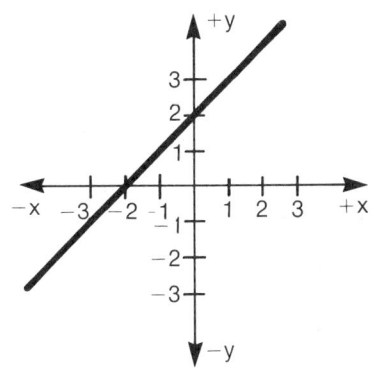

Figure 6-a

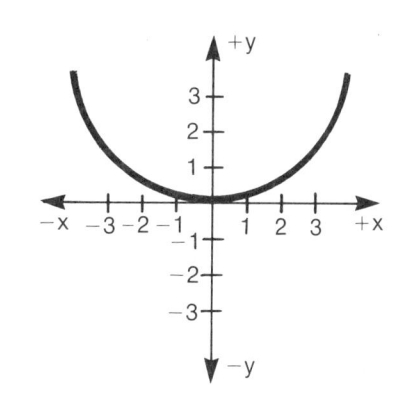

Figure 6-b

The relations illustrated in Figures 5-a and 5-b are not functions because it is possible to graph a vertical line on the plane so that it will contain more than one point of the relation. Note that two such points will have the same first coordinate.

Figures 6-a and 6-b are graphs of relations that are functions.

The relations illustrated in Figures 6-a and 6-b are functions since it is not possible to graph a vertical line on the plane so that it will contain more than one point of the relation. If no vertical line on the plane will contain two points of the graph of the relation, then the relation is a function. We shall call this method for determining whether a relation is a function the vertical line method or the vertical line test.

EXAMPLE 9

In these graphs of relations, a, b, and c, determine which are graphs of functions.

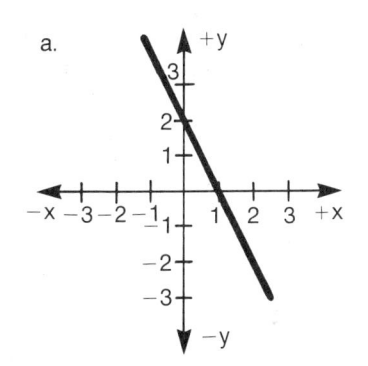

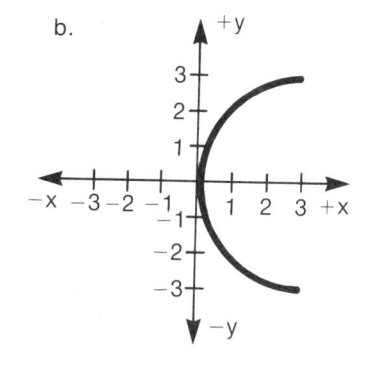

a.

b.

c.

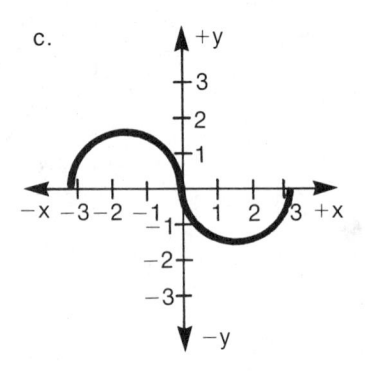

(graph showing an S-curve through the origin on a coordinate plane with x and y axes marked from -3 to 3)

Solution

a and c are graphs of functions and b is not. Using the vertical line method, it is not possible to draw a vertical line in a and c that will contain more than one point of the graph of the relation. In b, it is possible to draw such a line, therefore, it is not a graph of a function.

Consider the formula $C = \pi d$. This is the formula for finding the circumference of a circle where C represents the circumference, π is a constant, and d represents the diameter. If $d = 2$, then $C = 2\pi$ and if $d = 4$, then $C = 4\pi$. The value of C depends upon the value of d. We can say that C is a function of d. Since we can arbitrarily choose any value for d, we call d the *independent variable*. Because the value of C depends upon the value of d, (C is a function of d), we call C the *dependent variable*.

There are many different ways to express functions, and since we are dealing with ordered pairs of real numbers that represent points on the Cartesian plane, let us consider functions involving x and y. Consider the equation $y = x + 3$. In this equation, if $x = 0$, then $y = 3$ and if $x = 2$, then $y = 5$. Some ordered pairs that satisfy this equation are: (0, 3), (2, 5), (4, 7), (−3, 0). Note that x is the independent variable and y is the dependent variable, that is, the value of y depends upon the value of x. In other words, we can say that y is a function of x. The notation used in mathematics for the phrase "a function of x" is f(x). Note that f(x) does *not* mean f times x: it is an abbreviation for "a function of x" and is usually read as "f of x." Therefore, we can rewrite the statement "y is a function of x."

```
y is  a function of x.
↕    ↕
y =      f(x)
```

Using this notation, we can rewrite the equation $y = x + 3$ as $f(x) = x + 3$.

Relations and Functions

Instead of having a table of values such as that shown below on the left, we can also have that on the right:

x	y
0	3
2	5
4	7
−3	0

x	f(x)
0	3
2	5
4	7
−3	0

Instead of the ordered pair (x, y), we have (x, f(x)). Instead of writing y = 3 when x = 0, and y = 5 when x = 2, we can write f(x) = 3 when x = 0, and f(x) = 5 when x = 2. We can also have f(0) = 3 and f(2) = 5. In the first notation, f(0) = 3, this means that the value of the function f(x) is 3 when x = 0. The notation f(2) = 5 means that the value of the function f(x) is 5 when x = 2. Notation similar to this was discussed in Section 4.6.

Using this idea, what does f(4) mean? We want the value of the function f(x) when x = 4. Recall that the given function is y = x + 3 or f(x) = x + 3. The value of the function is 7 when x = 4, f(4) = 7. We obtain this value by substituting 4 for x in the equation. Therefore, f(x) = x + 3, and f(4) = 4 + 3 = 7. Similarly, f(−3) = −3 + 3 = 0, and f(5) = 5 + 3 = 8. In general, the notation f(a) means to find the value of the function f(x) when x = a. This is done by substituting a for x in the given equation.

EXAMPLE 10

If $f(x) = 3x - 1$, find:

a. f(1)   b. f(−1)   c. f(0)

Solution

a. The notation f(1) means to find the value of the function f(x) when x = 1. This is done by substituting 1 for x in the given equation:
$f(x) = 3x - 1$.
$f(1) = 3(1) - 1 = 3 - 1 = 2$.
b. $f(-1) = 3(-1) - 1 = -3 - 1 = -4$.
c. $f(0) = 3(0) - 1 = 0 - 1 = -1$.

It should be noted that other symbols can also be used to denote functions. For example, if we want to discuss two different functions of x, we can denote them by f(x) and g(x): $f(x) = 3x - 1$ and $g(x) = 4x + 2$ are two different functions of x. We can also write these functions as $y = 3x - 1$ and $y = 4x + 2$, but functions are normally expressed as f(x), g(x), and so on. Many other formulas also define functions. The area of a square whose sides is s is given by $A = s^2$. Here A is a function of s. The dependent variable is A and the independent variable is s. We can express the formula, $A = s^2$, in function notation. For example, $f(s) = s^2$, $g(s) = s^2$, $t(s) = s^2$. Special functions are represented by certain symbols. Some other examples of formulas that define functions are:

$$g(t) = 120t - 16t^2.$$
$$f(s) = s^2 - s.$$
$$Q(y) = y^2 + 3y + 2.$$
$$P(u) = u^3 - 2u^2.$$

EXAMPLE 11

If $g(t) = 120t - 16t^2$, find:

a. g(2)   b. g(0)   c. g(3)

Solution

Applying the function notation used in Example 10, we have

a. $g(2) = 120(2) - 16(2)^2 = 240 - 16(4) = 240 - 64 = 176.$
b. $g(0) = 120(0) - 16(0)^2 = 0 - 16 \cdot 0 = 0 - 0 = 0.$
c. $g(3) = 120(3) - 16(3)^2 = 360 - 16(9) = 360 - 144 = 216.$

Regardless of the notation used, we still evaluate a function in the same manner. Therefore, if $g(t) = 3t + 2$, then to find $g(x + 1)$ we must substitute $x + 1$ for t in the given equation,

$$g(t) = 3t + 2.$$
$$g(x + 1) = 3(x + 1) + 2$$
$$= 3x + 3 + 2$$
$$= 3x + 5.$$

EXAMPLE 12

If $f(s) = s^2 + 3s$, find:

a. f(3s)   b. f(s − 1)   c. f(a + b)

Relations and Functions

Solution

a. $f(3s) = (3s)^2 + 3(3s) = 9s^2 + 9s$.
b. $f(s-1) = (s-1)^2 + 3(s-1) = s^2 - 2s + 1 + 3s - 3$
    $= s^2 + s - 2$.
c. $f(a+b) = (a+b)^2 + 3(a+b) = a^2 + 2ab + b^2 + 3a + 3b$.

To evaluate a function f(x) at x = a, we denote this by f(a). We can think of f(a) as a command or instruction and it means to find the value of the function f(x) when x = a. Similarly f(a + b) means to find the value of the function f(x) when x = a + b.

EXERCISES FOR SECTION 8.2

(Answers given to exercises marked *)*

*1. Graph each of the given points on the coordinate plane:

A(4, 2), B(−3, 4), C(−2, −4), D(4, −2).

2. Graph each of the given points on the coordinate plane:

A(−3, −3), B(−4, 4), C(3, 4), D(4, −2).

*3. Write the coordinates of each point whose graph is shown below:

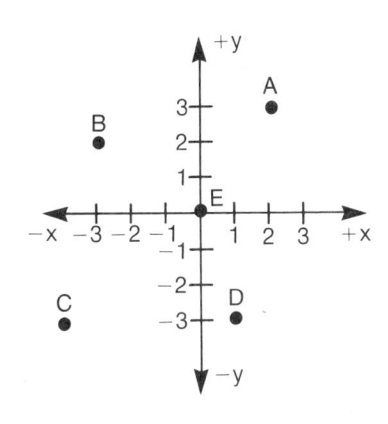

4. Write the coordinates of each point whose graph is shown below:

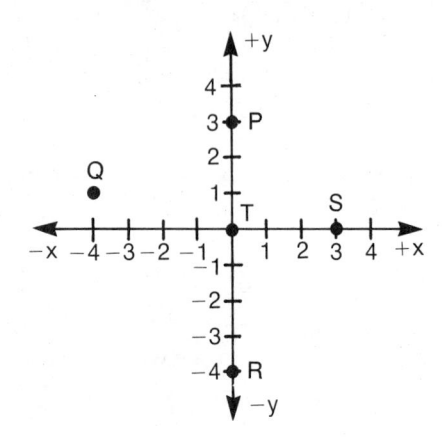

*5. Determine whether the following relations are functions or not:
   a. {(1, 1), (2, 2), (3, 3), (4, 4)}
   b. {(−2, 1), (1, −2), (3, −2), (4, 1)}
   c. {(−2, 2), (3, 2), (4, 2), (5, 2)}
   d. {(1, 2), (3, 4), (5, 7), (1, 3)}
   e. {(3, −2), (4, 1), (3, −1), (5, 1)}
   f. {(2, 4), (3, 1), (5, 2), (3, −2)}

6. Find the inverse of each relation in Exercise 5 and determine which of the inverse relations are functions.

*7. Determine which of these graphs of relations are graphs of functions.

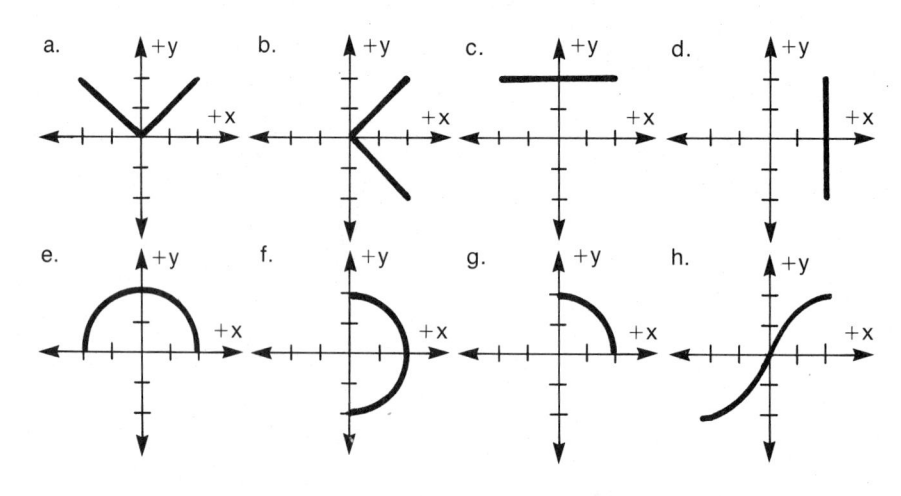

8. For each of the following equations, identify the independent variable and the dependent variable.

Relations and Functions

a. $y = 4x - 3$.　　　b. $A = \pi r^2$.　　　c. $C = 2\pi r$.
d. $g = s^2 + 2$.　　　e. $V = s^3$.　　　f. $A = s^2$.
g. $v = 4t^3 - t^2$.　　h. $u = v^2 + 2v$.　i. $C = \dfrac{5}{9}(F - 32)$.

*9. Solve each of the following equations for the indicated variable:
   a. $y = 2x + 5$, for x.　　　b. $A = \pi r^2$, for r.
   c. $60t - \dfrac{s}{2} = 8t^2$, for s.　　d. $C = \dfrac{5}{9}(F - 32)$, for F.
   e. $y = 2x^2 + 1$, for x.　　f. $C = 2\pi r$, for r.

10. If $f(x) = 2x + 5$, find:
    a. $f(1)$　　b. $f(2)$　　c. $f(-1)$　　d. $f(a)$

*11. If $f(x) = x^2 - 4$, find:
    a. $f(1)$　　b. $f(-1)$　　c. $f(2)$　　d. $f(0)$

12. If $g(x) = x^2 + 3x + 2$, find:
    a. $g(1)$　　b. $g(-1)$　　c. $g(-2)$　　d. $g(0)$

*13. If $s(t) = 120t - 16t^2$, find:
    a. $s(1)$　　b. $s(-1)$　　c. $s(2)$　　d. $s(-2)$

14. If $p(x) = 2x^2 - 3x + 2$, find:
    a. $p(0)$　　b. $p(-1)$　　c. $p(1)$　　d. $p(-2)$

*15. If $g(n) = n^3 - 2n^2$, find:
    a. $g(2)$　　b. $g(-2)$　　c. $g(1)$　　d. $g(-a)$

16. If $h(x) = 2\sqrt{x} - 1$, find:
    a. $h(1)$　　b. $h(4)$　　c. $h(0)$　　d. $h(9)$

*17. If $g(t) = 2t^2 - 3t$, find:
    a. $g(3)$　　b. $g(-2)$　　c. $g(a)$　　d. $g(a+1)$

18. If $f(x) = \dfrac{x+2}{x-3}$, find:
    a. $f(2)$　　b. $f(-2)$　　c. $f(4)$　　d. $f(3)$

*19. If $s(t) = 30t - 4t^2$, find:
    a. $s(-1)$　　b. $s(2)$　　c. $s(a)$　　d. $s(a+1)$

20. If $g(x) = \dfrac{x^2 - 1}{x + 1}$, find:

   a. $g(0)$ b. $g(3)$ c. $g(1)$ d. $g(-1)$

8.3 LINEAR FUNCTIONS

In the previous section, we graphed sets of ordered pairs on the coordinate plane (see Example 5, Section 8.2). In other words, we graphed relations. For example, consider the set of ordered pairs: $\{A(-5, -1), B(-4, 0), C(-2, 2), D(0, 4), E(2, 6)\}$. The relation containing this set of ordered pairs is pictured in Figure 7. Note that the points appear to form a straight line. If we were to graph more points that fit on the same line as A, B, C, etc., such as $\left(\dfrac{1}{2}, \dfrac{9}{2}\right), \left(\dfrac{3}{2}, \dfrac{11}{2}\right), \left(-\dfrac{1}{2}, \dfrac{7}{2}\right)$, the dotted line would appear to fill in to form a solid line. In fact, if we connect the points graphed in Figure 7 with a solid line, we will have the graph of a straight line. All points on the line correspond to an ordered pair and each pair will satisfy an equation that defines this particular relation. The equation that these ordered pairs satisfy is $y = x + 4$ and its graph is shown in Figure 8.

Any equation of the general form $ax + by = c$ where a, b, and c are real numbers is called a "linear equation," because when we graph the ordered pairs that satisfy such an equation, we get a straight line. Given a linear equation, we should be able to graph it on the coordinate plane. Consider the linear equation $x + y = 4$. If we want to determine the graph of this equation, we must find ordered pairs that will satisfy it. One way to do this is to let x be any real number and then solve for y. For example, if $x = 0$, then $y = 4$. If $x = 3$, then $y = 1$, and if $x = 4$, then $y = 0$. Three

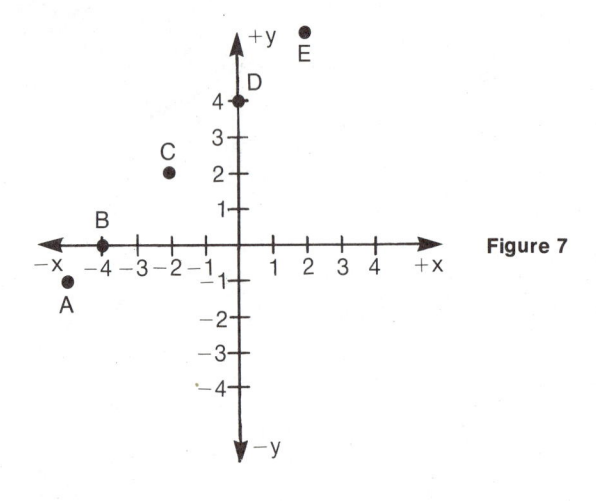

Figure 7

Relations and Functions

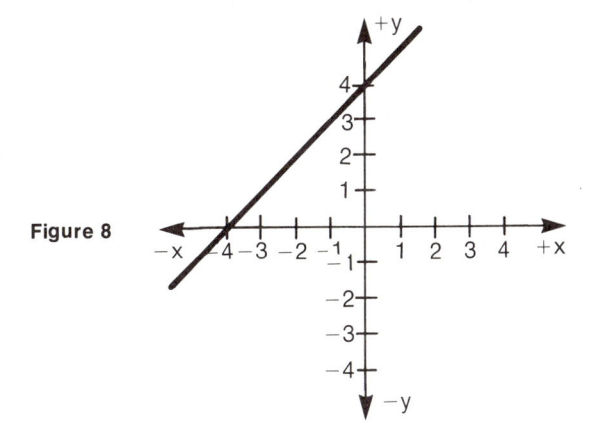

Figure 8

ordered pairs that satisfy the given equation, x + y = 4, are (0, 4), (3, 1), and (4, 0). We can plot these points on the Cartesian plane and connect them. Note that they all lie on the same straight line. This line is a picture or graph of the linear equation x + y = 4. See Figure 9.

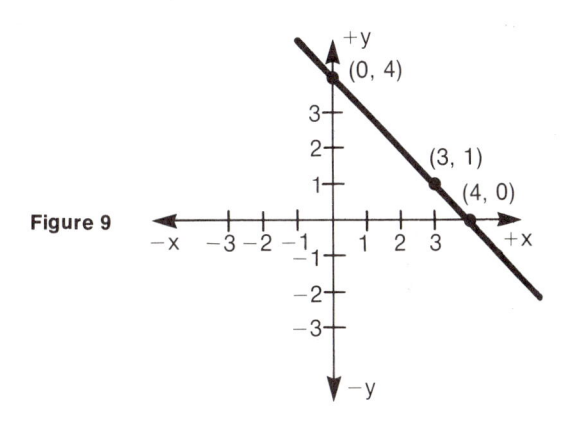

Figure 9

If the graph of a linear equation is always a straight line, then why do we use three points to determine the graph? It is true that two points determine a line, that is, through two given points one and only one straight line can be drawn. The third point is used as a check. It is located to check that all three points lie on the same straight line. If one of these points is not on the line, then we must go back and check for an error in determining the solutions to the given linear equation.

EXAMPLE 13

Graph $3x + 2y = 6$.

Solution

To graph $3x + 2y = 6$, we must first find three ordered pairs that are solutions to the equation.

a. Select a value for x, let $x = 2$.

$$3x + 2y = 6.$$
$$\text{Let } x = 2, \; 3(2) + 2y = 6.$$
$$6 + 2y = 6.$$
$$2y = 0.$$
$$y = 0 \quad \text{and } (2, 0) \text{ is a solution.}$$

b. Let $x = 4$, then

$$3(4) + 2y = 6.$$
$$12 + 2y = 6.$$
$$2y = -6.$$
$$y = -3 \quad \text{and } (4, -3) \text{ is a solution.}$$

c. Let $x = 0$, then

$$3(0) + 2y = 6.$$
$$0 + 2y = 6.$$
$$y = 3 \quad \text{and } (0, 3) \text{ is a solution.}$$

Now that we have three different ordered pairs, $(2, 0)$, $(4, -3)$, and $(0, 3)$, that are solutions to the equation $3x + 2y = 6$, we plot these points and draw the line that contains them. See Figure 10.

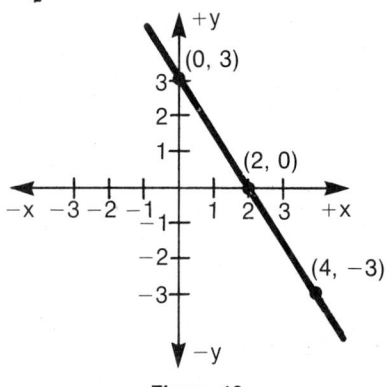

Figure 10

When graphing a linear equation, it is usually convenient to find the points where the graph crosses each axis. When a line crosses the y axis, $x = 0$ and when a line crosses the x axis, $y = 0$. In Example 13, the line

Relations and Functions

crossed the y axis at (0, 3) and it crossed the x axis at (2, 0). Where a line crosses the y axis is called the *y-intercept* and where it crosses the x axis is called the *x-intercept.* To determine the y-intercept, we let x = 0 and to find the x-intercept, we let y = 0.

EXAMPLE 14

Graph $x - 2y = 4$.

Solution

First, we shall find the intercepts. Therefore, the value we select for x is 0.

Let $x = 0$, then
$$0 - 2y = 4.$$
$$y = -2. \quad (0, -2) \text{ is a solution and it is the y-intercept.}$$

Next, let $y = 0$, then
$$x - 2(0) = 4.$$
$$x = 4. \quad (4, 0) \text{ is a solution and it is the x-intercept.}$$

Selecting a third point, in order to check for collinearity, let $x = 2$, then
$$2 - 2y = 4.$$
$$-2y = 2.$$
$$y = -1 \quad \text{and } (2, -1) \text{ is a solution.}$$

Using these points, we graph $x - 2y = 4$. See Figure 11.

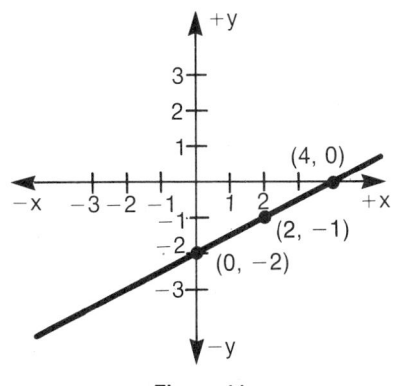

Figure 11

Using the vertical line method for determining whether a relation is a function or not, we see that the linear relation, $x - 2y = 4$ is also a function. Almost all linear equations are functions, but since the graph of a straight line could be a vertical line, we note that not all linear equations are functions. A function is a relation in which no two ordered pairs have the same first coordinate and a different second coordinate. A vertical line contains an infinite number of points with the same first coordinate, hence it cannot be a function.

EXAMPLE 15

Graph $2x - 3y = 7$.

Solution

First, find the intercepts.
Let $x = 0$, $2(0) - 3y = 7$.
$0 - 3y = 7$.
$y = -\frac{7}{3}$. $\left(0, -\frac{7}{3}\right)$ is a solution and it is the y-intercept.

Let $y = 0$, $2x - 3(0) = 7$.
$2x - 0 = 7$.
$x = \frac{7}{2}$. $\left(\frac{7}{2}, 0\right)$ is a solution and it is the x-intercept.

Note that the intercepts are ordered pairs that contain fractions. When this happens, it is usually more efficient to find solutions that contain only integers. It is more difficult (and probably less accurate) to locate a point such as $\left(0, -\frac{7}{3}\right)$ than a pair of integers. Therefore, we find three other solutions.

Let $x = 2$, then $2(2) - 3y = 7$.
$4 - 3y = 7$.
$-3y = 3$.
$y = -1$. $(2, -1)$ is a solution.

Let $x = -1$, then $2(-1) - 3y = 7$.
$-2 - 3y = 7$.
$-3y = 9$.
$y = -3$. $(-1, -3)$ is a solution.

Selecting a third point, in order to check for collinearity,

Relations and Functions

let x = 5, then 2(5) − 3y = 7.
10 − 3y = 7.
− 3y = −3.
y = 1.   (5, 1) is a solution.

Using these points, we graph 2x − 3y = 7. See Figure 12. Note that it is much easier to locate points such as (2, −1) and (−1, −3) compared to $\left(0, -\frac{7}{3}\right)$ and $\left(\frac{7}{2}, 0\right)$.

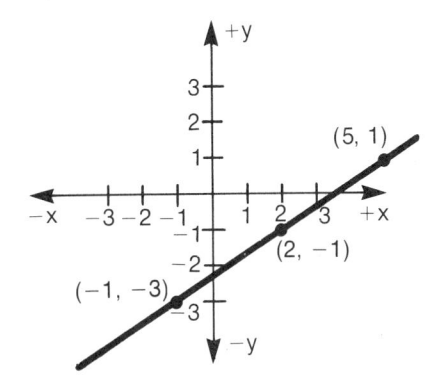

Figure 12

Earlier in this section, we stated that any equation of the general form ax + by = c (where a, b, and c are real numbers) is called a linear equation and the graph of a linear equation is a straight line. Sometimes it happens that a or b is equal to 0 in the general equation ax + by = c. Consider the case where b = 0; if b = 0 then we have ax + 0y = c or ax = c. A specific example that fits this general form is x = 3 (a = 1, b = 0, and c = 3). How can we graph x = 3? First, let's rewrite the equation x = 3 so that it fits the general form, that is, x = 3 is the same as x + 0y = 3. Now, let's find three solutions that satisfy this equation. Note that if y = 0, then x = 3. Similarly, if y = 2, then x = 3 and if y = −1, then x = 3. The ordered pairs (3, 0), (3, 2), and (3, −1) are all solutions of the equation x + 0y = 3.

Using these points, we graph x = 3. See Figure 13. The graph of any equation of the form x = c is a vertical line parallel to the y axis and passing through the point (c, 0). Note that the y-axis is the graph of the linear equation x = 0. It should also be noted that the graph of any equation of the form x = c is the only type of linear equation that is not a function. We can use a similar technique to graph an equation of the form y = c. Consider y = 3. Rewriting it in the general form, we have 0x + y = 3. Next, find three solutions that satisfy this equation. If x = 0, then y = 3. Similarly, if x = 2, then y = 3 and if x = −2, then y = 3. The

Figure 13

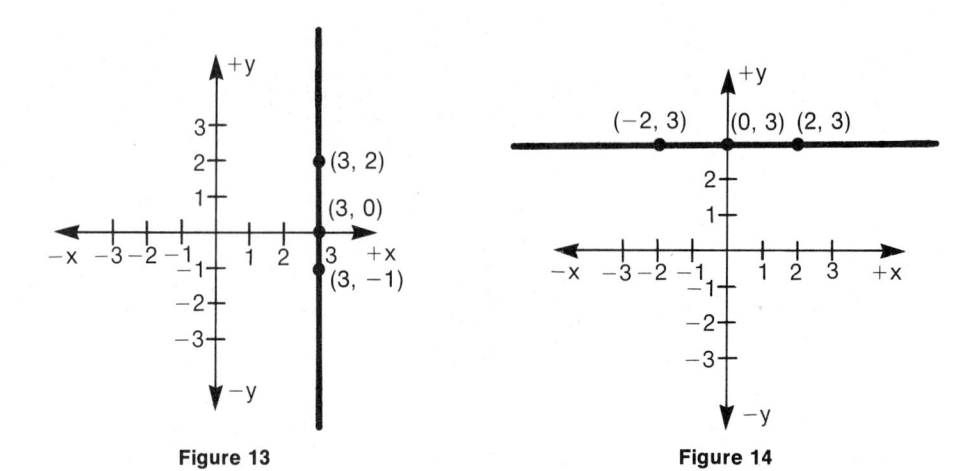

Figure 14

ordered pairs (0, 3), (2, 3), and (−2, 3) are all solutions of the equation 0x + y = 3. Using these points, we graph y = 3. See Figure 14. The graph of any equation of the form y = c is a horizontal line parallel to the x-axis and passing through the point (0, c). It has no x-intercept. Note that the x-axis is the graph of the linear equation y = 0.

EXAMPLE 16

Graph x = −3 and y = 2.

Solution

a. The graph of x = −3 is a vertical line parallel to the y-axis and passing through the point (−3, 0). It has no y-intercept. See Figure 15.
b. The graph of y = 2 is a horizontal line parallel to the x-axis and passing through the point (0, 2). It has no x-intercept. See Figure 15.

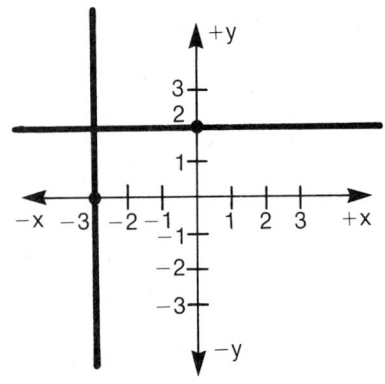

Figure 15

Relations and Functions

> The procedures discussed in this section enable us to graph any linear equation of the general form $ax + by = c$. When graphing a linear equation, it is usually convenient to find the x and y intercepts, that is, where the line crosses the x and y axes, respectively. Finding a third ordered pair that is a solution to the given equation is also recommended. The third point is located in order to check that all three points lie on the same straight line. The graph of a linear equation is a straight line passing through the points whose coordinates satisfy the equation.

EXERCISES FOR SECTION 8.3
(Answers given to exercises marked *)

Graph each of the following equations on the Cartesian plane.

*1. $x + y = 4$.
2. $x + y = 7$.
*3. $x - y = 4$.
4. $x - y = 2$.
*5. $x - y = 1$.
6. $2x + y = 8$.
*7. $2x - y = 4$.
8. $2x + y = -6$.
*9. $2x - y = -6$.
10. $-3x + y = 12$.
*11. $-2x - 3y = 6$.
12. $-3x + 2y = -12$.
*13. $2y = x - 6$.
14. $3x - 6 = y$.
*15. $y - 6 = -2x$.
16. $5y - 15 = -3x$.
*17. $x + y = 0$.
18. $2x + y = 0$.
*19. $x - y = 0$.
20. $x - 4y = 0$.
*21. $2x + 3y = 7$.
22. $3x - 2y = 5$.
*23. $-x + 3y = 1$.
24. $-2x - 3y = -1$.
*25. $x = -4$.
26. $y = 1$.
*27. $y = -2$.
28. $x = 4$.
*29. $x = 0$.
30. $y = 0$.

8.4 THE SLOPE OF A LINE

In the previous section, we graphed various types of lines. Some were vertical, others were horizontal. Those that were neither horizontal nor vertical can be called oblique lines. Now we shall consider the steepness or slope of lines. Consider the ramp pictured in Figure 16. Such a ramp could be found at the entrance to a building, or a stadium, or a parking garage. This particular ramp has a length of 50 feet and an elevation (or height) of 5 feet. For every 10 feet of horizontal distance, it rises 1 foot in vertical distance. Its steepness or slope is the ratio of the change in elevation to the change in horizontal distance. The ramp rises 1 foot for every change of 10 feet in horizontal distance. That is, its steepness or slope $= \frac{5}{50} = \frac{1}{10}$. We can use this same technique to measure the steepness or slope of a line. The slope of a line can be found by determining the ratio of the change in vertical distance to the change in the horizontal distance that the line makes, that is, we can compare the "rise" of the line to the "run" of the line.

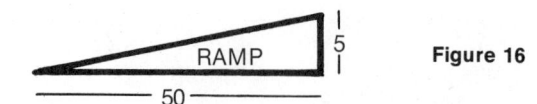

Consider the line pictured in Figure 17. It passes through the points (2, 0) and (4, 2) and we shall use these points as reference points. Note that as we proceed "up" the incline of the line (from left to right) the vertical distance (rise) changes one unit from every one unit that the horizontal distance (run) changes. Since the slope of a line is the ratio of the change in vertical distance to the change in horizontal distance (rise over run), we have

$$\text{slope} = \frac{\text{rise}}{\text{run}} = \frac{1}{1} = 1.$$

Note that the slope of the line pictured in Figure 17 is still 1 regardless of how we measure the change in vertical and horizontal distances. If we start at the point (2, 0) and proceed to the point (4, 2), the change in the vertical distance is 2 and the change in the horizontal distance is 2. Therefore, the ratio is $\frac{2}{2}$, which also equals 1. This indicates that the slope of

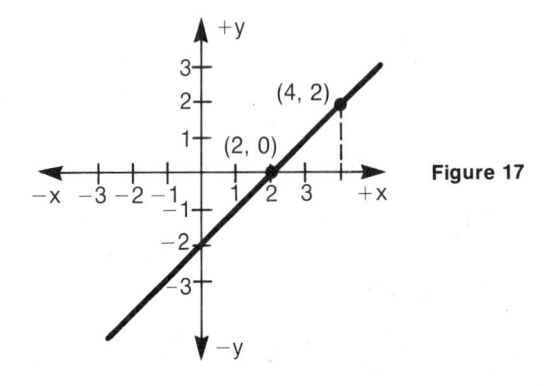

Relations and Functions 429

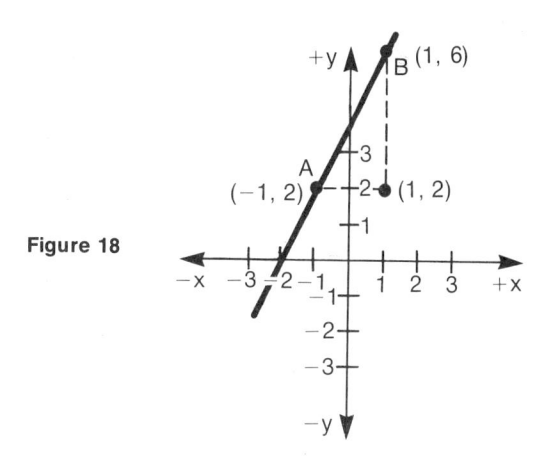

Figure 18

a line is the same, or constant, regardless of what two points on the line are used to determine the change in the vertical and horizontal distances.

Rather than actually count the changes in the vertical and horizontal distances of a line as we proceed from point to point, we can find these changes in a more efficient manner. Consider the line shown in Figure 18. Note that the points (−1, 2) and (1, 6) have been selected to form a "ramp" and the coordinates of the end of the base of the ramp are (1, 2). From this picture and these points, we can determine that the change in the vertical distance is 4 and the change in the horizontal distance is 2. Therefore, the slope of the line in Figure 18 is

$$\text{slope} = \frac{\text{rise}}{\text{run}} = \frac{4}{2} = \frac{2}{1} \text{ or } 2.$$

Note that the change in the vertical distance is the same as the change in the y coordinates and the change in the horizontal distance is the same as the change in the x coordinates. The change in the vertical distance is 4; the y coordinate of B is 6 and the y coordinate of A is 2: $6 - 2 = 4$. Similarly, the change in the vertical distance is 2, the x coordinate of B is 1, and the x coordinate of A is −1; $1 - (-1) = 1 + 1 = 2$.

To find the slope of a line containing two points, we can find the rise by subtracting the y values, and we can find the run by subtracting the x values, that is

> If a line contains the points (x_1, y_1) and (x_2, y_2), then the slope of the line is defined to be the rise of the line divided by the run of the line,
>
> $$\text{slope} = \frac{\text{rise}}{\text{run}} = \frac{y_2 - y_1}{x_2 - x_1}.$$

For example, consider the line shown in Figure 19. It passes through C(3, 4) and D(5, 7). To find the slope of the line, we designate one point

Figure 19

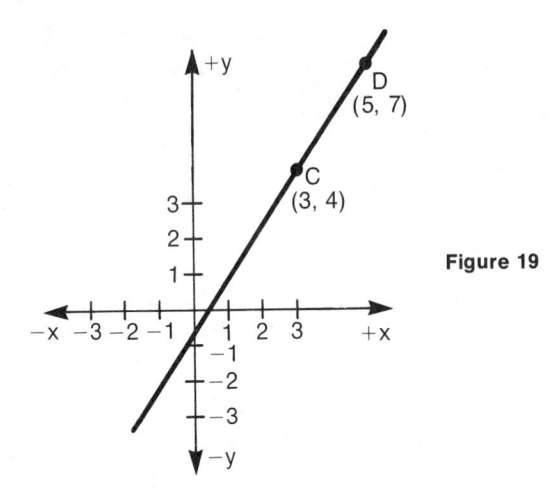

as a first point and the other as a second point (it doesn't matter which). In general, the coordinates of the first point are (x_1, y_1) and the coordinates of the second point are (x_2, y_2). Assume that C is the first point and D is the second point. Hence, $x_1 = 3$, $y_1 = 4$ and $x_2 = 5$, $y_2 = 7$. The slope of a line is often indicated by the letter "m." Therefore, the slope of the line in Figure 19 can be called m, and we have

$$\text{slope} = m = \frac{\text{rise}}{\text{run}} = \frac{y_2 - y_1}{x_2 - x_1} = \frac{7 - 4}{5 - 3} = \frac{3}{2}.$$

The slope of the line is the same if we choose D to be the first point and C to be the second point. In this case, $x_1 = 5$, $y_1 = 7$ and $x_2 = 3$, $y_2 = 4$ so we have

$$m = \frac{y_2 - y_1}{x_2 - x_1} = \frac{4 - 7}{3 - 5} = \frac{-3}{-2} = \frac{3}{2}.$$

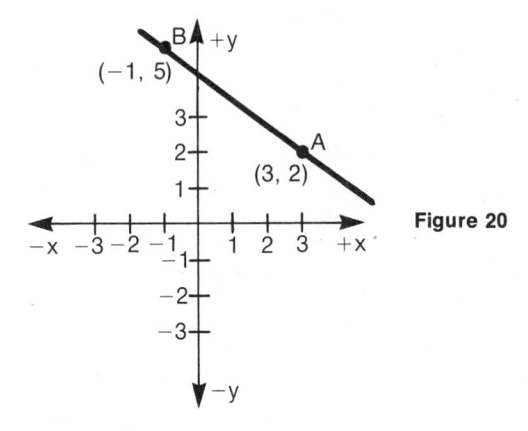

Figure 20

Relations and Functions

If a line slants upward to the right, then its slope is positive. If a line slants downward to the right, then its slope is negative. Consider the line shown in Figure 20. It passes through B$(-1, 5)$ and A$(3, 2)$. Let B be the first point and A be the second point, then $x_1 = -1$, $y_1 = 5$ and $x_2 = 3$, $y_2 = 2$. The slope of the line in Figure 20 is

$$m = \frac{y_2 - y_1}{x_2 - x_1} = \frac{2 - 5}{3 - (-1)} = \frac{-3}{3 + 1} = \frac{-3}{4} \text{ or } -\frac{3}{4}.$$

EXAMPLE 17

Find the slope of the line containing the points A$(2, 3)$ and B$(5, 6)$.

Solution

The slope of a line is the change in the y values divided by the change in the x values for any two different points on the line. Let A be the first point and B be the second point. Then $x_1 = 2$, $y_1 = 3$ and $x_2 = 5$, $y_2 = 6$. The slope of the line is

$$m = \frac{y_2 - y_1}{x_2 - x_1} = \frac{6 - 3}{5 - 2} = \frac{3}{3} = 1.$$

EXAMPLE 18

Find the slope of the line containing the points C$(-2, 3)$ and D$(-3, -2)$.

Solution

Let C be the first point and D be the second point. Then $x_1 = -2$, $y_1 = 3$ and $x_2 = -3$, $y_2 = -2$. The slope of the line is

$$m = \frac{y_2 - y_1}{x_2 - x_1} = \frac{-2 - 3}{-3 - (-2)} = \frac{-5}{-3 + 2} = \frac{-5}{-1} = 5.$$

Note: extreme care must be exercised in combining negative values such as $-3 - (-2)$. Also, remember that the slope will be the same regardless of which point is chosen first and which point is chosen second. In this example, if $x_1 = -3$, $y_1 = -2$, $x_2 = -2$, and $y_2 = 3$ then

$$m = \frac{y_2 - y_1}{x_2 - x_1} = \frac{3 - (-2)}{-2 - (-3)} = \frac{3 + 2}{-2 + 3} = \frac{5}{1} = 5,$$

which is exactly the answer we obtained previously.

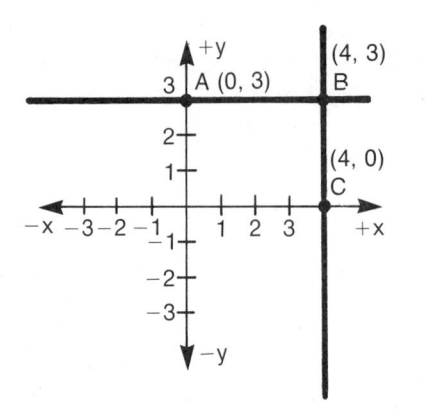

Figure 21

The slopes of horizontal and vertical lines deserve special consideration. Consider the lines pictured in Figure 21. The line containing the points A(0, 3) and B(4, 3) is a horizontal line parallel to the x-axis. If we let A be the first point and B be the second point, then $x_1 = 0$, $y_1 = 3$ and $x_2 = 4$, $y_2 = 3$. The slope of the horizontal line containing these points is

$$m = \frac{y_2 - y_1}{x_2 - x_1} = \frac{3 - 3}{4 - 0} = \frac{0}{4} = 0.$$

Note that the slope of any horizontal line parallel to the x-axis is 0. The line containing the points B(4, 3) and C(4, 0) is a vertical line parallel to the y-axis. If we let B be the first point and C be the second point, then $x_1 = 4$, $y_1 = 3$, $x_2 = 4$, and $y_2 = 0$. The slope of the vertical line containing these points is

$$m = \frac{y_2 - y_1}{x_2 - x_1} = \frac{0 - 3}{4 - 4} = \frac{-3}{0} = \text{undefined.}$$

Since division by 0 is undefined, this line has no slope. In general, the slope of a vertical line, one parallel to the y-axis, is undefined.

EXAMPLE 19

Find the slope of the lines pictured in Figure 22.

Solution

a. To find the slope of the line containing D and E, let D be the first point and E be the second point, then $x_1 = -2$, $y_1 = 3$ and $x_2 = -2$, $y_2 = -2$. The slope of the line is

Relations and Functions

$$m = \frac{y_2 - y_1}{x_2 - x_1} = \frac{-2 - 3}{-2 - (-2)} = \frac{-5}{-2 + 2} = \frac{-5}{0} = \text{undefined}.$$

The line containing D and E is a vertical line and the slope of a vertical line is undefined.

b. To find the slope of the line containing E and F, let E be the first point and F be the second point, then $x_1 = -2$, $y_1 = -2$ and $x_2 = 3$, $y_2 = -2$. The slope of the line is

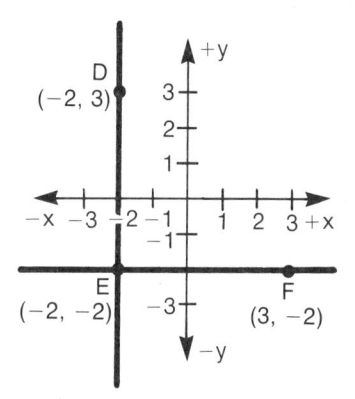

$$m = \frac{y_2 - y_1}{x_2 - x_1} = \frac{-2 - (-2)}{3 - (-2)} = \frac{-2 + 2}{3 + 2} = \frac{0}{5} = 0.$$

The line containing E and F is a horizontal line and the slope of a horizontal line is 0.

Thus far in our discussion of slopes we examined the slopes of oblique lines, vertical lines and horizontal lines. Before proceeding further we shall also consider the slopes of sets of lines. In particular, we shall examine the slopes of parallel and perpendicular lines. Consider the lines pictured in Figure 23. We are given that the line containing A and B is parallel to the line containing C and D. Also, the line containing E and F is perpendicular to the line containing C and D. How do the slopes of these lines compare? Let's find out. To find the slope of the line containing A and B, let $x_1 = -3$, $y_1 = 2$ and $x_2 = 1$, $y_2 = -3$. The slope of the line containing A and B, or the slope of line AB (denoted by $\overleftrightarrow{AB}$) is

$$m = \frac{y_2 - y_1}{x_2 - x_1} = \frac{-3 - 2}{1 - (-3)} = \frac{-5}{1 + 3} = \frac{-5}{4} = m\overleftrightarrow{AB}.$$

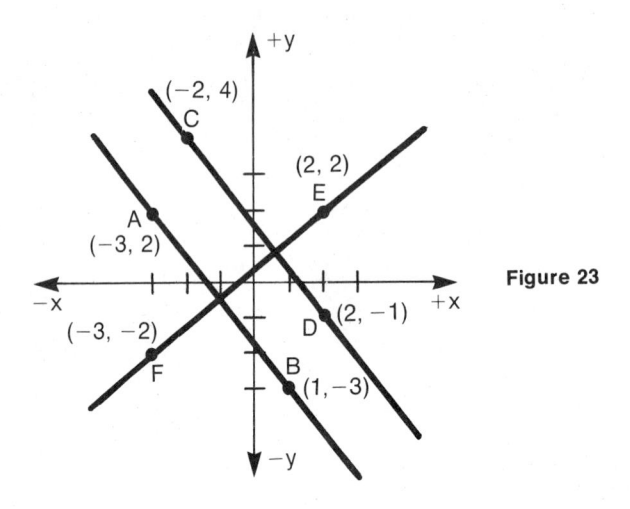

To find the slope of $\overleftrightarrow{CD}$, let $x_1 = -2$, $y_1 = 4$ and $x_2 = 2$, $y_2 = -1$. Then the slope of $\overleftrightarrow{CD}$ is

$$m = \frac{y_2 - y_1}{x_2 - x_1} = \frac{-1 - 4}{2 - (-2)} = \frac{-5}{2+2} = \frac{-5}{4} = m\overleftrightarrow{CD}.$$

We have shown that for these two particular parallel lines, their slopes are equal. It shall not be proved here, but it can be shown that, in general, if lines are parallel, then they must have the same slope. Also, lines that have the same slope are parallel (or they may be coincident).

To find the slope of $\overleftrightarrow{EF}$, let $x_1 = 2$, $y_1 = 2$ and $x_2 = -3$, $y_2 = -2$, then the slope of $\overleftrightarrow{EF}$ is

$$m = \frac{y_2 - y_1}{x_2 - x_1} = \frac{-2 - 2}{-3 - 2} = \frac{-4}{-5} = \frac{4}{5} = m\overleftrightarrow{EF}.$$

Recall that the slope of $\overleftrightarrow{CD}$ is $-\frac{5}{4}$. How does this compare with the slope of $\overleftrightarrow{EF}$? They are negative reciprocals of each other. We have shown for these two particular perpendicular lines that the slope of one line is the negative reciprocal of the slope of the other line. It also can be shown that, in general, if lines are perpendicular, then their slopes are negative reciprocals of each other. Also, if lines have slopes that are negative reciprocals of each other, then the lines are perpendicular. Are there any exceptions to this? The answer is yes. We must exclude the set of perpendicular lines where one line is vertical and the other is horizontal. The slope of a horizontal line is 0 and the slope of a vertical line is undefined — a vertical line has no slope. Therefore, we cannot apply the above rule to horizontal and vertical lines, even though they are perpendicular.

Relations and Functions 435

EXAMPLE 20

The coordinates of the vertices of a quadrilateral are: $A(-2, 2)$, $B(2, 6)$, $C(6, 4)$ and $D(1, -1)$. Determine which sides of the quadrilateral are

    a. parallel (if any).   b. perpendicular (if any).

Solution

First we plot the points on the Cartesian plane and connect them to obtain the appropriate diagram. See Figure 24.

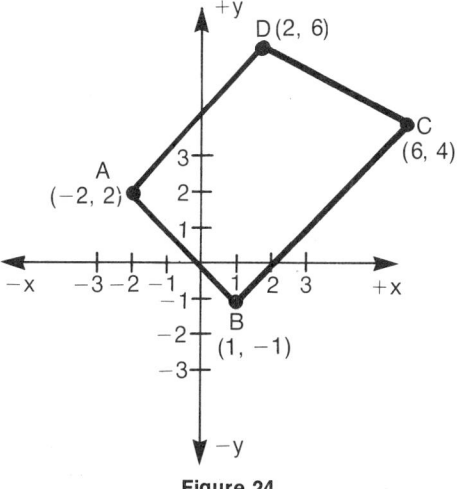

Figure 24

Next, we find the slopes of each side.

$$m\overline{AB} = \frac{y_2 - y_1}{x_2 - y_1} = \frac{-1 - 2}{1 - (-2)} = \frac{-3}{1 + 2} = \frac{-3}{3} = -1.$$

$$m\overline{BC} = \frac{y_2 - y_1}{x_2 - x_1} = \frac{4 - (-1)}{6 - 1} = \frac{4 + 1}{5} = \frac{5}{5} = 1.$$

$$m\overline{CD} = \frac{y_2 - y_1}{x_2 - x_1} = \frac{6 - 4}{2 - 6} = \frac{2}{-4} = -\frac{1}{2}.$$

$$m\overline{DA} = \frac{y_2 - y_1}{x_2 - x_1} = \frac{2 - 6}{-2 - 2} = \frac{-4}{-4} = 1.$$

    a. Sides BC and DA have the same slope, and lines that have the same slope are parallel. Therefore $\overline{BC}$ is parallel to $\overline{DA}$. (Note: we can also denote this by $\overline{BC} \parallel \overline{DA}$.)

b. The slope of side AB is −1 and the slope of side BC is 1. If lines have slopes that are negative reciprocals of each other, then the lines are perpendicular. Therefore, $\overline{AB}$ is perpendicular to $\overline{BC}$. Also, a straight line perpendicular to one of two parallel lines is perpendicular to the other. Since $\overline{BC} \parallel \overline{DA}$, we can also state that $\overline{AB}$ is perpendicular to $\overline{DA}$. (Note: we can also denote this by $\overline{AB} \perp \overline{DA}$ and $\overline{AB} \perp \overline{BC}$.)

EXERCISES FOR SECTION 8.4

(Answers given to exercises marked *)

Find the slope of the line containing the indicated pair of points.

*1. A(4, 2) and B(3, 4)
2. C(−1, 2) and D(3, 5)
*3. E(1, −2) and F(2, 3)
4. G(2, −3) and H(5, −2)
*5. I(−2, −3) and J(6, −1)
6. K(1, 0) and L(−2, 3)
*7. M(−2, 5) and N(−2, −3)
8. O(7, −1) and P(−2, −1)
*9. Q(3, −3) and R(−3, −3)
10. S(−3, 4) and T(2, −3)
*11. U(−3, −4) and V(−2, −1)
12. W(5, −3) and X(5, 7)

By means of the formula for the slope of a line, determine whether the three given points are collinear or not, that is, do the three points lie on the same line?

*13. A(1, 1), B(6, −2) and C(11, −5)
14. D(−1, −1), E(0, 0) and F(2, 3)
*15. G(1, −3), H(−1, −1) and I(−3, 2)
16. J(0, −3), K(1, −1) and L(3, 3)
*17. M(−1, 3), N(0, 1) and O(1, −1)
18. P(−1, −5), Q(0, −2) and R(2, 4)
*19. S(−2, −3), T(1, −2) and U(2, −1)
20. Given: A(0, −4), B(4, 4) and C(5, 6). Find the slope line segments AB and AC. What is your conclusion regarding the points A, B and C?
*21. The vertices of quadrilateral STUV are S(a, b), T(0, 0), U(c, 0) and V(a + c, b), in which a, b, and c are positive numbers. Find the slopes

Relations and Functions

of sides ST and UV. What is your conclusion regarding these two sides?

22. The vertices of quadrilateral PQRS are P(0, 0), Q(6a, 3b), R(3a, 4b) and S(a, 3b).

   a. Find the slopes of sides PQ and RS.
   b. What is your conclusion regarding these two sides?
   c. Is side PS parallel to side QR? Why?

*23. Sides PQ and SR of quadrilateral PQRS are parallel to each other. The coordinates of the vertices are: P(0, 0), Q(x, 5), R(7, −1), and S(x, −3).

   a. Find the slopes of sides PQ and SR (express in terms of x).
   b. Write an equation that can be used to solve for x and solve the equation for x.

24. Sides AB and DC of quadrilateral ABCD are parallel to each other. The coordinates of the vertices are: A(7, −3), B(2x, 2), C(x, 12) and D(3, 2).

   a. Find the slopes of sides AB and DC (express in terms of x).
   b. Write an equation that can be used to solve for x and solve the equation for x.

*25. The coordinates of the vertices of triangle ABC are A(0, −5), B(3, 4), and C(−3, −4). After determining the slopes of the sides of the triangle, show that triangle ABC is a right triangle.

26. A trapezoid is a quadrilateral that has two and only two sides parallel. The vertices of quadrilateral ABCD are A(a, 2b), B(4a, 4b), C(9a, 4b) and D(6a, 2b). After determining the slopes of sides AB, BC, CD, and DA, show that quadrilateral ABCD is *not* a trapezoid.

8.5 THE EQUATION OF A STRAIGHT LINE

Given any two points on a line, we can find the slope of that line. For example, if a line contains the points A(2, 1) and B(4, 4), then the slope of that line is $m = \frac{y_2 - y_1}{x_2 - x_1} = \frac{4 - 1}{4 - 2} = \frac{3}{2}$. But consider the line pictured in Figure 25. We note that it contains C(2, 1) and D(x, y). The values of x and y in point D are not given. We can still use the formula for the slope of a line to find this slope, but it will be expressed in terms of x and y. That is,

Figure 25

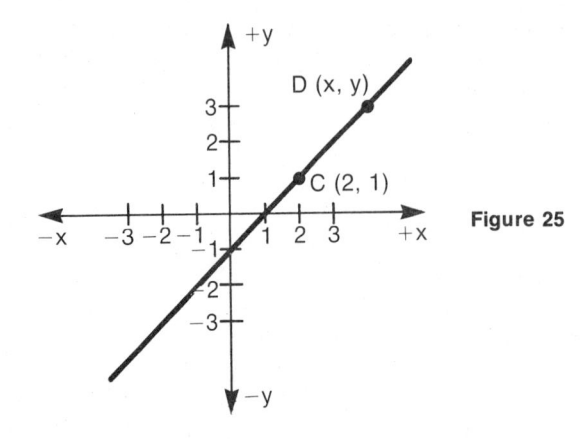

$m = \dfrac{y-1}{x-2}$. If we eliminate fractions from this equation, we have $m(x-2) = y - 1$, or we can rewrite the equation as $y - 1 = m(x - 2)$. Reviewing the process, we have the points $C(2, 1)$ and $D(x, y)$, which lie on the same straight line in Figure 25. The slope of the line is: $m = \dfrac{y-1}{x-2}$, and we rewrote the equation as $y - 1 = m(x - 2)$. This equation also represents the equation of the line pictured in Figure 25. Instead of a particular point such as (2, 1), we could have chosen some other point (different from D) such as $P(x_1, y_1)$ and then we would have the general equation

$$y - y_1 = m(x - x_1).$$

This equation is true whenever the points (x, y) and (x_1, y_1) lie on the same straight line. It is even true when (x, y) and (x_1, y_1) are the same point because then $y - y_1$ and $x - x_1$ are both 0, and $0 = 0$. What is the significance of the equation $y - y_1 = m(x - x_1)$? It is a general equation of the line. This formula can be used to find the equation of a line when we are given a point on the line and the slope of the line.

> $y - y_1 = m(x - x_1)$ is known as the *point-slope* formula for the equation of a line.

We shall use the point-slope formula to find the equation of a line. Consider the straight line whose slope is 3, which passes through the point (5, 2). This is the information we need to find the equation of the line. Using the point-slope formula, $y - y_1 = m(x - x_1)$, we have $m = 3$ and $x_1 = 5$, $y_1 = 2$. Therefore, we have

$y - y_1 = m(x - x_1)$.   point-slope formula
$y - 2 = 3(x - 5)$.   substituting for m, x_1, y_1

Relations and Functions

$$y - 2 = 3x - 15. \quad \text{simplifying}$$
$$y = 3x - 13. \quad \text{the desired equation}$$

The point-slope formula for the equation of a line can be used to find the equation of a line whenever we are given the slope of the line and a point on the line.

EXAMPLE 21

Find the equation of the line that passes through the point $(-3, 2)$ and has a slope of 2.

Solution

Using the point-slope formula $y - y_1 = m(x - x_1)$, we have $m = 2$ and $x_1 = -3$, $y_1 = 2$. Therefore,

$$y - 2 = 2(x - (-3)).$$
$$y - 2 = 2(x + 3).$$
$$y - 2 = 2x + 6.$$
$$y = 2x + 8.$$

EXAMPLE 22

Find the equation of the line that passes through the point $(-2, -4)$ and has a slope of $-\frac{1}{2}$.

Solution

$$y - y_1 = m(x - x_1). \qquad m = -\frac{1}{2},\ x_1 = -2,\ y_1 = -4.$$

$$y - (-4) = -\frac{1}{2}(x - (-2)).$$

$$y + 4 = -\frac{1}{2}(x + 2).$$

$$y + 4 = -\frac{1}{2} \cdot x - \frac{1}{2} \cdot 2.$$

$$y + 4 = -\frac{1}{2}x - 1.$$

$$y = -\frac{1}{2}x - 5.$$

We can also use the point-slope formula to find the equation of a line when we are given only two points that lie on the line, and not the slope. For example, if the points A(3, 2) and B(5, 6) lie on the same straight line, we can find the equation of the line. But first we must find the slope of the line. Using the formula for slope, we have

$$m = \frac{y_2 - y_1}{x_2 - x_1} = \frac{6 - 2}{5 - 3} = \frac{4}{2} = 2.$$

The slope of the line passing through A and B is 2, and now we can use either point, A or B, and the point-slope formula to find the equation of the line. If we choose A, then $m = 2$, $x_1 = 3$, $y_1 = 2$ and we have

$$y - y_1 = m(x - x_1).$$
$$y - 2 = 2(x - 3).$$
$$y - 2 = 2x - 6.$$
$$y = 2x - 4.$$

Since B also lies on the same straight line, we will get the same equation if we choose it. In this case $m = 2$, $x_1 = 5$, $y_1 = 6$ and we have

$$y - y_1 = m(x - x_1).$$
$$y - 6 = 2(x - 5).$$
$$y - 6 = 2x - 10.$$
$$y = 2x - 4.$$

We obtain the same equation that resulted when we used point A. This must be the case because both points lie on the line and hence, must satisfy the equation. Which point to use in determining the equation is a matter of personal preference.

EXAMPLE 23

Find the equation of the line that passes through the points C(3, 6) and D(7, 2).

Solution

Since we are not given the slope of the line, we must determine it.

$$m = \frac{y_2 - y_1}{x_2 - x_1} = \frac{6 - 2}{3 - 7} = \frac{4}{-4} = -1.$$

Next, we use the point-slope formula. Choosing point C, we have $m = -1$, $x_1 = 3$, $y_1 = 6$.

Relations and Functions

$$y - y_1 = m(x - x_1).$$
$$y - 6 = -1(x - 3).$$
$$y - 6 = -1x + 3.$$
$$y = -x + 9.$$

EXAMPLE 24

Find the equation of the line that passes through the points $E(-2, -2)$ and $F(-4, -1)$.

Solution

The slope of the line is

$$m = \frac{y_2 - y_1}{x_2 - x_1} = \frac{-2 - (-1)}{-2 - (-4)} = \frac{-2 + 1}{-2 + 4} = \frac{-1}{2} = -\frac{1}{2}.$$

Using the point-slope formula and choosing point E, we have $m = -\frac{1}{2}$, $x_1 = -2$, $y_1 = -2$.

$$y - y_1 = m(x - x_1).$$

$$y - (-2) = -\frac{1}{2}(x - (-2)).$$

$$y + 2 = -\frac{1}{2}(x + 2).$$

$$y + 2 = -\frac{1}{2}x - 1.$$

$$y = -\frac{1}{2}x - 3.$$

If we had chosen F instead, we have $m = -\frac{1}{2}$, $x_1 = -4$, $y_1 = -1$ and

$$y - y_1 = m(x - x_1).$$

$$y - (-1) = -\frac{1}{2}(x - (-4)).$$

$$y + 1 = -\frac{1}{2}(x + 4).$$

$$y + 1 = -\frac{1}{2}x - 2.$$

$$y = -\frac{1}{2}x - 3,$$ which is the same equation obtained with point E.

EXAMPLE 25

Find the equation of the line that passes through the points G(0, 3) and H(−1, 5).

Solution

The slope of the line is

$$m = \frac{y_2 - y_1}{x_2 - x_1} = \frac{5 - 3}{-1 - 0} = \frac{2}{-1} = -2.$$

Using the point-slope formula and choosing point G, we have $m = -2$, $x_1 = 0$, $y_1 = 3$.

$$y - y_1 = m(x - x_1).$$
$$y - 3 = -2(x - 0).$$
$$y - 3 = -2x + 0.$$
$$y = -2x + 3.$$

Note that point G(0, 3) is the y-intercept in Example 25. Also note that the slope of the line is −2 and the equation of the line is $y = -2x + 3$. For this particular example, the coefficient of x, (−2), is the slope of the line and the constant term, (+3), is the y-intercept. This is not a coincidence. Let's examine a more general case. Suppose a line whose coordinates are (0, b) crosses the y-axis at point B and the slope of the line is m. Using the point-slope formula, we can find the equation of this line,

$$y - y_1 = m(x - x_1).$$
$$y - b = m(x - 0).$$
$$y - b = mx - 0.$$
$$y = mx + b.$$

The equation of the line is $y = mx + b$. The coefficient x, (m), is the slope of the line and the constant term, (+b), is the y-intercept. Is this always the case? Let's consider another example. Suppose the slope of a line is m, but this time the line passes through some point other than the y-intercept. Consider point P, whose coordinates are (x_1, y_1). Using the point-slope formula, we have

$$y - y_1 = m(x - x_1).$$
$$y - y_1 = mx - mx_1.$$
$$y = mx - mx_1 + y_1.$$

Relations and Functions

Recall that the slope of a line is constant. Hence, m is a constant. Also, x_1 and y_1 are constants. Therefore, mx_1 is a constant, and $-mx_1 + y_1$ is a constant. We shall represent this sum by the letter b. The equation

$$y = mx - mx_1 + y_1$$
becomes $$y = mx + b.$$

The y-intercept of a line always has an abscissa whose x value is 0. If we substitute that value, $x = 0$, in the above equation, we obtain $y = m(0) + b$, or $y = b$. Therefore the y-intercept of the line is $(0, b)$. Again, we have the equation of a line, $y = mx + b$, where the coefficient of x is the slope of the line and the constant term is the y-intercept. Whenever an equation of a line is of this form, these conditions will hold. This equation, $y = mx + b$, is known as the *slope-intercept* form for the equation of a line.

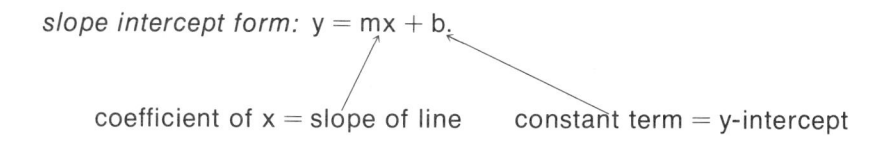

slope intercept form: $y = mx + b.$

coefficient of x = slope of line   constant term = y-intercept

---

EXAMPLE 26

Find the slope and y-intercept of the line whose equation is:

a. $y = 2x + 3.$
b. $y = -2x - 7.$
c. $2x - 5y = 10.$
d. $3x - 2y = 0.$

Solution

a. The equation $y = 2x + 3$ is of the general form $y = mx + b$. The coefficient of x is the slope of the line and the constant term is the y-intercept. Therefore, $m = 2$ and the y-intercept is $(0, 3)$.

b. $y = -2x - 7$, slope $= m = -2$, y-intercept is $(0, -7)$.

c. The equation $2x - 5y = 10$ is not of the general form $y = mx + b$. Therefore, we must transform the given equation into the general form. This is done by solving the given equation for y,

$$2x - 5y = 10.$$
$$-5y = -2x + 10.$$
$$y = \frac{2}{5}x - 2.$$

The equation is now of the general form, and slope $= m = \frac{2}{5}$, y-intercept is $(0, -2)$.

d. The equation $3x - 2y = 0$ is not of the general form $y = mx + b$. Solving it for y, we have

$$3x - 2y = 0.$$
$$-2y = -3x.$$
$$y = \frac{3}{2}x.$$

Note that we can also express the equation $y = \frac{3}{2}x$ as $y = \frac{3}{2}x + 0$. Therefore, slope $= m = \frac{3}{2}$ and the y-intercept $=$ (0, 0). Since the y-intercept is (0, 0) we also know that the line passes through the origin.

We can also use the slope-intercept form for the equation of a line to graph the line. Consider the line whose equation is $y = \frac{2}{3}x + 2$. Since the equation is of the general form, we note that the slope of the line is $\frac{2}{3}$ and the y-intercept is (0, 2). To graph the line, we locate the intercept and then sketch a triangle for the slope. Since the slope is $\frac{2}{3}$ and we can think of slope as rise over run, $\frac{2}{3} = \frac{\text{rise}}{\text{run}}$. We start at the point (0, 2) and move in a positive x direction three units (for the run) and then up in a positive y direction two units (for the rise) which brings us to the point (3, 4). Note that (3, 4) also satisfies the equation $y = \frac{2}{3}x + 2$. Since two points determine a line, we can graph the line as shown in Figure 26.

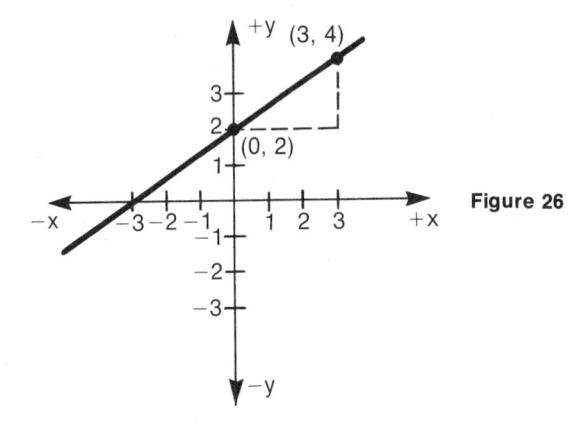

Figure 26

Relations and Functions   445

EXAMPLE 27

Draw graphs of each of the following equations:

a. $y = 3x + 1$.   b. $y = -\frac{1}{2}x - 1$.   c. $2x - 3y = 6$.

Solution

a. From the given equation, $y = 3x + 1$, we note that the slope is 3 and the y-intercept is (0, 1). A slope of 3 can also be expressed as $\frac{3}{1}$. We start at (0, 1) and move in a positive x direction one unit (run) and then up in a positive y direction three units (rise) which brings us to the point (1, 4), enabling us to graph the line. See Figure 27. Note that a line with a positive slope slants upward to the right.

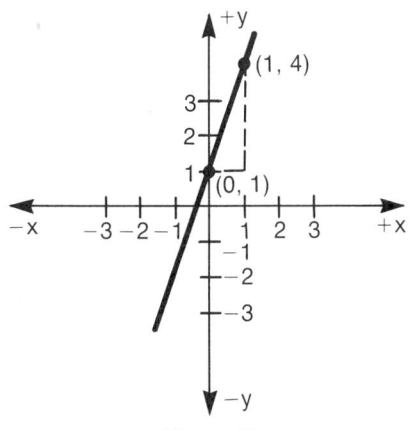

Figure 27

b. The same procedure is used to graph $y = -\frac{1}{2}x - 1$. The y-intercept is (0, −1). Note that the slope is $-\frac{1}{2}$. Therefore, we move in a positive x direction two units from (0, −1) and then down (−) in a negative y direction one unit. We also can determine a point by moving in a negative x direction two units from (0, −1) and then up in a positive y direction one unit, which enables us to graph the line. See Figure 28. Note that a line with a negative slope slants downward to the right.

c. Before proceeding, we transform $2x - 3y = 6$ into the general form $y = mx + b$,

$$2x - 3y = 6.$$
$$-3y = -2x + 6.$$
$$y = \frac{2}{3}x - 2.$$

Figure 28

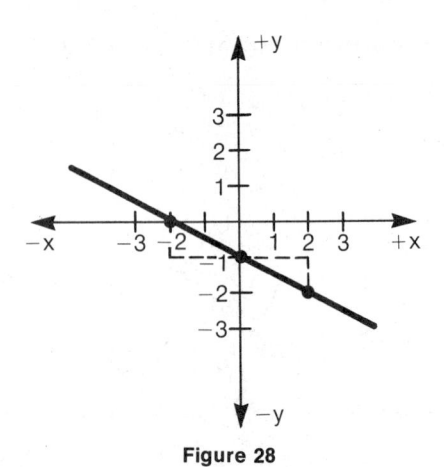

The slope is $\frac{2}{3}$ and the y-intercept is $(0, -2)$. From the point $(0, -2)$, we move in a positive x direction three units (run) and then up in a positive y direction two units (rise), which enables us to graph the line. See Figure 29.

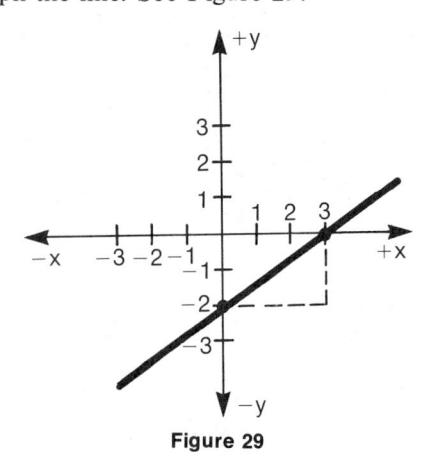

Figure 29

We shall not use the slope-intercept method to graph horizontal or vertical lines. In fact, we cannot use it to graph a vertical line since the slope of a vertical line is undefined and it has no y-intercept.* Recall (from Section 8.3) that the graph of any equation of the form $x = c$ is a vertical line parallel to the y-axis and passing through the point $(c, 0)$. The graph of any equation of the form $y = c$ is a horizontal line parallel to the x-axis and passing through the point $(0, c)$, which is the y-intercept. A horizontal line has a slope of 0.

---
*Except $x = 0$ (the y-axis), where every point on the y-axis is a y-intercept.

Relations and Functions

When we are given any two points that lie on a line, we can find the equation of the line by first finding the slope of the line and then substituting the coordinates of either point in the *point-slope formula*. But it should be noted that we can also use the *slope-intercept formula* to find the equation of a line when we are given two points. Let's find the equation of the line that passes through the points A(2, 4) and B(6, 2). First, we shall use the point-slope formula to find the equation of the line.

$$m = \frac{y_2 - y_1}{x_2 - x_1} = \frac{2 - 4}{6 - 2} = \frac{-2}{4} = -\frac{1}{2}.$$

Using the point-slope formula and choosing point A, we have $m = -\frac{1}{2}$, $x_1 = 2$, $y_1 = 4$.

$$y - y_1 = m(x - x_1).$$

$$y - 4 = -\frac{1}{2}(x - 2).$$

$$y - 4 = -\frac{1}{2} \cdot x + 1.$$

$$y = -\frac{1}{2}x + 5.$$

Now, we shall use the slope-intercept formula, $y = mx + b$, to obtain the same equation. We first determine the slope in the same manner as before. Therefore, $m = -\frac{1}{2}$; $y = mx + b$ becomes $y = -\frac{1}{2}x + b$. Now we must find a value for b and the equation will be complete. Since points A(2, 4) and B(6, 2) both lie on the line, their coordinates must satisfy the equation of the line. Hence, we substitute the coordinates of A or B in the equation $y = -\frac{1}{2}x + b$ and solve it for b. Choosing point A, we have $x = 2$ and $y = 4$. Therefore,

$$y = -\frac{1}{2}x + b.$$

$$4 = -\frac{1}{2} \cdot 2 + b. \quad \text{substituting 4 for y and 2 for x}$$

$$4 = -1 + b.$$
$$5 = b.$$

We have found the value of b and we can substitute it in the equation $y = -\frac{1}{2}x + b$. The equation of the line is

$$y = -\frac{1}{2}x + 5$$

which is precisely the same equation we obtained using the point-slope formula. We are not advocating one technique over the other, but you should be aware that there is more than one method for finding the equation of a line.

EXAMPLE 28

Using the slope-intercept formula, find the equation of the line that passes through the points A(3, 6) and B(7, 2).

Solution

We first determine the slope of the line:

$$m = \frac{y_2 - y_1}{x_2 - x_1} = \frac{2 - 6}{7 - 3} = \frac{-4}{4} = -1.$$

Using the slope-intercept formula, we have

$$y = mx + b \quad \text{and since } m = -1,$$
$$y = -1x + b.$$

Next, solve for b by substituting for x and y, using point A or B. If we choose B, we have x = 7, y = 2:

$$y = -1x + b.$$
$$2 = -1 \cdot 7 + b.$$
$$2 = -7 + b.$$
$$9 = b.$$

Therefore, the equation of the line is: $y = -x + 9$. Note that b would still equal 9 if we had chosen A: x = 3, y = 6.

$$y = -1 \cdot x + b.$$
$$6 = -1 \cdot 3 + b.$$
$$6 = -3 + b.$$
$$9 = b.$$

EXAMPLE 29

Using the slope-intercept formula, find the equation of the line that passes through the points C(−2, −2), and D(−4, −1).

Relations and Functions

Solution

We first determine the slope of the line.

$$m = \frac{y_2 - y_1}{x_2 - x_1} = \frac{-1 - (-2)}{-4 - (-2)} = \frac{-1 + 2}{-4 + 2} = \frac{1}{-2} = -\frac{1}{2}.$$

Hence, $y = mx + b$ becomes $y = -\frac{1}{2}x + b$.

Choosing point $D(-4, -1)$ and substituting for x and y, we have

$$y = -\frac{1}{2}x + b.$$

$$-1 = -\frac{1}{2}(-4) + b.$$

$$-1 = 2 + b.$$
$$-3 = b.$$

Therefore, the equation of the line is: $y = -\frac{1}{2}x - 3$.

A straight line is determined by two points, and if we are given two points that the line passes through, then we can determine the equation of the line. We can do this by means of the point-slope formula, $y - y_1 = m(x - x_1)$, or the slope-intercept formula, $y = mx + b$. If we are given the equation of a line, then we can draw a graph of the line by first solving the equation for y, so that it is in the form of $y = mx + b$. From this equation, we locate the y-intercept which is (0, b), then sketch a triangle from the intercept using the slope and the idea that slope $= m = \frac{\text{rise}}{\text{run}}$. This enables us to locate another point that lies on the line and since two points determine a line, we can graph the line.

EXERCISES FOR SECTION 8.5

(Answers given to exercises marked *)

Write the equation of the straight line that passes through the indicated point and has the indicated slope.

*1. $A(4, 2)$, $m = 3$.   2. $B(-2, 3)$, $m = -2$.

*3. $C(-2, -5)$, $m = -1$. 4. $D(-3, 7)$, $m = \frac{1}{2}$.

*5. $E(-3, 0)$, $m = \frac{2}{3}$. 6. $F(0, -2)$, $m = \frac{3}{2}$.

*7. $G(-3, -4)$, $m = -\frac{2}{5}$. 8. $H(-5, -2)$, $m = -\frac{3}{2}$.

*9. $I(0, -3)$, $m = -\frac{2}{3}$. 10. $J(4, 4)$, $m = \frac{1}{4}$.

Write the equation of the straight line that passes through the indicated points.

*11. $A(3, 2)$ and $B(4, 7)$ 12. $C(-2, 3)$ and $D(3, -2)$
*13. $E(-2, -3)$ and $F(-4, -5)$ 14. $G(-2, 0)$ and $H(0, -3)$
*15. $I(3, -1)$ and $J(4, -2)$ 16. $K(5, -2)$ and $L(6, 7)$
*17. $M(3, -3)$ and $N(-2, -2)$ 18. $O(2, 3)$ and $P(-1, 1)$
*19. $Q(-2, -2)$ and $R(-4, -7)$ 20. $S(-2, -3)$ and $T(-7, 0)$

Find the slope and y-intercept of the line whose equation is given and then graph the line.

*21. $y = 2x - 1$. 22. $y = -2x + 3$.
*23. $y = x + 2$. 24. $y = -x + 1$.
*25. $x + 2y = 4$. 26. $3x - 2y = 6$.
*27. $4x - 3y + 3 = 0$. 28. $-2x - 3y = 6$.
*29. $x + y = 0$. 30. $x - y = 0$.

*31. Determine the equation of the line passing through $(1, 1)$ and parallel to the line whose equation is $2x + 3y = 1$.

32. Determine the equation of the line passing through $(-2, -3)$ and parallel to the line whose equation is $2x - 5y = 6$.

*33. Determine the equation of the line passing through $(-3, -4)$ and parallel to the line whose equation is $x + 2y = 3$.

34. Determine the equation of the line passing through $(-2, 0)$ and perpendicular to the line whose equation is $2x + 3y = 1$.

*35. Determine the equation of the line passing through $(-1, -1)$ and perpendicular to the line whose equation is $x + 2y = 6$.

Relations and Functions

36. Determine the equation of the line passing through $(-2, -3)$ and perpendicular to the line whose equation is $x - y = 0$.

*37. Given the points $A(-3, 4)$, $B(3, 4)$ and $C(0, 0)$:
   a. What is the slope of $\overline{AB}$?
   b. Determine the equation of the straight line which is parallel to $\overline{AB}$ and which passes through the point $(5, 2)$.
   c. Determine the equation of the line that passes through A and C.
   d. Determine the equation of the line that passes through B and C.

38. Given the points $P(1, -1)$, $Q(4, 2)$ and $R(6, 4)$:
   a. Find the slopes of line segments $\overline{PQ}$ and $\overline{PR}$.
   b. From your answers to part a, what can you conclude regarding the points P, Q, and R?
   c. Determine the equation of the line that passes through P and R.
   d. Does point $S(10, 8)$ lie on the line PR?
   e. Find the coordinates of the point where line PR crosses the x-axis.

*39. Given that quadrilateral ABCD is a rectangle whose vertices are $A(-2, 2)$, $B(2, 6)$, $C(5, 3)$ and $D(1, -1)$:
   a. Determine the equation of side AB.
   b. Determine the equation of side AD.
   c. Determine the equation of diagonal BD.

40. Given the points $A(3, 2)$, $B(-2, -3)$, and $C(3, 0)$:
   a. Determine the equation of the line that is parallel to the x-axis and passes through A.
   b. Determine the equation of the line that is parallel to the y-axis and passes through B.
   c. Determine the equation of the line that is parallel to line $\overleftrightarrow{AB}$ and passes through C.
   d. Determine the equation of the line that is perpendicular to line $\overleftrightarrow{AB}$ and passes through C.
   e. Determine the equation of the line that is perpendicular to the x-axis and passes through C.

8.6 THE DISTANCE AND MIDPOINT FORMULAS

Many times we want to find the distance between two given points on the Cartesian plane. If the points are on the same vertical or horizontal line, we can simply count the number of units between the two given points. For example, in Figure 30 points A and B are on the same hori-

A (−2, 3)   B (5, 3)

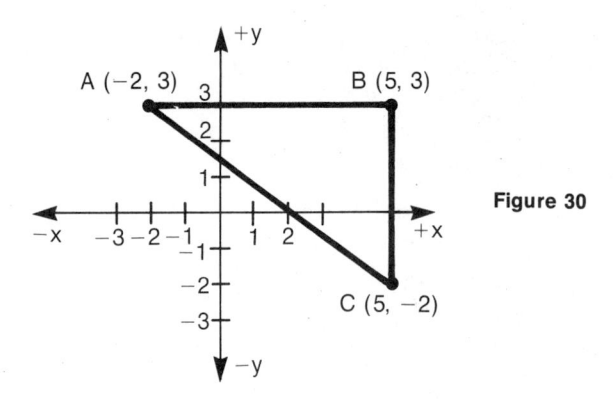

Figure 30

C (5, −2)

zontal line and points B and C are on the same vertical line. Note that the distance between A and B is seven units, while the distance between B and C is five units. We obtain these measures by counting the number of units between the given points. Note that if we regard A as the first point, and B as the second point, then $x_1 = -2$ and $x_2 = 5$, and the distance between these two points (A and B) is $x_2 - x_1 = 5 - (-2) = 7$, which is also the length of the horizontal line segment $\overline{AB}$. Similarly, if we regard C as the first point and B as the second point, then $y_1 = -2$ and $y_2 = 3$, and the distance between these points (B and C) is $y_2 - y_1 = 3 - (-2) = 3 + 2 = 5$, which is also the length of the vertical line segment $\overline{BC}$. The purpose of this discussion is to point out the fact that the length of a horizontal line segment can be determined by finding the difference of the abscissa, that is, $x_2 - x_1$. If the resulting value is negative, then we take the absolute value of it, since length cannot be negative. Similarly, the length of a vertical line segment can be determined by finding the difference of the ordinates, $y_2 - y_1$. If the resulting value is negative, then we take the absolute value of it, since length cannot be negative.

We can now find the lengths of line segments $\overline{AB}$ and $\overline{BC}$ in Figure 30. But what is the length of the line segment $\overline{AC}$? We cannot simply count the number of squares between A and C, nor can we subtract the

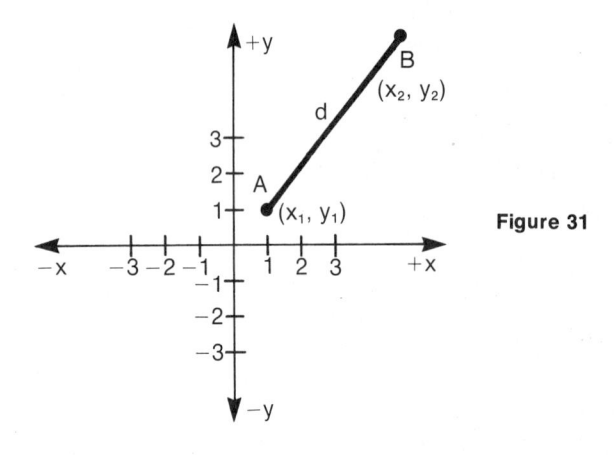

Figure 31

Relations and Functions

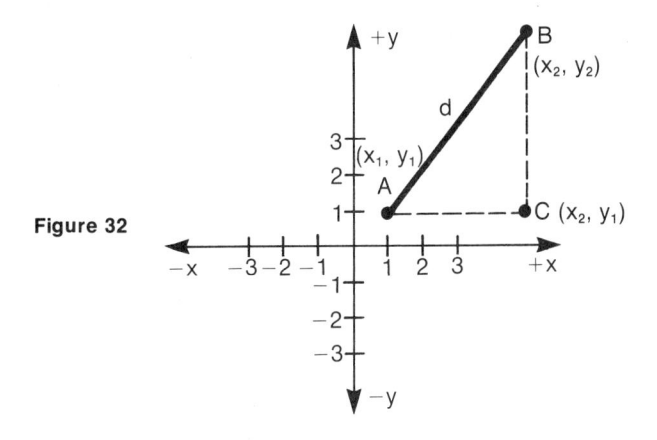

Figure 32

corresponding x or y values. That technique can only be used for horizontal or vertical lines. Line segment $\overline{AC}$ is an oblique line and, therefore, we must develop a method for finding the distance between any two points. Consider the line segment pictured in Figure 31. It is an oblique line segment whose end points are $A(x_1, y_1)$ and $B(x_2, y_2)$. We want to find its length—let's call it d. We can find d if we make it the hypotenuse of a right triangle. This can be done by constructing a line parallel to the x-axis passing through A, and constructing another line parallel to the y-axis passing through B. These lines will meet at a point (call it C) whose coordinates are (x_2, y_1). We now have a right triangle whose hypotenuse is $\overline{AB}$. See Figure 32.

Note that the length of the horizontal line segment $\overline{AC}$ is $x_2 - x_1$ and the length of the vertical segment $\overline{BC}$ is $y_2 - y_1$. Since triangle ABC is a right triangle, to find the length of the hypotenuse we can use the Pythagorean Theorem: The square of the hypotenuse of a right triangle is equal to the sum of the squares of the legs.* The theorem refers to the lengths of the sides of the right triangle. In Figure 32, triangle ABC is a right triangle. The length of side AC is $x_2 - x_1$, the length of side BC is $y_2 - y_1$, and we have indicated the length of the hypotenuse AB to be d. Therefore, we have

$$d^2 = \underbrace{(x_2 - x_1)^2 + (y_2 - y_1)^2}.$$

square of the hypotenuse   sum of the squares of the legs

Since we want to find d (the length of $\overline{AB}$ in Figure 32), we take the square root of both sides, and we have

$$\boxed{d = \sqrt{(x_2 - x_1)^2 + (y_2 - y_1)^2}.}$$

---

*It is called the Pythagorean Theorem since its proof is supposed to have been discovered by Pythagoras (approximately 550 B.C.), a famous Greek mathematician.

Figure 33

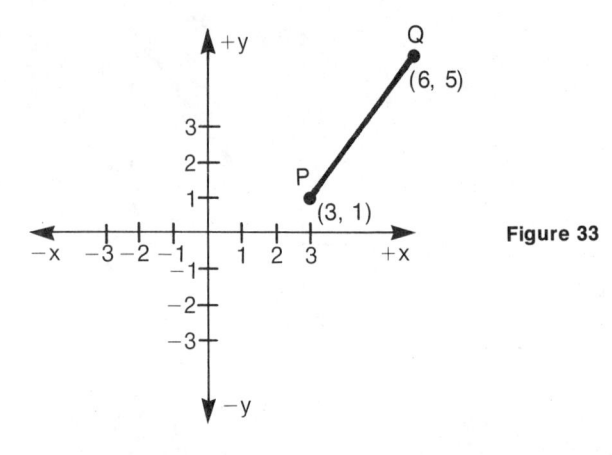

This equation or formula is called the *distance formula.* It is a general formula that works in all cases and we can use it to find the distance between any two points on the Cartesian plane. For example, consider the line segment pictured in Figure 33. The coordinates of P are (3, 1) and the coordinates of Q are (6, 5). To find the length of the line segment $\overline{PQ}$, we can use the distance formula. Let P be the first point and Q be the second point, then $x_1 = 3$, $y_1 = 1$ and $x_2 = 6$, $y_2 = 5$. Therefore we have

$$d = \sqrt{(x_2 - x_1)^2 + (y_2 - y_1)^2}.$$
$$d = \sqrt{(6 - 3)^2 + (5 - 1)^2}.$$
$$d = \sqrt{3^2 + 4^2}.$$
$$d = \sqrt{9 + 16}.$$
$$d = \sqrt{25}.$$
$$d = 5.$$

The length of line segment $\overline{PQ}$ is five units. We can also say that the distance between P and Q is five units. It does not matter which point is considered to be the first point and which is considered to be the second point. But note that the coordinates of the first point are (x_1, y_1) and the coordinates of the second point are (x_2, y_2). If we consider Q as the first point and P as the second point, then $x_1 = 6$, $y_1 = 5$ and $x_2 = 3$, $y_2 = 1$ and we have

$$d = \sqrt{(x_2 - x_1)^2 + (y_2 - y_1)^2}.$$
$$d = \sqrt{(3 - 6)^2 + (1 - 5)^2}.$$
$$d = \sqrt{(-3)^2 + (-4)^2}.$$
$$d = \sqrt{9 + 16}.$$
$$d = \sqrt{25}.$$
$$d = 5.$$

Relations and Functions

The distance formula provides us with a method for finding the distance between any two points on the Cartesian plane. The general formula is
$$d = \sqrt{(x_2 - x_1)^2 + (y_2 - y_1)^2}.$$

EXAMPLE 30

Find the length of the line segment joining the points whose coordinates are $A(0, 2)$ and $B(-4, 5)$.

Solution

Let A be the first point and B be the second point, then $x_1 = 0$, $y_1 = 2$ and $x_2 = -4$, $y_2 = 5$. The length of the line segment is

$$d = \sqrt{(x_2 - x_1)^2 + (y_2 - y_1)^2}.$$
$$d = \sqrt{(-4 - 0)^2 + (5 - 2)^2}.$$
$$d = \sqrt{(-4)^2 + (3)^2}.$$
$$d = \sqrt{16 + 9} = \sqrt{25}.$$
$$d = 5.$$

EXAMPLE 31

Find the length of the line segment joining the points whose coordinates are $C(-1, 3)$ and $D(2, 5)$.

Solution

Let C be the first point and D be the second point, then $x_1 = -1$, $y_1 = 3$ and $x_2 = 2$, $y_2 = 5$. The length of the line segment is

$$d = \sqrt{(x_2 - x_1)^2 + (y_2 - y_1)^2}.$$
$$d = \sqrt{(2 - (-1))^2 + (5 - 3)^2} = \sqrt{(2 + 1)^2 + 2^2}.$$
$$d = \sqrt{3^2 + 2^2} = \sqrt{9 + 4}.$$
$$d = \sqrt{13}.$$

Note that our answer is left in radical form. If we want a decimal value, then we can use the table in Appendix A and we would have $\sqrt{13} \approx 3.6$.

EXAMPLE 32

Find the distance between the points whose coordinates are $(-1, -1)$ and $(-4, -5)$.

Solution

Let $x_1 = -1$, $y_1 = -1$ and $x_2 = -4$, $y_2 = -5$. The distance between the points is

$$d = \sqrt{(x_2 - x_1)^2 + (y_2 - y_1)^2}.$$
$$d = \sqrt{(-4 - (-1))^2 + (-5 - (-1))^2}.$$
$$d = \sqrt{(-4 + 1)^2 + (-5 + 1)^2}.$$
$$d = \sqrt{(-3)^2 + (-4)^2}.$$
$$d = \sqrt{9 + 16} = \sqrt{25}.$$
$$d = 5.$$

EXAMPLE 33

The coordinates of the center of a circle are $(-3, 1)$. The circle passes through the point whose coordinates are $(5, -2)$. Find the length of the radius of the circle.

Solution

Since we are asked to find the length of the radius, we can use the distance formula. Let $x_1 = -3$, $y_1 = 1$ and $x_2 = 5$, $y_2 = -2$.

$$d = \sqrt{(x_2 - x_1)^2 + (y_2 - y_1)^2}.$$
$$d = \sqrt{(5 - (-3))^2 + (-2 - 1)^2}.$$
$$d = \sqrt{(5 + 3)^2 + (-3)^2} = \sqrt{8^2 + (-3)^2}.$$
$$d = \sqrt{64 + 9}.$$
$$d = \sqrt{73} \approx 8.5.$$

Therefore, the length of the radius is $\sqrt{73}$, or approximately 8.5.

Thus far we have discussed a method for finding the length of a line segment on the Cartesian plane when we are given the coordinates of the end points of the line segment. This also can be considered as finding

Relations and Functions

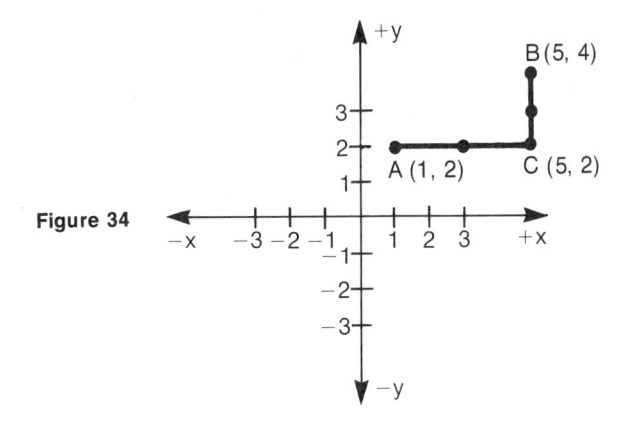

Figure 34

the distance between two points. Many times it is also necessary to find the coordinates of the *midpoint* of a line segment. If the line segment in question is a horizontal or vertical line segment, then it is not too difficult a task to find the midpoint. For example, consider the line segments pictured in Figure 34. Line segment $\overline{AC}$ is a horizontal line segment. Its length is four units, so its midpoint must divide it into two equal segments whose lengths are 2. What point does this? Counting two units from A to C, we see that it is the point whose coordinates are (3, 2). Since the point has to be on $\overline{AC}$, its y coordinate must be the same as A and C, that is, 2. But note that we could have also obtained the abscissa by means of the following computation: $\frac{1+5}{2} = \frac{6}{2} = 3$. This technique can be used in general for any horizontal line segment. That is, if a horizontal line segment (a line segment parallel to the x-axis) has as its end points $A(x_1, y_1)$ and $B(x_2, y_1)$ then its midpoint is $\left(\frac{x_1 + x_2}{2}, y_1\right)$.

A similar technique can be used to find the midpoint of a vertical line segment. Note in Figure 34 that the length of $\overline{BC}$ is 2. Therefore, its midpoint must divide it into two equal segments whose lengths are 1. The point that does this is (5, 3). Since the point has to be on $\overline{BC}$, its x coordinate must be the same as B and C, or 5. Note that the ordinate (3) can be obtained by means of the following computation: $\frac{4+2}{2} = \frac{6}{2} = 3$. This technique can be used in general for any vertical line segment. That is, if a vertical line segment (a line segment parallel to the y-axis) has as its end points $B(x_2, y_1)$ and $C(x_2, y_2)$, then its midpoint is $\left(x_2, \frac{y_1 + y_2}{2}\right)$.

Next, we demonstrate a method for finding the midpoint of an oblique line segment. Consider the line segment pictured in Figure 35. Note that we are considering the general case and we have also included point C to form right triangle ABC. Line segment $\overline{AC}$ is a horizontal line segment and its midpoint is E, whose coordinates are $\left(\frac{x_1 + x_2}{2}, y_1\right)$. Line segment $\overline{BC}$ is a vertical line segment whose midpoint is D and its coordinates are

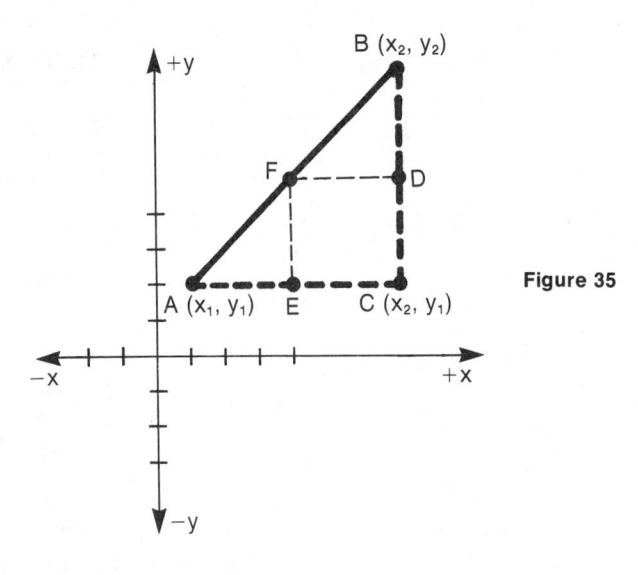

$\left(x_2, \frac{y_1+y_2}{2}\right)$. We have constructed line segment $\overline{FD}$ (in Figure 35) parallel to $\overline{AC}$. What is the ordinate of F? Since $\overline{FD}$ is a horizontal line parallel to $\overline{AC}$, point F must have D at the ordinate, that is, $\frac{y_1+y_2}{2}$. $\overline{FE}$ is parallel to $\overline{BC}$ and it is a vertical line segment. Therefore, point F must have the same abscissa as E, $\frac{x_1+x_2}{2}$. Hence, the coordinates of F are $\left(\frac{x_1+x_2}{2}, \frac{y_1+y_2}{2}\right)$. Is this the midpoint of $\overline{AB}$? Yes, because a line parallel to one side of a triangle and intersecting the other two sides divides these sides proportionally. Whether we consider $\overline{FD}$ or $\overline{FE}$, we have a line parallel to one side of a triangle. Consider $\overline{FD}$—it divides $\overline{BC}$ in half, hence it must also divide $\overline{AB}$ in half. Similarly, $\overline{FE}$ is parallel to $\overline{BC}$ and it divides $\overline{AC}$ in half, so it must divide $\overline{AB}$ in half. The midpoint of $\overline{AB}$ is F and the midpoint of $\overline{AB}$ is $\left(\frac{x_1+x_2}{2}, \frac{y_1+y_2}{2}\right)$.

If we have line segment $\overline{CD}$ whose endpoints are C(6, 5) and D(4, 7), then the midpoint of line segment $\overline{CD}$ is determined by finding one-half the sum of the abscissas and one-half the sum of the ordinates of its end points,

$$\text{midpoint of } \overline{CD} = \left(\frac{6+4}{2}, \frac{5+7}{2}\right) = \left(\frac{10}{2}, \frac{12}{2}\right) = (5, 6).$$

EXAMPLE 34

Given points A(-2, 7) and B(8, -5), find the coordinates of the midpoint of line segment $\overline{AB}$.

Relations and Functions

Solution

Let A be the first point and B be the second point, then $x_1 = -2$, $y_1 = 7$ and $x_2 = 8$, $y_2 = -5$. The midpoint of line segment $\overline{AB}$ is

$$\left(\frac{x_1 + x_2}{2}, \frac{y_1 + y_2}{2}\right)$$

$$\left(\frac{-2 + 8}{2}, \frac{7 + (-5)}{2}\right)$$

$$\left(\frac{6}{2}, \frac{2}{2}\right)$$

$$(3, 1).$$

EXAMPLE 35

The coordinates of the end points of a line segment are $(-3, -6)$ and $(7, 6)$. Find the coordinates of the midpoint of the segment.

Solution

Let $x_1 = -3$, $y_1 = -6$ and $x_2 = 7$, $y_2 = 6$. The midpoint of the segment is

$$\left(\frac{x_1 + x_2}{2}, \frac{y_1 + y_2}{2}\right)$$

$$\left(\frac{-3 + 7}{2}, \frac{-6 + 6}{2}\right)$$

$$\left(\frac{4}{2}, \frac{0}{2}\right)$$

$$(2, 0).$$

EXAMPLE 36

The coordinates of the midpoint of a line segment are $(1, 6)$. If the coordinates of one end point are $(-3, 4)$, find the coordinates of the other end point.

Solution

Again, we have to make use of the midpoint formula, but in a slightly different manner. The midpoint of a line segment is determined by finding one-half of the sum of the abscissas and one-half the sum of the ordinates of its end points. We can rewrite the midpoint formula

$$(x, y) = \left(\frac{x_1 + x_2}{2}, \frac{y_1 + y_2}{2}\right)$$

as $x = \frac{x_1 + x_2}{2}$ and $y = \frac{y_1 + y_2}{2}$. Let $x_1 = -3$, $y_1 = 4$ (an end point). We are finding x_2 and y_2. Therefore, $x = 1$, $y = 6$ and we have the equations

$$1 = \frac{-3 + x_2}{2} \quad \text{and} \quad 6 = \frac{4 + y_2}{2}.$$

Solving each equation,

$$2 = -3 + x_2, \qquad 12 = 4 + y_2,$$
$$5 = x_2, \qquad\qquad 8 = y_2.$$

The end point of the line segment is (5, 8).

EXAMPLE 37

The coordinates of the end points of the diameter of a circle are $(-3, 1)$ and $(-5, -7)$. Find the coordinates of the center of a circle.

Solution

The diameter of a circle passes through the center of the circle and the center of the circle is also the midpoint of the diameter. Therefore, we use the midpoint formula to find the center of the circle.

Let $x_1 = -3$, $y_1 = 1$ and $x_2 = -5$, $y_2 = -7$. The midpoint of the diameter is

$$\left(\frac{x_1 + x_2}{2}, \frac{y_1 + y_2}{2}\right)$$
$$\left(\frac{-3 + (-5)}{2}, \frac{1 + (-7)}{2}\right)$$
$$\left(\frac{-8}{2}, \frac{-6}{2}\right)$$
$$(-4, -3).$$

Relations and Functions

EXERCISES FOR SECTION 8.6

(Answers given to exercises marked *)

Find the distance between the following pairs of points. (Answers may be left in radical form.)

*1. A(4, 3) and B(7, 5)
2. C(6, 2) and D(−2, 6)
*3. E(−2, 5) and F(−8, 2)
4. G(−5, 6) and H(−6, −3)
*5. I(0, −3) and J(−4, 8)
6. K(−4, 0) and L(0, −5)
*7. M(−3, −4) and N(−8, −6)
8. O(−4, 7) and P(3, −5)
*9. Q(−8, −3) and R(−6, −2)
10. S(−5, −4) and T(−6, 3)

*11. Find the length of the longest side of the triangle whose vertices are P(1, 1), Q(9, 7), and R(9, 16).

12. The coordinates of the center of a circle are (−3, 4). The circle passes through the point whose coordinates are (5, −2). Find the length of the radius of the circle.

*13. The vertices of a triangle are A(0, 0), B(4, 4) and C(2, −2).
   a. Find the lengths of the three sides of the triangle.
   b. Show that triangle ABC is a right triangle.

14. The vertices of a triangle are A(−6, 9), B(−4, −2) and C(4, 4).
   a. Find the lengths of three sides of the triangle.
   b. Is triangle ABC an isosceles triangle? (Hint: an isosceles triangle has two sides equal in length.)

*15. The vertices of a quadrilateral are P(5, 2), Q(9, 4), R(11, 8), and S(7, 6).
   a. Find the lengths of the four sides of the quadrilateral.
   b. Find the slopes of the four sides of the quadrilateral.
   c. Quadrilateral PQRS is a rhombus. (True or false?)

The following pairs of points are end points of line segments. Find the coordinates of the midpoint of each line segment.

16. A(4, 6) and B(0, 2)
*17. C(3, 5) and D(7, 1)
18. E(−2, 6) and F(6, 2)
*19. G(−6, −3) and H(−4, −1)
20. I(5, −3) and J(−7, −5)
*21. K(−6, 1) and L(−2, 3)
22. M(5, 7) and N(−2, −4)
*23. O(3, 2) and P(6, 1)
24. Q(−3, −2) and R(−2, −3)
*25. S(0, −3) and T(5, −6)

26. Find the coordinates of the midpoint of the line segment whose end points are C(3a, −b) and D(a, b).

*27. The coordinates of the midpoint of a line segment are (−2, 5). If the coordinates of one end point are (−3, 9), find the coordinates of the other end point.

28. $\overline{RS}$ is a diameter of the circle whose center is the point (4, −2). If the coordinates of R are (6, 4), find the coordinates of S.

*29. The vertices of quadrilateral ABCD are A(0, 0), B(3, 3), C(−3, 3), and D(−6, 0). Find the coordinates of the midpoints of each side.

30. The vertices of triangle ABC are A(0, 0), B(2a, 2b), and C(4a, 0).

    a. Find the coordinates of the midpoints of each side.
    b. Find the lengths of the line segments connecting these midpoints.

*31. The vertices of triangle RST are R(0, 0), S(2r, 2s) and T(2t, 2u).

    a. Find the coordinates of the midpoints of $\overline{RS}$ and $\overline{RT}$.
    b. Find the slope of the line segment connecting the midpoints of $\overline{RS}$ and $\overline{RT}$.
    c. Is the line segment (in b above) parallel to side TS? Why or why not?

8.7 QUADRATIC FUNCTIONS

After examining all of the properties of a linear function, the next logical step is to examine the properties of quadratic functions. Recall that any equation of the general form ax + by = c or y = mx + b where a, b, and c are real numbers is called a linear equation. A quadratic equation is an equation containing second degree terms. For example, $y = x^2$, $y = x^2 - 9$, and $y = x^2 - 6x + 9$ are quadratic equations. The following are also quadratic equations: $x = y^2$, $x = y^2 - 9$ and $x = y^2 - 6y + 9$. In this section we shall only consider those quadratic equations of the general form

$$y = ax^2 + bx + c$$

because equations of this form also are *quadratic functions.* A function is a relation in which no two ordered pairs have the same first coordinate and a different second coordinate. Consider $y = x^2$. Some of the ordered pairs that satisfy this equation are: (0, 0), (1, 1), (−1, 1), (2, 4) and (−2, 4). Note that no two ordered pairs have the same first coordinate and a different second coordinate. Now, consider $x = y^2$. Some of the ordered pairs

Relations and Functions

that satisfy this equation are (0, 0), (1, 1), (4, 2), (4, −2). Some of the ordered pairs do have the same first coordinate and a different second coordinate. Hence, $x = y^2$ is not a function. In this section, we shall only consider quadratic functions, that is, equations of the general form $y = ax^2 + bx + c$.

Let's consider a specific example, $y = x^2$. This is a general quadratic equation of the form $y = ax^2 + bx + c$, and in this case $a = 1$, $b = 0$, $c = 0$. To graph $y = x^2$, we can find some ordered pairs that satisfy the given equation. Since an equation of this form is a function, we note that for every value of x, there corresponds a unique value of y. Given $y = x^2$, then if $x = 0$, $y = 0$. Similarly, if $x = 1$, $y = 1$; if $x = -1$, $y = 1$; if $x = 2$, $y = 4$; if $x = -2$, $y = 4$. These points can be summarized in a table of values, as in Table 1.

TABLE 1

x	0	1	−1	2	−2
y	0	1	1	4	4

If we plot these points and connect them by means of a smooth curve, we obtain the curve shown in Figure 36. Note that the graph cannot be a straight line, since $y = x^2$ is not a linear equation. We assume that all those points lying on the curve also have coordinates that satisfy $y = x^2$. For example, $\left(\frac{1}{2}, \frac{1}{4}\right)$, $\left(-\frac{1}{2}, \frac{1}{4}\right)$, $\left(\frac{3}{2}, \frac{9}{4}\right)$, and $\left(-\frac{3}{2}, \frac{9}{4}\right)$. The smooth curve pictured in Figure 36 is called a *parabola*. The study of parabolas is interesting and useful. For example, the path of most projectiles is that of a parabola. Suspension bridges have cables hanging from them whose shape is that of a parabola.

Thus far we have seen that a parabola can be graphed by finding points whose coordinates satisfy the given equation, plotting these points on the Cartesian plane, and drawing a smooth curve through these points. For example, to graph $y = x^2 - 4$, we find points whose coordinates satisfy the equation. See Table 2.

TABLE 2

x	0	1	−1	2	−2	3	−3
y	−4	−3	−3	0	0	5	5

Plotting these points and connecting them by means of a smooth curve, we obtain the parabola shown in Figure 37. Note that the point (0, −4) is a turning point of the parabola. That is, as we proceed from left to right along the x-axis, the y values decrease to −4 and then at precisely (0, −4) the y values begin to increase. The turning point of a parabola is formally

Figure 36

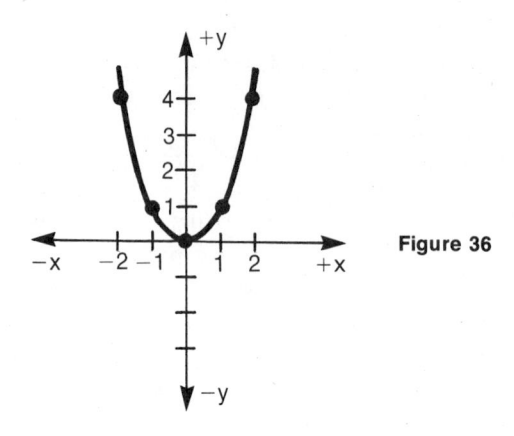

called the *vertex* of the parabola. In Figure 36, we see that the vertex of $y = x^2$ is (0, 0). For $y = x^2$ and $y = x^2 - 4$, the vertex of the parabola is also the minimum point because it is at this point (the vertex) that y has the lowest or least value of any point on the curve. The vertex of a parabola can also be a maximum point. Consider $y = -x^2 + 4$. Its graph is shown in Figure 38, and its vertex is (0, 4). Note that in this case y has the greatest value of any point on the curve. Therefore, the vertex of a parabola can be either a minimum point or a maximum point, depending upon whether the parabola opens upward or the parabola opens downward.

We can determine whether a parabola opens upward or downward by inspection. Recall that $y = x^2 - 4$ opens upward, while $y = -x^2 + 4$ opens downward. In general, the parabola $y = ax^2 + bx + c$ opens upward if $a > 0$, and it opens downward if $a < 0$. Note that for $y = x^2 - 4$, $a = 1$ which is greater than 0 and the parabola opens upward. In the equation $y = -x^2 + 4$, $a = -1$ which is less than 0 and the parabola opens down-

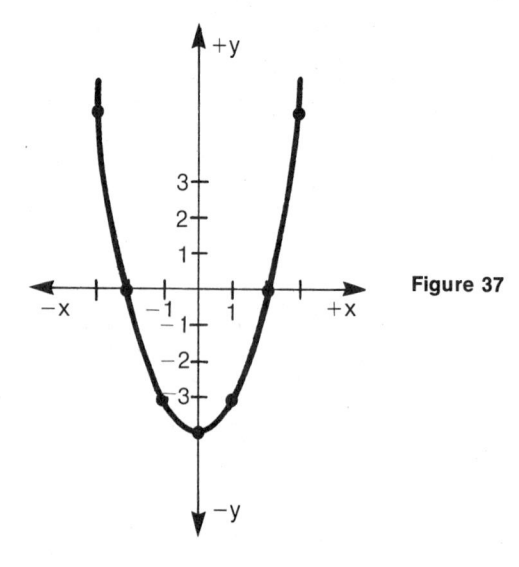

Figure 37

Relations and Functions

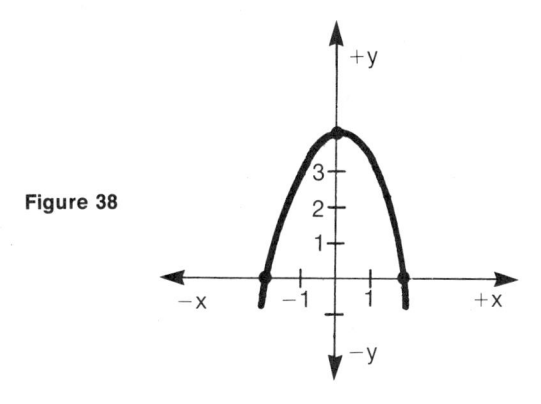

Figure 38

ward. We can use this information to help us graph the correct form of a given parabola.

The graph of $y = x^2$ is *symmetric* to the y-axis, that is, it is symmetric to the line $x = 0$ (see Figure 36). For every point on $y = x^2$ to the right of $x = 0$, there exists a "mirror image" to the left of $x = 0$ (the axis of symmetry). In other words, if we folded a piece of graph paper along the line $x = 0$, the points of $y = x^2$ would coincide. Formally, we say that a parabola has symmetry with respect to a line (the axis of symmetry) if for any point A on the parabola there is one and only one corresponding point B located in such a manner that a line (the axis of symmetry) is the perpendicular bisector of line segment $\overline{AB}$. Note that the axis of symmetry for the parabola $y = x^2 - 4$ is the line $x = 0$, as it is also for the parabola $y = -x^2 + 4$ (see Figures 37 and 38). The reader should be cautioned that the axis of symmetry of a parabola is *not* always the line $x = 0$ (the y-axis). A parabola may be graphed anywhere on the Cartesian plane and consequently the parabola will be symmetric to some other line.

There exists a formula for finding the axis of symmetry for a parabola. If we are given the parabola $y = ax^2 + bx + c$, then its *axis of symmetry* may be found by the formula

$$x = -\frac{b}{2a}.$$

For example, in the equation for the parabola, $y = x^2$, $a = 1$, and $b = 0$. Therefore, the axis of symmetry is

$$x = -\frac{b}{2a} = -\frac{0}{2(1)} = -\frac{0}{2} = 0.$$

As we noted before, the axis of symmetry for $y = x^2$ is $x = 0$.

The axis of symmetry can also be used to determine other useful information regarding a parabola. Regardless of what parabola we consider, the axis of symmetry will always pass through its vertex. Hence, $x = -\frac{b}{2a}$ must also be the abscissa of the vertex. We can find the y value of the vertex by substituting the determined x value in the equation of the parabola. Consider $y = x^2 - 4$. The value of a is 1 and $b = 0$. The axis of symmetry is $x = -\frac{b}{2a} = -\frac{0}{2 \cdot 1} = -\frac{0}{2} = 0$. Therefore, the abscissa of the vertex is also $x = 0$. To find the y value, we substitute in the given equation, $y = x^2 - 4$. If $x = 0$, then $y = 0^2 - 4 = 0 - 4 = -4$. Therefore, the vertex is at the point $(0, -4)$, as in Figure 37.

EXAMPLE 38

Given the equation of the parabola $y = x^2 + 4x + 4$, find (a) the equation of the axis of symmetry; (b) the coordinates of the vertex.

Solution

a. The formula for finding the axis of symmetry is $x = -\frac{b}{2a}$.
In the equation $y = x^2 + 4x + 4$, $a = 1$ and $b = 4$. Hence, the axis of symmetry is

$$x = -\frac{b}{2a} = -\frac{4}{2 \cdot 1} = -\frac{4}{2} = -2.$$

b. The abscissa of the vertex is $x = -2$. We find the y value by substituting in the original equation, $y = x^2 + 4x + 4$.

If $x = -2$, then $y = (-2)^2 + 4(-2) + 4$
$= 4 - 8 + 4$
$= 0.$

The vertex is $(-2, 0)$.

EXAMPLE 39

Given the equation of the parabola, $y = -2x^2 + 8x$, find (a) the equation of the axis of symmetry; (b) the coordinates of the vertex.

Solution

a. For $y = -2x^2 + 8x$, $a = -2$, $b = 8$. The equation of the axis of symmetry is

Relations and Functions

$$x = -\frac{b}{2a} = -\frac{8}{2(-2)} = -\frac{8}{-4} = 2.$$

b. The abscissa of the vertex is x = 2. We substitute in the equation $y = -2x^2 + 8x$ to find the ordinate.

If x = 2, then $y = -2(2)^2 + 8(2)$
$= -2(4) + 16 = -8 + 16$
$= 8.$

The vertex is (2, 8).

Recall from our previous discussions that the parabola $y = -2x^2 + 8x$ will open downward, since a (−2) is negative. Therefore, the vertex (2, 8) is the turning point of the parabola, and it is also the maximum on the graph of the parabola. That is, the value of y (8) is the maximum value of y on the graph of $y = -2x^2 + 8x$.

Sometimes we can determine other information that will also help us to graph a parabola. Recall that the parabola $y = x^2 - 4$ crossed the x-axis at the points (2, 0) and (−2, 0). Points of this type are often called the x-intercepts of the parabola. More formally, they are called the *zeros* of the function. Where a function crosses the x-axis is the x-intercept and that is where the value of y = 0. Therefore, the points (2, 0) and (−2, 0) are the x-intercepts or zeros of the parabola $y = x^2 - 4$. We can find the zeros of the parabola by letting y = 0 and solving the resulting quadratic equation. For $y = x^2 - 4$, we have $0 = x^2 - 4$ and the roots of this equation are $x = \pm 2$. It should be noted that the graphs of all parabolas will not necessarily cross the x-axis. For example, $y = x^2 + 1$ will not have any x-intercepts. The parabola $y = x^2 + 1$ will open upward since a > 0 (a = 1) and the axis of symmetry is x = 0, which means that the vertex is (0, 1). In this case the vertex is a minimum point which is above the x-axis, and since the parabola opens upward it will not cross the x-axis. We can also determine that the parabola $y = x^2 + 1$ does not have any x-intercepts (or zeros) by letting y = 0 and solving the quadratic equation $0 = x^2 + 1$. By means of the quadratic formula, we have a = 1, b = 0, c = 1 and

$$x = \frac{-b \pm \sqrt{b^2 - 4ac}}{2a} = \frac{-0 \pm \sqrt{0^2 - 4(1)(1)}}{2 \cdot 1},$$

$$x = \frac{0 \pm \sqrt{-4}}{2} = \frac{\pm 2i}{2} = \pm i.$$

The roots of $0 = x^2 + 1$ are imaginary. This indicates that the parabola $y = x^2 + 1$ has no points in common with the x-axis, that is, it has no real zeros.

EXAMPLE 40

Find the zeros of the parabola $y = x^2 - x - 2$.

Solution

To find the zeros of a parabola (if they exist), let $y = 0$ and solve the resulting quadratic equation. Therefore, we have

$$0 = x^2 - x - 2.$$
$$0 = (x - 2)(x + 1).$$
$$x - 2 = 0 \quad \text{or} \quad x + 1 = 0.$$
$$x = 2. \qquad\qquad x = -1.$$

Hence, the zeros are $(2, 0)$ and $(-1, 0)$.

EXAMPLE 41

Find the zeros of $y = x^2 - 6x + 9$.

Solution

Let $y = 0$ and solve the resulting quadratic equation.

$$0 = x^2 - 6x + 9.$$
$$0 = (x - 3)(x - 3).$$
$$x - 3 = 0 \quad \text{or} \quad x - 3 = 0.$$
$$x = 3. \qquad\qquad x = 3.$$

The equation has a double root, $x = 3$, and this indicates that the parabola has only one zero, $(3, 0)$.

When a parabola has only one 0 as indicated in Example 41, this implies that the zero point is also the vertex of the parabola. For the parabola $y = x^2 - 6x + 9$, the axis of symmetry is $x = -\frac{b}{2a} = -\frac{-6}{2 \cdot 1} = \frac{6}{2} = 3$. Therefore, the y value of the vertex is $y = 3^2 - 6 \cdot 3 + 9 = 9 - 18 + 9 = 0$. The vertex is $(3, 0)$, which is also an x-intercept.

EXAMPLE 42

Find the zeros of $y = x^2 + x + 1$.

Relations and Functions

Solution

Let y = 0 and solve the resulting quadratic equation.

$$0 = x^2 + x + 1, a = 1, b = 1, c = 1.$$

$$x = \frac{-b \pm \sqrt{b^2 - 4ac}}{2a}.$$

$$x = \frac{-1 \pm \sqrt{1^2 - 4(1)(1)}}{2 \cdot 1} = \frac{-1 \pm i\sqrt{3}}{2}.$$

The equation $0 = x^2 + x + 1$ has no real roots. Therefore, $y = x^2 + x + 1$ has no points in common with the x-axis. It has no real zeros.

Thus far we have seen that a parabola can be graphed by finding points whose coordinates satisfy the given equation, plotting those points on the Cartesian plane, and drawing a smooth curve through these points. But our task will be much easier if we make use of the information gained from examining the axis of symmetry, the vertex (turning point), and the zeros of the parabola (if it has any). Granted, a parabola can always be graphed by finding points whose coordinates satisfy the given equation, plotting these points, and drawing a smooth curve through them. In fact, even after determining the above information we should still find a few points whose coordinates satisfy the given equation in order to determine a more accurate graph.

EXAMPLE 43

Draw the graph of $y = x^2 + 4x + 3$.

Solution

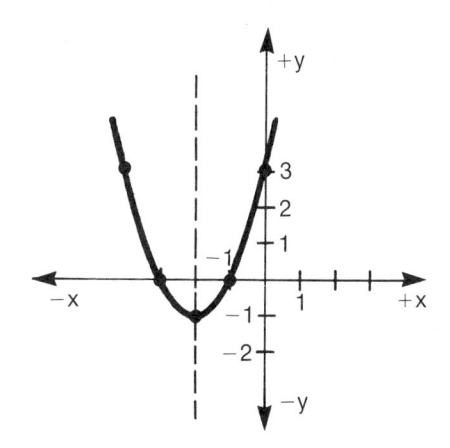

Analysis:

In the equation $y = x^2 + 4x + 3$, $a = 1$, $b = 4$, $c = 3$. Since $a > 0$, we know that the parabola opens upward. The axis of symmetry is $x = -\frac{b}{2a} = -\frac{4}{2 \cdot 1} = -\frac{4}{2} = -2$. If $x = -2$, then $y = (-2)^2 + 4(-2) + 3 = 4 - 8 + 3 = -1$. The vertex (minimum point) is $(-2, -1)$. Let $y = 0$, $0 = x^2 + 4x + 3$.

$$0 = (x + 1)(x + 3).$$
$$x + 1 = 0 \quad \text{or} \quad x + 3 = 0.$$
$$x = -1. \qquad\qquad x = -3.$$

The zeros of the function are $(-1, 0)$ and $(-3, 0)$. Some other points whose coordinates satisfy the equation are:

x	0	−4
y	3	3

EXAMPLE 44

Draw the graph of $y = -x^2 + 4x$.

Solution

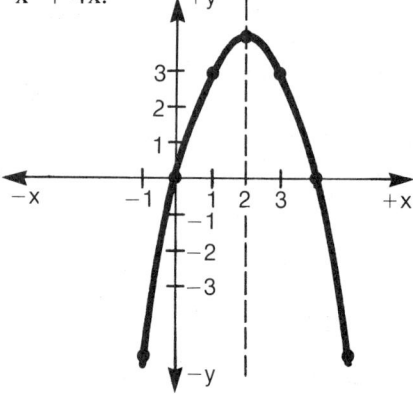

Analysis:

In the equation $y = -x^2 + 4x$, $a = -1$, $b = 4$, $x = 0$. Since $a < 0$, we know that the parabola opens downward. The axis of symmetry is $x = -\frac{b}{2a} = -\frac{4}{2(-1)} = \frac{4}{-2} = 2$. If $x = 2$, $y = -2^2 + 4 \cdot 2 = -4 + 8 = 4$.
The vertex (maximum point) is $(2, 4)$. Let $y = 0$, $0 = -x^2 + 4x$.

$$0 = x(-x + 4).$$
$$x = 0 \quad \text{or} \quad -x + 4 = 0.$$
$$\qquad\qquad\qquad x = 4.$$

The zeros of the function are (0, 0) and (4, 0). Some other points whose coordinates satisfy the equation are:

x	1	3	−1	5
y	3	3	−5	−5

EXAMPLE 45

Draw the graph of $y = x^2 - 4x + 6$.

Solution

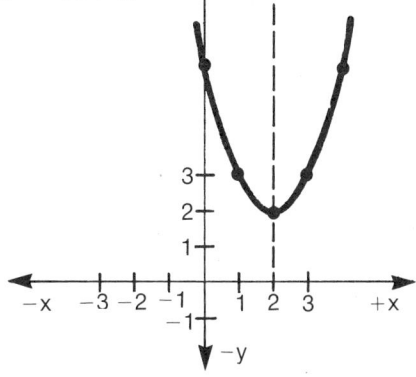

Analysis:

In the equation $y = x^2 - 4x + 6$, $a = 1$, $b = -4$, $c = 6$. Since $a > 0$, we know that the parabola opens upward. The axis of symmetry is $x = -\frac{b}{2a} = -\frac{-4}{2 \cdot 1} = \frac{4}{2} = 2$. If $x = 2$, $y = 2^2 - 4 \cdot 2 + 6 = 4 - 8 + 6 = 2$. The vertex (minimum point) is (2, 2). Let $y = 0$, $0 = x^2 - 4x + 6$.

$$x = \frac{-b \pm \sqrt{b^2 - 4ac}}{2a} = \frac{-(-4) \pm \sqrt{(-4)^2 - 4 \cdot 1 \cdot 6}}{2 \cdot 1}$$

$$= \frac{4 \pm \sqrt{16 - 24}}{2} = \frac{4 \pm \sqrt{-8}}{2} = \frac{4 \pm i\sqrt{8}}{2}.$$

The equation $0 = x^2 - 4x + 6$ has no real roots. Hence, $y = x^2 - 4x + 6$ has no points in common with the x-axis. Note that we could also determine this from the fact that (2, 2) is a minimum point and it is above the x-axis. Some other points whose coordinates satisfy the equation are:

x	0	1	3	4
y	6	3	3	6

EXERCISES FOR SECTION 8.7

(Answers given to exercises marked *)

Graph each of the given parabolas. Determine for each parabola, (a) the axis of symmetry, (b) the coordinates of the vertex, (c) the real zeros of the parabola (if they exist).

*1. $y = x^2 - 1$.
*3. $y = -x^2 + 4$.
*5. $y = x^2 + 2x + 1$.
*7. $y = -x^2 + 4x$.
*9. $y = x^2 + 2x$.
*11. $y = x^2 - 4x + 3$.
*13. $y = -x^2 + 6x - 9$.
*15. $y = x^2 - 6x + 5$.
*17. $y = 2x^2 + 4x - 6$.
*19. $y = x^2 + 3x + 2$.
*21. $y = x^2 + 5x + 6$.
*23. $y = x^2 + x + 1$.

2. $y = -x^2 + 1$.
4. $y = x^2 - 2x + 1$.
6. $y = x^2 - 2x$.
8. $y = -x^2 + 6x$.
10. $y = x^2 + 2x - 3$.
12. $y = 2x^2 + 2x - 4$.
14. $y = -2x^2 + 4x - 2$.
16. $y = x^2 + 2x - 8$.
18. $y = 2x^2 + 6x + 4$.
20. $y = x^2 - x - 2$.
22. $y = x^2 - 5x + 6$.
24. $y = -x^2 - x - 1$.

8.8 VARIATION

As we discovered in the beginning of this chapter, often pairs of variables such as x and y are related in some manner. Later in this section we shall consider some specific relationships between variables. But first we must discuss certain situations that will result in these relationships. The *ratio* of two quantities such as a and b is the quotient or indicated quotient obtained by dividing the value of a by the value of b, $\frac{a}{b}$ or a:b, which means a divided by b. We can think of ratios as fractions, and hence the value of a ratio will remain constant whenever the numerator and denominator of a ratio are multiplied or divided by the same number. For example, if we multiply the ratio $\frac{4}{8}$ by $\frac{2}{2}$, we obtain $\frac{8}{16}$ and the value of both ratios, $\frac{4}{8}$ and $\frac{8}{16}$, is $\frac{1}{2}$ or 0.5.

A *proportion* is the equality of two ratios. If the ratios a:b and c:d are equal, then we can form a proportion by writing

$$a:b = c:d \quad \text{or} \quad \frac{a}{b} = \frac{c}{d}.$$

Relations and Functions

We can describe this proportion by saying that "a is to b as c is to d." Note that b and c are called the *means* of the proportion and a and d are called the *extremes* of the proportion. Sometimes we can have the case where $b = c$, that is, the means of the proportion are equal. Therefore, we would have $a:b = b:d$ or $a:c = c:d$. We can have other situations where the means are equal, such as $a:b = b:c$. Whenever the means of a proportion are equal, then we describe either mean as the *mean proportional* between the first and last terms of the proportion.

If x is the mean proportional between 2 and 72, then what is the value of x? Since x is the mean proportional, we know that the second and third terms of the proportion are equal and we have

$$2:x = x:2 \quad \text{or} \quad \frac{2}{x} = \frac{x}{72}.$$

We can find the value of x by applying a well known property of proportions. It states

> In any proportion, the product of the means equals the product of the extremes, or, if $\frac{a}{b} = \frac{c}{d}$ then $ad = bc$.

Therefore, we can solve $\frac{2}{x} = \frac{x}{72}$ for x by determining the product of the means and extremes. In other words we can cross multiply

$$\frac{2}{x} = \frac{x}{72},$$

$$x^2 = 144,$$

$$x = \pm\sqrt{144} = \pm 12.$$

In order to have a proportion, the ratios must be equal. If $x = 12$, we have $\frac{2}{12}$ and $\frac{12}{72}$. Both ratios are equal to $\frac{1}{6}$. If $x = -12$, we have $\frac{2}{-12}$ and $\frac{-12}{72}$. Both ratios are equal to $-\frac{1}{6}$.

EXAMPLE 46

Find the value of x in each of the following proportions:

a. $2:4 = 3:x$. b. $3:7 = x:14$. c. $x:3 = 10:6$. d. $1:x = x:4$.

Solution

To find the value of x in each of the proportions, we can use the property that in any proportion the product of the means equals the product of the extremes.

a. $2:4 = 3:x$, $\dfrac{2}{4} = \dfrac{3}{x}$.

$$2 \cdot x = 4 \cdot 3.$$
$$2x = 12.$$
$$x = 6.$$

b. $3:7 = x:14$, $\dfrac{3}{7} = \dfrac{x}{14}$.

$$3 \cdot 14 = 7 \cdot x.$$
$$42 = 7x.$$
$$6 = x.$$

c. $x:3 = 10:6$, $\dfrac{x}{3} = \dfrac{10}{6}$.

$$6 \cdot x = 3 \cdot 10.$$
$$6x = 30.$$
$$x = 5.$$

d. $1:x = x:4$, $\dfrac{1}{x} = \dfrac{x}{4}$.

$$x^2 = 4.$$
$$x = \pm 2.$$

Direct Variation

Consider the proportion $\dfrac{y}{x} = \dfrac{k}{1}$, where y and x are variables and k is a constant. If we solve this proportion for y, we have

$$\dfrac{y}{x} = \dfrac{k}{1}.$$
$$y \cdot 1 = k \cdot x.$$
$$y = kx.$$

If $x = 2$ and $y = 4$, then the value of k in this proportion is 2. Since we placed a restriction on k, that is, it is a constant, what happens if we change the value of x? Suppose $x = 5$, then from the equation $y = kx$, we have $y = 2 \cdot 5$, $y = 10$. If $x = 7$, then $y = kx$ gives us $y = 2 \cdot 7 = 14$. Similarly, if $x = 3$ then $y = 2 \cdot 3 = 6$. Note that as x increases, y also increases and as x decreases, y also decreases. We can say that y and x *vary* in the same

Relations and Functions

manner. That is, y = kx means that y *varies directly* as x does, or y = kx represents *direct variation*.

If x and y are variables and k is a constant, then direct variation can be represented by the equation y = kx. In an equation of this type, k is called the *constant of proportionality,* or the variation constant. Sometimes the constant of proportionality may be known and other times it may not. If it is not known, then it can be determined, provided we know the values of x and y for a certain situation.

Given that y varies directly as x, and y = 10 when x = 5, find y when x = 7. To find the value of y, we first write an equation representing the variation,

$$y = kx.$$

Since y = 10 when x = 5, we can find the value of the constant of proportionality (k).

$$y = kx, \text{ let } x = 5, y = 10.$$
$$10 = k \cdot 5.$$
$$2 = k.$$

Now, we again use y = kx to find the value of y when x = 7,

$$y = kx.$$
$$y = 2 \cdot 7 \quad \text{substituting 2 for k and 7 for x.}$$
$$y = 14.$$

EXAMPLE 47

If y varies directly as x, and y = 6 when x = 2, find y when x = 5.

Solution

The variation statement for this problem is y = kx. Next, we determine the value of k. y = 6 when x = 2 gives us y = kx.

$$6 = k \cdot 2.$$
$$3 = k.$$

Now we substitute the value for k in the equation. That is, y = kx becomes y = 3x. To find y, when x = 5, y = 3 · 5,

$$y = 15.$$

EXAMPLE 48

If x varies directly as the square of y, and x = 12 when y = 4, find x when y = 6.

Solution

The variation statement for this problem is $x = ky^2$.
Determining the value of k, x = 12 when y = 4, we have

$$x = ky^2.$$
$$12 = k \cdot 4^2.$$
$$12 = k \cdot 16.$$
$$\frac{12}{16} = k.$$
$$\frac{3}{4} = k.$$

The equation can now be rewritten as

$$x = \frac{3}{4} y^2.$$

To find x, when y = 6, $\quad x = \frac{3}{4} \cdot 6^2.$

$$x = \frac{3}{4} \cdot 36.$$
$$x = 27.$$

EXAMPLE 49

The circumference of a circle varies directly as its radius. If the circumference of a circle is 6π when its radius is 3, find the circumference of a circle when its radius is 5.

Solution

The variation statement for this problem is $c = kr$.
Determining the value of k, $c = 6\pi$ when r = 3, we have

$$c = kr.$$
$$6\pi = k \cdot 3.$$
$$2\pi = k.$$

Relations and Functions

The equation can now be written as

$$c = 2\pi r.$$

To find c, when $r = 5$, $c = 2\pi \cdot 5$,

$$c = 10\pi.$$

Inverse Variation

If x and y are variables and k is constant, then direct variation can be represented by the equation $y = kx$. This means that y and x vary in the same manner. That is, as x increases, y also increases and if x decreases, then y also decreases. We can also have a relationship between two variables such that when one variable increases, the other variable decreases, or when one variable decreases, the other increases. In this situation, we say that one variable varies *inversely* as the other. For example, the statement "y varies inversely as x" can be interpreted to mean that as x increases in value, then y decreases in value, and when x decreases in value, the value of y increases. The statement "y varies inversely as x" can be symbolized as

$$y = \frac{k}{x} \quad \text{(where k is the constant of proportionality)}.$$

Given that y varies inversely as x, and $y = 6$ when $x = 2$, find the value of y when $x = 4$. First we write an equation representing the variation,

$$y = \frac{k}{x}.$$

Next, we find the value of k. We are given $y = 6$ when $x = 2$. Therefore,

$$y = \frac{k}{x}. \quad \text{Let } y = 6, x = 2.$$

$$6 = \frac{k}{2}.$$

$$12 = k.$$

The equation can now be rewritten as $y = \frac{12}{x}$. Now we can find y when $x = 4$, and we have

$$y = \frac{12}{x}. \qquad y = \frac{12}{4}. \qquad y = 3.$$

EXAMPLE 50

If x varies inversely as y, and x = 8 when y = 9, find y when x = 18.

Solution

The variation statement for this problem is $x = \dfrac{k}{y}$.

Determining the value of k, x = 8 when y = 9, we have

$$x = \frac{k}{y}.$$

$$8 = \frac{k}{9}.$$

$$72 = k.$$

The equation can now be rewritten as

$$x = \frac{72}{y}.$$

To find y when x = 18, $x = \dfrac{72}{y}$ becomes

$$18 = \frac{72}{y}.$$

$$18y = 72.$$
$$y = 4.$$

Note that in this inverse variation problem, the value of y decreases as the value of x increases.

EXAMPLE 51

Boyle's law states that under a constant temperature, the volume of a gas varies inversely as its pressure. If a certain gas has a volume of 100 cubic inches under a pressure of 5 pounds, what will be its volume when the pressure is 25 pounds?

Relations and Functions

Solution

The variation statement for this problem is $v = \dfrac{k}{p}$.
Determining the value of k, v = 100 when p = 5, we have

$$v = \frac{k}{p}.$$

$$100 = \frac{k}{5}.$$

$$500 = k.$$

The equation can now be rewritten as

$$v = \frac{500}{p}.$$

To find v when p = 25, we have

$$v = \frac{500}{25}.$$

$$v = 20.$$

Joint Variation

Consider the area of a rectangle. From previous discussions we know that the area of a rectangle can be found by multiplying its length times its width, A = l · w. In other words, the area of a rectangle depends upon the dimensions of its length and width. We can say that the area of a rectangle varies directly as the product of its length and width. We can also say that the area of a rectangle varies jointly as its length and width, that is,

$$A = kl \cdot w. \quad \text{(Note: in this case, k = 1.)}$$

The statement that A varies jointly as l and w means that A varies jointly as the product of l and w. We say that a quantity varies jointly with two or more quantities whenever it varies directly as their product. For example, the area of a triangle varies jointly as its base and height, A = kbh.

Given that the area of a triangle is 6 square metres when its base measures 4 metres and its height measures 3 metres, find the area of a triangle when its base measures 12 metres and its height measures 5 metres. The variation statement for this problem is

$$A = kbh.$$

Determining the value of k, A = 6 when b = 4 and h = 3, we have

$$6 = k \cdot 4 \cdot 3.$$
$$6 = k \cdot 12.$$
$$\frac{6}{12} = k.$$
$$\frac{1}{2} = k.$$

The equation can now be rewritten as $A = \frac{1}{2} bh$, the formula for the area of a triangle. To find A when b = 12 and h = 5, we have

$$A = \frac{1}{2} (12)5.$$
$$A = 30 \text{ square metres} \quad \text{or} \quad A = 30 m^2.$$

EXAMPLE 52

If z varies jointly as x and y, and z = 24 when x = 3 and y = 4, find z when x = 4 and y = 5.

Solution

The variation statement for this problem is z = kxy. Determining the value of k, z = 24 when x = 3 and y = 4, we have

$$z = kxy.$$
$$24 = k \cdot 3 \cdot 4.$$
$$24 = k \cdot 12.$$
$$2 = k.$$

The equation can now be expressed as z = 2xy. To find z when x = 4 and y = 5, we have

$$z = 2 \cdot 4 \cdot 5.$$
$$z = 40.$$

We have examined three different types of variation in this section: *direct variation, inverse variation* and *joint variation*. Often these different types of variations are combined to express the relationship that exists between different variables. For example, the safe load L for a

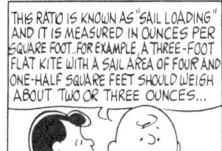

Relations and Functions

horizontal beam, supported at both ends, varies jointly as the width w and the square of the depth d, and inversely as the length l between supports. From this statement, we know that L varies jointly as w and d², and inversely as l. Therefore, the variation equation for this problem is

$$L = \frac{kwd^2}{l}.$$

To solve any variation problem, we must first write an equation that represents the type of variation stated. Next, we must use the given information in the problem to find the value of k, the constant of proportionality. Then rewrite the variation equation using the value of k. Now we can solve the problem for the desired variable, using the rest of the given information in the problem.

EXERCISES FOR SECTION 8.8

(Answers given to exercises marked *)

*1. If y varies directly as x, and y = 8 when x = 2, find y when x = 3.

2. If y varies directly as x, and y = 12 when x = 6, find y when x = 3.

*3. If x varies directly as y and x = 16 when y = 12, find x when y = 21.

4. If y varies directly as the square of x and if y = 100 when x = 5, find the value of y when x = 4.

*5. If x varies directly as y and if x = 13 when y = 52, find y when x = 100.

6. If y varies directly as x and x = 3 when y = 32, find x when y = 80.

*7. If y varies directly as x² and if y = 45 and x = 3, find y when x = 6.

8. If x varies directly as y and if x = 4 when y = 2, find x when y = $\frac{1}{2}$.

*9. If s varies directly as t and if s = 8 when t = 4, find s when t = −5.

10. If s varies directly as the square of r and if s = 18 when r = 3, find s when r = 4.

*11. If x varies inversely as y and if x = 2 when y = 20, find x when y = 8.

12. If y varies inversely as x and if y = 2 when x = 3, find y when x = 6.

*13. If s varies inversely as t and if s = 3 when t = 20, find t when s = 3.

14. If y varies inversely as x and if y = 6 when x = 2, find y when x = 4.

*15. If v varies inversely as z and if v = 8 when z = 9, find z when v = 12.

16. If s varies inversely as t and if t = 4 when s = 21, find t when s = 6.

*17. If y varies inversely as the square of x and if y = 4 when x = 3, find y when x = 2.

18. If y varies inversely as the square root of x and if y = 4 when x = 9, find x when y = 6.

*19. If x varies inversely as the cube root of y and if x = 2 when y = 8, find x when y = 27.

20. If s varies inversely as the square of t and if s = 5 when t = 2, find s when t = 5.

*21. If y varies jointly as x and z, and y = 30 when x = 2 and z = 3, find y when x = 3 and z = 5.

22. If s varies jointly as t and v and s = 36 when t = 3 and v = 4, find s when t = 4 and v = 5.

*23. If i varies jointly as p, r, and t, and i = 9 when p = 100, r = 0.06, and $t = \frac{1}{2}$, find i when p = 200, r = 0.06 and $t = \frac{1}{3}$.

24. If y varies jointly as x and the square of z, and if y = 24 when x = 3 and z = 2, find y when x = 4 and z = 3.

*25. If m varies jointly as p and q, and if m = 60 when p = 5 and q = 3, find q when m = 120 and p = 6.

26. If x varies jointly as z and the cube of y, and if x = 108 when z = 2 and y = 3, find x when z = 3 and y = 4.

Relations and Functions

*27. If v varies jointly as s and t, and inversely as x; and if v = 24 when s = 3, t = 4, and x = 2; find v when s = 6, t = 7 and x = 4.

28. If x varies jointly as y and the square of z, and inversely as the square of p, and if x = 64 when y = 3, z = 4 and p = 9, find x when y = 5, z = 7, and p = 16.

*29. If p varies jointly as q, r, and s, and if p = 48 when q = 2, r = 3 and s = 4, find p when q = 5, r = 7 and s = 11.

30. If y varies directly as the cube of x and inversely as the square of z, and if y = 144 when x = 4 and z = 6, find y when x = 5 and z = 9.

8.9 CONIC SECTIONS

In Section 8.7, we discussed the properties of a particular quadratic function, the parabola. The general equation of the parabolas that we examined is $y = ax^2 + bx + c$. Note that the equation contains two different variables, x and y, and also it contains the second power of x; it is a quadratic function. In this section, we shall briefly examine the graphs and properties of some other quadratic equations in two variables. The graphs of these equations will be figures called *circles, ellipses* and *hyperbolas.* Parabolas, circles, ellipses and hyperbolas are called *conic sections* because these curves can all be obtained by cutting a cone with a plane, as illustrated in Figure 39.

A circle is a set of points on the Cartesian plane that is a given distance r from a fixed point P. The fixed point P is the center of the circle and the given distance r is the radius of the circle. The distance formula can be used to determine the equation of a circle. Consider Figure 40, where we have circle P. The center is point P whose coordinates are (h, k).

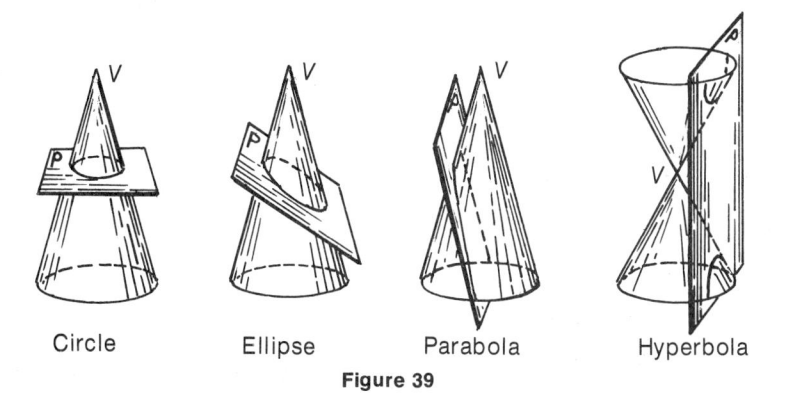

Circle   Ellipse   Parabola   Hyperbola

Figure 39

Figure 40

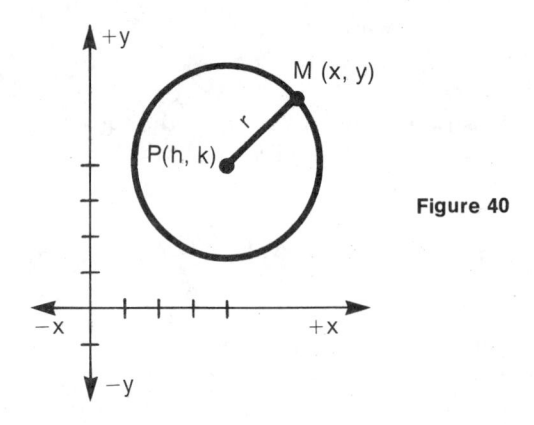

Point M is any point on the circle and the coordinates of M are (x, y). The length of the radius is r. Using the distance formula,

$$d = \sqrt{(x_2 - x_1)^2 + (y_2 - y_1)^2},$$

we have $\quad r = \sqrt{(x - h)^2 + (y - k)^2},$

$$r^2 = (x - h)^2 + (y - k)^2.$$

Therefore, the *equation of the circle* whose center is at (h, k) and whose radius is r is

$$(x - h)^2 + (y - k)^2 = r^2.$$

If the center of the circle is at the origin and the radius is r, then h = 0 and k = 0 and the general equation becomes

$$(x - 0)^2 + (y - 0)^2 = r^2.$$
$$x^2 + y^2 \qquad\quad = r^2.$$

It should be noted that the graph of a conic section can occur anywhere on the Cartesian plane. But in this section we shall consider almost exclusively those conic sections whose center is at the origin.

If we want to determine the equation of a circle whose center is at the origin and whose radius is 2, we use the general form $x^2 + y^2 = r^2$ and substitute for r, that is,

$$x^2 + y^2 = 4 \qquad \text{is the desired equation.}$$

If the center of the circle is at (2, 3) and the radius is 4, then we must use the general form $(x - h)^2 + (y - k)^2 = r^2$. In this case, h = 2, k = 3, and r = 4. Therefore, we have $(x - 2)^2 + (y - 3)^2 = 16$. If we simplify this equation, we have $x^2 - 4x + 4 + y^2 - 6y + 9 = 16$, or $x^2 + y^2 - 4x - 6y - 3 = 0$. Normally we would express the answer in the indicated form, $(x - 2)^2 + (y - 3)^2 = 16$.

Relations and Functions

EXAMPLE 53

Determine the equation for the following circles.

a. Center at (0, 0) and radius = 2.
b. Center at (3, 2) and radius = 4.
c. Center at (−2, −3) and radius = 3.

Solution

The general equation of a circle is $(x-h)^2+(y-k)^2=r^2$, whose center is at the point (h, k) and whose radius is r.

a. h = 0, k = 0 and r = 2.

$$(x-0)^2+(y-0)^2=2^2 \text{ or } x^2+y^2=4.$$

b. h = 3, k = 2 and r = 4.

$$(x-3)^2+(y-2)^2=16.$$

c. h = −2, k = −3, and r = 3.

$$(x-(-2))^2+(y-(-3))^2=3^2 \text{ or } (x+2)^2+(y+3)^2=9.$$

EXAMPLE 54

Determine the center and radius of each of the following circles.

a. $x^2+y^2=1$.
b. $x^2+y^2=4$.
c. $(x-2)^2+(y-3)^2=4$.
d. $(x+1)^2+(y-2)^2=1$.

Solution

a. A circle whose center is at the origin and whose radius is r has the general form $x^2+y^2=r^2$. The circle $x^2+y^2=1$ is of this form. Therefore, its center is at (0, 0) and $r^2=1$ and r = 1.
b. $x^2+y^2=4$, center is at (0, 0), $r^2=4$ and r = 2.
c. A circle whose center is at the point (h, k) and whose radius is r has the general form $(x-h)^2+(y-k)^2=r^2$. The circle $(x-2)^2+(y-3)^2=4$ is of this form. Hence, h = 2, k = 3 and the center is at (2, 3), $r^2=4$ and r = 2.
d. $(x+1)^2+(y-2)^2=1$, h = −1, k = 2 and the center is at (−1, 2), $r^2=1$ and r = 1.

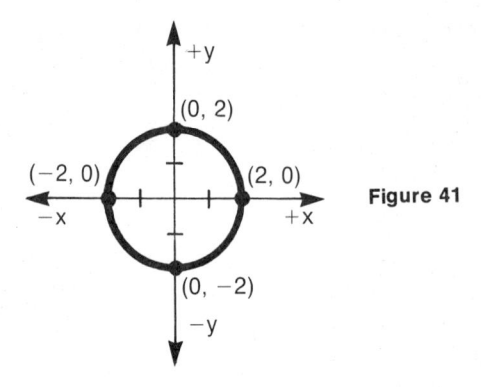

[Figure 41: circle centered at origin with radius 2, points (0,2), (2,0), (0,-2), (-2,0)]

Figure 41

Whenever the equation of a circle is of the form $x^2 + y^2 = r^2$, its graph is a circle whose center is at the origin and whose radius is r. Therefore the graph of $x^2 + y^2 = 4$ is a circle whose center is at the origin and whose radius is 2. See Figure 41. The equation $2x^2 + 2y^2 = 8$ is not of the general form $x^2 + y^2 = r^2$. But we can transform the given equation into such a form by multiplying both sides of the equation by $\frac{1}{2}$. Therefore, we have

$$\frac{1}{2}(2x^2 + 2y^2) = \frac{1}{2} \cdot 8.$$

$$\frac{1}{2} \cdot 2x^2 + \frac{1}{2} \cdot 2y^2 = 4.$$

$$x^2 + y^2 = 4.$$

This is the equation of a circle whose center is at the origin and whose radius is 2. Its graph is the same as that shown in Figure 41.

EXAMPLE 55

Graph each of the following circles.

a. $x^2 + y^2 = 9$.   b. $3x^2 + 3y^2 = 27$.

Solution

a. $x^2 + y^2 = 9$ is a circle whose center is at the origin and whose radius is 3. Its graph is shown in Figure 42.
b. $3x^2 + 3y^2 = 27$ is not of the general form $x^2 + y^2 = r^2$. Hence, we multiply both sides of the equation by $\frac{1}{3}$.

$$\frac{1}{3}(3x^2 + 3y^2) = \frac{1}{3} \cdot 27.$$

$$x^2 + y^2 = 9.$$

The center is at the origin and its radius is 3. Its graph is the same as that shown in Figure 42.

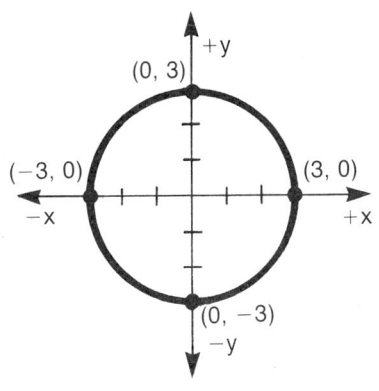

Figure 42

HISTORICAL NOTE

It was Apollonius who gave us the different names for the conic sections—parabola, ellipse, and hyperbola. It is interesting to note that Apollonius made all of his observations regarding the conic sections from oblique cones so that his propositions would hold in all cases. An oblique cone is one whose vertex is not directly over the center of its base. For his contributions to the study of conics, Apollonius became known as "The Great Geometer."

Apollonius (255–170 B.C.) was born in Perga, in Asia Minor. He studied at the famous school in Alexander, where Euclid had taught, and there he developed into one of the great mathematicians of his time. His best known contribution is a set of works titled Conic Sections, which consisted of eight volumes containing approximately 400 propositions pertaining to the different conic sections.

We have seen that an equation such as $3x^2 + 3y^2 = 27$ is of a circle whose center is at the origin and whose radius is 3. A general form of this type of equation is $ax^2 + by^2 = c$ where $a = b$, and a, b, c all have the same sign. We can also have the general equation $ax^2 + by^2 = c$ and $a \neq b$. For example, consider $9x^2 + 4y^2 = 36$. Here we have the case where $a \neq b$. The graph of such an equation is called an *ellipse*. How do we sketch a graph of such a figure? To begin with, we should first determine the intercepts of the ellipse, where the curve crosses the axes. We can determine the intercepts by letting each variable equal 0, and solve the resulting equation for the other variable.

[Figure 43: ellipse with intercepts (0, 3), (0, −3), (2, 0), (−2, 0)]

Figure 43

Given: $9x^2 + 4y^2 = 36$.

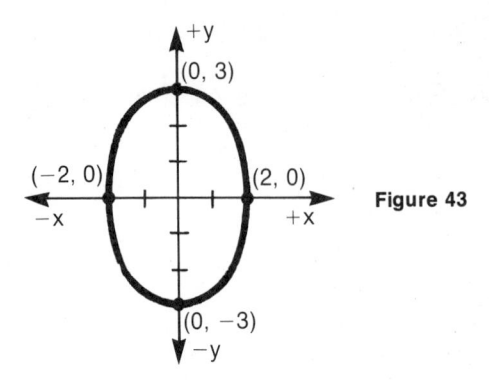

If $x = 0$, the resulting equation is	If $y = 0$, we have $9x^2 = 36$.
$4y^2 = 36$.	$x^2 = 4$.
$y^2 = 9$.	$x = \pm 2$.
$y = \pm 3$.	The intercepts are (2, 0) and
Hence, (0, 3) and (0, −3)	(−2, 0).
are intercepts.	

Now, by plotting these points (the intercepts), we can sketch the graph of the ellipse $9x^2 + 4y^2 = 36$. See Figure 43.

EXAMPLE 56

Graph the ellipse $x^2 + 4y^2 = 16$.

Solution

To aid us in sketching the graph of this ellipse, we find its intercepts. To find the y-intercepts, let $x = 0$ and we have

$0^2 + 4y^2 = 16$.
$4y^2 = 16$.
$y^2 = 4$.   The y-intercepts are (0, 2) and (0, −2).
$y = \pm 2$.

To find the x-intercepts, let $y = 0$ and we have

$x^2 + 4(0)^2 = 16$.
$x^2 = 16$.   The x-intercepts are (4, 0) and (−4, 0).
$x = \pm 4$.

By plotting these intercepts, we can sketch the graph of the ellipse, $x^2 + 4y^2 = 16$, as shown in Figure 44.

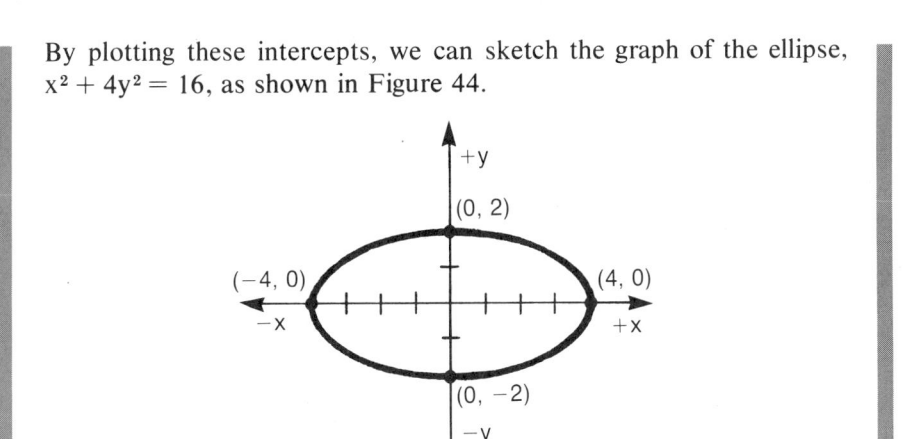

Figure 44

Thus far we have considered equations of the form $ax^2 + by^2 = c$ where $a = b$, and $ax^2 + by^2 = c$ where $a \neq b$ and a, b, c all have the same sign. The former is the equation of a circle and the latter is the equation of an ellipse. Next we consider the general equation $ax^2 - by^2 = c$ where $a = b$ or $a \neq b$. Note that the main difference between this equation and the others is that not all the signs of the terms are alike. In this case, the squared terms are separated by a minus sign. The graph of equations of the form $ax^2 - by^2 = c$ and $-ax^2 + by^2 = c$ are called *hyperbolas*.

Consider the equation $x^2 - y^2 = 4$. It is the equation of a hyperbola; note that $a = 1$, $b = -1$ and they have opposite signs. We can sketch its graph by determining its intercepts and finding a sufficient number of points whose coordinates are solutions of the equation $x^2 - y^2 = 4$. First, we find the intercepts. If we let $y = 0$, $x^2 = 4$ and $x = \pm 2$. The points (2, 0) and (−2, 0) are the x-intercepts. If we let $x = 0$, we have $-y^2 = 4$, $y^2 = -4$ and $y = \pm 2i$. This indicates that the y-intercepts are imaginary, that is, the graph does not cross the y-axis. Next, we select some arbitrary values for x and find the corresponding values of y. The table below lists such a set of points correct to the nearest tenth.

x	2	−2	3	3	−3	−3	4	4	−4	−4
y	0	0	2.2	−2.2	2.2	−2.2	3.5	−3.5	3.5	−3.5

By plotting these points, we can sketch the graph of the hyperbola, $x^2 - y^2 = 4$. See Figure 45.

Another equation whose graph is also a hyperbola is $xy = c$, where $c \neq 0$. Note that neither x nor y can assume a value of 0 since $xy = c$ can also be expressed as $x = \frac{c}{y}$ or $y = \frac{c}{x}$. Therefore, a hyperbola of the form

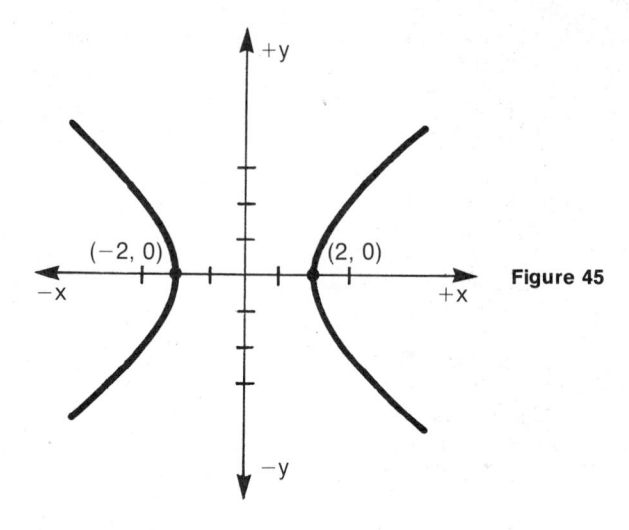

[Figure 45: hyperbola with vertices (−2, 0) and (2, 0)]

$xy = c$ has no intercepts. We can sketch the graph of such a hyperbola by generating a table of values that are solutions to the given equation.
Consider $xy = 6$. A table of coordinates for this equation is

x	1	2	3	−1	−2	−3
y	6	3	2	−6	−3	−2

Note that we can also have points such as $\left(12, \frac{1}{2}\right)$ and $\left(\frac{1}{2}, 12\right)$. Recall that a hyperbola of the form $xy = c$ has no intercepts, but points of this type indicate the hyperbola does approach the axes. That is, it gets closer and closer to the axes without ever touching them. The graph of $xy = 6$

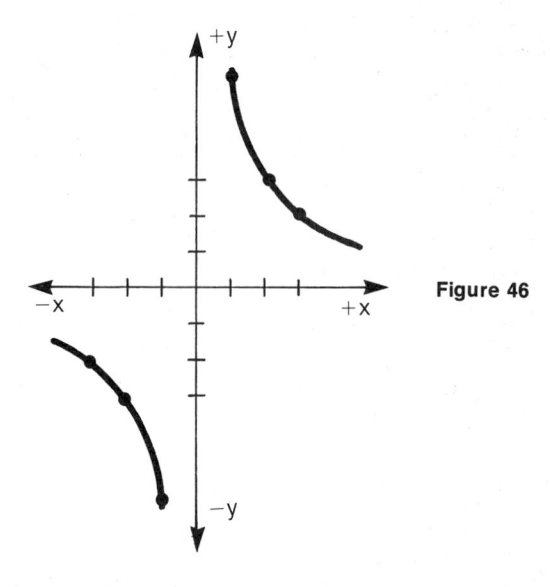

[Figure 46: hyperbola $xy = 6$ with branches in first and third quadrants]

Relations and Functions 491

is shown in Figure 46. Note that the graph is located in the first and third quadrants, since c > 0. If c < 0, then the graph of xy = c will occur in the second and fourth quadrants.

EXAMPLE 57

Graph the hyperbola $4y^2 - x^2 = 4$.

Solution

First we find the intercepts.
If x = 0, $4y^2 - 0^2 = 4$.
$4y^2 = 4$.
$y^2 = 1$.   The y-intercepts are
$y = \pm 1$.   (0, 1), (0, −1).

The hyperbola $4y^2 - x^2 = 4$ has no x-intercepts, because when we let y = 0, the roots of $-x^2 = 4$ are imaginary. Next, we select some arbitrary values for y and find the corresponding values of x. The table below lists such a set of points correct to the nearest tenth.

x	2	2	−2	−2	4	4	−4	−4
y	1.4	−1.4	1.4	−1.4	2.2	−2.2	2.2	−2.2

By plotting these points, we can sketch the graph of the hyperbola, $4y^2 - x^2 = 4$. See Figure 47.

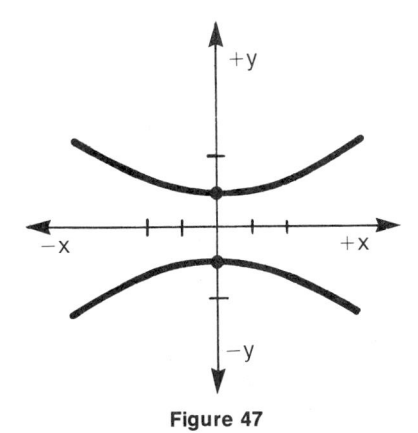

Figure 47

EXAMPLE 58

Graph the hyperbola $xy = -12$.

Solution

A table of coordinates for this equation is:

x	2	3	4	6	−2	−3	−4	−6
y	−6	−4	−3	−2	6	4	3	2

By plotting these points, we can sketch the graph of the hyperbola $xy = -12$. See Figure 48.

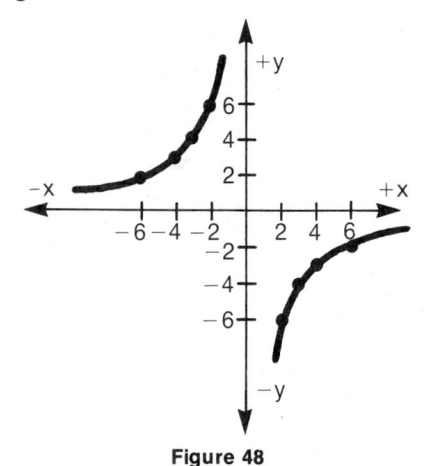

Figure 48

Equations of the form $ax^2 - by^2 = c (c \neq 0)$ and $-ax^2 + by^2 = c (c \neq 0)$ are equations of hyperbolas. It should be noted that if $c = 0$, then the graph of $ax^2 - by^2 = c$ is a set of two intersecting straight lines. For example, the graph of $x^2 - y^2 = 0$ is two straight lines that bisect the quadrants and intersect at the origin. A table of values is shown below and the graph of this equation is given in Figure 49. The graph is often referred to as a *degenerate hyperbola*.

x	0	1	1	2	2	−1	−1	−2	−2
y	0	1	−1	2	−2	1	−1	2	−2

Relations and Functions

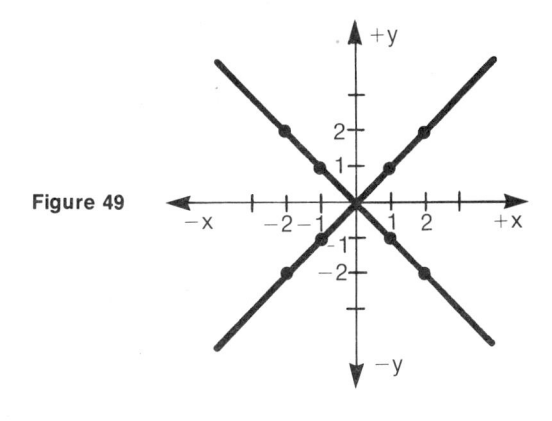

Figure 49

EXERCISES FOR SECTION 8.9

(Answers given to exercises marked *)

Find the equation for the following circles:

*1. Center at (0, 0) and radius = 1.
2. Center at (0, 0) and radius = 3.
*3. Center at (0, 0) and radius = 5.
4. Center at (0, 1) and radius = 1.
*5. Center at (2, 3) and radius = 1.
6. Center at (3, 1) and radius = 2.
*7. Center at (−2, 3) and radius = 2.
8. Center at (−2, −1) and radius = 1.
*9. Center at (−3, −1) and radius = 3.
10. Center at (−3, −4) and radius = 4.

Determine the center and radius of each of the following circles.

*11. $x^2 + y^2 = 9$.
12. $x^2 + y^2 = 25$.
*13. $x^2 + y^2 = 1$.
14. $x^2 + (y − 1)^2 = 1$.
*15. $(x − 1)^2 + y^2 = 4$.
16. $(x − 1)^2 + (y − 1)^2 = 1$.
*17. $(x − 2)^2 + (y − 3)^2 = 4$.
18. $(x − 3)^2 + (y + 1)^2 = 9$.
*19. $(x + 3)^2 + (y + 1)^2 = 1$.
20. $(x + 2)^2 + (y + 5)^2 = 4$.

Graph each of the given circles.

*21. $x^2 + y^2 = 1$.
22. $x^2 + y^2 = 9$.

*23. $x^2 + y^2 = 4$.
24. $x^2 + y^2 = 25$.
*25. $2x^2 + 2y^2 = 8$.
26. $3x^2 + 3y^2 = 12$.
*27. $(x-1)^2 + (y-1)^2 = 4$.
28. $(x-2)^2 + (y-1)^2 = 9$.
*29. $(x-3)^2 + (y-2)^2 = 4$.
30. $(x+1)^2 + (y+2)^2 = 4$.

Graph each of the given ellipses.

*31. $4x^2 + 9y^2 = 36$.
32. $9x^2 + 4y^2 = 36$.
*33. $4x^2 + y^2 = 36$.
34. $25x^2 + 4y^2 = 100$.
*35. $4x^2 + 25y^2 = 100$.
36. $4x^2 + 16y^2 = 144$.
*37. $16x^2 + 4y^2 = 144$.
38. $x^2 + 4y^2 = 16$.
*39. $9x^2 + 4y^2 = 9$.
40. $49x^2 + 25y^2 = 1225$.

Graph each of the given hyperbolas.

*41. $x^2 - y^2 = 9$.
42. $y^2 - x^2 = 9$.
*43. $2x^2 - y^2 = 4$.
44. $9y^2 - x^2 = 9$.
*45. $y^2 - 9x^2 = 25$.
46. $xy = 4$.
*47. $xy = -4$.
48. $xy = -8$.
*49. $4x^2 - 9y^2 = 0$.
50. $25y^2 - 16x^2 = 0$.

Identify each curve defined by the given equation.

*51. $x^2 + y^2 = 16$.
52. $4x^2 + 9y^2 = 36$.
*53. $9x^2 - 4y^2 = 36$.
54. $xy = 16$.
*55. $\frac{x^2}{4} + \frac{y^2}{9} = 1$.
56. $\frac{y^2}{4} - \frac{x^2}{9} = 1$.
*57. $x^2 + y^2 - 4x - 2y + 4 = 0$.
58. $x^2 - y^2 = 0$.
*59. $4x^2 - 16 + y = 0$.
60. $y - x^2 - 4x + 12 = 0$.

8.10 SUMMARY

A *relation* is a set of ordered pairs. An ordered pair contains a first element and a second element. The first element in an ordered pair is often called the first coordinate and the second element, the second coordinate. The *domain* of a relation is the set of all first coordinates of the given relation and the *range* of a relation is the set of all second coordinates. The *inverse* of a given relation can be formed by interchanging the first and second coordinates in all ordered pairs of it. Relations whose sets of ordered pairs contain only real numbers can be located on a coordinate system. The coordinate system contains the x and y axes and they intersect at a common point, the origin. Whenever we are given an ordered pair of real numbers for the coordinate system, such as (3, 2) the first coordinate is always the x value (abscissa) and the second coordinate is always the y value (ordinate). For every point on the Cartesian plane there corresponds an ordered pair of real numbers, and for every ordered pair of real numbers there corresponds a point on the plane.

A *function* is a relation in which no two ordered pairs have the same first coordinate and a different second coordinate. Whenever the graph of a relation has more than one point on the same vertical line, the relation is not a function. If no vertical line can contain two points of the graph of a relation, then the relation is a function. The notation commonly used in mathematics for the phrase "a function of x" is f(x). Hence, we can write the statement "y is a function of x" as y = f(x). To evaluate a function f(x) at x = a, we denote this by f(a). Similarly, f(a + b) means to find the value of the function f(x) when x = a + b.

Any equation of the general form ax + by = c where a, b, and c are real numbers is called a *linear* equation since its graph is a straight line. When graphing a linear equation it is usually convenient to find the x and y *intercepts*, where the line crosses the x and y axes, respectively. It is also recommended finding a third ordered pair that is a solution to the given equation as a check that all three points lie on the same straight line. The graph of a linear equation is a straight line passing through the points whose coordinates satisfy the given equation.

If a straight line contains the points (x_1, y_1) and (x_2, y_2), then the *slope* of the line is defined to be the "rise" of the line divided by the "run" of the line. That is, slope $= \frac{\text{rise}}{\text{run}} = \frac{y_2 - y_1}{x_2 - x_1}$. If a line slants upward to the right, its slope is positive, and if a line slopes downward to the right, its slope is negative. The slope of a horizontal line is 0 and the slope of a vertical line is undefined.

Given any two points on a line, we can find the slope of that line. We can also determine the equation of the line by means of the *point-slope* formula, which states that $y - y_1 = m(x - x_1)$. We can also use another general formula to determine the equation of a line, and it is y = mx + b, which is called the *slope-intercept* form for the equation of a line. The x coefficient (m) represents the slope of the line and the constant term (b) is the y-intercept of the line. We can also use the slope-intercept form for the equation of a line to graph the line. First we transform the given equation into the form y = mx + b. From this equation we locate the y-intercept, which is (0, b), then sketch a triangle from the intercept using the slope and the idea that slope $= m = \frac{\text{rise}}{\text{run}}$. This enables us to locate another point on the line and since two points determine a line, we can graph the line. The distance between two points on the Cartesian plane can be determined by means of the *distance formula,* d =

$\sqrt{(x_2 - x_1)^2 + (y_2 - y_1)^2}$. We can also find the length of any line segment on the plane by means of the distance formula. Many times it is also necessary to find the coordinates of the *midpoint* of a line segment. If the end points of a line segment are (x_1, y_1) and (x_2, y_2), then the midpoint of the line segment is determined by the formula $\left(\dfrac{x_1 + x_2}{2}, \dfrac{y_1 + y_2}{2}\right)$.

After examining the properties of a linear function, we next examined the properties of quadratic functions of the form $y = ax^2 + bx + c$. The graphs of quadratic functions of this form are called *parabolas*. A parabola can be graphed by finding points whose coordinates satisfy the given equation, plotting these points on the Cartesian plane, and drawing a smooth curve through these points. But we can also use other clues in helping us determine the graph of a parabola. For example, the parabola $y = ax^2 + bx + c$ opens upward if $a > 0$, and it opens downward if $a < 0$. The *axis of symmetry* is also useful in determining the graph of a parabola. The axis of symmetry for the parabola $y = ax^2 + bx + c$ is $x = -\dfrac{b}{2a}$. This x value is also the abscissa of the *vertex*. We find the y value by substituting the x value in the given equation and evaluating it for y. Sometimes we can also find the *zeros* of the parabola (if they exist), where the parabola crosses the x-axis. We can find the zeros of the parabola by letting $y = 0$ and solving the resulting quadratic equation.

Given the relationship that y and x vary in the same manner, we can say that y *varies directly* as x. The equation $y = kx$ represents direct variation where k is the constant of proportionality. The statement y *varies inversely* as x can be interpreted to mean that as x increases in value, then y decreases in value and when x decreases in value, then the value of y increases. The statement "y varies inversely as x" can be symbolized as $y = \dfrac{k}{x}$. We say that a quantity *varies jointly* with two or more quantities whenever it varies directly as their product. For example, the area of a triangle varies jointly as its base and height, $A = kbh$. To solve any variation problem, we must first write an equation that represents the type of variation stated in the given problem. Next, we must use the given information in the problem to find the value of k, the constant of proportionality. After finding the value of k, rewrite the variation equation using the value of k. Finally, solve the problem for the desired variable, using the rest of the given information in the problem.

Parabolas, circles, ellipses and hyperbolas are called conic sections because these curves can all be obtained by cutting a cone with a plane. The equation of the *circle* whose center is at (h, k) and whose radius is r is $(x - h)^2 + (y - k)^2 = r^2$. If the center of the circle is at the origin, then $h = 0$ and $k = 0$ and the general equation becomes $x^2 + y^2 = r^2$. Whenever the equation of a circle is of the form $x^2 + y^2 = r^2$, its graph is a circle whose center is at the origin and whose radius is r. For example, the graph of $x^2 + y^2 = 4$ is a circle whose center is at the origin and whose radius is 2. The equation of a circle whose center is at the origin can also

Relations and Functions

be represented by the general equation $ax^2 + by^2 = c$ where a, b, and c all have the same sign and $a = b$. If we have the case where $a \neq b$, then the graph of $ax^2 + by^2 = c$ is called an *ellipse*. For example, $x^2 + 4y^2 = 16$ is the equation of an ellipse. To aid us in sketching the graph of an ellipse, we find its intercepts. By plotting these intercepts, we can sketch the graph of the ellipse.

Next, we considered the general equations $ax^2 - by^2 = c$ and $-ax^2 + by^2 = c$ where $a = b$ or $a \neq b$. The graphs of equations of this form are called *hyperbolas*. To graph a hyperbola, we first find its intercepts (if any) and then select some arbitrary values for one variable and find the corresponding values of the other variable. By plotting these points, we can sketch the graph of the hyperbola. Another equation for a hyperbola is of the general form $xy = c$. Hyperbolas of this form do not have any intercepts as they do not cross the axes.

Here are some formulas (equations) that you should be familiar with as a result of the topics covered in this chapter.

Slope: $m = \dfrac{y_2 - y_1}{x_2 - x_1}$.

Point slope formula, for the equation of a line:
$y - y_1 = m(x - x_1)$.

Slope-intercept form, for the equation of a line:
$y = mx + b$.

Equation of a line: $ax + by = c$.
Parabola: $y = ax^2 + bx + c$.
Circle: $(x - h)^2 + (y - k)^2 = r^2$.
$x^2 + y^2 = r^2$.
$ax^2 + by^2 = c$,
if $a = b$ and they have the same sign.
Ellipse: $ax^2 + by^2 = c$,
if $a \neq b$ and they have the same sign.
Hyperbola: $ax^2 + by^2 = c$,
if a and b are opposite in sign,
$xy = c$.

REVIEW EXERCISES FOR CHAPTER 8
(Answers in Appendix B)

1. Given $A = \{a, b, c, d, e, f, g, h, i\}$, determine the relation R such that the first coordinate is a vowel in the English alphabet and the second coordinate is not.

    a. What is the domain of R?
    b. What is the range of R?
    c. Determine the inverse of R.
    d. What is the domain of the inverse relation?
    e. What is the range of the inverse relation?

2. Determine whether the following relations are functions or not.

    a. $\{(-2, 1), (1, -2), (3, -2), (4, 1)\}$

b. {(1, 2), (3, 4), (5, 7), (1, 3)}
c. {(−2, 2), (3, 2), (4, 2), (5, 2)}
d. {(2, 4), (3, 1), (5, 2), (3, −2)}

3. If $f(x) = 2x^2 - 3x + 2$, find:

   a. $f(0)$ b. $f(-1)$ c. $f(1)$ d. $f(-2)$ e. $f(a)$ f. $f(a+b)$

4. The vertices of quadrilateral PQRS are P(0, 0), Q(6a, 3b), R(3a, 4b) and S(a, 3b).

   a. Find the slopes of sides PQ and RS.
   b. What is your conclusion regarding these two sides?
   c. Is side PS parallel to side QR? Why?

5. Write the equation of the straight line that passes through the indicated point and has the indicated slope.

   a. $A(-2, 3), m = -2$.
   b. $B(-3, 7), m = \frac{1}{2}$.
   c. $C(0, -2), m = \frac{3}{2}$.
   d. $D(-5, -2), m = -\frac{3}{2}$.

6. Write the equation of the straight line that passes through the indicated points.

   a. $A(-2, 3)$ and $B(3, -2)$
   b. $C(-2, 0)$ and $D(0, -3)$
   c. $E(5, -2)$ and $F(6, 7)$
   d. $G(2, 3)$ and $H(-1, 1)$

7. Find the slope and y-intercept of the line whose equation is given.

   a. $y = -2x + 3$.
   b. $x + y = 1$.
   c. $3x - 2y = 6$.
   d. $x - 2y + 4 = 0$.

8. Given the points $P(1, -1)$, $Q(4, 2)$ and $R(6, 4)$:

   a. Find the slopes of line segments $\overline{PQ}$ and $\overline{PR}$.
   b. What can you conclude regarding points P, Q, and R?
   c. Determine the equation of the line that passes through P and R.
   d. Find the coordinates of the point where line PR crosses the x-axis.

9. The vertices of triangle ABC are A(0, 0), B(2a, 2b) and C(4a, 0).

   a. Find the coordinates of the midpoints of each side.
   b. Find the lengths of the line segments connecting these midpoints.

10. If x varies directly as y and if $x = 4$ when $y = 2$, find x when $y = \frac{1}{2}$.

11. If s varies inversely as t and if $t = 4$ when $s = 21$, find t when $s = 6$.

Relations and Functions

12. If y varies jointly as x and the square of z, and if y = 24 when x = 3 and z = 2, find y when x = 4 and z = 3.

13. If y varies directly as the cube of x and inversely as the square of z, and if y = 144 when x = 4 and z = 6, find y when x = 5 and z = 9.

Identify each curve defined by the given equation and draw the graph of each equation.

14. $-2x - 3y = 6$.
15. $y = x^2 - 1$
16. $9x^2 + 4y^2 = 36$.
17. $x^2 + y^2 = 9$.
18. $y = x^2 - 2x$.
19. $y - 6 = 2x$.
20. $x^2 + y^2 = 1$.
21. $y = -x^2 + 4$.
22. $x = 0$.
23. $y = x^2 + 3x + 2$.
24. $25x^2 + 4y^2 = 100$.
25. $y = 0$.
26. $4x^2 + 25y^2 = 100$.
27. $(x - 1)^2 + (y - 2)^2 = 4$.
28. $9y^2 - x^2 = 9$.
29. $y = x^2 + 2x - 3$.
30. $3x + 2y - 6 = 0$.
31. $x^2 - y^2 = 9$.
32. $(x - 3)^2 + (y + 1)^2 = 9$.
33. $x^2 - y^2 = 0$.
34. $xy = 2$.
35. $x^2 + 4y^2 = 16$.
36. $4x^2 - 9y^2 = 0$.
37. $xy = 4$.

CUMULATIVE REVIEW
(Answers in Appendix B)

Factor each of the given expressions completely.

1. $4y^2 - 2y$
2. $x(y + 3) + 5(y + 3)$
3. $(x + y)x - (x + y)3$
4. $ax + by + ay + bx$
5. $2ax + by + 4ax + 2by$
6. $st + uv + sv + ut$
7. $4x^2 - 9$
8. $y^{10} - 4$
9. $y^4 - 81$
10. $x^2 - y^2 - 2x - 2y$
11. $4y^4 + 8y^3 + 4y^2$
12. $x^2 + 11x - 12$
13. $y^4 - 7y^2 + 12$
14. $6x^2 + 11x + 5$
15. $6x^3 + 26x^2 + 24x$
16. $x^3 + y^3$
17. $x^6 - 1$
18. $x^4 - x^3 + 8x - 8$

Chapter 8

Perform the indicated operations and simplify, if possible.

19. $\dfrac{y+2}{y+5} \cdot \dfrac{y^2-25}{2y^2+3y+1}$

20. $\dfrac{xy+y^2}{x^2-xy} \div \dfrac{3x^2+3xy}{x^2-xy}$

21. $\dfrac{3x^2-x-2}{2x^2-x-1} \div \dfrac{3x^2+5x+2}{2x^2+3x+1}$

22. $\dfrac{y}{y^2-4} + \dfrac{2}{y^2-4}$

23. $\dfrac{2y-1}{3y+1} - \dfrac{y+1}{y-1}$

24. $\dfrac{2}{y-1} + \dfrac{3}{y+1} - \dfrac{1}{y-2}$

25. $\dfrac{\dfrac{1}{x^2}-\dfrac{1}{y^2}}{\dfrac{1}{x}-\dfrac{1}{y}}$

26. $\dfrac{\dfrac{y-1}{y+1} + \dfrac{y+1}{y-1}}{\dfrac{y-1}{y+1} - \dfrac{y+1}{y-1}}$

Find the solution set for each of the following equations.

27. $\dfrac{2x}{3} + \dfrac{3x}{4} = x + 5.$

28. $\dfrac{x+2}{x-3} = \dfrac{x+1}{x-1}.$

29. $\dfrac{4}{5x-10} - \dfrac{2}{3x-6} = \dfrac{1}{15}.$

30. $\dfrac{x+1}{x+2} - \dfrac{x-2}{x-3} = \dfrac{2-x}{x^2-x-6}.$

Perform the indicated operations and simplify, if possible.

31. $\sqrt{-4} + \sqrt{-9}$

32. $2i\sqrt{-27}$

33. $\sqrt[3]{-8} \cdot \sqrt[3]{-27}$

34. $(-3+2i)+(2-3i)$

35. $(4-3i)-(-2-3i)$

36. $3i(4+5i)$

37. $(3+2i)(4-3i)$

38. $\dfrac{4+2i}{-2+3i}$

39. $\dfrac{2-3i}{-2-3i}$

Solve the given equations by a method of your choice.

40. $3x^2 = 4x+4.$

41. $8y^2 - 6y + 1 = 0.$

42. $4y^2 = 3y - 8.$

43. $2y^2 = 3y - 9.$

44. $\sqrt{3x+1} - x - 1 = 0.$

45. $y - \sqrt{6-5y} = 0.$

46. Find the solution set for the given inequality and graph its solution set on the real number line:

$$6x^2 + 5x - 6 > 0.$$

47. When two snow plows are in operation together, they can plow a certain parking lot in three hours. The first snow plow alone takes twice as long

as the second snow plow alone to plow the parking lot. How long does it take the first snow plow to do the job alone?

48. A large oil tank can be filled by an inlet pipe in 10 hours, and emptied by an outlet pipe in 15 hours. The inlet pipe is open for 5 hours and then (by mistake) the outlet pipe is also opened. How much more time will be required to fill the oil tank?

49. The length of a rectangular piece of sheet metal is three more than twice its width. An open box is formed by cutting three-inch squares from the corners and bending up the sides. If the volume of the box is 135 cubic inches, find the dimensions of the original piece of sheet metal.

50. It takes Jamie three hours longer to paint the living room than it does her sister Pam. The sisters worked together for three hours, then Jamie left and Pam finished painting the living room alone in one hour. How many hours would it have taken Jamie to do the job alone?

chapter 9

Systems of Equations

As a result of this chapter, the student will be able to:

1. Solve a system of two linear equations with two variables by graphing;
2. Solve a system of two linear equations with two variables using the substitution method;
3. Solve a system of two linear equations with two variables using the addition method;
4. Solve a system of three linear equations with three variables by the addition method;
5. Determine whether a system of equations has a unique solution, or whether the system of equations is inconsistent or dependent;
6. Translate word problems into equations whose solutions require the solution of systems of linear equation in two or three unknowns;
7. Find the value of a second order determinant,

$$\text{e.g., } \begin{vmatrix} 3 & 2 \\ 1 & 4 \end{vmatrix} = 3 \cdot 4 - 1 \cdot 2 = 12 - 2 = 10;$$

8. Find the value of a third order determinant,

$$\text{e.g., } \begin{vmatrix} -2 & 1 & 3 \\ 1 & 2 & 4 \\ 5 & -1 & 2 \end{vmatrix} = -8 + 20 - 3 - (30 + 8 + 2) = 9 - 40 = 31;$$

9. Determine the solution for a system of linear equations containing two or three variables by means of Cramer's Rule;
10. Solve linear inequalities by graphing their solutions on the Cartesian plane, e.g., $x + y < 1$;
11. Solve systems of linear inequalities by graphing their solutions on the Cartesian plane, e.g., $x + y < 1$, $x - y < 2$.

9.1 INTRODUCTION

In Chapter 8, we examined the properties of various functions and their graphs. One that we examined in great detail was the linear function—the straight line. The general equation $ax + by = c$ represents the equation of a line. Formally, we say that it is a linear equation in two unknowns. A solution to the equation $ax + by = c$ is some ordered pair whose coordinates satisfy the equation. For example, (3, 2) and (6, −1) are solutions of the equation $x + y = 5$, since the coordinates of these points satisfy the given equation.

Systems of Equations

In this chapter we shall concern ourselves with finding the solutions of sets of linear equations. Consider the equations $x+y=5$ and $2x+y=7$. To find the solution for this set of equations is to determine those points whose coordinates satisfy both of the given equations at the same time. How many points will have coordinates that satisfy the two equations simultaneously? The answer is usually one, since the given equations are equations of straight lines and they normally intersect at a point. But if the lines are parallel then they will not intersect and their solution set will be empty. That is, they will have no points in common. We could also have the case where the equations represent the same straight line, and therefore the set of solutions for one equation will also be solutions for the other equation.

Let's determine the solution for the system of linear equations $x+y=5$, and $2x+y=7$. A solution for this set of equations could be any ordered pair whose coordinates satisfy both equations simultaneously. If we graph both equations on the same set of axes, then we can analyze their solution. Figure 1 shows the graph of $x+y=5$ and $2x+y=7$. Note that the graphs of the two equations intersect at one point, and the coordinates of that point are (2, 3). This point of intersection is a solution to the system of linear equations $x+y=5$ and $2x+y=7$. We can verify that this point satisfies both equations simultaneously by checking it in each equation.

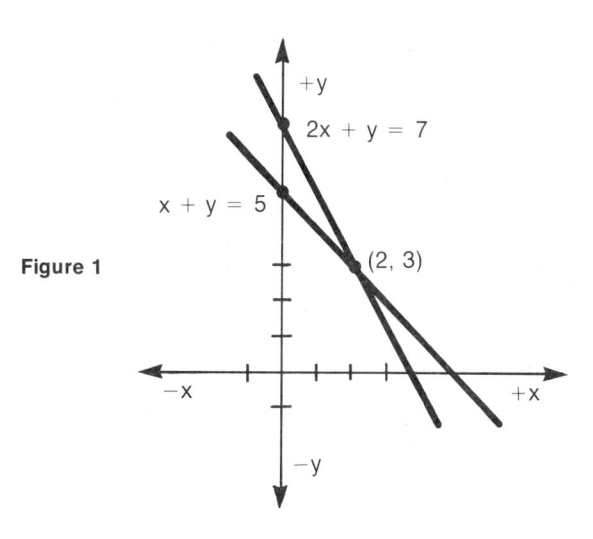

Figure 1

Check: Substitute (2, 3) in both equations.

$$x+y=5,\qquad 2x+y=7,$$
$$2+3 \stackrel{?}{=} 5,\qquad 2\cdot 2+3 \stackrel{?}{=} 7,$$
$$5=5.\qquad 4+3=7,$$
$$\qquad\qquad 7=7.$$

The ordered pair (2, 3) satisfies both equations.

Figure 2

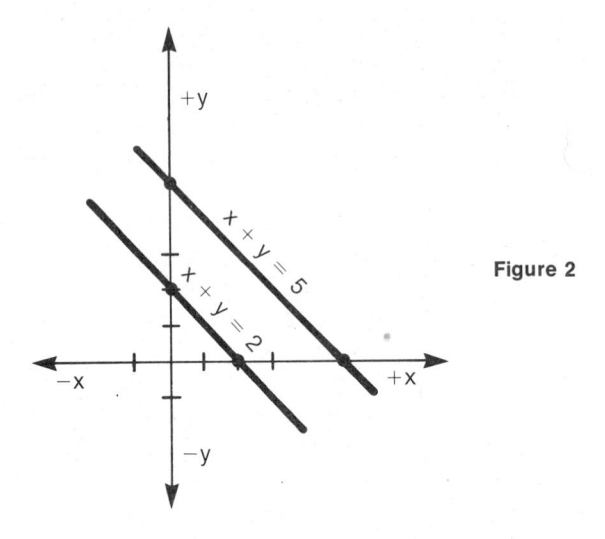

This example indicates that we can determine the solution for a set of linear equations by graphing them on the same set of axes and finding the coordinates of the point of intersection. Recall that there are three possibilities that could exist when determining the solution for a system of linear equations. (1) The two graphs may intersect at a point and the coordinates of the point represent a solution to the system of linear equations. The equations are *consistent.* This case occurred in the example that we just discussed. (2) The two graphs may be parallel lines. When this occurs, there is no solution to the system of equations. Or we can say that the solution set is the empty set ∅. When the solution for a system of equations is the empty set, the equations are called *inconsistent.* Consider the equations $x + y = 5$ and $x + y = 2$. If we graph these equations on the same set of axes, we have the solution shown in Figure 2. The graphs are parallel lines and therefore have no common point of intersection. The equations $x + y = 5$ and $x + y = 2$ are said to be *in-*

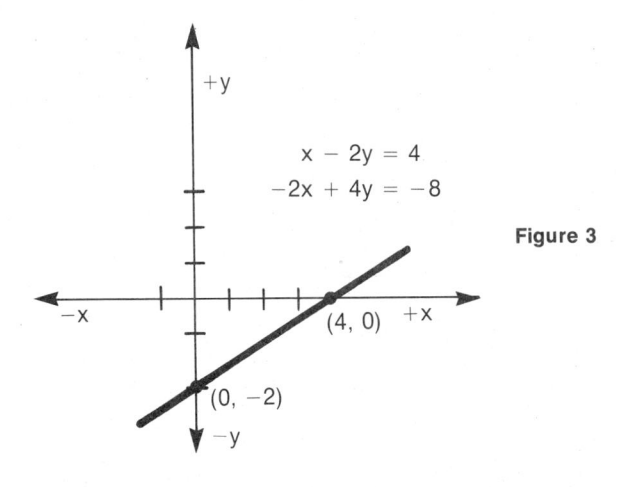

Figure 3

Systems of Equations

consistent. (3) The two graphs may be the same line. When this occurs, any solution of one equation is also a solution of the other. In this case the equations are called *dependent*. Consider the equations $x - 2y = 4$ and $-2x + 4y = -8$. Note that the intercepts for $x - 2y = 4$ are $(0, -2)$ and $(4, 0)$. These are also the intercepts for $-2x + 4y = -8$. If we graph these equations on the same set of axes, we have the situation shown in Figure 3. The graphs of the two equations are the same line. The equations $x - 2y = 4$ and $-2x + 4y = -8$ are said to be dependent. Rather than list a certain number of ordered pairs as solutions for dependent equations, we simply state that they are dependent.

EXAMPLE 1

Solve the given system of equations:

$$2x + y = -3.$$
$$-x + 2y = 4.$$

Solution

We graph each equation on the same set of axes, as in Figure 4.

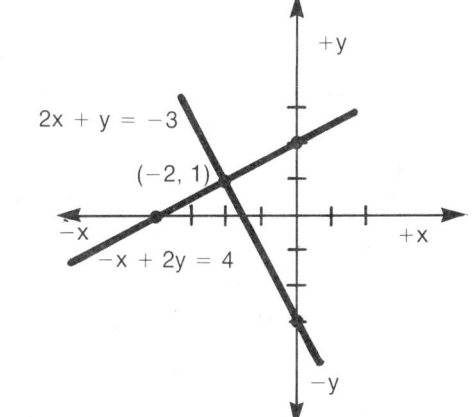

Figure 4

The two graphs intersect at a point, the coordinates of which are $(-2, 1)$. The solution of the system of equations is $(-2, 1)$. We verify this by checking the solution in both equations.

Check: Substitute $(-2, 1)$ in both equations.

$$2x + y = -3, \qquad -x + 2y = 4,$$
$$2(-2) + 1 \stackrel{?}{=} -3, \qquad -(-2) + 2 \cdot 1 \stackrel{?}{=} 4,$$
$$-4 + 1 = -3, \qquad 2 + 2 = 4,$$
$$-3 = -3. \qquad 4 = 4.$$

EXERCISES FOR SECTION 9.1

(Answers given to exercises marked *)

Solve each system of equations graphically.

1. $x + y = 5.$
   $x - y = 1.$

2. $x + y = 3.$
   $2x + y = 3.$

*3. $2x + 3y = -2.$
    $-x + 4y = 1.$

4. $x + y = 1.$
   $2x + y = 3.$

*5. $2x + y = 0.$
    $3x + 2y = 1.$

6. $x + y = 3.$
   $x + 2y = 5.$

*7. $x + y = 6.$
    $x + y = 1.$

8. $x - y = 1.$
   $x + y = -5.$

*9. $x + 2y = 3.$
    $-x + y = -6.$

10. $x + y = -4.$
    $x - y = 0.$

*11. $x - 2y = 6.$
     $-3x + 6y = 0.$

12. $x - 2y = 4.$
    $-2x + 4y = -8.$

*13. $-2x + y = 4.$
     $6x - 3y = -12.$

14. $2x - y = -3.$
    $-x + 2y = 0.$

*15. $2x + 3y = 0.$
     $x - y = -5.$

16. $2x - y = 4.$
    $x - 2y = -4.$

*17. $2x + 3y = 7.$
     $-3x - y = -7.$

18. $x + 2y = 0.$
    $3x - y = -3.$

*19. $2x - 3y = 6.$
     $-4x + 6y = 18.$

20. $3x + 3y = 6.$
    $-4x - 2y = -6.$

*21. $-2x + 2y = -2.$
     $-3x - 3y = 15.$

9.2 SOLUTION OF SYSTEMS OF EQUATIONS BY SUBSTITUTION

Here and in the next few sections, we shall examine algebraic solutions of systems of two linear equations in two unknowns. We just saw that systems of linear equations can be solved by means of graphing. But the graphical method of solution is not always accurate. Many times it is possible only to approximate the coordinates of the point of intersection. For example, if we were to determine the solution for the pairs of equations $x + 2y = 3$ and $2x - y = 4$ by means of graphing, it is highly unlikely that we would obtain the correct solution which is $\left(\frac{11}{5}, \frac{2}{5}\right)$. Coordinates such as these can only be approximated on a piece of graph paper. It would be even more difficult to determine the coordinates of a point such as $\left(-\frac{11}{17}, \frac{13}{9}\right)$. Therefore, we must consider other methods for solving systems of linear equations in two unknowns.

One technique is that of *substitution*. To solve a system of linear

Systems of Equations

equations by the substitution method, we first solve one of the given equations for either x or y. We next substitute the resulting expression for the chosen variable in the *other* equation and solve it. After determining a value for one of the coordinates, we can find the value of the other coordinate by substituting back into one of the original equations. Let's consider an example. To solve the system of equations

$$2x + 3y = -2,$$
$$-x + 4y = 1,$$

by the substitution method, we first solve one of the given equations for x or y. The best choice is to solve the second equation for x, since this will involve the least amount of computation. Hence, we have

$$-x + 4y = 1,$$
$$-x = 1 - 4y,$$
$$x = -1 + 4y.$$

Next, we substitute the resulting expression for the chosen variable in the other equation and solve it. We have

$$2x + 3y = -2.$$
$$2(-1 + 4y) + 3y = -2, \quad \text{substituting } (-1 + 4y) \text{ for } x$$
$$-2 + 8y + 3y = -2, \quad \text{simplifying}$$
$$-2 + 11y = -2,$$
$$11y = 0.$$
$$y = 0.$$

Now, if $y = 0$ we can find the value of x by substituting the determined value of y in either of the original equations and solving it for x. Choosing the first equation, we have

$$2x + 3y = -2.$$
$$\text{If } y = 0, \; 2x + 3 \cdot 0 = -2,$$
$$2x + 0 = -2,$$
$$2x = -2,$$
$$x = -1.$$

Therefore, the solution is $(-1, 0)$. We can verify that this point satisfies both equations simultaneously by checking it in each equation.

Check: Substitute $(-1, 0)$ in both equations.

$$2x + 3y = -2. \qquad\qquad -x + 4y = 1.$$
$$2(-1) + 3 \cdot 0 \stackrel{?}{=} -2, \qquad -(-1) + 4 \cdot 0 \stackrel{?}{=} 1,$$
$$-2 + 0 = -2, \qquad\qquad 1 + 0 = 1,$$
$$-2 = -2. \qquad\qquad\qquad 1 = 1.$$

The ordered pair (−1, 0) satisfies both equations.

How do we decide which equation to choose and what variable to solve for? Choose the one that involves the least amount of computation. If the coefficient of x or y is 1, then we should choose this variable since we need only to add or subtract terms to solve the equation for the chosen variable. If none of the variables has a coefficient of 1, then the choice is a matter of personal preference.

EXAMPLE 2

Solve the given system of equations by the substitution method.

$$2x + y = 0.$$
$$3x + 2y = 1.$$

Solution

We solve the first equation for y since its coefficient is 1.

$$2x + y = 0.$$
$$y = 0 - 2x,$$
$$y = -2x.$$

Next, substitute the resulting expression for y in the other equation.

$$3x + 2y = 1.$$
$$3x + 2(-2x) = 1, \quad \text{substituting } (-2x) \text{ for } y$$
$$3x - 4x = 1,$$
$$-x = 1,$$
$$x = -1.$$

Replace x by −1 in either of the original equations and solve it for y.

$$2x + y = 0.$$
$$\text{If } x = -1, \; 2(-1) + y = 0,$$
$$-2 + y = 0,$$
$$y = 2.$$

Therefore, the solution is (−1, 2).

Check: Substitute (−1, 2) in both equations.

$$2x + y = 0. \qquad\qquad 3x + 2y = 1.$$
$$2(-1) + 2 \stackrel{?}{=} 0, \qquad 3(-1) + 2(2) \stackrel{?}{=} 1,$$
$$-2 + 2 = 0, \qquad\qquad -3 + 4 = 1,$$
$$0 = 0. \qquad\qquad\qquad 1 = 1.$$

Systems of Equations   511

EXAMPLE 3

Solve the given system of equations by the substitution method.

$$2x + 4y = 6.$$
$$6x + 2y = 2.$$

Solution

Since neither of the equations contains a variable whose coefficient is 1, we can choose either and solve it for x or y. We shall pick the first equation and solve it for x.

$$2x + 4y = 6.$$
$$2x = 6 - 4y,$$
$$\frac{1}{2}(2x) = \frac{1}{2}(6 - 4y),$$
$$x = 3 - 2y.$$

Next, substitute the resulting expression for x in the other equation.

$$6x + 2y = 2.$$
$$6(3 - 2y) + 2y = 2, \quad \text{substituting } (3 - 2y) \text{ for } x$$
$$18 - 12y + 2y = 2,$$
$$18 - 10y = 2,$$
$$-10y = -16,$$
$$y = \frac{-16}{-10} = \frac{8}{5}.$$

Replace y by $\frac{8}{5}$ in either of the original equations and solve it for x.

$$2x + 4y = 6.$$
$$\text{If } y = \frac{8}{5}, \quad 2x + 4\left(\frac{8}{5}\right) = 6,$$
$$2x + \frac{32}{5} = 6,$$
$$2x = 6 - \frac{32}{5},$$
$$2x = \frac{30}{5} - \frac{32}{5},$$
$$2x = -\frac{2}{5},$$

$$\frac{1}{2}(2x) = \frac{1}{2}\left(-\frac{2}{5}\right),$$

$$x = -\frac{1}{5}.$$

Therefore, the solution is $\left(-\frac{1}{5}, \frac{8}{5}\right)$.

Check: Substitute $\left(-\frac{1}{5}, \frac{8}{5}\right)$ in both equations.

$$2x + 4y = 6. \qquad\qquad 6x + 2y = 2.$$

$$2\left(-\frac{1}{5}\right) + 4\left(\frac{8}{5}\right) \stackrel{?}{=} 6, \qquad 6\left(-\frac{1}{5}\right) + 2\left(\frac{8}{5}\right) \stackrel{?}{=} 2,$$

$$-\frac{2}{5} + \frac{32}{5} = 6, \qquad\qquad -\frac{6}{5} + \frac{16}{5} = 2,$$

$$\frac{30}{5} = 6, \qquad\qquad\qquad \frac{10}{5} = 2,$$

$$6 = 6. \qquad\qquad\qquad 2 = 2.$$

EXAMPLE 4

Solve the given system of equations by the substitution method.

$$2x + 3y = 1.$$
$$3x + 4y = 1.$$

Solution

Neither of the equations contains a variable whose coefficient is 1. We can choose either of the equations and solve it for x or y. We shall pick the first equation and solve it for y. (Note: The choice is arbitrary; we could solve it for x.)

$$2x + 3y = 1.$$
$$3y = 1 - 2x,$$
$$\frac{1}{3}(3y) = \frac{1}{3}(1 - 2x),$$
$$y = \frac{1 - 2x}{3}.$$

Next, substitute the resulting expression for x in the other equation.

Systems of Equations

$$3x + 4y = 1.$$
$$3x + 4\left(\frac{1-2x}{3}\right) = 1, \quad \text{substituting } \left(\frac{1-2x}{3}\right) \text{ for y}$$
$$3x + \frac{4-8x}{3} = 1,$$
$$3\left(3x + \frac{4-8x}{3}\right) = 3 \cdot 1, \quad \text{multiplying both sides of the equation by 3 to eliminate fractions.}$$
$$9x + 4 - 8x = 3,$$
$$4 + x = 3,$$
$$x = -1.$$

Replace x by -1 in either of the original equations and solve it for y.

$$2x + 3y = 1.$$
$$\text{If } x = -1, \; 2(-1) + 3y = 1,$$
$$-2 + 3y = 1,$$
$$3y = 3,$$
$$y = 1.$$

Therefore, the solution is $(-1, 1)$.

Check: Substitute $(-1, 1)$ in both equations.

$$2x + 3y = 1. \qquad\qquad 3x + 4y = 1.$$
$$2(-1) + 3 \cdot 1 \stackrel{?}{=} 1, \qquad 3(-1) + 4 \cdot 1 \stackrel{?}{=} 1,$$
$$-2 + 3 = 1, \qquad\qquad -3 + 4 = 1,$$
$$1 = 1. \qquad\qquad\qquad 1 = 1.$$

In the three examples we have just examined, each of the systems of linear equations had an ordered pair as a solution. That is, in each case the pair of lines intersected at a point. But recall that the two lines can be parallel and then there will be no solution for the given system of equations, or, the solution is the empty set ∅. When solving such a system by the substitution method we will obtain a *false statement of equality*. This is our clue that the equations are *inconsistent*. For example, consider the equations $x + y = 1.$
$$x + y = 3.$$

Solving the first equation for x, we have

$$x + y = 1.$$
$$x = 1 - y.$$

Substituting this expression for x in the second equation gives us

$$x + y = 3.$$
$$1 - y + y = 3,$$
$$1 = 3. \quad \text{(a false statement)}$$

Therefore, the equations are inconsistent. We could have also determined this by examining the slopes of the lines. Recall that parallel lines have equal slopes and different y-intercepts. If we write each equation in the slope-intercept form (y = mx + b), we note that the equations have equal slopes and are therefore parallel.

$$x + y = 1. \qquad x + y = 3.$$
$$y = -x + 1. \qquad y = -x + 3.$$

Hence, they have no common point of intersection and we say that the equations are inconsistent.

Two linear equations can also be *dependent*. That is, they are equations of the same line. When solving such a system of equations by the substitution method, we will obtain a *true statement of equality.* This is our clue that the equations are dependent. For example, consider the equations

$$x - 2y = 3.$$
$$-3x + 6y = -9.$$

Solving the first equation for x, we have

$$x - 2y = 3.$$
$$x = 3 + 2y.$$

Substituting this expression for x in the second equation gives us

$$-3x + 6y = -9.$$
$$-3(3 + 2y) + 6y = -9,$$
$$-9 - 6y + 6y = -9,$$
$$-9 = -9. \quad \text{(a true statement)}$$

Therefore, the equations are dependent.

EXAMPLE 5

Solve the given system of equations by the substitution method.

$$6x - 8y = 8.$$
$$-3x + 4y = 8.$$

Systems of Equations

Solution

Choosing the second equation and solving it for y, we have

$$-3x + 4y = 8.$$
$$4y = 8 + 3x,$$
$$\frac{1}{4} \cdot 4y = \frac{1}{4}(8 + 3x),$$
$$y = \frac{8 + 3x}{4}.$$

Substituting the resulting expression for y in the other equation,

$$6x - 8y = 8.$$
$$6x - 8\left(\frac{8 + 3x}{4}\right) = 8,$$
$$6x - 2(8 + 3x) = 8,$$
$$6x - 16 - 6x = 8,$$
$$-16 = 8. \quad \text{(a false statement)}$$

Therefore, the equations are inconsistent. That is, the graphs do not intersect and the solution is the empty set ∅.

EXERCISES FOR SECTION 9.2

(Answers given to exercises marked *)

Solve each system of equations by the substitution method.

*1. $2x - y = 5.$
   $2x + 3y = 1.$

2. $3x - y = 5.$
   $x + 2y = 4.$

*3. $2x + y = 6.$
   $x - 3y = 10.$

4. $2x - y = 3.$
   $3x + 3y = 9.$

*5. $2x - y = 5.$
   $x + y = -2.$

6. $x + y = 29.$
   $x - 3y = 5.$

*7. $x + y = 6.$
   $x + y = 1.$

8. $x - y = 1.$
   $x + y = -5.$

*9. $x + 2y = 3.$
   $-x + y = -6.$

10. $3x + 3y = 6.$
    $-4x - 2y = -6.$

*11. $2x - 3y = 6.$
    $-4x + 6y = 18.$

12. $-2x + 2y = -2.$
    $-3x - 3y = 15.$

*13. $-2x + y = 4.$
    $6x - 3y = -12.$

14. $2x - y = -3.$
    $-x + 2y = 0.$

*15. $x + 2y = 6.$
    $3x - y = -3.$

16. $x - 2y = 6.$
    $-3x + 6y = 0.$

*17. $x + y = 1.$
     $x - y = 1.$

18. $x - 2y = 4.$
    $-2x + 4y = -8.$

*19. $2x + 10y = 1.$
     $x + 4y = 0.$

20. $2x + y = 13.$
    $x + 3y = 7.$

*21. $3x + y = 2.$
     $5x - 3y = -13.$

22. $x + 3y = 9.$
    $4x - 2y = 1.$

*23. $2x + y = 1.$
     $6x + 2y = 1.$

24. $4x + 2y = 3.$
    $5x - 2y = 1.$

9.3 SOLUTION OF SYSTEMS OF EQUATIONS BY ADDITION

Thus far we have examined two methods for solving systems of linear equations. The first method is by graphing the given equations and finding the coordinates of the point of intersection (if it exists). This technique does not always produce the most accurate solutions. The second method is by means of substitution and although precise, it can be time consuming and cumbersome. In this section, we shall examine a third technique for solving systems of linear equations, the *addition* method.

Consider the following system of equations:

$$2x + y = 3.$$
$$x + 2y = 3.$$

The solution for this system is the ordered pair (1, 1). Now we combine the two equations by means of addition. We have $3x + 3y = 6$, and note that (1, 1) is also a solution of this new equation. This example is merely an application of a theorem which states that if an ordered pair (x, y) is a solution for the equations $ax + by + c = 0$ and $dx + ey + f = 0$, then the ordered pair (x, y) is also a solution for the sum of the two equations, $ax + by + c + dx + ey + f = 0$. In other words, if an ordered pair satisfies a system of equations, then it satisfies the sum of the equations.

How can we use this idea to solve systems of linear equations? Consider the following system:

$$x + y = 2.$$
$$x - y = 4.$$

If we add these two equations, we have $2x = 6$. Note that we have eliminated one of the variables and now we can solve the resulting equation.

$$2x = 6.$$
$$x = 3.$$

Systems of Equations

Now, if x = 3, we can replace x by 3 in either of the original equations and solve it for y.

$$x + y = 2.$$
$$\text{If } x = 3, 3 + y = 2,$$
$$y = -1.$$

Therefore, the solution is (3, −1). We check the solution by substituting (3, −1) in both equations.

$x + y = 2.$	$x - y = 4.$	The order pair (3, −1) satisfies both equations.
$3 + (-1) \stackrel{?}{=} 2,$	$3 - (-1) \stackrel{?}{=} 4,$	
$2 = 2.$	$3 + 1 = 4,$	
	$4 = 4.$	

We have seen that the solution to a system of linear equations also satisfies the sum of the equations. The reason for adding the two equations is to *eliminate* one of the variables and solve the resulting equation for the remaining variable. In order to apply this technique, we must have corresponding variables with the same numerical coefficient and opposite signs. That is, they must be additive inverses of each other.

EXAMPLE 6

Solve the given system of equations by the addition method.

$$3x + 2y = 6,$$
$$5x - 2y = 2.$$

Solution

If we add the two equations, the y variable is eliminated and we have

$$8x = 8.$$

Solving for x, x = 1.

Replace x by 1 in either of the original equations and solve it for y.

$$3x + 2y = 6.$$
$$\text{If } x = 1, 3 \cdot 1 + 2y = 6,$$
$$3 + 2y = 6,$$
$$2y = 3,$$
$$y = \frac{3}{2}.$$

Therefore, the solution is $\left(1, \frac{3}{2}\right)$.

Check: Substitute $\left(1, \frac{3}{2}\right)$ in both original equations.

$$3x + 2y = 6. \qquad\qquad 5x - 2y = 2.$$
$$3 \cdot 1 + 2 \cdot \frac{3}{2} \stackrel{?}{=} 6, \qquad\qquad 5 \cdot 1 - 2 \cdot \frac{3}{2} \stackrel{?}{=} 2,$$
$$3 + 3 = 6, \qquad\qquad 5 - 3 = 2,$$
$$6 = 6. \qquad\qquad 2 = 2.$$

EXAMPLE 7

Solve the given system of equations by the addition method.

$$3x + 2y = 6.$$
$$x - y = 2.$$

Solution

Note: We cannot use the addition method (yet) for this system. If we add the two equations, we have $4x + y = 8$ and one of the variables is not eliminated. Therefore, we cannot determine a value for x and y that will satisfy both the equations simultaneously.

We must now expand upon our addition technique so that it will work in all cases. Note that at this particular time we can solve the given system of linear equations in Example 7 only by graphing or substitution.

In Chapter 2, we solved equations in one unknown by transforming the given equation into an equivalent equation. One of the properties that we made use of was the Multiplication Property of Equality. It states

> If both members of an equation are multiplied by the same expression (provided the expression does not equal 0), then the resulting equation is equivalent to the given equation.

That is, if a, b, and c are real numbers ($c \neq 0$) and $a = b$, then $a = b$ is equivalent to $a \cdot c = b \cdot c$. Therefore, after examining the system of equations

$$3x + 2y = 6,$$
$$x - y = 2$$

Systems of Equations

and noting that no variable will be eliminated if we add the two equations directly, we can transform one of the equations into an equivalent equation by means of the Multiplication Property of Equality. It appears that a wise choice is to multiply the second equation by 2. If we multiply the second equation by 2, one of the variables, y, will be eliminated in the sum.

$3x + 2y = 6.$
$x - y = 2.$ multiply both members of the equation by 2

The system now appears as
$$3x + 2y = 6.$$
$$2x - 2y = 4.$$

If we add the two equations, the y variable is eliminated and we have
$$5x = 10.$$

Solving for x, $x = 2.$

Replace x by 2 in either of the original equations and solve it for y.

If $x = 2$, $3 \cdot 2 + 2y = 6,$
$6 + 2y = 6,$
$2y = 0,$
$y = 0.$

Therefore, the solution is (2, 0) and it can be verified by a check. It should be noted that we could have multiplied the second equation by -3. This would enable us to eliminate the x variable instead of the y variable. The process is illustrated below.

$3x + 2y = 6.$
$x - y = 2.$ multiply both members of the equation by -3

The system now appears as $3x + 2y = 6.$
$-3x + 3y = -6.$

Adding the two equations, we have $5y = 0$ and $y = 0$. Replace y by 0 in either of the original equations and solve it for x.

If $y = 0$, $3x + 2y = 6.$
$3x + 2 \cdot 0 = 6,$
$3x + 0 = 6,$
$3x = 6,$
$x = 2.$

Therefore, the solution is (2, 0) which is exactly the same solution that we obtained previously when we first eliminated the y variable. Normally, we will always have a choice as to which variable we want to eliminate first.

> To eliminate a variable in a system of equations by the addition method, we must multiply both members of an equation by the same real number so that a set of corresponding variables will have numerical coefficients that are additive inverses of each other. After performing this multiplication, we can add the two equations and eliminate the variable whose coefficients are additive inverses of each other.

EXAMPLE 8

Solve the given system of equations by the addition method.

$$3x - y = 5.$$
$$x + 2y = 4.$$

Solution

If we add the two equations immediately, then neither of the variables will be eliminated. Hence, we must multiply one of the equations by some number so that the coefficients of one of the variables will be additive inverses of each other. Choosing the first equation, we multiply it by 2.

$3x - y = 5.$ multiply both members by 2
$x + 2y = 4.$

The system now appears as

$$6x - 2y = 10.$$
$$x + 2y = 4.$$

Adding the two equations, we have $7x = 14$ and $x = 2$. Replace x by 2 in either of the original equations and solve it for y.

$$x + 2y = 4.$$
$$\text{If } x = 2, \ 2 + 2y = 4,$$
$$2y = 2,$$
$$y = 1.$$

Therefore, the solution is (2, 1).

Systems of Equations

Check: Substitute (2, 1) in both original equations.

$$3x - y = 5. \qquad x + 2y = 4.$$
$$3 \cdot 2 - 1 \stackrel{?}{=} 5, \qquad 2 + 2 \cdot 1 \stackrel{?}{=} 4,$$
$$6 - 1 = 5, \qquad 2 + 2 = 4,$$
$$5 = 5. \qquad 4 = 4.$$

Alternate Solution

Multiply both members of the second equation by -3 in order to eliminate the x's.

$3x - y = 5.$
$x + 2y = 4.$
The system now appears as:
$3x - y = 5.$
$-3x - 6y = -12.$

Adding the two equations, we have

$$-7y = -7 \text{ and } y = 1.$$

Replace y by 1 in either of the original equations and solve it for x.

$$3x - y = 5.$$
$$\text{If } y = 1, \ 3x - 1 = 5,$$
$$3x = 6,$$
$$x = 2.$$

Therefore, the solution is (2, 1).

EXAMPLE 9

Solve the given system of equations by the addition method.

$$2x + 4y = 6.$$
$$6x + 2y = 2.$$

Solution

Multiply the first equation by -3 so that the x terms will have numerical coefficients that are additive inverses of each other.

$2x + 4y = 6$
$6x + 2y = 2$
The system now appears as:
$-6x - 12y = -18.$
$6x + 2y = 2.$

Adding the two equations, we have $-10y = -16$,

$$y = \frac{-16}{-10},$$

$$y = \frac{8}{5}.$$

Replace y by $\frac{8}{5}$ in either of the original equations and solve it for x.

$$2x + 4y = 6.$$

If $y = \frac{8}{5}$, $2x + 4 \cdot \frac{8}{5} = 6$,

$$2x + \frac{32}{5} = 6,$$

$$2x = 6 - \frac{32}{5},$$

$$2x = \frac{30}{5} - \frac{32}{5},$$

$$2x = -\frac{2}{5},$$

$$\frac{1}{2}(2x) = \frac{1}{2}\left(-\frac{2}{5}\right),$$

$$x = -\frac{1}{5}.$$

Therefore, the solution is $\left(-\frac{1}{5}, \frac{8}{5}\right)$ and it can be verified by a check. Note that we could have chosen to multiply the second equation by −2, which would have eliminated the y variables first. This is strictly a matter of personal preference.

To eliminate a variable in a system of equations by the addition method, it is often necessary to multiply *both equations* by real numbers so that a set of corresponding variables will have numerical coefficients that are additive inverses of each other. For example, consider the following system of equations:

$$3x + 2y = 5.$$
$$4x + 3y = 7.$$

We multiply both equations by the appropriate real numbers to have the desired situation. If we multiply the first equation by 4, then we should

Systems of Equations

multiply the second equation by −3 to eliminate the x's by the addition method.

$3x + 2y = 5.$ multiply by 4 $12x + 8y = 20.$
                          which gives us
$4x + 3y = 7.$ multiply by −3 $-12x - 9y = -21.$

Note that we also could have multiplied the first equation by −4 and the second equation by 3. If we choose to eliminate the y's, then we can multiply the first equation by 3 and the second equation by −2. That is

$3x + 2y = 5.$ multiply by 3 $9x + 6y = 15.$
                          which gives us
$4x + 3y = 7.$ multiply by −2 $-8x - 6y = -14.$

Or we can multiply the first equation by −3 and the second equation by 2. Following is a complete solution:

Given: $3x + 2y = 5.$ multiply by 4
         $4x + 3y = 7.$ multiply by −3

which gives us $12x + 8y = 20.$ adding the $-y = -1.$
                $-12x - 9y = -21.$ two equations,
                                    we have $y = 1.$

Replace y by 1 in either of the original equations and solve it for x.

                $3x + 2y = 5.$
If $y = 1$, $3x + 2 \cdot 1 = 5,$
            $3x + 2 = 5,$
            $3x = 3,$
            $x = 1.$ Therefore, the solution is (1, 1).

Check: Substitute (1, 1) in both original equations.

          $3x + 2y = 5.$ $4x + 3y = 7.$
          $3 \cdot 1 + 2 \cdot 1 \stackrel{?}{=} 5,$ $4 \cdot 1 + 3 \cdot 1 \stackrel{?}{=} 7,$
          $3 + 2 = 5,$ $4 + 3 = 7,$
          $5 = 5.$ $7 = 7.$

EXAMPLE 10

Solve the given system of equations by the addition method.

$$3x + 4y = 6.$$
$$2x + 3y = 5.$$

Solution

To eliminate one of the variables, we must multiply both equations by the appropriate real numbers. We choose to eliminate the y's. Therefore,

$3x + 4y = 6.$ multiply by 3 $9x + 12y = 18.$
                                which gives us
$2x + 3y = 5.$ multiply by -4 $-8x - 12y = -20.$

Adding the two equations, we have $x = -2$. Replace x by -2 in either of the original equations and solve it for y.

$$3x + 4y = 6.$$
If $x = -2$, $3(-2) + 4y = 6$,
$$-6 + 4y = 6,$$
$$4y = 12,$$
$$y = 3.$$ Therefore, the solution is $(-2, 3)$.

Check: Substitute $(-2, 3)$ in both original equations.

$$3x + 4y = 6. \qquad 2x + 3y = 5.$$
$$3(-2) + 4 \cdot 3 \stackrel{?}{=} 6, \qquad 2(-2) + 3 \cdot 3 \stackrel{?}{=} 5,$$
$$-6 + 12 = 6, \qquad -4 + 9 = 5,$$
$$6 = 6. \qquad 5 = 5.$$

EXAMPLE 11

Solve the given system of equations by the addition method.

$$-2x + 3y = -9.$$
$$-3x - 2y = 19.$$

Solution

If we choose to eliminate the x's then we can multiply the first equation by 3 and the second equation by -2.

$-2x + 3y = -9.$ multiply by 3 $-6x + 9y = -27.$
                                which gives us
$-3x - 2y = 19.$ multiply by -2 $6x + 4y = -38.$

Adding the two equations, we have $13y = -65,$
$$y = -5.$$

Replace y by -5 in either of the original equations and solve for x.

Systems of Equations

$$-2x + 3y = -9.$$
If $y = -5$, $-2x + 3(-5) = -9$,
$$-2x - 15 = -9,$$
$$-2x = -9 + 15 = 6,$$
$$x = -3. \quad \text{Therefore, the solution is } (-3, -5).$$

Check: Substitute $(-3, -5)$ in both original equations.

$-2x + 3y = -9.$ $\qquad\qquad -3x - 2y = 19.$
$-2(-3) + 3(-5) \stackrel{?}{=} -9,$ $\qquad -3(-3) - 2(-5) \stackrel{?}{=} 19,$
$6 - 15 = -9,$ $\qquad\qquad\qquad 9 + 10 = 19,$
$-9 = -9.$ $\qquad\qquad\qquad\qquad 19 = 19.$

EXAMPLE 12

Solve the given system of equations by the addition method.

$$24x + 30y = 22.$$
$$18x + 12y = 13.$$

Solution

If we choose to eliminate the y's then we can multiply the first equation by 12 and the second equation by -30. But it is much easier to multiply the first equation by 2 and the second equation by -5 so that the corresponding coefficients will be 60 and -60. Therefore,

$24x + 30y = 22.$ multiply by 2 $48x + 60y = 44.$
                    which gives us
$18x + 12y = 13.$ multiply by -5 $-90x - 60y = -65.$

Adding the two equations, we have $\quad -42x = -21,$

$$x = \frac{-21}{-42} = \frac{1}{2}.$$

Replace x by $\frac{1}{2}$ in either of the original equations and solve it for x.

$$24x + 30y = 22.$$
If $x = \frac{1}{2}$, $24 \cdot \frac{1}{2} + 30y = 22,$
$$12 + 30y = 22,$$
$$30y = 10,$$

$$y = \frac{10}{30} = \frac{1}{3}.$$ Therefore, the solution set is $\left(\frac{1}{2}, \frac{1}{3}\right)$.

Check: Substitute $\left(\frac{1}{2}, \frac{1}{3}\right)$ in both original equations.

$$24x + 30y = 22. \qquad\qquad 18x + 12y = 13.$$
$$24 \cdot \frac{1}{2} + 30 \cdot \frac{1}{3} \stackrel{?}{=} 22, \qquad 18 \cdot \frac{1}{2} + 12 \cdot \frac{1}{3} \stackrel{?}{=} 13,$$
$$12 + 10 = 22, \qquad\qquad 9 + 4 = 13,$$
$$22 = 22. \qquad\qquad\qquad 13 = 13.$$

Recall that systems of linear equations can be *dependent* or *inconsistent.* If we solve a system of equations and obtain a true statement of equality, then the equations are dependent. If we solve a system of equations and obtain a false statement of equality, then the equations are inconsistent. These situations are true in solving systems by the addition method. For example, consider the equations

$$2x - 3y = -2.$$
$$-6x + 9y = 6.$$

Regardless of which variable we choose to eliminate, we can multiply the first equation by 3.

$$\begin{array}{l} 2x - 3y = -2. \\ -6x + 9y = 6. \end{array} \quad \text{multiply by 3} \quad \text{which gives us} \quad \begin{array}{l} 6x - 9y = -6. \\ -6x + 9y = 6. \end{array}$$

Adding the two equations, we obtain $0 + 0 = 0$ or $0 = 0$ which is a true statement of equality. Therefore, the equations are dependent. They represent the same line.

Next, consider the equations

$$2x - 3y = 7.$$
$$-8x + 12y = 1.$$

Regardless of which variable we choose to eliminate, we can multiply the first equation by 4.

$$\begin{array}{l} 2x - 3y = 7. \\ -8x + 12y = 1. \end{array} \quad \text{multiply by 4} \quad \text{which gives us} \quad \begin{array}{l} 8x - 12y = 28. \\ -8x + 12y = 1. \end{array}$$

Systems of Equations

Adding the two equations, we obtain $0 + 0 = 29$ or $0 = 29$ which is a false statement of equality. Therefore, the equations are inconsistent. That is, the graphs do not intersect and the solution is the empty set $\emptyset$.

EXERCISES FOR SECTION 9.3

(Answers given to exercises marked *)

Solve each system of equations by the addition method.

*1. $2x - y = 5$.
    $x + y = -2$.

2. $-2x - 3y = -1$.
    $2x - y = 5$.

*3. $x + 2y = 3$.
    $-x + y = -6$.

4. $3x - y = 5$.
    $x + 2y = 4$.

*5. $2x + y = 6$.
    $x - 3y = 10$.

6. $2x - y = 3$.
    $3x + 3y = 9$.

*7. $-2x + y = 4$.
    $6x - 3y = -12$.

8. $2x - y = -3$.
    $-x + 2y = 0$.

*9. $x + 2y = 6$.
    $3x - y = -3$.

10. $2x + 10y = 1$.
    $x + 4y = 0$.

*11. $2x + y = 13$.
    $x + 3y = 8$.

12. $3x + y = 2$.
    $5x - 3y = -13$.

*13. $3x + 2y = 11$.
    $5x - 3y = 31$.

14. $5x - 4y = 60$.
    $4x + 5y = 7$.

*15. $-2x + y = -2$.
    $3x + 2y = -25$.

16. $3x - 4y = 6$.
    $-5x + 8y = -14$.

*17. $5x + y = 10$.
    $-3x + 2y = -6$.

18. $3x - 4y = -5$.
    $x + 2y = 5$.

*19. $6x + 3y = 21$.
    $2x - 4y = 12$.

20. $6x - 2y = -14$.
    $2x + y = 2$.

*21. $3x - 6y = 9$.
    $-6x + 3y = 9$.

22. $3x + 3y = 12$.
    $8x + 4y = 40$.

*23. $4x + 8y = 32$.
    $9x - 3y = -33$.

24. $2x + 12y = 56$.
    $3x - 6y = -12$.

*25. $-2x + 4y = 8$.
    $6x + 3y = -9$.

26. $6x - 3y = 9$.
    $5x + 5y = -30$.

*27. $5x + 10y = 35$.
    $6x - 3y = -3$.

28. $4x + 8y = 16$.
    $9x - 3y = 15$.

*29. $2x - 6y = 10$.
    $x + y = 29$.

30. $10x - 5y = 25$.
    $4x + 4y = -8$.

*31. $6x - 3y = 9$.
$4x + 4y = 12$.

32. $8x + 4y = 24$.
$3x - 9y = 30$.

*33. $12x - 4y = 20$.
$5x + 10y = 20$.

9.4 APPLICATIONS

Previously we examined applications (word problems) whose solutions involved fractional equations and quadratic equations. We shall now examine some applications whose solution can be determined by using a system of equations. For example, consider the following problem:

> Find two positive numbers whose sum is 24 and whose difference is 2.

We begin the solution by letting x equal one number and y equal the other number. Since their sum is 24, we have the equation $x + y = 24$. Since their difference is 2, we also have the equation $x - y = 2$. Thus we have the following system of equations:

$$x + y = 24.$$
$$x - y = 2.$$

We can solve this system by the addition method. Adding the two equations, we have

$$2x = 26,$$
$$x = 13.$$

Replace x by 13 in either of the original equations and solve it for y.

$x + y = 24$.
If $x = 13$, $13 + y = 24$,
$y = 11$.

Therefore, the two numbers are 13 and 11.

Check: If $x = 13$ and $y = 11$, $x + y \stackrel{?}{=} 24$ and $x - y \stackrel{?}{=} 2$.
$13 + 11 = 24$, $13 - 11 = 2$,
$24 = 24$. $2 = 2$.

EXAMPLE 13

The sum of two positive numbers is 11. When the smaller number is subtracted from twice the larger number, the result is 10. Find the numbers.

Systems of Equations 529

Solution

Let $x =$ the larger number. Let $y =$ the smaller number. Therefore, we have the following system of equations:

$$x + y = 11.$$
$$2x - y = 10.$$

Adding the two equations, we have $3x = 21,$
$x = 7.$

Replace x by 7 in either of the original equations,

$x + y = 11.$
If $x = 7$, $7 + y = 11,$ Therefore, the two numbers are 7 and 4.
$y = 4.$

Check: If $x = 7$ and $y = 4$, $x + y \stackrel{?}{=} 11$ and $2x - y \stackrel{?}{=} 10.$
$7 + 4 = 11,$ $2 \cdot 7 - 4 = 10,$
$11 = 11.$ $14 - 4 = 10,$
$10 = 10.$

To solve any word problem using a system of equations, we must use care in designating each unknown quantity by a particular variable. The relations between the variables usually can be determined directly from the problem. If the relationships cannot be determined directly, then there are implied relationships that we are assumed to know. Examples of relationships that are assumed to be known are: distance equals rate times time; the area of a triangle is one-half the product of its base and height.

EXAMPLE 14

A clerk in a store wants to mix some hickory nuts selling at 65 cents a pound with some walnuts selling at 95 cents a pound. How many pounds of each are needed to make a mixture of 30 pounds which will sell for 75 cents a pound?

Solution

Let $x =$ number of pounds of hickory nuts.
$y =$ number of pounds of walnuts.

Using the given information, one of the equations is x + y = 30. If hickory nuts cost $0.65 a pound, then two pounds cost .65(2) and three pounds cost .65(3). Hence x pounds would cost .65x. Similarly, y pounds of walnuts would cost .95y. The total cost of the new mixture of 30 pounds at $.75 a pound is .75(30) or $22.50. Therefore, the cost of the new mixture is

$$.65x + .95y = 22.50$$

and if we multiply the equation by 100 to eliminate decimals, we have

$$65x + 95y = 2250$$

and the system of equations to be solved is

$$x + y = 30.$$
$$65x + 95y = 2250.$$

Solving the first equation for x, we have

$$x = 30 - y.$$

Substitute the resulting expression for x in the second equation,

$$65x + 95 = 2250.$$
$$65(30 - y) + 95y = 2250,$$
$$1950 - 65y + 95y = 2250,$$
$$1950 + 30y = 2250,$$
$$30y = 300,$$
$$y = 10.$$

If x + y = 30 and y = 10, then x + 10 = 30, and x = 20. Hence, the new mixture should contain 20 pounds of hickory nuts and 10 pounds of walnuts.

Check: x + y = 30.   65x + 95y = 2250.
        20 + 10 ≟ 30,   65(20) + 95(10) ≟ 2250,
        30 = 30.   1300 + 950 = 2250,
                               2250 = 2250.

Note: We must also check in the original problem as all conditions must be satisfied.

Systems of Equations   531

EXAMPLE 15

The sum of the digits of a two-digit number is 11. If the digits are reversed, the new number is 9 less than the original number. Find the original number.

Solution

Any two-digit number is composed of a tens digit and a units digit. For example, consider 32. The tens digit is 3 and the units digit is 2. We can express 32 as $10 \cdot 3 + 2$. Therefore, let $t =$ the tens digit of the number. Let $u =$ the units digit of the number. Hence,

$10t + u =$ the original number.
$10u + t =$ the number with the digits reversed.

Since the sum of the digits is 11, we have $t + u = 11$. Since the number with the digits reversed is 9 less than the original number, we have $10u + t = 10t + u - 9$. Simplifying this equation, we have $-9t + 9u = -9$ and the system of equations to be solved is

$$t + u = 11.$$
$$-9t + 9u = -9.$$

$t + u = 11.$ multiply by 9   which gives us $9t + 9u = 99.$
$-9t + 9u = -9.$ $-9t + 9u = -9.$

Adding the two equations, we obtain $18u = 90$,
$$u = 5.$$

If $t + u = 11$ and $u = 5$, then $t + 5 = 11$ and $t = 6$. Therefore, the original number is 65.

Check: $t + u = 11.$ $10u + t = 10t + u - 9.$
         $6 + 5 \stackrel{?}{=} 11,$ $10 \cdot 5 + 6 \stackrel{?}{=} 10 \cdot 6 + 5 - 9,$
         $11 = 11.$ $50 + 6 = 60 + 5 - 9,$
                                     $56 = 56.$

EXAMPLE 16

It takes one hour for Lewis to travel upstream in his boat, a distance of five miles. Traveling the same distance downstream in his boat, it only takes Lewis 30 minutes. Find the rate of the boat in still water, and the rate of the current.

Solution

Some uniform motion problems are easier to solve when two variables are used, but we still must make use of the formula d = rt, where d represents the distance traveled, r represents the rate of speed, and t represents the time. Let x = rate of the boat in still water. Let y = rate of the current. Then x + y = rate of the boat downstream; x − y = rate of the boat upstream.

Using the formula d = rt, we have the following system of equations:

$$5 = (x - y)1.$$

$$5 = (x + y)\frac{1}{2}.$$ Note: 30 minutes = $\frac{1}{2}$ hour, since r and t must be expressed in the same unit of time.

Simplifying: $5 = x - y$ and $\qquad 5 = x - y.$

$$5 = \frac{1}{2}x + \frac{1}{2}y, \qquad 10 = x + y.$$

Adding the two equations, we obtain $15 = 2x$.

$$\frac{15}{2} = x,$$

$$7\frac{1}{2} = x.$$

If $5 = (x - y)1$, then $5 = \left(\frac{15}{2} - y\right)1.$

$$5 = \frac{15}{2} - y,$$

$$\frac{10}{2} - \frac{15}{2} = -y, \qquad \text{Therefore, } x = 7\frac{1}{2} \text{ mph.}$$

$$-\frac{5}{2} = -y, \qquad\qquad y = 2\frac{1}{2} \text{ mph.}$$

$$\frac{5}{2} = y.$$

Check: $\qquad 5 = (x - y)1. \qquad\qquad 5 = (x + y)\frac{1}{2}.$

$$5 \stackrel{?}{=} \left(\frac{15}{2} - \frac{5}{2}\right)1, \qquad 5 \stackrel{?}{=} \left(\frac{15}{2} + \frac{5}{2}\right)\frac{1}{2},$$

$$5 = \frac{10}{2} \cdot 1, \qquad\qquad 5 = \frac{20}{2} \cdot \frac{1}{2},$$

$$5 = 5. \qquad\qquad\qquad 5 = 5.$$

Systems of Equations

To solve an application (word) problem by means of a system of equations, let each unknown be represented by a variable. Determine the relationships between the variables from the problem, or from the relationship that is assumed to be known, to generate a system of equations. Solve the system and check the solution.

EXERCISES FOR SECTION 9.4

(Answers given to exercises marked *)

*1. The sum of two positive numbers is 12. When the larger number is subtracted from twice the smaller number, the result is 3. Find the numbers.

2. Find two numbers whose sum is 0 and whose difference is 2.

*3. The difference between two positive numbers is 4. When twice the larger number is subtracted from three times the smaller number, the result is −2. Find the numbers.

4. If each of two positive numbers is doubled, then their sum is 48. When three times the smaller number is added to the larger number, the result is 46. Find the numbers.

*5. One positive number is 5 less than twice another positive number. Three times the smaller number is equal to twice the larger number. Find the numbers.

6. The Delta Psi Sorority sponsored a concert. The total number of tickets sold was 400. The price per ticket was $2.00 at the door and $1.50 in advance. If the receipts from ticket sales totaled $700, how many tickets were purchased at the door?

*7. A clerk mixes hard candy worth 40 cents a pound with chocolate candy worth 90 cents a pound to obtain a mixture worth 75 cents a pound. How many pounds of hard candy and how many pounds of chocolate candy should the clerk mix together to obtain 100 pounds of the 75-cent mixture?

8. A clerk in a store wants to mix 75 pounds of hickory nuts that sell for 60 cents a pound with some walnuts that sell for 50 cents a pound. How many pounds of walnuts should she mix with the hickory nuts so that she can sell the new mixture for 52 cents a pound?

*9. A certain type of latex semi-gloss paint worth $6.00 a gallon is mixed with a thinner worth $0.80 a gallon to form a cheaper-grade paint worth $2.36 a gallon. How many gallons of paint and how many gallons of thinner are needed to make 50 gallons of the new mixture?

10. Tickets for a concert cost $2.00 each if purchased at the advance sale before the day of the concert, and $2.75 each if purchased on the day of the concert. For this particular concert, 1600 tickets were sold and the total receipts were $3500. How many tickets were purchased at the advance sale and how many were purchased on the day of the concert?

*11. How much water must be added to 30 litres of a 10 per cent solution of salt and water in order to reduce it to a solution that is 6 per cent salt?

12. How much water must be evaporated from 40 litres of an 8 per cent solution of salt and water to produce a solution that is 15 per cent salt?

*13. The sum of the digits of a positive two-digit number is 15. If the digits are reversed, the new number is 9 less than the original number. Find the original number.

14. The sum of the digits of a positive two-digit number is 13. If the digits are reversed, the new number is 9 more than the original number. Find the original number.

*15. The sum of the digits of a positive two-digit number is 10. If the digits are reversed, the new number is 1 less than twice the original number. Find the original number.

16. The units digit of a positive two-digit number is 3 more than the tens digit. If the digits are reversed, the new number is 20 less than twice the original number. Find the original number.

*17. The sum of the digits of a positive two-digit number is 10. If the digits are reversed, the new number is 2 less than 3 times the original number. Find the original number.

18. If the units digit is subtracted from the tens digit for a certain positive two-digit number, the result is -6. If the digits are reversed, the new number is 15 more than twice the original number. Find the original number.

*19. The units digit of a positive two-digit number is 4 more than the tens digit. If the number is divided by the tens digit, the quotient is 12. Find the number.

20. A small single engine plane flies at 150 miles per hour in still air. On a certain day it took the same amount of time to fly 250 miles with the

wind as it did to return a distance of 200 miles against the same wind. Find the rate of the wind. (Express your answer to the nearest tenth.)

*21. A small single engine plane flies at 100 miles per hour in still air. On a certain day it took the same amount of time to fly 300 miles with the wind as it did to return a distance of 250 miles against the same wind. Find the rate of the wind. (Express your answer to the nearest tenth.)

22. On a certain day when Cal was practicing, he flew 240 miles with the wind in 80 minutes, and 180 miles against the wind in 90 minutes. Find the rate of the plane in still air, and the rate of the wind.

*23. It takes one hour for Benny to travel downstream in his boat, a distance of 12 miles. Traveling the same distance upstream in his boat, it takes Benny 90 minutes. Find the rate of the boat in still water, and the rate of the current.

24. A certain fraction is equivalent to $\frac{4}{7}$. If the numerator of the fraction is increased by 2 and the denominator is decreased by 1, then the resulting fraction is equivalent to 1. Find the original fraction.

*25. The numerator of a fraction is 4 less than the denominator. If the numerator of the fraction is decreased by 3 and the denominator is increased by 1, then the resulting fraction is equivalent to $\frac{1}{2}$. Find the original fraction.

26. The denominator of a fraction is 4 less than the numerator. If the numerator of the fraction is increased by 3 and the denominator is decreased by 5, then the resulting fraction is equivalent to $\frac{5}{2}$. Find the original fraction.

9.5 SYSTEMS OF LINEAR EQUATIONS WITH THREE VARIABLES

In the beginning of this chapter we discussed the solution of sets of linear equations with two variables. We have examined linear equations that contained one or two variables. For example, $x = 2$, $y = 3$, and $x + y = 1$ are all linear equations. The common characteristic of these equations is that they are all first degree equations. That is, they contain only first degree terms. In general, a linear equation can contain one, two, three, four, and so on, different variables as long as all the terms are of the first degree. Therefore, an equation such as $x+y+z=1$ is called a linear equation. However, it should be noted that such an equation would be graphed on a coordinate system containing x, y, and z axes, and its graph would be a plane, not a line. But we still can refer to the equation $x+y+z=1$ as a linear equation.

A linear equation has an infinite number of solutions. For example, (0, 1), (1, 0) and (2, −1) are solutions of $x+y=1$. These solutions are called ordered pairs. The solutions of linear equations with three variables are called *ordered triples*, and they are of the general form (x, y, z). For example, (1, 0, 0), (0, 1, 0) and (0, 0, 1) are solutions of the linear equation $x+y+z=1$. The systems of linear equations with three variables that we shall consider in this section are those that contain three equations; we shall solve three equations in three unknowns. Following is such a system:

$$3x - 2y + z = 4.$$
$$2x + 4y - 3z = 9.$$
$$-x + 8y - 2z = 4.$$

To solve a system of this nature, we want to find an ordered triple that is a common solution to all three of the given equations simultaneously. A graphical solution of such a system is practically impossible. Recall that the graph of a linear equation with three variables is a plane on the three dimensional coordinate system. Hence, we could have three planes intersecting at a common point, or intersecting at a common line. Also, the three planes could possibly have no points in common, or they might all coincide, that is, be the same plane.

Since a graphical solution is impractical, we shall develop an algebraic technique for solving three equations in three unknowns. This technique is merely an extension of the methods that we developed to solve systems containing two linear equations in two unknowns. You will recall that to solve a system of two linear equations in two unknowns, we eliminated a variable using the addition method. We shall apply this technique to solve the system given above.

Our choice is to eliminate the z variable from the first and second equation, although we could have chosen to eliminate either the x or y variable. Proceeding, we have

$3x - 2y + z = 4.$ multiply by 3 which gives us $9x - 6y + 3z = 12.$
$2x + 4y - 3z = 9.$ $\qquad\qquad\qquad\qquad\qquad\qquad\qquad\qquad$ $2x + 4y - 3z = 9.$

Adding the two equations, we have $11x - 2y = 21$. Now we have one equation in two unknowns. To obtain another equation containing only x and y, we can eliminate z from the first and third equations or from the second and third equations. But, we must be sure to use the third equation as it has not yet been involved in the solution process. Choosing the first and third equations, we have

Systems of Equations

$3x - 2y + z = 4$.   multiply by 2   which gives us $6x - 4y + 2z = 8$.
$-x + 8y - 2z = 4$. $-x + 8y - 2z = 4$.

Adding these two equations, we have $5x + 4y = 12$. We now have two equations in two unknowns.

$$11x - 2y = 21,$$
$$5x + 4y = 12,$$

and we can solve this system in the same manner that we used in Section 9.3.

$11x - 2y = 21$.   multiply by 2, which gives us $22x - 4y = 42$.
$5x + 4y = 12$. $5x + 4y = 12$.

Adding the two equations, we have $27x = 54$.
$$x = 2.$$

If $x = 2$, then from either $11x - 2y = 21$ or $5x + 4y = 12$, we see that $y = \frac{1}{2}$. If $x = 2$ and $y = \frac{1}{2}$, we replace x by 2 and y by $\frac{1}{2}$ in any of the three original equations. Choosing the first equation, we have

$$3x - 2y + z = 4.$$

If $x = 2$, $y = \frac{1}{2}$, $3(2) - 2\left(\frac{1}{2}\right) + z = 4,$ Therefore, the solution is the ordered triple:

$$6 - 1 + z = 4,$$ $\left(2, \frac{1}{2}, -1\right).$
$$5 + z = 4,$$
$$z = -1.$$

Check: Substitute $\left(2, \frac{1}{2}, -1\right)$ in each of the original equations.

$3x - 2y + z = 4.$ $2x + 4y - 3z = 9.$

$3(2) - 2\left(\frac{1}{2}\right) + (-1) \stackrel{?}{=} 4,$ $2(2) + 4\left(\frac{1}{2}\right) - 3(-1) \stackrel{?}{=} 9,$

$6 - 1 - 1 = 4,$ $4 + 2 + 3 = 9,$
$4 = 4.$ $9 = 9.$

$-x + 8y - 2z = 4.$

$-2 + 8\left(\frac{1}{2}\right) - 2(-1) \stackrel{?}{=} 4,$

$-2 + 4 + 2 = 4,$
$4 = 4.$

To solve a system of three linear equations with three variables, we proceed as follows:
1. Eliminate a variable (your choice) from two of the given equations by means of the addition method and obtain a resulting equation with two variables.
2. Using the same procedure outlined in Step 1, eliminate the *same variable* from two of the given equations. One must be the third equation not used in Step 1.
3. Solve the system of two equations in two unknowns (obtained from Steps 1 and 2) to find the value for one of the variables.
4. Replace the value of the variable (obtained in Step 3) in either of the equations obtained in Steps 1 and 2, and solve it for the remaining variable.
5. To find the value of the third variable, replace the two variables by their values (obtained in Steps 3 and 4) in one of the original equations, and solve it for the remaining variable.
6. Check the solution, the order triple, (x, y, z) in *each* of the original equations to verify that it is the simultaneous solution.

EXAMPLE 17

Solve the following system of equations:

$$x + 2y - 3z = 11.$$
$$2x + y - 3z = 13.$$
$$3x + 2y + z = 9.$$

Solution

It does not matter what variable we choose to eliminate for Steps 1 and 2 as long as we eliminate the *same* variable and use *all three* equations. We choose to eliminate x. For the first and second equations we have

$$x + 2y - 3z = 11. \qquad\qquad -2x - 4y + 6z = -22.$$
$$\text{multiply by } -2, \text{ which gives us}$$
$$2x + y - 3z = 13. \qquad\qquad 2x + y - 3z = 13.$$

Adding the two equations, we obtain $-3y + 3z = -9$. Next, we eliminate x from the first and third equations.

$$x + 2y - 3z = 11. \qquad\qquad -3x - 6y + 9z = -33.$$
$$\text{multiply by } -3, \text{ which gives us}$$
$$3x + 2y + z = 9. \qquad\qquad 3x + 2y + z = 9.$$

Systems of Equations

Adding the two equations, we obtain $-4y + 10z = -24$. Steps 1 and 2 have been completed and we now solve the resulting system of two equations in two unknowns.

$-3y + 3z = -9.$ multiply by 4 $-12y + 12z = -36$ } add
$-4y + 10z = -24.$ multiply by -3 $\underline{12y - 30z = 72.}$
$\phantom{-4y + 10z = -24.\ \text{multiply by} -3\ } -18z = 36,$
$\phantom{-4y + 10z = -24.\ \text{multiply by} -3\ \ \ \ } z = -2.$

If $-3y + 3z = -9$ and $z = -2$, then $-3y + 3(-2) = -9$,
$$-3y - 6 = -9,$$
$$-3y = -3,$$
$$y = 1.$$

Use one of the original equations to find the value of x. If $x + 2y - 3z = 11$, and $y = 1$, $z = -2$, then

$x + 2(1) - 3(-2) = 11$,
$\quad x + 2 + 6 = 11,$ Therefore, the solution is the
$\quad\quad x + 8 = 11,$ ordered triple: $(3, 1, -2)$.
$\quad\quad\quad x = 3.$

Check: Substitute $(3, 1, -2)$ in each of the original equations.

$x + 2y - 3z = 11.$ $2x + y - 3z = 13.$
$3 + 2(1) - 3(-2) \stackrel{?}{=} 11,$ $2(3) + 1 - 3(-2) \stackrel{?}{=} 13,$
$\quad 3 + 2 + 6 = 11,$ $\quad 6 + 1 + 6 = 13,$
$\quad\quad 11 = 11.$ $\quad\quad 13 = 13.$

$3x + 2y + z = 9.$
$3(3) + 2(1) + (-2) \stackrel{?}{=} 9,$
$\quad 9 + 2 - 2 = 9,$
$\quad\quad 9 = 9.$

Systems of linear equations can be *dependent* or *inconsistent*. If we solve a system of equations and obtain a true statement of equality, then the equations are dependent. If we solve a system of equations and obtain a false statement of equality, then the equations are inconsistent. This is also true for a system of linear equations with three variables.

EXAMPLE 18

Solve the following system of equations.

$$2x + y - 3z = 13.$$

$$x + 2y + z = 5.$$
$$-4x - 2y + 6z = 1.$$

Solution

First, we choose to eliminate the x from the first and second equations.

$2x + y - 3z = 13.$ $\qquad\qquad 2x + y - 3z = 13.$
$\qquad\qquad$ multiply by -2, which gives us
$x + 2y + z = 5.$ $\qquad\qquad -2x - 4y - 2z = -10.$

Adding the two equations, we obtain $-3y - 5z = 3$.
Next, we eliminate x from the second and third equations.

$x + 2y + z = 5.$ $\qquad\qquad 4x + 8y + 4z = 20.$
$\qquad\qquad$ multiply by 4, which gives us
$-4x - 2y + 6z = 1.$ $\qquad\qquad -4x - 2y + 6z = 1.$

Adding the two equations, we obtain $6y + 10z = 21$. We now solve the resulting system of two equations in two unknowns.

$-3y - 5z = 3.$ $\qquad\qquad -6y - 10z = 6.$
$\qquad\qquad$ multiply by 2, which gives us
$6y + 10z = 21.$ $\qquad\qquad 6y + 10z = 21.$

Adding the two equations, we have $\quad 0 + 0 = 27,$
a false statement of equality $\qquad\qquad 0 = 27.$

Therefore, the equations are inconsistent. There does not exist an ordered triple that satisfies the three given equations simultaneously. The solution is the empty set ∅.

EXAMPLE 19

Solve the following system of equations:

$$3x + 3y - 2z = 0.$$
$$4x - 9y + 4z = 6.$$
$$5x - 6y + 6z = 6.$$

Solution

First, we choose to eliminate y from the first and second equations.

Systems of Equations

$3x + 3y - 2z = 0.$ $\qquad\qquad\qquad 9x + 9y - 6z = 0.$
$\qquad\qquad$ multiply by 3, which gives us
$4x - 9y + 4z = 6.$ $\qquad\qquad\qquad 4x - 9y + 4z = 6.$

Adding the two equations, we obtain $13x - 2z = 6$. Next, we eliminate y from the first and third equations.

$3x + 3y - 2z = 0.$ $\qquad\qquad\qquad 6x + 6y - 4z = 0.$
$\qquad\qquad$ multiply by 2, which gives us
$5x - 6y + 6z = 6.$ $\qquad\qquad\qquad 5x - 6y + 6z = 6.$

Adding the two equations, we obtain $11x + 2z = 6$. We now solve the resulting system of two equations in two unknowns.

$13x - 2z = 6.$
$11x + 2z = 6.$
$\qquad$ Adding the two equations, we obtain $\qquad 24x = 12,$

$$x = \frac{12}{24} = \frac{1}{2}.$$

If $11x + 2z = 6$ and $x = \frac{1}{2}$ then $11 \cdot \frac{1}{2} + 2z = 6.$

$$\frac{11}{2} + 2z = \frac{12}{2},$$

$$2z = \frac{1}{2},$$

$$z = \frac{1}{4}.$$

Use one of the original equations to find the value of y.

If $4x - 9y + 4z = 6$, and $x = \frac{1}{2}$, $z = \frac{1}{4}$, then

$4 \cdot \frac{1}{2} - 9y + 4 \cdot \frac{1}{4} = 6,$

$\qquad 2 - 9y + 1 = 6,$ $\qquad$ Therefore, the solution is the
$\qquad\quad -9y + 3 = 6,$ $\qquad$ ordered triple, $\left(\frac{1}{2}, -\frac{1}{3}, \frac{1}{4}\right).$
$\qquad\qquad\; -9y = 3,$

$\qquad\qquad\quad y = -\frac{3}{9} = -\frac{1}{3}.$

Check: Substitute $\left(\frac{1}{2}, -\frac{1}{3}, \frac{1}{4}\right)$ in each of the original equations.

$$3x + 3y - 2z = 0. \qquad\qquad 4x - 9y + 4z = 6.$$

$$3\left(\frac{1}{2}\right) + 3\left(-\frac{1}{3}\right) - 2\left(\frac{1}{4}\right) \stackrel{?}{=} 0, \qquad 4\left(\frac{1}{2}\right) - 9\left(-\frac{1}{3}\right) + 4\left(\frac{1}{4}\right) \stackrel{?}{=} 6,$$

$$\frac{3}{2} - 1 - \frac{1}{2} = 0, \qquad\qquad 2 + 3 + 1 = 6,$$

$$\frac{2}{2} - 1 = 0, \qquad\qquad\qquad 6 = 6.$$

$$0 = 0.$$

$$5x - 6y + 6z = 6.$$

$$5\left(\frac{1}{2}\right) - 6\left(-\frac{1}{3}\right) + 6\left(\frac{1}{4}\right) \stackrel{?}{=} 6,$$

$$\frac{5}{2} + 2 + \frac{3}{2} = 6,$$

$$\frac{8}{2} + 2 = 6,$$

$$6 = 6.$$

EXAMPLE 20

Solve the following system of equations:

$$3x + 2y = 5.$$
$$4x - 3z = 7.$$
$$6y - 6z = -5.$$

Solution

Note that we again have a system of linear equations with three variables, but that each equation contains only two variables. Hence, we still eliminate a variable in the same manner as we have done previously and the resulting equations will contain the same two variables as one of the original equations. First, we choose to eliminate z from the second and third equations.

$4x - 3z = 7.$ multiply by -2, which gives us $\quad -8x + 6z = -14.$
$6y - 6z = -5.$ $\qquad\qquad\qquad\qquad\qquad\qquad\qquad 6y - 6z = -5.$

Systems of Equations

Note that when we add the two equations, we obtain $-8x + 6y = -19$. This equation contains the same two variables as the original first equation: $3x + 2y = 5$. We now solve the resulting system of two equations in two unknowns.

$3x + 2y = 5$. multiply by -3, which gives us $\quad -9x - 6y = -15$.
$-8x + 6y = -19$. $\hspace{5cm} -8x + 6y = -19$.

Adding the two equations, we obtain $-17x = -34$.
$$x = 2.$$

Since each of the original equations contains only two variables, we use them to find the values of y and z.

If $3x + 2y = 5$ and $x = 2$, then $3 \cdot 2 + 2y = 5$,
$$6 + 2y = 5,$$
$$2y = -1,$$
$$y = -\frac{1}{2}.$$

If $4x - 3z = 7$ and $x = 2$, then $4 \cdot 2 - 3z = 7$,
$$8 - 3z = 7,$$
$$-3z = -1,$$
$$z = \frac{1}{3}.$$
Therefore, the solution is the ordered triple $\left(2, -\frac{1}{2}, \frac{1}{3}\right)$.

Check: Substitute $\left(2, -\frac{1}{2}, \frac{1}{3}\right)$ in each of the original equations.

$\hspace{1cm} 3x + 2y = 5. \hspace{2cm} 4x - 3z = 7.$
$3 \cdot 2 + 2\left(-\frac{1}{2}\right) \stackrel{?}{=} 5, \hspace{1cm} 4 \cdot 2 - 3\left(\frac{1}{3}\right) \stackrel{?}{=} 7,$
$\hspace{1.5cm} 6 - 1 = 5, \hspace{2.3cm} 8 - 1 = 7,$
$\hspace{1.8cm} 5 = 5. \hspace{2.8cm} 7 = 7.$

$$6y - 6z = -5.$$
$$6\left(-\frac{1}{2}\right) - 6\left(\frac{1}{3}\right) \stackrel{?}{=} -5,$$
$$-3 - 2 = -5,$$
$$-5 = -5.$$

EXERCISES FOR SECTION 9.5

(Answers given to exercises marked *)

Solve the following systems of equations:

1. $x + y + z = -4.$
   $2x + 3y - z = 2.$
   $3x - y + 2z = -13.$

2. $2x + 3y - z = -3.$
   $3x - 2y + z = -2.$
   $-5x + 6y + 2z = 7.$

*3. $3x + 2y + z = 1.$
   $2x + 3y - z = 4.$
   $3x - y + 2z = -1.$

4. $x + y + z = 3.$
   $x - 2y - 3z = 0.$
   $-2x + 3y + 2z = -1.$

*5. $2x - 3y + 4z = 8.$
   $x + y + 2z = 9.$
   $x + 2y + z = 8.$

6. $2x - 3y - z = -4.$
   $3x - 2y + 4z = 8.$
   $-5x + 3y + z = 10.$

*7. $x - 2y - 3z = 12.$
   $-x + 3y - 2z = 1.$
   $2x - 5y + z = 5.$

8. $2x + y + z = 0.$
   $-3x - y + 2z = 7.$
   $x + 2y - 3z = 7.$

*9. $4x - 6y + 2z = -4.$
   $3x + 2y - z = 5.$
   $-2x + 3y - z = 2.$

10. $x - y + 2z = 2.$
    $4x - 4y + 8z = 8.$
    $-3x + 3y - 6z = -6.$

*11. $x + 2y - z = 5.$
    $3x - 4y - 2z = 7.$
    $2x + 2z = -1.$

12. $3x - 2y + 2z = 8.$
    $2x + 4y + 6z = -5.$
    $2x + 6y - 4z = -3.$

*13. $3x - 2y - 3z = -1.$
    $6x + y + 2z = 7.$
    $9x + 3y + 4z = 9.$

14. $2x + 3y - z = -1.$
    $4x + 3z = 2.$
    $x - y + 2z = 2.$

*15. $3x + 2y = 7.$
    $4x - 3z = -5.$
    $6y - 6z = -6.$

16. $3y - 2z = 4.$
    $3x - 5z = -5.$
    $5x + 7y = 14.$

*17. $2x - 3y = 2.$
    $4x - 4z = 1.$
    $3y - 8z = -3.$

18. $4y + 5z = 2.$
    $3x - 10z = -1.$
    $-6x - 8y = -4.$

*19. Find three positive numbers whose sum is 17, given that the sum of the first number and twice the second number is equal to the third number, and that if the sum of the first two numbers is subtracted from the third number then the result is 5.

Systems of Equations 545

20. Find the number of degrees contained in each angle of triangle ABC, given that the sum of the measures of angle A and angle B is equal to the measure of angle C, and angle B is equal to the measure of angle A.

*21. Find the number of degrees contained in each angle of triangle ABC, given that the measure of angle C is equal to the sum of the measures of angle A and angle B, and that twice the measure of angle A minus twice the measure of angle B is two-thirds the measure of angle C.

22. Find three numbers whose sum is 5, given that the sum of the first number and the second number is 1 less than the third number, and that twice the third number minus three times the first number equals the second number.

*23. Joey counted the coins in his bank. Upon examining the contents, he discovered that it contained only pennies, nickels and dimes, whose total value was $2.75. If the total number of coins was 62 and the number of pennies contained in the bank was twice the number of nickels, then how many of each kind of coin was in Joey's bank?

24. The sum of the digits of a three-digit number is 8. The sum of the units digit and the hundreds digit is equal to the tens digit. The number formed by reversing the digits is 198 more than the given number. Find the original number.

*25. The sum of the digits of a three-digit number is 10. The sum of the tens digit and the units digit is equal to the hundreds digit. The number formed by reversing the digits is 297 less than the given number. Find the original number.

9.6 DETERMINANTS

A rectangular array of numbers is called a *matrix*. For example, each of the following is a rectangular array of numbers, or matrix:

$$\begin{pmatrix} 1 & 2 \\ 3 & 4 \end{pmatrix} \begin{pmatrix} 1 & 1 & 2 \\ -1 & 0 & 3 \end{pmatrix} \begin{pmatrix} 3 & 1 \\ 0 & -1 \\ 2 & 3 \end{pmatrix} \begin{pmatrix} 3 & 1 & 2 & 0 \\ -1 & 2 & 3 & 4 \\ 0 & 1 & 2 & 1 \end{pmatrix}$$

A matrix can be described or named according to the number of rows and columns it contains. We can describe each of the above matrices in the following manner (proceeding from left to right): the first matrix contains two rows and two columns, and it is called a "two by two" matrix (denoted by 2 × 2); the second contains two rows and three columns, and it is called 2 × 3 matrix (read "two by three"); the third contains three rows and two columns, and is called a 3 × 2 matrix; the last matrix contains three rows and four columns, and is called a 3 × 4 matrix.

If a matrix contains the same number of rows and columns, then it is called a *square matrix*. Following are examples of square matrices. The first is a 2 × 2 matrix and the second is a 3 × 3 matrix. The number of rows is equal to the number of columns.

$$\begin{pmatrix} 1 & 2 \\ 0 & 3 \end{pmatrix} \begin{pmatrix} 0 & 1 & 3 \\ 1 & 2 & 4 \\ -1 & 2 & -1 \end{pmatrix}$$

Associated with each square matrix is a real number called the *determinant* of the matrix. For the matrix $\begin{pmatrix} 1 & 2 \\ 0 & 3 \end{pmatrix}$, the determinant is written as $\begin{vmatrix} 1 & 2 \\ 0 & 3 \end{vmatrix}$ and defined as $1 \cdot 3 - 0 \cdot 2 = 3 - 0 = 3$. In general terms, we have the 2 × 2 matrix $\begin{pmatrix} a & b \\ c & d \end{pmatrix}$ whose determinant is written $\begin{vmatrix} a & b \\ c & d \end{vmatrix}$ and is defined as $ad - cb$. The value of a 2 × 2 determinant is found by finding the product of the elements of the arrow slanting downward to the right and subtracting the product of the elements of the arrow slanting upward to the right, as illustrated in Figure 5.

$\begin{vmatrix} a & b \\ c & d \end{vmatrix} = ad - cb$.   **Figure 5**

To evaluate the determinant $\begin{vmatrix} 3 & 2 \\ 2 & 1 \end{vmatrix}$ we have

$\begin{vmatrix} 3 & 2 \\ 2 & 1 \end{vmatrix} = 3 \cdot 1 - 2 \cdot 2 = 3 - 4 = -1$

To evaluate the determinant $\begin{vmatrix} -1 & 4 \\ -3 & 2 \end{vmatrix}$ we have

$\begin{vmatrix} -1 & 4 \\ -3 & 2 \end{vmatrix} = -1 \cdot 2 - (-3)4 = -2 - (-12) = -2 + 12 = 10$

EXAMPLE 21

Evaluate each of the following determinants.

a. $\begin{vmatrix} 1 & 2 \\ 3 & 4 \end{vmatrix}$ b. $\begin{vmatrix} -1 & 2 \\ -3 & 4 \end{vmatrix}$ c. $\begin{vmatrix} -2 & 3 \\ -4 & -5 \end{vmatrix}$

Systems of Equations

Solution

The value of a 2×2 determinant $\begin{vmatrix} a & b \\ c & d \end{vmatrix}$ is defined as $ad - cb$.

a. $\begin{vmatrix} 1 & 2 \\ 3 & 4 \end{vmatrix} = 1 \cdot 4 - 3 \cdot 2 = 4 - 6 = -2.$

b. $\begin{vmatrix} -1 & 2 \\ -3 & 4 \end{vmatrix} = -1 \cdot 4 - (-3)2 = -4 - (-6) = -4 + 6 = 2.$

c. $\begin{vmatrix} -2 & 3 \\ -4 & -5 \end{vmatrix} = -2(-5) - (-4)(3) = 10 - (-12) = 10 + 12 = 22.$

In the next section we may encounter expressions such as $\dfrac{\begin{vmatrix} 2 & 1 \\ 3 & 4 \end{vmatrix}}{\begin{vmatrix} 2 & -1 \\ 3 & 5 \end{vmatrix}}$

This is a ratio of determinants, or a fraction. To evaluate such an expression, we evaluate the determinants of the numerator and denominator. That is,

$$\frac{\begin{vmatrix} 2 & 1 \\ 3 & 4 \end{vmatrix}}{\begin{vmatrix} 2 & -1 \\ 3 & 5 \end{vmatrix}} = \frac{2 \cdot 4 - 3 \cdot 1}{2 \cdot 5 - 3(-1)} = \frac{8 - 3}{10 + 3} = \frac{5}{13}.$$

EXAMPLE 22

Evaluate each of the following ratios:

a. $\dfrac{\begin{vmatrix} 3 & 4 \\ 2 & 4 \end{vmatrix}}{\begin{vmatrix} 3 & 1 \\ 4 & 2 \end{vmatrix}}$ b. $\dfrac{\begin{vmatrix} -1 & 2 \\ -3 & 4 \end{vmatrix}}{\begin{vmatrix} -5 & -3 \\ 4 & -1 \end{vmatrix}}$

Solution

a. $\dfrac{\begin{vmatrix} 3 & 4 \\ 2 & 4 \end{vmatrix}}{\begin{vmatrix} 3 & 1 \\ 4 & 2 \end{vmatrix}} = \dfrac{3 \cdot 4 - 2 \cdot 4}{3 \cdot 2 - 4 \cdot 1} = \dfrac{12 - 8}{6 - 4} = \dfrac{4}{2} = 2.$

b. $\dfrac{\begin{vmatrix} -1 & 2 \\ -3 & 4 \end{vmatrix}}{\begin{vmatrix} -5 & -3 \\ 4 & -1 \end{vmatrix}} = \dfrac{-1 \cdot 4 - (-3)2}{-5(-1) - (4)(-3)} = \dfrac{-4 - (-6)}{5 - (-12)}$

$= \dfrac{-4 + 6}{5 + 12} = \dfrac{2}{17}.$

Below is a 3 × 3, or third order, determinant.

$$\begin{vmatrix} a_1 & b_1 & c_1 \\ a_2 & b_2 & c_2 \\ a_3 & b_3 & c_3 \end{vmatrix}$$

The value of a third order determinant is defined to be $a_1b_2c_3 + b_1c_2a_3 + c_1a_2b_3 - (a_3b_2c_1 + b_3c_2a_1 + c_3a_2b_1)$. Rather than try to memorize this formula, you can obtain these values quickly and easily by rewriting the first and second columns to the right of the determinant and drawing the arrows as indicated in Figure 6.

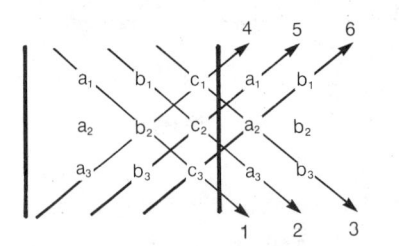

Figure 6

Now, the terms that have been defined previously for a 3 × 3 determinant can be obtained by finding the sum of the products of the elements of the arrows slanting downward to the right (arrows 1, 2, 3) and subtracting the sum of the products of the elements of the arrows slanting upward to the right (arrows 4, 5, 6). Therefore, the value of a 3 × 3 determinant is

$$a_1b_2c_3 + b_1c_2a_3 + c_1a_2b_3 - (a_3b_2c_1 + b_3c_2a_1 + c_3a_2b_1)$$

↑ ↑

sum of the products of the elements of arrows 1, 2, 3

sum of the products of the elements of arrows 4, 5, 6

To evaluate a determinant such as

$$\begin{vmatrix} 3 & 1 & 2 \\ 2 & -1 & 3 \\ 4 & 0 & 2 \end{vmatrix}$$

we rewrite the first and second columns to the right of the determinant and draw the arrows as indicated.

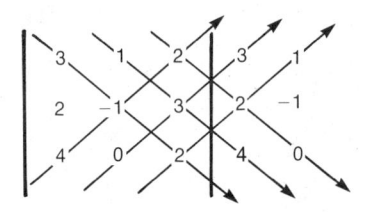

Systems of Equations 549

Next, we find the sum of the products of the elements of the arrows slanting downward to the right, and subtract the sum of the products of the elements of the arrows slanting upward to the right.

$$3(-1)(2) + 1 \cdot 3 \cdot 4 + 2 \cdot 2 \cdot 0 - (4(-1)(2) + 0 \cdot 3 \cdot 3 + 2 \cdot 2 \cdot 1)$$
$$-6 + \quad 12 + \quad 0 - (\quad -8 \quad + \quad 0 \quad + \quad 4)$$
$$6 - (\ -4)$$
$$6 + 4$$

The value of the determinant is: 10.

EXAMPLE 23

Evaluate the given determinant.

$$\begin{vmatrix} -2 & 1 & 3 \\ 1 & 2 & 4 \\ 5 & -1 & 2 \end{vmatrix}$$

Solution

Rewrite the first and second columns to the right of the determinant and draw the arrows to help determine the correct products. The value of the determinant is the sum of the products of the elements of the arrows slanting downward minus the sum of the products of the elements of the arrows slanting upward.

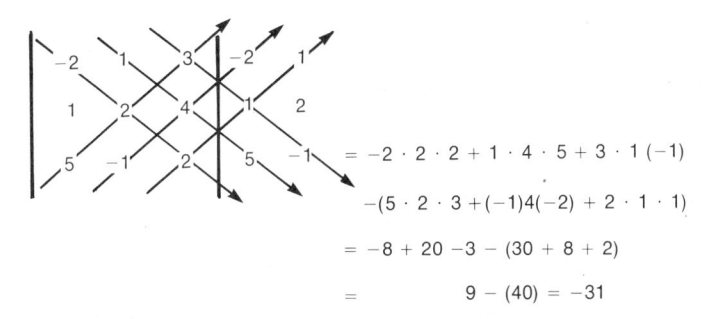

$$= -2 \cdot 2 \cdot 2 + 1 \cdot 4 \cdot 5 + 3 \cdot 1 \cdot (-1)$$
$$-(5 \cdot 2 \cdot 3 + (-1)4(-2) + 2 \cdot 1 \cdot 1)$$
$$= -8 + 20 - 3 - (30 + 8 + 2)$$
$$= 9 - (40) = -31$$

EXAMPLE 24

Evaluate the following ratio:

$$\frac{\begin{vmatrix} 1 & 3 & 5 \\ 1 & -1 & 2 \\ 2 & 0 & -2 \end{vmatrix}}{\begin{vmatrix} 3 & 2 & 1 \\ -1 & 3 & 2 \\ 0 & 2 & -2 \end{vmatrix}}$$

Solution

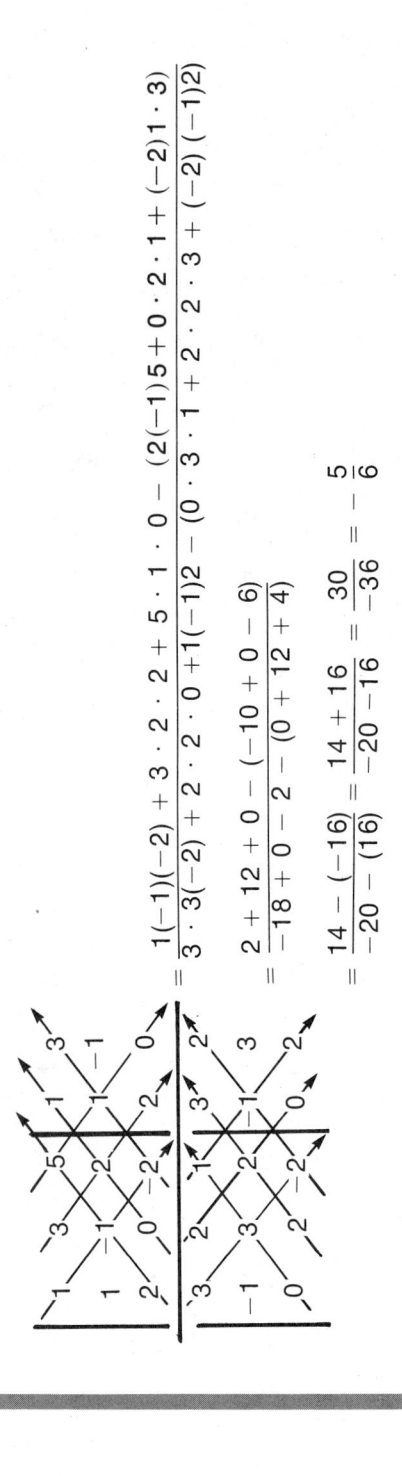

$$= \frac{1(-1)(-2) + 3 \cdot 2 \cdot 2 + 5 \cdot 1 \cdot 0 - (2(-1)5 + 0 \cdot 2 \cdot 1 + (-2)1 \cdot 3)}{3 \cdot 3(-2) + 2 \cdot 2 \cdot 0 + 1(-1)2 - (0 \cdot 3 \cdot 1 + 2 \cdot 2 \cdot 3 + (-2)(-1)2)}$$

$$= \frac{2 + 12 + 0 - (-10 + 0 - 6)}{-18 + 0 - 2 - (0 + 12 + 4)}$$

$$= \frac{14 - (-16)}{-20 - (16)} = \frac{14 + 16}{-20 - 16} = \frac{30}{-36} = -\frac{5}{6}$$

Systems of Equations

It should be noted at this point that we have only discussed and illustrated techniques for evaluating 2×2 and 3×3 determinants. The procedure used for evaluating a 3×3 matrix does *not* apply for determinants of higher orders. That is, it does not work for a 4×4 or larger determinant. Procedures for evaluating higher order determinants are usually discussed in advanced algebra courses.

EXERCISES FOR SECTION 9.6

(Answers given to exercises marked *)

Evaluate each of the following determinants:

*1. $\begin{vmatrix} 3 & 2 \\ 4 & 4 \end{vmatrix}$

2. $\begin{vmatrix} 5 & 3 \\ 4 & 6 \end{vmatrix}$

*3. $\begin{vmatrix} -1 & 2 \\ -1 & 3 \end{vmatrix}$

4. $\begin{vmatrix} 2 & 3 \\ -1 & -2 \end{vmatrix}$

*5. $\begin{vmatrix} -2 & -3 \\ -1 & -2 \end{vmatrix}$

6. $\begin{vmatrix} 0 & 1 \\ 1 & 0 \end{vmatrix}$

*7. $\begin{vmatrix} 4 & 3 \\ -3 & 4 \end{vmatrix}$

8. $\begin{vmatrix} -2 & 3 \\ 5 & -1 \end{vmatrix}$

*9. $\begin{vmatrix} -2 & 0 \\ 0 & 1 \end{vmatrix}$

10. $\begin{vmatrix} 3 & -1 \\ 5 & -2 \end{vmatrix}$

*11. $\begin{vmatrix} 3 & 0 \\ -2 & 0 \end{vmatrix}$

12. $\begin{vmatrix} -2 & -5 \\ 7 & -9 \end{vmatrix}$

*13. $\begin{vmatrix} 2 & 1 & 3 \\ 0 & -1 & 2 \\ 3 & -2 & 2 \end{vmatrix}$

14. $\begin{vmatrix} -1 & 2 & 1 \\ 0 & 1 & 2 \\ 3 & 2 & 1 \end{vmatrix}$

*15. $\begin{vmatrix} -2 & -1 & 3 \\ 4 & 2 & 1 \\ 0 & 2 & 3 \end{vmatrix}$

16. $\begin{vmatrix} 2 & -1 & -3 \\ 4 & -2 & 5 \\ 3 & 2 & 2 \end{vmatrix}$

*17. $\begin{vmatrix} 2 & -1 & 2 \\ 3 & 4 & 5 \\ -4 & 2 & -4 \end{vmatrix}$

18. $\begin{vmatrix} 3 & 4 & 2 \\ -1 & -1 & -1 \\ 2 & 1 & 3 \end{vmatrix}$

*19. $\begin{vmatrix} 0 & 1 & 0 \\ 2 & -3 & 2 \\ 4 & 2 & 1 \end{vmatrix}$

20. $\begin{vmatrix} -2 & 1 & 2 \\ -3 & 0 & 4 \\ -2 & 1 & 1 \end{vmatrix}$

*21. $\begin{vmatrix} -1 & -2 & -3 \\ -2 & -3 & -1 \\ 0 & -2 & 0 \end{vmatrix}$

22. $\begin{vmatrix} -2 & -3 & 4 \\ -5 & -2 & 0 \\ -1 & -2 & 0 \end{vmatrix}$

*23. $\begin{vmatrix} 1 & -1 & 1 \\ -1 & 1 & -1 \\ -2 & 2 & -2 \end{vmatrix}$

24. $\begin{vmatrix} 3 & 4 & 5 \\ 1 & 2 & 6 \\ 7 & 8 & 9 \end{vmatrix}$

Evaluate the following ratios:

*25. $\dfrac{\begin{vmatrix} 1 & 2 \\ 2 & 1 \end{vmatrix}}{\begin{vmatrix} 3 & 2 \\ 1 & 2 \end{vmatrix}}$

26. $\dfrac{\begin{vmatrix} -1 & 2 \\ -3 & 1 \end{vmatrix}}{\begin{vmatrix} 2 & 6 \\ -6 & 2 \end{vmatrix}}$

*27. $\dfrac{\begin{vmatrix} 3 & 2 \\ -3 & 4 \end{vmatrix}}{\begin{vmatrix} 4 & -5 \\ 2 & 1 \end{vmatrix}}$

28. $\dfrac{\begin{vmatrix} 1 & 2 & 1 \\ 3 & 1 & 2 \\ 2 & 1 & 0 \end{vmatrix}}{\begin{vmatrix} 1 & -1 & 0 \\ 2 & 0 & 1 \\ 3 & 2 & 2 \end{vmatrix}}$ *29. $\dfrac{\begin{vmatrix} 1 & 2 & 3 \\ 0 & 1 & 2 \\ -1 & -2 & 3 \end{vmatrix}}{\begin{vmatrix} -2 & -3 & 1 \\ 0 & 1 & 3 \\ 4 & 2 & 1 \end{vmatrix}}$ 30. $\dfrac{\begin{vmatrix} 3 & 4 & 1 \\ -2 & -3 & 1 \\ 0 & 2 & 1 \end{vmatrix}}{\begin{vmatrix} 2 & 3 & 5 \\ -1 & 0 & 2 \\ -3 & 2 & 4 \end{vmatrix}}$

9.7 SOLUTION OF SYSTEMS OF LINEAR EQUATIONS BY DETERMINANTS

We shall now see how determinants can be used to solve systems of linear equations. Consider the following system of two general linear equations where x and y are the variables, and the letters, a, b, c, d, e, f are real numbers.

$$ax + by = e.$$
$$cx + dy = f.$$

We shall solve this system by the addition method, and we choose to eliminate the x's. Therefore, we can multiply the first equation by $-c$ and the second equation by a.

$ax + by = e.$ multiply by $-c$ which gives us $-acx - cby = -ec.$

$cx + dy = f.$ multiply by a $acx + ady = af.$

Adding the two equations, we have: $ady - cby = af - ec.$

To solve this equation for y, we first factor the left member of the equation, $y(ad - cb) = af - ec$. Multiplying both sides by $\dfrac{1}{ad - cb}$, we obtain $y = \dfrac{af - ec}{ad - cb}$. Recall that in the previous section, the determinant $\begin{vmatrix} a & b \\ c & d \end{vmatrix}$ was defined as $ad - cb$. Note that this is exactly the same expression as is contained in the denominator of the solution for y! Similarly, the expression $af - ec$ can be written as the determinant $\begin{vmatrix} a & e \\ c & f \end{vmatrix}$. Therefore, the solution for y,

$$y = \frac{af - ec}{ad - cb}, \text{ is equivalent to } y = \frac{\begin{vmatrix} a & e \\ c & f \end{vmatrix}}{\begin{vmatrix} a & b \\ c & d \end{vmatrix}}.$$

Systems of Equations

Since the given system contained general equations, we can conclude that to solve such a system for y by means of determinants, the denominator is a determinant whose elements are coefficients of x and y, and the numerator is a similar determinant whose elements are coefficients of x while the constant terms replace the coefficients of y. It can be shown in a like manner that, for the given system, $x = \dfrac{\begin{vmatrix} e & b \\ f & d \end{vmatrix}}{\begin{vmatrix} a & b \\ c & d \end{vmatrix}}$. Therefore, to solve the system of linear equations

$$ax + by = e,$$
$$cx + dy = f \quad \text{(where } ad - cb \neq 0\text{)}$$

by means of determinants, we have $x = \dfrac{\begin{vmatrix} e & b \\ f & d \end{vmatrix}}{\begin{vmatrix} a & b \\ c & d \end{vmatrix}}$, $y = \dfrac{\begin{vmatrix} a & e \\ c & f \end{vmatrix}}{\begin{vmatrix} a & b \\ c & d \end{vmatrix}}$.

This method of solving a system of linear equations with two variables is often called *Cramer's Rule,* named after Gabriel Cramer (1704–1752), a Swiss mathematician. In solving a system of linear equations by means of the determinants, for the sake of convenience we shall denote the determinant in the denominator of the fractions by the capital letter D, whose elements are coefficients of x and y. When solving for x, the numerator will be denoted by D_x, which is the determinant consisting of the coefficients of y, and the constant terms replace the coefficients of x. Similarly, D_y is the determinant consisting of the coefficients of x, and the constant terms replace the coefficients of y.

To solve the following system of equations using Cramer's Rule,

$$-2x - 3y = -1.$$
$$2x - y = 5.$$

we have $x = \dfrac{D_x}{D} = \dfrac{\begin{vmatrix} -1 & -3 \\ 5 & -1 \end{vmatrix}}{\begin{vmatrix} -2 & -3 \\ 2 & -1 \end{vmatrix}}$ and $y = \dfrac{D_y}{D} = \dfrac{\begin{vmatrix} -2 & -1 \\ 2 & 5 \end{vmatrix}}{\begin{vmatrix} -2 & -3 \\ 2 & -1 \end{vmatrix}}$. Note that the elements of D are the coefficients of x and y; −2 and −3 are from the first equations while 2 and −1 are from the second equation. D_x is similar to D, but the x coefficients have been replaced by the constant terms. D_y is similar to D, but the y coefficients have been replaced by the constant terms. Completing the solution, we have

$$x = \dfrac{D_x}{D} = \dfrac{\begin{vmatrix} -1 & -3 \\ 5 & -1 \end{vmatrix}}{\begin{vmatrix} -2 & -3 \\ 2 & -1 \end{vmatrix}} = \dfrac{(-1)(-1) - (5)(-3)}{(-2)(-1) - (2)(-3)} = \dfrac{1-(-15)}{2-(-6)} = \dfrac{1+15}{2+6} = \dfrac{16}{8} = 2$$

and

$$y = \frac{D_y}{D} = \frac{\begin{vmatrix} -2 & -1 \\ 2 & 5 \end{vmatrix}}{\begin{vmatrix} -2 & -3 \\ 2 & -1 \end{vmatrix}} = \frac{(-2)(5) - (2)(-1)}{8} = \frac{-10 - (-2)}{8}$$

$$= \frac{-10 + 2}{8} = \frac{-8}{8} = -1.$$

Note: we have already evaluated this determinant in solving for x.

The solution is $(2, -1)$.

Check:

If $x = 2$, $y = -1$, then
$-2x - 3y = -1,$ $\qquad\qquad 2x - y = 5,$
$-2(2) - 3(-1) \stackrel{?}{=} -1,$ $\qquad 2 \cdot 2 - (-1) \stackrel{?}{=} 5,$
$-4 + 3 = -1,$ $\qquad\qquad\qquad 4 + 1 = 5,$
$-1 = -1.$ $\qquad\qquad\qquad\qquad 5 = 5.$

EXAMPLE 25

Use Cramer's Rule to solve the following system of equations:

$$3x + 4y = 6.$$
$$2x + 3y = 5.$$

Solution

According to Cramer's Rule, $x = \dfrac{D_x}{D}$ and $y = \dfrac{D_y}{D}$.

$$D = \begin{vmatrix} 3 & 4 \\ 2 & 3 \end{vmatrix} = 3 \cdot 3 - 2 \cdot 4 = 9 - 8 = 1.$$

$$D_x = \begin{vmatrix} 6 & 4 \\ 5 & 3 \end{vmatrix} = 6 \cdot 3 - 5 \cdot 4 = 18 - 20 = -2.$$

$$D_y = \begin{vmatrix} 3 & 6 \\ 2 & 5 \end{vmatrix} = 3 \cdot 5 - 2 \cdot 6 = 15 - 12 = 3.$$

Therefore, $x = \dfrac{D_x}{D} = \dfrac{-2}{1} = -2$, $y = \dfrac{D_y}{D} = \dfrac{3}{1} = 3.$

The solution is $(-2, 3)$.

Systems of Equations

Check:

If $x = -2$, $y = 3$, then
$3x + 4y = 6,$ $\qquad$ $2x + 3y = 5,$
$3(-2) + 4 \cdot 3 \stackrel{?}{=} 6,$ $\qquad$ $2(-2) + 3 \cdot 3 \stackrel{?}{=} 5,$
$-6 + 12 = 6,$ $\qquad$ $-4 + 9 = 5,$
$6 = 6.$ $\qquad$ $5 = 5.$

EXAMPLE 26

Use Cramer's Rule to solve the following system of equations.

$$2x + y = 3.$$
$$x - y = 1.$$

Solution

$$x = \frac{D_x}{D}, \; y = \frac{D_y}{D}.$$

$$D = \begin{vmatrix} 2 & 1 \\ 1 & -1 \end{vmatrix} = 2(-1) - 1 \cdot 1 = -2 - 1 = -3.$$

$$D_x = \begin{vmatrix} 3 & 1 \\ 1 & -1 \end{vmatrix} = 3(-1) - 1 \cdot 1 = -3 - 1 = -4.$$

$$D_y = \begin{vmatrix} 2 & 3 \\ 1 & 1 \end{vmatrix} = 2 \cdot 1 - 1 \cdot 3 = 2 - 3 = -1.$$

Therefore, $x = \dfrac{D_x}{D} = \dfrac{-4}{-3} = \dfrac{4}{3}$, $y = \dfrac{D_y}{D} = \dfrac{-1}{-3} = \dfrac{1}{3}$.

The solution is $\left(\dfrac{4}{3}, \dfrac{1}{3}\right)$.

Check:

If $x = \dfrac{4}{3}$, $y = \dfrac{1}{3}$, then
$2x + y = 3,$ $\qquad$ $x - y = 1,$
$2 \cdot \dfrac{4}{3} + \dfrac{1}{3} \stackrel{?}{=} 3,$ $\qquad$ $\dfrac{4}{3} - \dfrac{1}{3} \stackrel{?}{=} 1,$
$\dfrac{8}{3} + \dfrac{1}{3} = 3,$ $\qquad$ $\dfrac{3}{3} = 1,$
$\dfrac{9}{3} = 3,$ $\qquad$ $1 = 1.$
$3 = 3.$

EXAMPLE 27

Use Cramer's Rule to solve the following system of equations:

$$2x - 3y = 1.$$
$$-4x + 6y = 1.$$

Solution

$$x = \frac{D_x}{D}, \; y = \frac{D_y}{D}.$$

$$D = \begin{vmatrix} 2 & -3 \\ -4 & 6 \end{vmatrix} = 2 \cdot 6 - (-4)(-3) = 12 - 12 = 0.$$

$$D_x = \begin{vmatrix} 1 & -3 \\ 1 & 6 \end{vmatrix} = 1 \cdot 6 - 1(-3) = 6 + 3 = 9.$$

$$D_y = \begin{vmatrix} 2 & 1 \\ -4 & 1 \end{vmatrix} = 2 \cdot 1 - (-4)1 = 2 + 4 = 6.$$

Therefore, $x = \frac{D_x}{D} = \frac{9}{0}, \; y = \frac{D_y}{D} = \frac{6}{0}.$

Note: We cannot divide by 0. The solution is the empty set ∅. The equations are inconsistent. If we examine the slopes of the lines that the equations represent, we note that they are equal. Hence, the lines are parallel and do not intersect.

The preceding example illustrates the fact that for any system of linear equations, if $D = 0$ while $D_x \neq 0$ and $D_y \neq 0$, then the equations are *inconsistent*. Similarly, if $D = 0$, $D_x = 0$, and $D_y = 0$, then the equations are *dependent*.

Cramer's Rule can also be applied to solve systems of linear equations with three variables. That is, the solution of a system of three linear equations in three unknowns that is written in the form

$$a_1 x + b_1 y + c_1 z = d_1,$$
$$a_2 x + b_2 y + c_2 z = d_2,$$
$$a_3 x + b_3 y + c_3 z = d_3,$$

can be denoted by $x = \frac{D_x}{D}, \; y = \frac{D_y}{D},$ and $z = \frac{D_z}{D},$ where

$$D = \begin{vmatrix} a_1 & b_1 & c_1 \\ a_2 & b_2 & c_2 \\ a_3 & b_3 & c_3 \end{vmatrix} \neq 0, \; D_x = \begin{vmatrix} d_1 & b_1 & c_1 \\ d_2 & b_2 & c_2 \\ d_3 & b_3 & c_3 \end{vmatrix}$$

Systems of Equations

$$D_y = \begin{vmatrix} a_1 & d_1 & c_1 \\ a_2 & d_2 & c_2 \\ a_3 & d_3 & c_3 \end{vmatrix} \qquad D_z = \begin{vmatrix} a_1 & b_1 & d_1 \\ a_2 & b_2 & d_2 \\ a_3 & b_3 & d_3 \end{vmatrix}$$

Each of these determinants is a third order determinant, and a method for evaluating such a determinant was discussed in the previous section. Note that the denominator D is a determinant whose elements are coefficients of x, y, and z, respectively. D_x is the determinant consisting of the coefficients of y and z, and the constant terms replace the coefficients of x. D_y is the determinant consisting of the coefficients of x and z, and the constant terms replace the coefficients of y. Similarly, D_z is the determinant consisting of x and y, and the constant terms replace the coefficients of z. Once, we have constructed D, any of the other determinants — D_x, D_y, D_z — can be constructed by replacing the appropriate column by the constant terms. For example, D_x is similar in content to D except that we replace the x coefficients by the constant terms.

EXAMPLE 28

Use Cramer's Rule to solve the following system of equations:

$$2x + 3y - z = -3.$$
$$3x - 2y + z = -2.$$
$$-5x + 6y + 2z = 7.$$

Solution

$$x = \frac{D_x}{D}, \quad y = \frac{D_y}{D}, \quad z = \frac{D_z}{D}.$$

$$D = \begin{vmatrix} 2 & 3 & -1 \\ 3 & -2 & 1 \\ -5 & 6 & 2 \end{vmatrix} = -8 - 15 - 18 - (-10 + 12 + 18)$$
$$= -41 - (20) = -61.$$

$$D_x = \begin{vmatrix} -3 & 3 & -1 \\ -2 & -2 & 1 \\ 7 & 6 & 2 \end{vmatrix} = 12 + 21 + 12 - (14 - 18 - 12)$$
$$= 45 - (-16)$$
$$= 45 + 16 = 61.$$

$$D_y = \begin{vmatrix} 2 & -3 & -1 \\ 3 & -2 & 1 \\ -5 & 7 & 2 \end{vmatrix} = -8 + 15 - 21 - (-10 + 14 - 18)$$
$$= -14 - (-14)$$
$$= -14 + 14 = 0.$$

$$D_z = \begin{vmatrix} 2 & 3 & -3 \\ 3 & -2 & -2 \\ -5 & 6 & 7 \end{vmatrix} \begin{matrix} 2 & 3 \\ 3 & -2 \\ -5 & 6 \end{matrix} \quad \begin{aligned} &= -28 + 30 - 54 - (-30 - 24 + 63) \\ &= -52 - (+9) = -61. \end{aligned}$$

Therefore, $x = \dfrac{D_x}{D} = \dfrac{61}{-61} = -1$, $y = \dfrac{D_y}{D} = \dfrac{0}{-61} = 0$,

$$z = \dfrac{D_z}{D} = \dfrac{-61}{-61} = 1.$$

The solution is $(-1, 0, 1)$.

Check: Substitute $(-1, 0, 1)$ in each of the original equations.

$$\begin{aligned} 2x + 3y - z &= -3. \\ 2(-1) + 3 \cdot 0 - 1 &\stackrel{?}{=} -3, \\ -2 + 0 - 1 &= -3, \\ -3 &= -3. \end{aligned} \qquad \begin{aligned} 3x - 2y + z &= -2. \\ 3(-1) - 2 \cdot 0 + 1 &\stackrel{?}{=} -2, \\ -3 + 0 + 1 &= -2, \\ -2 &= -2. \end{aligned}$$

$$\begin{aligned} -5x + 6y + 2z &= 7. \\ -5(-1) + 6 \cdot 0 + 2 \cdot 1 &\stackrel{?}{=} 7, \\ 5 + 0 + 2 &= 7, \\ 7 &= 7. \end{aligned}$$

EXAMPLE 29

Use Cramer's Rule to solve the following system of equations:

$$\begin{aligned} 3x + 2y - 7 &= 0, \\ 4x - 3z + 5 &= 0. \\ 6y - 6z + 6 &= 0. \end{aligned}$$

Solution

Note that each equation does not contain three variables (but the system does contain three variables) and that the constant terms are placed on the the left side of the equals sign. Therefore, to use Cramer's Rule, the system must be rewritten as

$$\begin{aligned} 3x + 2y + 0z &= 7. \\ 4x + 0y - 3z &= -5. \\ 0x + 6y - 6z &= -6. \end{aligned}$$

If variables are missing, then their coefficients are 0 and 0's must be placed in the correct positions in the determinants. Hence,

Systems of Equations

$$D = \begin{vmatrix} 3 & 2 & 0 \\ 4 & 0 & -3 \\ 0 & 6 & -6 \end{vmatrix} \begin{matrix} 3 & 2 \\ 4 & 0 \\ 0 & 6 \end{matrix}$$

$= 0 + 0 + 0 - (0 - 54 - 48)$

$= 0 - (-102) = 102.$

$$D_x = \begin{vmatrix} 7 & 2 & 0 \\ -5 & 0 & -3 \\ -6 & 6 & -6 \end{vmatrix} \begin{matrix} 7 & 2 \\ -5 & 0 \\ -6 & 6 \end{matrix}$$

$= 0 + 36 + 0 - (0 - 126 + 60)$

$= 36 - (-66)$

$= 36 + 66 = 102.$

$$D_y = \begin{vmatrix} 3 & 7 & 0 \\ 4 & -5 & -3 \\ 0 & -6 & -6 \end{vmatrix} \begin{matrix} 3 & 7 \\ 4 & -5 \\ 0 & -6 \end{matrix}$$

$= 90 + 0 + 0 - (0 + 54 - 168)$

$= 90 - (-114)$

$= 90 + 114 = 204.$

$$D_z = \begin{vmatrix} 3 & 2 & 7 \\ 4 & 0 & -5 \\ 0 & 6 & -6 \end{vmatrix} \begin{matrix} 3 & 2 \\ 4 & 0 \\ 0 & 6 \end{matrix}$$

$= 0 + 0 + 168 - (0 - 90 - 48)$

$= 168 - (-138)$

$= 168 + 138 = 306.$

Therefore, $x = \dfrac{D_x}{D} = \dfrac{102}{102} = 1$, $y = \dfrac{D_y}{D} = \dfrac{204}{102} = 2$, $z = \dfrac{D_z}{D} = \dfrac{306}{102} = 3$.

The solution is (1, 2, 3).

Check: Substitute (1, 2, 3) in each of the original equations.

$3x + 2y - 7 = 0.$ $4x - 3z + 5 = 0.$ $6y - 6z + 6 = 0.$
$3 \cdot 1 + 2 \cdot 2 - 7 \stackrel{?}{=} 0,$ $4 \cdot 1 - 3 \cdot 3 + 5 \stackrel{?}{=} 0,$ $6 \cdot 2 - 6 \cdot 3 + 6 \stackrel{?}{=} 0,$
$3 + 4 - 7 = 0,$ $4 - 9 + 5 = 0,$ $12 - 18 + 6 = 0,$
$0 = 0.$ $0 = 0.$ $0 = 0.$

In each of the illustrative examples, we used Cramer's Rule to find the complete solution for the given system of linear equations. It should be noted that the last variable to be found in a solution can also be determined by replacing the known value(s) in either of the original equations and solving it for the remaining variable.

EXERCISES FOR SECTION 9.7

(Answers given to exercises marked *)

Use Cramer's Rule to solve the following systems of equations:

*1. $4x - 2y = 10.$
   $2x + 3y = 1.$

2. $4x - 2y = 6.$
   $3x + 3y = 9.$

*3. $2x + y = 6.$
   $x - 3y = 10.$

4. $x + 2y = 3.$
   $-x + y = -6.$

*5. $x + y = -5.$
   $x - y = 1.$

6. $x + 2y = 6.$
   $3x - y = -3.$

*7. $3x + y = 2.$
   $5x - 3y = -13.$

8. $x + y = 6.$
   $x + y = 1.$

*9. $-2x + y = 4.$
   $6x - 3y = -12.$

10. $2x + 10y = 1.$
    $x + 4y = 0.$

*11. $x + y = 1.$
    $x - y = 1.$

12. $4x + 2y = 3.$
    $5x - 2y = 1.$

*13. $3x + 2y + z = 1.$
    $2x + 3y - z = 4.$
    $3x - y + 2z = -1.$

14. $x + y + z = 3.$
    $x - 2y - 3z = 0.$
    $-2x + 3y + 2z = -1.$

*15. $2x - 3y + 4z = 8.$
    $x + y + 2z = 9.$
    $x + 2y + z = 8.$

16. $x - 2y - 3z = 12.$
    $-x + 3y - 2z = 1.$
    $2x - 5y + z = 5.$

*17. $x - y + 2z = 2.$
    $4x - 4y + 8z = 8.$
    $-3x + 3y - 6z = -6.$

18. $3x - 2y + 2z = 8.$
    $2x + 4y + 6z = -5.$
    $2x + 6y - 4z = -3.$

*19. $3x - 2y - 3z = -1.$
    $6x + y + 2z = 7.$
    $9x + 3y + 4z = 9.$

20. $2x + 3y - z = -1.$
    $x - y + 2z = 2.$
    $4x + 3z = 2.$

*21. $3x + 2y = 7.$
    $4x - 3z = -5.$
    $6y - 6z = -6.$

22. $3y - 2z - 4 = 0.$
    $3x - 5z + 5 = 0.$
    $5x + 7y - 14 = 0.$

*23. $4y + 5z - 2 = 0.$
    $3x - 10z + 1 = 0.$
    $-6x - 8y + 4 = 0.$

24. $2x - 3y - 2 = 0.$
    $4x - 4z - 1 = 0.$
    $3y - 8z + 3 = 0.$

Systems of Equations

9.8 SYSTEMS OF INEQUALITIES

In Section 8.3, we discussed equations of the form $ax + by = c$ where a, b, and c are real numbers. An equation of this type is called a linear equation. Statements of the form $ax + by < c$ and $ax + by > c$ are called *linear inequalities*. We have already discussed inequalities and graphed their solution sets on the real number line (see Chapters 2 and 7). Now we shall examine the solution of linear inequalities and their solution will be graphed on the Cartesian plane.

Consider the equation $x + y = 1$. Points such as (0, 1) and (1, 0) are solutions of the equation. When we substitute these values for x and y in the given equation, we obtain a statement of equality, that is, they satisfy the given equation. Now consider the point (0, 0). If we evaluate the equation $x + y = 1$ for (0, 0) we have

$$0 + 0 = 1.$$
$$0 = 1. \quad \text{(false)}$$

This is a false statement of equality and the value on the left side of the equals sign is "less than" the value on the right. Similarly, if we evaluate the given equation for (3, 0) we have

$$3 + 0 = 1.$$
$$3 = 1. \quad \text{(false)}$$

This is a false statement of equality and the value on the left side of the equals sign is "greater than" the value on the right. We have found some points on the plane that satisfy the given equation, such as those points where $x + y = 1$. We have also found points such that $x + y < 1$ and $x + y > 1$. A line on the Cartesian plane divides the plane into three sets of points: those that are on the line, those that are in the half-plane such that $x + y < 1$, and those that are in the half-plane such that $x + y > 1$.

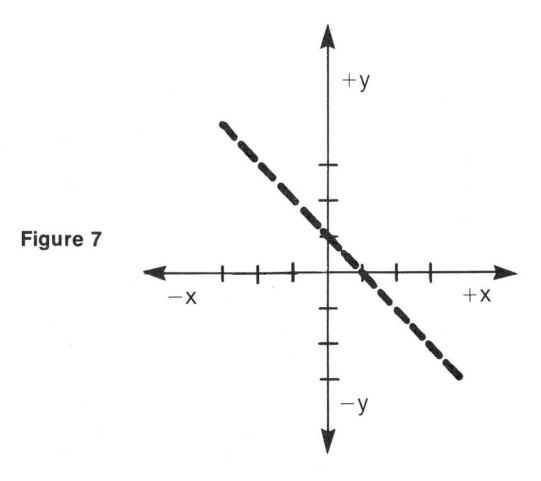

Figure 7

To graph a statement such as $x + y < 1$, we must first locate the boundary of the half-plane whose points satisfy the inequality. The boundary is the line $x + y = 1$. See Figure 7. Note that the line is dotted, since the points on the line $x + y = 1$ will not satisfy the inequality $x + y < 1$. Which half-plane contains points whose coordinates satisfy the inequality? We test a point on either side of the boundary to determine whether it satisfies the given condition. Let's try (3, 2). Therefore, we have

$$x + y < 1,$$
$$3 + 2 < 1,$$
$$5 < 1. \quad \text{(false)}$$

It should be noted that if a point does not satisfy the given condition, then none of the points in that half-plane satisfy the given condition. Therefore, the correct half-plane should be on the other side of the boundary. Let's verify this. A convenient point to try is the origin, (0, 0). Therefore, we have

$$x + y < 1,$$
$$0 + 0 < 1,$$
$$0 < 1. \quad \text{(true)}$$

If a point satisfies the given condition, then all of the points in that half-plane must also satisfy the given condition. Therefore, to indicate the solution of the inequality $x + y < 1$, we shade the half-plane that contains this point. In this case, we shade the half-plane that is below the boundary $x + y = 1$. (See Figure 8.) Note that the solution of the inequality $x + y \leq 1$ would have the same solution with one exception. The exception is that the boundary would be a solid line instead of a dotted line. The reason for this is that the points that satisfy the inequality $x + y \leq 1$ are those points in the half-plane $x + y < 1$, or those points on the line $x + y = 1$.

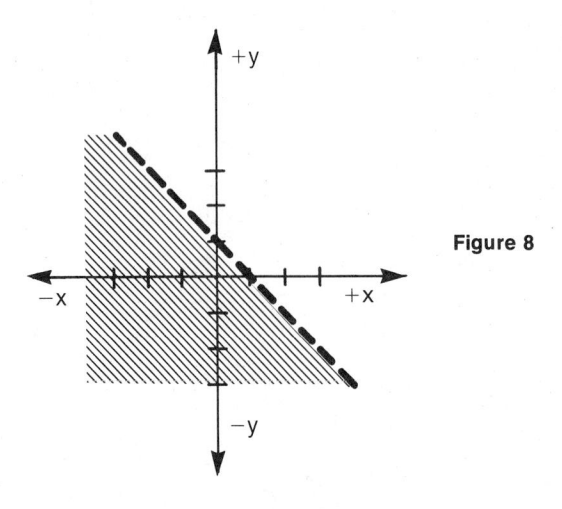

Figure 8

Systems of Equations 563

Recall that "or" is the same as "union" and therefore we unite the half-plane $x+y<1$ and the line $x+y=1$.

> To graph inequalities in two variables, we must first find the boundary of the half-plane that contains those points that satisfy the given inequality. We do this by graphing the equation derived from the inequality. If the given inequality is strict, then the boundary is a dotted line. The boundary is a solid line for statements containing ($\geq$) or ($\leq$). Next we test a point in a half-plane. If the coordinates of a point satisfy the given condition, then all the points in that half-plane also satisfy the given condition. We indicate the solution by shading the half-plane that contains those points that satisfy the given inequality.

EXAMPLE 30

Graph $x - 2y \leq 4$.

Solution

To graph inequalities in two variables, we must first find the boundary of the half-plane that contains those points that satisfy the given inequality. We locate the boundary by graphing the equation $x - 2y = 4$. Note that this boundary will be a solid line because of the relationship ($\leq$), that is, less than or equal to. Next, we test a point to determine if its coordinates satisfy the given inequality. A convenient point to try is the origin, (0, 0). If $x = 0$ and $y = 0$, we have

$$0 - 2(0) \leq 4,$$
$$0 - 0 \leq 4,$$
$$0 \leq 4. \quad \text{(true)}$$

If the coordinates of a point satisfy the given condition, then all of the points in that half-plane also satisfy the given condition. Hence, we shade the half-plane that contains the point (0, 0), the half-plane above the line $x - 2y = 4$.

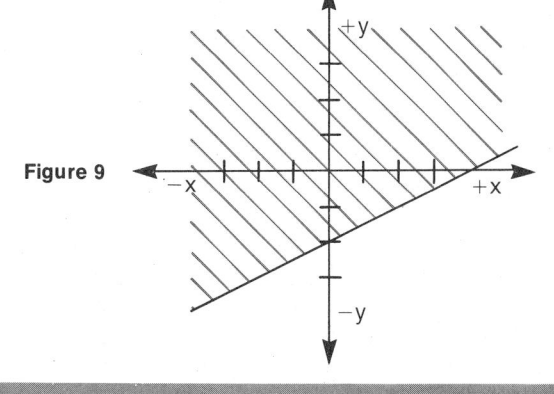

Figure 9

EXAMPLE 31

Graph $2x + y < -2$.

Solution

We locate the boundary of the half-plane by graphing the equation $2x + y = -2$. This boundary will be a dashed line because the relationship is a strict inequality, ($<$). Next, we test a point. A convenient point to try is the origin. Let $x = 0$, $y = 0$ and we have

$$2 \cdot 0 + 0 < -2,$$
$$0 + 0 < -2,$$
$$0 < -2. \quad \text{(false)}$$

The coordinates of the origin do not satisfy the given inequality. Then a point on the other side of the boundary should give us a true statement. Let's try $(-2, -2)'$ If $x = -2$ and $y = -2$, we have

$$2(-2) + (-2) < -2$$
$$-4 + -2 < -2$$
$$-6 < -2. \quad \text{(true)}$$

If the coordinates of a point satisfy the given condition, then all of the points in that half-plane also satisfy the given condition. Therefore, we shade the half-plane that contains the point $(-2, -2)$, the half-plane below the line $2x + y = -2$.

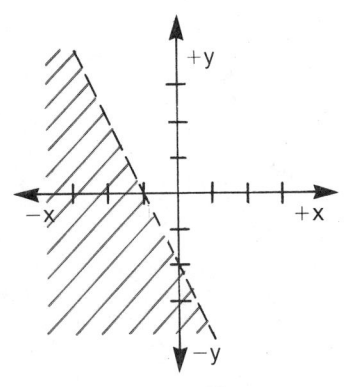

Figure 10

It is sometimes necessary to graph two linear inequalities on the same set of axes. For example, if we consider the following inequalities together, then we have a system of inequalities:

Systems of Equations

$$x - 2y \leq 4.$$
$$2x + y < 2.$$

The solution to such a system of inequalities is the intersection of the solutions for each inequality. That is, we want to find the set of all points (x, y) such that $x - 2y \leq 4$ *and* $2x + y < 2$. The solution must satisfy both of the given inequalities simultaneously. To obtain this solution, we draw both graphs on the same Cartesian plane. The solution is where the two half-planes (regions) intersect. The coordinates of all the points in the intersection will satisfy both of the given conditions and form the solution set. The solution is usually indicated by "crosshatching" or is more heavily shaded. The graphs of $x - 2y \leq 4$ (Fig. 11) and $2x + y < 2$ (Fig. 12) are shown separately, and then the solution for the system of inequalities is shown in the third graph, Figure 13.

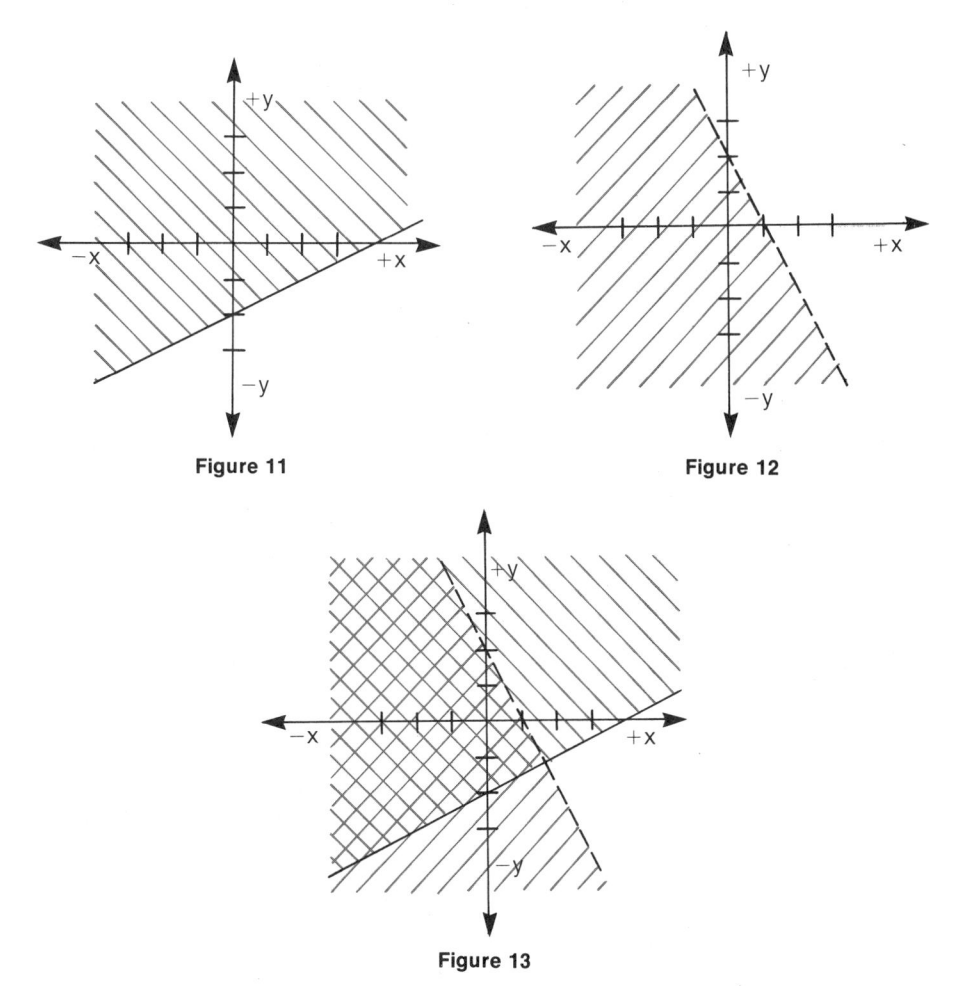

Figure 11

Figure 12

Figure 13

Note that the solution for the system of linear inequalities $x - 2y \leq 4$ and $2x + y < 2$ is the intersection of the two solutions.

EXAMPLE 32

Solve the following system of inequalities.

$$x + y < 3.$$
$$x - y \leq 2.$$

Solution

We graph each linear inequality on the same set of axes. The solution is where the two half-planes intersect. The coordinates of all points in the intersection will satisfy both of the inequalities simultaneously.

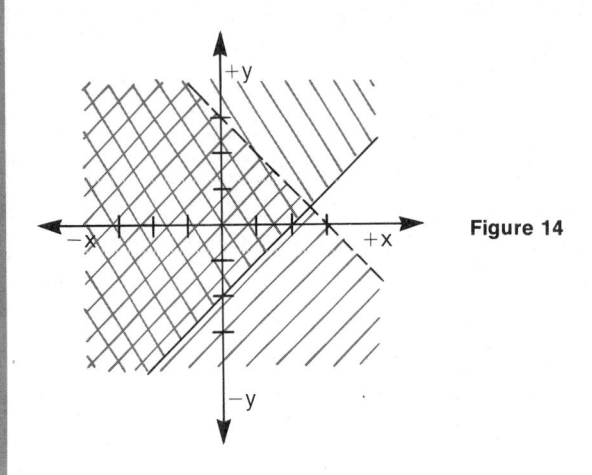

Figure 14

The solution is the set of all (x, y) such that $x + y < 3$ and $x - y \leq 2$. Note that one boundary line is solid, while the other is dashed.

EXERCISES FOR SECTION 9.8

(Answers given to exercises marked *)

Graph each of the linear inequalities on the Cartesian plane.

*1. $x + y > 4$.   2. $x + y < 2$.   *3. $x - y < 2$.

4. $x - y > 3$.   *5. $x - y \leq 2$.   6. $2x - 3y \geq 6$.

*7. $-2x + y \geq 6$.   8. $x - y \leq 0$.   *9. $y - x > 0$.

10. $x - 4y \leq 0$.   *11. $-3x - 3y > 0$.   12. $5x - 3y < 15$.

*13. $-5x + 3y \leq 15$.   14. $-3x + 2y \geq 6$.   *15. $-3x + 2y < -6$.

Systems of Equations

Solve the given systems of inequalities.

16. $x + y < 4.$
    $x - y < 4.$

*17. $x + y < 1.$
    $x - y < 1.$

18. $2x + 2y \geq 6.$
    $x - y < 3.$

*19. $x + y \geq 0.$
    $x - y < 0.$

20. $2x - 3y > 6.$
    $x + 2y \leq 4.$

*21. $-3x + 2y > 6.$
    $2x + 4y \leq 1.$

22. $x - 2y < -2.$
    $2x + 4y > 4.$

*23. $x - 4y \leq 0.$
    $x - y > 0.$

24. $5x - 3y < 15.$
    $-5x + 3y \leq 15.$

9.9 SUMMARY

In this chapter we discussed systems of equations (and inequalities) and various methods to solve such systems. A system of linear equations can contain any number of equations and these equations can contain any number of variables, as long as all the terms are of the first degree. We have considered methods for solving systems of linear equations containing two equations in two unknowns, and three equations in three unknowns.

A solution for a pair of linear equations can be obtained by graphing both equations on the same set of axes. If the graphs of the two equations intersect at a point, the equations are *consistent,* and the coordinates of this point represent the solution for the given system. The coordinates of the point of intersection must satisfy both equations simultaneously. The graphs of the two equations for a given system may be parallel lines. If this occurs, then there is no solution for the system, or we can say that the solution set is the empty set $\emptyset$. Two linear equations whose solution is the empty set are called *inconsistent.* We can also have the case where the graphs of the two equations are the same line. When this occurs, any solution of one equation is also a solution of the other equation, and we say that the equations are *dependent.*

Since the graphical method of solution is not always accurate, we also examined algebraic solutions of systems of linear equations. One technique for solving a system of two linear equations in two unknowns is by substitution. To solve a system of linear equations by the substitution method, we first solve one of the given equations for x or y. After solving the chosen equation for the chosen variable, substitute the resulting expression for the chosen variable in the other equation and solve it. After determining a value for one of the coordinates, we can find the value of the other coordinate by substituting back into one of the original equations.

A third technique for solving systems of linear equations is called the addition method. The reason for adding the two equations is to eliminate one of the variables and solve the resulting equation for the remaining variable. To eliminate a variable in a system of equations by the addition method, it is often necessary to multiply one or both equations by real numbers so that a set of corresponding variables will have numerical coefficients that are additive inverses of each other. After performing this multiplication, add the two equations to eliminate the variable whose coefficients are additive inverses of each other. Solve the resulting equation to find a value for one of the variables. Find the value of the other variable by substituting back into one of the original equations. If the equations in the given system are inconsistent, then a false statement of equality is obtained when the equations are added. If the equations are dependent, then a true statement of equality is obtained when the equations are added. When solving any system of linear equations, the solution should always be verified by checking it in each of the original equations.

To solve an application problem by means of a system of equations, let each unknown be represented by a variable. Determine the relationships between the variables from the problem, or from the relationship that is assumed to be known, to generate a system of equations. Solve the system and check the solution. The system generated may contain two equations with two unknowns, or it may contain three equations with three unknowns. To solve a system of three linear equations with three variables, we proceed as follows:

1. Eliminate a variable from two of the given equations by means of the addition method, and obtain a resulting equation with two variables.
2. Use the same procedure stated in Step 1 to eliminate the same variable from two of the given equations. One of these equations must be the third equation not used in Step 1.
3. Solve the system of two equations in two unknowns (obtained from Steps 1 and 2) to find the value of one of the variables.
4. Replace the value of the variable (obtained in Step 3) in either of the equations obtained in Steps 1 and 2, and solve it for the remaining variable.
5. To find the value of the third variable, replace the two variables by their respective values (obtained in Steps 3 and 4) in one of the original equations, and solve it for the remaining variable.
6. Check the solution in each of the original equations to verify that it is the simultaneous solution.

A rectangular array of numbers is called a matrix. If a matrix contains the same number of rows and columns, it is called a square matrix. Associated with each square matrix is a real number called the determinant of the matrix. In this chapter we evaluated 2×2 and 3×3 determinants. The value of the 2×2 determinant $\begin{vmatrix} a & b \\ c & d \end{vmatrix}$ is $ad - bc$. To evaluate a third order determinant, rewrite the first and second columns to the right of the determinant and draw in the arrows, as indicated.

Systems of Equations

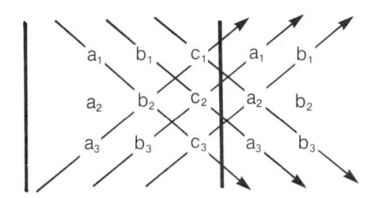

The value of the determinant can be obtained by finding the sum of the products of the elements of the arrows slanting downward to the right and subtracting the sum of the products of the elements of the arrows slanting upward to the right. Hence, the value of the given third order determinant is

$$a_1 b_2 c_3 + b_1 c_2 a_3 + c_1 a_2 b_3 - (a_3 b_2 c_1 + b_3 c_2 a_1 + c_3 a_2 b_1)$$

Determinants can be used to solve systems of linear equations. This technique is commonly referred to as *Cramer's Rule.* For example, to solve the system of linear equations

$$ax + by = e,$$
$$cx + dy = f,$$

(where $ad - cb \neq 0$)

by means of Cramer's Rule, we have

$$x = \frac{\begin{vmatrix} e & b \\ f & d \end{vmatrix}}{\begin{vmatrix} a & b \\ c & d \end{vmatrix}} = \frac{D_x}{D}, \quad y = \frac{\begin{vmatrix} a & e \\ c & f \end{vmatrix}}{\begin{vmatrix} a & b \\ c & d \end{vmatrix}} = \frac{D_y}{D}.$$

Note that D is a determinant whose elements are coefficients of the variables in respective order. D_x is a similar determinant, but the constant terms replace the coefficient of y. D_y is the determinant consisting of the coefficients of x and the constant terms replace the coefficients of y. Cramer's Rule can also be applied to solve systems of linear equations with three variables. The solution of a system of three linear equations in three unknowns that is written in the form

$$a_1 x + b_1 y + c_1 z = d_1,$$
$$a_2 x + b_2 y + c_2 z = d_2,$$
$$a_3 x + b_3 y + c_3 z = d_3,$$

can be denoted by $x = \dfrac{D_x}{D}$, $y = \dfrac{D_y}{D}$, and $z = \dfrac{D_z}{D}$

where

$$D = \begin{vmatrix} a_1 & b_1 & c_1 \\ a_2 & b_2 & c_2 \\ a_3 & b_3 & c_3 \end{vmatrix} \neq 0 \qquad D_x = \begin{vmatrix} d_1 & b_1 & c_1 \\ d_2 & b_2 & c_2 \\ d_3 & b_3 & c_3 \end{vmatrix}$$

$$D_y = \begin{vmatrix} a_1 & d_1 & c_1 \\ a_2 & d_2 & c_2 \\ a_3 & d_3 & c_3 \end{vmatrix} \qquad D_z = \begin{vmatrix} a_1 & b_1 & d_1 \\ a_2 & b_2 & d_2 \\ a_3 & b_3 & d_3 \end{vmatrix}$$

Statements of the form ax + by < c and ax + by > c are called linear inequalities. To graph a linear inequality such as ax + by < c, we must first locate the boundary of the half-plane that contains the points whose coordinates satisfy the inequality. The boundary of this half-plane is the line ax + by = c. The boundary is a dashed line if we are graphing a strict inequality, that is < or >. If the inequality contains part of an equals sign such as ≤ or ≥, then the boundary is a solid line. Next we test a point in a half-plane. If the coordinates of a point satisfy the given condition, then all of the points in that half-plane satisfy the given condition. We indicate the solution by shading the half-plane that contains those points that satisfy the given inequality. It is sometimes necessary to graph two linear inequalities on the same set of axes. When this is done, a system of linear inequalities is being solved. The solution for a system of linear inequalities is the set of (x, y) that satisfies both of the given inequalities simultaneously. That is, the solution for a system of inequalities is the intersection of the solutions for each inequality. The solution is usually indicated by crosshatching or is more heavily shaded.

REVIEW EXERCISES FOR CHAPTER 9

(Answers in Appendix B)

Solve each system of equations graphically.

1. x + y = 2.
   2x + y = 3.

2. x + y = 1.
   2x + y = 3.

3. x + y = 3.
   x + 2y = 5.

4. x − y = 1.
   x + y = −5.

5. x − 2y = 4.
   −2x + 4y = −8.

6. 2x − y = 4.
   x − 2y = −4.

Solve each system of equations by the substitution method.

7. 3x − y = 5.
   x + 2y = 4.

8. x − y = 1.
   x + y = 5.

9. 4x + 2y = 3.
   5x − 2y = 1.

10. 2x + y = 13.
    x + 3y = 7.

11. 2x − y = −3.
    −x + 2y = 0.

12. 6x + 2y = 4.
    5x − 3y = −13.

Solve each system of equations by the addition method.

13. 2x + y = 6.
    x − 3y = 10.

14. 5x − 4y = 60.
    4x + 5y = 7.

15. 3x − 4y = 6.
    −5x + 8y = −14.

Systems of Equations

16. $6x - 2y = -14.$
    $2x + y = 2.$

17. $3x + 3y = 12.$
    $8x + 4y = 40.$

18. $8x + 4y = 24.$
    $3x - 9y = 30.$

Solve the following systems of equations.

19. $2x + 3y - z = -3.$
    $3x - 2y + z = -2.$
    $-5x + 6y + 2z = 7.$

20. $2x + y + z = 0.$
    $-3x - y + 2z = 7.$
    $x + 2y - 3z = 7.$

21. $3y - 2z = 4.$
    $3x - 5z = -5.$
    $5x + 7y = 14.$

Evaluate each of the following determinants.

22. $\begin{vmatrix} 5 & 3 \\ 4 & 6 \end{vmatrix}$

23. $\begin{vmatrix} 2 & 3 \\ 4 & 0 \end{vmatrix}$

24. $\begin{vmatrix} -2 & 3 \\ 5 & -1 \end{vmatrix}$

25. $\begin{vmatrix} -2 & -3 & -2 \\ -4 & -1 & 0 \\ 0 & 2 & 3 \end{vmatrix}$

26. $\begin{vmatrix} -1 & 2 & 1 \\ 0 & 1 & 2 \\ 3 & 2 & 1 \end{vmatrix}$

27. $\begin{vmatrix} 3 & 4 & 5 \\ 1 & 2 & 6 \\ 7 & 8 & 9 \end{vmatrix}$

Use Cramer's Rule to solve the following systems of equations:

28. $4x - 2y = 6.$
    $3x + 3y = 9.$

29. $4x + 2y = 3.$
    $5x - 2y = 1.$

30. $x + 2y = 6.$
    $3x - y = -3.$

31. $3x - 2y + 2z = 8.$
    $2x + 4y + 6z = -5.$
    $2x + 6y - 4z = -3.$

32. $x - 2y - 3z = 12.$
    $-x + 3y - 2z = 1.$
    $2x - 5y + z = 5.$

33. $2x - 3y = 2.$
    $4x - 4z = 1.$
    $3y - 8z = -3.$

Graph each of the linear inequalities on the Cartesian plane.

34. $x + y < 6.$

35. $x - y \leq 0.$

36. $2x - 3y \geq 6.$

Solve the given systems of inequalities.

37. $2x + 2y \geq 6.$
    $x - y < 3.$

38. $2x - 3y > 6.$
    $x + 2y \leq 4.$

39. $5x - 3y < 15.$
    $-5x + 3y \leq 15.$

40. The denominator of a fraction is 4 less than the numerator. If the numerator of the fraction is increased by 3 and the denominator is decreased by 5, then the resulting fraction is equivalent to $\frac{5}{2}$. Find the original fraction.

41. A special showing of the movie "Jaws" was held at Monroe Community College. Tickets for the movie cost $2.00 each, if purchased at the advance sale and $2.75 each, if purchased on the day of the movie. For this special showing, 1600 tickets were sold and the total receipts were $3500. How many tickets were purchased at the advance sale and how many were purchased on the day of the movie?

42. The units digit of a positive two-digit number is 4 more than the tens digit. If the digits are reversed, the new number is 23 less than twice the original number. Find the original number.

43. Find three numbers whose sum is 18, given that the sum of the first two numbers equals the third number, and that twice the third number minus three times the first number is 2 more than the second number.

44. Scott has a collection of change consisting of nickels, dimes, and quarters, whose total value is $4.00. If the total number of coins is 44 and the number of quarters is 6 less than the number of dimes, then how many of each kind of coin does Scott have?

45. The sum of the digits of a three-digit number is 16. The units digit is equal to the sum of the hundreds digit and the tens digit. The number formed by reversing the digits is 693 more than the given number. Find the original number.

chapter 10

Exponential and Logarithmic Functions

As a result of this chapter, the student will be able to:

1. Identify an exponential function as an equation of the form $y = b^x$ (or $f(x) = b^x$), where $b > 0$ and $b \neq 1$;
2. Identify the inverse of an exponential function as a logarithmic function, that is, if $y = b^x$ is the exponential function, then $x = b^y$ is the logarithmic function;
3. Sketch the graphs of exponential and logarithmic functions;
4. Identify logarithms as exponents and for any logarithmic equation of the form $x = b^y$, express y in terms of x, thus $x = b^y$ is equivalent to $y = \log_b x$;
5. Express exponential equations in logarithmic form and express logarithmic equations in exponential form;
6. Solve logarithmic equations of the form $\log_b x = a$, $\log_b a = x$ and $\log_x b = a$;
7. Apply the properties of logarithms in performing computations:

$$\log_b MN = \log_b M + \log_b N,$$

$$\log_b \frac{M}{N} = \log_b M - \log_b N,$$

$$\log_b M^n = n \log_b M;$$

8. Determine the logarithm (characteristic + mantissa) for any three-digit number, and also determine the number that corresponds to a given logarithm (the antilogarithm);
9. Find the logarithm of a four-digit number by means of interpolation, and also use interpolation to find the antilogarithm of a number;
10. Use logarithms to perform calculations such as multiplication, division, raising to a power, and extracting roots;
11. Use logarithms to solve exponential equations.

10.1 INTRODUCTION

In earlier chapters, we have thoroughly discussed various types of algebraic functions—in particular, linear and quadratic functions. In this chapter, we shall examine two other kinds of functions: exponential and logarithmic. There is a certain formula used in finance that can be considered an exponential function, which you may have encountered in another course. It is the compound interest formula,

$$A = P(1 + r)^n$$

Exponential and Logarithmic Functions 575

where A = the compound amount, P = the principal, r = the rate of interest per conversion period, and n = the number of conversion periods. In this formula, if n is unknown, then we can consider the formula to be an exponential function. For example consider the following problem:

> How long will it take $1000 to amount to $2000 if invested at 6% compounded annually?

Using the formula $A = P(1 + r)^n$, A = $2000, P = $1000, r = 6%, or

$$2000 = 1000 (1 + .06)^n,$$
$$2000 = 1000 (1.06)^n.$$

At the present time we *cannot* solve this equation for n since it occurs as an exponent in the equation. We shall solve it later in the chapter. Until this point we have not encountered a problem of this nature. It has been discussed at this time to point out that an unknown (or variable) can occur as an exponent in an equation.

Next, consider the equation $y = 2^x$, where x is the independent variable and y is the dependent variable. Generating a table of values for this equation, we have:

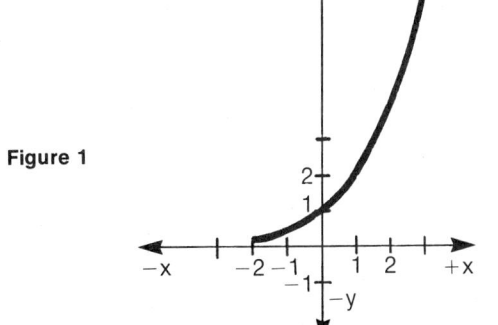

if $x = 0$, $y = 2^0 = 1$,   if $x = -1$, $y = 2^{-1} = \frac{1}{2^1} = \frac{1}{2}$,

if $x = 1$, $y = 2^1 = 2$,
if $x = 2$, $y = 2^2 = 4$,   if $x = -2$, $y = 2^{-2} = \frac{1}{2^2} = \frac{1}{4}$.

if $x = 3$, $y = 2^3 = 8$.

x	−2	−1	0	1	2	3
y	$\frac{1}{4}$	$\frac{1}{2}$	1	2	4	8

Figure 1

Plotting the corresponding points on the Cartesian plane and connecting them by means of a smooth curve, we obtain the graph shown in Figure 1. This represents the graph of $y = 2^x$. It should be noted that points such as $(\sqrt{2}, 2^{\sqrt{2}})$ also lie on this curve. As we have only considered rational numbers as exponents in our previous discussions, it is beyond the scope of this text to show that $(\sqrt{2}, 2^{\sqrt{2}})$ is actually a point on the graph. But we shall assume that the domain is the set of all real numbers (including irrational numbers) and that the graph is smooth and unbroken, hence this relation is a function.

An equation of the form $y = b^x$ (or $f(x) = b^x$), where $b > 0$ and $b \neq 1$, is called an *exponential function*. Note that for an exponential function the independent variable is the exponent. Also, we do not consider those cases where $b < 0$, $b = 0$, or $b = 1$.

Let's examine another exponential function. Consider the equation $y = 3^x$. Generating a table of values for this equation, we have:

if $x = 0$, $y = 3^0 = 1$,   if $x = -1$, $y = 3^{-1} = \dfrac{1}{3^1} = \dfrac{1}{3}$,

if $x = 1$, $y = 3^1 = 3$,

if $x = 2$, $y = 3^2 = 9$,   if $x = -2$, $y = 3^{-2} = \dfrac{1}{3^2} = \dfrac{1}{9}$.

x	−2	−1	0	1	2
y	$\dfrac{1}{9}$	$\dfrac{1}{3}$	1	3	9

The graph of $y = 3^x$ is shown in Figure 2. After examining the graphs of $y = 2^x$ and $y = 3^x$, we note some common characteristics of these graphs of exponential functions.

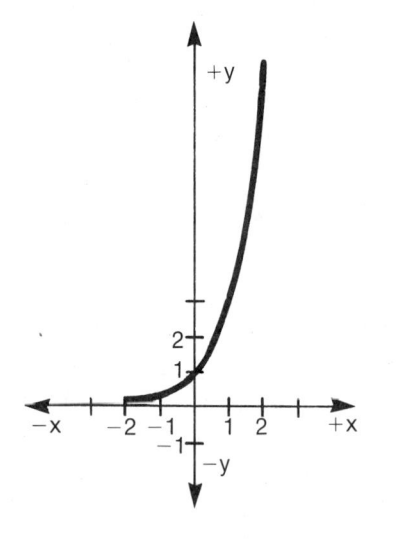

Figure 2

Exponential and Logarithmic Functions 577

> (1) Regardless of the equation, the y-intercept is always 1, because $b^0 = 1$ (b $\neq$ 0).
> (2) The graphs of these exponential functions lie completely above the x-axis. That is, the value of y is never negative since b is always greater than 0.
> (3) The graph can be extended indefinitely to the left and right of the y-axis. If we extend it to the right, when b > 1, the graph rises more and more. If we extend it to the left, when b > 1, the graph approaches the x-axis (closer and closer), but it never touches or intersects the x-axis.

EXAMPLE 1

Sketch the graph of $y = 4^x$.

Solution

We generate a table of values for the given equation and plot the corresponding points on the Cartesian plane. We connect them by means of a smooth curve and have the graph shown in Figure 3.

If $x = 0$, $y = 4^0 = 1$, if $x = -1$, $y = 4^{-1} = \dfrac{1}{4^1} = \dfrac{1}{4}$,

if $x = \dfrac{1}{2}$, $y = 4^{1/2} = 2$, if $x = -2$, $y = 4^{-2} = \dfrac{1}{4^2} = \dfrac{1}{16}$.

if $x = 1$, $y = 4^1 = 4$,

if $x = 2$, $y = 4^2 = 16$,

x	-2	-1	0	$\dfrac{1}{2}$	1	2
y	$\dfrac{1}{16}$	$\dfrac{1}{4}$	1	2	4	16

Figure 3

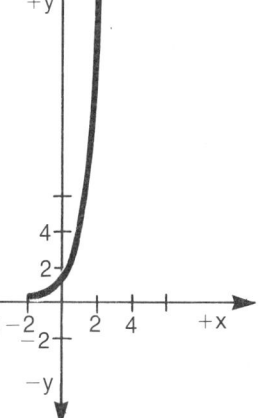

EXAMPLE 2

Sketch the graph of $y = \left(\dfrac{1}{2}\right)^x$.

Solution

Note that in this equation, $0 < b < 1$. Hence, we must use care when generating a table of values and plotting the corresponding points. For example,

$$\text{if } x = -3, \; y = \left(\dfrac{1}{2}\right)^{-3} = \dfrac{1}{\left(\dfrac{1}{2}\right)^3} = \dfrac{1}{\dfrac{1}{8}} = 8.$$

The corresponding table is shown and the graph of $y = \left(\dfrac{1}{2}\right)^x$ is plotted in Figure 4.

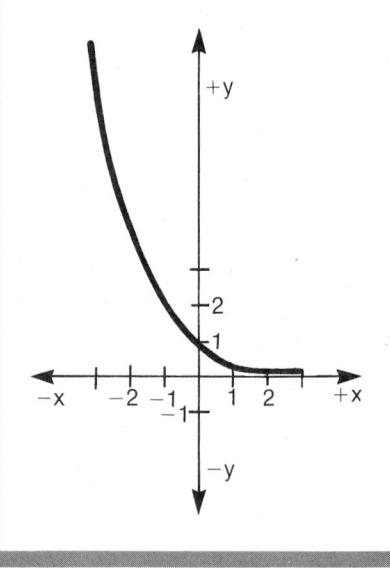

x	−3	−2	−1	0	1	2	3
y	8	4	2	1	$\dfrac{1}{2}$	$\dfrac{1}{4}$	$\dfrac{1}{8}$

Figure 4

Note that the graph of $y = \left(\dfrac{1}{2}\right)^x$ is similar in nature to the other exponential functions that we discussed previously. That is, the y-intercept is 1, and the graph is completely above the x-axis (y is always positive.). Also, the graph can be extended indefinitely to the right and left of the

Exponential and Logarithmic Functions

y-axis. But, in this case, where $0 < b < 1$, the graph rises to the left and approaches the x-axis on the right. It is still an exponential function. Rather than sketch the graph of $y = \left(\frac{1}{2}\right)^x$, we could have sketched the graph of $y = 2^{-x}$. These two equations are equal. Using the properties of negative exponents, we have

$$y = \left(\frac{1}{2}\right)^x,$$

$$y = (2^{-1})^x,$$

$$y = 2^{-x}.$$

Therefore, $y = \left(\frac{1}{2}\right)^x = 2^{-x}$ and the graph of $y = \left(\frac{1}{2}\right)^x$ is also the graph of $y = 2^{-x}$.

EXERCISES FOR SECTION 10.1
(Answers given to exercises marked *)

Make a table of values and sketch the graph of each equation:

*1. $y = 2^x$.
2. $y = 3^x$.
*3. $y = 5^x$.
4. $y = 2^{-x}$.
*5. $y = 3^{-x}$.
6. $y = 5^{-x}$.
*7. $y = \left(\frac{1}{2}\right)^x$.
8. $y = \left(\frac{1}{3}\right)^x$.
*9. $y = \left(\frac{1}{5}\right)^x$.
10. $y = \left(\frac{1}{2}\right)^{-x}$.
*11. $y = \left(\frac{1}{3}\right)^{-x}$.
12. $y = \left(\frac{1}{5}\right)^{-x}$.
*13. $y = 2^{2x}$.
14. $y = 3^{x+1}$.
*15. $y = 2^{x+1}$.
16. $y = 2^{-x-1}$.
*17. $y = 2 \cdot 2^x$.
18. $y = 2^{-x+1}$.

10.2 LOGARITHMIC FUNCTIONS

In the previous section, we examined various exponential functions of the form $y = b^x$, where x is the independent variable and y is the dependent variable. Now we want to see what happens when we interchange the variables. For example, we have sketched the graph of $y = 2^x$, but let us now consider $x = 2^y$. Generating a table of values for this equation, we have:

if $y = 0$, $x = 2^0 = 1$, \qquad if $y = -1$, $x = 2^{-1} = \frac{1}{2^1} = \frac{1}{2}$, if y = 1, x = 2^1 = 2,

if y = 2, x = 2^2 = 4,

if y = 3, x = 2^3 = 8,

if y = −2, x = $2^{-2} = \frac{1}{2^2} = \frac{1}{4}$.

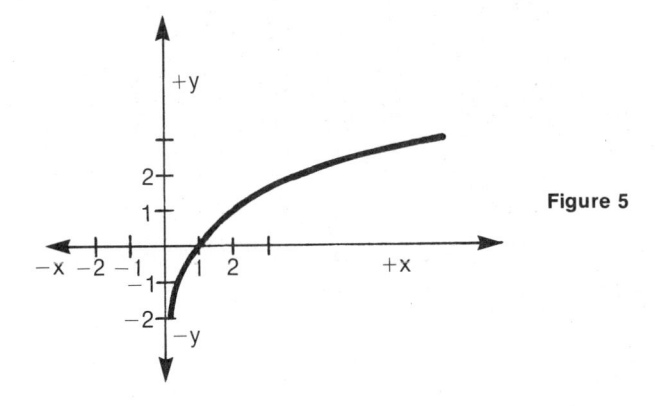

y	−2	−1	0	1	2	3
x	$\frac{1}{4}$	$\frac{1}{2}$	1	2	4	8

Figure 5

Plotting the corresponding points on the Cartesian plane and connecting them by means of a smooth curve, we obtain the graph shown in Figure 5. This represents the graph of x = 2^y. Figure 6 shows the graphs of y = 2^x and x = 2^y on the same set of coordinate axes. The dashed line represents the graph of y = x.

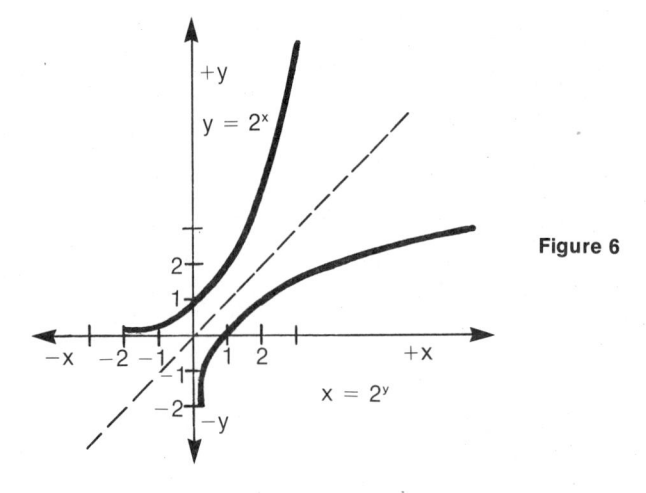

Figure 6

The equation x = 2^y is called the *inverse* of the exponential function y = 2^x. It is obtained by interchanging the variables. Recall that the inverse relation is formed by interchanging the coordinates of each ordered

Exponential and Logarithmic Functions 581

pair of the given relation (see Section 8.1). Note that in Figure 6 the graph of $y = 2^x$ and the graph of $x = 2^y$ are symmetric with respect to the line $y = x$. A function and its inverse will always be symmetric to the line $y = x$, and an exponential function and its inverse are also both functions.

The inverse of an exponential function is called a *logarithmic function*. That is, the logarithmic function is the inverse of the exponential function. Given $y = 2^x$, the inverse is $x = 2^y$ (interchanging the variables) and we can also say that the logarithmic function is $x = 2^y$. In general, if $y = b^x$ is the exponential function, then $x = b^y$ is the logarithmic function. Since it is convenient (and sometimes necessary) to express y in terms of x, that is, to state explicitly that y is a function of x, we must introduce a new notation and it is called logarithmic notation. Therefore, we note that

$$\boxed{x = b^y \text{ is equivalent to } y = \log_b x.}$$

The notation, $y = \log_b x$, is read as "y is the logarithm of x to the base b." We can also say "y equals the log of x to the base b." From the equation $y = \log_b x$, we know that y is the logarithm of x to the base b. Since $y = \log_b x$ is equivalent to $x = b^y$, we can say that *logarithms are exponents*. That is, the logarithm of x to the base b is the exponent of the base to which b is raised in order to obtain x. Again, we state that

$$\boxed{y = \log_b x \text{ is equivalent to } x = b^y.}$$

Using this definition, we can rewrite an exponential statement in logarithmic form. For example, we know that $2^3 = 8$ or $8 = 2^3$. Using the general equation $x = b^y$, we note that 8 corresponds to x, 2 corresponds to b (the base), and 3 corresponds to y (the exponent). Therefore, $8 = 2^3$ is equivalent to $3 = \log_2 8$ or $\log_2 8 = 3$. In writing logarithmic equations, remember that a logarithm is an exponent. In the equation $\log_2 8 = 3$, 3 is the exponent of the base, 2. This enables us to also write exponential equations from logarithmic equations. For example, in the equation $\log_{10} 100 = 2$, 10 is the base and 2 is the exponent. Hence,

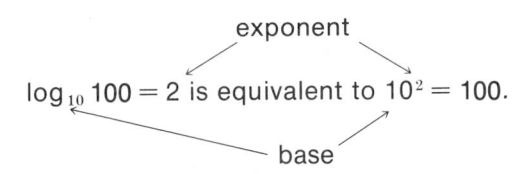

$$\underset{\text{base}}{\log_{10}} 100 = \overset{\text{exponent}}{2} \text{ is equivalent to } 10^2 = 100.$$

It is important that we be very familiar with the two different equations, exponential and logarithmic, so that we can change from $x = b^y$ to $\log_b x = y$, and vice versa. We must be able to change back and forth from the exponential form to the logarithmic form.

Listed below are some examples that illustrate the two equivalent forms.

Exponential Form	Logarithmic Form	Logarithmic Form	Exponential Form
$4^2 = 16.$	$\log_4 16 = 2.$	$\log_2 32 = 5.$	$2^5 = 32.$
$5^3 = 125.$	$\log_5 125 = 3.$	$\log_3 81 = 4.$	$3^4 = 81.$
$6^2 = 36.$	$\log_6 36 = 2.$	$\log_4 2 = \frac{1}{2}.$	$4^{1/2} = 2.$
$4^3 = 64.$	$\log_4 64 = 3.$	$\log_{10} 1 = 0.$	$10^0 = 1.$
		$\log_4 1 = 0.$	$4^0 = 1.$

EXAMPLE 3

Write each of the following equations in logarithmic form.

a. $5^2 = 25.$ b. $6^3 = 216.$ c. $8^{1/3} = 2.$

Solution

An exponential equation of the form $x = b^y$ is equivalent to the logarithmic equation, $\log_b x = y$. Therefore,

a. $5^2 = 25$ is equivalent to $\log_5 25 = 2.$
b. $6^3 = 216$ is equivalent to $\log_6 216 = 3.$
c. $8^{1/3} = 2$ is equivalent to $\log_8 2 = \frac{1}{3}.$

EXAMPLE 4

Write each of the following equations in exponential form.

a. $\log_3 243 = 5.$ b. $\log_9 3 = \frac{1}{2}.$ c. $\log_4 4 = 1.$

Solution

Recalling that logarithms are exponents and that the base is the same in either form, we have

a. $\log_3 243 = 5$ is equivalent to $3^5 = 243.$

b. $\log_9 3 = \frac{1}{2}$ is equivalent to $9^{1/2} = 3.$

c. $\log_4 4 = 1$ is equivalent to $4^1 = 4.$

Exponential and Logarithmic Functions 583

Many times a logarithmic equation will contain an unknown, and we have to solve it for the unknown variable. Basically, a logarithmic equation contains three different parts and therefore the unknown variable can occur in any one of three different places. Each of the following logarithmic equations contains an unknown variable, x, in a different place.

$$\log_3 x = 2. \qquad \log_3 9 = x. \qquad \log_x 9 = 2.$$

The first type of logarithmic equation can be solved rather quickly. To find the value of x in the equation, $\log_3 x = 2$, we only need to write the given equation in exponential form and evaluate it. Hence,

$\log_3 x = 2$ is equivalent to $3^2 = x$, and $x = 9$.

The second and third types of logarithmic equations can be solved using certain properties of exponents. To solve the second equation, $\log_3 9 = x$, for x, we again write the given equation in exponential form. That is,

$\log_3 9 = x$ is equivalent to $3^x = 9$.

To solve the equation $3^x = 9$, we make use of the fact that if two powers (with the same bases) are equal, then their exponents are equal. That is, if $x^m = x^n$ where $x > 0$ and $x \neq 1$, then $m = n$. Therefore, we must express $3^x = 9$ as an equation with the same bases. Since $9 = 3^2$, we have $3^x = 9$ is equivalent to $3^x = 3^2$ and $x = 2$.

To solve the third type of logarithmic equation, $\log_x 9 = 2$ where the unknown occurs in the base, we use the property that if $a = b$ then $a^m = b^m$. That is, if two quantities are equal, then they are still equal when they are both raised to the same exponent. Writing the given equation in exponential form, we have

$\log_x 9 = 2$ is equivalent to $x^2 = 9$.

In this case it may be apparent that $x = 3$, but we can use a standard procedure that will work in all cases. (Note: x cannot equal -3 since the base must be > 0.) We can solve the equation $x^2 = 9$ by solving it for x. One way to do this is to raise x^2 to the $\frac{1}{2}$ power in order to obtain x^1. That is,

$$(x^2)^{1/2} = x^{2/2} = x^1 \text{ or } x.$$

If we raise the left side of the equation to the $\frac{1}{2}$ power, then we maintain the equality by raising the right side of the equation to the $\frac{1}{2}$ power.

Therefore, we have

$$x^2 = 9,$$
$$(x^2)^{1/2} = 9^{1/2},$$
$$x = 3.$$

Note: Recall that
$$a^{m/n} = \sqrt[n]{a^m} = (\sqrt[n]{a})^m.$$

EXAMPLE 5

Solve each of the following equations.

a. $\log_2 x = 3$.   b. $\log_4 x = 0$.   c. $\log_9 x = \dfrac{3}{2}$.

d. $\log_4 x = -\dfrac{1}{2}$.

Solution

To find the value of x in each of the given equations, we express each as an equivalent exponential equation and evaluate it.

a. $\log_2 x = 3$ is equivalent to $2^3 = x$ and $x = 8$.
b. $\log_4 x = 0$ is equivalent to $4^0 = x$ and $x = 1$.
c. $\log_9 x = \dfrac{3}{2}$ is equivalent to $9^{3/2} = x$, $9^{3/2} = (\sqrt[2]{9})^3 = 3^3 = 27$.
   Hence $x = 27$.
d. $\log_4 x = -\dfrac{1}{2}$ is equivalent to $4^{-1/2} = x$ and $x = \dfrac{1}{4^{1/2}} = \dfrac{1}{\sqrt{4}} = \dfrac{1}{2}$.

EXAMPLE 6

Solve each of the following equations.

a. $\log_3 81 = x$.   b. $\log_5 25 = x$.   c. $\log_4 8 = x$.

d. $\log_7 \dfrac{1}{49} = x$.

Solution

First, we express each of the given equations as an equivalent exponential equation. Next, we express the members of the exponential equation so that they have the same base, and then solve the resulting equation that is obtained from the exponents.

a. $\log_3 81 = x$ is equivalent to $3^x = 81$. Expressing both mem-

Exponential and Logarithmic Functions 585

bers so that they have the same base, we have $3^x = 81$ is equivalent to $3^x = 3^4$ and $x = 4$.

b. $\log_5 25 = x$ is equivalent to $5^x = 25$, which is equivalent to $5^x = 5^2$ and $x = 2$.

c. $\log_4 8 = x$ is equivalent to $4^x = 8$. Both members can be expressed with bases of 2. $4^x = 8$ is equivalent to $(2^2)^x = 2^3$, which is equivalent to $2^{2x} = 2^3$. From this, we obtain the equation $2x = 3$, and $x = \dfrac{3}{2}$.

d. $\log_7 \dfrac{1}{49} = x$ is equivalent to $7^x = \dfrac{1}{49}$, which is equivalent to $7^x = \dfrac{1}{7^2}$, and since $a^{-m} = \dfrac{1}{a^m}$, we have $7^x = 7^{-2}$ and $x = -2$.

EXAMPLE 7

Solve each of the following equations.

a. $\log_x 8 = 3$.   b. $\log_x 4 = \dfrac{1}{2}$.   c. $\log_x 8 = \dfrac{3}{2}$.

d. $\log_x \dfrac{1}{4} = -2$.

Solution

First, we express each of the given equations as equivalent exponential equations. Next, we raise the power containing x as a base to the appropriate power to obtain x^1 and raise the other side of the equation to the same power. We evaluate the resulting expression to obtain the value of x.

a. $\log_x 8 = 3$ is equivalent to $x^3 = 8$. To obtain x, we raise x^3 to the one-third power (and also do the same to 8). That is,

$$x^3 = 8,$$
$$(x^3)^{1/3} = 8^{1/3},$$
$$x = 2.$$

b. $\log_x 4 = \dfrac{1}{2}$ is equivalent to $x^{1/2} = 4$.

$$(x^{1/2})^2 = 4^2,$$
$$x = 16.$$

c. $\log_x 8 = \dfrac{3}{2}$ is equivalent to $x^{3/2} = 8$.

$$(x^{3/2})^{2/3} = 8^{2/3},$$
$$x = (\sqrt[3]{8})^2 = 2^2 = 4.$$

d. $\log_x \frac{1}{4} = -2$ is equivalent to $x^{-2} = \frac{1}{4}$.

$$(x^{-2})^{-1/2} = \left(\frac{1}{4}\right)^{-1/2} = (4^{-1})^{-1/2},$$

$$x = 4^{1/2} = 2.$$

A logarithmic function is the inverse of an exponential function. For example, if $y = 2^x$ is the given exponential function, then its inverse is $x = 2^y$. Expressing y in terms of x, we use logarithmic notation and we have $y = \log_2 x$ or $\log_2 x = y$. Therefore, the logarithm of a number is the exponent to which the base must be raised in order to obtain the given number. If we have $\log_9 81 = 2$, then 2 is the logarithm of 81 to the base 9. Similarly, for $\log_2 8 = 3$, 3 is the logarithm of 8 to the base 2.

EXERCISES FOR SECTION 10.2

(Answers given to exercises marked *)

*1. Using the same set of axes, and the same scale, sketch the graphs of $y = 3^x$ and $x = 3^y$.

2. Using the same set of axes, and the same scale, sketch the graphs of $y = 4^x$ and $x = 4^y$.

*3. Write each of the following equations in logarithmic form.

    a. $4^2 = 16$.   b. $2^3 = 8$.   c. $3^3 = 27$.
    d. $2^4 = 16$.   e. $7^2 = 49$.   f. $5^3 = 125$.
    g. $4^{1/2} = 2$.   h. $8^{1/3} = 2$.   i. $9^{1/2} = 3$.
    j. $3^{-2} = \frac{1}{9}$.   k. $2^{-3} = \frac{1}{8}$.   l. $5^{-2} = \frac{1}{25}$.

4. Write each of the following equations in exponential form.

    a. $\log_4 16 = 2$.   b. $\log_2 8 = 3$.   c. $\log_3 27 = 3$.
    d. $\log_2 16 = 4$.   e. $\log_7 49 = 2$.   f. $\log_5 125 = 3$.
    g. $\log_4 2 = \frac{1}{2}$.   h. $\log_8 2 = \frac{1}{3}$.   i. $\log_9 3 = \frac{1}{2}$.
    j. $\log_3 \frac{1}{9} = -2$.   k. $\log_2 \frac{1}{8} = -3$.   l. $\log_5 \frac{1}{25} = -2$.

Exponential and Logarithmic Functions

*5. Solve each of the following equations:

a. $\log_5 x = 3$.
b. $\log_5 x = 2$.
c. $\log_5 x = 4$.
d. $\log_4 x = 3$.
e. $\log_7 x = 3$.
f. $\log_6 x = 3$.
g. $\log_2 x = 4$.
h. $\log_2 x = -3$.
i. $\log_9 x = -2$.
j. $\log_4 x = -\frac{1}{2}$.
k. $\log_9 x = -\frac{3}{2}$.
l. $\log_8 x = -\frac{2}{3}$.

6. Solve each of the following equations:

a. $\log_2 8 = x$.
b. $\log_3 27 = x$.
c. $\log_4 1 = x$.
d. $\log_8 2 = x$.
e. $\log_9 3 = x$.
f. $\log_2 32 = x$.
g. $\log_2 \frac{1}{8} = x$.
h. $\log_3 \frac{1}{9} = x$.
i. $\log_3 \frac{1}{81} = x$.
j. $\log_4 2 = x$.
k. $\log_9 27 = x$.
l. $\log_8 4 = x$.

*7. Solve each of the following equations:

a. $\log_x 8 = 3$.
b. $\log_x 16 = 4$.
c. $\log_x 125 = 3$.
d. $\log_x 2 = \frac{1}{2}$.
e. $\log_x 2 = \frac{1}{3}$.
f. $\log_x 2 = \frac{1}{4}$.
g. $\log_x \frac{1}{8} = -3$.
h. $\log_x \frac{1}{9} = -2$.
i. $\log_x 4 = \frac{2}{3}$.
j. $\log_x \frac{1}{2} = -\frac{1}{2}$.
k. $\log_x \frac{1}{27} = -\frac{3}{2}$.
l. $\log_x \frac{1}{8} = -\frac{3}{2}$.

8. Solve each of the following equations.

a. $\log_{10} 10 = x$.
b. $\log_{10} 1 = x$.
c. $\log_{10} 100 = x$.
d. $\log_{10} 0.1 = x$.
e. $\log_{10} 0.01 = x$.
f. $\log_{10} 1000 = x$.
g. $\log_{10} 0.001 = x$.
h. $\log_{10} 0.0001 = x$.
i. $\log_{10} 10000 = x$.

10.3 PROPERTIES OF LOGARITHMS

For a long time the most important use of logarithms was to aid in performing various complex calculations. For example, at one time the only way to find or compute various roots of numbers (other than square root and cube root) was by means of logarithms. To find or approximate the fifth root of 3, $\sqrt[5]{3}$, a person had to use logarithms. Admittedly, computers and attractively priced electronic calculators usually perform such calculations now, but logarithms are still essential as they enable us to solve many complex problems in a simple manner. They are invaluable in solving exponential and logarithmic equations. To use them in performing certain computations and solving equations, we must be familiar with their properties. At this time we shall examine three of these properties:

the logarithm of a product, the logarithm of a quotient, and the logarithm of a power.

Recall that logarithms are actually exponents. We can use this fact to find the logarithm of a product. Consider the following: if $x = \log_b M$ and $y = \log_b N$, then $b^x = M$ and $b^y = N$. If we multiply M times N, we have

$$MN = b^x b^y \quad \text{or} \quad MN = b^{x+y}. \quad \text{(adding exponents)}$$

Writing this exponential equation in logarithmic form, we have

$$\log_b MN = x + y.$$

Recall that $x = \log_b M$ and $y = \log_b N$; we can substitute for x and y in this equation. Therefore, $\log_b MN = x + y = \log_b M + \log_b N$ or

$$\boxed{\log_b MN = \log_b M + \log_b N.}$$

> The logarithm of a product of any positive real numbers M and N is the sum of the logarithms of the factors.

Note that for this property and the others that follow, it is assumed that $b \neq 0$ and $b > 1$. If this were not the case, then b could not be a base in the system of logarithms.

Using a procedure similar to that used for products, we can find the logarithm of a quotient. If $x = \log_b M$ and $y = \log_b N$, then $b^x = M$ and $b^y = N$. If we divide M by N, we have

$$\frac{M}{N} = \frac{b^x}{b^y} = b^{x-y}. \quad \text{(subtracting exponents)}$$

From the exponential equation, $\frac{M}{N} = b^{x-y}$, we obtain

$$\log_b \frac{M}{N} = x - y.$$

Substituting for x and y, we have

$$\boxed{\log_b \frac{M}{N} = \log_b M - \log_b N}$$

Exponential and Logarithmic Functions

The logarithm of a quotient of any positive real numbers M and N is the logarithm of the dividend minus the logarithm of the divisor. Also, the logarithm of a quotient of any positive real numbers $\frac{M}{N}$ is the logarithm of the numerator minus the logarithm of the denominator.

To obtain the logarithm of a power, we start with the fact that $x = \log_b M$ and then $b^x = M$. If we raise both members of this equation to the nth power, we have

$$(b^x)^n = M^n \quad \text{or} \quad b^{nx} = M^n.$$

From this exponential equation, we obtain

$$\log_b M^n = nx.$$

Substituting for x, we have

$$\boxed{\log_b M^n = n \log_b M}$$

> The logarithm of a power of a positive number M is equal to the exponent of that power times the logarithm of the number.

To summarize, we have developed the following properties of logarithms:

1. $\log_b MN = \log_b M + \log_b N$
2. $\log_b \frac{M}{N} = \log_b M - \log_b N$
3. $\log_b M^n = n \log_b M$

We can use each of these properties to simplify computations that involve multiplication, division, raising to powers, and finding roots of numbers. Here are some illustrative examples that apply these properties of logarithms:

$$\log_{10} 4 \cdot 5 = \log_{10} 4 + \log_{10} 5.$$

$$\log_b xyz = \log_b x + \log_b y + \log_b z.$$

$$\log_{10} \frac{4}{5} = \log_{10} 4 - \log_{10} 5.$$

$$\log_b \frac{xy}{z} = \log_b xy - \log_b z = \log_b x + \log_b y - \log_b z.$$

$$\log_{10} 3^2 = 2 \log_{10} 3.$$

$$\log_{10} \sqrt{3} = \log_{10} 3^{1/2} = \frac{1}{2} \log_{10} 3.$$

EXAMPLE 8

For each of the following, write equivalent expressions in terms of $\log_b x$, $\log_b y$, and $\log_b z$:

a. $\log_b x^2 y$ b. $\log_b \dfrac{x^2}{yz}$ c. $\log_b \sqrt{x^3 y^5 z}$

Solution

a. $\log_b x^2 y = \log_b x^2 + \log_b y = 2 \log_b x + \log_b y.$

b. $\log_b \dfrac{x^2}{yz} = \log_b x^2 - \log_b yz = 2 \log_b x - (\log_b y + \log_b z).$

c. $\log_b \sqrt{x^3 y^5 z} = \log_b (x^3 y^5 z)^{1/2} = \log_b (x^{3/2} y^{5/2} z^{1/2})$
$= \log_b x^{3/2} + \log_b y^{5/2} + \log_b z^{1/2}$
$= \dfrac{3}{2} \log_b x + \dfrac{5}{2} \log_b y + \dfrac{1}{2} \log_b z.$

EXAMPLE 9

Determine whether each of the following statements is true or false.

a. $\log_b x + \log_b y = \log_b x \cdot \log_b y.$

b. $\log_b x + \log_b y = \log_b (x + y).$

c. $\log_b \dfrac{x}{y} = \dfrac{\log_b x}{\log_b y}.$

d. $\log_b 3x = 3 \log_b x.$

e. $\log_b x^2 = \log_b x \cdot \log_b x.$

Solution

Each of the statements is false. Listed below are the corrected statements.

a. $\log_b x + \log_b y = \log_b xy.$ d. $\log_b 3x = \log_b 3 + \log_b x.$
b. $\log_b x + \log_b y = \log_b xy.$ e. $\log_b x^2 = 2 \log_b x.$
c. $\log_b \dfrac{x}{y} = \log_b x - \log_b y.$

Exponential and Logarithmic Functions

Now we are ready to do some computations using logarithms. Keep in mind that the following examples are to help us develop confidence in using logarithms for computations. Admittedly, they can be done without logarithms. Consider the following table of powers of ten:

TABLE 1

$1 = 10^{0.0000}$	$6 = 10^{0.7782}$
$2 = 10^{0.3010}$	$7 = 10^{0.8451}$
$3 = 10^{0.4771}$	$8 = 10^{0.9031}$
$4 = 10^{0.6021}$	$9 = 10^{0.9542}$
$5 = 10^{0.6990}$	$10 = 10^{1.0000}$

It should be noted that these powers of ten are approximate, not exact. If a greater degree of accuracy were desired, we would have to increase the number of decimal places in the exponents of 10. Each of the equations in the table is in the form of a logarithmic equation. For example, we see that $2 = 10^{0.3010}$, which we can express in the form of $\log_{10} 2 = 0.3010$. So we can convert the table containing powers of ten to a table of logarithms:

TABLE 2

$\log_{10} 1 = 0.0000$	$\log_{10} 6 = 0.7782$
$\log_{10} 2 = 0.3010$	$\log_{10} 7 = 0.8451$
$\log_{10} 3 = 0.4771$	$\log_{10} 8 = 0.9031$
$\log_{10} 4 = 0.6021$	$\log_{10} 9 = 0.9542$
$\log_{10} 5 = 0.6990$	$\log_{10} 10 = 1.0000$

From this table that we have developed, we can find the logarithm of the numbers 1 through 10 using 10 as a base. Many times we may know the logarithm of a number, but not the number itself. For example, if we have $\log_{10} N = 0.3010$, we can use the table to find the number corresponding to the logarithm. This is called the *antilogarithm* of a number. Therefore, from $\log_{10} N = 0.3010$, we write:

$$\text{antilog}_{10} 0.3010 = N,$$
$$N = 2. \quad \text{(using the Table)}$$

Now we shall perform some calculations, using logarithms to illustrate the different properties. For example, to compute 2×4 using base 10 logarithms, we have

$$2 \times 4,$$
$$\log_{10}(2 \times 4), \quad \text{using logarithm notation}$$
$$\log_{10}(2 \times 4) = \log_{10} 2 + \log_{10} 4,$$
$$\log_{10}(2 \times 4) = 0.3010 + 0.6021, \quad \text{(from the Table)}$$
$$\log_{10}(2 \times 4) = 0.9031.$$

Now we must find the antilogarithm and we write

$$\text{antilog}_{10} 0.9031 = ?,$$
$$\text{antilog}_{10} 0.9031 = 8. \quad \text{(from the Table)}$$
$$\text{Hence, } 2 \times 4 = 8.$$

For an example of division, consider $8 \div 4$. We proceed in the following manner:

$$\frac{8}{4},$$

$$\log_{10}\left(\frac{8}{4}\right), \quad \text{using logarithm notation}$$

$$\log_{10}\left(\frac{8}{4}\right) = \log_{10} 8 - \log_{10} 4,$$

$$\log_{10}\left(\frac{8}{4}\right) = 0.9031 - 0.6021, \quad \text{(from the Table)}$$

$$\log_{10}\left(\frac{8}{4}\right) = 0.3010,$$

$$\text{antilog}_{10} 0.3010 = ?,$$
$$\text{antilog}_{10} 0.3010 = 2. \quad \text{(from the Table)}$$

$$\text{Hence, } \frac{8}{4} = 2.$$

For an example of the properties of logarithms involving powers, consider 3^2. We proceed in the following manner:

$$3^2$$
$$\log_{10} 3^2, \quad \text{using logarithm notation}$$
$$\log_{10} 3^2 = 2 \log_{10} 3,$$
$$\log_{10} 3^2 = 2(0.4771), \quad \text{(from the Table)}$$
$$\log_{10} 3^2 = 0.9542,$$
$$\text{antilog}_{10} 0.9542 = ?,$$
$$\text{antilog}_{10} 0.9542 = 9. \quad \text{(from the Table)}$$
$$\text{Therefore, } 3^2 = 9.$$

EXAMPLE 10

Perform each of the following computations, using base 10 logarithms:

    a. 2×5 b. $10 \div 5$ c. $\sqrt{9}$

Exponential and Logarithmic Functions

Solution

a. $\log_{10}(2 \times 5) = \log_{10} 2 + \log_{10} 5$
$= 0.3010 + 0.6990$
$= 1.0000.$
$\text{antilog}_{10}\ 1.0000 = 10.$
Therefore, $2 \times 5 = 10.$

b. $\log_{10}\left(\dfrac{10}{5}\right) = \log_{10} 10 - \log_{10} 5$
$= 1.0000 - 0.6990$
$= 0.3010.$
$\text{antilog}_{10}\ 0.3010 = 2.$
Therefore, $\dfrac{10}{5} = 2.$

c. $\log_{10} \sqrt{9} = \log_{10} 9^{1/2} = \dfrac{1}{2} \log_{10} 9$
$= \dfrac{1}{2}(0.9542)$
$= 0.4771.$
$\text{antilog}_{10}\ 0.4771 = 3.$
Therefore, $\sqrt{9} = 3.$

EXERCISES FOR SECTION 10.3

*(Answers given to exercises marked *)*

Write equivalent expressions in terms of $\log_b x$, $\log_b y$, and $\log_b z$.

*1. $\log_b xyz^2$ 2. $\log_b (xyz)^2$ *3. $\log_b x^2 y^3$

4. $\log_b \dfrac{x^2 y}{z}$ *5. $\log_b \dfrac{xy^2}{z}$ 6. $\log_b \dfrac{x^3}{y^2 z}$

*7. $\log_b \sqrt{xy}$ 8. $\log_b \sqrt[3]{x^2 y}$ *9. $\log_b \sqrt{x}\, \sqrt[3]{z}$

10. $\log_b \sqrt{\dfrac{xy}{z^3}}$ *11. $\log_b \sqrt{\dfrac{6x}{yz^3}}$ 12. $\log_b \sqrt[3]{\dfrac{3x^2}{6yz}}$

Write each expression as a single logarithm.

*13. $2\log_b x + 3\log_b y - \log_b z$ 14. $3\log_b y - (2\log_b x + \log_b z)$

*15. $\dfrac{1}{2}\log_b x + \dfrac{3}{2}\log_b y + \dfrac{5}{2}\log_b z$ 16. $2\log_b y + \dfrac{1}{2}\log_b z$

*17. $3 \log_b x - 2 \log_b y - \frac{1}{2} \log_b z$ 18. $\frac{1}{2} \log_b x - \frac{3}{2} \log_b z$

*19. $\frac{1}{2} \log_b x + \frac{1}{2} \log_b y - \frac{3}{2} \log_b z$ 20. $\frac{1}{2} \log_b 6 + \frac{1}{2} \log_b x - \frac{3}{2} \log_b z$

Given the following logarithms: $\log_{10} 2 = 0.3010$; $\log_{10} 3 = 0.4771$; $\log_{10} 5 = 0.6990$; $\log_{10} 10 = 1.0000$, find each of the following:

*21. $\log_{10} 6$ 22. $\log_{10} 50$ *23. $\log_{10} 15$

24. $\log_{10} 30$ *25. $\log_{10} \frac{10}{5}$ 26. $\log_{10} \frac{10}{6}$

*27. $\log_{10} \frac{10}{2}$ 28. $\log_{10} \frac{10}{3}$ *29. $\log_{10} 5^2$

30. $\log_{10} 27$ *31. $\log_{10} 16$ 32. $\log_{10} 100$

*33. $\log_{10} 1000$ 34. $\log_{10} \sqrt{2}$ *35. $\log_{10} \sqrt[3]{5}$

Given the following logarithms: $\log_{10} 12 = 1.0792$; $\log_{10} 14 = 1.1461$; $\log_{10} 15 = 1.1761$, and using the logarithms from Table 2, perform each of the following computations, using base 10 logarithms:

36. 3×4 *37. 2×7 38. 3×5

*39. $12 \div 6$ 40. $14 \div 2$ *41. $15 \div 5$

42. $\sqrt{12} \cdot \sqrt{3}$ *43. $\sqrt[3]{8}$ 44. $\frac{3^2 \cdot 2^3}{6^2}$

10.4 COMMON LOGARITHMS

In almost all cases, logarithms used in computations have a base of 10. These are called *common logarithms.* Why should we use 10 as a base for logarithms? One of the main reasons is that 10 is the base of our number system and it is usually used to perform most computations. Previously, we have used notation such as $\log_{10} 2 = 0.3010$ and $\log_{10} 5 = 0.6990$. These are common logarithms. Now we shall adopt the convention of dropping or omitting the base 10 notation for common logarithms. For example, we shall now write that $\log 2 = 0.3010$ and $\log 5 = 0.6990$. When no other base is indicated, it is understood to be base 10.

Table 2 in the previous section lists the common logarithms for the numbers 1 through 10. Note that the common logarithm of any number between 1 and 10 is a real number between 0 and 1. What about the logarithm of a number between 10 and 100? Since $\log 10 = 1$ and $\log 100 = 2$ (see Exercise 32, Section 10.3), the logarithm of any number between 10 and 100 is a real number between 1 and 2. For example, con-

Exponential and Logarithmic Functions

sider the logarithm of 20. We can find log 20 by using Table 2 and the properties of logarithms:

$$\begin{align*}
\log 20 &= \log 10 \cdot 2 \\
&= \log 10 + \log 2 \\
&= 1 + 0.3010. \\
\log 20 &= 1.3010.
\end{align*}$$

Next, consider the logarithm of 2000:

$$\begin{align*}
\log 2000 &= \log 1000 \cdot 2 \\
&= \log 10^3 + \log 2 \\
&= 3 \log 10 + 0.3010 \\
&= 3 \cdot 1 + 0.3010 \\
&= 3 + 0.3010. \\
\log 2000 &= 3.3010.
\end{align*}$$

Note that the logarithms of 2, 20, and 2000 all have something in common. The decimal or fractional parts of the logarithms are identical, but the numbers in the units place are different. If we express each of these numbers in scientific notation, we can see why this is the case. Recall that scientific notation consists of $n \times 10^m$ where $1 \leq n < 10$ and m is an integer (see Section 3.3). Therefore, $2 = 2.0 \times 10^0$, $20 = 2.0 \times 10^1$, and $2000 = 2.0 \times 10^3$. Hence, we can state that

$$\begin{align*}
\log N &= \log(n \times 10^m) \\
&= \log n + \log 10^m \\
&= \log n + m \log 10 \quad \text{(Note: log 10 = 1.)} \\
&= \log n + m.
\end{align*}$$

From this derivation, we can see why the logarithms of 2, 20, and 2000 all have the same decimal or fractional parts and the number preceding the decimal point is different. The fractional or decimal portion of the logarithm (log n, where $1 < n < 10$) is called the *mantissa* of the logarithm. The integer preceding the decimal point, or the integral part of the logarithm (exponent of the power of 10) is called the *characteristic* of the logarithm. Therefore, we state that the logarithm of a number, N, is the sum of the characteristic and mantissa. That is,

$$\boxed{\log N = \text{characteristic} + \text{mantissa}}$$

We can determine the characteristic of a logarithm for any number by writing the given number in scientific notation ($N = n \times 10^m$ where $1 \leq n < 10$) and m represents the characteristic of the logarithm. For example, to find the characteristic of the logarithm of 187, we write the given number in scientific notation: $187 = 1.87 \times 10^2$. The exponent of 10 is the characteristic and it is 2.

EXAMPLE 11

Determine the characteristic of the logarithms of each of the following numbers:

a. 32   b. 423   c. 2030   d. 0.25   e. 0.013

Solution

We write each of the numbers in scientific notation and the exponent of 10 is the characteristic.

a. $32 = 3.2 \times 10^1$; the characteristic is 1.
b. $423 = 4.23 \times 10^2$; the characteristic is 2.
c. $2030 = 2.03 \times 10^3$; the characteristic is 3.
d. $0.25 = 2.5 \times 10^{-1}$; the characteristic is -1.
e. $0.013 = 1.3 \times 10^{-2}$; the characteristic is -2.

The logarithm of a number is the sum of the characteristic and mantissa. We have determined how to find the characteristic of a number — now we are ready to find the mantissa. Table II in Appendix A is a table of mantissas. With it, we can find the mantissa of any three-digit number. Note: All values in Table II are between 0 and 1, but the decimal points have been omitted. The column with the heading of "n" contains the first two digits of any three-digit number. To find the third digit, we must read across. Since "n" contains two digits, we must consider a number such as 2 to be 2.00. Similarly, 25 is treated as 25.0 and 99 is treated as 99.0.

Let's practice using Table II. To find log 2, we first determine the characteristic of 2. From previous discussions, we know that it is 0. Now we use Table II to find the mantissa. Under the "n" column, we find 20 and then move across under the 0 column. Recall that 2 is treated as 2.00. The mantissa in this position is .3010. Note that the decimal point is omitted; we must put it in. Therefore,

$$\log 2 = 0 + .3010 = 0.3010.$$

Next, consider log 354. We must find the characteristic and the mantissa. $354 = 3.54 \times 10^2$, so the characteristic is 2. Under the "n" column, we find 35, and then move across until we find the number under the 4 column. We find that the mantissa is .5490. Therefore, $\log 354 = 2 + .5490 = 2.5490$.

Exponential and Logarithmic Functions

Consider log 80.7. Since $80.7 = 8.07 \times 10^1$, the characteristic is 1. Under the "n" column, we find 80, and then move across until we find the number under the 7 column. We find that the mantissa is .9069. Therefore, log $80.7 = 1 + .9069 = 1.9069$.

EXAMPLE 12

Using Table II in Appendix A, find:

a. log 554   b. log 73   c. log 1.27

Solution

a. $554 = 5.54 \times 10^2$; the characteristic is 2. The mantissa for 5.54 is .7435. Therefore, log $554 = 2 + .7435 = 2.7435$.
b. $73 = 7.3 \times 10^1$; the characteristic is 1. The mantissa for 7.30 is .8633. Therefore, log $73 = 1 + .8633 = 1.8633$.
c. $1.27 = 1.27 \times 10^0$; the characteristic is 0. The mantissa for 1.27 is .1038. Therefore, log $1.27 = 0 + .1038 = 0.1038$.

It is advisable to keep handy a straight edge of some sort for finding the mantissa in Table II. It is easy to make an error in reading the Table.

Many times the characteristic of a logarithm will be negative. For example, consider log 0.002. Expressing 0.002 in scientific notation, we have 2.0×10^{-3}. The characteristic is -3. The mantissa for 2.0 is .3010. Therefore, log $0.002 = -3 + .3010$. Since all of the mantissas in Table II are non-negative, we must keep them non-negative, and we *do not* perform the indicated addition to obtain log 0.002. Instead, we express it by adding a multiple of 10 to it. But if we add a multiple of 10, then we must subtract the multiple of 10 to maintain the equality. (Recall the identity property of 0.) Therefore, $-3 = -3 + 10 - 10 = 7 - 10$. We can also express the characteristic -3 as $17 - 20$, $27 - 30$, $37 - 40$, and so on. Normally, we would express it as $7 - 10$, but sometimes it may be convenient to express it in terms of some other multiple of 10. Now we can state that log $0.002 = -3 + .3010$

$$= \overbrace{7 + .3010} - 10$$
$$= 7.3010 - 10.$$

Remember, the reason for doing this is to keep the mantissa non-negative. By using this technique, the mantissa for log 0.002 is still .3010 and the characteristic is -3 as indicated by $7 - 10$. Let's consider one more example: log 0.314. Expressing this number in scientific notation, we have 3.14×10^{-1}. The characteristic is -1. The mantissa for 3.14 is .4969. Therefore, log $0.314 = -1 + .4969$ and -1 can be expressed as $9 - 10$.

Hence,

$$\log 0.314 = -1 + .4969 = 9.4969 - 10.$$

EXAMPLE 13

Using Table II in Appendix A, find:

a. log 0.0123   b. log 0.524   c. log 0.00042

Solution

a. $0.0123 = 1.23 \times 10^{-2}$; the characteristic is -2. The mantissa for 1.23 is .0899. Therefore, $\log 0.0123 = -2 + .0899 = 8.0899 - 10$.
b. $0.524 = 5.24 \times 10^{-1}$; the characteristic is -1. The mantissa for 5.24 is .7193. Therefore, $\log 0.524 = -1 + .7193 = 9.7193 - 10$.
c. $0.00042 = 4.2 \times 10^{-4}$. The characteristic is -4. The mantissa for 4.2 is .6232. Therefore, $\log 0.00042 = -4 + .6232 = 6.6232 - 10$.

In each of the examples above, the characteristics could have been expressed in terms of some other multiple of 10. For example, $\log 0.0123 = 8.0899 - 10$ can also be expressed as $18.0899 - 20$ or $28.0899 - 30$.

In Section 10.3, we briefly discussed *antilogarithms.* When we find the antilogarithm of a number, we are finding the number which corresponds to a given logarithm. For example, if log N = 0.3010, then we want to find N. That is, we want to find the number which corresponds to a given logarithm or we want to find the antilog of 0.3010. Note that the characteristic is 0. Next, we look in Table II to find the mantissa, .3010, and what number corresponds to it. It is 2.00 and since the characteristic is 0, we have $N = 2.00 \times 10^0 = 2.00$. Therefore, the antilog of 0.3010 is 2.00 or 2.

To find the antilog of $7.8998 - 10$, we first note that the characteristic is $7 - 10 = -3$. The mantissa is .8998. Checking Table II, we note the number whose mantissa is .8998 and it is 7.94. Since the characteristic is -3, we have $N = 7.94 \times 10^{-3} = 0.00794$. Therefore, the antilog of $7.8998 - 10$ is 0.00794.

EXAMPLE 14

Using Table II in Appendix A, find the numbers whose logarithms are given:

a. 3.0170   b. $9.7832 - 10$ c. 0.9518   d. $8.5172 - 10$

Exponential and Logarithmic Functions

Solution

First, we determine the characteristic and then find the number that corresponds to the given mantissa. By means of scientific notation, we determine the number.

a. 3.0170: the characteristic is 3, the mantissa is .0170, and the number that corresponds to it is 1.04. Therefore, $N = 1.04 \times 10^3 = 1040$.

b. 9.7832 − 10: the characteristic is $9 - 10 = -1$, the mantissa is .7832, and the number that corresponds to it is 6.07. Therefore, $N = 6.07 \times 10^{-1} = 0.607$.

c. 0.9518: the characteristic is 0, the mantissa is .9518 and the number that corresponds to it is 8.95. Therefore, $N = 8.95 \times 10^0 = 8.95$.

d. 8.5172 − 10: the characteristic is $8 - 10 = -2$, the mantissa is .5172 and the number that corresponds to it is 3.29. Therefore, $N = 3.29 \times 10^{-2} = 0.0329$.

EXERCISES FOR SECTION 10.4

(Answers given to exercises marked *)

Determine the characteristic of the logarithm of each of the following numbers:

*1. 44
2. 108
*3. 437
4. 3
*5. 0.12
6. 0.00324
*7. 1.01
8. 1.472
*9. 3.14159
10. 0.00127
*11. 0.01203
12. 0.01010
*13. 340.7
14. 30,430
*15. 1,000,000

Use Table II in Appendix A to find:

16. log 2.12
*17. log 343
18. log 47.1
*19. log 147
20. log 0.012
*21. log 0.034
22. log 0.147
*23. log 0.345
24. log 0.003
*25. log 0.0090
26. log 0.0301
*27. log 30.9
28. log 996
*29. log 0.000807
30. log 0.0765

Use Table II in the appendix. Find the numbers whose logarithms are given:

*31. 2.7324
32. 1.8463
*33. 9.9795 − 10

34. 8.7782 − 10   *35. 8.9340 − 10   36. 0.9053

*37. 0.5551   38. 1.3160   *39. 7.4048 − 10

40. 2.3010   *41. 4.8439   42. 9.6085 − 10

*43. 18.0334 − 20   44. 29.6730 − 30   *45. 4.4900

10.5 INTERPOLATION

By means of Table II in Appendix A, we are able to find the logarithm of any three-digit number. We can also use it to find the logarithm of a four-digit number using a process called *interpolation*. For example, consider log 7.365. The log of 7.365 is between log 7.360 and log 7.370. Since 7.365 is halfway between 7.360 and 7.370, it is assumed that its logarithm is also halfway between log 7.360 and log 7.370. To find their relative positions, we list the numbers and their logarithms in the following manner:

```
                    ┌──────── log 7.370 = 0.8675. ────────┐
 .010  │     ┌── log 7.365 =    ?    ──┐                  │
       │.005 │                          │  x              │ .0006
       │     └──                       ──┘                │
       └──────── log 7.360 = 0.8669. ────────┘
```

We can determine the difference between the relative positions (as indicated) and set up the following proportion:

$$\frac{.005}{.010} = \frac{x}{.0006}.$$

Simplifying this equation, we have

$$\frac{5}{10} = \frac{x}{.0006}.$$

We can solve this proportion, recalling that in a proportion the product of the means equals the product of the extremes (see Section 8.4). Therefore,

$$10x = 5(.0006),$$
$$10x = .0030,$$
$$x = .0003.$$ (rounded off to four decimal places)

Therefore, log 7.365 = 0.8669 + .0003 = 0.8672.

Exponential and Logarithmic Functions

EXAMPLE 15

Find log 594.3.

Solution

The logarithm of 594.3 is between log 594.0 and log 595.0. To find the relative positions, we list the numbers and their logarithms.

$$1 \left[.3 \left[\begin{array}{l} \log 595.0 = 2.7745. \\ \log 594.3 = \;\; ? \\ \log 594.0 = 2.7738. \end{array} \right] x \right] .0007$$

We determine the position of log 594.3 by means of the following proportion:

$$\frac{.3}{1} = \frac{x}{.0007}, \quad \text{Simplifying,} \quad \frac{3}{10} = \frac{x}{.0007}.$$

$$10x = .0021,$$
$$x = .00021 = .0002. \quad \text{(rounded off to four decimal places)}$$

Therefore, log 594.3 = 2.7738 + .0002 = 2.7740.

EXAMPLE 16

Find log 0.01237.

Solution

The logarithm of 0.01237 is between log 0.01230 and log 0.01240. Hence, we have

$$.0010 \left[.0007 \left[\begin{array}{l} \log 0.01240 = 8.0934 - 10. \\ \log 0.01237 = \;\; ? \\ \log 0.01230 = 8.0899 - 10. \end{array} \right] x \right] .0035$$

We determine the position of log 0.01237 by means of the following proportion:

$$\frac{.0007}{.0010} = \frac{x}{.0035}. \quad \text{Simplifying,} \quad \frac{7}{10} = \frac{x}{.0035}.$$

$$10x = .0245,$$
$$x = .00245 = .0025. \quad \text{(rounded off to four decimal places)}$$

Therefore, log $0.01237 = 8.0899 - 10 + .0025 = 8.0924 - 10.$

add the decimals

We can use this same technique to find a four-digit number when we are given its logarithm. That is, we can find the antilogarithm of a number by means of interpolation. For example, consider that we are given log $N = 2.6752$. We want to find the number whose logarithm is 2.6752 or we want to find the antilog of 2.6752. We examine Table II to find those mantissas that are nearest in value to the given mantissa, .6752. We must locate the nearest mantissa that is smaller and the nearest that is larger than the given mantissa. They are: .6749, which appears opposite 47 and under the 3 column, and .6758, which appears opposite 47 and under the 4 column. Therefore, we can list the numbers and their logarithms in the following manner (note that the characteristic is 2):

```
          ┌──── log 474.0 = 2.6758. ────┐
          │                             │
1    ┌── log N    = 2.6752. ──┐        .0009
   x │                         │ .0003
     └──── log 473.0 = 2.6749. ┘
```

We can determine the difference between the relative positions (as indicated) and set up the following proportion:

$$\frac{x}{1} = \frac{.0003}{.0009}. \quad \text{Simplifying,} \quad \frac{x}{1} = \frac{3}{9} \quad \text{and } x \approx .3.$$

Therefore, $N = 473.0 + .3 = 473.3$. Note that x was rounded off to .3 since N must be between 473.0 and 474.0, and this will be the fourth figure of the antilogarithm.

EXAMPLE 17

Find the number whose logarithm is 0.4343.

Solution

The mantissa is .4343 and it lies between .4330 and .4346, which are logarithms of 2.71 and 2.72, respectively. Hence, we list the information as indicated (note that the characteristic is 0):

Exponential and Logarithmic Functions

```
             ┌──────── log 2.720 = 0.4346. ────────┐
.01     ┌──── log N      = 0.4343. ────┐              .0016
     x  │                                │  .0013
        └──── log 2.710 = 0.4330. ──────┘
```

We determine the value of x by means of the following proportion:

$$\frac{x}{.01} = \frac{.0013}{.0016}. \quad \text{Simplifying,} \quad \frac{x}{.01} = \frac{13}{16}.$$

$$x = \frac{.13}{16} \approx .008.$$

Therefore N = 2.710 + .008 = 2.718. Note that x was rounded off to .008 since N must be between 2.710 and 2.720 and this will be the fourth figure of the antilogarithm.

EXAMPLE 18

Find the number whose logarithm is 8.5974 − 10.

Solution

After locating the mantissas that the given mantissa is between, we list the information as indicated (note that the characteristic is 8 − 10 or −2):

```
               ┌──── log 0.03960 = 8.5977 − 10. ────┐
.0001    ┌──── log N       = 8.5974 − 10. ────┐       .0011
      x  │                                      │  .0008
         └──── log 0.03950 = 8.5966 − 10. ─────┘
```

We determine the value of x by means of the following proportion:

$$\frac{x}{.0001} = \frac{.0008}{.0011}. \quad \text{Simplifying,} \quad \frac{x}{.0001} = \frac{8}{11}.$$

$$x = \frac{.0008}{11} \approx .00007.$$

Therefore, N = 0.03950 + .00007 = 0.03957. Note that x was rounded off to .00007 since N must be between 0.03950 and 0.03960 and this will be the fourth figure of the antilogarithm.

EXAMPLE 19

Find the antilogarithm of $9.7353 - 10$.

Solution

After locating the mantissas that the given mantissa is between, we list the information as indicated (note that the characteristic is $9 - 10$ or -1):

$$.001 \left[x \left[\begin{array}{l} \log 0.5440 = 9.7356 - 10. \\ \log N = 9.7353 - 10. \\ \log 0.5430 = 9.7348 - 10. \end{array} \right] .0005 \right] .0008$$

We determine the value of x by means of the following proportion:

$$\frac{x}{.001} = \frac{.0005}{.0008}. \quad \text{Simplifying,} \quad \frac{x}{.001} = \frac{5}{8}.$$

$$x = \frac{.005}{8} \approx .0006.$$

Therefore, $N = 0.543 + .0006 = 0.5436$. Note that x was rounded off to .0006 since N must be between 0.5430 and 0.5440 and this will be the fourth figure of the antilogarithm.

To find the number that corresponds to a given logarithm, or to find the logarithm of a given number by means of interpolation requires practice. Make sure that you are thoroughly familiar with, and understand, the illustrative examples in this section before attempting the exercises.

EXERCISES FOR SECTION 10.5

*(Answers given to exercises marked *)*

Find the following:

*1. log 4275   2. log 8765   *3. log 72.17
 4. log 5.728   *5. log 0.4167   6. log 0.2356
*7. log 0.04123 8. log 456.6   *9. log 7634
10. log 2.973   *11. log 744.2   12. log 457.6

Exponential and Logarithmic Functions 605

*13. log 312.3   14. log 0.05523   *15. log 0.6383
16. log 8753   *17. log 55.94   18. log 314.2
*19. log 237.2   20. log 38.24   *21. log 2.753
22. log 0.08437   *23. log 0.6432   24. log 238.9

Find the number of each logarithm:

*25. 2.8958   26. 2.1650   *27. 3.2916
28. 0.4574   *29. 0.8174   30. 0.4752
*31. 9.5594 − 10   32. 9.8085 − 10   *33. 9.7432 − 10
34. 8.5531 − 10   *35. 8.4060 − 10   36. 8.6791 − 10
*37. 2.3085   38. 0.4343   *39. 3.8680
40. 0.6752   *41. 9.6318 − 10   42. 3.7004
*43. 9.1025 − 10   44. 1.5974   *45. 8.7353 − 10

HISTORICAL NOTE

The word logarithm means "ratio number." It was derived from the Greek words *logos* which means "ratio" and *arithmos* which means "number." John Napier (1550–1617), a Scottish mathematician, invented logarithms in 1614. Napier first published a table of logarithms for trigonometric functions—in particular, logarithms of the sines of angles. Later, an English mathematician, Henry Briggs (1561–1631), visited Napier to discuss the logarithms. As a result of this visit, Briggs introduced what are now known as common logarithms. Briggs also introduced the terms "mantissa" and "characteristic." It is interesting to note that logarithm tables included the characteristic and the mantissa until the eighteenth century.

Napier announced his discovery of logarithms in a book titled Mirifici Logarithmorum Canonis Descriptio (A Description of the Admirable Table of Logarithms). In a prefatory paragraph, he wrote, "Seeing there is nothing, right well-beloved students of mathematics, that is so troublesome to mathematical practice, nor doth more molest and hinder calculators, than the multiplication, division, square and cubical extractions of great numbers, which besides the tedious expense of time are for the most part subject to many slippery errors, I began therefore to consider in my mind by what certain and ready art I might remove those hinderances."

10.6 COMPUTATIONS USING LOGARITHMS

We can use logarithms to perform certain calculations. We can multiply, divide, raise to powers, and extract roots using logarithms. Logarithms simplify these computations, although, as we stated earlier, many of these complex calculations are now done by means of computers and electronic calculators. We shall perform some calculations using logarithms to illustrate their various properties and also to illustrate how logarithms do simplify computations. For example, consider the problem of finding the square root of five, $\sqrt{5}$. You may have been taught in an elementary algebra course a long, tedious process for finding square roots. We can compute $\sqrt{5}$ by means of logarithms:

$$N = \sqrt{5},$$

$$\log N = \log \sqrt{5},$$

$$\log N = \log 5^{1/2},$$

$$\log N = \frac{1}{2} \log 5,$$

$$\log N = \frac{1}{2}(0.6990),$$

$$\log N = 0.3495.$$

To find the antilog of 0.3495, we must use interpolation. Therefore, we have

$$.01 \left[x \left[\begin{array}{l} \log 2.240 = 0.3502. \\ \log N\ \ \ \ = 0.3495. \\ \log 2.230 = 0.3483. \end{array} \right] .0012 \right] .0019$$

$$\frac{x}{.01} = \frac{.0012}{.0019}. \quad \text{Simplifying,} \quad \frac{x}{.01} = \frac{12}{19}.$$

$$x = \frac{.12}{19} \approx .006.$$

Therefore, $N = 2.230 + .006 = 2.236 = \sqrt{5}.$

Recall that $\sqrt{5}$ is an irrational number, that is, a non-terminating, non-repeating decimal. Hence, when we state that $\sqrt{5} = 2.236$, we mean that this is an approximation with four-digit accuracy.

EXAMPLE 20

Find the indicated root by use of logarithms:

$$\sqrt[3]{83.7}.$$

Exponential and Logarithmic Functions 607

Solution

$$N = \sqrt[3]{83.7},$$
$$\log N = \log \sqrt[3]{83.7} = \log (83.7)^{1/3},$$
$$\log N = \frac{1}{3} \log 83.7,$$
$$\log N = \frac{1}{3} (1.9227),$$
$$\log N = 0.6409.$$

To find the antilog of 0.6409, we use interpolation.

$$.01 \begin{bmatrix} x \begin{bmatrix} \log 4.380 = 0.6415. \\ \log N = 0.6409. \\ \log 4.370 = 0.6405. \end{bmatrix} .0004 \end{bmatrix} .0010$$

$$\frac{x}{.01} = \frac{.0004}{.0010}. \quad \text{Simplifying,} \quad \frac{x}{.01} = \frac{4}{10} = \frac{2}{5}.$$

$$x = \frac{.02}{5} = .004.$$

Therefore, $N = 4.370 + .004 = 4.374 = \sqrt[3]{83.7}$. Note that 4.374 is an approximation, since $\sqrt[3]{83.7}$ is an irrational number.

Now, let's examine another problem that involves a little more computation. Consider:

$$\frac{0.345 \times 24.6}{4320}.$$

Again, we shall use logarithms to evaluate this expression. We proceed as follows:

$$N = \frac{0.345 \times 24.6}{4320},$$
$$\log N = \log \frac{0.345 \times 24.6}{4320},$$
$$\log N = \log (0.345 \times 24.6) - \log 4320,$$
$$\log N = \log 0.345 + \log 24.6 - \log 4320,$$
$$\log N = 9.5378 - 10 + 1.3909 - 3.6355.$$

Note: It is advisable to look up all of the mantissas at the same time to avoid confusion and errors. First, we combine $9.5378 - 10 + 1.3909$, which equals $10.9287 - 10$. We subtract 3.6355 from $10.9287 - 10$. Therefore, we now have

$$\log N = 10.9287 - 10 - 3.6355,$$
$$\log N = 7.2932 - 10.$$

To find the antilog of $7.2932 - 10$, we must use interpolation:

$$.00001 \left[x \left[\begin{array}{l} \log 0.00197 = 7.2945 - 10. \\ \log N = 7.2932 - 10. \\ \log 0.00196 = 7.2923 - 10. \end{array} \right] .0009 \right] .0022$$

$$\frac{x}{.00001} = \frac{.0009}{.0022}. \quad \text{Simplifying,} \quad \frac{x}{.00001} = \frac{9}{22}.$$

$$x = \frac{.00009}{22} \approx .000004.$$

Therefore, $N = 0.00196 + .000004 = 0.001964$.

EXAMPLE 21

Using logarithms, find to four digits the value of N:

$$N = \sqrt{\frac{462 \times 4.87}{8.56}}.$$

Solution

$$\log N = \log \sqrt{\frac{462 \times 4.87}{8.56}} = \log \left(\frac{462 \times 4.87}{8.56}\right)^{1/2},$$

$$\log N = \frac{1}{2} \log \left(\frac{462 \times 4.87}{8.56}\right),$$

$$\log N = \frac{1}{2} (\log 462 + \log 4.87 - \log 8.56),$$

$$\log N = \frac{1}{2} (2.6646 + 0.6875 - 0.9325),$$

$$\log N = \frac{1}{2} (3.3521 - 0.9325),$$

Exponential and Logarithmic Functions 609

$$\log N = \frac{1}{2}(2.4196),$$

$$\log N = 1.2098.$$

We must use interpolation to find the antilog of 1.2098.

$$.1 \begin{bmatrix} \begin{bmatrix} \log 16.30 = 1.2122. \\ x \begin{bmatrix} \log N = 1.2098. \\ \log 16.20 = 1.2095. \end{bmatrix} .0003 \end{bmatrix} .0027 \end{bmatrix}$$

$$\frac{x}{.1} = \frac{.0003}{.0027}. \quad \text{Simplifying,} \quad \frac{x}{.1} = \frac{3}{27} = \frac{1}{9}.$$

$$x = \frac{.1}{9} \approx .01. \quad \text{(rounded off)}$$

Therefore, $N = 16.20 + .01 = 16.21.$

EXAMPLE 22

Using logarithms, find to four digits the value of N:

$$N = \frac{45.6}{\sqrt[3]{.574}}.$$

Solution

$$\log N = \log \frac{45.6}{\sqrt[3]{.574}},$$

$$\log N = \log 45.6 - \log \sqrt[3]{.574},$$

$$\log N = \log 45.6 - \log (.574)^{1/3} = \log 45.6 - \frac{1}{3} \log .574,$$

$$\log N = 1.6590 - \frac{1}{3}(9.7589 - 10),$$

Note: We must find $\frac{1}{3}$ of $9.7589 - 10$, but to maintain consistency and keep the characteristic expressed as a multiple of 10, we express $9 - 10$ as $29 - 30$, since -30 is divisible by 3. Hence $\frac{1}{3}(9.7589 - 10) = \frac{1}{3}(29.7589 - 30) = 9.9196 - 10.$ The mantissa is rounded off in the fourth decimal place. Therefore,

$$\log N = 1.6590 - (9.9196 - 10).$$

To subtract 9.9196 − 10 from 1.6590, we must rewrite 1.6590 as 11.6590 − 10. Hence, log N = 11.6590 − 10 − (9.9196 − 10) = 1.7394. To find the antilog of 1.7394, we use interpolation.

$$.1 \begin{bmatrix} & \text{log } 54.90 = 1.7396. & \\ x \begin{bmatrix} \text{log N} = 1.7394. \\ \text{log } 54.80 = 1.7388. \end{bmatrix} .0006 \end{bmatrix} .0008$$

$$\frac{x}{.1} = \frac{.0006}{.0008}. \qquad \text{Simplifying,} \qquad \frac{x}{.1} = \frac{6}{8} = \frac{3}{4}.$$

$$x = \frac{.3}{4} \approx .08. \qquad \text{(rounded off)}$$

Therefore, N = 54.80 + .08 = 54.88.

EXAMPLE 23

Using logarithms, find to four digits the value of N:

$$N = 3.14 \sqrt{\frac{2.34}{0.451}}.$$

Solution

$$\log N = \log 3.14 \sqrt{\frac{2.34}{0.451}},$$

$$\log N = \log 3.14 + \log \left(\frac{2.34}{0.451}\right)^{1/2},$$

$$\log N = \log 3.14 + \frac{1}{2} \log \left(\frac{2.34}{0.451}\right),$$

$$\log N = \log 3.14 + \frac{1}{2} \log 2.34 - \frac{1}{2} \log 0.451,$$

$$\log N = 0.4969 + \frac{1}{2}(0.3692) - \frac{1}{2}(9.6542 - 10).$$

Note: We must find $\frac{1}{2}$ of 9.6542 − 10, and to express the resulting characteristic as a multiple of 10, we express 9 − 10 as 19 − 20 since −20 is divisible by 2. Therefore,

$$\log N = 0.4969 + \frac{1}{2}(0.3692) - \frac{1}{2}(19.6542 - 20),$$

$$\log N = 0.4969 + 0.1846 - (9.8271 - 10),$$

Exponential and Logarithmic Functions

$$\log N = 0.6815 - (9.8271 - 10).$$

To subtract $9.8271 - 10$ from 0.6815, we must rewrite 0.6815 as $10.6815 - 10$. Hence,

$$\log N = 10.6815 - 10 - (9.8271 - 10),$$
$$\log N = 0.8544.$$

To find the antilog of 0.8544, we use interpolation.

$$.01 \begin{bmatrix} \log 7.160 = 0.8549. \\ x \begin{bmatrix} \log N = 0.8544. \\ \log 7.150 = 0.8543. \end{bmatrix} .0001 \end{bmatrix} .0006$$

$$\frac{x}{.01} = \frac{.0001}{.0006}. \quad \text{Simplifying,} \quad \frac{x}{.01} = \frac{1}{6}.$$

$$x = \frac{.01}{6} \approx .002. \quad \text{(rounded off)}$$

Therefore, $N = 7.150 + .002 = 7.152$.

EXAMPLE 24

Using logarithms, find to four digits the value of N:

$$N = \sqrt[3]{\frac{-8.17}{0.326}}.$$

Solution

Note that the expression for N contains a negative number. We *cannot* find the logarithm of a negative number. Recall that a common logarithm is an exponent of 10 and no real number x exists that can be used as an exponent such that 10^x is a negative number. Therefore, to evaluate an expression containing negative numbers, we treat all numbers as positive and then affix the correct sign (+ or −) to the answer. The quotient inside the radical sign will be negative, since a negative number divided by a positive number is negative. The cube root of a negative number is negative. Hence, the answer will be a negative number.

$$\log N = \log \sqrt[3]{\frac{8.17}{0.326}} = \log \left(\frac{8.17}{0.326}\right)^{1/3}.$$

> Note: We are treating -8.17 as $+8.17$.
>
> $$\log N = \frac{1}{3} \log \left(\frac{8.17}{0.326}\right) = \frac{1}{3} \log 8.17 - \frac{1}{3} \log 0.326,$$
>
> $$\log N = \frac{1}{3}(0.9122) - \frac{1}{3}(9.5132 - 10),$$
>
> $$\log N = 0.3041 - \frac{1}{3}(29.5132 - 30),$$
>
> (rounded off) (rewritten)
>
> $$\log N = 0.3041 - (9.8377 - 10),$$
>
> $$\log N = 10.3041 - 10 - (9.8377 - 10),$$
>
> $$\log N = 0.4664.$$
>
> To find the antilog of 0.4664, we use interpolation.
>
> .01 ⎡ ⎡ log 2.930 = 0.4669. ⎤ ⎤
>   x ⎢ log N   = 0.4664. ⎥ .0010 ⎥ .0015
>   ⎣ log 2.920 = 0.4654. ⎦
>
> $$\frac{x}{.01} = \frac{.0010}{.0015}. \quad \text{Simplifying,} \quad \frac{x}{.01} = \frac{10}{15} = \frac{2}{3}.$$
>
> $$x = \frac{.02}{3} \approx .007. \quad \text{(rounded off)}$$
>
> Therefore, $-N = 2.920 + .007 = 2.927$ and $N = -2.927$. (Recall that our answer is to be a negative number.)

In the beginning of this chapter, we encountered the compound interest formula, $A = P(1 + r)^n$ where $A =$ the compound amount, $P =$ the principal, $r =$ the rate of interest per conversion period, and $n =$ the number of conversion periods. Following is an example where this formula is applied and logarithms are used to compute the answer:

How much money would a person have in 5 years if she deposited $1000 in a savings account that pays 8% interest compounded quarterly?

Using the formula $A = P(1 + r)^n$, we have $P = \$1000$. Note that $n = 20$ since the interest is compounded quarterly and there are 4 conversion periods per year, therefore $n = 4 \times 5 = 20$. Similarly, the rate of interest

Exponential and Logarithmic Functions 613

per conversion period would be 2%, i.e., 8% ÷ 4 = 2% = .02. Substituting in the formula, we have

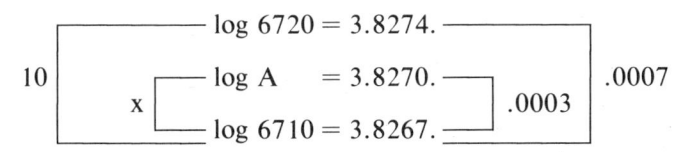

$$A = 1000(1 + .02)^{20},$$
$$A = 1000(1.02)^{20},$$
$$\log A = \log 1000(1.02)^{20},$$
$$\log A = \log 1000 + \log (1.02)^{20},$$
$$\log A = \log 1000 + 20 \log (1.02),$$
$$\log A = \quad 3 \quad + 20 \,(0.0086),$$
$$\log A = \quad 3 + 0.1720,$$
$$\log A = \quad 3.1720.$$

```
         ┌────── log 1490 = 3.1732. ──────┐
  10     │  ┌── log A    = 3.1720. ──┐    │  .0029
         │x │                        │.0017│
         │  └── log 1480 = 3.1703. ──┘    │
         └────────────────────────────────┘
```

$$\frac{x}{10} = \frac{.0017}{.0029}. \quad \text{Simplifying,} \quad \frac{x}{10} = \frac{17}{29}.$$

$$x = \frac{170}{29} \approx 6. \quad \text{Therefore, } A = 1480 + 6 = \$1486. \quad \text{(to the nearest dollar)}$$

EXAMPLE 25

Five years ago, $5000 was invested at 6% compounded semi-annually. Find the amount of this investment today.

Solution

We use the formula $A = P(1 + r)^n$. P = $5000, n = 5 (number of years) × 2 (the number of conversion periods per year) = 10, r = rate of interest per conversion period = 6% ÷ 2 = 3% = .03. Substituting in the formula, we have

$$A = 5000(1 + .03)^{10},$$
$$A = 5000(1.03)^{10},$$
$$\log A = \log 5000(1.03)^{10},$$
$$\log A = \log 5000 + \log (1.03)^{10},$$
$$\log A = \log 5000 + 10 \log (1.03),$$
$$\log A = 3.6990 + 10(0.0128),$$
$$\log A = 3.6990 + 0.1280,$$
$$\log A = 3.8270.$$

```
         ┌────── log 6720 = 3.8274. ──────┐
  10     │  ┌── log A    = 3.8270. ──┐    │  .0007
         │x │                        │.0003│
         │  └── log 6710 = 3.8267. ──┘    │
         └────────────────────────────────┘
```

$$\frac{x}{10} = \frac{.0003}{.0007}. \quad \text{Simplifying,} \quad \frac{x}{10} = \frac{3}{7}.$$

$$x = \frac{30}{7} \approx 4. \text{ Therefore } A = 6710 + 4 = \$6714. \quad \text{(to the nearest dollar)}$$

EXERCISES FOR SECTION 10.6

*(Answers given to exercises marked *)*

Find the indicated roots of the given numbers by means of logarithms to four digits:

1. $\sqrt{2}$
2. $\sqrt{3}$
*3. $\sqrt{7}$
4. $\sqrt[3]{5}$
*5. $\sqrt[3]{16}$
6. $\sqrt[3]{9}$
*7. $\sqrt[3]{100}$
8. $\sqrt[4]{5}$
*9. $\sqrt[4]{7}$
10. $\sqrt{50}$
*11. $\sqrt{125}$
12. $\sqrt{55.1}$
*13. $\sqrt[3]{0.698}$
14. $\sqrt{0.0354}$
*15. $\sqrt{0.0507}$
16. $(3.14)^{2/3}$
*17. $(3.14)^{3/2}$
18. $(0.123)^{2/3}$

Using logarithms, find to four digits the value of N:

*19. $N = \frac{321 \times 2.56}{7.21}$.
20. $N = \frac{432 \times 5.23}{3.14}$.
*21. $N = \sqrt{\frac{4.35 \times 273}{8.41}}$.

22. $N = \sqrt{\frac{0.821 \times 437}{9.72}}$.
*23. $N = \frac{49.2}{\sqrt[3]{3.14}}$.
24. $N = \frac{375}{\sqrt{21.2}}$.

*25. $N = \frac{586}{\sqrt{2.75 \times 0.812}}$.
26. $N = 42.1 \sqrt{\frac{3.14}{0.245}}$.

*27. $N = 2.37 \sqrt{\frac{3.14}{0.489}}$.
28. $N = \sqrt[3]{\frac{3.14 \times 0.986}{237}}$.

*29. $N = \sqrt{\frac{(-21.4) \times (3.14)}{-0.815}}$.
30. $N = \sqrt[3]{\frac{-4.37}{0.812}}$.

*31. $N = \frac{4.21 \times (-3.02) \times 3.14}{(-0.475) \times 9.02}$.
32. $N = \frac{237 \times 3.14 \times (-0.112)}{(-2.15) \times (-0.105)}$.

*33. $N = \sqrt[3]{\frac{(-3.14) \times 2.17}{0.814}}$.
34. $N = \sqrt[3]{\frac{-332 \times (-2.17)}{-0.902}}$.

*35. Ten years ago, $1000 was invested at 6% compounded semi-annually. Find the amount of this investment today.

Exponential and Logarithmic Functions   615

36. In 1976, the United States celebrated its 200th birthday. If Benjamin Franklin had deposited $1 in a bank on July 4, 1776, at 1% interest compounded annually, how much money would be in the bank account on July 4, 1976?

*37. If $2000 is invested at 6% compounded semi-annually for 15 years, what will be the compounded amount? How much of this will be interest?

38. If $10,000 is invested at 8% compounded quarterly for ten years, what will be the compounded amount? How much of this will be interest?

*39. What principal is needed to produce $10,000 in 15 years, if it is invested at 8% compounded quarterly?

40. The period, t, in seconds, for the complete oscillation of a pendulum whose length is L, where g is the acceleration of gravity, is determined by the formula

$$t = 2\pi\sqrt{\frac{L}{g}}.$$

Using logarithms, find to the nearest hundredth the value of t when L = 1.38, g = 32.2 and π = 3.14.

*41. Given the formula $R = \sqrt[3]{\frac{r^2 h}{1.33}}$, and using logarithms, find R to the nearest tenth when r = 49.3 and h = 37.2.

42. On a girl's second birthday, her parents placed $1000 in her name in an investment which pays 8% compounded quarterly. How much will she have on her 18th birthday?

10.7 EXPONENTIAL EQUATIONS

In Section 10.2, we encountered some exponential equations, which are equations in which variables appear as exponents. For example, $3^x = 9$ and $2^{2x-1} = 8$ are exponential equations. To solve them, we can express each member of the exponential equation so that it has the same base, and then solve the resulting equation that is obtained from the exponents. That is,

$3^x = 9$, $2^{2x-1} = 8$,
$3^x = 3^2$,   Check: $3^2 \stackrel{?}{=} 9$, $2^{2x-1} = 2^3$,   Check: $2^{2(2)-1} \stackrel{?}{=} 8$,
$x = 2$. $9 = 9$. $2x - 1 = 3$, $2^{4-1} = 8$,
                                  $2x = 4$, $2^3 = 8$,
                                  $x = 2$. $8 = 8$.

Both of these equations can be solved by means of logarithms. It is important to know how to solve exponential equations with logarithms since it is not always possible to express the members of the exponential equation so that they have the same base. For example, $2^x = 5$ cannot be written so that each member has the same base. Now we shall solve each of the two examples, $3^x = 9$ and $2^{2x-1} = 8$, by means of logarithms:

$$3^x = 9.$$
$$\log 3^x = \log 9,$$
$$x \log 3 = \log 9,$$
$$x = \frac{\log 9}{\log 3},$$
$$x = \frac{0.9542}{0.4771} = 2.$$

$$2^{2x-1} = 8.$$
$$\log 2^{2x-1} = \log 8,$$
$$(2x - 1)\log 2 = \log 8,$$
$$2x - 1 = \frac{\log 8}{\log 2},$$
$$2x - 1 = \frac{0.9031}{0.3010} \approx 3,$$
$$2x - 1 = 3,$$
$$2x = 4,$$
$$x = 2.$$

It should be noted that $\frac{\log 9}{\log 3}$ and $\frac{\log 8}{\log 2}$ are quotients of logarithms and not the logarithms of quotients. Take care when evaluating expressions such as this so there won't be any confusion.

EXAMPLE 26

Solve for x: $2^x = 5$.

Solution

Since each member of the given exponential equation cannot be written so that it has the same base, we use logarithms to solve for x.

$$2^x = 5.$$
$$\log 2^x = \log 5,$$
$$x \log 2 = \log 5,$$
$$x = \frac{\log 5}{\log 2} = \frac{0.6990}{0.3010},$$

$x \approx 2.3$ (to the nearest tenth). Note that the solution is approximate.

Exponential and Logarithmic Functions

EXAMPLE 27

Solve for x: $4^{2x-1} = 8^x$.

Solution

We can express both members of the equation so they have the same base, and then solve the resulting equation that is obtained from the exponents.

$$4^{2x-1} = 8^x.$$

Note that $4 = 2^2$ and $8 = 2^3$, hence

$$(2^2)^{2x-1} = (2^3)^x,$$
$$2^{4x-2} = 2^{3x}.$$

From this equation, we obtain

$$4x - 2 = 3x, \qquad \text{Check:} \quad 4^{2 \cdot 2 - 1} \stackrel{?}{=} 8^2.$$
$$x = 2. \qquad\qquad\qquad 4^{4-1} = 64,$$
$$\qquad\qquad\qquad\qquad\qquad 4^3 = 64,$$
$$\qquad\qquad\qquad\qquad\qquad 64 = 64.$$

Alternate Solution

It should be noted that any exponential equation can be solved by the use of logarithms. Therefore, we shall solve this same equation by means of logarithms:

$$4^{2x-1} = 8^x.$$
$$\log 4^{2x-1} = \log 8^x,$$
$$(2x - 1)\log 4 = x \log 8,$$

(expanding) $\quad 2x \log 4 - 1 \log 4 = x \log 8,$
$\qquad\qquad 2x \log 4 - x \log 8 = 1 \log 5,\quad$ (collecting like terms)
$\qquad\qquad x(2 \log 4 - \log 8) = \log 4.\quad$ (factoring)

$$x = \frac{\log 4}{2 \log 4 - \log 8},$$

$$x = \frac{0.6021}{2(0.6021) - 0.9031},$$

$$x = \frac{0.6021}{1.2042 - 0.9031},$$

$$x = \frac{0.6021}{0.3011} \approx 2.$$

Logarithms could have also been used to perform the indicated division to find the value of x. Note that x approximayely equals 2, since the mantissas have been rounded off to four decimal places.

EXAMPLE 28

Solve for x: $2^{x+1} = 9$.

Solution

The members of the given exponential equation cannot be written so that they have the same base, hence we use logarithms to solve for x.

$$2^{x+1} = 9.$$
$$\log 2^{x+1} = \log 9,$$
$$(x+1) \log 2 = \log 9,$$
(expanding) $\quad x \log 2 + 1 \log 2 = \log 9,$
$$x \log 2 = \log 9 - 1 \log 2. \quad \text{(collecting like terms)}$$

$$x = \frac{\log 9 - \log 2}{\log 2},$$

$$x = \frac{0.9542 - 0.3010}{0.3010} = \frac{0.6532}{0.3010},$$

$$x \approx 2.2. \quad \text{(to the nearest tenth)}$$

In the beginning of this chapter, we considered the following problem:

How long will it take $1000 to amount to $2000 if invested at 6% compounded annually?

Now we can solve it, using logarithms and the formula $A = P(1+r)^n$.

A = $2000, P = $1000, r = 6%, or

$$2000 = 1000 (1 + .06)^n,$$
$$2000 = 1000 (1.06)^n,$$
$$2 = (1.06)^n, \quad \text{(dividing both sides by 1000)}$$
$$\log 2 = \log (1.06)^n,$$
$$\log 2 = n \log 1.06,$$

$$\frac{\log 2}{\log 1.06} = n,$$

$$\frac{0.3010}{0.0253} = n,$$

$$11.9 \text{ years} = n. \quad \text{(to the nearest tenth)}$$

Exponential and Logarithmic Functions

EXAMPLE 29

How long will it take $5000 to amount to $10,000 if invested at 8% compounded annually?

Solution

Using the formula, $A = P(1 + r)^n$, we have $A = \$10,000$, $P = \$5000$, $r = .08$, and

$$10,000 = 5,000(1.08)^n,$$
$$2 = (1.08)^n, \quad \text{(dividing both sides by 5000)}$$
$$\log 2 = \log(1.08)^n,$$
$$\log 2 = n \log 1.08.$$

$$\frac{\log 2}{\log 1.08} = n,$$

$$\frac{0.3010}{0.0334} = n,$$

$$9 \text{ years} \approx n.$$

EXERCISES FOR SECTION 10.7

*(Answers given to exercises marked *)*

Solve the given exponential equations by expressing the members so that have the same base and solving the resulting equation obtained from the exponents:

*1. $2^{2x+1} = 16$.   2. $4^{2x-1} = 8$.   *3. $16^{3x+2} = 8$.

4. $27^{2x-2} = 9$.   *5. $9^{2x-1} = 81$.   6. $4^{3x-1} = 32^{2x}$.

*7. $64^{x-1} = 32^{2x+1}$.   8. $27^{3x+2} = 9^{x-1}$.   *9. $2^{4x-2} = 8^x$.

10. $25^{2x-1} = 125^x$.   *11. $4^{2x-1} = 8^x$.   12. $9^{2x+1} = 27^x$.

*13. $4^{-x-2} = 8^{2x+1}$.   14. $16^{-2x} = 8^{3x+1}$.   *15. $4^{-3x+1} = 32^{-2x}$.

Solve the given exponential equations by means of logarithms. Express your answer to the nearest tenth.

16. $3^x = 5$.   *17. $4^x = 9$.   18. $3^x = 10$.

*19. $2^x = 10$. 20. $3^{x+1} = 5$. *21. $2^{x+1} = 5$.

22. $4^{x-1} = 6$. *23. $3^{x-1} = 15$. 24. $5^{x-1} = 10$.

*25. $2.3^{x-1} = 4.75$. 26. $2.47^{x+1} = 3.14$. *27. $2^{x+1} = 3^x$.

28. $3^{x+1} = 5^x$. *29. $2^{x+1} = 5^{x-2}$. 30. $2^{2x+1} = 3^{x-1}$.

*31. How long will it take $1000 to amount to $5000 if invested at 8% compounded annually? (Express your answer to the nearest year.)

32. How long will it take $5000 to amount to $7500 if invested at 6% compounded annually? (Express your answer to the nearest year.)

*33. How long will it take $1000 to amount to $2000 if invested at 6% compounded semi-annually? (Express your answer to the nearest year.)

34. How long will it take $1000 to amount to $5000 if invested at 8% compounded quarterly? (Express your answer to the nearest year.)

*35. How long will it take $10,000 to amount to $20,000 if invested at 8% compounded quarterly? (Express your answer to the nearest year.)

10.8 SUMMARY

An equation of the form $y = b^x$ (or $f(x) = b^x$), where $b > 0$ and $b \neq 1$, is called an exponential function. An exponential function can be graphed on the Cartesian plane. Regardless of the equation, the y-intercept is always 1 and the graph of an exponential function lies completely above the x-axis. The value of y is never negative since b is always greater than 0 and the graph can be extended indefinitely to the left and right of the y-axis.

The equation $x = b^y$ is called the inverse of the exponential function $y = b^x$. It is obtained by interchanging the variables. We can also call the inverse of an exponential function a logarithmic function. In general, if $y = b^x$ is the exponential function, then $x = b^y$ is the logarithmic function. It is often necessary to express y in terms of x, that is, to state explicitly that y is a function of x, and logarithmic notation is used to do this. The logarithmic function $x = b^y$ is equivalent to $y = \log_b x$. This is read as "y equals the log of x to the base b." From this expression we note that logarithms are exponents.

Basically, a logarithmic equation contains three different parts, therefore, the unknown variable can occur in any one of three different places. For example, $\log_2 x = 3$, $\log_2 8 = x$, and $\log_x 8 = 3$ are all logarithmic equations and x occurs in a different position. Following are solutions for these three equations:

1. $\log_2 x = 3$ is equivalent to $2^3 = x$, and $x = 8$;
2. $\log_2 8 = x$ is equivalent to $2^x = 8$; since $8 = 2^3$, we have $2^x = 8$ is equivalent to $2^x = 2^3$, and $x = 3$;
3. $\log_x 8 = 3$ is equivalent to $x^3 = 8$; to solve for x, we raise x^3 and also 8 to the $\frac{1}{3}$ power. Therefore, $x^3 = 8$ is equivalent to $(x^3)^{1/3} = 8^{1/3}$, and $x = 8^{1/3} = 2$.

The logarithm of a product of any positive real numbers M and N is the sum of the logarithms of the factors. That is, $\log_b MN = \log_b M + \log_b N$. The logarithm of a quotient of any positive real numbers M and N is the logarithm of the dividend minus the logarithm of the divisor. That is, $\log_b \frac{M}{N} = \log_b M - \log_b N$. The logarithm of a power of a positive number M is equal to the exponent of that power times the logarithm of the number. That is, $\log_b M^n = n \log_b M$. These properties of logarithms can be used to simplify computations that involve multiplication, division, raising the powers, and finding roots of numbers.

Normally, logarithms that are used in computations have a base of 10, and logarithms that have a base of 10 are called *common logarithms*. When no other base is indicated for a logarithm, it is understood to be base 10. The decimal portion of the logarithm (log n, where $1 < n < 10$) is called the *mantissa* of the logarithm. The integer preceding the decimal point is called the *characteristic* of the logarithm. The logarithm of a number is the sum of the characteristic and mantissa. That is, log N = characteristic + mantissa. Table II in Appendix A is a table of mantissas. The characteristic of the logarithm of a number is determined by expressing the given number in scientific notation. The exponent of 10 is the characteristic. Many times the characteristic of a logarithm will be negative. We express a negative characteristic by adding and subtracting a multiple of 10 to it. For example, if the characteristic is -1, then we have $-1 + 10 - 10$ or $9 - 10$. Note that $9 - 10$ can also be expressed as $19 - 20$, $29 - 30$, etc. The reason for doing this is to keep the mantissa non-negative. For example, the mantissa for 2.0 and 0.2 is 0.3010, but log 2.0 = 0.3010 and log 0.2 = 9.3010 − 10. When we find the *antilogarithm* of a number, we are finding the number which corresponds to a given number. For example, if log N = 2.3010, then we want to find N or the antilog of 2.3010, which is 200. For a detailed discussion of logarithms, characteristics, mantissas, and antilogarithms, refer to Section 10.4.

To find the logarithm of a four-digit number, we use a process called *interpolation*. We can also use this same technique to find a four-digit number when we are given its logarithm. That is, we can find the antilogarithm of a number by means of interpolation. Following is an example using interpolation to find the logarithm of a four-digit number. To find log 68.34, we proceed as follows: The logarithm of 68.34 is between 68.30 and 68.40. To find the relative positions, we list the numbers and their logarithms.

$$.1 \left[.04 \left[\begin{array}{l} \text{log } 68.40 = 1.8351. \\ \text{log } 68.34 = ? \\ \text{log } 68.30 = 1.8344. \end{array} \right] x \right] .0007$$

We determine the position of log 68.34 by means of the proportion

$$\frac{.04}{.1} = \frac{x}{.0007}. \quad \text{Simplifying,} \quad \frac{4}{10} = \frac{x}{.0007},$$

$10x = .0028.$
$x = .00028 \approx .0003.$ (rounded off)

Therefore, log 68.24 = 1.8344 + .0003 = 1.8347.

Logarithms can be used to perform certain calculations, such as multiplication, division, raising to powers, and extracting or finding roots of numbers. In performing these calculations, we must apply the properties of logarithms. For example, to find the value of N in the following problem, we must first apply the properties of logarithms.

$$N = \sqrt{\frac{324 \times 2.17}{3.14}}.$$

$$\log N = \log \sqrt{\frac{324 \times 2.17}{3.14}} = \log \left(\frac{324 \times 2.17}{3.14} \right)^{1/2},$$

$$\log N = \frac{1}{2} \log \frac{324 \times 2.17}{3.14},$$

$$\log N = \frac{1}{2} (\log 324 + \log 2.17 - \log 3.14).$$

Next, we find the logarithms of the numbers, then perform the indicated addition and subtraction. After multiplying this number by one-half, we have the log of N. Interpolation is used to find N to four digits. These portions of the problem are left for the student as a review exercise. The value of N is 14.96.

An exponential equation is any equation in which variables appear as exponents. For example, $8^{2x-1} = 16$ is an exponential equation. Some exponential equations can be solved by expressing the members of the equation so that they have the same base, then solving the resulting equation that is obtained from the exponents. For example,

Exponential and Logarithmic Functions 623

$$8^{2x-1} = 16.$$
$$(2^3)^{2x-1} = 2^4,$$
$$2^{6x-3} = 2^4,$$

$$6x - 3 = 4,$$
$$6x = 7,$$
$$x = \frac{7}{6}.$$

Check: $8^{2 \cdot 7/6 - 1} \stackrel{?}{=} 16.$
$$8^{7/3 - 1} = 16,$$
$$8^{4/3} = 16,$$
$$16 = 16.$$

Exponential equations can also be solved by means of logarithms. For example, to solve $3^x = 5$, we first apply a property of logarithms and, after determining the logarithms of the numbers in the equation, perform the indicated division. That is,

$$3^x = 5.$$
$$\log 3^x = \log 5,$$
$$x \log 3 = \log 5,$$

$$x = \frac{\log 5}{\log 3} = \frac{0.6990}{0.4771} \approx 1.5. \quad \text{(to the nearest tenth)}$$

Logarithms are useful in performing many different complex calculations. Although many of these can now be done by computers and electronic calculators, logarithms are still useful in performing calculations that are not routine and, particularly, when calculating devices are not immediately available.

REVIEW EXERCISES FOR CHAPTER 10
(Answers in Appendix B)

Make a table of values and sketch the graph of each equation:

1. $y = 3^x$.
2. $y = 3^{-x}$.
3. $y = \left(\frac{1}{3}\right)^x$.
4. $y = \left(\frac{1}{2}\right)^{-x}$.
5. $y = \left(\frac{1}{2}\right)^x$.
6. $y = 3^{x+1}$.

7. Using the same set of axes, and the same scale, sketch the graphs of $y = 2^x$ and $x = 2^y$.

8. Write each of the following equations in logarithmic form:

   a. $2^3 = 8$.
   b. $5^3 = 125$.
   c. $2^{-3} = \frac{1}{8}$.

9. Write each of the following equations in exponential form:

   a. $\log_2 8 = 3$.
   b. $\log_5 125 = 3$.
   c. $\log_2 \frac{1}{8} = -3$.

Solve each of the following equations for x:

10. $\log_5 x = 2$.
11. $\log_2 x = -3$.
12. $\log_8 x = -\frac{2}{3}$.
13. $\log_3 27 = x$.
14. $\log_9 3 = x$.
15. $\log_8 4 = x$.
16. $\log_x 2 = \frac{1}{2}$.
17. $\log_x 125 = 3$.
18. $\log_x \frac{1}{8} = -\frac{3}{2}$.
19. $\log_{10} 1 = x$.
20. $\log_{10} 0.001 = x$.
21. $\log_x 1000 = 3$.

Write equivalent expressions in terms of $\log_b x$, $\log_b y$, and $\log_b z$:

22. $\log_b (xyz)^2$
23. $\log_b \sqrt[3]{x^2 y}$
24. $\log_b \sqrt[3]{\dfrac{6x}{yz^2}}$

25. Find the logarithm of 34.74.

26. Find the number whose logarithm is $7.6752 - 10$.

27. Find the number whose logarithm is $9.8680 - 10$.

28. Find the logarithm of 3.142.

Determine whether the given statement is true or false.

29. Log $3x^2$ is equal to 6 log x.

30. Log $(x^2 - y^2)$ equals 2 log x $-$ 2 log y.

31. The expression 2 log x $-$ log y is the logarithm of $\dfrac{2x}{y}$.

32. The expression log x^2 + log x^3 is equal to 5 log x.

33. If log N equals log x $-$ log y then $N = x - y$.

34. If log $x^3 = 0.9294$, then log $10x = 1.9294$.

35. If log $x^4 = 0.7404$, then log $x^2 = 0.3702$.

36. Log $10x^3$ equals 1 + log 3 + log x.

37. If log x $=$ y then log $10x = 1 + y$.

38. If log N = log x $-$ 3 log y + 2 log z, then $N = \dfrac{xz^2}{y^3}$.

39. Using logarithms, find $\sqrt{3}$, to four digits.

40. Using logarithms, find $\sqrt[3]{9}$, to four digits.

Exponential and Logarithmic Functions

41. Using logarithms, find to four digits the value of N:

$$N = \sqrt{\frac{0.821 \times 437}{9.72}}.$$

42. Using logarithms, find to four digits the value of N:

$$N = 42.1\sqrt{\frac{3.14}{0.245}}.$$

43. Using logarithms, find to four digits the value of N:

$$N = \sqrt[3]{\frac{-4.37 \times (-2.17)}{-0.902}}.$$

44. Ten years ago, $500 was invested at 6% compounded semi-annually. Find the amount of this investment today.

45. If $10,000 is invested at 8% compounded quarterly for five years, what will be the compounded amount? How much of this will be interest?

46. In how many years will $1000 amount to $2000 if interest is compounded annually at 8%? Express your answer to the nearest year.

47. In how many years will $5000 amount to $10,000 if interest is compounded semi-annually at 6%? Express your answer to the nearest year.

appendix A

Useful Tables and Formulas

Table 1 Powers and Roots

No.	Sq.	Sq. Root	Cube	Cube Root	No.	Sq.	Sq. Root	Cube	Cube Root
1	1	1.000	1	1.000	51	2,601	7.141	132,651	3.708
2	4	1.414	8	1.260	52	2,704	7.211	140,608	3.733
3	9	1.732	27	1.442	53	2,809	7.280	148,877	3.756
4	16	2.000	64	1.587	54	2,916	7.348	157,464	3.780
5	25	2.236	125	1.710	55	3,025	7.416	166,375	3.803
6	36	2.449	216	1.817	56	3,136	7.483	175,616	3.826
7	49	2.646	343	1.913	57	3,249	7.550	185,193	3.849
8	64	2.828	512	2.000	58	3,364	7.616	195,112	3.871
9	81	3.000	729	2.080	59	3,481	7.681	205,379	3.893
10	100	3.162	1,000	2.154	60	3,600	7.746	216,000	3.915
11	121	3.317	1,331	2.224	61	3,721	7.810	226,981	3.936
12	144	3.464	1,728	2.289	62	3,844	7.874	238,328	3.958
13	169	3.606	2,197	2.351	63	3,969	7.937	250,047	3.979
14	196	3.742	2,744	2.410	64	4,096	8.000	262,144	4.000
15	225	3.873	3,375	2.466	65	4,225	8.062	274,625	4.021
16	256	4.000	4,096	2.520	66	4,356	8.124	287,496	4.041
17	289	4.123	4,913	2.571	67	4,489	8.185	300,763	4.062
18	324	4.243	5,832	2.621	68	4,624	8.246	314,432	4.082
19	361	4.359	6,859	2.668	69	4,761	8.307	328,509	4.102
20	400	4.472	8,000	2.714	70	4,900	8.367	343,000	4.121
21	441	4.583	9,261	2.759	71	5,041	8.426	357,911	4.141
22	484	4.690	10,648	2.802	72	5,184	8.485	373,248	4.160
23	529	4.796	12,167	2.844	73	5,329	8.544	389,017	4.179
24	576	4.899	13,824	2.884	74	5,476	8.602	405,224	4.198
25	625	5.000	15,625	2.924	75	5,625	8.660	421,875	4.217
26	676	5.099	17,576	2.962	76	5,776	8.718	438,976	4.236
27	729	5.196	19,683	3.000	77	5,929	8.775	456,533	4.254
28	784	5.292	21,952	3.037	78	6,084	8.832	474,552	4.273
29	841	5.385	24,389	3.072	79	6,241	8.888	493,039	4.291
30	900	5.477	27,000	3.107	80	6,400	8.944	512,000	4.309
31	961	5.568	29,791	3.141	81	6,561	9.000	531,441	4.327
32	1,024	5.657	32,768	3.175	82	6,724	9.055	551,368	4.344
33	1,089	5.745	35,937	3.208	83	6,889	9.110	571,787	4.362
34	1,156	5.831	39,304	3.240	84	7,056	9.165	592,704	4.380
35	1,225	5.916	42,875	3.271	85	7,225	9.220	614,125	4.397
36	1,296	6.000	46,656	3.302	86	7,396	9.274	636,056	4.414
37	1,369	6.083	50,653	3.332	87	7,569	9.327	658,503	4.431
38	1,444	6.164	54,872	3.362	88	7,744	9.381	681,472	4.448
39	1,521	6.245	59,319	3.391	89	7,921	9.434	704,969	4.465
40	1,600	6.325	64,000	3.420	90	8,100	9.487	729,000	4.481
41	1,681	6.403	68,921	3.448	91	8,281	9.539	753,571	4.498
42	1,764	6.481	74,088	3.476	92	8,464	9.592	778,688	4.514
43	1,849	6.557	79,507	3.503	93	8,649	9.644	804,357	4.531
44	1,936	6.633	85,184	3.530	94	8,836	9.695	830,584	4.547
45	2,025	6.708	91,125	3.557	95	9,025	9.747	857,375	4.563
46	2,116	6.782	97,336	3.583	96	9,216	9.798	884,736	4.579
47	2,209	6.856	103,823	3.609	97	9,409	9.849	912,673	4.595
48	2,304	6.928	110,592	3.634	98	9,604	9.899	941,192	4.610
49	2,401	7.000	117,649	3.659	99	9,801	9.950	970,299	4.626
50	2,500	7.071	125,000	3.684	100	10,000	10.000	1,000,000	4.642

Appendix A

Table 2   Four-Place Logarithms of Numbers

n	0	1	2	3	4	5	6	7	8	9
10	0000	0043	0086	0128	0170	0212	0253	0294	0334	0374
11	0414	0453	0492	0531	0569	0607	0645	0682	0719	0755
12	0792	0828	0864	0899	0934	0969	1004	1038	1072	1106
13	1139	1173	1206	1239	1271	1303	1335	1367	1399	1430
14	1461	1492	1523	1553	1584	1614	1644	1673	1703	1732
15	1761	1790	1818	1847	1875	1903	1931	1959	1987	2014
16	2041	2068	2095	2122	2148	2175	2201	2227	2253	2279
17	2304	2330	2355	2380	2405	2430	2455	2480	2504	2529
18	2553	2577	2601	2625	2648	2672	2695	2718	2742	2765
19	2788	2810	2833	2856	2878	2900	2923	2945	2967	2989
20	3010	3032	3054	3075	3096	3118	3139	3160	3181	3201
21	3222	3243	3263	3284	3304	3324	3345	3365	3385	3404
22	3424	3444	3464	3483	3502	3522	3541	3560	3579	3598
23	3617	3636	3655	3674	3692	3711	3729	3747	3766	3784
24	3802	3820	3838	3856	3874	3892	3909	3927	3945	3962
25	3979	3997	4014	4031	4048	4065	4082	4099	4116	4133
26	4150	4166	4183	4200	4216	4232	4249	4265	4281	4298
27	4314	4330	4346	4362	4378	4393	4409	4425	4440	4456
28	4472	4487	4502	4518	4533	4548	4564	4579	4594	4609
29	4624	4639	4654	4669	4683	4698	4713	4728	4742	4757
30	4771	4786	4800	4814	4829	4843	4857	4871	4886	4900
31	4914	4928	4942	4955	4969	4983	4997	5011	5024	5038
32	5051	5065	5079	5092	5105	5119	5132	5145	5159	5172
33	5185	5198	5211	5224	5237	5250	5263	5276	5289	5302
34	5315	5328	5340	5353	5366	5378	5391	5403	5416	5428
35	5441	5453	5465	5478	5490	5502	5514	5527	5539	5551
36	5563	5575	5587	5599	5611	5623	5635	5647	5658	5670
37	5682	5694	5705	5717	5729	5740	5752	5763	5775	5786
38	5798	5809	5821	5832	5843	5855	5866	5877	5888	5899
39	5911	5922	5933	5944	5955	5966	5977	5988	5999	6010
40	6021	6031	6042	6053	6064	6075	6085	6096	6107	6117
41	6128	6138	6149	6160	6170	6180	6191	6201	6212	6222
42	6232	6243	6253	6263	6274	6284	6294	6304	6314	6325
43	6335	6345	6355	6365	6375	6385	6395	6405	6415	6425
44	6435	6444	6454	6464	6474	6484	6493	6503	6513	6522
45	6532	6542	6551	6561	6571	6580	6590	6599	6609	6618
46	6628	6637	6646	6656	6665	6675	6684	6693	6702	6712
47	6721	6730	6739	6749	6758	6767	6776	6785	6794	6803
48	6812	6821	6830	6839	6848	6857	6866	6875	6884	6893
49	6902	6911	6920	6928	6937	6946	6955	6964	6972	6981
50	6990	6998	7007	7016	7024	7033	7042	7050	7059	7067
51	7076	7084	7093	7101	7110	7118	7126	7135	7143	7152
52	7160	7168	7177	7185	7193	7202	7210	7218	7226	7235
53	7243	7251	7259	7267	7275	7284	7292	7300	7308	7316
54	7324	7332	7340	7348	7356	7364	7372	7380	7388	7396

Table 2 Four-Place Logarithms of Numbers (Continued)

n	0	1	2	3	4	5	6	7	8	9
55	7404	7412	7419	7427	7435	7443	7451	7459	7466	7474
56	7482	7490	7497	7505	7513	7520	7528	7536	7543	7551
57	7559	7566	7574	7582	7589	7597	7604	7612	7619	7627
58	7634	7642	7649	7657	7664	7672	7679	7686	7694	7701
59	7709	7716	7723	7731	7738	7745	7752	7760	7767	7774
60	7782	7789	7796	7803	7810	7818	7825	7832	7839	7846
61	7853	7860	7868	7875	7882	7889	7896	7903	7910	7917
62	7924	7931	7938	7945	7952	7959	7966	7973	7980	7987
63	7993	8000	8007	8014	8021	8028	8035	8041	8048	8055
64	8062	8069	8075	8082	8089	8096	8102	8109	8116	8122
65	8129	8136	8142	8149	8156	8162	8169	8176	8182	8189
66	8195	8202	8209	8215	8222	8228	8235	8241	8248	8254
67	8261	8267	8274	8280	8287	8293	8299	8306	8312	8319
68	8325	8331	8338	8344	8351	8357	8363	8370	8376	8382
69	8388	8395	8401	8407	8414	8420	8426	8432	8439	8445
70	8451	8457	8463	8470	8476	8482	8488	8494	8500	8506
71	8513	8519	8525	8531	8537	8543	8549	8555	8561	8567
72	8573	8579	8585	8591	8597	8603	8609	8615	8621	8627
73	8633	8639	8645	8651	8657	8663	8669	8675	8681	8686
74	8692	8698	8704	8710	8716	8722	8727	8733	8739	8745
75	8751	8756	8762	8768	8774	8779	8785	8791	8797	8802
76	8808	8814	8820	8825	8831	8837	8842	8848	8854	8859
77	8865	8871	8876	8882	8887	8893	8899	8904	8910	8915
78	8921	8927	8932	8938	8943	8949	8954	8960	8965	8971
79	8976	8982	8987	8993	8998	9004	9009	9015	9020	9025
80	9031	9036	9042	9047	9053	9058	9063	9069	9074	9079
81	9085	9090	9096	9101	9106	9112	9117	9122	9128	9133
82	9138	9143	9149	9154	9159	9165	9170	9175	9180	9186
83	9191	9196	9201	9206	9212	9217	9222	9227	9232	9238
84	9243	9248	9253	9258	9263	9269	9274	9279	9284	9289
85	9294	9299	9304	9309	9315	9320	9325	9330	9335	9340
86	9345	9350	9355	9360	9365	9370	9375	9380	9385	9390
87	9395	9400	9405	9410	9415	9420	9425	9430	9435	9440
88	9445	9450	9455	9460	9465	9469	9474	9479	9484	9489
89	9494	9499	9504	9509	9513	9518	9523	9528	9533	9538
90	9542	9547	9552	9557	9562	9566	9571	9576	9581	9586
91	9590	9595	9600	9605	9609	9614	9619	9624	9628	9633
92	9638	9643	9647	9652	9657	9661	9666	9671	9675	9680
93	9685	9689	9694	9699	9703	9708	9713	9717	9722	9727
94	9731	9736	9741	9745	9750	9754	9759	9763	9768	9773
95	9777	9782	9786	9791	9795	9800	9805	9809	9814	9818
96	9823	9827	9832	9836	9841	9845	9850	9854	9859	9863
97	9868	9872	9877	9881	9886	9890	9894	9899	9903	9908
98	9912	9917	9921	9926	9930	9934	9939	9943	9948	9952
99	9956	9961	9965	9969	9974	9978	9983	9987	9991	9996

Appendix A

Table 3*

METRIC CONVERSION FACTORS
Approximate Conversions to Metric Measures

Symbol	When You Know	Multiply by	To Find	Symbol
		LENGTH		
in	inches	*2.5	centimeters	cm
ft	feet	30	centimeters	cm
yd	yards	0.9	meters	m
mi	miles	1.6	kilometers	km
		AREA		
in^2	square inches	6.5	square centimeters	cm^2
ft^2	square feet	0.09	square meters	m^2
yd^2	square yards	0.8	square meters	m^2
mi^2	square miles	2.6	square kilometers	km^2
	acres	0.4	hectares	ha
		MASS (weight)		
oz	ounces	28	grams	g
lb	pounds	0.45	kilograms	kg
	short tons (2000 lb)	0.9	tonnes	t
		VOLUME		
tsp	teaspoons	5	milliliters	ml
Tbsp	tablespoons	15	milliliters	ml
fl oz	fluid ounces	30	milliliters	ml
c	cups	0.24	liters	l
pt	pints	0.47	liters	l
qt	quarts	0.95	liters	l
gal	gallons	3.8	liters	l
ft^3	cubic feet	0.03	cubic meters	m^3
yd^3	cubic yards	0.76	cubic meters	m^3
		TEMPERATURE (exact)		
°F	Fahrenheit temperature	5/9 (after subtracting 32)	Celsius temperature	°C

°F -40   0   32   40   80   98.6   120   160   200   212 °F
°C -40   -20   0   20   37 40   60   80   100 °C

*1 in = 2.54 cm (exactly).

*(From Superintendent of Documents, U. S. Government Printing Office, Washington, D.C. 20402. Available as a wallet-sized plastic card for 20 cents. Catalog No. C13.10:365/2.)

Table 3 Metric Conversion Factors (Continued)

Approximate Conversions from Metric Measures

Symbol	When You Know	Multiply by	To Find	Symbol
\multicolumn{5}{c}{LENGTH}
mm	millimeters	0.04	inches	in
cm	centimeters	0.4	inches	in
m	meters	3.3	feet	ft
m	meters	1.1	yards	yd
km	kilometers	0.6	miles	mi
\multicolumn{5}{c}{AREA}
cm^2	square centimeters	0.16	square inches	in^2
m^2	square meters	1.2	square yards	yd^2
km^2	square kilometers	0.4	square miles	mi^2
ha	hectares (10,000 m^2)	2.5	acres	
\multicolumn{5}{c}{MASS (weight)}
g	grams	0.035	ounces	oz
kg	kilograms	2.2	pounds	lb
t	tonnes (1000 kg)	1.1	short tons	
\multicolumn{5}{c}{VOLUME}
ml	milliliters	0.03	fluid ounces	fl oz
l	liters	2.1	pints	pt
l	liters	1.06	quarts	qt
l	liters	0.26	gallons	gal
m^3	cubic meters	35	cubic feet	ft^3
m^3	cubic meters	1.3	cubic yards	yd^3
\multicolumn{5}{c}{TEMPERATURE (exact)}
°C	Celsius temperature	9/5 (then add 32)	Fahrenheit temperature	°F

Appendix A

TABLE 4: USEFUL FORMULAS

Circles

Circumference: $C = 2\pi r = \pi d$.

Area: $A = \pi r^2 = \dfrac{\pi d^2}{4}$

Areas of Polygons

Rectangle	$A = bh$.
Square	$A = s^2 = \dfrac{1}{2} d^2$.
Parallelogram	$A = bh$.
Rhombus	$A = bh = \dfrac{1}{2} d_1 d_2$.
Trapezoid	$A = \dfrac{1}{2} h(b_1 + b_2)$.
Triangle	$A = \dfrac{1}{2} bh$.
Equilateral Triangle	$A = \dfrac{s^2 \sqrt{3}}{4}$.

Right Triangles

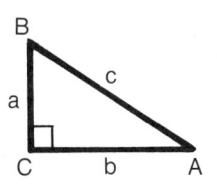

In a right triangle, the square of the hypotenuse equals the sum of the squares of the two legs, $c^2 = a^2 + b^2$.

Common Pythagorean triples: 3, 4, 5
5, 12, 13
8, 15, 17 } or multiples thereof

In a 30°, 60°, 90° triangle:

the leg opposite 30° angle $= \dfrac{1}{2}$ hypotenuse,

the leg opposite 60° angle = $\frac{1}{2}$ hypotenuse $\times \sqrt{3}$.

In a 45°, 45°, 90° triangle:

the leg opposite 45° angle = $\frac{1}{2}$ hypotenuse $\times \sqrt{2}$.

Coordinate Geometry

Length of a line segment (d) whose end points are (x_1, y_1) and (x_2, y_2)

$$d = \sqrt{(x_2 - x_1)^2 + (y_2 - y_1)^2}.$$

Midpoint of a line segment whose end points are (x_1, y_1) and (x_2, y_2):

$$\text{coordinates of midpoint} = \left(\frac{x_1 + x_2}{2}, \frac{y_1 + y_2}{2}\right).$$

The slope (m) of a line containing the points (x_1, y_1) and (x_2, y_2):

$$\text{slope (m)} = \frac{y_2 - y_1}{x_2 - x_1}.$$

The equation of a straight line:
 Point-slope form: $y - y_1 = m(x - x_1)$, where x_1 and y_1 are coordinates of any point on the line.
 Slope-intercept form: $y = mx + b$, where $m =$ slope, $b =$ y-intercept.
 Parallel to the x-axis: $y = k$, where k is the y-coordinate of any point on the line.
 Parallel to the y-axis: $x = c$, where c is the x-coordinate of any point on the line.

appendix B

Answers to Selected Exercises

Chapter 1. The Number System

Section 1.1

1. a. true c. true e. true g. true i. true k. true
   m. true o. true q. false s. true u. false w. true
   y. true

2. a. {1, 2, 3}, ∅
   {1} {1, 2}
   {2} {1, 3}
   {3} {2, 3}
   c. {m, a, t, h} {h}, {a, t} {m, a, h}
   {m} {m, a}, {a, h} {m, t, h}
   {a} {m, t}, {t, h} {a, t, h}
   {t} {m, h}, {m, a, t} ∅

3. a. {1, 2, 3, 4, 5, 6, 7} e. {1, 2, 3, 4, 5, 6}
   c. {a, b, c, d, ..., z} g. {a, e, i, o, u}

4. a. {x | x is an odd counting number}
   c. {x | x is a Great Lake}
   e. {x | x is a prime number}

5. a. {3, 4, 5}   c. {5, 6}   e. {1, 2, 3, 4, 5, 6, 7, 8}   g. ∅
   i. {3, 4, 5, 6, 7, 8}   k. {5, 6}

Section 1.2

	a.	c.	e.	g.	i.
1. Variables	x	x, y	s, t	r	x
Constants	7	3, 4, 6	4, 7, 5	2, π	a, b, c

2. a. 81   c. 14   e. 14   g. 10   i. $9a + 3b + c$

3. a. $3x^2, -2x$ c. $4x^2, +2y^2, +3(x-y)$

4. a. polynomial: trinomial   c. polynomial: monomial
   e. polynomial: binomial   g. no   i. polynomial: monomial

5. a. y^2 and $+2y^2$ c. none   e. $2x^3y^2$ and $-4x^3y^2$ g. none

6. a. $x = 2$ c. always   e. always   g. never   i. always

Section 1.3

1. a. commutative property of addition

636

Answers to Selected Exercises 637

    c. distributive property
    e. associative property of multiplication
    g. identity property of addition
    i. identity property of multiplication

2. a. true   c. false   e. false   g. false   i. false

3. a. 3x   c. $7(4+2)$ e. $(4+6)y$ g. $\pi(x+y)$
   i. $(x+y)(2+4)$ k. $x^2(x+3)$ m. $x^2(2xy^2z - x + 3)$
   o. $(x+y)(x+z+x+y)$

4. a. III   c. V   e. VIII   g. II

Section 1.4

1. a. -2 c. -100 e. 0

2. No, not all integers have multiplicative inverses that are integers. For example, the multiplicative inverse of 3 is $\frac{1}{3}$, which is not an integer.

3. a. true   c. true   e. false   g. true   i. false   k. false
   m. true   o. false

4. a. $-2 < 10.$ c. $-10 < 0.$ e. $-3 - 2 < -6 + 2.$
   g. $(-2)(-3) = 2 \cdot 3.$ i. $(-2)(-3)(-1) < 6.$ k. $(-6) \div (-2) = 3.$

5. a. -1 c. -3 e. 3   g. -1 i. -3 k. 13   m. 1
   o. 3   q. -1 s. -1 u. 1

6. a. -9 c. -6 e. 6   g. 0   i. 1   k. 10   m. 2   o. 1
   q. undefined

7. a. true   c. true   e. true   g. false   i. true

8. a. $x = -1.$ c. $x = 4.$ e. $x = 0.$ g. $x = -3.$ i. $x = 3.$

Section 1.5

1. a. 0.125   c. $0.8\overline{33}$ e. $0.2\overline{727}$ g. $0.0\overline{606}$ i. $0.\overline{384615}$
   k. $0.\overline{142857}$

2. a. 3/4   c. 1/40   e. 8/9   g. 5/11   i. 41/333   k. 67/495
   m. 1379/300   o. 5/2

3. a. 22/15 c. 3/8 e. −22/15 g. 8/15 i. 1/5 k. 5/6

4. a. $\frac{1}{2} > \frac{1}{3}$ c. $\frac{5}{11} < \frac{6}{12}$ e. $\frac{4}{5} = \frac{36}{45}$ g. $\frac{6}{13} < \frac{7}{11}$ i. $\frac{5}{3} > \frac{9}{7}$

5. a. true c. false e. true g. true i. false

6. a. 7/24 c. 13/84 e. 55/126 g. 41/99 i. 23/154

Section 1.6

1. a. rational c. rational e. irrational g. irrational
   i. irrational k. rational m. irrational o. irrational

2. a. terminating c. terminating e. non-terminating, non-repeating
   g. non-terminating, non-repeating i. repeating
   k. non-terminating, non-repeating

3. 3.14, 3.141141114..., $3.\overline{14}$, 3.1441444, $3.1\overline{4}$

5. a. false c. true e. true g. false i. true

6. a. commutative property for addition
   c. identity property for multiplication
   e. associative property for addition
   g. inverse property for multiplication
   i. identity property for multiplication
   k. inverse property for multiplication
   m. closure property for addition

7. a. $-\sqrt{2} + \sqrt{2} = 0$, and 0 is not an irrational number.

9. a. given b. equality axiom c. inverse property for addition
   d. substitution axiom e. identity property for addition

10. a. 8 c. −24 e. −18 g. −1 i. 4 k. 44 m. 8/13
    o. −18

Section 1.7

1. a. 1 c. 1 e. 1 g. 5 i. 1 k. 1 m. 1 o. 1

2. a. 4 c. 4 e. 16 g. 0 i. 0

Answers to Selected Exercises

3. a. 1   c. 7   e. 12   g. $\frac{4}{3}$ i. -1

4. a. 4, -4 c. $\emptyset$ e. 3, -5 g. $\emptyset$ i. $-2, -8$ k. 7, -1
    m. $\emptyset$ o. $-2, -4$

5. a. y   c. 0

Review Exercises for Chapter 1

1. a. $\{0, 1, 2\}$ b. $\{1, 2, 3\}$
    c. $\{1, 2\}$ d. $\{0, 1, 2, 3, \ldots\}$
    e. $\{-2, -1, 0, 1, 2, 3\}$ f. $\emptyset$
    g. $\{0, 1, 2, 3\}$ h. $\{0, 1, 2, 3, \ldots\}$
    i. $\{-2, -1, 0, 1, 2, 3\}$ j. $\{0, 1, 2\}$

2. a. $\{1, 2, 3, \ldots\}$ b. $\{0, 1, 2, 3, \ldots\}$
    c. $\{\ldots, -2, -1, 0, 1, 2, \ldots\}$ d. $\{1, 4, 9, 16, 25, \ldots\}$

3. a. 36   b. 16   c. 25π d. -6 e. -80 f. 120
    g. $9a - 3b + c$ h. $4a + 9b + c$

4. a. true   b. false   c. true   d. false   e. false
    f. false   g. false   h. true   i. true   j. false

5. a. -3 b. -7 c. -9 d. -1 e. 2   f. -2
    g. -8 h. 15   i. 2   j. -3 k. $-\frac{1}{2}$ l. 2

6. a. 0.375   b. $0.\overline{27}$ c. $0.\overline{142857}$ d. $0.\overline{384615}$

7. a. 4/9   b. 4/33   c. 313/90   d. 5072/990

8. a. 43/35   b. $-1/15$ c. 11/35   d. 2/15
    e. $-10/21$ f. 6/35   g. $-8/11$ h. 6/35
    i. 2/3

9. a. rational   b. rational   c. irrational   d. rational
    e. irrational   f. rational   g. rational   h. rational
    i. irrational   j. rational   k. rational   l. rational

10. 4.512, 4.5121121112…, $4.5\overline{12}$, 4.5121221222…, $4.\overline{512}$

11. a. identity property for multiplication
    b. commutative property for addition
    c. identity property for addition

d. inverse property for addition
e. commutative property for multiplication
f. identity property for multiplication
g. associative property for multiplication
h. inverse property for multiplication
i. associative property for addition
j. distributive property
k. inverse property for addition
l. commutative property for addition
m. closure property for addition
n. closure property for multiplication

12. a. -59 b. 24   c. $-3/5$ d. 10/7

13. a. 3   b. 7   c. 7   d. 3   e. 10
    f. 10   g. -3 h. 5/2   i. 5/2

Chapter 2. First Degree Equations and Inequalities in One Variable

Section 2.1

1. a. conditional   c. neither   e. identity   g. conditional
   i. conditional

2. a. yes   c. no   e. no   g. yes

3. a. yes   c. no   e. no   g. yes   i. yes

Section 2.2

1. $x = 7$.   3. $z = -2$.   5. $x = 3$.

7. $z = 8$.   9. $x = -27$.   11. $x = 2$.

13. $z = \frac{4}{3}$.   15. $y = \frac{20}{3}$.   17. $y = -4$.

19. $x = -5$.   21. $y = 15$.   23. $x = \frac{14}{8}$.

Answers to Selected Exercises

25. $y = -\dfrac{15}{4}$. 27. $z = -\dfrac{25}{42}$. 29. $y = -5$.

31. $x = 14$. 33. $z = 1$. 35. $y = 10$.

Section 2.3

1. $x = 30$. 3. $z = 12$. 5. $x = 3$.

7. $z = 9$. 9. $x = 4$. 11. $z = 12$.

13. $x = 10$. 15. $y = -1$. 17. $y = 5$.

19. $x = -1$. 21. $x = 1$. 23. $y = \dfrac{21}{10}$.

25. $x = \dfrac{17}{10}$. 27. $z = \dfrac{103}{10}$. 29. $y = 10$.

31. $x = 6$. 33. $x = -1$.

Section 2.4

1. $x = \dfrac{a+b}{2}$. 3. $x = \dfrac{c-b}{4a}$. 5. $x = 2b - a$.

7. $y = \dfrac{4a+2b}{3}$. 9. $y = \dfrac{3x-ab}{a}$. 11. $y = \dfrac{4c+2h+a}{2}$.

13. $t = \dfrac{d}{r}$. 15. $t = \dfrac{i}{pr}$. 17. $h = \dfrac{2A}{b}$.

19. $r = \dfrac{C}{2\pi}$. 21. $b = \dfrac{3v}{h}$. 23. $F = \dfrac{9}{5}C + 32$.

25. $h = \dfrac{2A}{b+B}$. 27. $h = \dfrac{S-\pi r^2}{2\pi r}$. 29. $m = \dfrac{y-b}{x}$.

Section 2.5

1. $10x$ 3. $3x + 5$ 5. $4x + 1$

7. $6(x + 8)$ 9. $5(x - 3)$ 11. $n = 2$

13. $n = 5$. 15. Bill is 36 17. width = 8 m
     Addie is 34   length = 24 m

Appendix B

19. 90°, 45°, 45°
21. 6
23. 190 mph
    310 mph

25. 4 nickels
    8 quarters
27. 12 nickels
    12 dimes
    22 quarters
29. 49, 50, 51

31. 33, 34, 35
33. 44°, 56°, 80°

Section 2.6

1. $\{x \mid x \geq 2\}$

3. $\{x \mid x < 3\}$

5. $\{x \mid x \geq -2\}$

7. $\{x \mid x > 2\}$

9. $\{x \mid x \leq 4\}$

11. $\{x \mid x < -2\}$

13. $\{x \mid x < -8\}$

15. $\{x \mid x \geq \frac{2}{3}\}$

17. $\{x \mid x \leq 10\}$

19. $\{x \mid x \leq -3\}$

21. $\{x \mid x < \frac{1}{2}\}$

23. $\{x \mid -2 \leq x < 0\}$

25. $\{x \mid -2 \leq x < 2\}$

27. $\{x \mid 1 \leq x < 3\}$

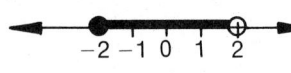

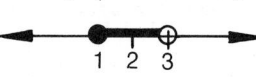

Section 2.7

1. $\{x \mid -1 < x < 2\}$

3. $\{x \mid -3 \leq x \leq 2\}$

5. $\{x \mid 0 \leq x \leq 3\}$

Answers to Selected Exercises 643

7. $\{x \mid 1 \leq x \leq 2\}$ 9. $\{x \mid -2 < x < -1\}$ 11. $\{x \mid -3 < x < 1\}$

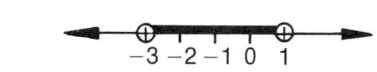

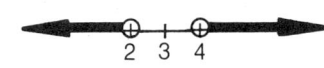

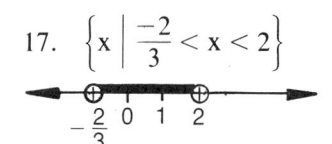

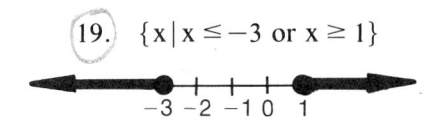

13. $\{x \mid x < 2 \text{ or } x > 4\}$ 15. $\{x \mid -1 \leq x \leq 4\}$

17. $\left\{x \mid \dfrac{-2}{3} < x < 2\right\}$ 19. $\{x \mid x \leq -3 \text{ or } x \geq 1\}$

21. $\{x \mid x < -1 \text{ or } x > 0\}$ 23. $\{x \mid x \leq -3 \text{ or } x \geq -2\}$

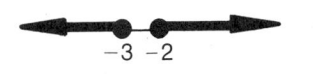

25. $\emptyset$ 27. $\emptyset$

Review Exercises for Chapter 2

1. a. conditional   b. identity   c. neither
   d. conditional   e. identity   f. neither

2. a. $y = 3$.   b. $x = -27$.   c. $z = 8$.
   d. $z = -2$.   e. $x = 1$.   f. $y = 4$.
   g. $y = \dfrac{16}{15}$.   h. $z = 18$.   i. $x = -\dfrac{25}{42}$.
   j. $z = 14$.   k. $x = 10$.   l. $y = 18$.
   m. $z = 2$.   n. $y = \dfrac{190}{63}$.   o. $y = 6$.

3. a. $r = \dfrac{d}{t}$.   b. $r = \dfrac{i}{pt}$.   c. $h = \dfrac{2A}{b}$.
   d. $C = \dfrac{5}{9}(F - 32)$.   e. $F = \dfrac{9}{5}C + 32$.   f. $h = \dfrac{S - 2\pi r^2}{2\pi r}$.
   g. $y = \dfrac{c - b}{a}$.   h. $W = \dfrac{p - 2L}{2}$.   i. $x = \dfrac{c - 2b}{a}$.

4. $8.60   5. 60 mph and 45 mph   6. length = 365 feet
                                       width = 135 feet

7. 5, 6   8. 350   9. 11, 13, 15

10. Scott = 8, Joe = 6, Father = 36.

11. a. $\{x \mid x > 2\}$ b. $\{x \mid x < -3\}$ c. $\{x \mid x \geq -5\}$

d. $\{x \mid x \geq 10\}$ e. $\{x \mid x > -3\}$ f. $\left\{x \mid x > \frac{1}{2}\right\}$

12. a. $\{x \mid x < 0 \text{ or } x > 2\}$ b. $\{x \mid 1 < x < 3\}$

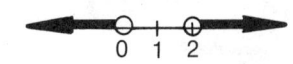

c. $\{x \mid x \leq -2 \text{ or } x \geq 3\}$ d. $\left\{x \mid -\frac{1}{3} < x < 1\right\}$

e. $\{x \mid -3 \leq x \leq 1\}$ f. $\emptyset$

13. 12 minutes

Chapter 3. Exponentials and Radicals

Section 3.1

1. a. x^3 c. $4^3 a^2$ e. $x^2 y z^3$ g. $2^3 x^2 y^3$ i. $1 b^3 c^3 d$

2. a. $x \cdot x \cdot x \cdot y \cdot y$ c. $2 \cdot 2 \cdot a \cdot a \cdot b \cdot b \cdot b$ e. $x \cdot x \cdot x \cdot x \cdot x \cdot x$

3. a. 8   c. 25   e. -25

Answers to Selected Exercises 645

    g. 16   i. 64   k. -512
    m. 3   o. 1

4. a. -6 c. 36   e. 144
    g. -108 i. 36   k. -216

5. a. x^5 c. z^6 e. 2^9
    g. $3^3 y^5$ i. x^{m+n} k. b^{7x}

6. a. x^3 c. y e. 4^2
    g. -4 i. $3xy$ k. 10^2

7. a. x^8 c. z^9 e. $2^2 a^4 b^6$
    g. $x^8 y^{12}$ i. $\dfrac{a^4}{b^6}$ k. $\dfrac{x^6 y^9}{z^6}$

    m. $\dfrac{a^{3m}}{b^{3m}}$ o. $\dfrac{x^{2z}}{y^{3z}}$ q. $\dfrac{2^4 x^6 y^4}{3^4 z^8}$

    s. $3^6 a^6 b^6$

Section 3.2

1. 1   3. 4   5. x^5

7. a^6 9. $4z^3$ 11. $4x^3$

13. $\dfrac{1}{m^3}$ 15. $\dfrac{3}{b^2}$ 17. $\dfrac{1}{9b^2}$

19. $\dfrac{1}{a^2 b^4}$ 21. $\dfrac{1}{x^3}$ 23. b^2

25. a 27. 1   29. $\dfrac{1}{512}$

31. $\dfrac{27}{y^2}$ 33. b^n 35. $\dfrac{1}{a^5}$

37. $\dfrac{1}{16}$ 39. x^4 41. 1

43. 1024   45. $\dfrac{1}{ab^2}$ 47. 1

49. a^8 51. $\dfrac{1}{4a^4 b^6}$ 53. $\dfrac{x^9 y^6}{64}$

55. $\dfrac{x^8}{y^6}$

57. $\dfrac{b^{15}}{a^{15}}$

59. $\dfrac{81y^4}{16x^6z^8}$

Section 3.3

1. 4.00×10^9
3. 9.00×10^{-28}
5. 6.00×10^{21}
7. 0.0041
9. $314{,}200$
11. $48{,}590{,}000$
13. 1.53×10^{-8}
15. 1.54×10^{18}
17. 5.00×10^3
19. 1.10×10^{-1}
21. 1.00×10^8
23. 4.00×10^8
25. 1.00×10^{17}

Section 3.4

1. a. $\sqrt{4}$ c. $\sqrt{3}$ e. $\sqrt[3]{x}$
   g. $\sqrt[3]{x^2}$ i. $\sqrt[4]{y}$ k. $\sqrt{3x}$
   m. $\sqrt{5^3 x^6}$ o. $\sqrt{x+y}$

2. a. $x^{1/2}$ c. $a^{1/3}$ e. $a^{3/4}$
   g. $8^{1/6}$ i. 2 k. $(2x)^{1/2}$
   m. $(3x^3)^{1/4}$ o. $x^2 y^3$

3. a. 2 c. 1 e. 3
   g. $\emptyset$ i. x^3 k. y^4

4. a. 2 c. -2 e. 8
   g. 16 i. $\dfrac{1}{2}$ k. $\dfrac{1}{4}$
   m. $\dfrac{1}{2}$ o. $\dfrac{2}{3}$ q. 4

5. a. x c. $y^{7/12}$ e. $2^{5/6}$
   g. $4^{5/6}$ i. $y^{1/2}$ k. $y^{1/8}$
   m. $2^{1/6}$ o. $5^{1/6}$ q. $m^{5/6}$
   s. $3^{1/6}$ u. $y^{23/12}$

Answers to Selected Exercises

Section 3.5

1. 2
3. -2
5. $5\sqrt{2}$
7. $2\sqrt[3]{2}$
9. $4\sqrt[3]{2}$
11. $2\sqrt[4]{2x^3}$
13. $2x^3y^2\sqrt{3xy}$
15. $2x^2y^2\sqrt{2x}$
17. $\dfrac{\sqrt{3}}{2}$
19. $\dfrac{\sqrt{15}}{5}$
21. $\dfrac{\sqrt[3]{12}}{3}$
23. $\dfrac{\sqrt[3]{2x^2}}{x}$
25. $\dfrac{\sqrt[3]{xy^2}}{y}$
27. $\dfrac{\sqrt{abc}}{bc}$
29. $\dfrac{\sqrt{5}}{2}$
31. $\dfrac{\sqrt{6xy}}{3y}$
33. $\dfrac{2a\sqrt{6ab}}{3b}$
35. $\dfrac{\sqrt[4]{27a^3b}}{3}$
37. 2
39. $\sqrt{5}$
41. 3
43. $\sqrt[4]{3}$
45. $\sqrt{2x}$
47. $3\sqrt{xy}$
49. $\sqrt[4]{4x^2y}$
51. $x\sqrt{3y}$

Section 3.6

1. $\sqrt{15}$
3. $\sqrt[3]{4}$
5. 30
7. $8\sqrt{6xy}$
9. $12x\sqrt{6}$
11. $30x$
13. $\sqrt[3]{9x^2}$
15. $\sqrt[6]{432}$
17. $\sqrt[6]{x^4y^3}$
19. $12\sqrt[6]{2x^4y^3}$
21. $3\sqrt[12]{16x^{10}y^7}$
23. $ab\sqrt[6]{108ab}$
25. $2\sqrt[6]{432x^3y^5}$
27. $6ab\sqrt[6]{4ab}$
29. $\sqrt{6}-\sqrt{10}$
31. $3y+2y\sqrt{x}$
33. $y+\sqrt[3]{xy}$
35. $12\sqrt{xy}-3\sqrt{2yz}$
37. $x\sqrt[3]{12y^2}+4y\sqrt[3]{x}$
39. $\sqrt[6]{72}-\sqrt[6]{32}$
41. $\sqrt[6]{x^5}-\sqrt[6]{x^2y^2}$

Section 3.7

1. $5\sqrt{2}$
3. $3\sqrt[3]{3}$
5. $\sqrt{2}+\sqrt{3}$
7. $2\sqrt{2}$
9. 0
11. $-2\sqrt{2}$
13. $7\sqrt{3}$
15. $5\sqrt[3]{2}$
17. $16\sqrt{2}$

19. $5\sqrt{2x}$ 21. $2.7x$ 23. $(4-x)\sqrt[3]{x^2}$

25. $\sqrt{x}+\sqrt[3]{x}$ 27. $\sqrt{x^3}+\sqrt[3]{x^4}$ 29. $5\sqrt{3}$

31. $6\sqrt{2}$ 33. $\sqrt{3}$

Section 3.8

1. $\sqrt{6}-\sqrt{2}$ 3. $6\sqrt{6}-4\sqrt{3}$ 5. $16-4\sqrt{6}$

7. $6x-10x\sqrt{y}$ 9. $15x+10\sqrt{xy}$ 11. $28+16\sqrt{3}$

13. $\sqrt{10}+\sqrt{15}-\sqrt{6}-3$ 15. $6\sqrt{6}+4\sqrt{10}-3\sqrt{15}-10$

17. $6x-3y+7\sqrt{xy}$ 19. $4\sqrt{15}-6\sqrt{10}+6\sqrt{6}-18$

21. 17 23. 15

25. $x+2\sqrt{3x}+3$ 27. $38-12\sqrt{10}$

29. $18x+48\sqrt{xy}+32y$ 31. $10+2\sqrt{6}+2\sqrt{10}+2\sqrt{15}$

33. $6+7\sqrt{x}+\sqrt{y}+2x+\sqrt{xy}-y$

Section 3.9

1. $\dfrac{\sqrt{3}}{3}$ 3. $\dfrac{\sqrt{6}}{2}$ 5. $\dfrac{\sqrt{xy}}{y}$

7. $\dfrac{\sqrt[3]{4}}{2}$ 9. $\dfrac{\sqrt[3]{6}}{3}$ 11. $\dfrac{x\sqrt{y}}{y}$

13. $\dfrac{\sqrt[3]{4}}{2}$ 15. $\dfrac{\sqrt[3]{yx^2}}{x}$ 17. $\dfrac{6+2\sqrt{2}}{7}$

19. $\dfrac{x\sqrt{2x}-x}{2x-1}$ 21. $\dfrac{3\sqrt{2}-3\sqrt{3}}{-1}$ 23. $-2\sqrt{5}-2\sqrt{3}$

25. $\dfrac{3+2\sqrt{2}}{1}$ 27. $4-\sqrt{15}$

29. $\dfrac{\sqrt{15}+\sqrt{6}+\sqrt{10}+2}{3}$ 31. $\dfrac{4\sqrt{15}+6\sqrt{10}+9\sqrt{6}+18}{-17}$

34. $\dfrac{x+2\sqrt{xy}+y}{x-y}$ 35. $\dfrac{9\sqrt{xy}+2y+9x}{4y-9x}$

38. $\dfrac{3\sqrt{2y}+3\sqrt{10}+2\sqrt{xy}+2\sqrt{5x}}{3y-15}$

Answers to Selected Exercises

39. $\dfrac{20\sqrt{xy} - 15x\sqrt{10} - 4y\sqrt{6} + 6\sqrt{15xy}}{8y - 45x}$

Section 3.10

1. 9
3. 4
5. ∅
7. 2
9. 8
11. ∅
13. 9
15. -14
17. 41
19. 2
21. ∅
23. 26
25. 9
27. -1
29. ∅

Review Exercises for Chapter 3

1. a. x^7 b. $2x^5$ c. $8x^3y^4$
   d. $3^3 x^5 y^2$ e. $-6x^3y^2$ f. $x^{5/6}$
   g. $10a^5 b^3$ h. $6a^2 bc$ i. a^{7x}
   j. $x^{13/6}$ k. 2^2 l. xy
   m. $x^{3/2}$ n. $a^2 b^2$ o. 3^4
   p. x^4 q. $2^3 x^3 y^3$ r. $3^4 a^2 b^2$

2. a. 1   b. 1   c. 1
   d. $x^4 y^6$ e. $\dfrac{1}{x^2}$ f. $3y^3$
   g. $\dfrac{1}{9b^2}$ h. $\dfrac{1}{x^6}$ i. $\dfrac{1}{x^5}$
   j. y^6 k. $x^5 y$ l. $\dfrac{x^4}{y^6}$
   m. $\dfrac{y^{10}}{x^{10}}$ n. $\dfrac{x^9}{2^9}$ o. $x^{10} y^2$

3. a. 2.84×10^9 b. 5.46×10^{-8} c. 1.00×10^7 d. 4.80×10^1

4. a. $\sqrt[3]{x^2}$ b. $\sqrt[3]{x^2 y}$ c. $\sqrt{x+y}$
   d. $\dfrac{\sqrt{xy}}{y}$ e. $\dfrac{3\sqrt{xy}}{x^2}$ f. $\dfrac{x\sqrt{y}}{8}$

5. a. $x^{2/3}$ b. $2^{1/4} x^{3/4}$ c. $x^2 y^3$
   d. $(x+y)^{1/2}$ e. $x^{2/3} y^{1/3}$ f. $(x^2 + y^2)^{1/2}$

6. a. $2\sqrt{3}$ b. $3\sqrt[3]{2}$ c. $-2\sqrt[3]{x}$

d. $2x^2y^3\sqrt{3xy}$ e. $\dfrac{\sqrt{2}}{3}$ f. $\dfrac{\sqrt{6}}{3}$

g. $\dfrac{\sqrt[3]{12}}{2}$ h. $\dfrac{2x\sqrt{6xy}}{3y}$ i. $\sqrt{2}$

j. $\sqrt{2}$ k. $\sqrt{2x}$ l. $\sqrt[4]{3xy^2}$

7. a. $7\sqrt{2}$ b. $-\sqrt[3]{3}$ c. $9\sqrt{3}$
   d. 0 e. $11\sqrt{3y}$ f. $6\sqrt{2}$

8. a. $\sqrt{21}$ b. $\sqrt{xy}$ c. $6\sqrt[3]{12ab}$
   d. $3y\sqrt{3y}$ e. $\sqrt{2xy} - 3x\sqrt{2y}$ f. $6y\sqrt{x} + 6\sqrt{2y}$
   g. 0 h. $\sqrt{6} - 2 + \sqrt{15} - \sqrt{10}$ i. $x - y$
   j. $17\sqrt{2x} + 10x + 6$ k. $\dfrac{\sqrt{xy}}{y}$ l. $\sqrt{3}$
   m. $\dfrac{\sqrt{2}}{2}$ n. $\dfrac{\sqrt[3]{12}}{2}$ o. $\dfrac{1-\sqrt{3}}{-1}$
   p. $\dfrac{3\sqrt{5}+6}{1}$ q. $\dfrac{2\sqrt{2}+6+\sqrt{6}+3\sqrt{3}}{-7}$ r. $\dfrac{5+2\sqrt{6}}{1}$
   s. $\dfrac{6x + 2y + 7\sqrt{xy}}{9x - 4y}$ t. $\dfrac{3\sqrt{2y} + 3\sqrt{6} + \sqrt{xy} + \sqrt{3x}}{2y - 6}$

9. a. 4 b. ∅ c. −1
   d. 9 e. ∅ f. ∅
   g. 4 h. 6 i. 5

Chapter 4. Polynomials

Section 4.1

1. a. $x, -2$ c. $4xy, -1$ e. $4y^2, -3y, 2$ g. ax^2, bx

2. a. binomial c. binomial e. trinomial g. binomial

3. a. 3, 2 c. 1, −2 e. 3, 2, −2

4. a. 2 c. 3 e. 6 g. 3 i. 6

5. a. $2 - 2xyz^2 + 3x^2y^3z^3 + x^3y^2z$ c. $3x^2y^3z^3 - 2xyz^2 + x^3y^2z + 2$

Answers to Selected Exercises 651

6. a. yes   c. no   e. yes

7. a. 9   c. -1 e. 102

Section 4.2

1. $12x^2 + 7x + 6$ 3. $-y^2 + 4y + 4$ 5. $3x + 2xy + 2y$

7. $10y + 1$ 9. $7 - 3x + 4x^2$ 11. $x^3 - 4x^2 - 2x$

13. $7x - 2$ 15. $3x^2 + 1$ 17. $x^3 + 2x^2 - 3$

19. $-10x^2 - 9x - 8$ 21. $-2x + 2y - 2$ 23. $5x + 6$

25. $-y^3 - 2y^2 - 1$ 27. $-5x^2 + 4x - 9$ 29. $2x^4 + 5x^2 + x - 7$

31. 0 33. $7y^2 - 4y + 2$ 35. $3x^2 + 5x - 7$

37. $7x^2 - 4x$ 39. $4x^2 - x - 3$ 41. $y^2 + 5y - 2$

43. $-5x - 5$

Section 4.3

1. $6x^5$ 3. $-6y^7$ 5. $3x^2 + 3xy$

7. $-6x^3y - 8x^4$ 9. $6x^3y^3 + 8x^3y^2$

11. $-6x^2y^3 - 9x^3y^2 - 12x^4y$

13. $x^2 - x - 2$ 15. $6y^2 - y - 2$

17. $6y^2 - 5y - 6$ 19. $2x^3 - x^2 + 6x - 3$ 21. $6y^4 + 7y^2 - 3$

23. $3y^3 - 5y^2 - 2y$ 25. $6x^2 + 7xy - 5y^2$ 27. $4xy - 3x^2 + 4y^2$

29. $3r^5 + 3r^2s^2 - 2s^3r^3 - 2s^5$ 31. $6t^2u^2 - 4t^3 + 3u^5 - 2u^3t$

33. $x^3 + 5x^2 + 5x - 2$ 35. $6y^4 + 5y^3 + 6y^2 - 8y$

37. $12r^5 + 14r^3s + 3r^2s^2 + 4rs^2 + 2s^3$ 39. $x^3 + 8y^3$

41. $2x^2 - xy + 7x - 6y^2 + 7y + 3$

43. $2x^4 + 4x^2y + 4x^2 + 3x^2y^2 + 6y^3 + 9y^2 - 2x^2 - 4y - 6$

45. $y^4 - 2y^3 - 4y^2 + 2y + 3$ 47. $r^3 + 2r^2 - r - 2$

Section 4.4

1. $6x^2 + 4xy$
3. $12x^3y^3 - 8x^2y^2$
5. $y^2 + 2y - 15$
7. $x^2 - 8x + 7$
9. $y^2 - 8y - 20$
11. $8x^2 + 2x - 3$
13. $10y^2 - y - 3$
15. $6x^2 - xy - 2y^2$
17. $6y^2 + 11yz - 10z^2$
19. $6x^4 + 17x^2y + 12y^2$
21. $6y^4 + 7xy^2 - 20x^2$
23. $5y^6 - 13y^3z^2 - 6z^4$
25. $x^2 + 2xy + y^2$
27. $x^2 + 4xy + 4y^2$
29. $z^2 + 10z + 25$
31. $9z^2 + 12xz + 4x^2$
33. $16y^2 + 40y + 25$
35. $16y^2 + 40y + 25$
37. $9x^4 - 24x^2y + 16y^2$
39. $16y^6 - 16y^3z^2 + 4z^4$
41. $4x^6 + 12x^3y^2 + 9y^4$
43. $x^2 - y^2$
45. $9x^2 - 4y^2$
47. $x^4 - 16y^2$
49. $25x^6 - 9y^2$
51. $9z^8 - y^6$
53. $25x^4y^2 - 9z^4$
55. 1596
57. 391
59. 384
61. 396
63. $2xy$
65. $4x^2$
67. $4y^2$

Section 4.5

1. $x^3 + 3x^2y + 3xy^2 + y^3$
3. $x^3 - 3x^2y + 3xy^2 - y^3$
5. $x^3 + 9x^2 + 27x + 27$
7. $x^3 - 12x^2 + 48x - 64$
9. $8x^3 + 12x^2 + 6x + 1$
11. $27x^3 - 27x^2y + 9xy - y^3$
13. $x^3 + 9x^2y + 27xy^2 + 27y^3$
15. $27x^3 - 108x^2y + 144xy^2 - 64y^3$
17. $8y^3 + 48y^2x + 96yx^2 + 64x^3$
19. $y^6 - 3y^4x + 3y^2x^2 - x^3$
21. $27x^6 - 54x^4y + 36x^2y^2 - 8y^3$
23. $x^3 - y^3$
25. $x^3 + 27$
27. $27x^3 - 8$
29. $8x^3 - 27y^3$
31. $8x^6 - 27y^3$

Answers to Selected Exercises   653

33. $8x^9 + 27y^6$

35. $2x^2 - xy + 7x - 6y^2 + 7y + 3$

37. $y^4 - 5y^3 + 5y^2 + 7y - 20$

39. $x^4 + 4x^3y + 4x^2y^2 - y^4$

Section 4.6

1. $2x^2$
3. $-2x^2z$
5. $3x^2yz$

7. $-4x$
9. $\dfrac{2y^2z}{x}$
11. $3ab^2c - 2b$

13. $2x^3 + x^2 - 2$
15. $4y^2 + 2y + 1$
17. $3x - 2y^2 - 5$

19. $-3tu + 2 - u^2$
21. $\dfrac{5}{x^3y^3} - \dfrac{4}{x^2y^4}$
23. $x + 1$

25. $x + 4$
27. $4y - 1$
29. $3x + 2$

31. $2x + 3$
33. $3y + 3 + \dfrac{9}{6y - 2}$
35. $y^2 + 2y - 3$

37. $3x^2 + 7x - 1$
39. $x^4 - x^3 + 4x^2 - 4x + 6 + \dfrac{-6}{x + 1}$

41. $x^2 + 2x + 4$
43. $2y^4 + 3y^3 + 6y^2 + 9y + 16 + \dfrac{48}{2y - 3}$

45. $8x^4 - 28x^2 + 2x + 112 + \dfrac{-9x - 445}{x^2 + 4}$

47. a. -5 b. 1 c. -59 d. 1

49. 63 51. 11 53. 2

Section 4.7

1. $x + 1$
3. $6x^2 + 8x + 11 + \dfrac{9}{x - 1}$

5. $x^2 - 4x + 9 + \dfrac{-13}{x + 1}$
7. $x^2 + 2x + 4$

9. $3x^2 + 3x + 4 + \dfrac{-1}{x - 1}$
11. $6x^3 - 6x^2 + 10x - 7 + \dfrac{7}{x + 1}$

13. $x^4 - 2x^3 + 5x^2 - 10x + 21 + \dfrac{-42}{x + 2}$

15. $x^5 - x^4 + x - 1 + \dfrac{1}{x+1}$

17. $x^5 - x^4 + x^3 - x^2 + x - 1 + \dfrac{2}{x+1}$

19. $x^5 + 3x^4 + 9x^3 + 27x^2 + 81x + 243 + \dfrac{702}{x-2}$

21. $P(1) = 0, P(2) = 0, P(-2) = 0, (x-1)(x-2)(x+2)$
    $= x^3 - x^2 - 4x + 4.$

Review Exercises for Chapter 4

1. a. $-6x^3 + 10x^2 - 5x$ b. $-4x^2 - 6x + 10xy$ c. $2y^3 + y^2 + y$

2. a. $7x^2 + 7x - 13$ b. $9x^2 - x + 1$ c. $4x^2 + y^2 + x - y + 5$

3. a. $2x - 4y - 11$ b. $x^2 - x + 1$ c. $-x^2 - 2x - 1$

4. a. $3x^2 + 6x - 1$ b. $y^2 + 6y - 3$ c. $x - 9$

5. a. $-8x^3 + 6x^2$ b. $6x^4y^4 - 8x^5y^3$ c. $x^2 + x - 6$
   d. $8x^2y + 12x^2 + 6y + 9$ e. $2y^3 + 3y^2 - 13y + 6$
   f. $6x^2 + 5xy + 2x - 6y^2 - 23y - 20$

6. a. $x^2 - x - 12$ b. $y^2 - 5y - 50$
   c. $6x^2 + 3x - 9$ d. $2x^2 - xy - 6y^2$
   e. $x^4 - x^2y - 6y^2$ f. $6y^4 + 5xy^2 - 6x^2$
   g. $x^2 + 2xy + y^2$ h. $x^2 - 6xy + 9y^2$
   i. $4x^2 + 12xy + 9y^2$ j. $x^4 + 4x^2y + 4y^2$
   k. $25x^4 - 30x^2y^3 + 9y^6$ l. $x^2 - y^2$
   m. $x^4 - 4y^2$ n. $9x^4y^2 - 1$
   o. $x^3 + 3x^2y + 3xy^2 + y^3$ p. $x^3 - 3x^2 + 3xy^2 - y^3$
   q. $8x^3 + 36x^2y + 54xy^2 + 27y^3$ r. $x^3 - 8$
   s. $x^3 + 8y^3$ t. $x^6 + y^3$
   u. $27x^3 - 1$

7. a. $2xy$ b. $4x^2$ c. $9y^2$

8. a. $8x^2 + 6x - 2$ b. $x^2yz^2 - 2xz$ c. $4x^2y + 2y - 3$
   d. $x - 3$ e. $2x - 1$ f. $3x - 2$
   g. $2x^2 + \dfrac{4}{4x^2 + 1}$ h. $x^2 + 4$ i. $y^2 - y + 1$

9. a. 91   b. -13

Answers to Selected Exercises 655

10. a. $x - 1 + \dfrac{-4}{x-1}$ b. $2x + 1 + \dfrac{3}{x+1}$ c. $4x^2 + 11x + 26 + \dfrac{53}{x-2}$

   d. $4x^3 - 8x^2 + 20x - 40 + \dfrac{84}{x+2}$

   e. $5x^4 + 15x^3 + 48x^2 + 144x + 432 + \dfrac{1297}{x-3}$

   f. $x^5 - 2x^4 + 4x^3 - 8x^2 + 16x - 32 + \dfrac{56}{x+2}$

11. none

Cumulative Review I (Chapters 1–4)

1. a. $\{4, 5\}$ b. $\{4, 5, 6, 7\}$
   c. $\{0, 1, 2, 3, 4, 5, 6, 7\}$ d. $\{3, 4, 5, 6, 7\}$
   e. $\{3, 4, 5\}$ f. $\{3, 4, 5, 6, 7, 8, 9\}$
   g. $\{0, 1, 2, 3, 4, 5, 6, 7, 8, 9\}$ h. $\emptyset$ i. $\emptyset$

2. a. false   b. false   c. false
   d. false   e. false   f. false
   g. true

3. 2.312, $2.3121121112\ldots$, $2.3\overline{12}$, $2.3122122212222\ldots$, $2.\overline{312}$

4. a. $x = \dfrac{14}{3}$ b. $x = 10$ c. $y = \dfrac{-25}{42}$
   d. $x = 18$ e. $x = 2$ f. $x = 6$

5. 35 cm., 35 cm., and 30 cm.

6. 11, 13, 15

7. 6:00 P.M.

8. a. $x > 5$.   b. $x \geq 10$.   c. $-3 \leq x \leq 1$.

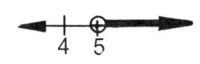

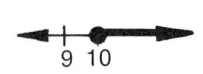

9. a. $x^{5/6}$ b. $x^{5/6}$ c. $2y^{1/3}$
   d. $\dfrac{1}{xy}$ e. $\dfrac{1}{x^5 y}$ f. $\dfrac{x}{y^7 z^3}$

10. a. 1.00×10^7 b. 4.80×10^1

11. a. $-12\sqrt{5}$ b. 0 c. $14\sqrt{2x} - 10x + 24$
    d. $\dfrac{2 - 2\sqrt{3}}{-2}$ e. $\sqrt[3]{4}$ f. $5 + 2\sqrt{6}$

12. a. $y = -1.$ b. $x = 12.$ c. $\emptyset$

13. a. $x^2 + 2x - 3$ b. $-x^2 + 5x - 5$
    c. $2x^2 + 11xy + 15y^2$ d. $2x^3 + 7x^2 - 21x + 9$
    e. $x^2 - 4xy + 4y^2$ f. $x^3 + 3x^2y + 3xy^2 + y^3$
    g. $x^2 + 2xy + x + y^2 + y - 6$ h. $2x^2 + 4x + 11 + \dfrac{18}{x - 2}$
    i. $5x^4 - 5x^3 + 8x^2 - 6x + 6 + \dfrac{-5}{x + 1}$

Chapter 5. Factoring

Section 5.1

1. $2(x + 5)$ 3. $5(x + 2)$ 5. $3(2y + 1)$

7. $2(x - 3)$ 9. $4(2y - 1)$ 11. $y(y + 3)$

13. $y(3y + 5)$ 15. $2x(x + 3)$ 17. $xy(y + 1)$

19. $5xy(-2 + x)$ 21. $2y(2y^2 + 4y + 1)$

23. $2z(2xy + 4xz - y)$ 25. $-3x^2y^2(1 - 2x^2 + 3y^2)$

27. $(x + 5)(y + 2)$ 29. $(x + 3)(y - 3)$

31. $(3x - 2)(y - 2)$ 33. $(x + y)(a - b)$

35. $2(2a + b)(y + x)$ 37. $(u + s)(v - t)$

39. $(y + a)(y + 4)$ 41. $(y + x)(y + 8)$

43. $2(x^2 - 2)(x + 2)$ 45. $(3a + 2b - 2c)(x - 2y)$

Section 5.2

1. $(x + 3)(x - 3)$ 3. $(y + 4)(y - 4)$

Answers to Selected Exercises

5. $(3 + x)(3 - x)$
7. $(1 + y)(1 - y)$
9. $(y^5 + 2)(y^5 - 2)$
11. $(3x + 4)(3x - 4)$
13. $(7 + 4y)(7 - 4y)$
15. $(4z + 7)(4z - 7)$
17. $(4x + 1)(4x - 1)$
19. $4xy^2(x + y)(x - y)$
21. $3x(3x^3 + y^2)(3x^3 - y^2)$
23. $(x + y)(x - y - 2)$
25. $(x + y)(x - y + 2)$
27. $(3x + y)(3x - y + 4)$
29. $(2x + 3 + y)(2x + 3 - y)$

Section 5.3

1. $(x + 2)^2$
3. $(y + 1)^2$
5. $(z - 4)^2$
7. $(1 - x)^2$
9. $(x - y)^2$
11. $(3x - 5y)^2$
13. $(5y - 2x)^2$
15. $(ax - 5y)^2$
17. $(2x^2 + 3y^2)^2$
19. $-1(2x - 3y)^2$
21. $3y(y + 1)^2$
23. $2z(2x + 5y)^2$
25. $(y - 3 + x)(y - 3 - x)$
27. $(x - y + 2z)(x - y - 2z)$

Section 5.4

1. $(x + 1)^2$
3. $(x - 6)(x - 2)$
5. $(x + 12)(x - 1)$
7. $(x - 4)(x - 3)$
9. $(x + 4)(x - 3)$
11. $(y - 6)(y + 3)$
13. $(y - 18)(y + 1)$
15. $(z + 7)(z - 4)$
17. $(y + 4)(y - 1)$
19. $3y(y - 7)(y + 5)$
21. $3y^2z^2(x + 4)(x - 2)$
23. $(y - 4)(y + 1)$
25. $-5(x + 3)(x - 2)$
27. $(x + 1)^2(x - 1)^2$
29. $(y^2 - 3)(y + 2)(y - 2)$
31. $(y^n - 4)(y^n - 3)$
33. $(a + 6b)(a - 4b)$

Section 5.5

1. $(2x + 1)(x + 2)$
3. $(3y + 2)(y + 1)$
5. $(5y - 1)(y - 1)$

7. $3(3x+1)(x+3)$ 9. $(5z+6)(z-1)$ 11. $(3y+2)(2y+3)$

13. $(5z+4)(3z+2)$ 15. $(5y-4)(3y-2)$ 17. $(2-3x)(1-7x)$

19. $(3+x)(-1+2x)$ 21. $(3y^2+1)(2y^2+1)$

23. $(3y^2-1)(2y^2-3)$

25. $2x(3x+4)(x+3)$ 27. $2y(3x+1)(2x-1)$

29. $(2x+3y)(x+y)$ 31. $(4x-3y)(3x-5y)$

33. $(y-x)(y+2x)$ 35. $(x-y)(x+2y-3)$

37. $(x-y)(x+3y-2)$ 39. $(a+5b)(a+b+2)$

Section 5.6

1. $(y+4)(y^2-4y+16)$ 3. $(2z+5)(4z^2-10z+25)$

5. $(3y^2+1)(9y^4-3y^2+1)$ 7. $(2x+3y)(4x^2-6xy+9y^2)$

9. $(5y^3+4x^4)(25y^6-20x^4y^3+16x^8)$ 11. $(y-1)(y^2+y+1)$

13. $(3x^2-y)(9x^4+3x^2y+y^2)$ 15. $(4z-2x^3)(16z^2+8x^3z+4x^6)$

17. $(5z^3-4x^2)(25z^6+20x^2z^3+16x^4)$

19. $(xy^2+2z^3)(x^2y^4-2xy^2z^3+4z^6)$

21. $(2x^2y-3z)(4x^4y^2+6x^2yz+9z^2)$

23. $(y+2)(y^2-2y+4)(y-2)(y^2+2y+4)$

25. $(y^3-2)(y+1)(y^2-y+1)$

27. $(z+2)(z-1)(z^2+z+1)$

29. $(x+3)(x-1)(x^2+x+1)$

Review Exercises for Chapter 5

1. $x(x+3)$ 2. $2y(2y-1)$

3. $xy(y-1)$ 4. $5xy(-2+x)$

Answers to Selected Exercises

5. $2xy(xy + 2x^2 + 3xy^2)$
6. $(x + 5)(y + 3)$
7. $(x + y)(x + z)$
8. $(x + y)(x - 3)$
9. $(2x + 3)(y - 4)$
10. $(a + b)(x + y)$
11. $3(2ax + by)$
12. $(s + u)(t + v)$
13. $(y + 3)(y - 3)$
14. $(2x + 3)(2x - 3)$
15. $(4 + y)(4 - y)$
16. $(3x + 4)(3x - 4)$
17. $(5 + 2x)(5 - 2x)$
18. $(y^5 + 2)(y^5 - 2)$
19. $(x^2 + 1)(x + 1)(x - 1)$
20. $(y^2 + 9)(y + 3)(y - 3)$
21. $(4 + x^2)(2 + x)(2 - x)$
22. $3x(3x^3 + u^2)(3x^3 - u^2)$
23. $4xy^2(x + y)(x - y)$
24. $2x^2(4x^2 + 5y)(4x^2 - 5y)$
25. $(x - y)(x + y - 2)$
26. $(x + y)(x - y + 2)$
27. $(3x + y)(3x - y + 4)$
28. $(2x + 1 + y)(2x + 1 - y)$
29. $(x + y + a + b)(x + y - a - b)$
30. $(x - 4)^2$
31. $(4y - 3x)^2$
32. $(2x^2 + 3y^2)^2$
33. $4y^2(y + 1)^2$
34. $(x^3 + y^4)^2$
35. $(x + 12)(x - 1)$
36. $(x - 9)(x - 2)$
37. $(x - 11)(x + 5)$
38. $(8 + y)(6 + y)$
39. $2x(x + 7)(x - 5)$
40. $(x^2 - 6)(x^2 + 2)$
41. $(y^2 - 3)(y + 2)(y - 2)$
42. $(3y + 1)(2y - 1)$
43. $(3z + 4)(z + 3)$
44. $(6x + 5)(x + 1)$
45. $(5z - 4)(3z - 2)$
46. $2y(3y + 4)(y + 3)$
47. $3x(5x - 1)(x - 1)$
48. $(y - x)(y + 2x)$
49. $(3x - 2y)(x + y)$
50. $(x - y)(x + 2y - 3)$
51. $(x + 2y)(x + y + 3)$
52. $(x + y)(x^2 - xy + y^2)$

53. $(2z + 5)(4z^2 - 10z + 25)$ 54. $(5y^3 + 4x^4)(25y^6 - 20x^4y^3 + 16x^8)$

55. $(2z - 3)(4z^2 + 6z + 9)$ 56. $(2y^2 - 3x^3)(4y^4 + 6x^3y^2 + 9x^6)$

57. $(x + 1)(x^2 - x + 1)(x - 1)(x^2 + x + 1)$

58. $(z + 2)(z^2 - 2z + 4)(z - 2)(z^2 + 2z + 4)$

59. $(x - 1)(x + 2)(x^2 - 2x + 4)$

60. $(z + 2)(z - 1)(z^2 + z + 1)$

Chapter 6. Rational Expressions (Algebraic Fractions)

Section 6.2

1. yes   3. yes   5. no   7. yes   9. yes

11. no   13. $\dfrac{7}{22}$ 15. $\dfrac{b}{c}$ 17. $\dfrac{2z}{3xy}$ 19. $\dfrac{-2}{3st}$

21. $\dfrac{-a}{3bc}$ 23. $\dfrac{(x - 2y)}{4xy}$ 25. $\dfrac{(1 - 2y)}{6y}$

27. $\dfrac{3(x + 2y)}{(y - 2)}$ 29. $\dfrac{(x - y)}{(x + y)}$ 31. $\dfrac{(x + 4)}{(x - 3)}$

33. $\dfrac{(x^2 + y^2)}{(x^2 - y^2)}$ 35. $-\dfrac{1}{2}$ 37. $\dfrac{(3x - 2y)}{(x + y)}$

39. $\dfrac{(x - y)}{(x + y)}$ 41. $\dfrac{(x + y)}{(-2x - y)}$ 43. $\dfrac{(x + y)}{(-2y - x)}$

45. $\dfrac{-1}{(x^2 + y)}$

Section 6.3

1. $\dfrac{-3}{16}$ 3. $\dfrac{5}{9}$ 5. $\dfrac{2}{11}$

Answers to Selected Exercises

7. $\dfrac{2x^2}{3y}$
9. $\dfrac{bd}{a^2c}$
11. s^2uv^3

13. 1
15. $\dfrac{(z-3)(z-3)}{(z-1)(z+3)}$
17. $3x + 2y$

19. $\dfrac{3y+5}{y+1}$
21. $\dfrac{(x-2)(2-x)}{(x-1)}$
23. $\dfrac{(3y+2)(4y+1)(2y+1)}{-1(y+3)(2+y)}$

25. $\dfrac{(x+y)(x-y)}{(c+d)(c-d)}$
27. $\dfrac{-3(x-3)}{(x^2+1)(x-5)}$
29. $\dfrac{3y(x+2)}{8(-x+2)}$

31. $\dfrac{3x^2}{-y^2}$
33. $\dfrac{y+1}{y+3}$
35. -1

37. $\dfrac{2y+1}{(3y+2)(y+2)}$
39. -1

Section 6.4

1. $\dfrac{4+x}{y}$
3. $\dfrac{2+y}{x}$
5. $\dfrac{2x-3}{x-2}$

7. $\dfrac{1}{x+y}$
9. $\dfrac{1}{x+3}$
11. $\dfrac{1}{x+1}$

13. $\dfrac{7}{2x}$
15. $\dfrac{1}{x(x-1)}$
17. $\dfrac{7y+2}{y^2-4}$

19. $\dfrac{x-y}{x+y}$
21. $\dfrac{-9y}{(y+2)(y-1)}$
23. $\dfrac{3yz - 2xz - 5xy}{xyz}$

25. $\dfrac{21}{8x}$
27. $\dfrac{31}{18y}$
29. $\dfrac{3y+1}{60}$

31. $\dfrac{5x^2 - x - 3}{(3x+2)(x+1)}$
33. $\dfrac{-y^2 - 7y}{(3y+1)(y-1)}$
35. $\dfrac{3x+5}{4(x-3)}$

37. $\dfrac{2x}{(x-1)(x+1)}$
39. $\dfrac{-y^2 - 5y - 12}{y^2 - 9}$
41. $\dfrac{3}{y-1}$

43. $\dfrac{2y}{x-y}$
45. $\dfrac{2}{y(y-3)(y-1)}$

47. $\dfrac{5x^2 + 5x + 2}{(x-2)(x-2)(x+2)}$
49. $\dfrac{5x+4}{(x+2)(x-1)(x-1)}$

51. $\dfrac{6y}{(y+1)(y-2)(y-2)}$ 53. $\dfrac{2x^2+3x-3}{(x-3)(x+1)(x-1)}$

55. $\dfrac{6x^2+7x-1}{(x+1)(x-1)(x+2)}$ 57. $\dfrac{-2x^2+2x-18}{(x-1)(x+2)(x-3)}$

Section 6.5

1. $\dfrac{27}{44}$ 3. $\dfrac{9}{10}$ 5. $\dfrac{258}{35}$ 7. $\dfrac{2x+1}{3x-1}$ 9. $\dfrac{x}{y}$

11. $\dfrac{3y+2x}{2y+3x}$ 13. -1 15. $\dfrac{xy}{x+y}$

17. $\dfrac{x}{2x-y}$ 19. x^2+x+1 21. $\dfrac{2x^2-y^2}{x^2-xy+x+y}$

23. 1 25. $\dfrac{(2x^2+4x-1)}{(x^2+x-6)} \cdot \dfrac{(x^2+7x+10)}{(-3x+3)}$

27. $\dfrac{6y^2-20y+16}{-4y^2+17y-15}$ 29. $\dfrac{1}{x-1}$ 31. $\dfrac{x+3}{x-1}$

33. $\dfrac{y^2+2y}{y+1}$ 35. $\dfrac{8}{5}$

Section 6.6

1. $\{6\}$ 3. $\{-15\}$ 5. $\{-23\}$

7. $\{6\}$ 9. $\{1\}$ 11. $\emptyset$

13. $\left\{\dfrac{-8}{3}\right\}$ 15. $\{2\}$ 17. $\{6\}$

19. $\{-11\}$ 21. $\left\{\dfrac{-1}{2}\right\}$ 23. $\{4\}$

25. $\emptyset$ 27. $\left\{\dfrac{-1}{5}\right\}$ 29. $\{-2\}$

31. $\{-1\}$ 33. $\emptyset$

Answers to Selected Exercises

Section 6.7

1. freight train, 30 mph
   express train, 60 mph

3. 4.5 mph

5. 16.7 mph

7. 1.2 hours

9. 7.5 hours

11. 7 days

13. $3\frac{1}{3}$ hours

15. 15 hours

17. $\frac{5}{9}$

19. $\frac{3}{8}$

21. $3\frac{1}{3}$ hours

23. Rate of boat in still water = 8 mph.
    Rate of current = 2 mph.

25. $\frac{5}{13}$

Review Exercises for Chapter 6

1. no
2. no
3. yes
4. no
5. $\frac{13}{19}$
6. $\frac{2x}{y}$
7. $\frac{2}{3xy}$
8. $\frac{2xy}{(x-3y)}$
9. $\frac{y-2}{2(x+2)}$
10. $\frac{5}{x-2y}$
11. $\frac{x^2+xy+2}{x+y}$
12. $\frac{2x-3}{2x-1}$
13. $\frac{y+2}{y-5}$
14. $\frac{-1}{3}$
15. $\frac{x-3y}{3x+y}$
16. $\frac{2x-y}{-2x-y}$
17. $\frac{4y}{5}$
18. $\frac{a^2d^2}{bc}$
19. $\frac{x-5}{2x-1}$
20. $\frac{3x(x+y)}{x-y}$
21. $\frac{1}{y+3}$
22. $\frac{(3y-1)(y+1)}{3y^2+5y+1}$
23. $\frac{4y}{y+5}$
24. $\frac{y}{3x}$
25. $\frac{3(x-3)}{2(x+1)}$
26. 1
27. $\frac{(x-y)(2x-3y)(x+y)}{(2x+y)(x-3y)}$
28. $\frac{(3x-2y)(x+y)}{(2y+x)(y-3x)}$
29. $\frac{1}{x-2}$
30. $\frac{1}{y-2}$
31. $\frac{5x^2+7x}{(x+3)(x-1)}$
32. $\frac{10x+1}{132}$
33. $\frac{x^2-7x}{(x-1)(3x+1)}$

34. $\dfrac{8x + 2}{3(x + 2)}$ 35. $\dfrac{x^2 + 3x + 3}{x^2 - 4}$ 36. $\dfrac{2y^2 + y + 1}{y(y + 1)(y + 2)}$

37. $\dfrac{3y^2 + 5y + 10}{(y + 5)(y - 5)(y - 5)}$ 38. $\dfrac{6y + 2}{(y + 2)(y + 3)(y - 1)}$

39. $\dfrac{2y^2 - 5y - 9}{(y + 1)(y - 1)(y + 2)}$ 40. $\dfrac{4x^2 - 11x + 3}{(x - 1)(x + 1)(x - 2)}$

41. $\dfrac{308}{345}$ 42. $\dfrac{xy}{y + x}$ 43. $\dfrac{x^2 + 1}{-2x}$

44. $\dfrac{2y^3 + y^2 - y - 1}{y(y^2 - 1)}$ 45. $\{12\}$ 46. $\left\{\dfrac{10}{13}\right\}$

47. $\{7\}$ 48. $\left\{-\dfrac{1}{3}\right\}$ 49. $\{4\}$

50. $\{-1\}$ 51. A = 200 mph. 52. 3.5 hours
B = 175 mph.

53. 24 hours 54. $\dfrac{1}{2}$ and 5 55. Rate of boat = 2.5 mph.
Rate of current = 0.5 mph.

Chapter 7. Complex Numbers and Quadratic Equations

Section 7.1

1. i 3. $4i$ 5. $7i$ 7. $-3i\sqrt{3}$ 9. $6i\sqrt{3}$

11. $6i\sqrt{3}$ 13. i 15. i 17. 1 19. $-8i$

21. -2 23. -16 25. -2 27. -64 29. $2i$

31. $10i$ 33. $-11i$ 35. $7i$ 37. $5 + 3i$ 39. $2\sqrt{2} - i\sqrt{6}$

41. $20i$ 43. -6 45. 30 47. $-6\sqrt{2}$ 49. $-\sqrt{10}$

51. $5i$ 53. $-6\sqrt{6}$ 55. $-4i$ 57. 2 59. $\sqrt{3}$

61. $i\sqrt{3}$ 63. $-i\sqrt{3}$ 65. $\dfrac{-i\sqrt{30}}{5}$ 67. $3i$ 69. $\dfrac{3\sqrt{6}}{2}$

Answers to Selected Exercises 665

Section 7.2

1. $5 + 8i$
3. $3 + 6i$
5. $2 + 2i$
7. $-6 - 4i$
9. 0
11. $2 - 2i$
13. $1 + 7i$
15. $-2 + 2i$
17. -4
19. $6 - 15i$
21. $-10 - 15i$
23. $4 + 17i$
25. $-4 + 7i$
27. $16 - 11i$
29. $15 + 5i$
31. $-5 + 12i$
33. $-5 + 12i$
35. $2i$
37. 2
39. 10
41. 13
43. $15 + 10i$
45. $2 - 2i$
47. $2 - 3i$
49. $\dfrac{3 - 3i}{2}$
51. $\dfrac{1 + i}{2}$
53. i
55. $\dfrac{-8 + 15i}{17}$
57. $2 + i$
59. $\dfrac{-5 - i}{13}$
61. $\dfrac{2 - i}{5}$
63. $\dfrac{-5 - 12i}{169}$
65. $\dfrac{1 + 3i}{2}$
67. $-2 + 2i$
69. $2 + 2i$

Section 7.3

1. $\{0\}$
3. $\{0\}$
5. $\{\pm 1\}$
7. $\{\pm\sqrt{2}\}$
9. $\{\pm 3\}$
11. $\{\pm\sqrt{5}\}$
13. $\{\pm i\sqrt{2}\}$
15. $\{\pm 2i\}$
17. $\left\{\pm\dfrac{\sqrt{15}}{3}\right\}$
19. $\left\{\dfrac{\pm 2i\sqrt{6}}{3}\right\}$
21. $\left\{\pm\dfrac{i\sqrt{10}}{5}\right\}$
23. $\left\{\pm\dfrac{i\sqrt{2}}{2}\right\}$
25. $\left\{0, \dfrac{2}{3}\right\}$
27. $\left\{0, \dfrac{5}{3}\right\}$
29. $\{0, -2\}$
31. $\{0, 6\}$
33. $\{0, 6\}$
35. $\left\{0, \dfrac{b}{a}\right\}$

Section 7.4

1. $\{-2, -1\}$
3. $\{-3, -2\}$
5. $\{1, 3\}$
7. $\{-3, 1\}$
9. $\{7, -3\}$
11. $\left\{1, -\frac{1}{3}\right\}$
13. $\left\{\frac{3}{2}, 2\right\}$
15. $\left\{-\frac{7}{2}, 1\right\}$
17. $\left\{\frac{1}{3}, -\frac{1}{2}\right\}$
19. $\left\{-\frac{2}{3}\right\}$
21. $\left\{\frac{4}{3}\right\}$
23. $\{\pm 2\}$
25. $\left\{\pm \frac{3}{2}\right\}$
27. $\left\{\pm \frac{1}{5}\right\}$
29. $\left\{-\frac{1}{2}, 1\right\}$
31. $\left\{\frac{1}{2}\right\}$
33. $\left\{\frac{1}{2}, 1\right\}$
35. $\{3, 1\}$
37. $\left\{\frac{4}{3}\right\}$
39. $\left\{\frac{5}{2}\right\}$
41. $\left\{\frac{1}{5}, 1\right\}$
43. $\left\{-\frac{1}{3}, 2\right\}$
45. $\{0, 2\}$
47. $\left\{\frac{2}{3}, 2\right\}$
49. $\{1\}$
51. $\{1, 8\}$
53. $x^2 - 2x - 3 = 0$.
55. $x^2 + 4x - 5 = 0$.
57. $6x^2 - 7x + 2 = 0$.
59. $35x^2 - 11x - 10 = 0$.
61. $x^2 - 2x - 1 = 0$.
63. $x^2 - 4x + 1 = 0$.
65. $x^2 - 2x + 5 = 0$.

Section 7.5

1. $\{-2, -4\}$
3. $\{-3, 1\}$
5. $\{-5, -3\}$
7. $\{6, 2\}$
9. $\{5, -2\}$
11. $\{5, 2\}$
13. $\{4, -1\}$
15. $\{-3, 1\}$
17. $\{-2, -1\}$
19. $\left\{-\frac{1}{2}, 3\right\}$
21. $\left\{\frac{-1 \pm i\sqrt{3}}{2}\right\}$
23. $\left\{\frac{-1 \pm i\sqrt{7}}{2}\right\}$
25. $\{-1 \pm i\}$
27. $\left\{\frac{1 \pm i\sqrt{71}}{6}\right\}$
29. $\left\{\frac{-3 \pm \sqrt{17}}{4}\right\}$

Answers to Selected Exercises

31. $\left\{\dfrac{1 \pm 2i\sqrt{6}}{5}\right\}$
33. $\left\{\dfrac{-5 \pm i\sqrt{59}}{6}\right\}$

Section 7.6

1. $\left\{-\dfrac{3}{2}, -1\right\}$
3. $\left\{-\dfrac{2}{3}, -\dfrac{1}{2}\right\}$
5. $\{3, 2\}$

7. $\{4, -1\}$
9. $\{-3, 2\}$
11. $\left\{\dfrac{5}{3}, -\dfrac{1}{2}\right\}$

13. $\left\{\dfrac{1}{3}, -\dfrac{1}{2}\right\}$
15. $\left\{\dfrac{-3 \pm \sqrt{41}}{8}\right\}$
17. $\left\{\dfrac{3}{5}, -2\right\}$

19. $\left\{\dfrac{-1 \pm i\sqrt{3}}{2}\right\}$
21. $\left\{1, -\dfrac{1}{2}\right\}$
23. $\left\{\dfrac{-3 \pm \sqrt{13}}{2}\right\}$

25. $\{1 \pm \sqrt{3}\}$
27. $\{1 \pm \sqrt{6}\}$
29. $\left\{\dfrac{-1 \pm i\sqrt{7}}{2}\right\}$

31. $\{-1 \pm i\}$
33. $\left\{\dfrac{1 \pm i\sqrt{71}}{6}\right\}$
35. $\left\{\dfrac{-3 \pm \sqrt{17}}{4}\right\}$

37. $\left\{\dfrac{1 \pm 2i\sqrt{6}}{5}\right\}$
39. $\left\{\dfrac{-5 \pm i\sqrt{59}}{6}\right\}$
41. $\emptyset$

43. $\{2, 1\}$
45. $\emptyset$
47. $\left\{\dfrac{3 + \sqrt{5}}{2}\right\}$

49. $\{-1\}$
51. $\{0\}$

Section 7.7

1. $\{-11, -10\}$ or $\{10, 11\}$
3. $\{-15, -13\}$ or $\{13, 15\}$

5. $\{-11, -9\}$ or $\{9, 11\}$
7. length = 11 metres
   width = 4 metres

9. length = 10 inches
   width = 10 inches
   height = 2 inches
11. length = 20 inches
    width = 8 inches
    height = 2 inches

13. length = 8 centimetres
    width = 3.75 centimetres
15. 8 centimetres and
    15 centimetres

17. Larry = 32 mph
    Benny = 40 mph
19. 38 mph

21. 21 hours
23. 30 minutes
25. Pam = 30 minutes
    Angie = 40 minutes

Section 7.8

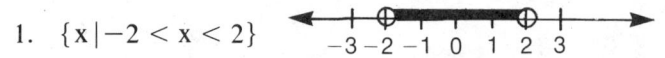

1. $\{x \mid -2 < x < 2\}$
3. $\{x \mid -1 < x < 1\}$
5. $\{x \mid -3 < x < 3\}$
7. $\emptyset$ 9. $\emptyset$
11. $\{x \mid x \leq -5 \text{ or } x \geq 1\}$
13. $\{x \mid x \text{ is a real number}\}$
15. $\{x \mid x \text{ is a real number}\}$
17. $\emptyset$
19. $\{x \mid -2 \leq x \leq -1\}$
21. $\{x \mid x \leq -2 \text{ or } x \geq 1\}$
23. $\{x \mid -1 < x < 4\}$
25. $\{x \mid x < -4 \text{ or } x > 1\}$
27. $\{x \mid x \leq -5 \text{ or } x \geq -2\}$
29. $\{x \mid -5 \leq x \leq 2\}$
31. $\left\{x \mid -\dfrac{1}{2} \leq x \leq 2\right\}$
33. $\left\{x \mid -\dfrac{1}{3} < x < \dfrac{1}{2}\right\}$
35. $\left\{x \mid -\dfrac{4}{5} \leq x \leq 1\right\}$

Answers to Selected Exercises

Review Exercises for Chapter 7

1. $2i$
2. $5i$
3. $2i\sqrt{2}$
4. $-3i\sqrt{2}$
5. $6i\sqrt{5}$
6. $15i\sqrt{5}$
7. i
8. i
9. i
10. 1
11. $-32i$
12. -2
13. $3i$
14. $4i$
15. $-6\sqrt{3}$
16. $3i$
17. $-6i\sqrt{6}$
18. $-6i\sqrt{2}$
19. $-i\sqrt{6}$
20. $\dfrac{\sqrt{10}}{2}$
21. $\dfrac{3i\sqrt{10}}{2}$
22. $-6i$
23. $-20+5i$
24. 0
25. $15+6i$
26. $8-i$
27. $-16-30i$
28. i
29. -1
30. $\dfrac{5-i}{2}$
31. $\left\{\pm\dfrac{\sqrt{3}}{3}\right\}$
32. $\{\pm 1\}$
33. $\{0, 4\}$
34. $\left\{\pm\dfrac{\sqrt{3}}{3}\right\}$
35. $\{0, -3\}$
36. $\left\{0, -\dfrac{b}{a}\right\}$
37. $\left\{-\dfrac{1}{2}, 1\right\}$
38. $\left\{-\dfrac{1}{4}, -1\right\}$
39. $\left\{\dfrac{1}{5}, 1\right\}$
40. $\left\{\dfrac{2}{3}, 3\right\}$
41. $\left\{\dfrac{3}{2}, 2\right\}$
42. $\left\{\pm\dfrac{\sqrt{34}}{2}\right\}$
43. $\{5, -3\}$
44. $\{-5, 2\}$
45. $\left\{\dfrac{2}{3}, -1\right\}$
46. $\left\{\dfrac{-1\pm i\sqrt{15}}{4}\right\}$
47. $\left\{\dfrac{1\pm i\sqrt{11}}{6}\right\}$
48. $\left\{\dfrac{3\pm i\sqrt{119}}{8}\right\}$
49. $\left\{-\dfrac{2}{3}, 2\right\}$
50. $\left\{\dfrac{1}{4}, \dfrac{1}{2}\right\}$
51. $\left\{\dfrac{1\pm\sqrt{5}}{2}\right\}$
52. $\{-1\pm\sqrt{3}\}$
53. $\left\{\dfrac{-3\pm i\sqrt{3}}{2}\right\}$
54. $\left\{\dfrac{3\pm i\sqrt{119}}{8}\right\}$
55. $\left\{\dfrac{3\pm 3i\sqrt{7}}{4}\right\}$
56. $\{1\}$
57. $\{0, 1\}$
58. $\{x \mid -3 < x < 3\}$

59. {x | x is a real number}

60. {x | −2 < x < 1}

61. {x | −1 < x < 2}

62. $\left\{x \mid -\dfrac{3}{2} < x < \dfrac{2}{3}\right\}$

63. $\left\{x \mid x \leq -\dfrac{5}{2} \text{ or } x \geq -\dfrac{7}{3}\right\}$

64. {−9, −7} or {7, 9}

65. {−9, −8} or {8, 9}

66. length = 21 inches
    width = 9 inches

67. 42 mph

68. 21 hours

69. 5 centimetres and 12 centimetres

Chapter 8. Relations and Functions

Section 8.1

1. {(1, 4), (1, 5), (1, 6), (2, 4), (2, 5), (2, 6), (3, 4), (3, 5), (3, 6)}

3. a. {(0, 2), (0, 4), (0, 6), (2, 4), (2, 6), (4, 6)}
   b. {(6, 4), (6, 2), (6, 0), (4, 2), (4, 0), (2, 0)}
   c. {(0, 0), (2, 2), (4, 4), (6, 6)}
   d. {(0, 2), (0, 4), (0, 6), (6, 2), (4, 2), (2,2), (4, 4), (6, 6)}
   e. {(2, 0), (4, 0), (6, 0), (2, 6), (2, 4), (2, 2), (4, 4), (6, 6)}

5. {(1, 3), (1, 5), (1, 7), (3, 5), (3, 7), (5, 7)}
   a. {1, 3, 5}
   b. {3, 5, 7}
   c. {(3, 1), (5, 1), (7, 1), (5, 3), (7, 3), (7, 5)}

Answers to Selected Exercises

7. {(−2, −1), (−2, 0), (−2, 1), (−2, 2), (−1, 0), (−1, 1), (−1, 2), (0, 1), (0, 2), (1, 2)}
   a. {−2, −1, 0, 1}
   b. {−1, 0, 1, 2}
   c. {(−1, −2), (0, −2), (1, −2), (2, −2), (0, −1), (1, −1), (2, −1), (1, 0), (2, 0), (2, 1)}
   d. {−1, 0, 1, 2}
   e. {−2, −1, 0, 1}

9. {(−2, −1), (−2, 1), (−2, 2), (−1, 1), (0, −2), (0, −1), (0, 1), (0, 2), (1, −1), (2, 1), (2, −1), (2, −2)(−2, −2), (−1, −1), (1, 1), (2, 2)}
   a. {−2, −1, 0, 1, 2}
   b. {−2, −1, 1, 2}
   c. {(−1, −2), (1, −2), (2, −2), (1, −1), (−2, 0), (−1, 0), (1, 0), (2, 0), (−1, 1), (1, 2), (−1, 2), (−2, 2)(−2, −2), (−1, −1), (1, 1), (2, 2)}
   d. {−2, −1, 1, 2}
   e. {−2, −1, 0, 1, 2}

Section 8.2

1.

3. A(2, 3), B(−3, 2), C(−4, −3), D(1, −3), E(0, 0)

5. a. yes   b. yes   c. yes   d. no   e. no   f. no

7. a. yes   b. no   c. yes   d. no
   e. yes   f. no   g. yes   h. yes

9. a. $x = \dfrac{y-5}{2}$ b. $r = \sqrt{\dfrac{A}{\pi}}$. c. $s = -16t^2 + 120t$.

   d. $F = \dfrac{9}{5}C + 32$. e. $x = \sqrt{\dfrac{y-1}{2}}$. f. $r = \dfrac{C}{2\pi}$.

11. a. $f(1) = -3$. b. $f(-1) = -3$.
    c. $f(2) = 0$. d. $f(0) = -4$.

13. a. $s(1) = 104$. b. $s(-1) = -136$.
    c. $s(2) = 176$. d. $s(-2) = -304$.

15. a. $g(2) = 0$. b. $g(-2) = -16$.
    c. $g(1) = -1$. d. $g(-a) = -a^3 - 2a^2$.

17. a. $g(3) = 9$. b. $g(-2) = 14$.
    c. $g(a) = 2a^2 - 3a$. d. $g(a+1) = 2a^2 + a - 1$.

19. a. $s(-1) = -34$. b. $s(2) = 44$.
    c. $s(a) = 30a - 4a^2$. d. $s(a+1) = -4a^2 + 22a + 26$.

Section 8.3

1.

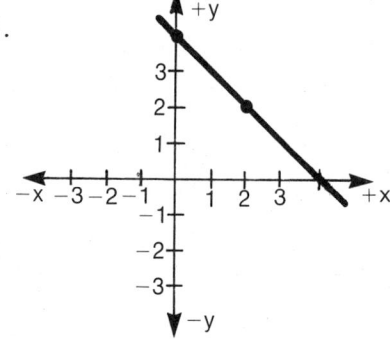

3.

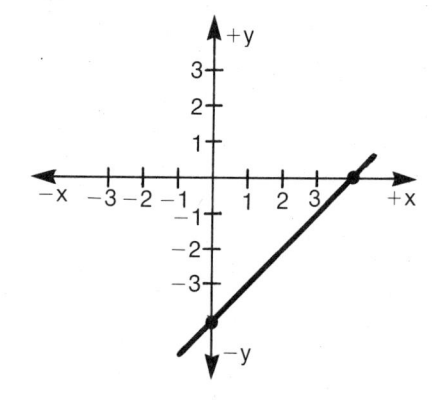

5.

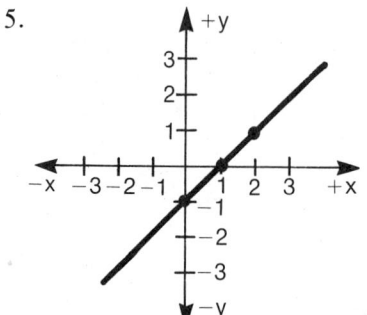

7.

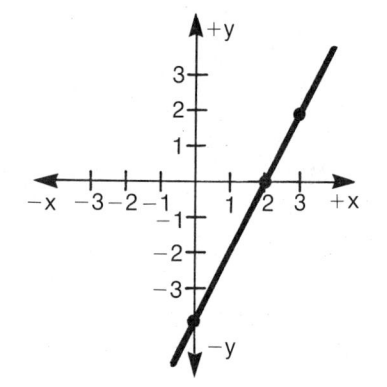

Answers to Selected Exercises

9.

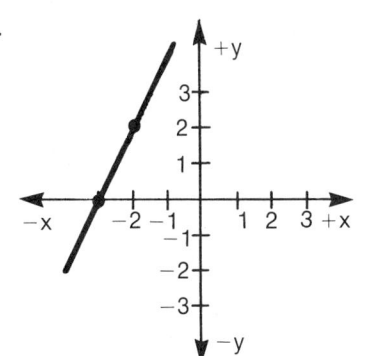

11.

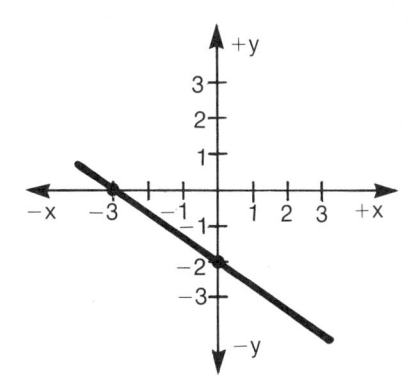

13.

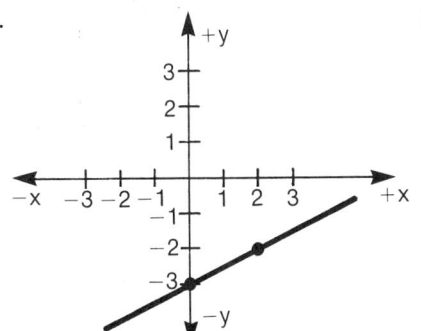

15.

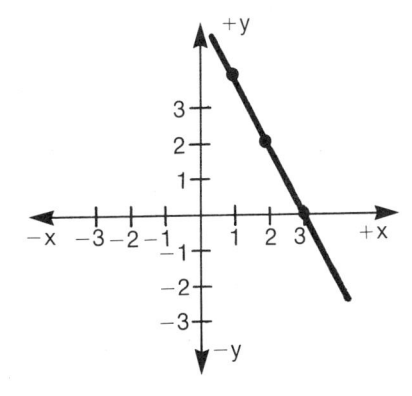

17.

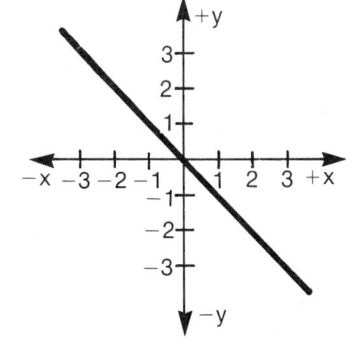

19.

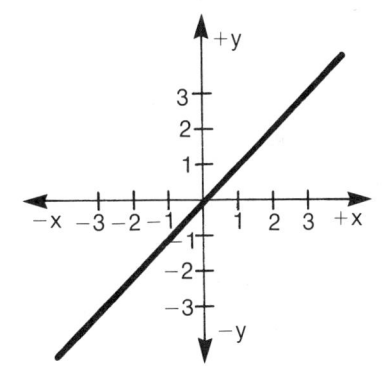

21.

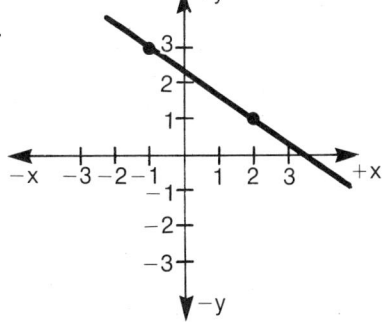

23.

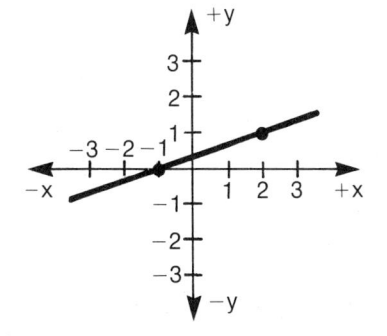

25.

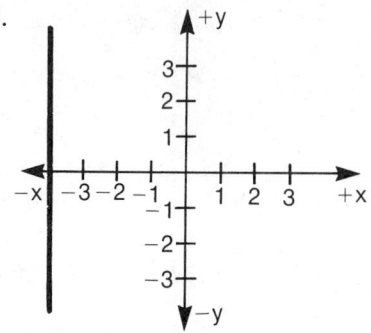

27.

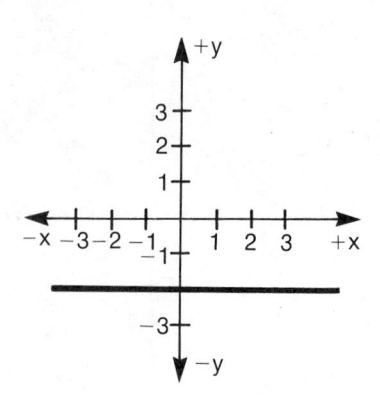

29.

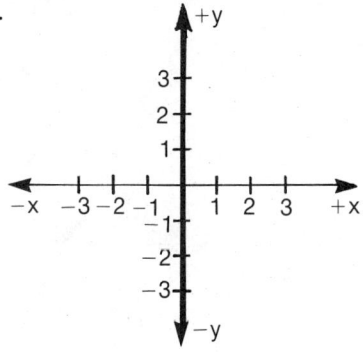

Section 8.4

1. -2

3. 5

5. $\dfrac{1}{4}$

7. undefined

9. 0

11. 3

13. yes

15. no

17. yes

19. no

21. $m\overline{ST} = \dfrac{b}{a}$, $m\overline{UV} = \dfrac{b}{a}$; the two sides are parallel.

23. a. $m\overline{PQ} = \dfrac{5}{x}$, $m\overline{SR} = \dfrac{2}{(7-x)}$.

  b. $\dfrac{5}{x} = \dfrac{2}{(7-x)}$; $x = 5$.

25. $m\overline{AB} = 3$, $m\overline{BC} = \dfrac{4}{3}$, $m\overline{CA} = -\dfrac{1}{3}$. The slope of AB is the negative reciprocal of AC, therefore the sides are perpendicular. If two sides of a triangle are perpendicular then the triangle is a right triangle.

Answers to Selected Exercises

Section 8.5

1. $y = 3x - 10$.
3. $y = -x - 7$.
5. $3y = 2x + 6$.
7. $5y = -2x - 26$.
9. $3y = -2x - 9$.
11. $y = 5x - 13$.
13. $y = x - 1$.
15. $y = -x + 2$.
17. $5y = -x - 12$.
19. $2y = 5x + 6$.

21. $m = 2$, y-intercept $= -1$.

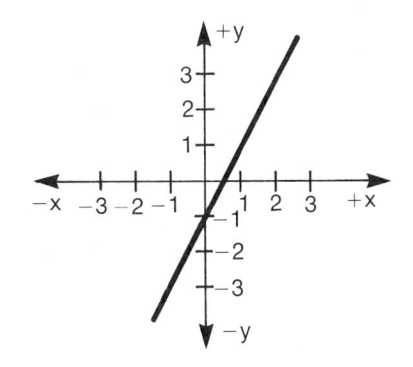

23. $m = 1$, y-intercept $= 2$.

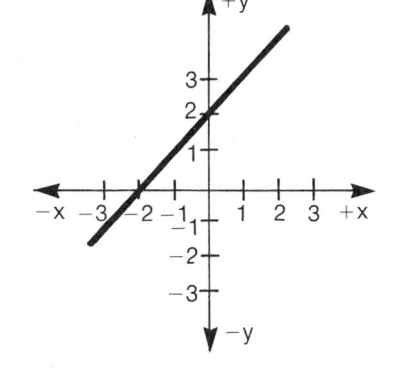

25. $m = -\dfrac{1}{2}$, y-intercept $= 2$.

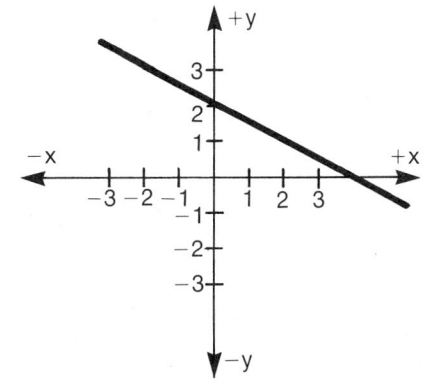

27. $m = \frac{4}{3}$, y-intercept = 1.

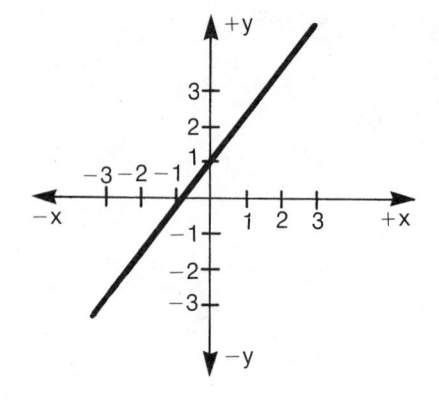

29. $m = -1$, y-intercept = 0.

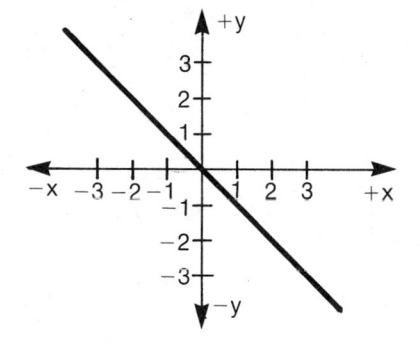

31. $3y = -2x + 5$.

33. $2y = -x - 11$.

35. $y = 2x + 1$.

37. a. $m\overline{AB} = 0$.
    b. $y = 2$.
    c. $3y = -4x$.
    d. $3y = 4x$.

39. a. $y = x + 4$.   b. $y = -x$.   c. $y = 7x - 8$.

Section 8.6

1. $\sqrt{13}$ 3. $\sqrt{45}$ 5. $\sqrt{137}$

7. $\sqrt{29}$ 9. $\sqrt{5}$ 11. $\overline{PR} = 17$.

13. a. $\overline{AC} = \sqrt{8}$, $\overline{CB} = \sqrt{40}$, $\overline{BA} = \sqrt{32}$.
    b. If the square of the length of one side of a triangle equals the sum of the squares of the lengths of the other two sides, then the triangle is a right triangle.

Answers to Selected Exercises 677

We have $(\sqrt{40})^2 = (\sqrt{32})^2 + (\sqrt{8})^2$.

$40 = 32 + 8$.
$40 = 40$. Triangle ABC is a right triangle.

15. a. $\overline{PQ} = \sqrt{20}$, $\overline{QR} = \sqrt{20}$, $\overline{RS} = \sqrt{20}$, $\overline{SP} = \sqrt{20}$.

b. $m\overline{PQ} = \dfrac{1}{2}$, $m\overline{QR} = 2$, $m\overline{RS} = \dfrac{1}{2}$, $m\overline{SP} = 2$.

c. true

17. (5, 3) 19. (−5, −2) 21. (−4, 2)

23. $\left(\dfrac{9}{2}, \dfrac{3}{2}\right)$ 25. $\left(\dfrac{5}{2}, -\dfrac{9}{2}\right)$ 26. (2a, o)

27. (−1, 1)

29. midpoint $\overline{AB} = \left(\dfrac{3}{2}, \dfrac{3}{2}\right)$. midpoint $\overline{CD} = \left(-\dfrac{9}{2}, \dfrac{3}{2}\right)$.

midpoint $\overline{BC} = (0, 3)$. midpoint $\overline{DA} = (-3, 0)$.

31. a. midpoint $\overline{RS} = (r, s)$.
midpoint $\overline{RT} = (t, u)$.

b. $m = \dfrac{s - u}{r - t}$. c. Yes, they have equal slopes.

Section 8.7

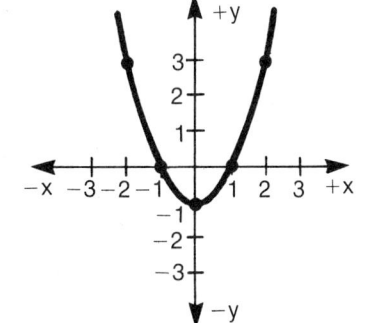

1. Axis of symmetry, $x = 0$;
vertex $= (0, -1)$;
zeros $= (1, 0), (-1, 0)$.

3. Axis of symmetry, x = 0;
   vertex = (0, 4);
   zeros = (2, 0), (−2, 0).

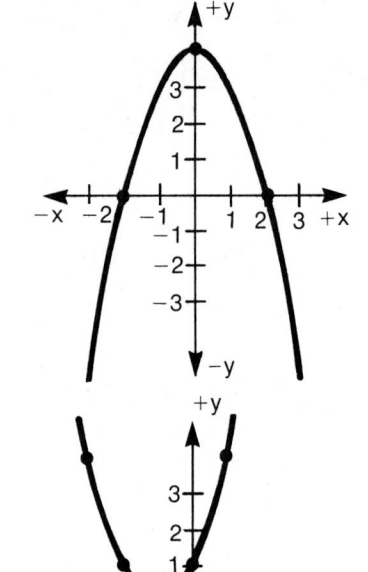

5. Axis of symmetry, x = −1;
   vertex = (−1, 0);
   zero = (−1, 0).

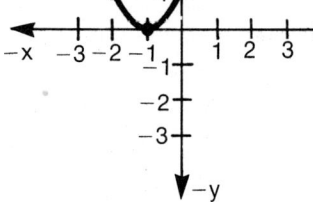

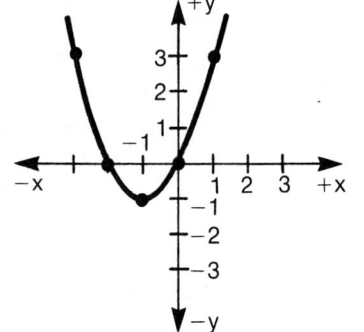

7. Axis of symmetry, x = 2;
   vertex = (2, 4);
   zeros = (0, 0), (4, 0).

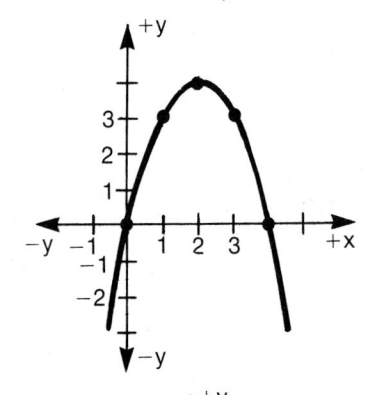

9. Axis of symmetry, x = −1;
   vertex = (−1, −1);
   zeros = (0, 0), (−2, 0).

Answers to Selected Exercises 679

11. Axis of symmetry, $x = 2$;
    vertex $= (2, -1)$;
    zeros $= (3, 0), (1, 0)$.

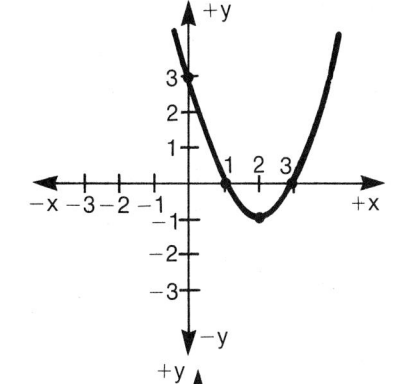

13. Axis of symmetry, $x = 3$;
    vertex $= (3, 0)$;
    zero $= (3, 0)$.

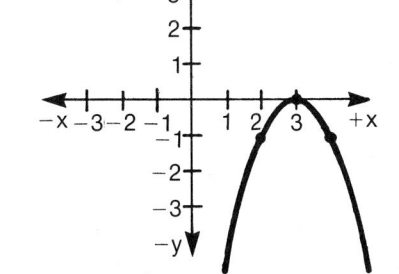

15. Axis of symmetry, $x = 3$;
    vertex $= (3, -4)$;
    zeros $= (5, 0), (1, 0)$.

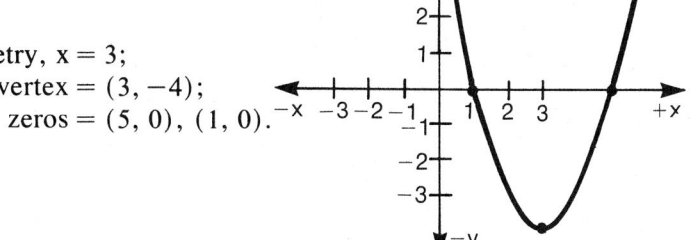

17. Axis of symmetry, $x = -1$;
    vertex $= (-1, -8)$;
    zeros $= (-3, 0), (1, 0)$.

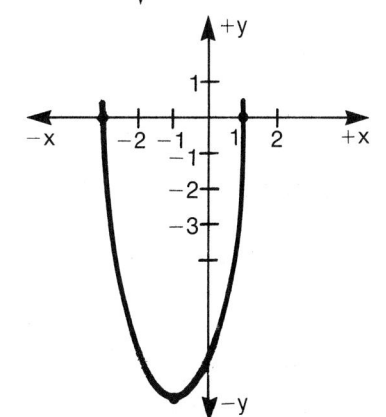

19. Axis of symmetry, $x = -\dfrac{3}{2}$;

   vertex $= \left(-\dfrac{3}{2}, -\dfrac{1}{4}\right)$;

   zeros $= (-2, 0), (-1, 0)$.

21. Axis of symmetry, $x = -\dfrac{5}{2}$;

   vertex $= \left(-\dfrac{5}{2}, -\dfrac{1}{4}\right)$;

   zeros $= (-2, 0), (-3, 0)$.

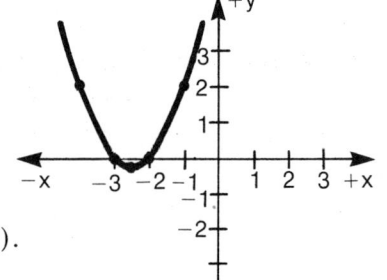

23. Axis of symmetry, $x = -\dfrac{1}{2}$;

   vertex $= \left(-\dfrac{1}{2}, \dfrac{3}{4}\right)$;

   zeros $= \emptyset$.

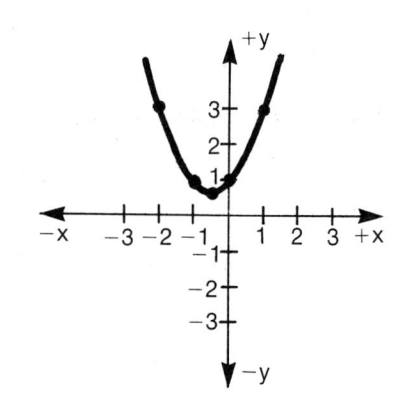

Section 8.8

1. $y = 12$. 3. $x = 28$. 5. $y = 400$. 7. $y = 180$. 9. $s = -10$.

11. $x = 5$. 13. $t = 20$. 15. $z = 6$. 17. $y = 9$. 19. $x = \dfrac{4}{3}$.

21. $y = 75$. 23. $i = 12$. 25. $q = 5$. 27. $v = 42$. 29. $p = 770$.

Answers to Selected Exercises 681

Section 8.9

1. $x^2 + y^2 = 1$.
3. $x^2 + y^2 = 25$.
5. $(x - 2)^2 + (y - 3)^2 = 1$.
7. $(x + 2)^2 + (y - 3)^2 = 4$.
9. $(x + 3)^2 + (y + 1)^2 = 9$.
11. Center at $(0, 0)$, radius $= 3$.
13. Center at $(0, 0)$, radius $= 1$.
15. Center at $(1, 0)$, radius $= 2$.
17. Center at $(2, 3)$, radius $= 2$.
19. Center at $(-3, -1)$, radius $= 1$.

21.

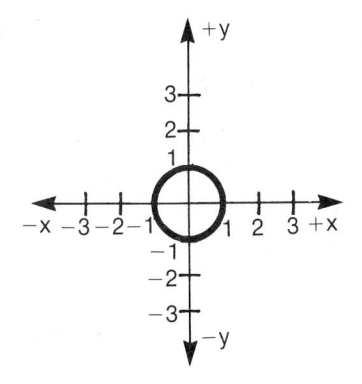

23.

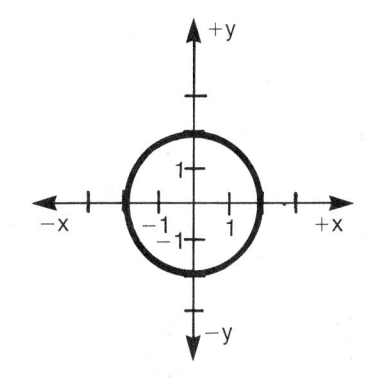

25.

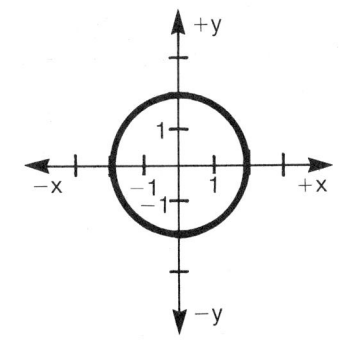

27.

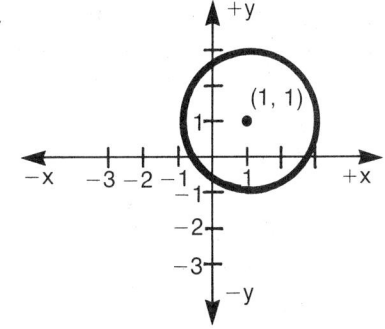

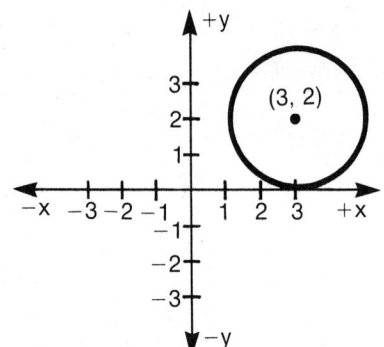

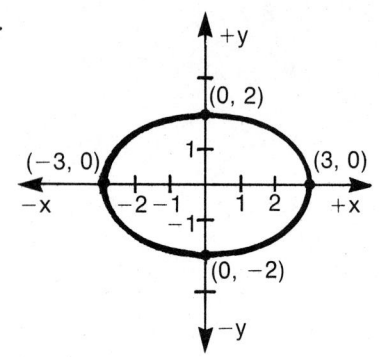

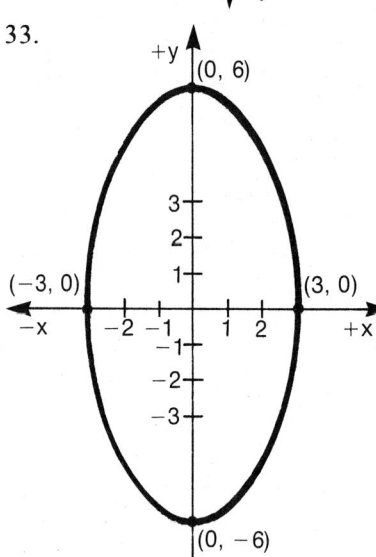

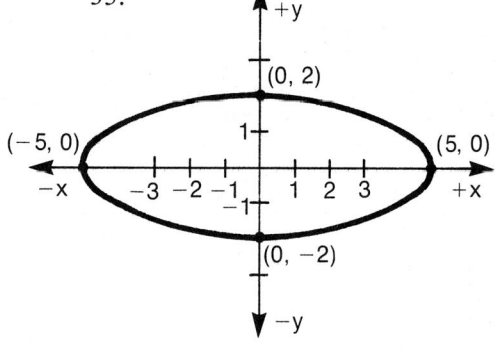

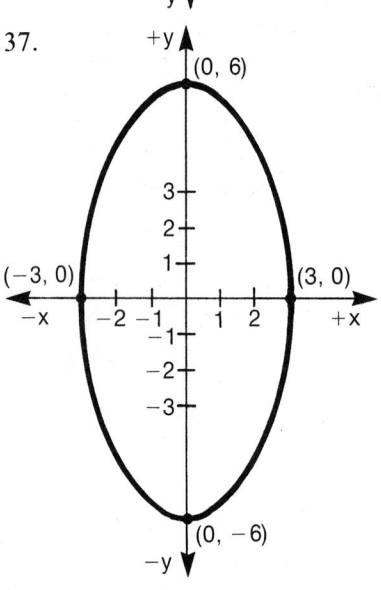

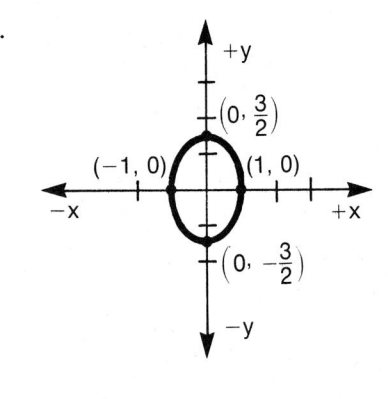

29.

31.

33.

35.

37.

39.

Answers to Selected Exercises

41. Graph showing hyperbola with vertices at $(0, -3)$ and $(3, 0)$.

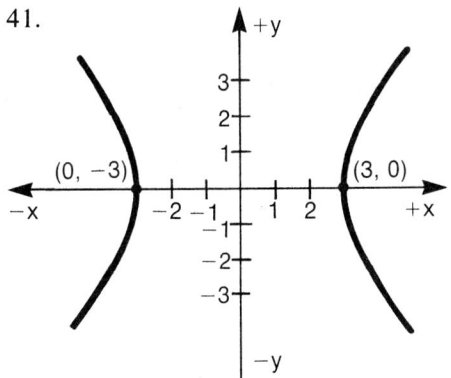

43. Graph showing hyperbola with vertices at $(-\sqrt{2}, 0)$ and $(\sqrt{2}, 0)$.

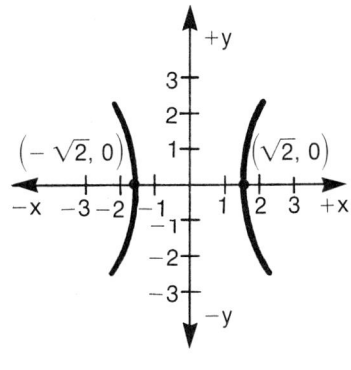

45. Graph showing curves with points $(0, 5)$ and $(0, -5)$.

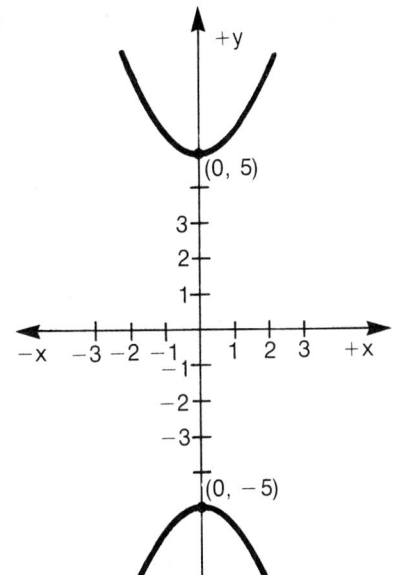

47. Graph showing curves through $(-2, 2)$ and $(2, -2)$.

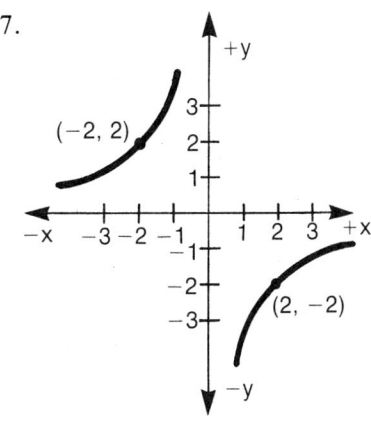

49. Graph showing three lines through the origin.

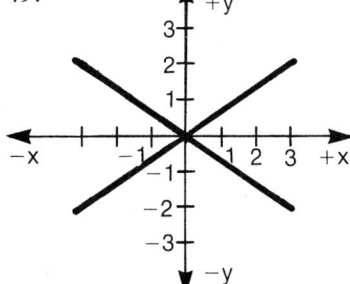

51. circle **53.** hyperbola **55.** ellipse **57.** circle **59.** parabola

Review Exercises for Chapter 8

1. R = {(a, b), (a, c), (a, d), (a, f), (a, g), (a, h), (e, b), (e, c), (e, d), (e, f), (e, g), (e, h), (i, b), (i, c), (i, d), (i, f), (i, g), (i, h)}
    a. {a, e, i}
    b. {b, c, d, f, g, h}
    c. {(b, a), (c, a), (d, a), (f, a), (g, a), (h, a), (b, e), (c, e), (d, e), (f, e), (g, e), (h, e), (b, i), (c, i), (d, i), (f, i), (g, i), (h, i)}
    d. {b, c, d, f, g, h}   e. {a, e, i}

2. a. yes   b. no   c. yes   d. no

3. a. $f(0) = 2$.   b. $f(-1) = 7$.   c. $f(1) = 1$.
    d. $f(-2) = 16$.   e. $f(a) = 2a^2 - 3a + 2$.
    f. $f(a + b) = 2a^2 + 4ab + 2b^2 - 3a - 3b + 2$.

4. a. $m\overline{PQ} = \dfrac{b}{2a}$, $m\overline{RS} = \dfrac{b}{2a}$.
    b. Side PQ is parallel to side RS.
    c. No, slopes of sides are not equal.

5. a. $y = -2x - 1$.   b. $2y = x + 17$.
    c. $2y = 3x - 4$.   d. $2y = -3x - 19$.

6. a. $y = -x + 1$.   b. $2y = -3x - 6$.
    c. $y = 9x - 47$.   d. $3y = 2x + 5$.

7. a. $m = -2$, y-intercept = $(0, 3)$.
    b. $m = -1$, y-intercept = $(0, 1)$.
    c. $m = \dfrac{3}{2}$, y-intercept = $(0, -3)$.
    d. $m = \dfrac{1}{2}$, y-intercept = $(0, 2)$.

8. a. $m\overline{PQ} = 1$, $m\overline{PR} = 1$.
    b. They are collinear.
    c. $y = x - 2$.
    d. $(2, 0)$.

9. a. midpoint of $\overline{AC} = (2a, 0)$.   b. $\sqrt{a^2 + b^2}$
    midpoint of $\overline{CB} = (3a, b)$. $\sqrt{a^2 + b^2}$
    midpoint of $\overline{AB} = (a, b)$. $2a$

10. $x = 1$.   11. $t = 14$.   12. $y = 72$.   13. $y = 125$.

Answers to Selected Exercises   685

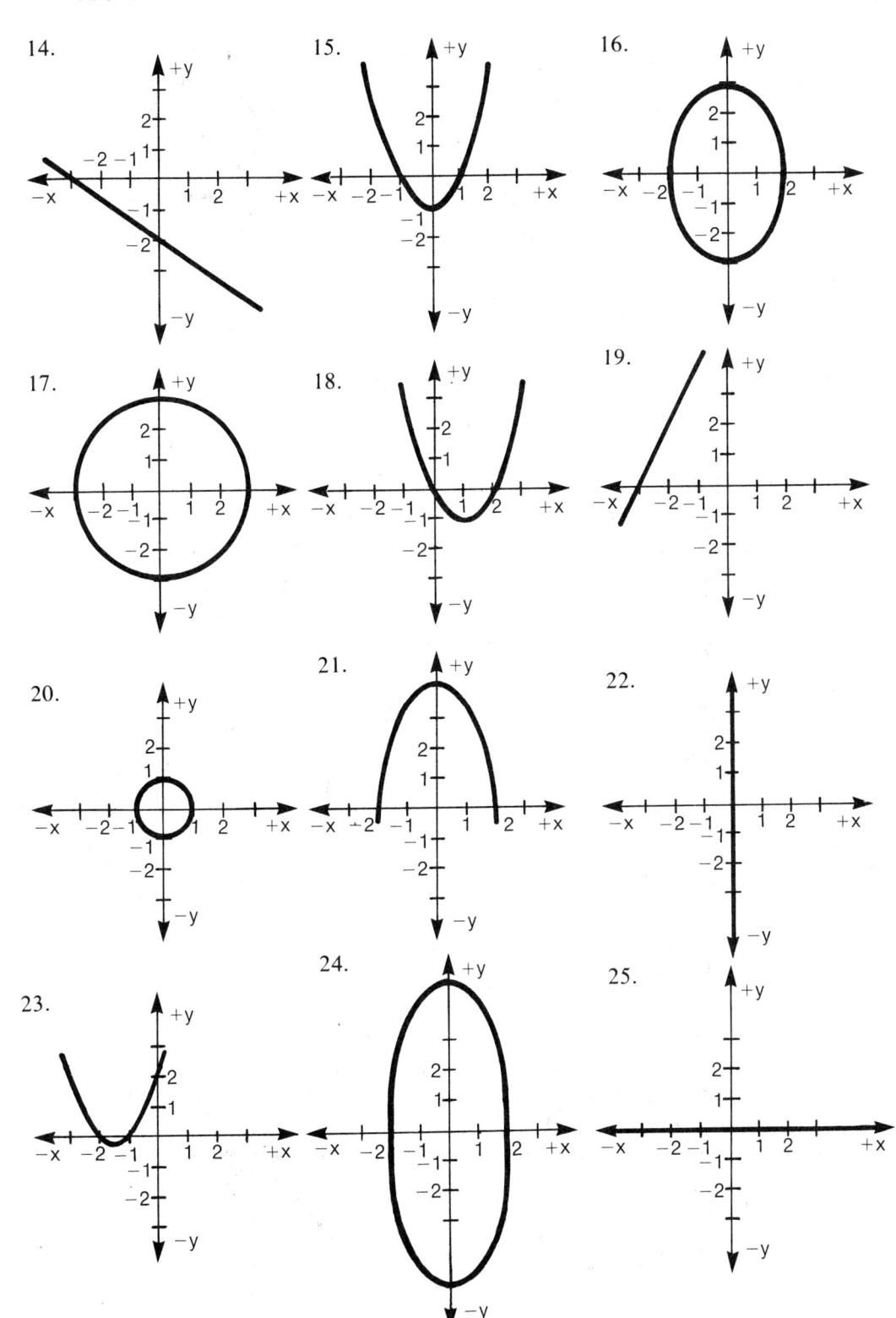

14.
15.
16.
17.
18.
19.
20.
21.
22.
23.
24.
25.

686 Appendix B

26.

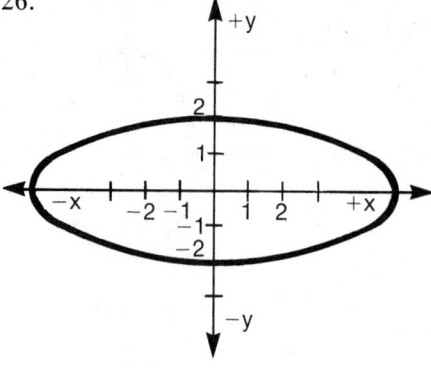

27.

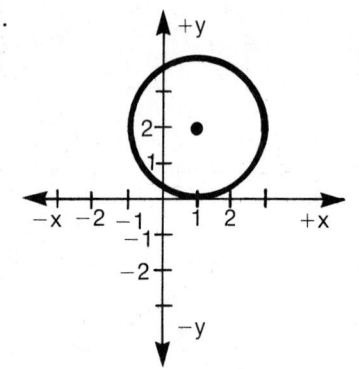

28.

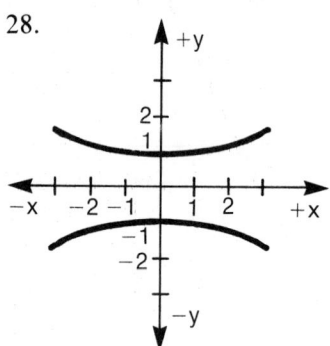

29.

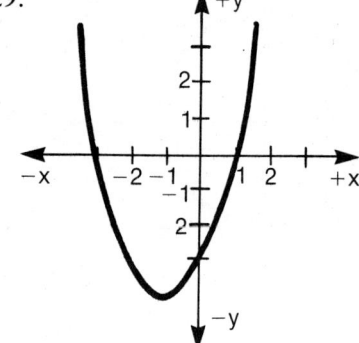

30.

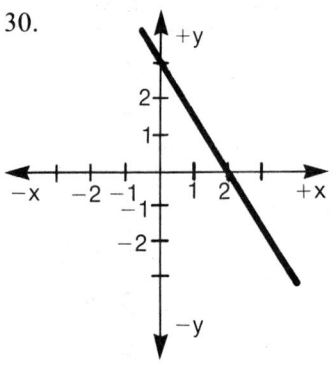

31.

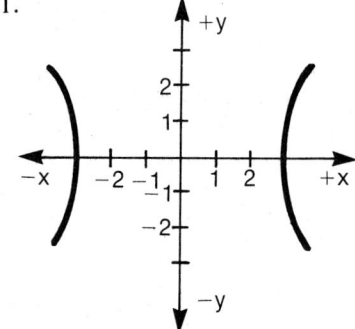

32.

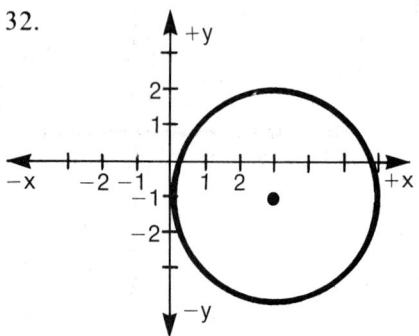

33.

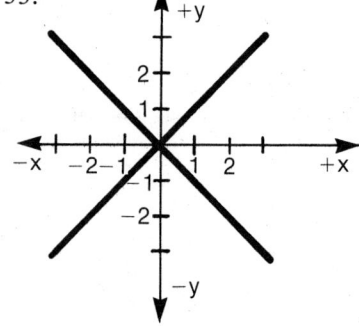

Answers to Selected Exercises 687

34.

35.

36.

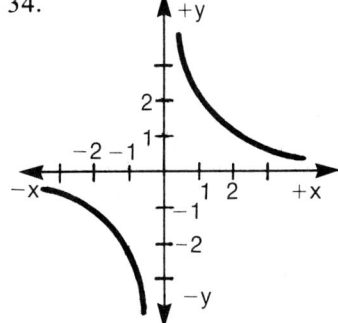

37.

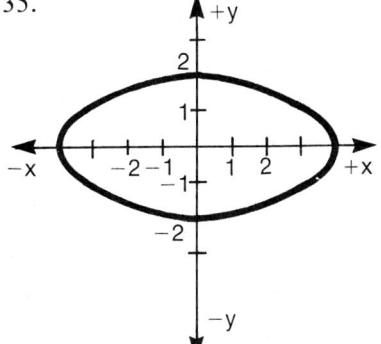

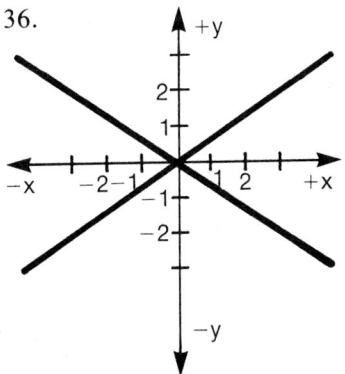

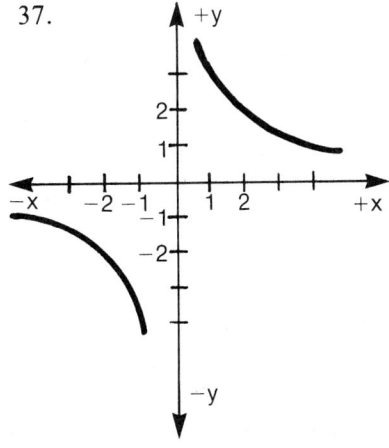

Cumulative Review II (Chapters 5-8)

1. $2y(2y - 1)$
2. $(x + 5)(y + 3)$
3. $(x + y)(x - 3)$
4. $(a - b)(x + y)$
5. $3(2ax + by)$
6. $(s + u)(t + v)$
7. $(2x + 3)(2x - 3)$
8. $(y^5 + 2)(y^5 - 2)$
9. $(y^2 + 9)(y + 3)(y - 3)$
10. $(x - y)(x + y - 2)$
11. $4y^2(y + 1)(y + 1)$
12. $(x + 12)(x - 1)$
13. $(y^2 - 3)(y + 2)(y - 2)$
14. $(6x + 5)(x + 1)$
15. $2x(3x + 4)(x + 3)$
16. $(x + y)(x^2 - xy + y^2)$
17. $(x + 1)(x^2 - x + 1)(x - 1)(x^2 + x + 1)$

18. $(x + 2)(x^2 - 2x + 4)(x - 1)$ 19. $\dfrac{(y + 2)(y - 5)}{(2y + 1)(y + 1)}$

20. $\dfrac{y}{3x}$ 21. 1 22. $\dfrac{1}{y - 2}$

23. $\dfrac{-y^2 - 7y}{(3y + 1)(y - 1)}$ 24. $\dfrac{4y^2 - 11y + 3}{(y - 1)(y + 1)(y - 2)}$

25. $\dfrac{y + x}{xy}$ 26. $\dfrac{y^2 + 1}{-2y}$ 27. $\{12\}$

28. $\left\{-\dfrac{1}{3}\right\}$ 29. $\{4\}$ 30. $\{-1\}$

31. $5i$ 32. $-6\sqrt{3}$ 33. 6

34. $-1 - i$ 35. 6 36. $-15 + 12i$

37. $18 - i$ 38. $\dfrac{-2 - 16i}{13}$ 39. $\dfrac{5 + 12i}{13}$

40. $\left\{-\dfrac{2}{3}, 2\right\}$ 41. $\left\{\dfrac{1}{4}, \dfrac{1}{2}\right\}$ 42. $\left\{\dfrac{3 \pm i\sqrt{119}}{8}\right\}$

43. $\left\{\dfrac{3 \pm 3i\sqrt{7}}{4}\right\}$ 44. $\{0, 1\}$ 45. $\{1\}$

46. $\left\{x \,\Big|\, x < -\dfrac{3}{2} \text{ or } x > \dfrac{2}{3}\right\}$

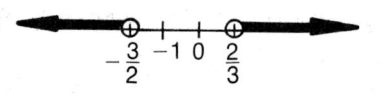

47. 9 hours   48. 15 hours

49. length = 21 inches   50. 9 hours
    width = 9 inches

Chapter 9. Systems of Equations

Section 9.1

1. $(3, 2)$ 3. $(-1, 0)$ 5. $(-1, 2)$

Answers to Selected Exercises

7. ∅
9. (5, −1)
11. ∅
13. dependent
15. (−3, 2)
17. (2, 1)
19. ∅
21. (−2, −3)

Section 9.2

1. (2, −1)
3. (4, −2)
5. (1, −3)
7. ∅
9. (5, −1)
11. ∅
13. dependent
15. (0, 3)
17. (1, 0)
19. $\left(-2, \frac{1}{2}\right)$
21. $\left(-\frac{1}{2}, \frac{7}{2}\right)$
23. $\left(-\frac{1}{2}, 2\right)$

Section 9.3

1. (1, −3)
3. (5, −1)
5. (4, −2)
7. dependent
9. (0, 3)
11. $\left(\frac{31}{5}, \frac{3}{5}\right)$
13. (5, −2)
15. (−3, −8)
17. (2, 0)
19. (4, −1)
21. (−3, −3)
23. (−2, 5)
25. (−2, 1)
27. (1, 3)
29. (23, 6)
31. (2, 1)
33. (2, 1)

Section 9.4

1. {5, 7}
3. {10, 6}
5. {15, 10}
7. 30 pounds of hard candy
   70 pounds of chocolate candy
9. 15 gallons of paint
   35 gallons of thinner
11. 20 litres
13. 87
15. 37
17. 28
19. 48
21. 9.1 mph
23. rate of boat = 10 mph
    rate of current = 2 mph
25. $\frac{11}{15}$

Section 9.5

1. $(-2, 1, -3)$
3. $(1, 0, -2)$
5. $(1, 2, 3)$
7. $(-1, -2, -3)$
9. dependent
11. $\left(\dfrac{5}{3}, \dfrac{7}{12}, -\dfrac{39}{18}\right)$
13. $\left(\dfrac{2}{3}, -3, 3\right)$
15. $(1, 2, 3)$
17. $\left(\dfrac{1}{2}, -\dfrac{1}{3}, \dfrac{1}{4}\right)$
19. $(1, 5, 11)$
21. m ∠ A = 60°
    m ∠ B = 30°
    m ∠ C = 90°
23. 30 pennies
    15 nickels
    17 dimes
25. 532

Section 9.6

1. 4
3. -1
5. 1
7. 25
9. -2
11. 0
13. 19
15. 28
17. 0
19. 6
21. -10
23. 0
25. $-\dfrac{3}{4}$
27. $\dfrac{9}{7}$
29. $\dfrac{6}{-30}$

Section 9.7

1. $(2, -1)$
3. $(4, -2)$
5. $(-2, -3)$
7. $\left(-\dfrac{1}{2}, \dfrac{7}{2}\right)$
9. dependent
11. $(1, 0)$
13. $(1, 0, -2)$
15. $(1, 2, 3)$
17. dependent
19. $\left(\dfrac{2}{3}, -3, 3\right)$
21. $(1, 2, 3)$
23. $\left(\dfrac{1}{3}, \dfrac{1}{4}, \dfrac{1}{5}\right)$

Answers to Selected Exercises

Section 9.8

1. 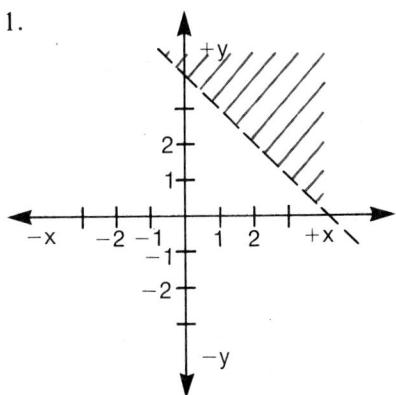
3. 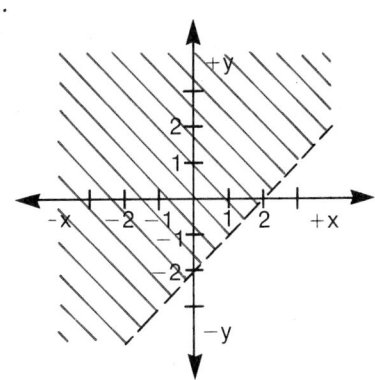
5. 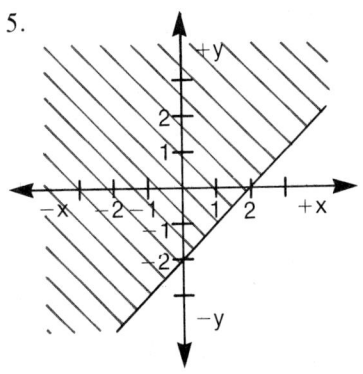
7. 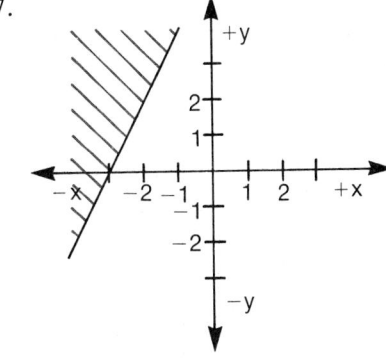
9. 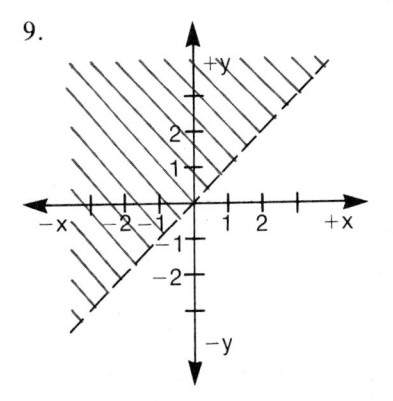
11.

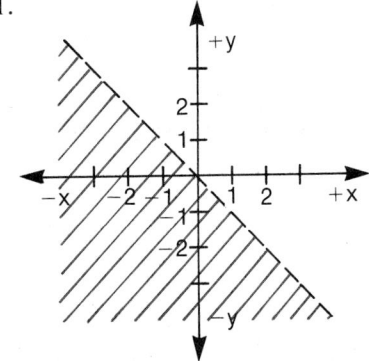

13.

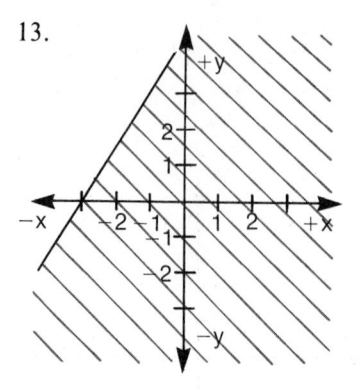

15.

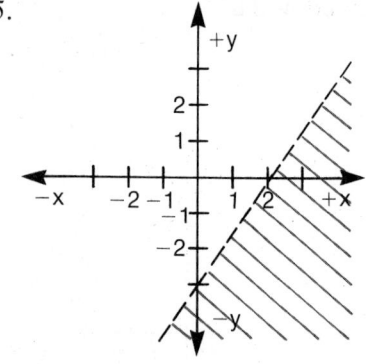

17.

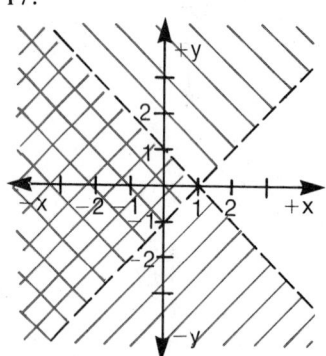

19.

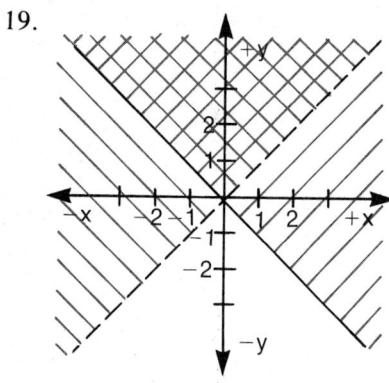

21.

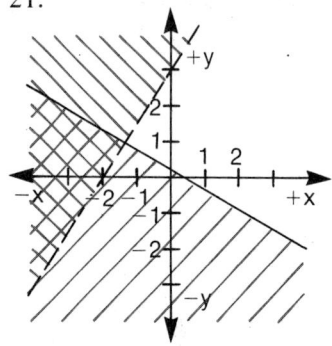

23.

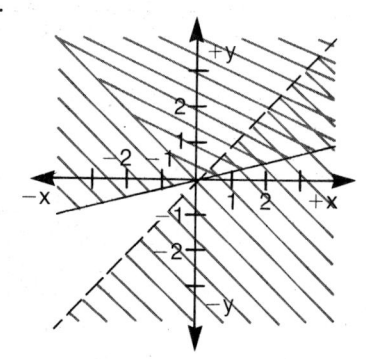

Review Exercises for Chapter 9

1. (1, 1)
2. (2, −1)
3. (1, 2)
4. (−2, −3)
5. dependent
6. (4, 4)
7. (2, 1)
8. (3, 2)
9. $\left(\dfrac{4}{9}, \dfrac{11}{18}\right)$

Answers to Selected Exercises

10. $\left(\dfrac{32}{5}, \dfrac{1}{5}\right)$
11. $(-2, -1)$
12. $\left(-\dfrac{1}{2}, \dfrac{7}{2}\right)$
13. $(4, -2)$
14. $(8, -5)$
15. $(-2, -3)$
16. $(-1, 4)$
17. $(6, -2)$
18. $(4, -2)$
19. $(-1, 0, 1)$
20. $(-4, 7, 1)$
21. $(0, 2, 1)$
22. 18
23. -12
24. -13
25. -14
26. 12
27. 12
28. $(2, 1)$
29. $\left(\dfrac{4}{9}, \dfrac{11}{18}\right)$
30. $(0, 3)$
31. $\left(2, -\dfrac{3}{2}, -\dfrac{1}{2}\right)$
32. $(-1, -2, -3)$
33. $\left(\dfrac{1}{2}, -\dfrac{1}{3}, \dfrac{1}{4}\right)$

34.

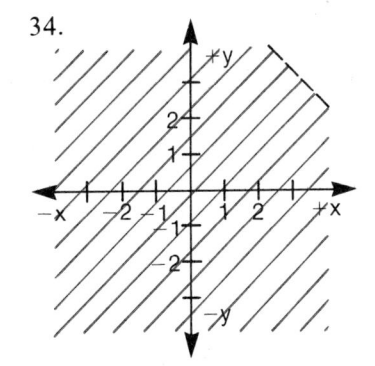

35.

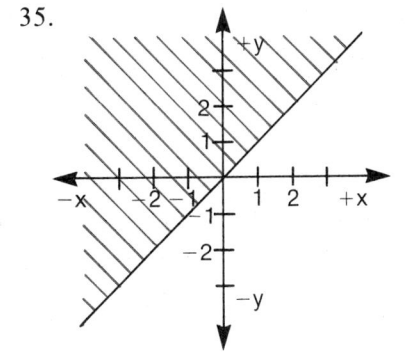

36.

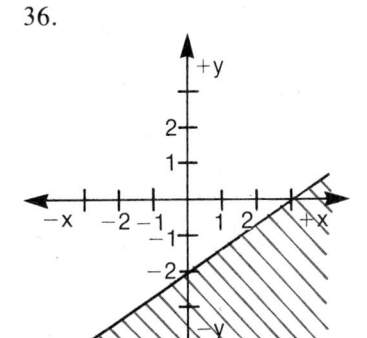

37.

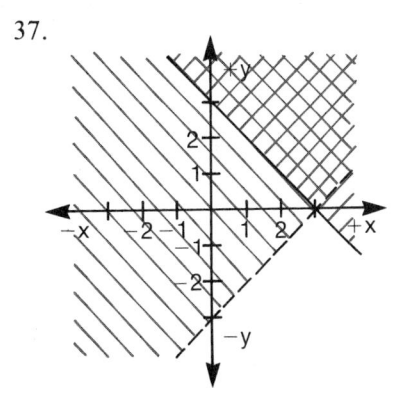

38.

39.

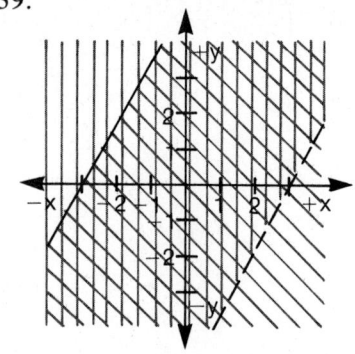

40. $\dfrac{17}{13}$

41. 1200 tickets at $2.00
    400 tickets at $2.75

42. 59

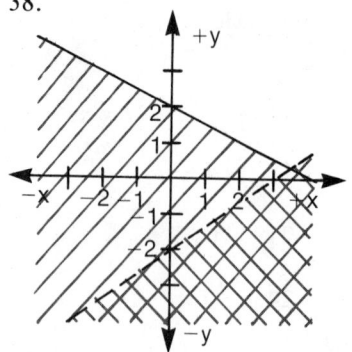

43. $\left(\dfrac{7}{2}, \dfrac{11}{2}, 9\right)$

44. 26 nickels
    12 dimes
    6 quarters

45. 178

Chapter 10. Exponential and Logarithmic Functions

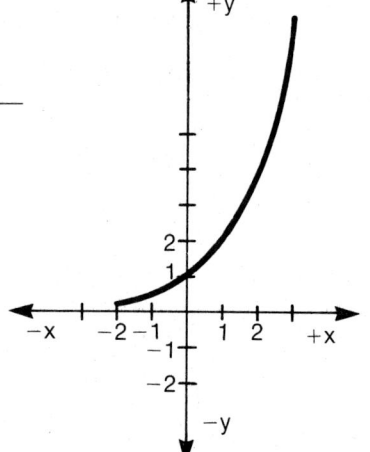

Section 10.1

1.

x	−2	−1	0	1	2	3
y	$\dfrac{1}{4}$	$\dfrac{1}{2}$	1	2	4	8

Answers to Selected Exercises

3.

x	−2	−1	0	1	2
y	$\frac{1}{25}$	$\frac{1}{5}$	1	5	25

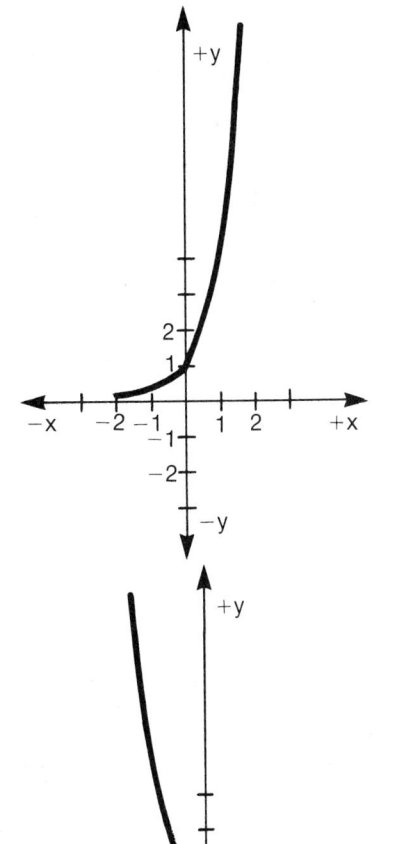

5.

x	−2	−1	0	1	2
y	9	3	1	$\frac{1}{3}$	$\frac{1}{9}$

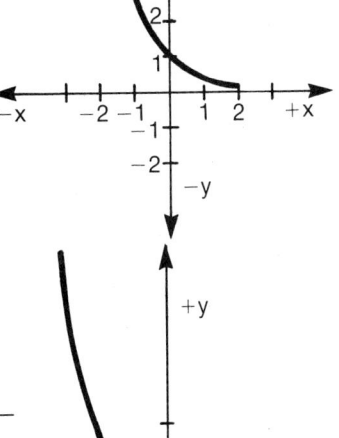

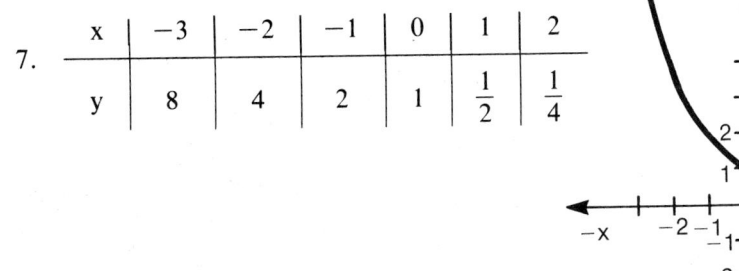

7.

x	−3	−2	−1	0	1	2
y	8	4	2	1	$\frac{1}{2}$	$\frac{1}{4}$

9.

x	−2	−1	0	1	2
y	25	5	1	$\frac{1}{5}$	$\frac{1}{25}$

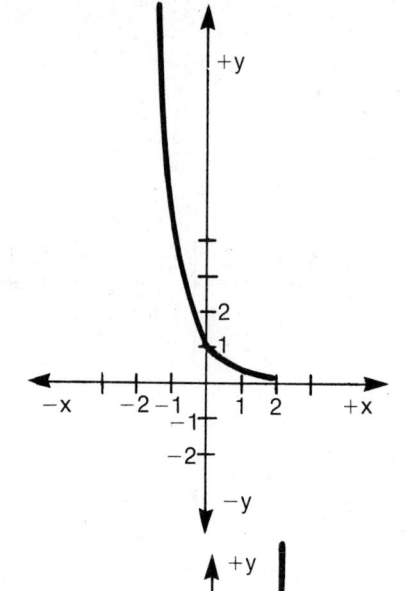

11.

x	−2	−1	0	1	2
y	$\frac{1}{9}$	$\frac{1}{3}$	1	3	9

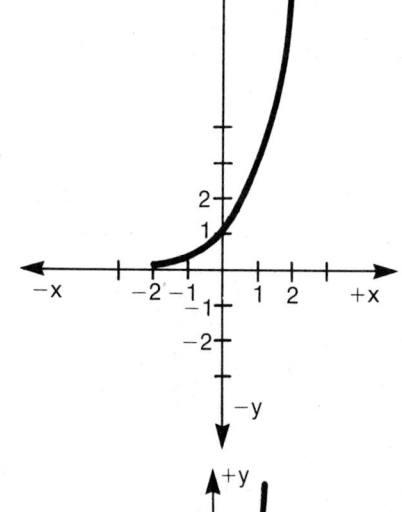

13.

x	−2	−1	0	1	2
y	$\frac{1}{16}$	$\frac{1}{4}$	1	4	16

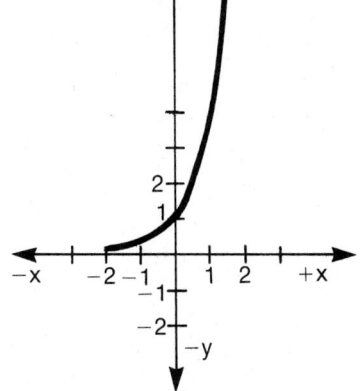

Answers to Selected Exercises

15.

x	−2	−1	0	1	2
y	$\frac{1}{2}$	1	2	4	8

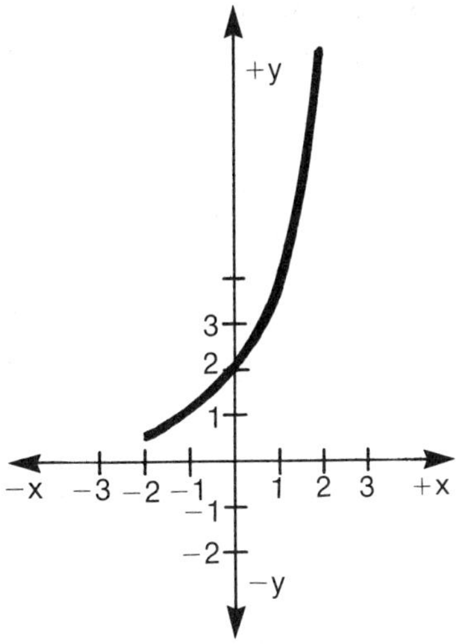

17.

x	−2	−1	0	1	2
y	$\frac{1}{2}$	1	2	4	8

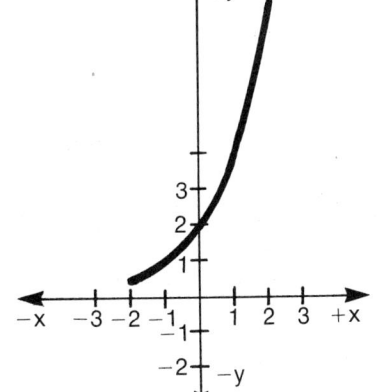

Section 10.2

1. $y = 3^x$.

x	−2	−1	0	1	2
y	$\frac{1}{9}$	$\frac{1}{3}$	1	3	9

$x = 3^y$.

x	$\frac{1}{9}$	$\frac{1}{3}$	1	3	9
y	−2	−1	0	1	2

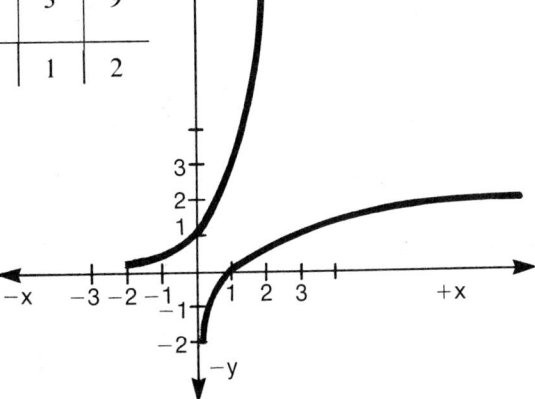

3. a. $\log_4 16 = 2$. b. $\log_2 8 = 3$. c. $\log_3 27 = 3$.
   d. $\log_2 16 = 4$. e. $\log_7 49 = 2$. f. $\log_5 125 = 3$.
   g. $\log_4 2 = \frac{1}{2}$. h. $\log_8 2 = \frac{1}{3}$. i. $\log_9 3 = \frac{1}{2}$.

   j. $\log_3 \frac{1}{9} = -2$. k. $\log_2 \frac{1}{8} = -3$. l. $\log_5 \frac{1}{25} = -2$.

5. a. $x = 125$. b. $x = 25$. c. $x = 625$.
   d. $x = 64$. e. $x = 343$. f. $x = 216$.

   g. $x = 16$. h. $x = \frac{1}{8}$. i. $x = \frac{1}{81}$.

   j. $x = \frac{1}{2}$. k. $x = \frac{1}{27}$. l. $x = \frac{1}{4}$.

7. a. $x = 2$. b. $x = 2$. c. $x = 5$. d. $x = 4$.
   e. $x = 8$. f. $x = 16$. g. $x = 2$. h. $x = 3$.
   i. $x = 8$. j. $x = 4$. k. $x = 9$. l. $x = 4$.

Section 10.3

1. $\log_b x + \log_b y + 2 \log_b z$
3. $2 \log_b x + 3 \log_b y$
5. $\log_b x + 2 \log_b y - \log_b z$
7. $\frac{1}{2} \log_b x + \frac{1}{2} \log_b y$
9. $\frac{1}{2} \log_b x + \frac{1}{3} \log_b z$
11. $\frac{1}{2} \log_b 6 + \frac{1}{2} \log_b x - \frac{1}{2} \log_b y - \frac{3}{2} \log_b z$

13. $\log_b \frac{x^2 y^3}{z}$ 15. $\log_b \sqrt{xy^3 z^5}$ 17. $\log_b \frac{x^3}{y^2 \sqrt{z}}$

19. $\log_b \sqrt{\frac{xy}{z^3}}$ 21. 0.7781 23. 1.1761

25. 0.3010 27. 0.6990 29. 1.3980

31. 1.2040 33. 3.0 35. 0.2330

37. 14 39. 2 41. 3

43. 2

Answers to Selected Exercises

Section 10.4

1. 1
3. 2
5. −1
7. 0
9. 0
11. −2
13. 2
15. 6
17. 2.5353
19. 2.1673
21. 8.5315 − 10
23. 9.5378 − 10
25. 7.9542 − 10
27. 1.4900
29. 6.9069 − 10
31. 540
33. 0.954
35. 0.0859
37. 3.59
39. 0.00254
41. 69800
43. 0.0108
45. 30900

Section 10.5

1. 3.6309
3. 1.8583
5. 9.6198 − 10
7. 8.6152 − 10
9. 3.8827
11. 2.8717
13. 2.4946
15. 9.8050 − 10
17. 1.7477
19. 2.3751
21. 0.4398
23. 9.8083 − 10
25. 786.7
27. 1957
29. 6.567
31. 0.3626
33. 0.5536
35. 0.02547
37. 203.5
39. 7378
41. 0.4284
43. 0.1266
45. 0.05436

Section 10.6

1. 1.414
3. 2.646
5. 2.520
7. 4.641
9. 1.627
11. 11.18
13. 0.8872
15. 0.2252
17. 5.565
19. 114.0
21. 11.89
23. 33.61
25. 392.1
27. 6.004
29. 9.080
31. 9.318

33. -2.030 35. $\$1803$ 37. $\$4842, \$2842$

39. $\$3048$ 41. 40.8

Section 10.7

1. $x = \frac{3}{2}$.   3. $x = -\frac{5}{12}$.   5. $x = \frac{3}{2}$.

7. $x = -\frac{11}{4}$.   9. $x = 2$.   11. $x = 2$.

13. $x = -\frac{7}{8}$.   15. $x = -\frac{1}{2}$.   17. $x = 1.6$.

19. $x = 3.3$.   21. $x = 1.3$.   23. $x = 3.5$.

25. $x = 2.9$.   27. $x = 1.7$.   29. $x = 4.3$.

31. $n = 21$.   33. 12 years   35. 9 years

Review Exercises for Chapter 10

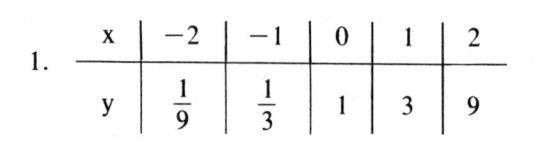

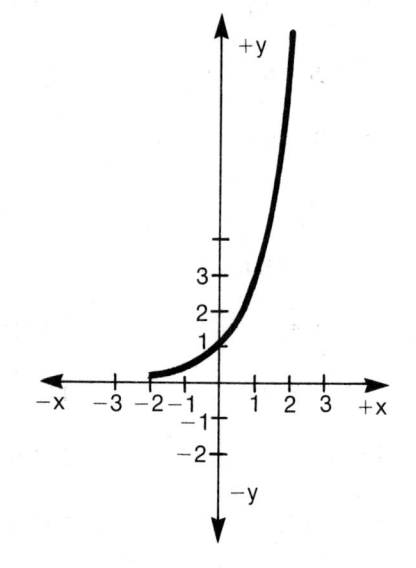

1.

x	-2	-1	0	1	2
y	$\frac{1}{9}$	$\frac{1}{3}$	1	3	9

Answers to Selected Exercises

2.

x	−2	−1	0	1	2
y	9	3	1	$\frac{1}{3}$	$\frac{1}{9}$

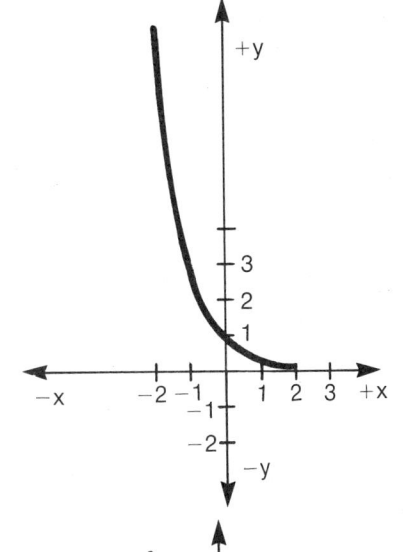

3.

x	−2	−1	0	1	2
y	9	3	1	$\frac{1}{3}$	$\frac{1}{9}$

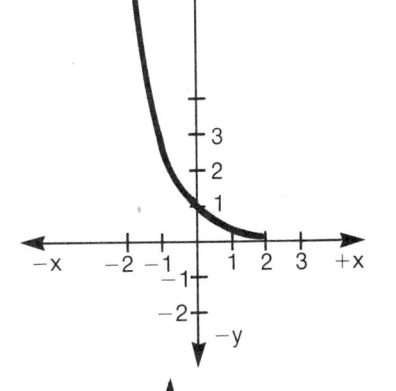

4.

x	−3	−2	−1	0	1	2
y	$\frac{1}{8}$	$\frac{1}{4}$	$\frac{1}{2}$	1	2	4

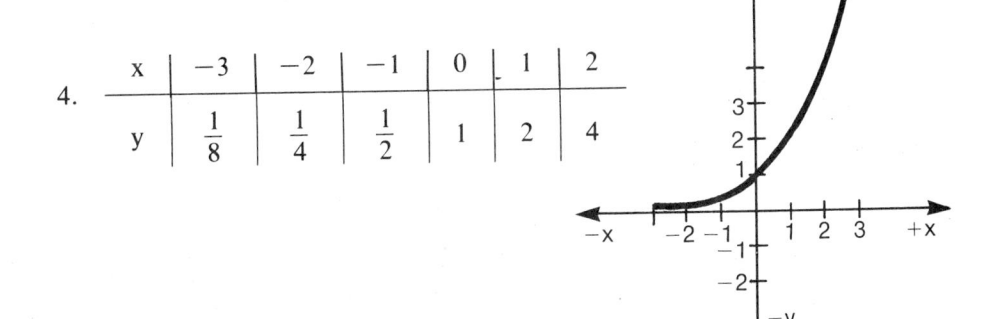

5.

x	−3	−2	−1	0	1	2
y	8	4	2	1	$\frac{1}{2}$	$\frac{1}{4}$

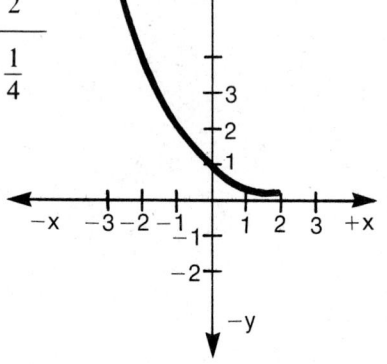

6.

x	−2	−1	0	1
y	$\frac{1}{3}$	1	3	9

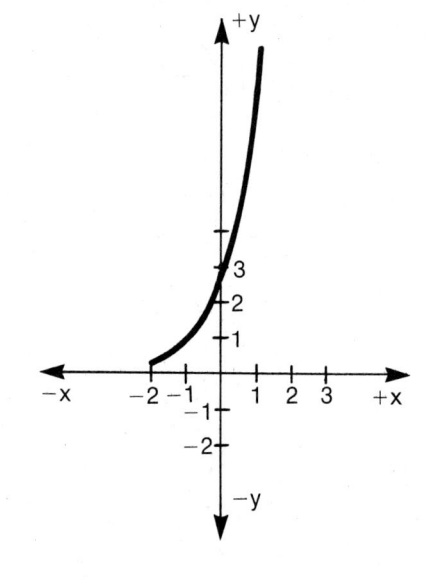

7. $y = 2^x$.

x	−2	−1	0	1	2	3
y	$\frac{1}{4}$	$\frac{1}{2}$	1	2	4	8

$x = 2^y$.

x	$\frac{1}{4}$	$\frac{1}{2}$	1	2	4	8
y	−2	−1	0	1	2	3

Answers to Selected Exercises

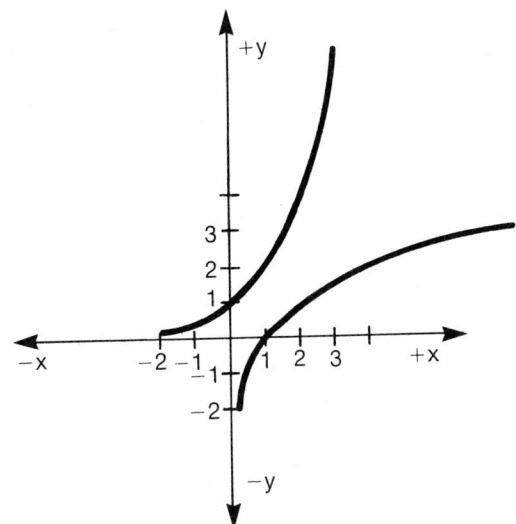

[Graph showing exponential curve in upper right and logarithmic curve in lower right, with axes labeled +x, -x, +y, -y, marked from -2 to 3]

8. a. $\log_2 8 = 3$. b. $\log_5 125 = 3$. c. $\log_2 \frac{1}{8} = -3$.

9. a. $2^3 = 8$. b. $5^3 = 125$. c. $2^{-3} = \frac{1}{8}$.

10. $x = 25$ 11. $x = \frac{1}{8}$. 12. $x = \frac{1}{4}$. 13. $x = 3$.

14. $x = \frac{1}{2}$. 15. $x = \frac{2}{3}$. 16. $x = 4$. 17. $x = 5$.

18. $x = 4$. 19. $x = 0$. 20. $x = -3$. 21. $x = 10$.

22. $2 \log_b x + 2 \log_b y + 2 \log_b z$ 23. $\frac{2}{3} \log_b x + \frac{1}{3} \log_b y$

24. $\frac{1}{3} \log_b 6 + \frac{1}{3} \log_b x - \frac{1}{3} \log_b y - \frac{2}{3} \log_b z$

25. 1.5408 26. 0.004733 27. 0.7378 28. 0.4972

29. false 30. false 31. false 32. true

33. false 34. false 35. true 36. false

37. true 38. true 39. 1.732 40. $N = 2.080$.

41. $N = 6.076$. 42. $N = 150.7$. 43. $N = -2.191$.

44. $901.60 45. $14,859, $4859.

46. 9 years 47. 12 years

index

Abscissa, 409
Absolute value, 63, 68, 120
  definition of, 63
  equations, 64
  inequalities, 121
Addition, 57, 198
  associative property of, 57, 68
  commutative property of, 57, 68
  of complex numbers, 337
  of fractions, 44
  of polynomials, 198
  of radicals, 170
  of rational numbers, 44
Addition property of equality, 75, 125
Additive identity element, 57
Additive inverse element, 27
Ahmes, 14, 100
Algebra, 2
Antilogarithm, 591
Apollonius, 487
Applications, 314, 381, 528
Associative property, 2, 20, 57
  of addition, 20, 57, 68
  of multiplication, 20, 57, 68
Axiom, 58
  of equality, 58
  of inequality, 58
  reflexive, 58
  symmetry, 58
  transitive, 58
  trichotomy, 58
Axis of symmetry, 465

Base, 130
  of exponential, 582
  of logarithm, 581
Binomial, 16, 196
  cube of, 214
  definition of, 16, 196
  expansion of, 215
  product of, 205
  square of, 210
Boyle's law, 478
Braces, 59
Brackets, 59
Briggs, Henry, 605

Cartesian coordinates, 409
Cartesian plane, 402
Characteristic, 574
Circle(s), 484
  useful formulas for, 633
Closure property, 2, 23, 57
Coefficient, 12, 196
Combining, 28
Common factors, 351
Common logarithm, 594
Commutative property, 2, 19
  of addition, 19
  of multiplication, 19
Completing the square, 330, 363

Complex fractions, 47, 300
  definition of, 47
  simplifying, 48, 302
Complex numbers, 330, 336
  addition of, 337
  definition of, 336
  division of, 341
  equality of, 337
  multiplication of, 338
  standard form, 336
  subtraction of, 337
Compound interest, 618
Conic sections, 483
  circle, 484
  ellipse, 487
  hyperbola, 489
  parabola, 463
Conjugates, 184, 279
Consistent equations, 506
Constant, 12, 414
Contradiction, proof by, 54
Conversion, metric, 631–632
Coordinate(s), 404
Coordinate geometry, useful formulas for, 634
Coordinate system, 402
Cramer, Gabriel, 553
Cramer's rule, 504, 553
Cross-product rule, 274, 291
Cubes, 215
  of a binomial, 215
  sum or difference of, 216, 270

Decimals, 53, 144
  non-terminating, 53
  repeating, 53
  terminating, 53
Dense, 50
Dependent equations, 504
Dependent variable, 414
Descartes, Rene, 14, 343
Determinants, 504, 545
  Cramer's rule, 553
  definition of, 546
  elements of, 546
  second order, 546
  of systems of linear equations, 552
  third order, 548
Difference of two squares, 212
Diophantas, 14, 84
Direct variation, 474
Disjoint sets, 8
Distance formula, 402, 451
Distributive property, 2, 22, 33
Division, 35
Domain, 402

Elements, 3
  of a determinant, 546
  of a set, 3
Ellipse, 487

705

Empty set, 5
Equations, 72, 402
  absolute value, 64
  of circle, 402
  conditional, 73
  consistent, 506
  dependent, 504
  of ellipse, 402
  exponential, 574
  first degree, 74
  fractional, 307, 357
  of hyperbola, 402
  inconsistent, 504
  linear, 74, 420
  literal, 96
  logarithmic, 583
  of parabola, 402
  quadratic, 345
  radical, 187
  solution set of, 188
  of straight line, 402
  systems of, 504
Equivalence, 73, 154
Equivalent equations, 73
Equivalent fractions, 275
Euler, Leonard, 343
Exponent(s), 13, 130
  definition of, 130, 131
  integral, 137
  negative, 138
  operations involving, 130
  positive, 131
  rational, 151
  zero, 137
Exponential equation, 574, 615
Exponential function, 574
Expressions, 13
Extraneous roots, 358, 377
Extremes, 473

Factor, 19, 240, 267
  binomial, 249, 255
  common, 241, 244
  prime, 240, 267
Factoring, 240, 270
  difference of two cubes by, 265
  difference of two squares by, 247
  equation solving by, 355
  by grouping, 245, 262
  perfect square trinomial, 251
  sum of two cubes by, 264
Field, 58
FOIL, 209, 338
Formula, quadratic, 371
Fractions, 275, 290
  addition of, 219, 274
  combining, 290, 298
  complex, 300
  division of, 286
  equal, 275
  multiplication of, 277
  simplifying, 276
  subtraction of, 290

Function(s), 402, 411
  definition of, 411
  domain of, 412
  exponential, 576
  graphs of, 413
  linear, 420
  logarithmic, 579
  notation, 415
  quadratic, 462
  range of, 412
Fundamental Principle of Fractions, 277
Fundamental Theorem of Algebra, 380

Gauss, Karl, 343
Geometry, coordinate, useful formulas for, 634
Greater than symbol, 28
Grouping, 270

Horizontal lines, 426
  slope of, 432
Horner's method, 230
Hyperbola, 489

i, definition of, 331
Identity, 2, 57, 68
  element, 24, 49
  property, 24, 57
Imaginary numbers, 331
Inconsistent equations, 504, 506
Independent variable, 414
Index of radical, 152
Indirect proof, 54
Inequalities, 112, 561
  absolute, 112
  compound, 118
  conditional, 112
  graphing, 116
  properties of, 113
  quadratic, 388
  solution of, 114
Integers, 26
Intercept, 423
Interest, 574
Interpolation, 600
Intersection, 7
Intervals, 389
Inverse, 2, 27, 68
  additive, 27, 57
  multiplicative, 57, 287
  variation, 477
Irrational number, 56

Joint variation, 479

Laws, 130
  of exponents, 130
  of logarithms, 589
Least common denominator (LCD), 44, 292

Abscissa, 409
Absolute value, 63, 68, 120
  definition of, 63
  equations, 64
  inequalities, 121
Addition, 57, 198
  associative property of, 57, 68
  commutative property of, 57, 68
  of complex numbers, 337
  of fractions, 44
  of polynomials, 198
  of radicals, 170
  of rational numbers, 44
Addition property of equality, 75, 125
Additive identity element, 57
Additive inverse element, 27
Ahmes, 14, 100
Algebra, 2
Antilogarithm, 591
Apollonius, 487
Applications, 314, 381, 528
Associative property, 2, 20, 57
  of addition, 20, 57, 68
  of multiplication, 20, 57, 68
Axiom, 58
  of equality, 58
  of inequality, 58
  reflexive, 58
  symmetry, 58
  transitive, 58
  trichotomy, 58
Axis of symmetry, 465

Base, 130
  of exponential, 582
  of logarithm, 581
Binomial, 16, 196
  cube of, 214
  definition of, 16, 196
  expansion of, 215
  product of, 205
  square of, 210
Boyle's law, 478
Braces, 59
Brackets, 59
Briggs, Henry, 605

Cartesian coordinates, 409
Cartesian plane, 402
Characteristic, 574
Circle(s), 484
  useful formulas for, 633
Closure property, 2, 23, 57
Coefficient, 12, 196
Combining, 28
Common factors, 351
Common logarithm, 594
Commutative property, 2, 19
  of addition, 19
  of multiplication, 19
Completing the square, 330, 363

Complex fractions, 47, 300
  definition of, 47
  simplifying, 48, 302
Complex numbers, 330, 336
  addition of, 337
  definition of, 336
  division of, 341
  equality of, 337
  multiplication of, 338
  standard form, 336
  subtraction of, 337
Compound interest, 618
Conic sections, 483
  circle, 484
  ellipse, 487
  hyperbola, 489
  parabola, 463
Conjugates, 184, 279
Consistent equations, 506
Constant, 12, 414
Contradiction, proof by, 54
Conversion, metric, 631–632
Coordinate(s), 404
Coordinate geometry, useful formulas for, 634
Coordinate system, 402
Cramer, Gabriel, 553
Cramer's rule, 504, 553
Cross-product rule, 274, 291
Cubes, 215
  of a binomial, 215
  sum or difference of, 216, 270

Decimals, 53, 144
  non-terminating, 53
  repeating, 53
  terminating, 53
Dense, 50
Dependent equations, 504
Dependent variable, 414
Descartes, Rene, 14, 343
Determinants, 504, 545
  Cramer's rule, 553
  definition of, 546
  elements of, 546
  second order, 546
  of systems of linear equations, 552
  third order, 548
Difference of two squares, 212
Diophantas, 14, 84
Direct variation, 474
Disjoint sets, 8
Distance formula, 402, 451
Distributive property, 2, 22, 33
Division, 35
Domain, 402

Elements, 3
  of a determinant, 546
  of a set, 3
Ellipse, 487

Index

Empty set, 5
Equations, 72, 402
  absolute value, 64
  of circle, 402
  conditional, 73
  consistent, 506
  dependent, 504
  of ellipse, 402
  exponential, 574
  first degree, 74
  fractional, 307, 357
  of hyperbola, 402
  inconsistent, 504
  linear, 74, 420
  literal, 96
  logarithmic, 583
  of parabola, 402
  quadratic, 345
  radical, 187
  solution set of, 188
  of straight line, 402
  systems of, 504
Equivalence, 73, 154
Equivalent equations, 73
Equivalent fractions, 275
Euler, Leonard, 343
Exponent(s), 13, 130
  definition of, 130, 131
  integral, 137
  negative, 138
  operations involving, 130
  positive, 131
  rational, 151
  zero, 137
Exponential equation, 574, 615
Exponential function, 574
Expressions, 13
Extraneous roots, 358, 377
Extremes, 473

Factor, 19, 240, 267
  binomial, 249, 255
  common, 241, 244
  prime, 240, 267
Factoring, 240, 270
  difference of two cubes by, 265
  difference of two squares by, 247
  equation solving by, 355
  by grouping, 245, 262
  perfect square trinomial, 251
  sum of two cubes by, 264
Field, 58
FOIL, 209, 338
Formula, quadratic, 371
Fractions, 275, 290
  addition of, 219, 274
  combining, 290, 298
  complex, 300
  division of, 286
  equal, 275
  multiplication of, 277
  simplifying, 276
  subtraction of, 290

Function(s), 402, 411
  definition of, 411
  domain of, 412
  exponential, 576
  graphs of, 413
  linear, 420
  logarithmic, 579
  notation, 415
  quadratic, 462
  range of, 412
Fundamental Principle of Fractions, 277
Fundamental Theorem of Algebra, 380

Gauss, Karl, 343
Geometry, coordinate, useful formulas for, 634
Greater than symbol, 28
Grouping, 270

Horizontal lines, 426
  slope of, 432
Horner's method, 230
Hyperbola, 489

i, definition of, 331
Identity, 2, 57, 68
  element, 24, 49
  property, 24, 57
Imaginary numbers, 331
Inconsistent equations, 504, 506
Independent variable, 414
Index of radical, 152
Indirect proof, 54
Inequalities, 112, 561
  absolute, 112
  compound, 118
  conditional, 112
  graphing, 116
  properties of, 113
  quadratic, 388
  solution of, 114
Integers, 26
Intercept, 423
Interest, 574
Interpolation, 600
Intersection, 7
Intervals, 389
Inverse, 2, 27, 68
  additive, 27, 57
  multiplicative, 57, 287
  variation, 477
Irrational number, 56

Joint variation, 479

Laws, 130
  of exponents, 130
  of logarithms, 589
Least common denominator (LCD), 44, 292

Index

Less than symbol, 28
Like terms, 16, 199
Line, 438
   point-slope formula of, 438
   slope-intercept formula of, 443
Linear equation, 420, 504
   graph of, 444
Linear inequality, 504, 561
Linear systems of equations, solution of, 504
   by addition, 516
   by determinants, 552
   by elimination, 517
   by graphing, 507
   by substitution, 508
Logarithm(s), 581
   base of, 582
   characteristics of, 595
   common, 594
   computations with, 605
   definition of, 581
   four place, table of, 629–630
   mantissa of, 595
   properties of, 589
Logarithmic functions, 574, 579
Lowest common denominator, 44

Mantissa, 574
Matrix, 545
Means, 473
Metric conversion, 631–632
Midpoint, 402, 451
Mixture, 107, 530
Monomial(s), 15, 196
Multiplication, 19, 33
   associative properties of, 57, 68
   commutative properties of, 19
   complex numbers, 338
Multiplication property of equality, 75, 125
Multiplicative identity, 34, 49
Multiplicative inverse, 49, 287

Napier, John, 605
Natural numbers, 2
Negative numbers, 26
Null set, 5
Number(s), 2, 11
   absolute value of, 63
   additive inverse of, 27
   complex, 330
   counting, 19, 56
   imaginary, 331
   integers, 2, 26
   irrational, 2, 56
   multiplicative inverse of, 49
   natural, 2, 18
   prime, 19
   rational, 38, 56
   real, 2, 56
   whole, 18, 56
Number line, 28
Numeral, 2, 12

Open sentence, 72
Ordered pair, 536
Ordered triple, 536
Ordinate, 409
Origin (zero values), 407–408

Parabola, 463
   equation of, 463
   graph of, 464
   vertex of, 464
Parallel lines, 434
Parentheses, 59
Pascal, Blaise, 216
Pascal's triangle, 216
Perfect square, 251
Perpendicular lines, 434
Point, coordinates of, 404
Point-slope form, 438
Polygons, areas of, 633
Polynomial(s), 16, 196
   addition of, 198
   definition of, 16
   division of, 219
   factoring, 240
   multiplication of, 204
   subtraction of, 198
Positive square root, 54, 152
Powers, 130
   table of, 628
Prime factors, 240
Prime numbers, 19
Principal n^{th} root, 335, 395
Proof, 59
Properties, 2
Proportion, 472
Pythagoras, 55
Pythagorean theorem, 453

Quadrant, 408
Quadratic equations, 330, 345
Quadratic formula, 330, 371
Quadratic inequalities, 388
Quotient, 35
   of complex numbers, 342
   logarithm of, 589
   of polynomials, 219
   of radical expressions, 181

Radical(s), 152
   addition and subtraction of, 170
   definition of, 152
   division of, 181
   equations containing, 187
   index of, 152
   like, 170
   product of, 164, 175
   rationalizing, 181
   simplifying, 157, 161
Radicand, 152
Range, 402
Ratio, 428, 472

Index

Rational expressions, 274, 312
Rational number, 38
Real number, 56
Recorde, Robert, 307
Rectangular coordinates, 409
Reflexive axiom, 158
Relation, 402
  definition of, 403
  domain of, 404
  inverse of, 405
  range of, 404
Remainder theorem, 225
Repeating decimals, 39
Right triangles, useful formulas for, 633
Roots, table of, 628
Roster form, 3

Scientific notation, 144
Sense, 113
Set(s), 2, 3
  builder notation, 6
  disjoint, 8
  elements of, 3
  empty, 5
  equality of, 3
  finite, 4
  infinite, 4
  of integers, 26
  intersection of, 2, 7
  of irrational numbers, 56
  of natural numbers, 2
  null, 5
  ordered pairs, 402
  of rational numbers, 38
  of real numbers, 56
  solution, 356, 389
  subset, 5
  union of, 2, 7
  of whole numbers, 18
Signed numbers, 26
Slope, 402, 427
Slope-intercept form, 443
Square(s), 212, 251
  binomial, 340
  completing the, 330
  difference of two, 212
  perfect, 251
Square root, 53
Standard form, 374
Subset, 5
Subtraction, 30, 32
  of complex numbers, 337
  definition of, 32
  of polynomials, 198
  of signed numbers, 33
Sum, 57
  of complex numbers, 337

Sum (Continued)
  of polynomials, 198
  of radical expressions, 170
Synthetic division, 228
Systems of equations, 516
  consistent, 506
  dependent, 504
  inconsistent, 506
  linear, 516, 535

Term, 15, 196
Terminating decimal, 39
Transitive property, 58
Triangles, right, useful formulas for, 633
Trichotomy, 58
Trinomial(s), 16, 196
  definition of, 16, 196
  factoring, 251

Uniform motion, 106, 532
Union of sets, 7, 8

Value, absolute, 63
Variable(s), 12
  definition of, 12
  dependent, 414
  independent, 414
Variation, 402, 472
  combined, 480
  constant of, 475
  direct, 474
  inverse, 477
  joint, 479
Vertex, 465
Vertical lines, 413, 426

Whole numbers, 18, 27
Word problems, 101, 316, 381

X-axis, 407
X-intercept, 423

Y-axis, 407
Y-intercept, 423

Zero, 27
  addition of, 57
  division by, 35, 250, 309
  as exponent, 137
  multiplication by, 33